D0778955

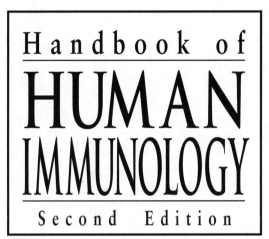

Handbook of
HUMAN
IMMUNOLOGY
Second Edition

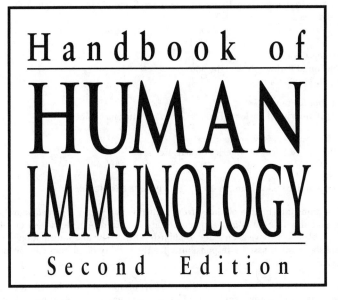

Handbook of HUMAN IMMUNOLOGY

Second Edition

Edited by

Maurice R. G. O'Gorman, Ph.D.

The Children's Memorial Hospital
Chicago, Illinois

Albert D. Donnenberg, Ph.D.

Hillman Cancer Center
Pittsburgh, Pennsylvania

CRC Press
Taylor & Francis Group
Boca Raton   London   New York

CRC Press is an imprint of the
Taylor & Francis Group, an informa business

Cover art by Yoju from the original artwork "Ao ni Aru Saibo"-525A. Copyright Yoju's Studio. Used with permission.

CRC Press
Taylor & Francis Group
6000 Broken Sound Parkway NW, Suite 300
Boca Raton, FL 33487-2742

© 2008 by Taylor & Francis Group, LLC
CRC Press is an imprint of Taylor & Francis Group, an Informa business

No claim to original U.S. Government works
Printed in the United States of America on acid-free paper
10 9 8 7 6 5 4 3 2 1

International Standard Book Number-13: 978-0-8493-1984-6 (Hardcover)

Library of Congress Cataloging-in-Publication Data

Handbook of human immunology / editors, Maurice R. G. O'Gorman and Albert
D. Donnenberg -- 2nd ed.
    p. ; cm.
"A CRC title."
Includes bibliographical references and index.
  ISBN 978-0-8493-1984-6 (alk. paper)
    1. Immune system. 2. Immunology. I. O'Gorman, Maurice R. G. II.
Donnenberg, Albert David. III. Title.
    [DNLM: 1. Immunity. 2. Immune System Diseases. 3. Immunologic
Techniques. 4. Immunologic Tests. QW 540 H236 2008]

QR181.H27 2008
616.07'9--dc22
                                                    2007035313

Visit the Taylor & Francis Web site at
http://www.taylorandfrancis.com

and the CRC Press Web site at
http://www.crcpress.com

Contents

Preface

Since the publication of the first edition of the *Handbook of Human Immunology* in 1997, major scientific achievements have contributed directly to an increased understanding of the complexities of the human immune system in health and disease. Whether directly as a result of the sequencing of the entire human genome or as a result of the technological advancements in the completion of the latter, several new components of the immune system have been revealed, and new technologies for their measurement and evaluation have been developed. In the past decade, the number of recognized clusters of differentiation (CD) on the surface of leukocytes and associated cells has more than doubled; an entire new chemokine and chemokine receptor nomenclature system has been established; the number of "lymphokines" (now greater than 30) and humanized monoclonal antibody therapy have become a staple of our pharmacologic armamentarium (to mention only a few of the significant immunological developments of the past decade). The discovery of all of these immune system components has been accompanied by the development of new and improved methods for their detection as well as the recognition of the roles that these components play in health and disease. These major breakthroughs are reflected in the present edition of the *Handbook of Human Immunology*.

There are six new chapters for a total of 20 (compared to 14 in the previous edition) written by 12 new authors covering all the major components of the innate and adaptive human immune system, descriptions of the specific human conditions characterized by their assessment, the specific technologies and methods currently used for their measurement, and finally the relevance and potential pitfalls involved in the interpretation of their specific measurements.

As in the past, the book is introduced with an overview of the immune system, which is immediately followed by a new, practical, and fun-to-read chapter on "Statistics of Immunological Testing" that is invaluable both in interpreting test results and also in the validation of new tests and the development of reference ranges. Throughout this edition, readers will find "normal ranges" including serum immunoglobulins, complement components, cytokines, and new to this edition—age-associated normal ranges for lymphocyte subsets. The first edition's chapter on flow cytometry has been expanded into six completely revised chapters including a thorough treatise on general flow cytometry principles and practice, quality control and theory, leukemia and lymphoma immunophenotyping analyses along with cytogenetic abnormalities, an update on CD4 measurement guidelines in HIV-infected persons and two new chapters covering (1) the flow cytometry–based diagnosis of primary immunodeficiency disease and (2) a history of the human leukocyte differentiation antigen workshops (now renamed the human cell differentiation antigen workshops) along with two comprehensive and very informative tables on CD.

The chapters on clinical laboratory disease monitoring include laboratory methods for monitoring new biological therapies in the context of clinical trials and routine practice, an entire chapter devoted specifically to the immunological assessment of gastrointestinal diseases, and a global review of the laboratory assessment of autoimmune disease. Infectious disease monitoring is discussed from the serological perspective, as well as the timely review of the technology and applications of new and now well-established molecular techniques. Finally, the laboratory's role in bone marrow and solid organ transplant is presented from the perspectives of antibody screening and crossmatching as well as molecular-based human leukocyte antigen (HLA) typing, followed by Chapter 20, which presents novel methods for monitoring functional immunosuppression of alloreactivity in patients posttransplant.

This edition of the handbook provides a practical reference of the important immunological parameters along with up-to-date descriptions of the methods used for their assessment and applications in health and disease. The text is particularly relevant to practicing clinicians, clinical laboratory professionals, and students interested in human immunology both from laboratory and applied clinical perspectives.

Editors

Maurice R.G. O'Gorman, PhD, is a professor of pediatrics and pathology at the Feinberg School of Medicine of Northwestern University, Chicago, Illinois. He is also the vice chair of the department of pathology and laboratory medicine and the director of the diagnostic immunology and flow cytometry laboratories at Children's Memorial Hospital.

Dr. O'Gorman was born in Windsor, Ontario, Canada and graduated in 1981 from the University of Western Ontario, London, Ontario, with a BSc with honors in microbiology and immunology. He then attended the University of British Columbia and earned an MSc in pathology in 1985 and 3 years later, a PhD in academic pathology from the same university. Following his doctorate, Dr. O'Gorman was awarded a 1-year fellowship in the division of neurology at UBC to develop and implement functional immune monitoring assays for clinical trials of new biological treatments for multiple sclerosis (MS). The following year Dr. O'Gorman was awarded a 2-year fellowship in clinical and diagnostic laboratory immunology in the department of pathology at the University of North Carolina in Chapel Hill. His research as a postdoctoral fellow evolved around the development, validation, and standardization of new clinical assays, many of which were flow cytometry based.

In his present position at the Children's Memorial Hospital in Chicago, Illinois, Dr. O'Gorman has established the Diagnostic Immunology and Flow Cytometry Laboratory at the Children's Memorial Hospital as an internationally recognized laboratory for the evaluation of primary immune deficiency disorders, leukemia and lymphoma immunophenotyping, and autoimmune disease. His clinical and developmental research program encompasses broad areas in immunology, including the establishment of normal reference ranges for the pediatric population, development of surrogate markers for monitoring disease activity and response to therapy, and the development of new diagnostic tests. He is also the codirector of the CPEP accredited postdoctoral program in clinical and diagnostic laboratory immunology. His main research interests are the investigation of immunopathogenetic mechanisms and immune regulation in autoimmunity and primary immunodeficiency diseases. Dr. O'Gorman has over 100 scholarly publications, has been an invited lecturer at numerous national and international venues, and has organized and chaired several national and international conferences and training programs in immunology and flow cytometry.

Dr. O'Gorman's leadership positions have included among others, president of the Association of Medical Laboratory Immunologists (AMLI, 2001–2003); president of the Great Lakes International Image Analysis and Flow Cytometry Association (2004); editor-in-chief of *Clinical Immunology Newsletter* (1994–2000); editor-in-chief of *Clinical and Applied Immunology Reviews* (2000–2006); and reviews editor for the *Journal of Immunological Methods* (2007). In 2005, Dr. O'Gorman earned an MBA from the executive MBA program at the Kellogg School of Management. He is married and has two sons, and enjoys a broad variety of sports and recreational activities.

Albert D. Donnenberg was born in New York, in 1951. As an undergraduate student, he studied philosophy at the University of Colorado, Boulder. He received his PhD in infectious disease epidemiology at the Johns Hopkins University in 1980, studying cellular immunity to herpes simplex virus. On graduation, he was elected to Delta Omega, the honorary Public Health Society. After a postdoctoral fellowship under the direction of bone marrow transplant pioneer Dr. George Santos at the Johns Hopkins Oncology Center, Dr. Donnenberg was appointed instructor of oncology in 1982, assistant professor in 1983, and associate professor in 1989. He worked on adoptive transfer of donor immunity during allogeneic bone marrow transplantation, and the development and clinical use of T-cell depletion of bone marrow to prevent graft-versus-host disease, an early implementation of engineered cellular therapy. He also performed early studies on cellular immunity in human immunodeficiency virus (HIV) infection, and codeveloped the concept of T-cell homeostasis.

In 1991, Dr. Donnenberg was recruited to the University of Pittsburgh to serve as the director of laboratory research for the Bone Marrow Transplant Program. He has also served as program codirector, and as interim director. He has directed the UPMC Hematopoietic Stem Cell Laboratory and the University of Pittsburgh Cancer Institute's Flow Cytometry Facility since 1998. He was promoted to professor of medicine in 2001.

His current research interests are in stem cell therapy and graft engineering, immunity after hematopoietic stem cell transplantation, and the role of stem cells in neoplasia—a project he pursues with his wife Dr. Vera Donnenberg. He is internationally recognized for his work in somatic cell therapy and flow cytometry. Dr. Donnenberg has authored more than 130 scholarly publications. He is the proud father of four daughters and one son, and lives on Pittsburgh's Southside with hobbies such as winemaking and art collecting.

Contributors

Ali Abdullah
Department of Pathology
University of Pittsburgh
Pittsburgh, Pennsylvania

Robert A. Bray
Department of Pathology and
  Laboratory Medicine
Emory University
  School of Medicine
Atlanta, Georgia

Nydia Chien
Department of Pathology
University of Pittsburgh
Pittsburgh, Pennsylvania

Barbara Detrick
Department of Pathology
Johns Hopkins
  School of Medicine
Baltimore, Maryland

Albert D. Donnenberg
Department of Medicine
Division of Hematology and Oncology
University of Pittsburgh
Pittsburgh, Pennsylvania

Vera S. Donnenberg
University of Pittsburgh Schools of
  Medicine and Pharmacy
UPMC Heart Lung and Esophageal
  Surgery Institute
Pittsburgh, Pennsylvania

Marcelo A. Fernández-Viña
Department of Pathology and
  Laboratory Medicine
MD Anderson Cancer Center
University of Texas
Houston, Texas

James D. Folds
Department of Pathology and
  Laboratory Medicine
University of North Carolina at
  Chapel Hill
Chapel Hill, North Carolina

Howard M. Gebel
Department of Pathology and
  Laboratory Medicine
Emory University
  School of Medicine
Atlanta, Georgia

Patricia C. Giclas
Department of Pediatric
  Allergy and Immunology
National Jewish
  Medical and Research
  Center
Denver, Colorado

Jennifer S. Goodrich
University of North Carolina
  Hospitals
Chapel Hill, North Carolina

Stefano Guandalini
Department of Pediatrics
Comer Children's Hospital
University of Chicago
Chicago, Illinois

Robert G. Hamilton
Department of Medicine
Division of Allergy and
  Clinical Immunology
Johns Hopkins
  School of Medicine
Baltimore, Maryland

Denise L. Heaney
Department of Pathology and
  Laboratory Medicine
Emory University
  School of Medicine
Atlanta, Georgia

John J. Hooks
Laboratory of Immunology
National Eye Institute
National Institutes of Health
Bethesda, Maryland

Bana Jabri
Department of Pathology
Comer Children's Hospital
University of Chicago
Chicago, Illinois

Andrés Jaramillo
Department of Pathology
Rush University Medical Center
Chicago, Illinois

Chethan Ashok Kumar
Department of Pathology
University of Pittsburgh
  School of Medicine
Pittsburgh, Pennsylvania

Alison Logar
Department of Pathology
University of Pittsburgh
  School of Medicine
Pittsburgh, Pennsylvania

Holden T. Maecker
BD Biosciences
Becton, Dickinson and Company
San Jose, California

Susana G. Marino
Department of Pathology
University of Chicago
Chicago, Illinois

Melissa B. Miller
Department of Pathology and
  Laboratory Medicine
University of North Carolina at
  Chapel Hill
Chapel Hill, North Carolina

Chandrasekharam N. Nagineni
Laboratory of Immunology
National Eye Institute
National Institutes of Health
Bethesda, Maryland

Maurice R.G. O'Gorman
Department of Pathology
  and Pediatrics
Feinberg School of Medicine
Northwestern University
Chicago, Illinois

Maria A. Proytcheva
Department of Pathology
Feinberg School of Medicine
Northwestern University
Chicago, Illinois

Noel R. Rose
Departments of Pathology and Molecular
  Microbiology and Immunology
Johns Hopkins University
Baltimore, Maryland

Chee L. Saw
Department of Pathology and
  Laboratory Medicine
Emory University
  School of Medicine
Atlanta, Georgia

John L. Schmitz
Department of Pathology and
  Laboratory Medicine
University of North Carolina at
  Chapel Hill
Chapel Hill, North Carolina

Mala Setty
Department of Pediatrics
Comer Children's Hospital
University of Chicago
Chicago, Illinois

Rakesh Sindhi
Department of Surgery
University of Pittsburgh
   School of Medicine
Pittsburgh, Pennsylvania

Mandal Singh
Department of Pathology
University of Pittsburgh
   School of Medicine
Pittsburgh, Pennsylvania

Anjan Talukdar
Department of Surgery
University of Pittsburgh
   School of Medicine
Pittsburgh, Pennsylvania

Gulbu Uzel
National Institute of Allergy and
   Infectious Diseases
National Institutes of Health
Bethesda, Maryland

Theresa L. Whiteside
Department of Pathology
University of Pittsburg
   School of Medicine
Pittsburgh, Pennsylvania

Patrick Wilson
Children's Hospital of Pittsburg
   of UPMC
Pittsburgh, Pennsylvania

1 Overview of Immunity

James D. Folds

CONTENTS

1.1   INTRODUCTION

The events leading to the development of immunity directed against pathogens are exceedingly complex. There are two distinct systems—innate and adaptive—that act in concert as well as separately in the development of immunity. The innate system provides a first line of defense against a foreign substance. It is nonspecific, rapid, lacks immunologic memory, and is usually of short duration. The adaptive system has exquisite specificity, is slower in development, exhibits immunological memory, and is long lasting.

Innate and adaptive immune systems are distinct systems but interact at several levels to develop a complete defense against invading pathogens. Both systems have mechanisms for distinguishing self from nonself, therefore, under normal situations they are not directed against the host's tissues and cells.

This chapter is intended to provide an introduction to the development of immunity. The purpose is to provide an overview of the different cells involved in both systems and the interactions occurring between the cells. The vigorous signaling mechanisms and the regulatory cells and enhancer and suppressor substances will be discussed in the context of developing immunity. This chapter serves as a broad overview.

1.2 OVERVIEW OF INNATE IMMUNITY

Elements of the innate immune system have been known for many years. However, in the past few years there has been a greater focus on innate immunity and its role in protection against infection and tissue injury [1] and its role in tolerance to self-antigens. Innate immunity defines a collection of protective mechanisms the host uses to prevent or minimize infection. The innate immune system operates in the absence of the specific adaptive immune system but is tied to adaptive immunity in many ways. The innate immune system is characterized by a rapid response to an invading pathogen or foreign or effete cells. In addition to the rapid response, it is also nonspecific and usually of a short duration. Innate immunity lacks immunological memory and there is no clonal expansion of lymphocytes as seen in the adaptive immune response. The innate immune response is also important in directing the specific, long-lived adaptive immune response.

The host defense mechanisms associated with innate immunity consist of a number of physical barriers (intact skin) and secretions accompanied by a number of serum factors such as complement, certain cytokines, and natural immunoglobulins [2]. The cellular components of innate immunity include a number of cell types, many of which are found at potential points of entry of pathogens [3]. Examples of these cells include natural killer (NK) cells, polymorphonuclear neutrophils (PMNs), macrophages, and dendritic cells (DCs).

The intact skin and mucosal tissues provide considerable protection against invading infectious agents. However, once the agents pass through the skin a number of important events take place. This includes activation of the complement cascade that triggers the development of a number of substances to attract phagocytes to the area. A number of antimicrobial peptides are produced at epithelial cell surfaces. These antimicrobial peptides play an important role in local defense mechanisms, disrupt bacterial cell membranes, and probably play a role in preventing skin infections.

1.2.1 ANTIMICROBIAL PEPTIDES

Human β-defensins are produced by epithelial cells in the mucous membranes of the airways and intestinal tract [4]. Defensins are small cationic peptides that have broad antimicrobial activities against a number of microbial agents [4] including Gram-positive and Gram-negative bacteria, fungi, and enveloped viruses. Defensins are nonglycosylated peptides containing approximately 35 amino acid residues, and β-defensins have six cysteine residues that provide a distinct structure.

Stimulation of the epithelium by certain cytokines can induce defensin production. The exact mode of action of defensins' antimicrobial activity is unknown. It is likely that defensins cause membrane disruption resulting from electrostatic interaction with the polar head groups of membrane lipids [5].

There are three defensin subfamilies: α-defensins, β-defensins, and θ-defensins [5]. The α- and β-defensins are products of distinct gene families and are structurally different from the θ-defensins. The θ-defensins are not seen in humans and probably represent a mutated form of the α-defensins.

The α-defensins were first purified from azurophilic granules of PMNs [6]. Several species, including rabbits, humans, and some rodents have α-defensins in their PMNs. Human monocytes and NK cells produce α-defensins that are similar to the α-defensins of PMNs [6]. The α-defensins play a role in the oxygen-independent killing of microorganisms after phagocytosis by PMNs [5].

The β-defensins are expressed in epithelial cells and leukocytes throughout the body. The β-defensins work in concert with a number of other components of the innate system to provide an important defense against microorganisms. There are at least four human β-defensins (HBD-1–HBD-4) [7]. Production of β-defensins may be constitutive or inducible. HBD-2, -3, and -4 are inducible. There is evidence that certain epithelial cells can be stimulated to produce HBD-2 in response to activation by bacterial products and toll-like receptors (TLRs) found on the epithelial cells [7].

Defensins also appear to have immunoregulatory properties in addition to their antimicrobial properties [8]. Immunoregulatory properties include chemoattractants for PMNs, immature DCs (iDCs), mast cells, and some memory T cells. Defensins may also stimulate iDCs to undergo maturation [8].

Several other antimicrobial peptides have been described in epithelial cells and PMNs. Lysozyme was described as an antimicrobial peptide found in human neutrophils and is known to attack the peptidoglycan cell walls of bacteria. Cathelicidin is expressed in human cells such as epithelial cells, PMNs, monocytes, and T, B, and NK cells [9]. It is usually expressed by cells lining the respiratory and gastrointestinal tracts. Cathelicidin is a well-known chemoattractant for various cells including PMNs, mast cells, monocytes, and thymus-derived lymphocytes (T lymphocytes). Cathelicidin has antimicrobial activity against most Gram-positive and Gram-negative bacteria.

Histatins are a family of cationic peptides (MW = 3–4 kDa) that are present in human saliva [6]. Histatins probably play an important role in oral health by providing potent antibacterial and antifungal actions. Of the several known histatins, histatin 5 is the most potent antifungal agent and is secreted by human parotid and submandibular glands [10].

The human skin, when intact, is refractory to most pathogens. This natural resistance is reportedly due to the presence of constitutively produced and inducible antimicrobial peptides. These peptides are cathelicidins, defensins, and dermicidins [9,7,11]. These antimicrobial peptides appear to act by directly inhibiting pathogen growth and enhancing other components of the immune responses. Psoriasin is another antimicrobial peptide found in the skin, especially in areas where bacterial invasion is likely to occur [12]. Psoriasin shows bactericidal activity preferentially against *Escherichia coli*, and also shows activity against other organisms that may colonize the skin.

Together, these antimicrobial peptides and proteins contribute significantly by providing a "chemical barrier" to reenforce the physical barriers of the intact skin and mucous membranes.

1.2.2 THE COMPLEMENT SYSTEM

The complement system is another important component of innate immunity. The system consists of 30 proteins found in serum or on the surface of certain cells [13]. Activation of the complement system results in a cascade of biochemical reactions that ultimately ends in lysis and disruption of foreign or effete cells. Without activation, the components of the complement system exist as proenzymes in body fluids. As a by-product of the activation of the cascade, a number of biologically reactive complement fragments are generated. The complement fragments can modulate other parts of the immune system by binding directly to T lymphocytes and bone marrow–derived lymphocytes (B lymphocytes) of the adaptive immune system and also stimulate the synthesis and release of cytokines.

As shown in Figure 1.1, there are three activation pathways for the complement system. Although the activation pathways are different, they all act at the microbial surface to assemble an enzyme convertase that cleaves C3 to form C3b that binds to a microbial surface where it activates C5 and the other components of the cascade.

The three pathways are the classical, mannan-binding lectin (MBL), and the alternative. Each of the pathways has its own recognition mechanism and is activated through different mechanisms, but all result in the formation of a membrane attack complex (MAC) and lysis of a target cell.

The classical pathway is activated by either IgM or IgG attached to a microbial surface antigen. The recognition molecule for the classical pathway is complement component C1q. A conformational change occurs in C1q, which results in activation of C1r and C1s, which, in turn, activates C4 and C2, which leads to the formation of the C4b2a complex (C3 convertase). The C3 covertase acts on C3, which ultimately leads to the formation of the MAC.

Activation of the MBL pathway begins after the recognition of mannose-binding lectin on various carbohydrate ligands [14]. MBLs and ficolins are found in serum and are structurally similar to C1q. MBLs and ficolins bind to mannose-containing carbohydrates on the surface of microbes. The MBLs and ficolins are considered to be typical pattern recognition molecules and as such attach to the MBL-associated serine proteases. On activation, the MBL-associated serine proteases cleave C4 and C2 to generate the C3 convertase C4bC2a and activate the remainder of the cascade [15].

The alternative pathway is important in innate immunity because it does not require specific antibodies for activation of C3. There are low levels of C3 present in body fluids at all times. C3 undergoes hydrolysis to produce $C3(H_2O)$, which is an activated form. $C3(H_2O)$ can bind to factor B that is then cleaved by the factor D to form the fluid-phase C3 convertase $C3(H_2O)Bb$. Small amounts of C3b are needed to activate the alternative pathway at microbial surfaces [16]. C3b on the microbial surface binds to factor B, which is cleaved by factor D to form C3bBb, the C3 convertase. Properdin serves to stabilize the convertase whose role is to cleave C5, which activates the remainder of the cascade. There are several agents that can activate the alternative pathway: bacterial cells, tumor cells, enveloped viruses, and damaged mast cells.

The complement system and its by-products serve to facilitate opsonization and may ultimately remove or destroy invading microorganisms. Tissue and circulating PMNs and macrophages are the cells that are most often involved in the ingestion

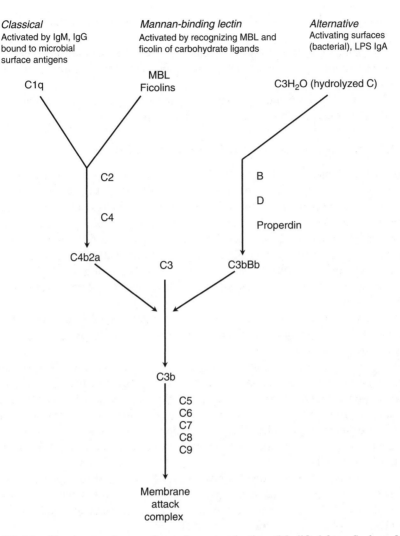

Classical
Activated by IgM, IgG
bound to microbial
surface antigens

Mannan-binding lectin
Activated by recognizing MBL and
ficolin of carbohydrate ligands

Alternative
Activating surfaces
(bacterial), LPS IgA

C1q

MBL
Ficolins

$C3H_2O$ (hydrolyzed C)

C2

B

C4

D

Properdin

C4b2a   C3   C3bBb

C3b

C5
C6
C7
C8
C9

Membrane
attack
complex

FIGURE 1.1 The three pathways of complement activation. (Modified from Seeleen, M., Roos, A., Daha, M.R., *J. Nephrol.*, 18, 642–653.)

of intracellular pathogens and killing of the invading microbes. Surface-bound C3b and iC3b (on the microbes) facilitate the attachment of the microbes to phagocyte complement receptors, which activates the ingestion and intracellular killing by the phagocytes [17].

In addition to enhancing opsonization in the presence or absence of antibodies, complement components have other important biological functions [17]. For example, free cleavage fragments of C3 and C5 are known to promote host inflammatory responses. C3a and C5a stimulate the bone marrow to release additional PMNs (C3b) and to serve as strong chemoattractants (C3a) for PMNs, monocytes, and eosinophils. Complement components C4a and C5a behave as anaphylotoxins to induce histamine release, which, in turn, causes increased vascular dilatation and permeability.

As mentioned earlier, the complement system helps modulate the adaptive immune response by enhancing antigen recognition and by stimulating the synthesis and release of cytokines.

Together, the complement system is another important factor in defense against invading microbes and it functions to provide a rapid response.

1.2.3 NATURAL ANTIBODIES (OR IMMUNOGLOBULINS)

Natural antibodies have been recognized for some time but recently they were described as a component of the innate immune system [18]. Natural antibody is defined as an antibody that is found in normal, healthy individuals who have no evidence of exogenous antigenic stimulation. Natural antibodies are believed to develop in a highly regulated manner; they are usually found in low titer in serum and are low-affinity antibodies [17]. A high percentage of the natural antibodies found in serum are of the IgM class. These antibodies are produced by a primitive B lymphocyte, called the B-1 lymphocytes [18]. B-1 cells are usually CD5[+] and considered to be long-lived and self-replicating.

Natural antibodies play an important role as a first line of defense against pathogens and other types of cells, including precancerous, cancerous, cell debris, and some self-antigens [19].

The cells of innate immunity apparently rely on an array of nonclonally expressed "pattern-recognition receptors (PRRs)" found in the target cells [20]. This response does not recognize specific single antigenic structures or epitopes, as in adaptive immunity, but respond to specific patterns, which are expressed independent of mutational events. This recognition system allows the innate immune mechanism to respond rapidly by focusing on structures most likely found in pathogens or effete cells.

The B-1 cells are positioned at possible sites of entry of pathogens, along with monocytes, to mediate a rapid response to pathogens. The B-1 lymphocytes differ from the usual B lymphocyte (B-2) in terms of phenotype, anatomic location, and mechanisms of activation and signaling [21].

Thus, innate immunity uses an inherited set of receptors found on NK cells, $\gamma\delta$ T cells, and CD5[+] B cells to recognize and interact with a broad spectrum of different antigens.

1.2.4 TOLL-LIKE RECEPTORS

TLRs are found on phagocytic cells, including mononuclear phagocytes, circulating monocytes, tissue macrophages, and endothelial cells, and are important components of the innate immune system [22]. TLRs make up a family of cell surface protein receptors present on several cell types that function to recognize certain conserved molecular components of microorganisms and signal that microbes have breached the body's barrier defenses [23]. TLRs serve as first responders in a mammalian host to recognize the presence of an invading pathogen. They also generate an inflammatory response to attempt to remove the invading agent. There are at least

TABLE 1.1

Toll-Like Receptors in Mammals

Receptors	Functions and Ligands
TLR-1	Triacyl lipopeptides—from *Mycobacteria* TLR-1 is a coreceptor with TLR-2
TLR-2	Peptidoglycans, lipoproteins, lipoteichoic acids, *Mycobacteria*, and spirochetes
TLR-4	Lipopolysaccharide, heat shock protein
TLR-5	Bacterial flagellin
TLR-6	Lipoteichoic acid, zymosan diacyl lipopeptides TLR-6 is a coreceptor with TLR-2
TLR-7	Single-stranded RNA from viruses
TLR-8	Single-stranded RNA from viruses
TLR-9	Cytidine–phosphate–guanosine nucleotides (CpG)—natural or synthetic unmethylated
TLR-10	Unknown

Source: Adapted from Sigal, L.H., *J. Clin. Rheumatol.*, 10, 353, 2004.

10 TLRs in humans and they are capable of detecting a broad range of microbial ligands (see Table 1.1).

The primary role of TLRs, as mentioned earlier, is to recognize and control bacterial infection. The mechanism of recognition is based on the receptor binding to a structurally conserved and unique pathogen-associated molecular pattern (PAMP) [24]. PAMPs are structural components of microbes that are important to them physiologically and are expressed on the pathogen but not on the host cells. TLRs consist of a family of "PRRs," which are inherited molecules that exist as transmembrane proteins that detect the presence of pathogenic agents. The recognition mechanism between the TLRs and PAMPs provides an efficient method for self/nonself discrimination. The interaction between the TLR and the microbial PAMPs triggers host cell activation [24].

Despite intensive investigations, the molecular details of the interaction between TLRs and pathogens is still unclear. Binding sites for the different PAMPs are known to contain different extracellular leucine-rich repeat units. However, binding and recognition of many diverse bacterial ligands is not understood.

After a TLR binds to a PAMP, the TLR dimerizes either with itself or, in other cases, it binds with a different TLR to induce an intracellular conformational change, resulting in the recruitment of certain other proteins (adaptor proteins) in the cytoplasm [25]. These adaptor proteins, such as MyD88, CD14, and others, transmit the message that TLR activation has occurred and initiate an intracellular cascade resulting in induction of an inflammatory response [26].

Activation of some TLRs induces the expression of a costimulatory molecule, B7 (CD80), which is found on antigen-presenting cells (APCs) and is needed for activation of naïve T lymphocytes [27].

Table 1.1 gives a listing of the TLRs found in mammals. TLRs of humans primarily recognize microbial structures and PAMPs and go on to trigger host cell activation and inflammation.

In summary, TLRs function to curb an acute infection by activation and regulation of rapid effector responses in the innate immune system. TLR activation regulates a number of systems known to be important in innate immunity. These include the release of inflammatory cytokines and chemokines, the oxidative burst in phagocytic cells, as well as the activation of various cationic peptides. TLRs serve to discriminate between self and nonself through the recognition and reactivity to PAMPs that are not expressed on host cells. TLRs also impact and moderate the adaptive immune response system through the induction of costimulatory molecular and cytokines that are involved in T- and B-lymphocyte reactivity.

1.2.5 PHAGOCYTOSIS

Polymorphonuclear neutrophilic leukocytes have been well-known components of the innate immune system for many years. Detailed studies of PMN phagocytosis and intracellular killing of microorganisms have led to a better understanding of important defense mechanisms against invasion by pathogenic bacteria, fungi, and enveloped viruses. PMNs are attracted to the site of microbial invasion, recognize the microbe, become activated, kill the microorganisms, resolve the infection, undergo apoptosis, and are then ingested and removed by either macrophages or neighboring endothelial cells to resolve the inflammatory response.

PMNs arise as myeloid progenitors in the bone marrow. Specific growth factors and cytokines mediate the differentiation of myeloid precursors into mature PMNs [28]. After entering the circulation, the PMNs have a half-life of about 8–12 h before undergoing a programmed cell death (apoptosis) and are reabsorbed through endothelial walls. The PMN turnover is about 10^{11} cells per day [29].

PMNs are actively recruited to the site of an infection by a complicated multistep process whereby they are mobilized both from the circulation and bone marrow. PMNs are constantly rolling along walls of the postcapillary venules where they readily detect the presence of a chemoattractant signal generated either by endothelial cells or the microbes themselves [30]. The chemokine, interleukin-8 (IL-8), is an important chemoattractant produced by multiple host cell types during an inflammatory event [31]. Bacteria also produce substances that are chemoattractants for PMNs. Once the PMNs reach the site, they become primed and develop enhanced functional capabilities [32].

Phagocytes are actively recruited to the site of microbial invasion as a response to a number of lectin glycoproteins called selectins [33]. Selectins are found on the surface of endothelial cells lining the blood vessels at the site of infection. P-selectin (from activated platelets) is upregulated on the surface of the endothelial cells and the selectins facilitate the "rolling and tethering" of PMNs [30]. PMNs secrete several molecules to support their migration from the circulation through the endothelium into the tissues. A number of neutrophil chemoattractants have been reported, including C5a, N-formyl bacterial oligopeptides, and leukotriene B4 [34]. As PMNs move toward the site, they become activated or primed to produce their antibacterial substances [32].

Phagocytosis is the process whereby PMNs recognize, bind, and ingest the microorganisms that stimulated the inflammatory reaction. Phagocytosis is greatly enhanced by opsonization of the bacteria. Attachment of a specific IgG or complement fragment, C3, can greatly enhance the efficiency of phagocytosis, although phagocytosis can occur without opsonization [35]. Complement receptors, CR1 and CR3, are the primary receptors for opsonization by the complement. The PMNs also express receptors for IgG fragment Fc (FcγRs) facilitates phagocytosis [36]. The two most prominently expressed Fcγ receptors on circulating PMNs are known as FcγRII (CD32) and FcγIII (CD16), and binding to these receptors triggers the oxidative burst in PMNs. The binding of antibody and complement receptors at the PMNs surface activates phagocytic process.

Activation of phagocytosis causes changes in the cytoskeletal contractile elements, which leads to an invagination of the cell membrane of the PMNs. This occurs at the site of the attachment of the opsonized microorganism [37]. Pseudopods extend from the PMNs and fuse around the invagination encasing the microorganism inside the phagolysosomal vacuole [37].

The PMNs have two broad types of killing mechanisms. One is oxygen-dependent and the other is oxygen-independent [38]. Phagocytosis of microbes stimulates the production of superoxide radicals and other reactive oxygen species. These are potent microbicidal agents and include, among others, hydrogen peroxide and chloramines. The enzyme NADPH oxidase is found in the cell membrane of the PMNs and generates the superoxide (called the respiratory burst) [38]. The superoxide is unstable and quickly dismutates to hydrogen peroxide and other substances that are microbicidal. These reactions take place inside the phagolysosome (also called phagosome).

The PMNs contain two types of cytoplasmic granules, azurophilic and specific (also referred to as secondary and tertiary granules) [39]. Each type of granule houses a number of proteins and peptides with microbicidal properties. Lysozyme is found in both types of granules and cleaves peptidoglycans of bacterial cell walls to disrupt the microbes [40]. The azurophilic granules contain a number of small cationic proteins with microbicidal activity. Azurophilic granules also contain myeloperoxidase (MPO), which during PMN activation is directed to phagosomes where it catalyzes a reaction with chloride and hydrogen peroxide to form hypochlorous acid [37], which is extremely microbicidal. The beta2 integrin CD11b/CD18 is present in the plasma membrane and secondary granules of neutrophils, and functions as a major adhesion molecule. On PMN activation, there is translocation of intracellular pools of CD11b/CD18 to the plasma membrane in concert with enhanced cellular adhesion. Although much is known about the function of CD11b/CD18, how this protein is transported within the cell is less well defined.

α-Defensins are also found in azurophilic granules. These are small cationic peptides that can interact with negatively charged molecules at the pathogen surface to change the permeability of the bacterial cell membranes and cause death [41]. Lactoferrin is an iron-binding protein found in azurophilic granules and is capable of binding iron, which is needed for bacterial growth [42]. Iron is also needed by the PMN to form other antibacterial compounds.

After death of the invading microorganisms, the PMNs also die through a process known as apoptosis. This process is affected by both proinflammatory host reactions and microbes and is important in the resolution of the inflammatory process.

Phagocytosis is well known and extremely important to host defenses as noted by the severe infections seen in patients with phagocytic defects.

1.2.6 CYTOKINES AND CHEMOKINES

Cytokines and chemokines are small, secreted polypeptides that regulate essentially all functions of the immune system. Cytokines participate in determining the nature of the immune response by regulating or controlling cell growth, differentiation, activation, immune cell trafficking, and the location of immune cells within the lymphoid organs [43].

Cytokines are a group of "intercellular messengers" that contribute to inflammatory responses through activation of the host's immune cells. Cytokines are host-derived products that enhance the recruitment of circulating leukocytes as a response to the presence of pathogens [44]. Cytokines also play important roles in leukocyte attraction by inducing the production of chemokines, which are known to be potent mediators of chemoattractant activity for inflammatory cells. Chemokines and cytokines provide a complex network of signals that can either activate or suppress inflammatory responses [44].

Cytokine secretions can alter the behavior or properties of that cell itself as well as the behavior of surrounding tissue cells. Cytokines are produced by many different cells. One cell type is capable of making many different cytokines and a particular cytokine may be secreted by multiple cell types. A particular cytokine may have many different effects on different cells depending on the environment. In some cases, the target cell itself may produce a cytokine that influences itself as well as the neighboring cells of the target cell. Some cytokines require cell-to-cell interaction to exert their effects [45].

Chemokines are another family of small structurally similar polypeptides that can regulate the trafficking of subsets of leukocytes [46]. Chemokines differ from other cytokines because all chemokines are ligands for the G-protein-coupled receptors [44]. Chemokines are potent cell activators capable of inducing migration of immune and inflammatory cells. Most cells in the immune system express receptors for at least one chemokine. Inflammatory cells may release a variety of chemokines and there is some evidence that infection with certain bacteria and viruses can stimulate the host cells to produce characteristic sets of immune cells.

Cytokines have multiple, broad properties. For example, cytokines may work effectively in concert with or in competition with both immune and nonimmune cells [47]. In addition to their impact on innate immunity, cytokines support B- and T-lymphocyte maturation and proliferation, differentiation of T helper cells into Th1 and Th2 subsets, maturation and polarization of DC subtypes and memory cell development [48]. These are important facets of adaptive or acquired immunity.

There have been more than 30 cytokines described [49]. Most cells of the immune system and many other host cells release cytokines. In some cases, the same cell type may respond to cytokines through specific cytokine receptors and may release and respond to the same cytokines it just produced.

Two cytokines, IL-1 and tumor necrosis factor-α (TNF-α) play important roles in responses to bacterial infection [49,50]. They are both small polypeptides that exert a broad range of effects on multiple host reactions, including immunologic responses, inflammation, and hematopoiesis. Experimentally, IL-1 and TNF-α if injected into mice may produce many of the features of Gram-negative sepsis in the absence of infection with the microorganisms [49]. In endotoxic shock caused by

Gram-negative organism, the cytokines, IL-1 and TNF-α are secreted by mononuclear phagocytes in response to activation by TLRs by the bacterial lipopolysaccharides [49]. This reaction results in the secretion of other cytokines and chemokines, which exacerbate the reaction.

Other cytokines have important roles in cell development, for example, they impact myeloid cell progenitor cell development, enhance IgG4 subclass development, regulate Th2 responses, cause mast cell proliferation *in vitro*, stimulate B-cell production, activate and stimulate growth of eosinophils, stimulate survival and expansion of immature precursors that are committed to T- and B-cell lineages, stimulate endothelial cells to produce adhesion molecules, enhance PMN recruitment, and many other functions. For review, see Refs. 44 and 51 and Figure 1.1.

1.2.7 Natural Killer Cells

NK cells were first reported in the 1970s. Initially, NK cells were referred to as nonspecific lymphocytes because NK cells could kill certain virally infected and malignant cells without known prior sensitization. NK cells were known to resemble large lymphocytes morphologically and were referred to as large granular lymphocytes. Approximately, 10–15% of the lymphocytes circulating in peripheral blood are NK cells. NK cells are distinct from T- and B lymphocytes because they express neither immunoglobulin receptors nor T-cell antigen receptors. There are other distinctions including phenotype and function. NK cells have receptors that recognize major histocompatibility complex (MHC) class I antigens. Because NK cells have cytotoxic properties, their function is highly regulated in their interactions in both the innate and adaptive immune systems [52,53].

NK cells develop from a common lymphoid progenitor cell in the bone marrow. The NK cells diverge from other lymphocyte lineages and acquire specific cell surface markers to guide them through their developmental stages. After developing and maturing in the bone marrow, the NK cells migrate and circulate in the peripheral blood and may be found in various organs including the lung, liver, spleen, and uterus. On antigenic stimulation, NK cells rapidly "home" to the lymph nodes and lymphatics.

NK cells play important roles in innate immune responses and immune regulation. They communicate with other cells through a complex of both activation and inhibitory signals through cell surface receptors.

There are many recognizable NK cell subsets [40–50] found in peripheral blood [54]. NK cells were first defined by a lack of B- and T-cell surface markers but the NK cells are now identified into subsets based on the expression of certain phenotypic surface markers. Most NK cells express a neural cell adhesion marker called CD56. Staining with a monoclonal antibody to CD56 permits division into two major subsets, CD56[bright] and CD56[dim]. CD56[bright] NK cells are characterized by having an expression of many CD56 surface molecules [55]. These cells have lower levels of some of the cytotoxic molecules such as perforin and express high levels of cytokines. CD56[bright] NK cells are thought to be important in inflammatory responses and probably play a role in immune regulation. NK cells are distinct from NKT cells that express CD3, and rearrange their germline DNA T-cell receptor (TCR) genes [55].

In contrast, CD56^{dim} NK cells are the most effective killer lymphocytes. The CD56^{dim} NK cells make up about 90–95% of NK cells in peripheral blood and they also express CD16 (Fcγ receptor) on their surface. In contrast to CD56^{bright} NK cells, these NK cells express large amounts of perforin that mediates cytoxicity. Perforin-dependent cytoxicity is the major mechanism of NK cell lysis of target cells [55].

NK cells are programmed to kill target cells and they are inherently capable of killing autologous cells. They are actively inhibited from killing "self" cells by inhibitory receptors and signals. The MHC defines "self" and it has been suggested that "self-MHC" surface receptors engage the inhibitory receptors on the NK cell and prevent lysis of "self" cells [56]. Viral infection of a cell causes a change in the MHC class I expression, which in turn removes the normal inhibitory signal and NK cell activation occurs, resulting in cytotoxicity and death of the viral-infected cells.

There are two families of inhibitory receptors affecting NK cells [57]. The best-described inhibitory signals are those transduced by HLA-specific receptors and are members of the inhibitory killer immunoglobulin-like receptor family (KIR). There is another family of inhibitors that are lectinlike receptors identified as NKG2A/3. It is believed that the net sum of activation and inhibition signals tightly regulates the function of NK cells [55].

NK cells recognize and lyse pathogen-infected cells and malignant cells [58]. They also play an important immunoregulatory role. There are several mechanisms used by NK cells to remove cells. NK cells are effective killers by releasing large number of cytolytic granules at the site of interaction with the target. A major component of the NK cell lysosomal granules is perforin, which, as mentioned earlier, is the major cytolytic substance [58].

The cytokine and chemokine secretions of NK cells are involved in the death of target cells [52]. NK cells produce IFN-γ, TNF-α, GM-CSF, IL-5, and IL-13 among other active substances. Some antiviral activity can be attributed to cytokine production by NK cells.

NK cells are known to express costimulatory molecules for T- and B-lymphocytes and activate the adaptive immune system [59]. TLRs are also expressed on NK cells and these receptors participate in the early detection of an impending infection. NK cells are activated by cytokines produced by virally infected APCs, which may result in cell lysis. At the same time, NK cells may interact with DCs to participate in generating an adaptive immune response. NK cells and DCs interact to induce DCs maturation through cytokines produced by NK cells [52].

NK cell biology is very complex and appears to be directly or indirectly involved in establishing and maintaining immunity. By expression of cell surface receptors, NK cells may go through several stages of maturation and the by-products of maturation affect most components of immunity. The balance between activation and inhibition is closely regulated in the environment where NK cells and other types of immune cells exist.

Patients have been described with various NK cell abnormalities. The most prevalent observation is unusual susceptibility to certain types of viral infections primarily (herpes) [60,61]. In any case, NK cell deficiency must be rigorously documented. Most NK cell enumeration studies are performed using peripheral blood, which may not give an accurate reflection of the numbers of functional NK cells available. Functional assays are difficult to perform and quantify but it is possible to determine cytolytic function and cytokine secretory properties of peripheral blood NK cells. Patients have been reported with normal numbers of peripheral blood NK cells but with a deficiency of perforin, which would impact cytoxicity.

1.2.7.1 Natural Killer T Cells

Natural killer T (NKT) cells are a subset of T lymphocytes that share some properties of NK cells and conventional T cells. Classical NKT (or type 1) cells are CD1d-restricted T cells that express a semi-invariant TCR $V\alpha24$-$J\alpha18$, which distinguishes them from CD1d-dependent T cells that do not express this semi-invariant TCR (type 2 NKT cells) [62]. Most NKT cells express both an invariant TCR and the NK receptor NK1.1 type 1; CD1d-restricted NKT cells are found primarily in the liver, thymus, spleen, and bone marrow [63]. Cells with type 1 TCR can be activated by a synthetic ligand α-galactosylceramide (α-GalCer) presented on CD1d [64]. Type 2 or nonclassical NKT cells fail to be activated by α-GalCer [65]. An endogenous ligand for type 1 NKT cells was identified as a lysosomal glycosphingolipid [66].

There are distinct subsets of CD1d-restricted T cells. NK cell associated markers were expressed primarily within the $CD4^-$ $CD8^-$ $V\alpha$ 24 NKT subset. However, both $CD4^+$- and $CD4^-$ $CD8^-$ $V\alpha$ 24 NKT cells were capable of Th1 cytokine production. This includes IFNγ and TNF-α. The Th2 cytokines, such as IL-4 and IL-13, were secreted by the $CD4^+$ subset [67]. Both the $CD4^+$ and the $CD4^-$ $CD8^-$ $V\alpha$ 24 NKT exist wherever other T lymphocytes are found.

The fact that NKT cells recognize glycolipid antigens in association with CD1d (a nonclassical antigen-presenting molecule) sets the NKT cells apart from conventional T cells. Type 1 NKT cells recognize both foreign and self-glycolipids. Recently, it has been shown that type 1 NKT cells recognize various types of glycolipids and related compounds found in a number of parasites and bacteria but the specific ligands have not been identified. This suggests that the classical NK cells focus activity on viruses and viral infection while NKT cells, however, are primarily involved in detection of parasite and bacterial pathogens [68].

On activation, NK cells respond within a few hours with vigorous production of cytokines [69]. NKT cells release Th1-type cytokines including IFNγ and TNF-α as well as the Th2-type cytokines IL-4 and IL-13 [70]. Individual NKT cells are able to produce both Th1- and Th2-type cytokines at the same time following stimulation *in vivo* [71]. This is unusual because it is possible that Th1 and Th2 antagonize each other. The implications of this observation and mechanisms are unknown.

NKT cells appear to be involved in immediate immune responses, tumor rejection, control of autoimmune disease, and immune surveillance [63]. NKT cells may act as effector cells as seen with their cytotoxic activity; they may also act as regulators. The important question concerns how they determine which way to go. The NKT cells may produce either pro- or anti-inflammatory cytokines. This depends on the type of signal they receive. This is probably related to cytokine profiles produced. The cytokine profile is dependent on the type of TCR stimulation the NKT cells receive.

The study of NKT cells is an important topic of investigation today and probably will be for some time. It is clear that NKT cells are involved in a number of pathological conditions and they appear to regulate a number of others. NKT cells appear to have both protective and harmful roles in disease progression of certain allergic and autoimmune disorders and they may modulate viral infections and have a role in tumor growth and progression.

1.2.8 GAMMA/DELTA T LYMPHOCYTES

Gamma/delta T lymphocytes ($\gamma\delta$ T cells) are a relatively recent discovery within the T-cell population. It is difficult to categorize them but it is becoming clear they are an important component of host defense and may represent a different parallel immune system component [72]. It is likely that $\gamma\delta$ T-cell functions fall somewhere "in-between" the innate and adaptive immune systems. A major characteristic of the $\gamma\delta$ T-cell population is that they have a TCR consisting of a $\gamma\delta$ heterodimer rather than the more prevalent $\alpha\beta$ TCR [73].

$\gamma\delta$ T cells make up a small percentage of T cells (1–5%) in peripheral blood and other lymphoid organs. However, they are found in higher concentrations in the skin, gastrointestinal tract, and the genitourinary system [73]. These locations may be related to the types of antigens they encounter and the immunological responses they deliver. $\gamma\delta$ T cells may be important in preventing infection with organisms such as *Mycoplasma penetrans*, an organism capable of causing urethritis and respiratory diseases in immunocompromised individuals [74]. In addition, *Mycobacterium tuberculosis* has been shown to elicit a $\gamma\delta$ T-cell response [75].

$\gamma\delta$ T cells may also recognize and express cytoxic activity against certain types of tumors, including both hematopoietic and solid tumors [73].

It has been suggested that $\gamma\delta$ T cells recognize ligands, which are different from the short peptides that are detected by $\alpha\beta$ T cells in the context of MHC class I or II molecules [76]. At least one subclass of $\gamma\delta$ T cells recognizes lipidlike antigens from pathogens. Another functional subtype recognizes stress-inducible MHC-related molecules and a number of other ligands [76]. $\gamma\delta$ T cells are active producers of cytokines, which are cytotoxic for many tumor cells. Activated $\gamma\delta$ T cells, through cytokine production, may modulate conventional immune responses by acting on macrophages and DCs.

It has been shown that $\gamma\delta$ T cells function as both activators and inhibitors of immune reactions through surface-bound receptors [73]. Most $\gamma\delta$ T cells possess the NKG2D receptor, which may provide a costimulatory signal that is essential for the $\gamma\delta$ T-cell response against certain types of tumor cells.

Through their interactions, both directly and indirectly, $\gamma\delta$ T cells appear to supplement the cellular immune response by recognizing and responding to antigens that may not be detected by the more prevalent $\alpha\beta$ T cells. These $\gamma\delta$ T cells appear to have an influence on both innate and adaptive responses through patternlike recognition systems, cytoxicity against tumor cells, and providing protection against intracellular pathogens.

1.2.9 DENDRITIC CELLS

DCs have been known since 1973; however, the importance of DCs in both innate and adaptive immunity has been defined more clearly in the past few years. The DCs develop in the bone marrow and are found in the circulating blood and tissues such as the spleen, lungs, gut mucosa, and other places where they may play a role in immunosurveillance. The DCs develop in the bone marrow from hematopoietic pluripotential stem cells [77]. Precursor DCs are constantly generated in the bone marrow and are released into the peripheral blood. After leaving the bone marrow, the precursor DCs "home" to a number of different tissues where they reside as sentinels waiting to interact with antigen. The precursor DCs express low-density MHC class II antigens and after encountering a proper stimulus differentiate into highly endocytic and phagocytic iDCs [78].

DCs probably make up a heterogeneous population of cells. However, precursor DCs, iDCs, and mature DCs play different roles in the immune system. Precursor DCs circulate in the environment and on contacting a pathogen produce cytokines, that is, γ-interferon, and undergo maturation to iDCs. The iDCs acquire new properties, that is, markedly increased phagocytic and endocytic capabilities that lead to binding antigen by the iDCs and then maturation to mature DCs [78]. The mature DCs have specialized properties (receptors) to bind foreign or effete cells through lectin and Fcγ receptors. Once the antigens enter the mature DCs, they are processed for antigen presentation and the mature DCs become an APC and lose their phagocytic properties. The mature DCs are the only cells capable of activation of naïve T cells and are defined by this characteristic [79]. A variety of different stimuli can initiate DCs maturation including pathogens, damaged tissue-derived antigens, and ultraviolet light. The mature DCs, carrying the antigen, then migrates to the secondary lymphoid tissues [78].

1.2.9.1 Immature Dendritic Cells

iDCs are activated upon exposure to the so called danger signals, including PAMPs and become actively phagocytic for the infectious agent. As a part of their maturation to mature DCs, the iDCs become less phagocytic but more mobile once the DC has taken up the antigen and begun processing it. Activation and maturation of DCs occur through the NF-κB signaling pathway [80]. The iDC neither provides T-cell stimulation nor cosignaling.

1.2.9.2 Mature Dendritic Cells

After maturation from iDC, the mature DC migrates to the secondary lymphoid tissues where it begins processing the antigenic material. This migration occurs because of the expression of specific chemokine receptors, especially CCR-7. Mature DCs have highly developed antigen-processing cell capabilities and are capable of loading endocytosed antigenic peptides on both MHC class I and II molecules thereby permitting presentation to both CD8$^+$ and CD4$^+$ T lymphocytes [81]. Mature DCs have high-density costimulatory molecules for presenting processed antigen to T cells [79]. Maturation of DCs is associated with the up-regulation of the costimulatory molecules. Costimulatory molecules CD40, CD80, and CD86 enhance the stability of DCs interactions with naïve antigen-specific CD4$^+$ and CD8$^+$ T lymphocytes and the secretion of cytokines such as INFα, IL6, IL-10, and IL-12. The mature DCs are potent activators of T-cell responses. The DCs transmit information about the danger signals to the T cell and help define the T-cell responses [82].

There are two subsets of DCs in the blood based on the expression of CD11c (β_2 integrin). The subsets named, myeloid DCs (M-DCs) and lymphoid DCs or plasmacytoid DCs (P-DCs), differ in morphology and expression of markers and function [83]. They do, however, share some surface markers for adhesion, activation, costimulation, and coinhibition.

M-DCs express myeloid surface markers, that is, CD13, CD33, and CD11c. The M-DCs also express large numbers of mannose receptors and rapidly take up polysaccharide antigens [84]. The M-DCs readily capture antigen in peripheral tissues by phagocytosis and migrate as immature M-DCs to lymph nodes where stimulation with CD40L (CD40 ligand) induces maturation to mature M-DCs. The M-DCs are potent inducers of both Th1 and Th2 cytokines in naïve CD4$^+$ T lymphocytes. The local microenvironments bias the development of either Th1 or Th2 types of reactions [84].

The P-DCs have morphology similar to plasma cells and are derived from a lymphoid lineage. Instead of myeloid markers, the P-DCs express high levels of CD123 and MHC molecules. CD40L (ligand) causes stimulation of P-DCs and results in DCs maturation. These DCs support Th2 cytokines (primarily). The P-DCs populate T-cell areas of lymph nodes and may have a special ability to recognize self-antigens or viruses [85].

The cell surface receptor CD40 plays an important role in humoral and cell-mediated immune responses. CD40 is expressed on many cell types including B cells, epithelial and endothelial cells, and all APC. The ligand for CD40 is CD40L and is a trimeric TNF-α-like molecule. CD40L is expressed primarily on T helper cells [86].

Ligation of CD40 on the surface of the DCs induces maturation that is detected by markedly enhanced T-cell stimulatory capacity. This maturation of the DCs causes it to activate naïve CD8$^+$ cytotoxic T-cells (CTL) that are important in developing immunity against certain pathogens and tumors [86].

The types of T-cell subsets induced by DCs is dependent on several factors including the DCs subset involved, the nature and dose of the antigen, and the types of cytokines present in the microenvironment where the interaction between DCs and pathogen occurs. DCs play an important role in how a host responds to foreign antigens, effete, or other cells that have initiated its maturation. The DCs help direct the T cell to respond—how it should respond, and where to go to respond. The DC provides a number of sequential signals to the responding T cells. The first signal consists of the interaction of the TCR with the specific antigen in the context of the MHC protein on the surface of the DC. This determines the antigenic specificity of the response and triggers the differentiation of naïve CD4$^+$ and CD8$^+$ T cells into T helper and CTL, respectively. This antigen-specific T-cell activation requires the engagement of the TCR/CD3 complex. The antigenic peptide is presented by the MHC and the engagement of appropriate costimulatory reception by costimulatory ligands on the DCs [87].

The DCs also provide the costimulatory signaling that T cells require to respond to antigen. The cosignaling can be either positive or negative. Cosignaling may be provided by a number of different molecules including CD80, CD86, and CD28. The activation and maturation signals may be diverse, but all involve activation of the NF-κB signaling pathway [88]. In the presence of a negative cosignal (coinhibitory) or in the absence of a positive costimulatory signal, T cells will fail to respond and may not be capable of reacting to that specific antigen in the future. Some pathogens have evolved mechanisms to help evade this component of the innate immune response. These pathogens possess compounds capable of arresting DCs in their immature state where they cannot produce costimulating molecules. Without costimulatory activity, the DCs are unable to completely stimulate T cells to respond [89] and the pathogens to avoid an immune response directed against them.

DCs also direct the functional polarization of CD4$^+$ T cells into Th1, Th2, and Treg cells [90]. The nature of the "danger signal" or PAMP defines the DCs response as Th1-, Th2-, or Treg-type responses, which result in the DCs producing certain cytokines to induce T-cell differentiation into Th1, Th2, or Treg CD4$^+$ T lymphocytes [90]. It is unclear whether or not the DCs are restricted to one polarization type or if they are flexible depending on the nature of the stimulus and this microenvironment.

The nature of the binding of the stimulating agent (type of PAMP) to certain types of TLRs induces the DC maturation, which results in the development of Th1 T cells [91]. For example, binding of microbial double strand RNA (ds RNA) to TLR3 triggers the formation of the Th1 T cells. Other TLRs associated binding signals produce either Th1 or Th2 types of responses depending on the TLR type and the danger signal. It appears that the nature of the PAMP is important in defining the T-cell response. In general, Th1-type responses are directed to cell-mediated types of responses and the nature of the cytokines produced drives these responses.

A Th2 T-cell response may result from antigens of a parasites-inducing type 2 DCs, which go on to produce a Th2 type of response. Th2 responses are usually antibody or humoral immune responses [89,92].

Treg-type T cells are CD4$^+$ T cells that can suppress responses of other T cells and probably play an important role in regulating self-tolerance. Multiple subtypes of Treg cells have been identified and each has its own specific phenotype, cytokine profile, and mechanism of activation for suppressing immune responses. The most frequently described phenotype includes CD25 as a surface marker [93,94]. The Treg cells have been shown to interfere with tumor immunity and a number of parasites are known to induce regulatory DCs that induce Treg responses. For example, a hemagglutin from *Bordetella pertusis* serves as a ligand for TLR2 and this induces the development of regulatory DCs and ultimately Treg T cells.

It is important that immune responses are initiated against foreign substances but immune responses should not be directed at self-antigens. DCs are responsible for the induction of peripheral and central tolerance. A major role of the DCs is to recognize "danger signals" and activate the appropriate response by passing information to T cells through cytokines and chemokines to direct the T-cell response. As expected, cytokines production by DCs is tightly regulated. The DCs possess PRRs that can detect concerned motifs on invading pathogens and distinguish them from self-antigens. There is constant communication between DCs and T cells with information being shared in both directions. The exact mechanisms involved in how DCs and T cells combine to distinguish self from non-self are not understood completely. However, iDCs are involved in the maintenance of tolerance to self-antigens by constantly defining self in the periphery. The iDCs constantly sample different self-antigens and present them to T cells under noninflammatory conditions permitting the detection of autoreactive T cells. The iDCs may also induce tolerance to self by stimulating naïve CD4$^+$ and CD8$^+$ T cells to differentiate into Treg cells that produce IL-10, which in turn causes these cells to inhibit Th1 T-cell differentiation and suppress CD8$^+$ memory cell responses. Figure 1.2 summarizes the pivotal role of DCs in immunity.

1.2.10 OVERVIEW OF ADAPTIVE IMMUNITY

In contrast to innate immunity, adaptive immunity is flexible, specific, and has immunological memory, that is, it can respond more rapidly and vigorously on a second exposure to an antigen. Immunologic memory provides a more powerful response to a repeated exposure to the same foreign substance or antigen. Adaptive immunity is more complex because it provides the ability to respond very specifically. Innate and adaptive immunity responses interact effectively to enhance the body's defense mechanisms against foreign or damaged host cells. Inherent in both innate and adaptive immune responses are the mechanisms to distinguish self from nonself.

The primary blood cell elements of the adaptive immune system are T lymphocytes and B lymphocytes. These T- and B-cells provide the unique specificity for their target antigens by virtue of the antigen-specific receptors expressed on their surfaces. The B- and T-lymphocyte antigen-specific receptors develop by somatic rearrangement of germline gene elements to form the TCR genes and the immunoglobulin receptor genes. This recombination mechanism provides unique antigen receptors capable of recognizing almost any antigen encountered, and provides the specific immunological memory for a rapid, vigorous, and specific response to a later exposure to the same antigen. It is estimated that millions of different antigen receptors may be formed from a collection of a few hundred germline-encoded gene elements.

For many years, innate and adaptive immune responses were studied as separate systems because of their different mechanisms of action. However, it is now understood that synergy between the two systems is required to provide adequate immune reactivity against invading pathogens. Innate immune responses, through their barrier and relatively broad types of actions, represent the first line of defense against pathogens. At the time the innate system is getting activated, the adaptive system becomes activated also. The adaptive response becomes evident a few days later because it requires time for sufficient antigen-specific receptors to be generated through clonal expansion/proliferation. There are multiple interactions occurring between the two systems, which results in the coamplification of each respective response and leads to the ultimate destruction and elimination of the invading pathogen.

1.2.11 B LYMPHOCYTES

The primary function of B lymphocytes is the production of antibodies that are specific for a given antigenic component of an invading pathogen. Antibodies are encoded by the heavy (H)- and light (L)-chain immunoglobulin genes. Antibodies may be secreted or cell surface–bound on B lymphocytes. There are five classes of immunoglobulins: IgM, IgG, IgA, IgD, and IgE; and the classification is based on the isotypes of the H chain. B lymphocytes represent roughly 10–15% of the

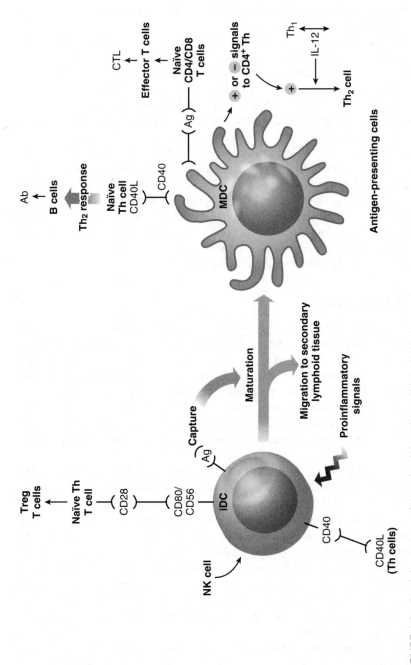

FIGURE 1.2 Partial summary of the role of dendritic cell in cellular immunity.

peripheral blood lymphocyte population and free immunoglobulins make up a considerable proportion of serum proteins. After an encounter with a specific pathogen and an antibody response is generated, the level of specific antibodies to that antigen decreases in serum over a relatively short period of time. However, immunological memory persists in the B-cell population, which is capable of rapid clonal expansion upon reexposure to that same antigen [95].

Protective immunity for a specific antigen requires cooperation between B and T lymphocytes. In many instances, helper T-cell interaction is required for the development of high-affinity antibodies and complete protection. For naïve B lymphocytes to undergo proliferation and differentiation in response to most antigens, they require stimulation by a CD4+ T helper cell with the same antigenic specificity. B lymphocytes are capable of antigen presentation to T cells through their surface MHC class II proteins [96].

Antibodies recognize the tertiary structure of proteins and react with specific epitopes in the antigen structure [97]. During an immune response, immunoglobulin formation may undergo switching from one isotype of immunoglobulin to another, usually from IgM to IgG and IgA or IgE. This switching requires additional genetic recombination between the same variable region genes and new H-chain isotype genes. As the humoral immune response continues, antibodies with higher affinity for the antigen develops. Competition for antigen provides a selective advantage to B lymphocytes with the highest affinity for antigen.

1.2.12   T LYMPHOCYTES

Whereas B lymphocyte products recognize extracellular pathogens, T lymphocytes are adept at identifying and destroying cells that have been infected by intracellular pathogens. For T cells to recognize antigenic peptides, the peptide must be presented in the context of cell surface MHC class I or class II proteins [98]. In other words, T cells can only recognize molecular complexes consisting of the antigenic peptide and a self-structure, that is, the MHC. Depending on whether the antigenic peptide has been synthesized within the host cell or ingested by the cell and modified by proteolytic digestion, either MHC class I or class II proteins are required [99].

Proteins of the MHC are intimately tied to T-lymphocyte responses and recognition of antigenic peptides. The MHC class I proteins consist of three HLA classes: HLA-A, HLA-B, and HLA-C with hundreds of allelic variants of each. Structural studies have shown that class I molecules exist as cell surface heterodimers with a polytransmembrane α-chain associated (noncovalently) with a nonpolymorphic β_2 microglobulin protein [99]. The protein chains are folded in such a way as to form a physical groove capable of binding up to an 11 amino acid long peptide. Antigenic proteins are degraded by proteolytic enzymes to about this size for binding to the MHC class I proteins for antigenic presentation. Antigenic peptides are bound in the groove of the HLA molecule and expressed to the cell surface for presentation to initiate a T-cell response [99].

The large number of HLA class I alleles reflects structural polymorphism, which adds to the number of antigenic peptides that can be recognized. Most humans are heterozygous for HLA, further increasing the numbers of antigenic polypeptides bound and capable of activating T cells [100].

There are three major groups of MHC class II proteins, HLA-DR, HLA-DQ, and HLA-DP, each with large numbers of alleles. Class II proteins are also folded in such a way as to form a groove for binding peptides for presentation as a complex consisting of the protein fragment—HLA structure. Class II proteins consist of two polypeptide chains, which are MHC-encoded transmembrane proteins [101].

As mentioned earlier, the antigenic peptides bound to class II proteins are derived from exogenous antigens taken up by APCs and are degraded into peptides of a size appropriate for loading in the class II protein groove. The phagocytosed antigens were degraded by proteolytic enzymes found in the lysosomes of the APC and further processed by specialized proteins in the cytoplasm to generate peptides of the correct length for binding [101].

As with the MHC class I proteins, the large repertoire of the peptide-binding class II molecule is generated through polymorphisms of the class II proteins where there are multiple alleles of each HLA type.

Once again, class I and class II HLA proteins differ in the types of peptides recognized and where and how the antigenic peptides are generated.

The thymus is an organ dedicated to T-cell development. T cells originate in the bone marrow as progenitors and circulate in the blood stream before eventually homing to the thymus. The particular environment in the thymus provides the optimal milieu for thymic lineage development [102]. It is likely that interaction with the thymic stroma provides the signals required for the thymic progenitor cells to undergo proliferation and differentiation into mature naïve T cells [102]. This process is tightly regulated and is mediated by various transcription factors, cytokines, chemokines, and one or more selectins.

T lymphocytes make up the majority of lymphocytes in peripheral blood and are most readily defined by the expression of TCR molecules on their surface. The $\alpha\beta$ TCR molecules recognize specific peptide antigens that are presented in the context of the MHC class I or class II proteins (in the form of a complex). T cells expressing $\alpha\beta$ molecules differentiate into different subsets including the CD4$^+$ T cells and the CD8$^+$ T cells. The CD4$^+$ T-cell subsets are active as regulators of cellular and humoral immune responses. The other subset, CD8$^+$ T cells, is cytotoxic for cells infected with intracellular pathogens.

T-lymphocyte receptors are different from the B-lymphocyte immunoglobulin receptors in that they are never secreted and are able to recognize peptides produced after proteolytic breakdown of antigens as opposed to B-cell receptors that react to native proteins [103]. TCRs recognize the primary amino acid structure of the protein antigen. TCRs recognize antigenic peptides only when they are presented as cell surface–bound complexes with MHC class I or II proteins. Each T cell bears a TCR with a single antigenic specificity and these cells provide recognition for a very large number of possible pathogens. T-lymphocyte activation through the TCR is an important step in the initiation of an adaptive immune response. TCR activation requires two interactions [99]. One is the interaction between the TCR and the peptide MHC complex, which gives a partial signal for T-cell activation. Full activation requires both the TCR-peptide-MHC binding and the interaction with a costimulatory molecule,

CD28, on the T cell and CD80 or CD86 present on the APC [99]. This second signal stimulates proliferation and differentiation of the T cell. If T cells do not receive the second signal they may become unresponsive and possibly anergic.

There is considerable cross talk between T- and B-cells in an adaptive immune response. For proliferation and differentiation to occur, naïve B cells must be stimulated by a particular type of helper T cell, the CD4$^+$ T cell. Moreover, the B and the CD4$^+$ helper T cells must have receptors that are specific for epitopes present on the same antigenic molecule. The complex interaction between T helper cells and B cells (with matching specificities for a pathogen-specific antigen) and a second signal to the T cell (possibly provided by the B cell) results in the proliferation and differentiation of both B- and T-cells and ultimately in an adaptive immune response.

T cells are divided into a number of subsets based on their migration patterns and functional abilities. The naïve T lymphocytes tend to circulate between the blood and lymph nodes in response to the homing receptor L-selectin [104]. Recirculation of naïve T cells allow them to move in and out of areas of the body where they have the best chance of detecting a pathogen bearing their specific antigen receptor [104]. At least two types of memory T cells are active in adaptive immunity. One subtype, the effector memory cell, is short-lived and aggressively migrates to the site of the target tissue to destroy the pathogen. The second subset, called central memory cells, serves a role in immunologic memory and has migratory patterns that place them in environments where pathogens may enter [104]. Antigenic stimulation of these memory cells leads to a rapid proliferative response leading to the production of both effector and central memory T cells. There appears to be a tendency for memory T cells to migrate back to the tissue where they first encountered a specific antigen. This appears to be controlled by specific homing receptors on the surface of memory T cells.

As mentioned earlier, $\alpha\beta$ T cells, while in the thymus, differentiate into T-cell subpopulations expressing CD4 or CD8 cell surface markers. These cell populations were described first as phenotypic markers and their functions were described later. Originally, these T cells were considered to be helper cells (CD4) and suppressor cells (CD8) but later the CD8 cells were identified as cytotoxic T cells [105]. These associations are not absolute, since some CD4$^+$ T cells may also have cytotoxic properties.

CD4$^+$ molecules and CD8$^+$ molecules serve as coreceptors in the interaction between the T cells and APC. The CD4 molecules expressed on the surface of T cells bind to class II molecules expressed on the surface of APCs and serve to stabilize the interaction between that particular T cell and the APC. CD8 molecules, however, bind to class I molecules expressed on the surface of APCs and serve to stabilize the interaction between CD8 T cells and APCs.

When naïve CD4$^+$ or CD8$^+$ T cells are activated by APC, they undergo differentiation into distinct subsets with different functions. The effector functions of the T-cell subsets are determined by the nature of the costimulatory signals given and the cytokines secreted [106].

There are two main subsets of CD4$^+$ T cells generated or determined by the cytokines secreted by the APC, Th1, and Th2.

The Th1 subsets are generated by APCs secreting IL-2. The T cells differentiate into effector cells producing high levels of IFN-γ and IL-2. This subset, Th1, generally supports cell-mediated immunity by producing cytokines INF-γ and TNF-β

that are efficient in activating macrophages and cytotoxic T cells [107]. In contrast, naïve $CD4^+$ T cells activated by IL-4 stimulated DCs differentiate into effector T cells designated Th2. The Th2 subset actively supports the development of humoral or antibody responses [107]. Th2 lymphocytes secrete IL-4, IL-5, IL-9, and IL-13, which in turn are efficient at stimulating B lymphocytes to differentiate into antibody-forming cells (particularly IgE) and actively secrete antibody.

Generally, most immune responses show a combination of both features of Th1 and Th2 pathways and a prolonged immunization process may lead to one pathway becoming the most dominant. To a certain extent, Th1 and Th2 subsets secrete cytokines that can suppress one another, that is, Th1 secrete cytokines that can suppress Th2 responses and Th2 cells produce cytokines that can suppress Th1 responses.

Another family of $CD4^+$ T cells called Treg was recently shown to suppress the responses of other T cells [93]. It is likely the Treg cells play a role in regulating self-tolerance but may have a harmful effect on tumor immunity. Several subsets of Treg cells have been reported and each has a distinct surface phenotype, cytokine profile, and mechanism of action for suppressing immune responses. Initially, high CD25 expression was used to identify the majority of human Treg cells. The $CD4^+$ cells with the greatest regulatory activity had high levels of CD25 (i.e., $CD4^+$ CD25 high T cells). A considerable number of other surface markers have been shown on Treg cells but expression is not as great or as consistent as CD25 [108]. More recently, the intracellular protein FOXP3 has been identified as a key molecule involved in driving the activity of Treg and now serves as a marker to enumerate these cells.

1.2.13 Comments

This chapter serves as a background and provides a broad overview for the remainder of the chapters.

REFERENCES

1. Blach-Olszweska, Z., Innate immunity: cells, receptors and signaling pathways, *Arch. Immunol. Ther. Exp.*, 53, 245, 2005.
2. Janeway, C.A. Jr., Medzhitov, R., Innate immune recognition, *Ann. Rev. Immunol.*, 20, 197, 2002.
3. Banchereau, J., Steinman, R.M., Dendritic cells and the control of immunity, *Nature*, 392, 245, 1998.
4. Ganz, T., Defensins: antimicrobial peptides of innate immunity, *Nat. Rev. Immunol.*, 3, 710, 2003.
5. Yang, D., Chertov, O., Bykovskia, S.N., Chen, Q., Buffo, M.J., Shogan, J., Anderson, M., Schroeder, J.M., Wang, J.M., Howard, O.M., Oppenheim, J.J., β-defensins: linking innate and adaptive immunity through dendritic and T cell CCR6, *Science*, 286, 525, 1999.
6. De Smet, K., Contreras, R., Human antimicrobial peptides: defensins, cathelicidins and histantins, *Biotechnol. Lett.*, 18, 1337, 2005.
7. Pazgiera, M., Hoover, D.M., Yang, D., Lu, W., Lubkowski, J., Human β-defensins, *Cell. Mol. Life Sci.*, 63, 1294, 2006.

8. Boniotto, M., Jordan, W.J., Eskdale, J., Tossi, A., Antcheva, N., Crovella, S., Connell, N.D., Human β-defensin 2 induces a vigorous cytokine response in peripheral blood mononuclear cells, *Antimicrob. Agents. Chemother.*, 50, 1433, 2006.

9. Tjabringa, C.S., Ninaber, D.K., Drijfhout, J.W., Rabe, K.F., Hiemstra, P.S., Human cathelicidin LL-37 is a chemoattractant for eosinophils and neutrophils that acts via formyl-peptide receptors, *Int. Arch. Allergy Immunol.*, 140, 103, 2006.

10. Castagnola, M., Inzitari, R., Rossetti, D.V., Olmi, C., Cabras, T., Piras, V., Nicolussi, P., Sanna, M.T., Pellegrini, M., Giardina, B., Messana, I., A cascade of 24 histantins (histatin 3 fragments) in human saliva: suggestions for a pre-secretory sequential cleavage pathway, *J. Biol. Chem.*, 279, 41436, 2004.

11. Schittek, B., Hipfel, R., Sauer, B., Bauer, J., Stevanovic, S., Schrile, M., Schroeder, K., Blin, N., Meier, F., Rassner, G., Garbe, C., Dermicidin: a novel human antibiotic peptide secreted by sweat glands, *Nat. Immunol.*, 2, 1133, 2001.

12. Madsen, P., Rasmussen, H.H., Leffers, H., Honore, B., Dejgaard, K., Olsen, E., Kill, J., Walbum, E., Andersen, A.H., Basse, B., Molecular cloning, occurrence and expression of a novel partially secreted protein "psoriasin" that is highly up-regulated in psoriatic skin, *J. Invest. Dermatol.*, 97, 701, 1991.

13. Rus, H., Cudrici, C., Niculescu, F., The role of the complement system in innate immunity, *Immunol. Res.*, 33, 103, 2005.

14. Fujita, T., Endo, Y., Nonaka, M., Primitive complement system—recognition and activation, *Mol. Immunol.*, 41, 103, 2004.

15. Petersen, S.V., Thiel, S., Jensen, L., Vorup-Jensen, T., Koch, C., Jensenius, J.C., Control of the classical and the MBL pathway of complement activation, *Mol. Immunol.*, 37, 803, 2000.

16. Xu, Y., Narayana, S.V., Volanakis, J.E., Structural biology of the alternative pathway convertase, *Immunol. Rev.*, 180, 123, 2001.

17. Frank, M.M., Complement system, in *Samter's Immunologic Diesease*, Frank, M.M., Claman, H.N., Unanue, E.R., Eds., Boston, MA, Little, Brown & Company, 1995, 331p.

18. Baumgarth, N., Tung, J.W., Herzenberg, L.A., Inherent specificities in natural antibodies: a key to immune defense against pathogen invasion, *Springer Semin. Immunopathol.*, 26, 347, 2005.

19. Brändlein, S., Vollmers, H.P., Natural IgM antibodies, the ignored weapons in tumor immunity, *Histol. Histopathol.*, 19, 897, 2004.

20. Medzhitov, R., Janeway, C.A. Jr., Innate immunity: the virtues of a non-clonal system of recognition, *Cell*, 91, 295, 1997.

21. Sagaert, X., DeWolf-Peeters, C., Classification of B-cells according to their differentiation status, their micro-anatomical localization and their developmental lineage, *Immunol. Lett.*, 90, 179, 2003.

22. Christofaro, P., Opal, S.M., Role of toll-like receptors in infection and immunity: clinical implications, *Drugs*, 66, 15, 2006.

23. Kaisho, T., Akira, S., Toll-like receptor function and signaling, *J. Allergy Clin. Immunol.*, 117, 979, 2006.

24. Beutler, B., Hoebe, K., Du, X., Ulevitch, R.J., How we detect microbes and respond to them: the toll-like receptors and their transducers, *J. Leukocyte Biol.*, 74, 479, 2003.

25. Means, T.K., Golenbock, D.T., Fenton, M.J., The biology of Toll-like receptors, *Cytokine Growth Factor Rev.*, 11, 219, 2000.

26. Akira, S., Takeda, K., Toll-Like receptor signaling, *Nat. Rev. Immunol.*, 4, 499, 2004.

27. Freeman, G.J., Gray, G.S., Gimmi, C.D., Lombard, D.B., Zhou, L.J., White, M., Fingeroth, J.D., Gribben, J.G., Nadler, L.M., Structure, expression and T cell costimulatory activity of the murine homologue of the human B lymphocyte activation antigen B7, *J. Exp. Med.*, 174, 625, 1991.

28. Baume, C.M., Weissman, I.L., Tsukamoto, A.S., Buckle, A.M., Peault, B., Isolation of a candidate human hematopoietic stem-cell population, *Proc. Natl. Acad. Sci. USA*, 89, 2804, 1992.
29. Athens, J., Haab, O.P., Raab, S.O., Mauer, A.M., Ashenbrucker, H., Cartwright, G.E., Wintrobe, M.M., Leukokinetic studies. IV. The total blood, circulating and marginal granulocyte pools and the granulocyte turnover rate in normal subjects, *J. Clin. Invest.*, 40, 989, 1961.
30. Butcher, E.C., Leukocyte-endothelial cell recognition: three (or more) steps to specificity and activity and diversity, *Cell*, 67, 1033, 1991.
31. Ley, K., Integration of inflammatory signals by rolling neutrophils, *Immunol. Rev.*, 186, 8, 2002.
32. Doerfler, M.E., Danner, R.L., Shelhamer, J.H., Parrillo, J.E., Bacterial lipopolysaccharides prime human neutrophils for enhanced production of leukotriene B4, *J. Clin. Invest.*, 83, 970, 1989.
33. Lawrence, M.B., Springer, T.A., Leukocytes roll on a selectin at physiologic flow rates: distinction from and prerequisite for adhesion through integrins, *Cell*, 65, 859, 1991.
34. Rossi, D., Zlotnik, A., The biology of chemokines and their receptors, *Annu. Rev. Immunol.*, 18, 217, 2000.
35. Joiner, K.A., Brown, E.J., Frank, M.M., Complement and bacteria: chemistry and biology in host defense, *Annu. Rev. Immunol.*, 2, 461, 1984.
36. Unkeless, J.C., Shen, Z., Liu, C.W., DeBeus, E., Function of human Fc gamma RIIA and Fc gamma RIIIB, *Semin. Immunol.*, 7, 37, 1995.
37. Stossel, T.P., Phagocytosis, *Prog. Clin. Biol. Research*, 13, 87, 1977.
38. Klebanoff, S.J., Myeloperoxidase: friend or foe, *J. Leukoc. Biol.*, 77, 598, 2005.
39. Cohn, Z.A., Hirsch, J.G., Degranulation of polymorphonuclear leukocytes following phagocytosis of microorganisms, *J. Exp. Med.*, 112, 1005, 1960.
40. Spitznagel, J.K., Dalldorf, F.G., Leffell, M.S., Folds, J.D., Welsh, I.R., Clooney, M.H. et al., Character of azurophil and specific granules purified from human polymorphonuclear leukocytes, *Lab Invest.*, 30, 774, 1974.
41. Ganz, T., Defensins in the urinary track and other tissues, *Infect. Dis.*, 183, S41, 2001.
42. Klebanoff, S.J., Waltersdorph, A.M., Prooxidant activity of transferrin and lactoferrin, *J. Exp. Med.*, 172, 1293, 1990.
43. Gouwy, M., Struyf, S., Proost, P., van Damme, J., Synergy in cytokine and chemokine networks amplifies the inflammatory response, *Cytokine Growth Factor Rev.*, 16, 561, 2005.
44. Borish, L.G., Steinke, J.W., 2. Cytokines and chemokines, *J. Allergy Clin. Immunol.*, 111, S460, 2003.
45. Sigal, L.H., Basic science for the clinician 33: interleukins of current clinical relevance (part 1), *J. Clin. Rheumatol.*, 10, 353, 2004.
46. Charo, I.F., Ransohoff, R.M., Mechanisms of disease: the many roles of chemokines and chemokine receptors in inflammation, *N. Engl. J. Med.*, 354, 610, 2006.
47. Guimond, M., Fry, T., Mackall, C., Cytokine signals in T-cell homeostatis, *J. Immunother.*, 28, 289, 2005.
48. Romagnani, P., From basic science to clinical practice: use of cytokines and chemokines as therapeutic reagents in renal diseases, *J. Nephrol.*, 18, 229, 2005.
49. Oppenheim, J., Feldman, M., in *Cytokine References*, Oppenheim, J., Feldman, M., Durum, S.K., Hirano, T., Vicek, J., Nicola, N., Eds., San Diego, CA, Academic Press, 2001, pp. 3–20.
50. Bentley, B., Cerami, A., The biology of cachectin/TNFα-primary mediator of the host response, *Rev. Immunol.*, 7, 625, 1989.
51. Steinke, J.W., Borish, L., 3. Cytokines and chemokines, *J. Allergy Clin. Immunol.*, 117, S441, 2006.

52. O'Conner, G.M., Hart, O.M., Gardiner, C.M., Putting the natural killer cell in its place, *Immunology*, 117, 1, 2005.
53. Raulet, D.H., Interplay of natural killer cells and their receptor with the adaptive immune response, *Nat. Immunol.*, 5, 996, 2004.
54. Jonges, L.E., Albertsson, P., vanVlierberghe, R.L., Ensink, N.G., Johansson, B.R., van de Velde, C.J., Fleuren, G.J., Nannmark, U., Kuppen, P.J.K., The phenotypic heterogeneity of human natural killer cells: presence of at least 48 different subsets in the peripheral blood, *Scand. J. Immunol.*, 53, 103, 2001.
55. Orange, J.S., Ballas, Z.K., Natural killer cells in human health and disease, *Clin. Immunol.*, 118, 1, 2006.
56. Cooper, M.A., Fehniger, T.A., Caligiuri, M.A., The biology of human natural-killer-cell subsets, *Trends Immunol.*, 22, 633, 2001.
57. Sigal, L.N., Molecular biology and immunology for clinicians 22: natural killer cell receptors and activation mechanisms, *J. Clin. Rheumatol.*, 9, 55, 2003.
58. Biron, C.A., Nguyen, K.B., Dien, B.C., Counsens, L.P., Salazar-Mather, T.P., Natural killer cells in antiviral defense: function and regulation by innate cytokines, *Ann. Rev. Immunol.*, 17, 189, 1999.
59. Snyder, M.R., Weyand, C.M., Goronzy, J.J., The double life of NK receptors: stimulation or co-stimulation? *Trends Immunol.*, 25, 25, 2004.
60. Orange, J.S., Brodeur, S.R., Jain, A., Bonilla, F.A., Schneider, L.C., Kretschmer, R., Nurko, S., Rassmussen, W.L., Köhler, J.R., Gellis, S.E., Ferguson, B.M., Stominger, J.L., Zonana, J., Ramesh, N., Ballas, Z.K., Geha, R.S., Deficient natural killer cell cytotoxicity in patients with IKK-γ/NEMO mutations, *J. Clin. Invest.*, 109, 1501, 2002.
61. Orange, J.S., Human natural killer cell deficiencies and susceptibility to infection, *Microbes Infect.*, 4, 1545, 2002.
62. Kronenberg, M., Toward understanding of NKT cell biology: progress and paradoxes, *Annu. Rev. Immunol.*, 26, 877, 2005.
63. Jameson, J., Witherden, D., Havran, W., T cell effector mechanisms: $\gamma\delta$ and CD1d-restricted subsets, *Curr. Opin. Immunol.*, 15, 349, 2003.
64. Kawano, T., Cui, J., Koezuka, Y., Toura, I., Kaneko, Y., Motoki, K., Ueno, H., Nakagawa, R., Sato, H., Kondo, E., Koseki, H., Taniguchi, M., CD1d-restricted and TCR-mediated activation of V_a14NKT cells by glycosylceramides, *Science*, 278, 1626, 1997.
65. Solomon, M., Sarvetnick, N., The pathogenesis of diabetes in the NOD mouse, *Adv. Immunol.*, 84, 239, 2004.
66. Zhou, D., Mattner, J., Cantu, C. III., Schrantz, N., Yin, N., Gao, Y., Sagiv, Y., Hudspeth, K., Wu, Y.-P., Yamashita, T., Teneberg, S., Wang, D., Proia, R.L., Levery, S.B., Savage, P.B., Teyton, L., Bendelac, A., Lysosomal glycosphingolipid recognition by NKT cells, *Science*, 306, 1786, 2004.
67. Gumperz, J.E., Miyake, S., Yamamura, T., Brenner, M.B., Functionally distinct subsets of CD1d-restricted natural killer T cells revealed by CD1d tetramer staining, *J. Exp. Med.*, 195, 625, 2002.
68. Brigh, M., Brenner, M.B., CD1d: antigen presentation and T cell function, *Annu. Rev. Immunol.*, 22, 817, 2004.
69. Godfrey, D.I., Hammond, K.J., Poulton, L.D., Smyth, M.J., Baxter, A.G., NKT cells: facts, functions and fallacies, *Immunol. Today*, 21, 573, 2000.
70. Smyth, M.J., Godfrey, D.T., NKT cells and tumor immunity—a double-edged sword, *Nat. Immunol.*, 1, 459, 2000.
71. Matsuda, J.L., Naidenko, O.V., Gapin, L., Tracking the response of natural killer T cells to a glycolipid antigen using CD1d tetramers, *J. Exp. Med.*, 192, 741, 2000.
72. Segal, L.H., Basic science for the clinician 35: CD1, invariant NKT (iNKT) cells, and $\gamma\delta$ T-cells, *J. Clin. Rheumatol.*, 11, 336, 2005.

73. Champagne, E., Martinez, L.O., Vantourant, P., Collet, X., Barbaras, R., Role of apo-lipoproteins in $\gamma\delta$ and NKT cell-mediated innate immunity, *Immunol. Res.*, 3, 241, 2005.
74. Born, W.K., Reardon, C.L., O'Brien, R.L., The function of $\gamma\delta$ T cells in innate immunity, *Curr. Opin. Immunol.*, 18, 31, 2006.
75. Guperz, J.E., Brenner, M.B., CD1-specific T cells in microbial immunity, *Curr. Opin. Immunol.*, 13, 471, 2001.
76. Holtmeir, W., Kabelitz, D., Gamma/delta T cells link innate and adaptive immune responses, *Chem. Immunol. Allergy*, 86, 151, 2005.
77. Schuurhuis, D.H., Fu, N., Ossendorp, F., Melief, C.J.M., Ins and outs of dendritic cells, *Int. Arch. Allergy Immunol.*, 140, 53, 2006.
78. Chain, B.M., Current issues in antigen presentation—focus on the dendritic cell, *Immunol. Lett.*, 89, 237, 2003.
79. Banchereau, J., Steinman, R.M., Dendritic cells and the control of immunity, *Nature*, 392, 245, 1998.
80. Guermonprez, P., Valladeau, J., Zitvogel, L., Thery, C., Amigorena, S., Antigen presentation and T-cell stimulation of dendritic cells, *Annual Rev. Immunol.*, 20, 621, 2002.
81. Adams, S., O'Neill, B., Recent advances in dendritic cell biology, *J. Clin. Immunol.*, 25, 177, 2005.
82. Kapsenberg, M.L., Dendritic-cell control of pathogen-driven T-cell polarization, *Nat. Rev. Immunol.*, 3, 984, 2003.
83. Dallal, R.M., Lotze, M.T., The dendritic cell and human cancer vaccines, *Curr. Opin. Immunol.*, 12, 583, 2000.
84. Grouard, G., Rissoan, M.C., Filgueira, L., Durand, I., Banchereau, J., Liu, Y.J., The enigmatic plasmacytoid T-cells develop into dendritic cells with interleukin (IL)-3 and CD40 ligand, *J. Exp. Med.*, 185, 1101, 1997.
85. Banchereau, J., Pulendran, B., Steinman, R., Palucka, K., Will the making of plasmacytoid dendritic cells *in vitro* help unravel their mysteries? *J. Exp. Med.*, 192, F39, 2000.
86. VanKooten, C., Banchereau, J., Functions of CD40 on B cells, dendritic cells and other cells, *Curr. Opin. Immunol.*, 9, 330, 1997.
87. Baxter, A.G., Hodgkin, P.D., Activation rules: the two signal theories of immune activation, *Nat. Rev. Immunol.*, 2, 439, 2002.
88. Bonizzi, G., Karin, M., The two NF-κB activation pathways and their role in innate and adaptive immunity, *Trends Immunol.*, 25, 280, 2004.
89. DeJong, E.C., Vieira, P.C., Kalinski, P., Schuitemaker, J.H.N., Tanaka, Y., Wierenga, E.A., Yazdanbakhsh, M., Kapsenberg, M.L., Microbioal compounds selectively induce Th1 cell-promoting or Th2 cell-promoting dendritic cells *in vitro* with diverse Th cell-polarizing signals, *J. Immunol.*, 168, 1704, 2002.
90. Agrawal, S., Agrawal, A., Doughty, B., Gerwitz, A., Blenis, J., vanDyke, T., Puledndran, B., Cutting edge: different toll-like receptor agonists instruct dendritic cells to induce distinct Th responses via differential modulation of extracellular signal-regulated kinase-mitogen-activated protein kinase and c-fos, *J. Immunol.*, 171, 4984, 2003.
91. Alexopoulou, L., Holt, A.C., Medzhitov, R., Flavell, R.A., Recognition of double-stranded RNA and activation of NF-κB by toll-like receptor 3, *Nature*, 413, 732, 2001.
92. Whelan, M., Harnett, M.M., Houston, K.M., Patel, V., Harnett, W., Rigley, K.P., A filarial nematode-secreted product signals dendritic cells to acquire a phenotype that derives development of Th2 cells, *J. Immunol.*, 164, 6453, 2000.
93. Sigal, L.H., Basic science for the clinician 32: T-cells with regulatory function, *J. Clin. Rheumatol.*, 11, 286, 2005.
94. D'Ambrosio, D., Regulatory T cells: how do they find their space in the immunological arena? *Semin. Cancer Biol.*, 16, 91, 2006.

95. Cooper, M.D., Alder, M.N., The evolution of adaptive immune systems, *Cell*, 124, 815, 2006.
96. Delves, P.J., Roitt, I.M., Advances in immunology: the immune system—first of two parts, *NEJM*, 343, 37, 2000.
97. Garcia, K.C., Teyton, L., Wilson, I.A., Structural basis of T-cell recognition, *Annu. Rev. Immunol.*, 17, 369, 1999.
98. McCullough, K.C., Summerfield, A., Basic concepts of immune response and defense development, *ILAR J.*, 46, 230, 2005.
99. Alam, R., Gorska, M., 3. Lymphocytes, *J. Allergy Clin. Immunol.*, 111, S476, 2003.
100. Groothuis, T., Neefies, J., Ins and outs of intracellular peptides and antigen presentation by MHC class I molecules, *Curr. Top. Microbiol. Immunol.*, 300, 127, 2005.
101. Jones, E.Y., Fugger, L., Strominger, J.L., Siebold, C., MHC class II proteins and diseases: a structural perspective, *Nat. Rev. Immunol.*, 6, 271, 2006.
102. Sebzda, E., Mariathassan, S., Ohteki, T., Jones, R., Bachmann, M.F., Ohashi, P.S., Selection of T cell repertoire, *Annu. Rev. Immunol.*, 17, 829, 1999.
103. Nemazee, D., Receptor selection in B and T lymphocytes, *Annu. Rev. Immunol.*, 18, 19, 2000.
104. Obhrai, J.S., Oberbarnscheidt, M.H., Hand, T.W., Diggs, L., Chalasani, G., Lakkis, F.G., Effector T cell differentiation and memory T cell maintenance outside secondary lymphoid organs, *J. Immunol.*, 176, 4051, 2006.
105. Castellino, F., Germain, R.N., Cooperation between CD4$^+$ and CD8$^+$ T cells: when, where and how, *Annu. Rev. Immunol.*, 24, 519, 2006.
106. Shortman, K., Liu, Y.J., Mouse and human dendritic cell subtypes, *Nat. Rev. Immunol.*, 2, 151, 2002.
107. Neurath, M.F., Finotto, S., Glimcher, L.H., The role of Th1/Th2 polarization in mucosal immunity, *Nat. Med.*, 8, 567, 2002.
108. Graca, L., Silva-Santor, B., Coutinho, A., The blind-spot of regulatory T cells, *Eur. J. Immunol.*, 36, 802, 2006.
109. Seeleen, M., Roos, A., Daha, M.R., *J. Nephrol.* 18, 642–653, 2005.

2 Statistics of Immunological Testing

Albert D. Donnenberg

CONTENTS

2.1 OVERVIEW

Lord Kelvin* famously said, "Until you can measure something and express it in numbers, you have only the beginning of understanding." With few exceptions, interpreting immunological assays are all about interpreting numbers, and this can be aided (or not) by the application of statistical methods. This chapter is intended to help students of immunology, who have already had some exposure to statistics, navigate the most common statistical methods that are applied to tests of immunity. To quote the humorist Evan Esar, a statistician is one "who believes figures don't lie, but admits than under analysis some of them won't stand up either." By understanding a few basic statistical methods and avoiding the most common pitfalls of data analysis, the student of immunology increases the odds that his or her figures will, in fact, stand up.

Jim Deacon of the University of Edinburgh cites an unnamed Cambridge professor's take on statistics in Biology:

> In our Department we have a long corridor with a notice board at one end. I draw a histogram of my results, pin it to the notice board, then walk to the other end of the corridor. If I can still see a difference between the treatments then it's significant.

The don was advocating a rudimentary form of something that the great statistician and teacher John Tukey called exploratory data analysis (EDA) [1]. In our quest to understand our data, EDA is only a beginning, a method to discern trends and sniff out possible sources of variability *without making prior assumptions about the data.* Tukey drew a distinction between exploratory and confirmatory data analyses. The latter draws on a rich body of statistical methodology, which in the words of Tukey "[are] principles and procedures to look at a sample, and at what the sample has told us about the population from which it came, and assess the precision with which our inference from sample to population is made." Tukey advocated that exploratory and confirmatory data analyses proceed side by side.

2.2 EXPLORATORY DATA ANALYSIS IN THE COMPUTER AGE

The desktop computer did not exist when John Tukey wrote his seminal tract on EDA. Most of the methods that he invented or promoted were designed to be performed with a pencil, a ruler, and graph paper. Although this must have been somewhat cumbersome, EDA techniques are now implemented in most statistical packages. SYSTAT (Systat Software, Inc., San Jose, California), which I have used for the graphics and analyses in this chapter, has a powerful suite of EDA and graphic tools.

2.2.1 The Box Plot

Figure 2.1 is just what the don ordered. It is a dot plot of reciprocal antibody titers in nine healthy subjects before and one week after booster immunization with tetanus toxoid.

---

* William Thomson (1824–1907), First Baron Kelvin, who among other notable achievements gave us the concept of absolute zero, was fond of such pronouncements. He also said, "In science there is only physics, all the rest is only stamp collecting" and "Large increases in cost with questionable increases in performance can only be tolerated in race horses and women."

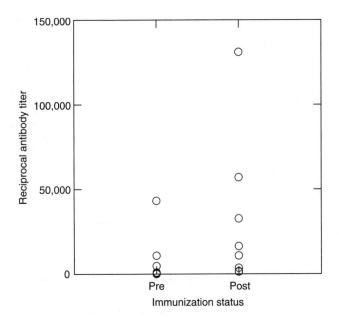

FIGURE 2.1   Dot plot of reciprocal antibody titers in nine subjects before and one week after immunization with tetanus toxoid vaccine.

It does not pass the don's test when I step to the back of my less than commodious office and look at the data on my computer screen. What I see is a lot of overlapping dots and a single dot in the *pre*group and several dots in the *post*group that are larger than the rest of the values. Did the immunization work? Perhaps not. Figure 2.2 shows the same data displayed using one of the most powerful tools of EDA, the box plot. Box plots display the sample range, median, and the spread of the central 50% of the data surrounding the median. Tukey saw the box as a *hinge* turned on its side. The outer bounds of the hinge give the location of the central 50% of the data (known as the hinge spread); where the two leaves of the hinge join (the *waist*) is the sample median (visible in the *post*group). The *whiskers* (another Tukey coinage) show the range (location of the smallest and largest values), exclusive of outliers, which are indicated by circles (explained in the figure legend). What one sees in the box plot is quite different from the raw dot plot. The median value of the *post*group is higher than that of the *pre*group, which on this scale appears to be near the origin. We also notice that the data is not symmetrically distributed, but is quite skewed toward high values. In the *pre*group, the first and second quartile are smashed against the origin, and the *post*group has a short whisker below the second quartile, and a long whisker above the third. Both the *pre*- and *post*groups have one high outlier. I might be changing my opinion about whether the immunization worked, or at least formulating some confirmatory tests that might be used to shed light on the question.

2.2.2   TRANSFORMATIONS IN EXPLORATORY DATA ANALYSIS

Tukey advocated transformation (the most common being logarithmic and square root) as the next logical step in EDA when one has *skewed* data. Figure 2.3 shows the

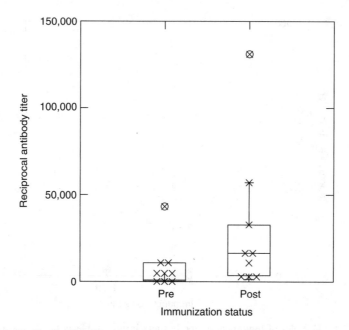

FIGURE 2.2   Classic Tukey box plots display nonparametric (distribution-free) descriptive statistics. Individual data points are marked by Xs. The waist indicates the group median and the hinges (upper and lower boundaries of the box) indicate interquartile distances. The whiskers (bars) give the ranges, exclusive of outliers. Outliers (more than 1.5 times the hinge spread from the median) are shown by asterisks (there are none in this illustration) and far outliers (more than three times the hinge spread from the median) are shown by circled Xs. The lower hinge cannot be seen in the pre group, because more than five of the nine data points lie close to the origin.

data of Figure 2.2, log transformed to the base 10. Now we can clearly see what was not apparent at all in Figure 2.1, and was only hinted in Figure 2.2. From the location of the median bars, we see that immunization led to a more than 10-fold increase in median antibody titer ($<10^3$ versus $>10^4$). Further, the data in both groups are more or less symmetrically distributed, one of the key requirements for the legitimate use of a variety of confirmatory statistical tests (explained in Section 2.6). The two far outlying values seen in Figure 2.2, now fit neatly into the distribution.

2.3   ESTIMATES OF DATA LOCATION AND DISPERSION

2.3.1   DATA LOCATION

The median (the midpoint of a series of ordered numbers, where half the values are above and half below) is the first data-location summary statistic to be evaluated in EDA because it does not require any assumptions about the distribution of the data. Additional summary statistics of data location are the arithmetic mean (the numerical average) and the mode (the most prevalent value within a distribution). Although the arithmetic mean is the most widely used statistic, we will see that

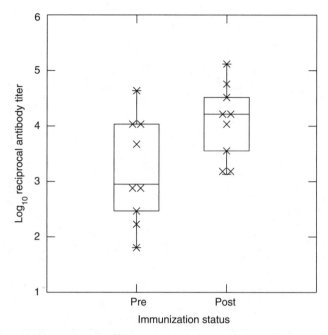

FIGURE 2.3 Box plot of log-transformed antibody data. The base 10 logarithm of the reciprocal titers shown in Figures 2.1 and 2.2 are plotted. Note that the distributions are fairly symmetrical about the waists (medians), and the two values that appeared as outliers in Figure 2.2 are now within 1.5 hinge spreads of the median.

this is not always to the best effect. When data are approximately symmetrically distributed, the median, mean, and mode are very close. As we saw in our EDA example of antibody titers, biological data, and especially immunological testing data, are often highly skewed. To continue with antibodies as an example, titers in a population may consist of individuals with no detectable antibodies, those with low titers, and a very small proportion with extremely high titers. If one were to accept the mode as a summary statistic to characterize the human immunodeficiency virus Enzyme-Linked ImmunoSorbent Assay (HIV ELISA) titers of university students, one would entirely discount the infrequent HIV positive subjects, because the most common result would be zero (or undetectable). The median would also be zero because more than half of the students are seronegative. Likewise, the arithmetic mean would barely register above zero, influenced as it is by the high proportion of subjects without detectable antibody. Although all these summary statistics are technically correct, they may not capture the biology that we seek to explain. Even if one confined the analysis to "positive" subjects, as is the case with tetanus toxoid because all individuals have been immunized some time in the past (Figure 2.4), the left-skewed distribution would result in mean and median values that differ markedly (the mean is higher than the median because the data distribution contains a tail of very high values). Thus, as in our EDA example, the choice of the most informative statistic to summarize data location is dependent on the data distribution. This will be discussed more in Section 2.3.2.

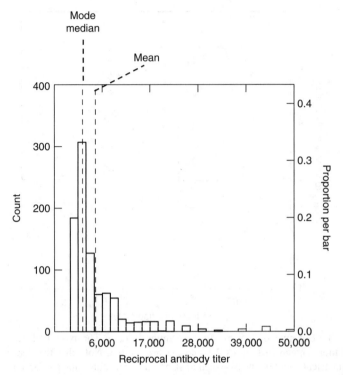

FIGURE 2.4 Distribution of antitetanus antibodies in 1000 subjects. Antibody titer was measured in serial dilutions of serum by ELISA and is expressed as a reciprocal titer (i.e., a value of 32 means that the serum the dilution curve crossed the threshold defining a positive result at a 1/32 dilution). The distributions of values is shown as a bar density histogram. Like many measures of immunity, antibody titer is not symmetrically distributed, but rather is left-skewed. Here, the median and modal values are seen to provide a better summary of the data, the mean being influenced by a few, very high values.

2.3.2 DATA DISPERSION

Although it is easy to find graphs in the scientific literature in which group means are compared without recourse to error bars, statistics of location alone do not provide an adequate summary of a set of observations, and certainly do not provide all the information required to compare results from different groups. In addition to location, we need an indication of the spread of individual observations within the group. In EDA, we chose the hinge spread (distance between the beginning of the second quartile and the end of the third) as a statistic that, along with the location of the median, provided useful information about the shape of the data distribution. In confirmatory data analysis, the sample standard deviation (SD) (see Technical Definition 1) is the most widely used and forms the basis for the many statistical tests based on the assumption of a normal or Gaussian distribution. In general, statistical tests, which assume that the data fit a particular distribution are called parametric tests.

Technical Definition 1

Sample Standard Deviation. Experimental data sets are collected to make generalizations about the population from which test data are drawn. The standard deviation is a measure of average absolute difference in individual measurements (x_i), compared to the sample average (x bar).

$$s = \sqrt{\frac{1}{N-1} \sum_{i=1}^{N} (x_i - \bar{x})^2}$$

The standard deviation tells us about the relative homogeneity or spread of a sample. It can be used, for example, to compute sample percentiles, or it can be normalized by dividing it by the sample mean to create a scale-independent measure of variability, the coefficient of variation.

Before we are ready to use parametric statistical tests (e.g., Student's t test*), there is something you must know about all tests relying on the normal distribution:

The normal model uses only two parameters (mean and standard deviation) to describe your entire data set.

As discussed, these are measures of location and dispersion, respectively. The validity of tests that assume normality depends entirely on how well the data can be described by these two parameters. Figure 2.4 shows more antibody data, this time from a group of more than 1000 subjects. As we have begun to appreciate from Figures 2.1–2.3, antibody distributions are usually left-skewed, and the data in Figure 2.4 are no exception. In particular, notice how the high outlying values cause the mean to be higher than the median or mode. Since the normal distribution is a family of bell-shaped curves that can be described by the mean and SD, we can create such a curve by plugging our sample mean and SD into an equation. Figure 2.5 shows what happens when we do this for the data in Figure 2.4. By definition, the resulting bell-shaped curve has its peak located at the sample mean, and its points of inflection at one SD below and above the mean. At this point, we have to question whether we want to entrust our data analysis to these two statistics. Considering the skewed nature of the distribution and the absence of zero or negative values, a logarithmic transformation looks promising. *In practice, a great deal of quantitative, continuous data from immunologic testing is lognormally distributed, so this is the first transformation that one would try.* Figure 2.6 shows the resulting distribution of the log-transformed data (bars) and the predicted normal distribution (curve) (see Technical Definifion 2). What an

---

* Student was the pseudonym of William Sealy Gosset, an industrial chemist for the Guinness Brewery in Dublin, Ireland. After discussions with the statistical pioneer Karl Pearson, he developed the t test to monitor the quality of batches of beer. Although he published the test in Pearson's statistical journal *Biometrika* [2], he was required by his employer to use a pseudonym, because they considered their monitoring process a trade secret.

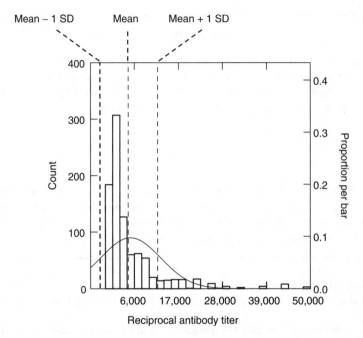

FIGURE 2.5 Application of the normal model to antibody data. The bars show the actual distribution of the data. The bell-shaped curve shows the normal distribution calculated from the sample mean and SD. One can see that the actual peak of the distribution is of a much lower value than the peak of the curve. It is also disturbing that the lower end of the distribution extrapolates to negative titers, and the upper end does not account for the highest values.

Technical Definition 2

Normal Distribution. The normal distribution forms the basis of many parametric statistics, such as Student's t test. It is specified by the equation

$$\frac{1}{\sigma\sqrt{2\pi}}e^{-(x-\mu)^2/(2\sigma^2)}$$

This equation has only two parameters, the population mean, μ, and the population standard deviation, σ. The entire family of normal curves is generated by varying μ and σ. If you plug $\mu = 0$, $\sigma = 1$, and x values of -3 through 3 into this equation, you will get the standard normal curve (which we will see in Section 2.5) spanning three standard deviations above and below a mean of 0. This curve forms the basis of all standard statistical tables.

improvement—the mean, median, and mode are virtually identical (3.26, 3.31, and 3.30, respectively). The data themselves appear to be symmetrically distributed about the mean, and even the extreme values are fairly well modeled. However, this is real-world data and there is also a degree of positive Kurtosis (sharper peak with fatter tails, compared to the normal distribution). This is not a disease, but rather a way to quantify

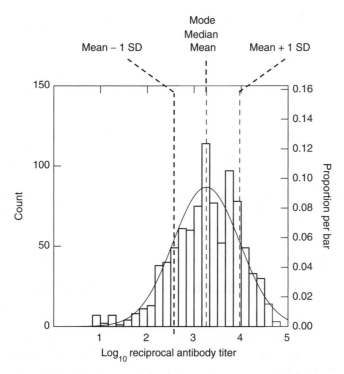

FIGURE 2.6 Distribution of log-transformed reciprocal titers. The mean, median, and mode are virtually identical and the superimposed normal distribution fits the data reasonably well. Although the data are in $\log_{10}$ units, you can get back to the original reciprocal titers by raising the numbers on the x-axis to the 10th power. Alternatively, one could simply plot the x-axis on a log scale and achieve the same effect.

(along with skewness) a departure from the normal distribution. Practically, experience tells that with antibody data, this is just about as good as it gets, and the assumption of normality is a reasonable one. We could then proceed to compare log-transformed titers. Going back to the data in Figures 2.1–2.3, which are naturally paired values (each subject was measured before and after immunization), Student's paired t test is the most appropriate confirmatory test (Figure 2.7). This test looks at the difference between the pre- and postimmunization values of the individual subjects. In our example, the average subject had a 7.9-fold rise in antibody titer following immunization (Student's two-tailed test, $p = 0.014$). This is as good a time as any for a discussion of one- and two-tailed p values. A p value can be put into words as the probability that a difference this great (i.e., 7.9-fold) would occur by chance alone (given the magnitude of the SD between individual differences and the number of paired observations). In the present case, a chance difference of 7.9-fold could be expected only 1 in 71 times ($p = 0.014$), well below the conventional probability traditionally considered statistically significant (1 in 20, or $p = 0.05$). Because this is a two-tailed value, we are considering the probability of a 7.9-fold increase *or* decrease in antibody titer. Some would favor a one-tailed t test (which would yield exactly half the probability, $p = 0.007$) arguing that we have a prior hypothesis that the only two expected outcomes

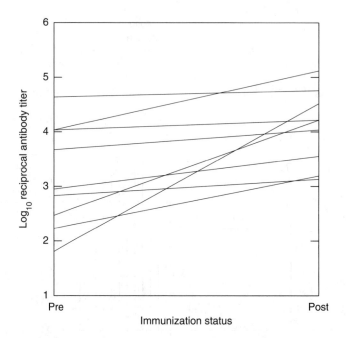

FIGURE 2.7 A line graph showing paired pre- and postimmunization log-transformed reciprocal titers. Student's paired t test calculates the difference between pre- and postvalues for each individual data pair. The mean difference is then tested to determine whether it is significantly different from zero (the famous null hypothesis). In this example, the mean difference was 0.896 ± 0.303 (SEM). Because these log units are not terribly intuitive, it is best to transform them back into something more intelligible. Taking the base 10 antilog of the mean difference (i.e., converting the number from the log domain, back into the linear domain), we get $10^{0.896} = 7.87$. This means that the subjects had, on average, a 7.9-fold increase in antibody titer. The t value associated with this value is 3.314, with a two-tailed p value of 0.014.

of immunization are that either the antibody titer remains the same or increases. I find such arguments to be an annoying example of p-value inflation. Although we do not expect immunization to result in a decrease in titer, this is certainly an intelligible possibility, and as such, deserves to be factored into our analysis by the two-tailed test.

2.3.3 DATA SAMPLING

Sampling has a special meaning in statistics. One of the tenets of the scientific method is that we can make inferences about an entire population by examining a sample drawn from that population. It is lost to history whether Student had a more convivial definition of sampling in mind when he invented his famous t test for the Guinness brewery [2]. To bring the subject back to immunology, the notion of sampling means that you do not have to measure the CD4 count of every living person to get a very good estimate of the distribution of CD4 counts in the population. Statisticians make a useful distinction between means and SDs calculated for entire populations, and those estimated from population samples. In educational testing, for example, it is possible to know the mean (location) and SD (dispersion) of the test scores for the entire population of students taking the examination. In clinical testing, a population can only be sampled to infer normal ranges (i.e., the distribution of scores in subjects without known pathology). In experimental testing, replicate experiments are thought of as representing a sample from an infinite population of identical experiments that could have theoretically been performed. Although this sounds quite incredible, we will see how, by increasing the number of experimental replicates, our estimated mean and SD converge on the mean and SD of this fictitious population of all possible experiments.

2.4 EFFECT OF SAMPLE SIZE ON PARAMETRIC SUMMARY STATISTICS

The effect of sampling can easily be seen if we start with a data set in which, like our educational testing example discussed earlier, the results can be known for the entire population. Figure 2.8 shows data for white blood cell (WBC) counts measured on all patients who underwent peripheral stem cell mobilization at the University of Pittsburgh Medical Center between January 1999 and December 2006. To move hematopoietic stem cells from the bone marrow into the blood, these patients are treated with the growth factor granulocyte colony stimulating factor (G-CSF), and their WBCs are monitored for an increase that usually coincides with the peripheralization of hematopoietic stem and progenitor cells. Thus, the counts have much more variability than those of normal subjects. Results from the entire population (1224 determinations) were randomly sampled using increasingly large samples, until the entire population was included. Up to an n of 300, the sample mean oscillates between

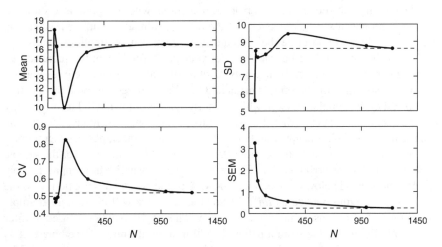

FIGURE 2.8 The figure depicts how the mean, sample SD, CV, and SEM change as a function of the number of observations (N). The data are peripheral WBCs (per microliter in 1000s) from patients receiving the hematopoietic growth factor G-CSF for mobilization of stem cells into their peripheral blood. Data were chosen at random from a set of 1224 observations. Mean, SD, CV, and SEM were calculated with an n of 3, 10, 30, 100, 300, 1000, and the full data set of 1224. The data points are connected by a spline fit. The statistics for the entire population are shown as a dashed line.

10.0 and 18.1, after which the mean stabilizes at about 16.5. The large SD indicates that there is a great deal of patient-to-patient variability. When expressed relative to the mean (in the coefficient of variation [CV]), the SD stabilizes at about 50%. Unlike the mean, SD, and CV, the estimates of which become more precise (not smaller) with an increasing number of observations, the standard error of the mean (SEM) is a measure of how precisely we have estimated the location of the population mean. Accordingly, it decreases as the number of observations grows, and theoretically, would be 0 when the entire population has been sampled.

2.5 STANDARD DEVIATIONS, STANDARD ERRORS, AND WHEN TO USE THEM

SDs and standard errors of the mean are the most commonly used statistics to summarize data dispersion in the immunology literature. Now we have seen how the sample mean, SD, and standard error change as the sample size increases, we can apply these statistics to our data with confidence. Bar graphs, line graphs, and dot plots are frequently shown with error bars to help us understand and compare data dispersion. Similarly, we often see numerical data summarized as a value ± another value (e.g. 197 ± 62). Most commonly, these errors are expressed as SDs or standard errors. As seen, these two useful statistics have very different meanings. Despite this, they are sometimes (incorrectly) used interchangeably. Equally incorrectly, error bars or intervals are often used in technical publications without specifying which statistic they represent.

The *sample SD* tells you something about the spread of the data distribution. Use it as a summary statistic (e.g., as an error bar in a graph) when the point is to show the relative homogeneity or heterogeneity of the data. As already mentioned, when you use the sample SD to describe the spread of your data, you are implicitly endorsing the use of the normal distribution to model your data. You accept the premise that your data are well described by a family of symmetrical bell-shaped curves that are completely specified by the sample mean and SD. Your inferences about the location and spread of the population from which your sample was drawn will be correct only to the extent that the normal distribution fits this population. If the model does fit, you can use the SD to infer the upper and lower bounds, which encompass a given proportion of the population data. For example, 95th percentiles are calculated as the sample mean ±1.96 SDs. Referring back to Figure 2.6, one can visualize how approximately two SDs above and below the mean include 95% of the normal distribution, with the remaining 5% lying outside as *tails*. In Section 2.7.2, we will provide an example where the SD of a measurement sampled from a healthy population is used to infer when an individual value from a patient is within normal limits. Remember that as the sample size (*n*) increases, the sample SD becomes a more precise estimate of the population SD. It does not become smaller. In this respect, it behaves something like the sample mean.

The *standard error of the sample mean* can be interpreted as a SD describing how means sampled from the entire population vary with respect to one another. Use the SEM (e.g., as an error bar in a graph) when you want to express the precision with which you know the location of the population mean or when comparing the means of two or more or experimental treatments or groups (to determine the probability that they were drawn from the same population). As we saw in Figure 2.8, the greater the number of observations, the more precise is our estimate of the population mean.

2.6 ASSUMPTIONS OF CONFIRMATORY TESTS

Some of the most useful statistical tests rely on the normal distribution. These include Student's t test, analysis of variance (a generalization of Student's t test for more than two groups), linear regression analysis, and certain power calculations that are very helpful for experimental design. In addition to the assumption of normality, there are two other requirements that bear some explanation. The first is homogeneity of variance. For comparison of groups, both groups must have approximately equal variance (or equal SDs, the square root of the variance). The sample variances are then *pooled* to obtain the best estimate of the population variance. For linear regression, the SD of the dependent variable must not change as a function of the independent variable. In practice, a variant of Student's t test, the unequal variance t test, does not require this assumption, and can be used, provided the data are normally distributed [3]. For linear regression, a dependent variable with a variance that grows as a function of the magnitude of the independent variable will probably behave better after transformation, as we will see in one of our case studies (Section 2.7.3).

The second assumption, known as continuity, requires that the universe of possible data values is continuous on the number line. That is, observations can be given to many decimal places arbitrarily. This is never actually the case in the real world where assay precision is limited by instrumentation and data are, in the strictest sense, discontinuous. To pick an extreme example, a hemagglutination assay may call for serum to be serially diluted in twofold increments. The titer is the highest dilution that causes the red blood cells to clump. Here, the only possible assay results (given as reciprocal titer) are 2, 4, 8, 16, etc. Given the assumptions of normality (hemagglutination titers are beautifully lognormally distributed) and homogeneity of variance, can we use Student's t test to compare data from two groups? Here is how John Tukey reconciled the "unrealism" of the assumption of continuity: "To the man with the data, such unrealism hardly looks helpful, but to the worker with a theory such an assumption turns out to make things simpler. Since all the man with the data can hope for is an approximation, and since an approximation to a few decimal places is often more than he deserves, he soon learns not to worry very often about this particular sort of unrealism." We may interpret this to mean that continuity is a formal requirement, and for us, the women and men with the data; it is sufficient that continuity of our measurement be intelligible, if not actual.

Why do I have to meet these requirements anyway?

- When you use a test like Student's t test to compare the data from groups A and B you are not really comparing the data itself.
- When you apply Student's t test to compare the results of group A and B, you are giving up the raw data in favor of a model that is *completely specified by only three numbers*: the mean of A, the mean of B, and a pooled SD that is supposed to accurately represent the spread of both groups.

- The test is invalid to the extent that the resulting normal distributions fail to accurately describe the experimental results.
- Remember that when the assumptions of normality or homogeneity of variance fail, transformations can often bring the data back into line.

2.6.1 A WORD ABOUT THE NULL HYPOTHESIS

In confirmatory statistics, the interpretation of a test is often related in terms of the null hypothesis. This elicits in most of us a mental cramp, since this double negative (trying to refute the hypothesis of no difference), seems at a first glance the equivalent of trying to prove the hypothesis of a difference (what we students of science usually think we should be doing). This is not the case, as disproving the null hypothesis provides supporting evidence for, but does not prove an alternative hypothesis. Forcing ourselves to see our experimental problems in terms of the null hypothesis, and formulating statistical tests to reject the null hypothesis brings us into line, philosophically, with the falsifiability model of scientific discovery proposed by the twentieth century philosopher Karl Popper [4]. Popper believed that the closest science could come to truth is to provide theories with ample opportunity to be proven wrong by counterexamples. In other words, we gain confidence in a scientific hypothesis by attempting to prove it wrong and failing. Let us return to Figure 2.7 and our paired t test of antibody titers of sera obtained before and after immunization. Because this is a paired t test, the null hypothesis states that there is no difference between titers of individual subjects before and after immunization. To perform the statistical test, we subtract the log of the *pre* titer from the log of the *post* titer. When we test the null hypothesis, we determine the probability that the distribution of resulting *post–pre* differences could have been drawn from a (hypothetical) population of individuals for whom the mean difference between *pre-* and *post* values is zero, and the data dispersion, though unknown, can be estimated from the SD of our sample. Because we have sampled only nine pairs of data from this hypothetical population, the mean difference is very likely, by chance alone, to be a nonzero value. But how different from zero does it have to be before we can reject the idea that our data pairs were drawn from a population of subjects in whom there is no difference before and after immunization? As we have already seen, the mean difference of our sample was 0.896 (not 0) with a SEM of 0.303. A fast and intuitive way of thinking about this problem is to approximate the least significant difference (LSD) at $p = 0.05$. Just take the null hypothesis mean difference (0), and add two times the SEM (0.606). Is the sample mean difference (0.896) greater? Then we can reject the null hypothesis. Note that this does not formally prove that the immunization worked. There could have been other factors responsible for the rise in titers after immunization, or it could just be bad luck (in Figure 2.7, we found that a difference this large could be expected by chance alone, 1 in 71 times, if the nine data pairs were drawn from a population with a mean difference of 0). It does, however, provide evidence consistent with the notion that immunization caused a rise in titer. Further unsuccessful attempts to falsify our hypothesis (such attempts may include performing similar experiments on additional subjects, or immunizing subjects with a sham immunogen) will give us greater confidence in our conclusion that the vaccine is efficacious.

2.6.2 A WORD ABOUT STATISTICAL SIGNIFICANCE

The word *significant* does two jobs in the language of biomedical science. When we are speaking of a *statistically significant* difference, we usually mean that we have rejected the null hypothesis because, for example, the probability of two groups of data drawn from the same population is less than 1 in 20 ($p < 0.05$). This is an arbitrary cutoff, and we may wish to set a more rigorous criterion for statistical significance (say, $p < 0.01$) if our problem demands that avoiding type I error (false positives, see Technical Definition 3) is more important than avoiding type II error (false negatives). Strictly speaking, this concept of statistical significance applies to comparisons that are performed in isolation. If one were to perform 100 comparisons, then 5 (1 in 20) should turn up spuriously significant at a p value of 0.05, even when no differences exist. In such cases, the Bonferroni correction for multiple comparisons is used to adjust the required p value [5]. Finally, we have the semantic problem of the word significance, and the double job that it does in the language of science. Statistical significance in no way implies significance in the ordinary sense of meaningfulness or importance. We have seen earlier how the SEM decreases with sample size. Thus, with increasing sample size, even small and biologically unimportant differences may grow to be statistically significant. In words attributed to Gertrude Stein, author and Johns Hopkins School of Medicine dropout, "A difference, to be a difference, must make a difference."

Technical Definition 3

A type I error (or alpha error) is a false positive. In other words, it is rejecting the null hypothesis when it is actually true. The *p value* is the probability of alpha error.

A type II error (or beta error) is a false negative. It means failing to observe a difference when in fact a difference exists (or failing to reject a false null hypothesis).

2.6.3 SENSITIVITY AND SPECIFICITY: POSITIVE AND NEGATIVE PREDICTIVE VALUES

When we are dealing with a test that provides dichotomous outcomes, such as a test to detect a viral infection (positive or negative), we can evaluate it in terms of sensitivity and specificity, which are defined in comparison to a reference or *gold standard*. The gold standard may be an existing and accepted test, or it may be an outcome that will eventually become known with time, such as the onset of clinical symptoms (e.g., HIV infection). Table 2.1 shows the relationship between a hypothetical-test result and the gold standard. We can see that sensitivity, the ability to identify true positives, is defined as the ratio of true positives to the sum of true positives plus false negatives. Similarly, specificity, the ability to detect true negatives is the ratio of true negatives to true negative plus false positives. Table 2.1 can also be used to conceptualize two other important parameters that depend not only on the test, but also on the prevalence of true positives in the population. These are the

TABLE 2.1

Sensitivity and Specificity Are Defined in Relation to a Gold Standard

Test Result	Gold Standard	
	Positive	Negative
Positive	TP	FP
Negative	FN	TN

Note: About the test—Sensitivity = TP/(TP + FN), ability to identify TP; Specificity = TN/(FP + TN), ability to identify TN.

About the population—Positive predictive value = TP/(TP + FP), will be low when prevalence of TP is low; Negative predictive value = TN/(FN + TN), will be low if prevalence of TP is very high.

positive and negative predictive values of the test. For a test of a given sensitivity and specificity, the positive predictive value will be low when the prevalence of true positives is low. Similarly, the negative predictive value is low when the prevalence of true positives is high. This can be very important in the interpretation of test results. For example, in a blood donor pool that has been prescreened to exclude subjects with risk behavior the prevalence of HIV infection may be very low (say 1/10,000). On screening a million blood products from this donor group, we would expect to find 100 positive samples. Using an assay with 99.9% sensitivity and specificity, we would correctly identify virtually all the 100 positive samples. Of the 999,900 products that are not infected, the test would also correctly identify 999,000 as negative, but would incorrectly find 1000 samples to be HIV positive. Despite the excellent sensitivity and specificity, the positive predictive value is only 9%.

We also sometimes use the term sensitivity in the context of quantitative testing, for example, nephelometry is more sensitive than radial immunodiffusion for the measurement of serum immunoglobulin. Here, the term is used to compare the lower limit of detection of the assays. We mean that nephelometry can quantify immunoglobulins down to a lower concentration than immunodiffusion. This is usually related to the signal-to-noise ratio, a term borrowed from electrical engineering. It is based on the assumption that a measurement is an admixture of signal (the parameter that we wish to measure) and noise (a random or irrelevant contribution that contaminates the signal). Sensitivity depends on the relative strengths of signal and noise. In the context of assays of immunity, noise is the intensity of the background or negative control measurement (e.g., optical density of the negative control in an ELISA assay and fluorescence intensity of an isotype control in flow cytometry) and the signal is inferred to be what remains when the background is subtracted from the test result. Sensitivity in this sense can be increased both by strengthening the signal (in flow cytometry using a fluorescent dye with a higher quantum efficiency) or reducing the noise (blocking nonspecific antibody binding). From a statistical perspective, the lower limit of detection can be defined as an upper limit (e.g., 95th percentile) of an appropriate negative control value. For example, the lower limit of detection of an

MHC class I-tetramer assay was defined by determining the mean and SD of the percent of log tetramer positive cells in an MHC mismatched setting (permitting only nonspecific binding). The upper 95th percentile of this negative control, determined to be about 1 in 10,000 cells, was considered to be the lower limit of detection [6].

2.6.4 PRECISION AND ACCURACY

Precision and accuracy are two additional terms that have a specialized meaning in the context of assays of immunity. Precision is closely allied to reproducibility, and reflects the closeness of replicate determinations. The CV (Figure 2.8) among replicate determinations provides an estimate of precision. Like sensitivity and specificity, accuracy relies on comparison to some gold standard. An intuitive example is a pipetting device that may aliquot the same amount of liquid each time within a very close tolerance, but is out of calibration. This device is precise but not accurate. It is entirely possible to be accurate without being precise, and vice versa (Figure 2.9).

An example in which the precision of a new assay was determined is shown in Figure 2.10, where a nuclear stain was used to quantify leukocytes contaminating filtered platelet products [7]. Each assay was performed in triplicate and the CV was determined and plotted as a function of the leukocyte count. The assay was very precise at high leukocyte concentrations, but the precision deteriorated rapidly at counts less than 100 cells/mL. Accuracy of the same assay was determined by carefully diluting leukocytes in a sample of known concentration (Figure 2.11). The cell count calculated from the dilution (*x*-axis) was taken as a gold standard, and the observed results (*y*-axis) were compared by linear regression analysis. The unexpected nonlinearity at the lower region of the curve may have come from the dilution error, but the model assumes that it is from counting error. Despite this, an intercept near 0 and a slope near 1, and small differences between observed and expected values indicate the accuracy of this assay.

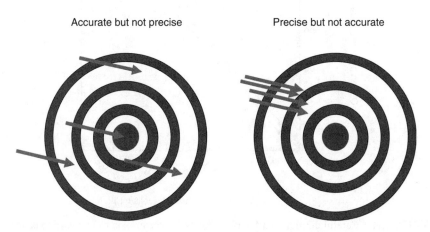

FIGURE 2.9   Accuracy and precision are independent attributes. The target on the left-hand side illustrates accuracy without precision (the position of the arrows average to a perfect bull's-eye), whereas, the arrows on the right-hand side form a tight cluster (high precision) but fail to hit the mark.

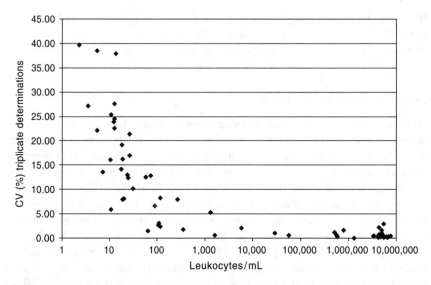

FIGURE 2.10 The precision of an assay that measures the leukocyte content of leukore-duced platelet products decreases at lower cell concentrations, as determined by the CV of triplicate determinations.

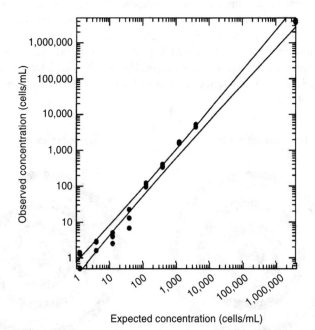

FIGURE 2.11 Careful dilution of a sample of known concentration can be used to create a gold standard (expected concentration) to judge the accuracy of an assay designed to count cells at very low concentration. The sample was diluted in half-$\log_{10}$ increments and assayed in triplicate. There is unexplained nonlinearity at some of the lower concentrations. The lines represent the 95% confidence intervals about the linear regression line of best fit.

2.6.5 STATISTICS AND EXPERIMENTAL DESIGN

The only complaint that statisticians have about experimentalists (and especially immunologists) is that we think about statistics only after getting the results of our experiments. Modern statistical packages contain tools to assist with experimental design, the most important of which are for power analysis. For a two-way comparison, power analysis forces one to think about one's experiment in terms of the smallest difference that one would care to detect between groups (remember Gertrude Stein), and the degree of data dispersion expected within these groups. In practice, it is like backing into a test of significance. For a real-world example, a graduate student came to me with a data set comparing the expression of an immunophenotypic marker in two groups. After doing a preliminary experiment with two groups, each containing three subjects, he noticed a difference in the hypothesized direction, but on performing a t test, failed to reject the null hypothesis. He decided that the minimal difference in marker expression that he wished to be powered to detect was 25% (a smaller difference seemed to him unlikely to be biologically important). In the example shown in Figure 2.12, his hypothesized results were a mean of 10% in group 1 versus 35% in group 2. From his preliminary data, he estimated the SD

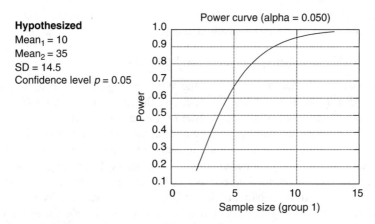

FIGURE 2.12 This is a true example from the doctoral dissertation work of Dr. Aki Hoji, an immunology graduate student at the University of Pittsburgh Graduate School of Public Health. He was comparing phenotypic markers on two T-cell subsets, therefore the data is expressed as percent positive for a given marker. He did a preliminary experiment with three subjects per group. At that time the data looked promising, but had not reached statistical significance. He decided that the minimal difference that he wished to be powered to detect was 25%. That is, a difference of less than 25% expression for a given marker was not deemed to be of biological importance. In the example shown, the hypothesized percent positive were 10 and 35% for the two subsets being compared. This is arbitrary and the same conclusion would have been reached for hypothesized values of 0% versus 25% and 50% versus 75%. The SD (14.5%) was calculated from his preliminary data. The results were encouraging. They indicated that at his present group size ($n = 3$), he had only a 37% probability of detecting a 25% difference between the groups (at $p = 0.05$), even if such a difference actually existed. Further, the power calculation indicated that increasing the group size to $n = 7$ would give him an 84% probability of detecting such a difference.

at 14.5%, and he chose a conventional alpha value (significance level) of 0.05. The results of the power curve, which examines the effect of sample size on power (1—beta error), showed that his preliminary data ($n = 3$) had only a 37% probability of detecting a 25% difference between the groups, if any existed, at a p value of 0.05. Increasing the sample size to $n = 7$ increased this probability to 84%. Knowing this, he went back to the bench. To be kosher, the power calculation should have been made before the experiment was initiated, or should have been made using data from an independent set of experiments. Performing interim analyses (taking peeks at the data) influences the meaning of a p value in much the same way that multiple comparisons do.

How many mice/patients/samples do you need to complete your experiment? All you need to know is

- The assay results conform to the requirements of the test
- The smallest between-group difference that you would call biologically meaningful
- An estimate of the SD of the assay itself
- The alpha level. The p value you will call significant (conventionally at $p = 0.05$ or less)
- How much power you need (usually 0.8, i.e., in 80% of such experiments you would detect such a difference, if it exists, at the stated confidence level)?

2.6.6 A Word about Nonparametric Statistics

Nonparametric statistical tests, such as the Wilcoxon signed rank test, the Mann–Whitney U test, the Kruskal Wallace test, and the Spearmen rank coefficient are confirmatory tests based on the rank order of data points rather than their magnitude. They, therefore, do not require the data to conform to a particular distribution, and are not sensitive to the magnitude of outlying values. They can also accommodate data points that are outside the dynamic range of the test (e.g., undetectable and too numerous to count). These tests can be very useful when comparing data that are drawn from a complex distribution, such as one with two or more peaks, or small data sets, where it is difficult to determine the distribution from which the data are drawn. As robust as nonparametric tests are, experience teaches that they are rarely necessary for the vast majority of immunological testing data. Given a data set that can legitimately be analyzed, raw, or in transformation, by tests that rely on normality, an equivalent nonparametric test will often have less statistical power (sensitivity).

2.6.7 Graphic Representation of Data

2.6.7.1 Words of Wisdom Concerning Data Graphics

A picture may be worth a thousand words. Here are a few hundred words about pictures.

- Pictures based on exploration of data should force their messages upon us. Pictures that emphasize what we already know—security blankets to reassure us—are frequently not worth the space they take. Pictures that have to be gone over with a reading glass to see the main point are wasteful of time and inadequate of effect. The greatest value of a picture is when it forces us to notice what we never expected to see.   **John Tukey**

- Less is more.   **Ludwig Mies Van der Rohe**
- Pie charts are among the most abused graphics icons.
- We cannot think of a single instance in which a perspective bar graph should be used for any application.
- A nasty relative of the perspective bar chart is the pseudo perspective bar chart. Illustrators frequently feel the need to make two dimensional bars look like blocks or sky scrapers.   **Leland Wilkinson [8]**
- Forego chartjunk, including moiré vibration, the grid, and the duck.*
- The task of the designer is to give visual access to the subtle and the difficult—that is, the revelation of the complex.   **Edward Tufte [9]**

2.6.7.2 Practical Advice on Data Presentation

- Watch your proportions. In the olden days, presentation data were shown using 35-mm slides that have an aspect ratio (width/height) of 1.55:1. Today, most of our graphics end up in PowerPoint, which has an aspect ratio of 1.33:1. Therefore, graphs, tables, and word charts should be about 30% wider than tall. A significant departure means wasted space and a graphic that will read smaller when projected on a screen.
- "Wider-than-tall shapes usually make it easier for the eye to follow from left to right ... smoothly-changing curves can stand being taller than wide, but a wiggly curve needs to be wider than tall" (John Tukey).
- Choose backgrounds carefully. White is fine for print, but creates a terrible glare when projected in a large auditorium. I personally prefer a light gray, but deep blue with white letters can also be effective in a retro kind of way, in imitation of diazo print photographic slides.
- For publication graphics, anticipate the final size of the printed graphic. Choose the type size and the line thickness accordingly. Letters should not clot when they are reduced (e.g., as when the upper half of the lower case "e" fills in). Likewise, lines should not become so thin that they fade.
- Choose fonts carefully. Sans serif fonts (e.g., Ariel and Helvetica) tend to look cleaner, but words are more easily recognized by sight when seriffed fonts (e.g., Times Roman and Georgia) are used.

---

* Tufte defines chartjunk as ink that does not tell the viewer anything new. This includes all forms of spurious or superfluous decoration and especially computer-generated grids and crosshatching patterns that cause the Moiré effect: "designs [that] interact with the physiological tremor of the eye to produce the distracting appearance of vibration and movement." To know more about the duck, see Ref. 8.

2.7   THREE CASE STUDIES

2.7.1   CASE I: THYMIDINE UPTAKE DATA

Uptake of radiolabeled thymidine is a tried and true method for measuring the proliferation of lymphocytes in response to mitogens or specific antigens [10]. Cells are cultured for a sufficient duration to enter the log phase of proliferation, after which they are given a short pulse (usually 4 h) with tritiated thymidine, which is taken up specifically by proliferating cells. The cells are trapped on a filter, washed, and the incorporated thymidine is measured by scintillation spectrometry. The results involve a comparison of signal (label incorporation in cultures exposed to a mitogenic stimulus) and noise (background incorporation of label in the absence of stimulation). In the present example, the data are from an antigen-specific lymphoproliferation test performed on peripheral blood from a series of 76 patients. The results are expressed as net counts per minute (net CPM), that is, the CPM incorporated in the presence of antigen minus the CPM incorporated in a no antigen control from the same sample (Figure 2.13). The net CPM were very skewed, with a median of 1750 and a mean of 11,125, the mean being strongly influenced by a tail of large values. The data were, therefore, log-transformed and evaluated. Zero or negative values, resulting from background counts equal to or greater than the experimental counts ($n = 6$) were assigned an arbitrary small value of 1 prior to transformation, because logarithms are undefined for these values. In this case, negative values represented less than 10% of the data. A preponderance of negative values would have steered us away from log transformations and assumptions of normality, and toward a nonparametric test. Log transformation went a long way to make our data fit the normal distribution, with a mean of 3.023 (1054 when back-transformed to a geometric mean by taking the antilog, and much closer to the median of 3.243). This can be appreciated in the horizontal notched box plots shown in Figure 2.13. Figure 2.13 also introduces the normal probability plot, another useful and commonly available tool for evaluating data distributions. Here, the ordered data (x-axis) are plotted against the value predicted by the normal distribution (normalized to a mean of 0 and a SD of 1). A linear relationship indicates that the data are well described by the normal distribution and is evidence of the appropriateness of using tests requiring normality. The EDA tools applied in Figure 2.13 point to the same conclusion. Although the log-transformed data are not perfect (there is a hump on the left resulting from negative values and a resulting flattening of the curve [negative Kurtosis]), the transformed data are decidedly better described by the normal distribution than are the raw data. This is about all one can expect from real-world data.

As mentioned earlier, the use of t tests, analysis of variance (a generalization of the t test for multiple groups), linear regression analysis, and several other statistical tests depend on the assumption that the data are normally distributed. Often immunological data are not. When data are not normally distributed, the mean and SD are poor summary statistics. The greater the departure from normality, the less sensitive the test is to detect real differences in the data. Many types of immunological data are log-normally distributed. Examples include radionuclide

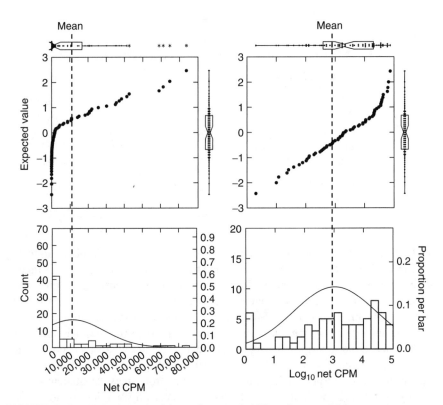

FIGURE 2.13 Use of notched box plots, probability plots, and histograms to assess the conformance of a data set to the normal distribution. In the top panels, net tritiated thymidine uptake (net CPM) and the log transform of net CPM are displayed as normal probability plots, where the *x*-axis shows ordered observed results, and the *y*-axis shows results predicted by the normal distribution. The linearity of the resulting relationship indicates the normality of the data. The probability plots are flanked by notched box plots, a variation of the box plot in which the notches at the waist show the 95% confidence interval about the median. The lower panels show histograms with superimposed normal distributions calculated from the data means and standard deviations, comparable to those shown in Figures 2.5 and 2.6. The CV of the raw data is 158% compared to a CV of 46% for the log-transformed data.

uptake, antibody titers, absolute blood counts, MHC tetramer binding frequencies, precursor frequencies, fluorescence intensity of cell surface immunophenotypic markers, and frequencies of rare populations detected by flow cytometry. When such data have been log transformed, the mean and SD better represent the data, tests depending on these parameters are more sensitive, and summary statistics are more appropriate.

Before we leave the effects of log transformation on radioactive thymidine uptake data, let us consider a comparison of the lymphoproliferative responses of two groups of 175 subjects each (Table 2.2). Samples of increasing size (*n*) were drawn from this data set at random. The sampled data were compared by Student's *t* test using the raw

TABLE 2.2

Student's t Test Performed on CPM Data from Two Groups

	Mean CPM (Groups 1 and 2)			p Value (Two Tailed)	
N	Arithmetic	Log	Geometric	Raw Data	Logged
3	2960, 6293	3.465, 3.703	2951, 5045	0.221	0.369
10	1890, 3180	3.239, 3.445	1734, 2784	0.063	0.040
30	1980, 4675	3.262, 3.546	1827, 3512	0.001	7×10^{-5}
175	1949, 4826	3.248, 3.550	1772, 3547	4×10^{-14}	6×10^{-23}

Note: Samples of increasing size were drawn at random from a dataset containing 175 observations per group.

data (arithmetic means) or geometric means. The geometric mean is obtained by log transforming the data, taking the average, and then back transforming the average to the linear domain by taking the antilog. At $n = 3$ (three subjects per group), group 2 appears to have a higher mean than group 1, but the comparison is not statistically significant. An independent sample of 10 subjects yields statistical significance using both raw and log-transformed data, but the p value is somewhat smaller using logs. In this data set, the superiority of log transformation is most evident when n is large. This case study demonstrates both the practical advantages of transforming the data to better to conform to normality, and the incredible robustness of Student's t test, which manages to provide the correct answer, albeit with some loss of sensitivity, even if the assumption of normality is flagrantly violated. It is worth noting here that despite our insistence on the importance of transformation, the central limit theorem states that no matter what the shape of the population distribution, the sampling distribution of the mean approaches normality as the sample size increases. However, a very large sample size is required before a highly skewed distribution converges to normality, and we rarely have this luxury with experimental data.

2.7.2 CASE II: T-CELL RECEPTOR HETEROGENEITY AS MEASURED BY THE V-BETA DISTRIBUTION

The antigen receptors on T-cell clones acquire their incredible diversity by a process involving gene rearrangement and mutation at splice sites (see details in Chapter 6). One of the elements contributing to receptor diversity is the 57-member v-beta gene family that codes for the variable portion of the T-cell receptor beta chain [11]. Only one v-beta specificity is expressed on a given T cell or its clonal descendants. Since 24 of the 57 members account for more than 75% of all TCR v-beta usage, a panel of 24 antibodies can be used to classify the majority of human T cells. v-beta usage within these 24 families follows a characteristic distribution, with some members representing a larger proportion of the T cell population than others. At most, a given specificity will represent 10% of the total T cells, but most are much less prevalent.

This is illustrated in Figure 2.14, which shows the v-beta distribution on CD4+ T cells from 21 healthy subjects. Applying a Gaussian model to our data (Figure 2.15), we can determine normal ranges for each v-beta specificity, arbitrarily defining individual v-beta deletions and expansions as outside the bounds of the 2.5th and 97.5th percentiles, respectively, of the normal distribution calculated from the means and SDs of our data. For each v-beta specificity, we expect, by definition, a frequency of 2.5% expansions and 2.5% deletions in our healthy population. We can check this by applying the calculated cutpoints to our control data set. Given that we have measured 24 v-beta specificities in 21 subjects, we would expect to find, on average 12.6 deletions and 12.6 expansions ($24 \times 21 \times 0.025 = 12.6$). However, when we apply our cutpoints, we get 25 expansions and only three deletions. What went wrong? To find out, let us examine our assumptions about normality. Lest you begin to think that all immunological data require transformation, let us

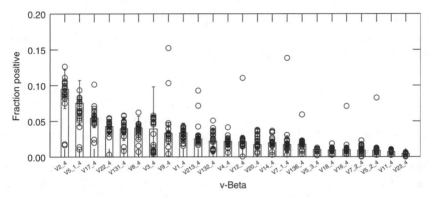

FIGURE 2.14 Distribution of v-beta specificities in peripheral blood CD4+ T cells from 21 healthy subjects. Individual v-beta specificities are ordered by prevalence. The circles show individual data points. The bars show the arithmetic mean values, and the error bars represent 1.96 standard deviations about the mean.

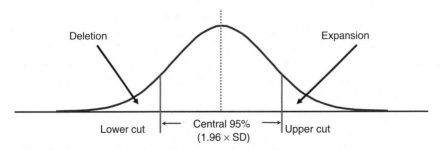

FIGURE 2.15 The Gaussian model can be used to define normal ranges of usage for each v-beta specificity. Deletions and expansions were defined as lying ouside the central 95% of the distribution.

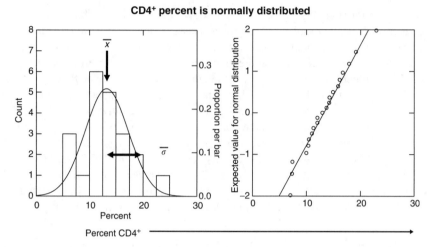

FIGURE 2.16   Distribution of CD4$^+$ cells among lymphocytes. The histogram shows the distribution of the actual data; the superimposed normal curve is calculated from the sample mean and standard deviation. The probability plot confirms the normality of the data.

begin by applying our EDA tools to the percent of CD4$^+$ cells in our 21 samples (Figure 2.16). Visual inspection of a bar histogram of the data, the superimposed normal distribution calculated from the sample mean and SD, and a probability plot all confirm that CD4 expression is normally distributed among lymphocytes of healthy subjects. In Figure 2.17, I have applied probability plots to both raw and log-transformed frequencies of usage for two v-beta specificities, v-beta 1 and v-beta 7.2, demonstrating quite conclusively that v-beta usage is lognormally distributed. In general, this appears to be the case for all measurements of rare events. Figure 2.18 is a reworking of Figure 2.14, using a logarithmic scale on the y-axis to reflect the lognormal distribution of TCR v-beta. Defining deletions and expansions as lying outside the central 95% of the log-transformed control data set, as we did previously on the raw data, we now calculate 12 expansions and 14 deletions among our controls. This is remarkably close to the 12.6 deletions and expansions predicted on the basis of the normal distribution. Unlike the thymidine uptake example, where failure to log transform merely reduced our power to detect a difference, failure to transform TCR v-beta usage data gave us completely erroneous normal ranges. Had we blindly applied the normal range cutpoints calculated from the raw control data to a set of patient data, we would have greatly overestimated the frequency of expansions and underestimated the frequency of deletions, and therefore seriously misread the underlying immunobiology.

The following happens if you fail to transform your data when

- The sample mean is greater than the sample median (lognormal data is asymmetrical and has a tail of high values).
- The SD is larger than it would be for log-transformed values. The example in Figure 2.13 shows that the CV was more than three times greater in the untransformed data.

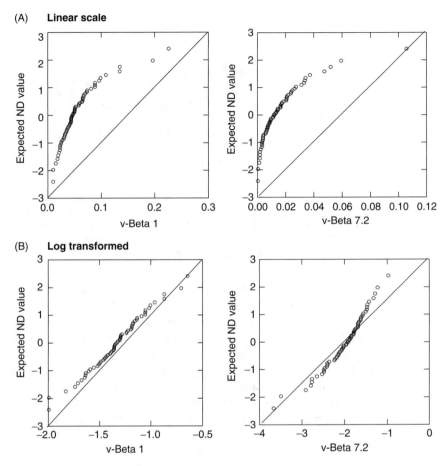

FIGURE 2.17 v-Beta usage is lognormally distributed. The distribution of two v-beta specificities, v-beta 1 and v-beta 7.2, are shown for both raw (A) and log-transformed (B) results from a larger dataset. The probability plots clearly demonstrate that v-beta usage is more closely modeled by the normal distribution after log transformation.

- The SD does not represent the distance from the mean to the point of inflection of the sample distribution (this is not possible since the distribution is asymmetrical). This greatly throws off percentile (probability) calculations based on the number of SDs above and below the mean.
- It takes a greater difference between the means of two groups to attain statistical significance, because the sample SDs are larger.

2.7.3   CASE III: LINEAR REGRESSION AND CORRELATION—PERIPHERAL BLOOD CD34 COUNTS PREDICT THE CD34 DOSE HARVESTED BY LEUKAPHERESIS

Simple linear regression analysis is designed to test a particular hypothesis concerning the relationship between two variables: The dependent variable increases in proportion to the independent variable. The distinction between the independent

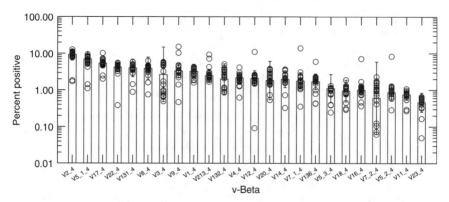

FIGURE 2.18 Distribution of log-transformed v-beta specificities in peripheral blood CD4$^+$ T cells from 21 healthy subjects. Individual v-beta specificities are ordered by prevalence. The circles show individual data points. The bars show the geometric mean values, and the error bars represent 1.96 standard deviations about the geometric mean.

variable, conventionally plotted on the x-axis, and the dependent variable, plotted on the y-axis, is an important one. The independent variable is like a gold standard; it is assumed to have no variance, whereas, the dependent variable is a measurement that entails inherent variability. An example might be the relationship between the measurement cell number (dependent variable) and time in culture (independent variable), or a response predicted by theory (independent variable), and the actual measured response (dependent variable). There is often the implicit assumption that the independent variable causes changes in the dependent variable. Linear regression assumes that the relationship is a straight line (Figure 2.19) which can be stated as

$$y = a + bx$$

where

> y = predicted value of the dependent variable
> a = y intercept of the line
> b = slope of the line
> x = value of the independent variable

You may have learned the same equation in algebra class as $y = mx + b$. Simple linear regression analysis uses the method of least squares to determine the line of best fit through a series of data points (pairs of observations) by minimizing the difference between the observed values and those predicted by the linear equation (residuals). Unless the line of best fit passes exactly through each pair of expected and observed values, the intercept (a) and the slope (b) will have variance. We can express this variance in terms of the SDs of the intercept and slope, or by calculating confidence intervals about these parameters. We can also determine the coefficient of correlation (r^2), a parameter that is estimated as 1—the sum of the squared differences between predicted and observed y values (regression sum

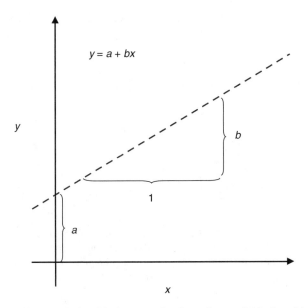

FIGURE 2.19 A linear relationship between the dependent variable (y-axis) and the independent variable (x-axis) is modeled by an intercept parameter (a) and a slope parameter (b).

of squares), divided by the sum of the squared differences between the predicted y values and the mean of the observed y values (total sum of squares). In other words, it is the fraction of the total variance in the variable y that is explained by the regression of y against the variable x. r^2 is a fraction between 0 and 1, where 0 indicates no correlation and 1 indicates that all the variance in y can be explained by the value of x (i.e., a perfect fit). Unlike regression analysis, there is no assumption that one variable is dependent and the other independent. The correlation coefficient r^2 can be computed between any two measurements, even if both have variance.

It is important to remember that a "line of best fit" with its intercept, slope, and r^2 can be calculated for any set of paired values. Even a slope that is significantly different from 0 does not guarantee that a linear relationship exists between x and y. Figure 2.20 shows the line of best fit through a hyperbola, defined by the equation $y = 1/x$. Despite the high coefficient of correlation ($r^2 = 0.915$) and a slope significantly different from 0 (-0.301, $p < 0.0005$), a straight line does not make a very compelling fit. The residual plot (a plot of the difference between observed and predicted values as a function of the estimated value) provides a very powerful EDA tool with which to judge the appropriateness of the linear model. In Figure 2.20, we can see that at increasing values of x, the linear model first underestimates, then overestimates, then underestimates the observed values.

Linear regression also assumes that data are normally distributed, continuous, and homoskedastic (here, this means that the variance of y does not change as a function of x). The residual plot can help you assess the homoskedasticity of the data, because the residuals should be evenly distributed across the estimated values in both sign and magnitude.

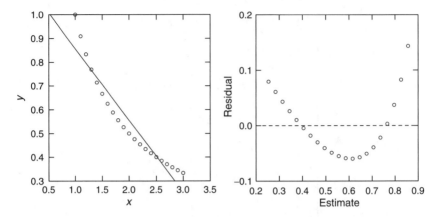

FIGURE 2.20 Fitting a linear model to hyperbolic data. The data points plotted on the left were generated from the hyperbolic function $y = 1/x$. Despite the obvious departure of these data from linearity, the coefficient of correlation was quite high (0.915). The nonrandom distribution of residuals (right-hand side), clearly show the shortcomings of the linear model.

Autologous or matched allogeneic hematopoietic stem cells can be used to rescue the bone marrow of patients treated with high-dose chemotherapy for a variety of hematologic malignancies. The hematopoietic stem cell graft is obtained by administering hematopoietic growth factors such as G-CSF [12]. This causes hematopoietic stem cells to mobilize from the bone marrow into the peripheral circulation. G-CSF is administered for several days after which leukapheresis, a technique that captures WBCs, while returning red cells, platelets, and plasma to the donor, is performed. The duration of treatment required to mobilize stem cells into the blood is variable, especially in patients who have had extensive previous chemotherapy. If you leukapherese the patient too soon, few stem cells are obtained necessitating further leukapheresis. Leukapherese too late and the patient will have to be rested and remobilized at a later time delaying the critical therapy. A biomarker predicting the results of leukapheresis would be most helpful.

To determine whether the peripheral CD34 count predicts the cell dose obtained on the same day by leukapheresis, we performed linear regression analysis of the CD34$^+$ cell dose obtained by leukapheresis (measured in CD34$^+$ cells/kilogram of patient body weight) as a function of the peripheral CD34 count (measured in CD34$^+$ cells/µL). Our first order of business would be to determine whether these variables are approximately normally distributed. As we have seen, there are many tools for this, probability plots, box plots, and density histograms among them. For the purpose of this exercise, we will blast ahead with the analysis at our peril, without EDA. Figure 2.21 shows the poor results. The data are crammed against the y-axis. A few very high values are dragging the regression downward (this is called leverage, and is not a good thing). The residual plot tells the complete story. The leftward skewing of the data is even more apparent (departure from normality), and the residuals are small for low values and get larger for high values (heteroskedasticity). The residuals seem pretty evenly distributed around 0 for low values, but are all negative and large for the four highest values (leverage). Had we started with probability plots, we

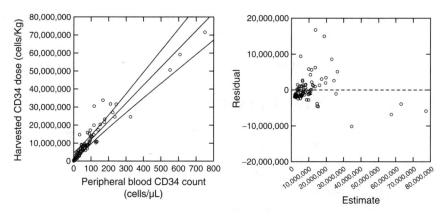

FIGURE 2.21 Linear regression of CD34 harvest dose as a function of peripheral CD34 count. The individual data points are shown as circles, the least squares line of best fit and its 95% confidence bounds are shown on the left-hand side. A plot of the residuals of regression is shown on the right-hand side.

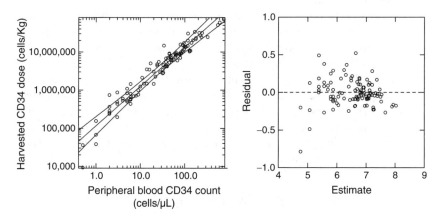

FIGURE 2.22 Linear regression of log CD34 harvest dose as a function of log peripheral CD34 count.

would have immediately recognized that both variables are log normal. Figure 2.22 shows what happens when we log transform the data and repeat the regression analysis. The data are fairly evenly spread, and the residuals are much better behaved, although now, at low values some negative residuals with high leverage have popped up (these are after all data from the real world). Most importantly, we have a regression equation of the form as follows:

$$y = 10^{5.1+1.03x}, \quad r^2 = 0.94$$

where y is the predicted CD34 harvest dose in cells per kilogram. This can be used to predict the cell dose that can be expected for a patient with a given peripheral CD34 count. In practice, we were able to greatly improve on this regression equation, using a technique called multiple linear regression, by factoring in other variables that carried independent information. These included how long the patient was on the leukapheresis machine, how many times the patient had been previously leukapheresed, and the sex of the patient. This has provided a practical tool that is used everyday in the clinic.

2.8   CAVEATS AND HANDY FACTS

- *Look at the data first.* Data analysis should always begin with EDA. You can tell a lot about your data just by plotting them. This includes the shape of the distributions, the presence of extreme values, and the nature of the relationships between variables. EDA tools such as probability plots can help you choose your models and your transformations.
- *The mean of differences.* It is not the same thing as the difference of two means. If data are naturally paired, you usually want to take the mean of individual differences. For example, if you have the data from which you subtract the "background," the average of these differences is not mathematically equivalent to taking the mean of the data and subtracting the mean background.
- *Beware of ratios.* Ratios are sometimes essential to the analysis of immunological data but they are rarely well behaved from a statistical standpoint. Also, remember that, as mentioned earlier, the average of ratios is not computationally equivalent to the ratio of averages.
- *Outliers.* Victor McKusick, a pioneer of medical genetics, exhorted his students "cherish your exceptions." Outliers often provide a key to normal biology. In contrast, if you know that technical problems encountered during an experiment explain the outlying values, it is a disservice to leave them in the analysis where they are a known source of noise. Let your conscience be your guide.

2.9   SOME HANDY PROPERTIES OF LOGS ADAPTED FROM TUKEY

- Logs in different bases differ only by multiplicative constants. You can convert a logarithm of base x to a logarithm of base y by dividing it by the base x log of y. For example, to convert a natural logarithm to the base 10, divide it by the natural log of 10. Because one base is as good as another, serial fourfold dilutions of serum, for example, can be treated as arbitrary whole number units: 0, 1, 2 for 1/8, 1/32, 1/128, respectively. After averaging these arbitrary units, the mean can be "back-converted" to a reciprocal titer by the formula: $4^x \times 8$, where 4 is the fold-dilution, x is the titer in arbitrary fourfold units, and 8 is the starting dilution. In the traditional serological literature, this is known as a geometric mean. Another way to get to the same place is to take the logs of the reciprocal titers, average them, and then take the antilog of the average.
- Engineers and mathematicians often express logarithms to the base e because e (2.71828183) crops up naturally in a variety of situations involving

derivatives and limits and simplifies calculations. Base *e* logarithms are known as natural logarithms (ln). In the biological sciences, base 10 logarithms are more common (in fact, they are known as common logarithms) because it is easier for most of us to think in powers of 10 rather than in powers of 2.7: the $\log_{10}$ of 10 is 1, of 100 is 2, of 1000 is 3, etc.

- The log of 1 is 0 (any base).
- The log of 0 is undefined. This sometimes poses a problem in the case of data such as CPM above background that can take on 0 or negative values. Most data realists agree that it is acceptable to add an arbitrary small number to data values before taking the log. Thus, radioisotopic data might be transformed as $\log_{10}$ (net CPM + 1).
- The log of a product is the sum of the logs of the factors: $\log (x \times y) = \log x + \log y$.
- The difference of log values is a ratio. For example, in base 10 logs, subtracting 3 from 6 is equal to 3. In the linear domain, this translates to $10^6/10^3 = 10^3$ or $1,000,000/1,000 = 1,000$.
- The geometric mean is obtained by logging the values, taking their average, and then taking the antilog of the average.

2.10   PRESENTING DATA THAT HAVE BEEN LOG TRANSFORMED

Although working in base 10 logarithms gets to be intuitive very quickly, it is always a good idea to plot or tabulate data in the linear domain. This is accomplished by taking the antilog of the logged values. Be aware of the properties of logarithms mentioned. For example, if you need to calculate error bars, the lower limit of your error bar should be $10^{(\text{mean}-\text{SEM})}$ and the upper limit $10^{(\text{mean}+\text{SEM})}$, not $10^{\text{mean}} \pm 10^{\text{SEM}}$. You will know that your error bars are correct if they are symmetrical about the mean when plotted on a log scale, but asymmetrical (with the lower bar being shorter) when plotted on a linear scale.

2.11   CLOSING REMARKS

This chapter discusses how a handful of exploratory and confirmatory statistical techniques can improve the design and interpretation of immunological experiments. Although immunology has always been a quantitative science, improvements in instrumentation and the direct interface of instruments to computers have allowed us to design more complex experiments in which more parameters are measured with higher resolution and greater dynamic range. Although this is certainly a blessing, it is a mixed one, because the interpretation of experimental results is commensurately more complex and, as we saw in the TCR v-beta example, the chance that rote application of statistical methods will lead to erroneous interpretations is magnified. The key to good data analysis then is to begin with techniques that allow one to visualize the lay of the data before one chooses confirmatory tests that require assumptions about it. One cannot overemphasize the importance of data transformation during the exploratory phase of data analysis. Finally, the end of scientific investigation is to communicate the findings as effectively as possible. In this regard, good graphic design is almost as critical as good data analysis. On recognizing the power and beauty of EDA data transformation, confirmatory data analysis, and graphical data display, we may find ourselves agreeing with George Bernard Shaw who said that, "It is the mark of a truly intelligent person to be moved by statistics."

ACKNOWLEDGMENTS

This chapter is written from the perspective of a biologist, with data to analyze and present, not a professional statistician. I am deeply indebted to my heroes of data analysis, John Tukey, Leland Wilkinson, and Edward Tufte, who are all statisticians.

REFERENCES

1. Tukey, JW. *Exploratory Data Analysis*. Addison-Wesley Publishing Company, Reading, MA, 1977.
2. Student (Gosset, WS). The probable error of a mean. *Biometrika* 6: 1–25, 1908.
3. Ruxton, GD. The unequal variance t-test is an underused alternative to student's *t*-test and the Mann–Whitney U test. *Behav. Ecol.* 17(4): 688–690, 2006.
4. Popper, K. *The Logic of Scientific Discovery*. Routledge Classic, London, 2002.
5. Shaffer, JP. Multiple hypothesis testing. *Annu. Rev. Psychol.* 46: 561–584, 1995.
6. Hoffmann, TK, Donnenberg, AD, Finkelstein, SD, Donnenberg, VS, Friebe-Hoffmann, U, Myers, EN, Appella, E, DeLeo, AB, Whiteside, TL. Frequencies of tetramer + T cells specific for the wild-type sequence p53(264–272) peptide in the circulation of patients with head and neck cancer. *Cancer Res.* 5; 62(12): 3521–3529, 2002.
7. Triulzi, DJ, Meyer, EM, Donnenberg, AD. Leukocyte subset analysis of leukoreduced platelet components. *Transfusion* 40: 771–780, 2000.
8. Wilkinson, L. *SYSTAT, Version 9*. SPSS Inc., Chicago, IL, 1999.
9. Tufte, ER. *The Visual Display of Quantitative Information*. Graphics Press, Cheshire, CT, 1983, 117p.
10. Schellekens, P, Eijsvoogel, VP. Lymphocyte transformation *in vitro*: Tissue culture conditions and quantitative measurements. *Clin. Exp. Immunol.* 3: 571–584, 1968.
11. Robinson, MA. The human T cell receptor beta-chain gene complex contains at least 57 variable gene segments. Identification of six V beta genes in four new gene families. *J. Immunol.* 46(12): 4392–4397, 1991.
12. To, LB, Haylock, DN, Simmons, PJ, Juttner, CA. The biology and clinical uses of blood stem cells. *Blood* 89: 2233–2258, 1997.

3 Human Immunoglobulins

Robert G. Hamilton

CONTENTS

3.1   INTRODUCTION

The human immunoglobulins are a family of proteins that confer humoral immunity and perform vital roles in promoting cellular immunity. Five distinct classes or *isotypes* of immunoglobulins (IgG, IgA, IgM, IgD, and IgE) have been identified in human serum on the basis of their structural, biological, and antigenic differences.[1–4] IgG and IgA have been further subdivided into subclasses IgG1, IgG2, IgG3, and IgG4 or subclasses IgA1 and IgA2 on the basis of unique antigenic determinants.[5,6] Multiple *allotypic* determinants in the constant region domains of human IgG and IgA molecules as well as kappa (κ) light chains indicate inherited genetic markers. Finally, there are several immunoglobulin-associated polypeptides such as secretory component (SC) and J chain that have no structural homology with the immunoglobulins, but serve important functions in immunoglobulin polymerization and transport across membranes into a variety of secretions (e.g., saliva, sweat, nasal secretions, breast milk, and colostrum). This diversity of the immunoglobulin components of the humoral immune system provides a complex network of protective and surveillance functions (e.g., see the role of immunoglobulins in the gastrointestinal tract in Chapter 13 by Guandalini).

From a clinical perspective, quantitative levels of these analytes in serum can aid in the diagnosis and management of immunodeficiency, abnormal protein metabolism, and malignant states (e.g., multiple myeloma). As such, they provide a differential diagnosis as to possible causes of recurrent infections and can indicate a strategy for subsequent therapeutic intervention. However, the reported target ranges of each immunoglobulin vary widely as a result of differences in the quantification methods, reagents employed, and populations studied. To date, no compendium of information is available that summarizes the levels of immunoglobulins in healthy pediatric and adult populations in an attempt to provide a consensus for reference intervals on which action levels can be based.

The goals of this chapter are threefold. First, primary structural and biological properties of human immunoglobulins are overviewed to highlight their antigenic diversity, which is the basis on which they are quantified. Second, the design and performance of the clinical laboratory methods that are used in the quantification of immunoglobulins are discussed within the context of available commercial and research reagents and their performance in interlaboratory proficiency surveys. Finally, a reference compendium has been prepared that summarizes the ranges of the immunoglobulins in the serum, urine, and cerebral spinal fluid (CSF) of healthy populations, where possible, as a function of age and sex and other demographic variables.

3.2 PROPERTIES OF HUMAN IMMUNOGLOBULINS

3.2.1 GENERAL IMMUNOGLOBULIN STRUCTURAL PROPERTIES

Immunoglobulins are functionally defined as glycoproteins that possess the ability to bind to substances (antigens) that have elicited their formation. Tables 3.1 and 3.2 summarize many of the known physical and biological properties of the human immunoglobulin heavy and light chains, SC, J chain, and the five classes of intact immunoglobulins. As a group, the immunoglobulins are composed of 82–96% polypeptides and 4–18% carbohydrate, and they account for approximately 20% of all proteins in plasma.[3-5] When placed in porous agarose gels along with other serum proteins under current and selected ionic conditions, most immunoglobulins along with C-reactive proteins migrate toward the cathode, forming a broad band that has been labeled the classical gamma-globulin region. Heterogeneity in their composition and net charge causes some immunoglobulins to also migrate more toward the anode, overlapping with hemopexin, transferrin, and a variety of other proteins in the beta-globulin region.[4]

As a family, the human immunoglobulins share a basic structural unit that is composed of four polypeptide chains, which are held together by both noncovalent forces and covalent disulfide bonds among their heavy chains, and with the exception of IgA2, also between their heavy and light chains.[1,3-5] Each four-chain unit is bilaterally symmetrical, containing two structurally identical *heavy* (H) chains and two identical *light* (L) chains (e.g., H_2L_2). Each polypeptide chain is composed of a number of domains comprising 100–110 amino acid residues, each forming a loop as a result of intrachain disulfide bonds. The N-terminal domain of each chain contains the area designated as the variable or V region. The V region contains several highly variable or hypervariable regions that together form the antigen-binding pocket that confers the property of antigen specificity on the immunoglobulin molecule. The COOH-terminal domains (CH1, hinge, CH2, CH3, and CH4) have been collectively defined as the constant region because the polypeptide backbone is generally invariant (with exception of allotypic differences) within a particular class of immunoglobulin.

Human IgG4 appears to be unique in the human immunoglobulin family with its apparent functional monovalency.[6] An *in vivo* exchange of IgG half-molecules (one H plus one L chain) occurs that reportedly results in bispecific antibodies that behave *in vivo* as functionally monovalent antibodies. The structural basis for this abnormal monovalent behavior has been attributed to the substitution of serine in the hinge region for a proline, which results in a marked shift in the equilibrium between inter- and intrachain disulfide bridges. The consequence is that 25–75% of the IgG4 molecules lack a covalent interaction among the heavy chains. Strong noncovalent interactions among the CH3 domains, however, cause IgG4 to remain stable four-chain molecules that do not easily exchange half-molecules under standard physiological conditions.[7]

From a clinical laboratory point of view, the constant region structure is vital to the design of immunoglobulin quantification methods. Assays are constructed using poly- and monoclonal antibody (MAb) reagents that bind to nonallotypic, invariant, isotype-unique determinants in the immunoglobulin constant region domains.

TABLE 3.1
Human Immunoglobulin Chain Characteristics

Property	Light Chains		Heavy Chains					Secretory Component	J Chain
Designation	κ	λ	α	δ	ε	γ	μ	SC	J
Associated isotypes	All	All	IgA	IgD	IgE	IgG	IgM	IgA	IgA–IgM
Subclasses or subtypes	—	1–4	1–2	—	—	1–4	1–2	—	—
Allotypes	Km(1)–(3)	—	A2m(1)–(2)	—	—	Gm(1)–(25)	—	—	—
Molecular weight (D)[a]	23,000	23,000	55,000	62,000	70,000	50,000–60,000 = g3	70,000	70,000	15,000
V-region subgroups	VKI-IV	VLI-VI	VHI-IV	VHI-IV	VHI-IV	VHI-IV	VHI-IV	—	—
Carbohydrate average %	0	0	8–10	18	18	3–4	12–15	16	8
No. of oligosaccharides	0	0	2–3	NA	5	1	5	NA	1

Note: NA = not available.

[a] Approximate MWs.

TABLE 3.2
Structural and Biological Properties of Human Immunoglobulins

	IgA1	IgA2	IgD	IgE	IgG1	IgG2	IgG3	IgG4	IgM
Heavy-chain class	$\alpha 1$	$\alpha 2$	δ	ε	$\gamma 1$	$\gamma 2$	$\gamma 3$	$\gamma 4$	μ
Light-chain type	κ and λ	κ and λ	κ and λ	κ and λ	κ and λ	κ and λ	κ and λ	κ and λ	κ and λ
Averaged Ig light-chain κ/λ ratio[15]	1.4	1.6			2.4	1.1	1.4	8.0	3.2
Molecular weight (D) of secreted form[a]	160,000m 300,000d	160,000m 350,000d	180,000m	190,000m	150,000m	150,000m	160,000m	150,000m	950,000p
Heavy-chain domain no.	4	4	4	5	4	4	4	4	5
Hinge (amino acids)	18	5	Yes	None	15	12	62	12	None
Interchain disulfide bonds per monomer					2	4	11	2	
pI range mean (SD)					8.6 (0.4)	7.4 (0.6)	8.3 (0.7)	7.2 (0.8)	
Tail piece	Yes	Yes	Yes	No	No	No	No	No	Yes
Allotypes	None	A2m(1), A2m(2)	None	Em1	G1m: a(1), x(2), f(3), z(17)	G2m: n(23)	G3m: b1(5), c3(6), b5(10), b0(11), b3(13), b4(14), s(15), t(16), g1(21), c5(24), u(26), v(27), g5(28)	G4m, Gm4a(i), Gm4b(i)	None
Distribution (% intravascular)[15]	55 ± 4	57 ± 2	75	50	45	45	45	45	80
Biological half-life (days)[3,15]	5.9 ± 0.5	4.5 ± 0.3	2–8	1–5	21–24	21–24	7–8	21–24	5–10
Fractional catabolic rate (% intravascular pool catabolized per day)	24 ± 2	32 ± 4	37	71	7	7	17	7	8.8
Synthetic rate (mg/kg/day)[3,15]	24 ± 5	4.3 ± 1	0.4	0.002	33	33	33	33	3.3

(continued)

TABLE 3.2 (continued)
Structural and Biological Properties of Human Immunoglobulins

	IgA1	IgA2	IgD	IgE	IgG1	IgG2	IgG3	IgG4	IgM
Approximate total Ig in adult serum (%)	11–14	1–4	0.2	0.004	45–53	11–15	0.03–0.06	0.015–0.045	10
Adult range: age 16–60 in serum (g/L)[b]	1.81	0.22			5–12	2–6	0.5–1	0.2–1	0.25–3.1
Functional valency	2,4	2,4	2	2	2	2	2	1–2	5–10
Transplacental transfer	0	0	0	0	++	+	++	++	0
Binding to phagocytic cells	0	0	0	0	++	+	++	±	0
Complement activation classical path	0	0	0	0	++	+	+++	0	+++

[a] Approximate MW (m = monomer, d = dimer, t = trimer of IgA are produced, p = pentamer).

[b] Ranges of immunoglobulins in serum are examined in detail in the text. These ranges are provided as a general indication of target levels. Immunoglobulins rates of metabolism were extracted from Waldmann, T.A., Strober, W., Blaese, R.M., Metabolism of immunoglobulins, in *Progress in Immunology*, Amos, B., Ed., Academic Press, New York, 1971, pp. 891–903. With permission.

3.2.1.1   Human Light Chains

Human light chains are approximately 23,000 Da proteins that contain no oligo-saccharides.[4] They have been classified into two types: kappa (κ) and lambda (λ) based on their unique antigenic determinants that result from structural differences in their constant region domains. Based on structural and antigenic differences λ light chains have been further subdivided into four subtypes. The combination of a light chain with a heavy chain is a random process, and thus a complete repertoire of light chains can be found bound to every immunoglobulin isotype heavy chain. It is the heavy chain, therefore, that determines the isotype and the biological effector functions of the immunoglobulin. The kappa to lambda (κ/λ) ratio in the serum of healthy humans is approximately 2:1; however, it can reportedly vary from 1.1 to 8.0 depending on the human immunoglobulin isotype (Table 3.2).

3.2.1.2   Human Heavy Chains

The principal structural characteristics of the five major classes of heavy chains are summarized in Table 3.1. Heavy chains vary in their molecular weight (MW) (50,000–70,000 Da), the percentage of carbohydrates (4–18) and number of oligo-saccharides (1–5), number of respective subclasses (IgG1-4 and IgA1-2) and allo-types, number of constant region domains, and number of interchain disulfide bonds. Further details about the structural aspects of the human heavy chains are beyond the scope of this chapter and are presented elsewhere.[3–5] Most important to this discussion, the unique antigenic determinants on the heavy chain define the isotype of an immunoglobulin, and thus form the basis on which the different classes of immunoglobulins are differentiated and quantified in the immunoassays discussed subsequently.

3.2.1.3   Secretory Component

Human SC is a 90,000-MW glycoprotein that is expressed as an integral protein on the basolateral membrane of mucosal epithelial cells.[3,8,9] SC is either released into mucosal secretions as a 70,000-MW soluble fragment or bound to polymers of IgA through strong noncovalent interactions. Structurally, it shares no homology with human heavy or light polypeptide chains or the human J chain. It contains a high carbohydrate content and serves as a receptor to transport IgA across mucosal tissues into various human secretions. Secreted human IgA is composed of two IgA monomers, a J-chain linker, and a molecule of SC. Mucosal secretions contain a mixture of secretory IgA and free SC.[10] Highly specific murine MAbs are available (Table 3.3) that bind to selected determinants on the SC and allow their quantification in serum and mucosal secretions.[11,12]

3.2.1.4   J Chain

The J chain is an elongated glycoprotein of approximately 15,000 Da that can be distinguished by its unusually high quantity of glutamic and aspartic acid residues.[3,13] It reportedly serves as a facilitator of polymeric immunoglobulin (e.g., IgM and IgA) polymerization. A single J chain has been identified in each pentameric IgM

TABLE 3.3
Human Immunoglobulin-Specific Murine MAbs

Clone	Mouse Isotype	Human Ig Specificity	Epitope Specificity[a]
HP6017	IgG2a	IgG-PAN	Fc-CH2
HP6043[b]	IgG2b	IgG-PAN	Fc-CH2
HP6046[b]	IgG1	IgG-PAN	Fd-CH1
HP6069[b]	IgG1	IgG1	Fc-CH2
HP6070	IgG1	IgG1	Fc-CH2
HP6002[b]	IgG2b	IgG2	Fc-CH2
HP6014[b]	IgG1	IgG2	Fd-CH1
HP6047[b]	IgG1	IgG3	Fd-hinge
HP6050	IgG1	IgG3	Fd-hinge
HP6023	IgG3	IgG4	Fc-CH3
HP6025[b]	IgG1	IgG4	Fc-CH3
HP6019[b]	IgG1	IgG1,3,4	Fc-CH2
HP6030[b]	IgG1	IgG2,3,4	Fc-CH2
HP6058[b]	IgG1	IgG1,2,3	Fc-CH2
HP6029	IgG1	IgE	Fc
HP6061	IgM	IgE	Fc
HP6081	IgG1	IgM	Fc
HP6083	IgG1	IgM	Fc
HP6086	IgG1	IgM	Fc
HP6111	IgG1	IgA-PAN	Fc
HP6123	IgG1	IgA-PAN	Fd
HP6054	IgG2a	λ Light chain	—
HP6062	IgG3	κ Light chain	—
HP6130	IgG1	SC	—
HP6141	IgG1	SC	—

[a] Designation of the domain to which the IgG-specific MAbs bind have been obtained from Matsson, P., Hamilton, R.G. and Diagnostic Allergy Techniques Working Group. *Analytical Performance Characteristics and Clinical Utility of Immunological Assays for Human IgE Antibody of Defined Allergen Specificities: Guideline*, Clinical and Laboratory Standards Institute (formally the National Committee on Clinical Laboratory Standards), Wayne, PA, 2007; Jefferis, R., Reimer, C.B., Skvaril, F., De Lange, G., Ling, N.R., Lowe, J., Walker, M.R., Phillips, D.J., Aloiso, C.H., Wells, T.W., Vaerman, J.P., Magnusson, C.G., Kubagawa, H., Cooper, M., Vartdal, F., Vandvik, B., Haaijnan, J.J., Makela, O., Sarnesto, A., Lando, Z., Gergely, J., Rajnavolgyi, E., Laszlo, G., Radl, J., Molinaro, G., *Immunol. Lett.*, 10, 223, 1985; Hamilton, R.G., *The Human IgG Subclasses: Molecular Analysis of Structure, Function and Regulation*, Pergamon Press, New York, 1990; Reimer, C.B., Phillips, D.J., Aloisio, C.H., Moore, D.D., Galland, G.G., Wells, T.W., Black, C.M., McDougal, J.S., *Hybridoma*, 3, 263, 1984; Phillips, D.J., Wells, T.W., Reimer, C.B., *Immunol. Lett.*, 17, 159, 1987; Reimer, C.B., Phillips, D.J., Aloisio, C.H., Black, C.M., Wells, T.M., *Immunol. Lett.*, 21, 209, 1989. With permission. The Fd refers to the heavy chain of the $F(ab')_2$.

[b] MAbs specific for human IgG that were selected as components for the polymonoclonal antihuman IgG reagent [43].

or polymeric IgA molecule, covalently bound to the penultimate cysteine residue of mu or alpha heavy chains. From a clinical point of view, the J chain is rarely quantified unless a structural or functional abnormality of this protein is suspected or it is required as a marker to distinguish multiple myeloma from benign gammopathy.[13]

3.2.2 ISOTYPE-UNIQUE STRUCTURAL AND BIOLOGICAL PROPERTIES

Immunoglobulins are bifunctional molecules that bind antigen through their V region. This binding process can elicit a variety of secondary effector functions (e.g., complement activation leading to bacteriolysis and augmentation of phagocyte chemotaxis and opsonization, and histamine release from mast cells), which are independent of the immunoglobulin's antigen specificity and depend on C-region determinants (Table 3.2). Although all the five major isotypes of human immunoglobulins share the common structural features of the four-chain monomer subunits discussed earlier, they vary in terms of minor structural aspects that confer some special biological properties.

3.2.2.1 Human IgA

Polymeric secretory IgA is composed of two four-chain basic units and one molecule each of SC and J chain (approximately 400,000 MW).[3,4,9] It is the predominant immunoglobulin in colostrum, saliva, tears, bronchial secretions, nasal mucosa, prostatic fluid, vaginal secretions, and mucous secretions of the small intestine.[8,9] In contrast, 10% of the circulating serum IgA is polymeric, whereas 90% is monomeric (160,000 MW). Together, they constitute approximately 15% of the total serum immunoglobulins. Trimers and higher polymeric forms can exist, but in small amounts. Two subclasses of IgA have been identified (IgA1 and IgA2), which differ by 22 of the 365 amino acids.[14] Apart from the 13 amino acid deletion in the IgA2 hinge region, IgA1 and IgA2 constant region domains vary by 20 amino acid substitutions. The IgA1 hinge contains a carbohydrate attachment site, whereas the IgA2 hinge does not. IgA2 is present in two allotypic forms: IgA2m(1) and IgA2m(2). The human IgA PAN-specific murine MAbs listed in Table 3.3 react either to the crystalizable fragment (Fc) or to the antigen binding fragment (Fab) fragment of both subclasses of IgA and both allotypic forms of IgA2. Approximately 80% of the serum IgA is IgA1. In secretions, however, IgA2 concentrations can approach 50% of the total IgA. IgA2 lacks proteolytically sensitive epitopes that makes it particularly resistant to cleavage by enzymes produced by a variety of bacteria (*Clostridium* spp., *Streptococcus pneumoniae*, *S. sanguis*, *Haemophilus influenzae*, *Neisseria gonorrhoeae*, and *N. meningitidis*) that otherwise readily cleave IgA1 into Fab and Fc fragments. The polymeric nature and presence of SC on IgA in secretions increases its resistance to bacterial proteolysis. The IgA2m(1) allotype lacks inter-heavy-chain disulfide bonds. The light chains of IgA2m(2) are linked by disulfide bonds rather than to their respective alpha heavy chain; however, no special biological function has been associated with this unique structural difference.[14,15]

In terms of complement activation, IgA poorly activates the classical pathway.[8] This process has been hypothesized as a host mechanism for attenuating inflammatory responses induced by IgG antibodies at the mucosal surface. In contrast, IgA reportedly activates the alternative pathway of complement to provide some direct protective functions. IgA, once bound to a bacterial or parasitic surface antigen, may bind CD89 (IgA receptor) on inflammatory cells (monocytes, macrophages, neutrophils, and eosinophils), leading to their destruction by means of antibody-dependent cell-mediated cytotoxicity (ADCC). Moreover, its binding to viral or microbial surface antigens may restrict the mobility of microorganisms and prevent their binding to mucosal epithelium. Finally, secretory IgA can play an important first line of defense in antigen clearance by binding to antigens that leak across an epithelium and transporting them back across to prevent their entry.[9] To summarize, IgA's unique structure resists proteolysis and it functions to block uptake of antigen, bacterial or viral attachment, limit inflammation induced by classical pathway complement activation, and promote microbial destruction through ADCC by binding to leukocyte receptors.

3.2.2.2   Human IgD

IgD is a four-chain monomer of approximately 180,000 MW with a long hinge region that increases its susceptibility for proteolytic cleavage. Although IgD is normally present in serum in trace amounts (0.2% of total serum immunoglobulin), it predominantly serves as a membrane-bound antigen receptor on the surface of human B lymphocytes.[3] Despite suggestions that IgD may be involved in B-cell differentiation, its principal function is as yet unknown. As such, IgD is rarely quantified in a general workup of an individual suspected of a humoral immune deficiency or a B-cell dyscrasia. Hyperimmunoglobulinemia D with serum IgD levels >100 U/mL, however, has been noted in conjunction with periodic fever syndrome.[16] This condition is a rare, autosomal recessive disorder that is characterized by recurrent episodes of fever accompanied by abdominal distress, lymphadenopathy, joint involvement, and skin lesions. It appears to be particularly responsive to anti-tumor necrosis factor (TNF) treatment. Mutations that lead to this disease occur in the mevalonate kinase gene, which encodes an enzyme involved in cholesterol and nonsterol isoprenoid biosynthesis.

3.2.2.3   Human IgE

IgE (190,000 MW) was identified in 1967 as a unique immunoglobulin that circulates in serum as a four-chain monomer.[17,18] Although IgE constitutes only 0.004% of the total serum immunoglobulins, it possesses a clinically significant biological function by binding through its Fc region to the alpha chain on high-affinity receptors (FcεR1) on mast cells and basophils.[19] On subsequent exposure to relevant protein allergens from trees, grasses, weeds, pet dander, molds, foods, or insect venoms, IgE antibodies on mast cells become cross-linked. This process triggers the production and release of vasoactive mediators (e.g., histamine, prostaglandins, and leukotrienes) that can induce mild to severe immediate type I hypersensitivity reactions in sensitized atopic individuals. Clinical diagnostic allergy laboratories focus on the quantitation of total serum IgE and allergen-specific IgE to identify an individual's propensity to develop a spectrum of type I reactions on exposure to a defined panel of 400–500 known allergenic substances (see also Chapter 13 for a discussion on food allergy).[20]

Total serum IgE is commonly expressed in international units per milliliter (IU/mL) or converted to mass units using 1 IU = 2.44 ng of protein. Recently, International System of Units have proposed units in which1 SI = 1 μg/L; however, these units have not been widely adopted in clinical immunology laboratories that perform allergy testing.

In 2003, omalizumab was licensed by the U.S. Food and Drug Administration (FDA) for use in the management of asthma. Omalizumab is a humanized IgG1 antihuman IgE Fc MAb that binds to the region on the epsilon heavy chain that interacts with the alpha chain of the FcεR1. This interaction blocks IgE binding to FcεR1 and downregulates the number of FcεR1 receptors on mast cells and basophils. Important to this chapter on immunoglobulin quantification is the observation that the presence of exogenously administered anti-IgE (omalizumab) in serum degrades the accuracy of some but not all total and allergen-specific IgE assays.[21] This issue is discussed in Section 3.7.

3.2.2.4 Human IgG

In healthy adults, the four polypeptide chain IgG monomer (150,000 MW) constitutes approximately 75% of the total serum immunoglobulins.[3,4] IgG is approximately equally distributed between intra- and extravascular serum pools. Moreover, IgG possesses the unique ability to cross the placenta, which provides protection for the fetus and newborn. Human IgG has been subdivided into four subclasses on the basis of unique antigenic determinants. Table 3.2 summarizes major structural and biological differences among the IgG subclasses. Relative subclass percentages of the total IgG in serum are IgG1, 60–70%; IgG2, 14–20%; IgG3, 4–8%; and IgG4, 2–6%.[3,5,22] IgG1, IgG2, and IgG4 possess an MW of approximately 150,000, whereas IgG3 is heavier (160,000 MW) as a result of an extended 62-amino acid hinge region that contains 11 interchain disulfide bonds. IgG3's highly rigid hinge region promotes accessibility of proteolytic enzymes to sensitive Fc cleavage sites, which results in an increased fractional catabolic rate and a shorter biological half-life (7–8 days) than has been observed for IgG1, IgG2, and IgG4 (21–24 days). In terms of complement activation, IgG1 and IgG3 are the most effective, whereas IgG4 due to its compact structure does not readily activate the classical pathway of complement. IgG4 antibodies are also unique in that they appear to be functionally monovalent due to *in vivo* exchange of IgG4 half-molecules.[6,7] As such, this is believed to lead to the formation of small IgG4 immune complexes that have a low potential for inducing immune inflammation. Moreover, IgG4 antibodies have the ability to interfere with immune inflammation caused by the interaction of complement-fixing IgG subclasses with antigen. Researchers in the field of allergy have speculated that IgG4 antibodies also scavenge antigen that prevents mast cell-bound IgE antibody from being cross-linked by antigen, and thus blocking IgE-mediated hypersensitivity reactions in atopic individuals who have undergone immunotherapy. Other important structural and biological differences among the human IgG subclasses relate to their Fc receptor binding, and the different binding sites on the constant region domains for rheumatoid factors, complement components, and bacterial proteins (protein A and protein G). The reader can refer to several reviews that discuss these differences in detail.[5,22,23]

3.2.2.5 Human IgM

IgM is a pentameric immunoglobulin of approximately 900,000 MW that is composed of a J chain and five IgM monomers. Pentameric IgM constitutes approximately 10% of serum immunoglobulins in healthy individuals. Along with IgD, monomeric IgM is also a major immunoglobulin that is expressed on the surface of B cells where it serves as an antigen receptor. The C-terminal portion of pentameric secreted IgM differs from that of its monomeric cell-bound form. Secreted IgM has a mu chain with a 20-amino acid hydrophilic tail and a penultimate cysteine that facilitates polymerization. Cell membrane-bound IgM has a 41-amino acid membrane tail that contains a hydrophobic 26-amino acid segment that anchors the IgM molecule in the B-cell membrane lipid bilayer. IgM antibodies are clinically important because they predominate as an antigen receptor in early immune responses to most antigens. With a theoretical functional valency of 10, IgM antibodies are highly efficient in activating the classical complement pathway. IgM's actual functional valency, however, is only 5 due to steric hindrance among its many antigen-binding sites.[3,24]

3.3 CLINICAL APPLICATIONS

An immunological workup of an individual who presents with the complaint of chronic or recurrent infections, sometimes with unusual infecting agents, commonly involves the examination for one or more defects in the patient's antibody-mediated (B-cell), cell-mediated (T-cell), phagocytic, or complement segments of their immune system. The level of serum immunoglobulins are commonly measured to identify an underlying defect in the humoral immune system.[25,26]

There are a variety of primary immunodeficiency disorders (reviewed in Chapter 9) that can produce immunoglobulin patterns ranging from a complete absence of all isotypes of immunoglobulins (e.g., hypogammaglobulinemia) to a selective decrease in a single isotype (selective IgA deficiency). Sometimes, a deficiency in one or several isotypes (e.g., IgG and IgA) can be associated with an elevated level of a third isotype (e.g., IgM). The immunoglobulin profiles of the major primary immunodeficiency diseases are presented in Table 3.4. In the case of hyper-IgE syndrome, levels of IgE in excess of 12,000 ng/mL can be accompanied by a general diminished antibody response following immunization. A spectrum of secondary causes of decreased serum immunoglobulin levels may include malignant neoplastic diseases (e.g., myeloma), protein-losing states (e.g., nephrotic syndrome and protein-losing enteropathy), and immunosuppressive treatment (e.g., transient decrease from corticosteroids). A detailed description of the common symptoms, laboratory findings, other immune markers used in the differential diagnosis of these, and other immunodeficiency disorders is presented elsewhere.[26,27]

At the other extreme from immunodeficiency are hematological diseases such as plasma-cell dyscrasias that can lead to gross elevations in one or several immunoglobulin isotypes as a result of malignant proliferation of one or several clones of B cells.[28] As a group, these conditions are often referred to as paraproteinemias or monoclonal gammopathies and they can be distinguished by the presence of a monoclonal immunoglobulin in the patient's serum or urine. The laboratory investigation

TABLE 3.4
Serum Immunoglobulin Levels in Primary Immunodeficiency Disorders

Disease	Total Ig	IgA	IgD	IgE	IgG	IgM
X-linked infantile hypogammaglobulinemia (Bruton's agammaglobulinemia)[a]	<2.5 g/L	<	<	<N	<2.0 g/L	<
Transient hypogammaglobulinemia of infancy	<3.0 g/L	<	<	<	<2.5 g/L	<
Common, variable immunodeficiency (acquired hypogammaglobulinemia: 15–35 years of age)	<3.0 g/L	<-N	<	<	<2.5 g/L	<N
Immunodeficiency with hyper-IgM	<>	<	<	<N	<	>
Selective IgA deficiency[b]	N>	<0.05 g/L	N>	N>	N>	N>
Selective IgM deficiency (rare)	N	N	N	N	N>	<
Severe combined B-/T-cell immunodeficiency[c]	<	<	<	<	<M	<
Cellular immunodeficiency with abnormal immunoglobulin synthesis (Nezelof's syndrome)[d]	<N>	<N>	<N>	<N>	<N>	<N>
Immunodeficiency with thrombocytopenia, eczema, and recurrent infection (Wiskott–Aldrich syndrome)	?	>	?	>	N	<

Note: < = below age-adjusted reference range or nondetectable, N = normal level for age group, > = above age-adjusted reference range, <N = below or at age-adjusted reference range, N> = at or above age-adjusted reference range, <N> = below, at or above age-adjusted reference range, <m = below maternal IgG level transferred through the placenta, ? = unknown.

a   Most infants go through a period of hypogammaglobulinemia at approximately 5–6 months of age as infant shifts from exogenous IgG (maternal through placenta) to endogenously produced immunoglobulin. During this period the infant can experience recurrent respiratory tract infections.

b   Selective IgA deficiency has been also associated with an IgG2 deficiency in some individuals; some patients have normal levels of serum IgA levels <0.05 g/L and either normal or low secretory IgA. Rarely, normal serum IgA levels can be accompanied by low secretory IgA, possibly as a result of an absence of SC.

c   Onset of symptoms by 6 months of age with recurrent viral, bacterial, fungal, and protozoal infection. The presence of placentally transferred material IgG can make diagnosis difficult.

d   Various degrees of B-cell immunodeficiency produce decreased, normal, or increased (<N>) immunoglobulin levels.

of paraproteinemias involves a variety hematologic (e.g., complete blood count with differential), routine clinical chemistry (e.g., total protein, albumin, globulin, calcium, phosphate, electrolytes, and uric acid), hemostatic (e.g., clotting time and platelet count), serum viscosity, radiological examination and renal function tests. Immunological tests are then performed, beginning with a total serum immunoglobulin and ending with a serum protein electrophoresis with immunofixation if a paraprotein is suspected.[28] A quantitative measurement of the ratio of serum κ/λ light-chain concentrations has been proposed as a simpler alternative method to electrophoresis–immunofixation for identifying monoclonal proteins. In theory, a serum κ/λ light-chain ratio that is above or below a reference range for healthy adults may indicate the presence of a paraprotein.[29] However, because serum levels of immunoglobulins other than the myeloma isotype are highly variable and commonly significantly lower in most myeloma patients than the adult reference ranges, the observed serum κ/λ ratio may be decreased, normal, or increased in individuals with known paraprotein. Thus, a normal serum κ/λ light-chain ratio does not guarantee the absence of a paraprotein. Abnormal serum κ/λ ratios are generally followed by protein electrophoresis with immunofixation to confirm the presence and the type of the paraprotein(s).[30] Bence Jones protein (light chains) can be detected in the urine of about half of all patients with multiple myeloma. About 20% of these myeloma patients have only Bence Jones proteinuria as the sole distinguishing feature. Waldenstrom's macroglobulinemia is a special disease state in which the patients experience hyperviscosity of the blood as a result of excess monoclonal IgM production. Although the monoclonal IgM is often a pentamer, monomeric IgM has also been observed in this abnormal immunological state.

IgE is a special immunoglobulin isotype in terms of its clinical utility.[20] A moderately elevated total serum IgE positively reinforces the differential diagnosis of atopic disorders such as allergic rhinitis, allergic asthma, and atopic dermatitis. Very high serum IgE levels are necessary for the diagnosis of hyper-IgE syndrome in patients with an increased susceptibility to infections and dermatitis. Many parasitic infections can produce extremely elevated total serum IgE levels, and thus a high IgE in the absence of other explanations strongly suggests the possibility of parasitism. A normal IgE level makes the diagnosis of parasitism less likely as a cause of eosinophilia, which is otherwise a common feature of nonallergic asthma. Normal total serum IgE levels can identify nonallergic or intrinsic asthma and exclude allergic bronchopulmonary aspergillosis.

Several notes of caution are warranted when measuring immunoglobulins in clinical specimens from individuals with disease. First, clinical immunoglobulin assays are designed to measure immunoglobulins at levels that are commonly observed in healthy children and adults. Some clinical assays may not have the analytical sensitivity required to detect low levels of immunoglobulins in pediatric sera. Second, paraproteins are often structurally atypical immunoglobulins that may produce inaccurate results in some clinically used immunoglobulin assays. This can be a problem for polyclonal antibody-based immunodiffusion or nephelometric assays where the size of the immunoglobulin (e.g., IgM pentamer or monomer) will affect its rate of migration or extent of reflectance of light. The resultant diameter of the immunoprecipitation ring or luminescence measured may not be an accurate reflection of the immunoglobulin's concentration. There may also be a problem with poly- or monoclonal antibody-based immunoassays that use antisera, which fail to recognize altered structural determinants on an atypical paraprotein. Finally, an individual may have an immunoglobulin level that varies about its norm for that individual. When it varies significantly from the individual's norm, it may still be within the population reference intervals and thus considered "normal" when it is actually abnormal for that individual. This has contributed to a heightened interest in distinguishing an *antibody deficiency* as distinct from an *immunoglobulin deficiency* in the identification of causes of recurrent infectious disease. In other words, an individual with a serum immunoglobulin within the reference range for an age-adjusted healthy population may be unable to mount a specific antibody response against a panel of protein or carbohydrate antigens.

3.4 METHODS OF QUANTITATION

3.4.1 SPECIMEN TYPES

Human immunoglobulins have been detected in a variety of body fluids. The most extensively studied and reproducible clinical specimen between individuals of different sexes and races is serum (plasma).[25] Urine is evaluated for the presence of heavy or light chains and occasionally for intact immunoglobulins if kidney dysfunction and plasma-cell dyscrasias are suspected. The level of IgG and the assessment of oligoclonal immunoglobulin bands in CSF is part of the workup for an individual suspected of having multiple sclerosis. Finally, immunoglobulins (e.g., secretory IgA, IgG, and IgE) are occasionally investigated on a research basis in other human body fluids such as tears, sweat, peritoneal fluid, colostrum, saliva, bronchial secretions, nasal mucosa, prostatic fluid, vaginal secretions, and mucous secretions of the small intestine.[31-33] The reference ranges presented in this chapter focus on immunoglobulin levels that have been reported in serum, urine, and CSF.

3.4.2 REAGENTS

A variety of immunological reagents are used in clinical assays to quantify human immunoglobulins. In the early years, polyclonal antibodies were extensively used with both immunoprecipitation-based and two-site immunometric assays. More recently, well-documented murine MAbs are used because of their exceptional specificity for human immunoglobulins. MAbs have been especially useful in the quantitation of the human IgG and IgA subclasses where maximal specificity and high avidity are requirements.

3.4.2.1 Polyclonal Antibodies

Assays involving the formation of an immune complex that is subsequently detected visually or by light-scatter techniques (e.g., immunodiffusion, nephelometry, and turbidimetry) almost always require the use of highly avid polyclonal antibody reagents to achieve sufficient analytical sensitivity. A majority of clinical laboratories measuring human immunoglobulins purchase an "FDA-cleared" assay that has been through a lengthy documentation process overseen by the FDA (e.g., 510k). In these cases, the specificity of antibody reagents used in the assay has been documented by cross-reactivity analyses that have been performed by the manufacturer. Cross-reactivity for heterologous immunoglobulin isotypes should be negligible (e.g., <0.001%) especially when IgE, IgM, or IgA are being quantified in fluids containing high levels of IgG.[34]

3.4.2.2 Monoclonal Antibodies

Clinical assays such as those that measure IgG and IgA subclass levels in serum require the exceptional specificity that is provided by MAbs. International collaborative studies have documented available murine MAbs for their specificity and utility in the detection of the human IgG and IgA subclasses and human SC.[12,35,36] Highly avid murine MAbs have become commercially available with specificity for human IgG and its four individual subclasses,[35–38] human IgA and its two individual subclasses,[12,39] human IgM, human IgE,[31,40] human secretory piece,[39] and the human κ and λ light chains.[35] The mouse isotype, human immunoglobulin specificity, and clone number of a selected panel of such MAbs are presented in Table 3.3 for illustration. HP6014, for example, describes a mouse IgG1 antihuman IgG2 Fab that is the hybridoma product of clone 6014. The term PAN has been used in the documentation studies to indicate that a particular MAb binds to all subclasses and allotypic forms of that particular human immunoglobulin isotype. Thus, HP6043 is a human IgG Fc PAN-reactive MAb that has been shown to react to the Fc region of all four subclasses and known allotypic forms of human IgG. Most of these MAbs have been purified from ascites by immunochemical techniques such as sequential anion exchange resin and hydroxylapatite chromatography. Affinity chromatography using protein G or another immunoglobulin-binding reagent is not encouraged since any release of protein G and anti-IgG into the final preparation may cause a loss of the purified MAb's specificity. Occasionally, the purified MAbs are labeled with biotin or an enzyme (horseradish peroxidase, alkaline phosphatase, and β-galactosidase) and then quality controlled by enzyme immunoassay or electrophoretic-blotting methods such as isoelectric focusing (IEF) immunoblot analysis.[41] The degree of immuno- and cross-reactivity can be studied with human paraproteins to confirm their consistency and restricted specificity.[35,36,41] Of the antibodies listed in Table 3.3, MAbs produced from clones HP6043 (antihuman IgG Fc PAN), HP6083 (antihuman IgM Fc), and HP6123 (antihuman IgA Fd PAN) have been identified by investigators in the United States as reference antibodies for the detection of human IgG, IgA, and IgM antibodies in infectious disease serological immunoassays.

3.4.2.3 International and National Reference Proteins and Serum Standards

3.4.2.3.1 Reference Proteins for Specificity Analysis
International collaborative studies of human immunoglobulin-specific MAbs have involved the use of well-characterized human reference immunoglobulins that are supplied by agencies such as the World Health Organization (WHO). Their use has been invaluable in documenting the restricted isotype specificity and lack of allotype selectivity of human immunoglobulin-specific immunochemical reagents.[12,35] The

majority of these reference immunoglobulins are myeloma paraproteins that have been isolated from human serum and characterized in terms of their light-chain type and heavy-chain isotype and concentration. As such, they must be considered atypical immunoglobulins. The use of multiple paraproteins of the same heavy- and light-chain type and allotype can minimize biases that may result from possible structural differences caused during transcription or translation of the paraprotein from the myeloma cell line.

Molecular engineering techniques have been used to produce human–mouse chimeric antibodies that possess human immunoglobulin constant region domains and a defined V-region specificity for haptens such as nitrophenyl (NP) or dansyl.[42,43] Although these are not internationally recognized reference proteins, they have been successfully applied to the documentation of the isotype-restricted specificity of the panel of human immunoglobulin-specific murine MAbs in Table 3.3.[44] In some cases, their ability to bind to a defined insolubilized antigen and present their C-region determinants in an orientation that would mimic human antibodies binding to their insolubilized antigen has made them candidates as calibration proteins for future human antibody standards.[45]

3.4.2.3.2 Human Serum Pools

A number of internationally recognized serum pools have been calibrated by value transfer or consensus procedures for use as reference sera to calibrate clinical human serum immunoglobulin assays. Historically, a number of primary reference sera have been used to calibrate human IgG, IgA, and IgM assays. These have included the WHO International Reference Preparation (WHO 67/86, WHO 67/97); the U.S. National Reference Preparation (USNRP-IS1644); Netherlands Red Cross Reference Preparation (NRCRP-H0002); International Federation of Clinical Chemistry (IFCC) Immunoglobulin Standard 71/4; and College of American Pathologists (CAP) Reference Preparation for Serum Proteins (CAP-RPSP-4).[46–48] The WHO International Reference Preparation for human IgD (WHO 67/37) and IgE (WHO 75/502) have been used to calibrate total serum IgD and IgE assays, respectively.[49,50] Unfortunately, most of these international reference preparations have exceeded their life span and are no longer available for use.

By 1993, a new reference preparation (CRM-740/RPPHS lot 5) had been prepared and the lengthy process began to establish it as a certified reference material (CRM).[51,52] Collaborative studies involved extensive cross-validation with the earlier reference preparations to verify the concentrations of 14 serum proteins in the CRM-740. The CRM-740 has become the current internationally recognized serum protein reference preparation for immunoglobulin assays.[53–55] A portion of the CRM-740 is maintained by the IFCC and the remainder by the CAP.

3.4.3 Assay Designs

Human IgG, IgA, IgM, IgD, IgE, and the light chains are routinely measured in the clinical immunology laboratory. SC and J chain are considered as research analytes. Three types of assays are currently used by clinical immunology laboratories to quantify human IgG, IgA, IgM, IgD, IgE, and the κ and λ light chains. These are immunodiffusion assays, nephelometric–turbidimetric assays, and immunoassays.[56]

3.4.3.1 Immunodiffusion

The radial immunodiffusion assay (RID) was originally described in 1965 by Mancini.[57] RID employs polyclonal antisera or in rare cases, a carefully constructed mixture of MAbs in a porous agarose gel into which a small quantity of serum (5–10 μL) is pipetted. As the serum proteins migrate through the gel, immune-complexed proteins form a visible white precipitin ring with a diameter that is proportional to the concentration of the particular analyte specific for the antiserum in the gel. The ring diameter is measured either at a defined time such as 18–20 h (Fahey–McKelvey technique)[58] or at maximal endpoint equivalence (Mancini technique), and interpolated from a dose–response curve constructed with multiple dilutions of a reference serum. Immunodiffusion assays are used in smaller clinical laboratories that have fewer specimens and can accept a 2–3 days turnaround time. In the 2006 Diagnostic Immunology Proficiency Survey conducted by the CAP, approximately 20% of participating laboratories measure human IgG1, IgG2, IgG3, and IgG4 using immunodiffusion assays.[59] Some laboratories also measure human IgD in serum by means of RID. Performance of these laboratories in the CAP survey in terms of accuracy and variance was equivalent (e.g., interlaboratory variation <18% coefficient of variation [CV]) to laboratories using other assay methods. Gel-based immunodiffusion methods tend to be limited in their analytical sensitivity (1 μg/mL),[56] and thus they are not clinically useful in the measurement of immunoglobulins that are normally in low concentrations in serum (e.g., IgE). Variance in the immunodiffusion assay is primarily dependent on the accuracy with which the test and reference sera are pipetted and the precipitin ring diameters are measured.

3.4.3.2 Nephelometric–Turbidimetric Assays

Both nephelometric and turbidimetric assays function on a similar principle in which serum (containing variable amounts of the analyte) is added to a reaction chamber containing a constant amount of optically clear, IgG-, IgG1-, IgG2-, IgG3-, IgG4-, IgA-, or IgM-specific antiserum. The extent of immune-complex formation varies as a function of the quantity of the particular immunoglobulin being measured. The rate or extent of immunoglobulin–anti-immunglobulin complex formation is then measured by the extent of light incident on the reaction chamber that is (a) scattered or reflected toward a detector that is not in the direct path of the transmitted light (nephelometry), or (b) attenuated (decreased) in intensity as measured by a detector in the direct path of the transmitted light as a result of scattering, reflectance, and absorption (turbidimetry).[60] Light scatter or turbidity increases immediately following the mixture of antigen and antibody to a maximum value (equivalence) and then decreases. The extent of scatter or absorption obtained with dilutions of a reference serum containing known quantities of IgG, IgG1, IgG2, IgG3, IgG4, IgA, IgD, or IgM allow construction of a reference serum from which response results obtained with test sera are interpolated. Earlier nephelometric assays generally exhibited a lower analytical sensitivity than turbidimetric assays due to difficulty in accurately and precisely measuring small changes in light absorbance in the forward direction.[61] However, stable, high-resolution photometric systems have insured that the two methods are competitive. Both assays suffer from inaccuracies caused when the immunoglobulin (antigen) is in a molar excess relative to the anti-immunoglobulin (antiserum); however, computer algorithms are designed to flag antigen excess automatically. Finally, any particle or solvent as well as serum macromolecules, can scatter light causing inaccuracies. The advantages of these two methods reside in their speed and relative simplicity. High levels of lipoproteins in lipemic serum, hemoglobulin concentrations >5.0 g/L in hemolyzed blood and bilirubin levels >0.15 g/L in icteric serum may cause interference in both nephelometric and turbidimetric assays.[62] Of the participating clinical laboratories in 2006 CAP Diagnostic Immunology Proficiency Survey, 33% used one of the six commercially available turbidimetric assays, whereas 67% used one of three commercially available nephelometric assays for the measurement of human IgG, IgA, and IgM in serum.[59]

3.4.3.3 Immunoassay

The human immunoglobulins in low concentrations in serum, urine, and other body fluids were measured by immunoassay that can achieve analytical sensitivities of 1 ng/mL.[56] More specifically, human IgG1, IgG2, IgG3, and IgG4,[63,64] human IgA1 and IgA2, SC,[11,14] J chain,[13] and IgE[65] can be effectively measured by MAb-based two-site immunometric assays. These assays use an insolubilized capture antibody to bind the immunoglobulin isotype of interest from serum, urine, or other body fluids and a biotin-, enzyme-, fluorophor-, or radio-labeled polyclonal, polymonoclonal, or monoclonal antibody specific for different immunoglobulin epitopes to detect bound immunoglobulin. Analysis of multiple dilutions of a reference serum permits the construction of a reference curve from which response values obtained with test specimens can be interpolated in mass per volume units of immunoglobulin. Owing to the assay's sensitivity, serum specimens are normally diluted 1:10 (IgE) to over 1:10,000 (IgG1-4) and thus hemolysis, bilirubin, and lipemia rarely cause interference. Immunoassays are technically more complex than immunodiffusion and immunoprecipitation (nephelometric and turbidimetric) assays and they generally require more replicates and dilutions of the unknown specimen to obtain accurate results. Nephelometric and turbidimetric assays have thus become the clinical laboratory methods of choice for quantifying immunoglobulins in human serum. This trend has relegated immunoassay methods primarily to the measurement of immunoglobulins in atypical research specimens where analytical sensitivity is critical.

3.5   REFERENCE VALUES

In an ideal world, laboratory personnel would select an immunoglobulin quantitation method, and then establish reference ranges for a population whose demographics closely resemble the expected patient population that will be tested. This, however, can be difficult to accomplish for clinical immunology laboratories that perform interstate commerce, and thus receive specimens from large geographical areas that contain a spectrum of individuals with varying ages and ethnic backgrounds. The alternative strategy has been to adopt published reference intervals, some of which are recommended by the manufacturer. A compendium of reported human immunoglobulin mean and 95 percentile reference intervals is thus presented in Tables 3.5 and 3.6 to aid the laboratorian in the interpretation of immunoglobulin

TABLE 3.5

Total Human IgG, IgA, and IgM Levels in Serum as a Function of Age

Age	N	IgG Mean (g/L)	IgG 95 Percentile Range	IgA Mean (g/L)	IgA 95 Percentile Range	IgM Mean (g/L)	IgM 95 Percentile Range	Total Ig Mean (g/L)	Total Ig 95 Percentile Range	Reference No.	Assay Method
Cord blood	22	10.31	6.31 14.31	0.02	0.00 0.08	0.11	0.01 0.21	10.44	6.42 14.46	53	RID
1–3 months	29	4.30	1.92 6.68	0.21	0.00 0.47	0.30	0.08 0.52	4.81	2.27 7.35	53	RID
4–6 months	33	4.27	0.55 7.99	0.28	0.00 0.64	0.43	0.09 0.77	4.98	0.90 9.06	53	RID
7–12 months	56	6.61	2.23 10.99	0.37	0.01 0.73	0.54	0.08 1.00	7.52	2.68 12.36	53	RID
13–24 months	59	7.62	3.44 11.80	0.50	0.02 0.98	0.58	0.12 1.04	8.70	3.54 13.86	53	RID
25–36 months	33	8.92	5.26 12.58	0.71	0.00 1.45	0.61	0.23 0.99	10.24	6.14 14.34	53	RID
3–5 years	28	9.29	4.73 13.85	0.93	0.39 1.47	0.56	0.20 0.92	10.78	5.88 15.68	53	RID
6–8 years	18	9.23	4.11 14.35	1.24	0.34 2.14	0.65	0.15 1.15	11.12	5.23 16.98	53	RID
9–11 years	9	11.24	6.54 15.94	1.31	0.11 2.51	0.79	0.13 1.45	13.34	8.26 18.42	53	RID
12–16 years	9	9.46	6.98 11.94	1.48	0.22 2.74	0.59	0.19 0.99	11.53	8.15 14.91	53	RID
Adults	30	11.58	5.48 17.68	2.00	0.78 3.22	0.99	0.45 1.53	14.57	7.51 21.63	53	RID
Cord blood	50	11.21	6.36 16.06	0.02	0.01 0.04	0.13	0.06 0.25	11.36	6.44 16.35	58	Neph
1 month	50	5.03	2.51 9.06	0.13	0.01 0.53	0.45	0.20 0.87	5.61	2.72 10.46	58	Neph
2 months	50	3.65	2.06 6.01	0.15	0.03 0.47	0.46	0.17 1.05	4.26	2.26 7.53	58	Neph

3 months	50	3.34	1.76	5.81	0.17	0.05	0.46	0.49	0.24	0.89	4.00	2.05	7.16	58	Neph
4 months	50	3.43	1.96	5.58	0.23	0.04	0.73	0.55	0.27	1.01	4.21	2.27	7.32	58	Neph
5 months	50	4.03	1.72	8.14	0.31	0.08	0.84	0.62	0.33	1.08	4.96	2.13	10.06	58	Neph
6 months	50	4.07	2.15	7.04	0.25	0.08	0.68	0.62	0.35	1.02	4.94	2.58	8.74	58	Neph
7–9 months	50	4.75	2.17	9.04	0.36	0.11	0.90	0.80	0.34	1.26	5.91	2.62	11.20	58	Neph
10–12 months	50	5.94	2.94	10.69	0.40	0.16	0.84	0.82	0.41	1.49	7.16	3.51	13.02	58	Neph
1 year	50	6.79	3.45	12.13	0.44	0.14	1.06	0.93	0.43	1.73	8.16	4.02	14.92	58	Neph
2 years	50	6.85	4.24	10.51	0.47	0.14	1.23	0.95	0.48	1.68	8.27	4.86	13.42	58	Neph
3 years	50	7.28	4.41	11.35	0.66	0.22	1.59	1.04	0.47	2.00	8.98	5.10	14.94	58	Neph
4–5 years	50	7.80	4.63	12.36	0.68	0.25	1.54	0.99	0.43	1.96	9.47	5.31	15.86	58	Neph
6–8 years	50	9.15	6.33	12.80	0.90	0.33	2.02	1.07	0.48	2.07	11.12	7.14	16.89	58	Neph
9–10 years	50	10.07	6.08	15.72	1.13	0.45	2.36	1.21	0.52	2.42	12.41	7.05	20.50	58	Neph
Adult	120	9.94	6.39	13.49	1.71	0.70	3.12	1.56	0.56	3.52	13.21	7.65	20.13	58	Neph

Note: Results are presented as the mean serum immunoglobulin level in gram per liter and the 95 percentile range (2.5–97.5 percentile: mean ± 2 SD).

Source: Data extracted from Hamilton, R.G., *Manual of Molecular and Clinical Laboratory Immunology.* American Society for Microbiology, Washington DC, 2006 and Stiehm, E.R., Fudenberg, H.H., *Pediatrics* 37, 717, 1966 that used immunodiffusion (RID) or nephelometry (Neph) assays, respectively. With permission.

TABLE 3.6
Adult Human Serum IgG, IgA, and IgM Levels in Serum Reference Intervals[a]

Author and Reference	N Adults	Age Range	IgG Mean (g/L)	IgG 95 Percentile Interval (g/L)		IgA Mean (g/L)	IgA 95 Percentile Interval (g/L)		IgM Mean (g/L)	IgM 95 Percentile Interval (g/L)		IgD Mean (g/L)	Total IgD 95 Percentile Interval (g/L)		Company Name	Assay Name	Assay Method
Weeke and Krasilnikoff[88]	200	15–93	10.40	6.80	16.00	1.37	0.54	3.50	0.55	0.21	1.44	NA	NA	NA	Dansk	Laurell	IEP
Stiehm and Fudenberg[66]	30	>16	11.60	5.60	17.60	2.00	0.78	3.22	0.99	0.47	1.50	NA	NA	NA	NA	None	RID
Bulletin[69]	300	>16	10.47	5.64	17.65	1.77	0.85	3.85	1.26	0.45	2.50	0.04	0.00	0.14	Sanofi	Quanti-clone	RID
Jolliff et al.[67]	120	16–62	9.94	6.39	13.49	1.71	0.70	3.12	1.56	0.56	3.52	NA	NA	NA	Beckman	ICS	Neph
Dati[70]	773 M	>16	12.00	8.00	17.00	2.30	1.00	4.90	1.40	0.50	3.20	NA	NA	NA	Behring	BN-100	Neph
Dati[70]	680 F	>16	M/F combined			2.10	0.85	4.50	1.55	0.60	3.70	NA	NA	NA	Behring	BN-100	Neph

												Abbott DuPont	TDx ACA	Neph Turb
Bulletin[71]	349	>16	NA	6.78	17.14	NA	0.73	4.22	NA	0.54	2.96	NA	NA	NA
Hutson[72]	213	17–65	NA	5.68	14.83	NA	0.57	4.14	NA	0.20	2.74	NA	NA	NA
Intermethod mean[a]			11			1.95			1.3			0.04	NA	NA
Intermethod 1SD			0.96			0.27			0.24			NA	NA	NA
Intermethod CV (%)			8.7			13.8			18.6			NA	NA	NA
Maximum of range			17.65			4.90			3.70					0.14
Minimum of range			5.6			0.57			0.2				0	
N			4			4			4			1	NA	NA

a IEP results not included in intermethod analyses.

b Population consisted of 149 M and 64 F with a race distribution of 91% Caucasian, 6% Black, and 3% Hispanic in study by Huston et al.[72] NA = data not reported, M = male, F = female, IEP = immunoelectrophoresis, Turb = turbidimetric assay, Neph = nephelometric assay, RID = radial immunodiffusion assay. Results are presented as the mean serum immunoglobulin level in gram per liter and the 95 percentile range (2.5–97.5 percentile; mean ± 2 SD). N = number of measurements used in computational means, standard deviation and CV.

measurements. One common denominator among these published studies has been the use of specimens from *healthy individuals* to establish the reference range. A healthy individual may be defined as one with "a state of complete physical, mental, and social well being and not merely the absence of disease or infirmity." Individuals satisfying this definition can be difficult to find because even blood bank donors may not meet this criterion. Thus, with all their potential flaws, the published reference intervals that are summarized in Tables 3.5 through 3.9 may be the best information available on which action levels can be assigned for immunoglobulin concentrations in serum, urine, and CSF for those laboratories that have not determined their own reference ranges.

3.5.1 FACTORS INFLUENCING IMMUNOGLOBULIN LEVELS

Multiple factors influence immunoglobulin levels in humans. Age is possibly the most important personal attribute that determines serum immunoglobulin levels. Other suggested factors include the subject's sex, ethnic background, history of allergies and recurrent infections, and whether the individual lives in a geographic region where parasites are endemic. For purposes of this report, the reference ranges extracted from the literature have been partitioned based on the study population's age, sex, and specimen type. In one earlier study using immunodiffusion methods to measure immunoglobulin levels in approximately 300 healthy individuals (1/3 black, 2/3 Caucausian/Hispanic), no racial differences were suggested by the data and thus results were grouped only according to age and sex.[66] In a similar manner, subsequently reported reference immunoglobulin ranges have also been grouped

TABLE 3.7
Human IgG Subclass Levels as a Function of Age

Age Group	Human IgG1 (g/L)	Human IgG2 (g/L)	Human IgG3 (g/L)	Human IgG4 (g/L)
Cord blood	4.35–10.84	1.43–4.53	0.27–1.46	<0.01–0.47
0–2 months	2.18–4.96	0.40–1.67	0.04–0.23	<0.01–0.33
3–5 months	1.43–3.94	0.23–1.47	0.04–1.00	<0.01–0.14
6–8 months	1.90–3.88	0.37–0.60	0.12–0.62	<0.01
9–24 months	2.86–6.80	0.30–3.27	0.13–0.82	<0.01–0.65
3–4 years	3.81–8.84	0.70–4.43	0.17–0.90	<0.01–1.16
5–6 years	2.92–8.16	0.83–5.13	0.08–1.11	<0.01–1.21
7–8 years	4.22–8.02	1.13–4.80	0.15–1.33	<0.01–0.84
9–10 years	4.56–9.38	1.63–5.13	0.26–1.13	<0.01–1.21
11–12 years	4.56–9.52	1.47–4.93	0.12–1.79	<0.01–1.68
13–14 years	3.47–9.93	1.17–4.40	0.23–1.17	<0.01–0.83
Adult	3.10–9.10	0.72–4.10	0.17–0.72	0.02–0.65

Source: Data extracted from Hutson, D.K., *DuPont Company Technical Bulletin E-61050*. With permission.

TABLE 3.8

Human Urine and CSF Immunoglobulin G/A/M Reference Intervals

Assay Name	Total Urinary Human IgG (mg/24 h volume)	Total Urinary Human IgA (mg/24 h volume)	Total Urinary Human IgM (mg/24 h volume)	Reference and Author
LC-Partigen	1.2–6.5	1.3–5.0	Undetectable	Ritzmann[82]

Assay Name	Total CSF Human IgG G/L	Total CSF Human IgG G/L	Total CSF Human IgM G/L	Reference and Author
LC-Partigen RID-Behring	UD–0.055	0.0015–0.006	<0.001	Ritzmann[82]
RIA	0.035–0.058 15–60 years	<0.002	<0.002	Tietz[91]
RID N = 93 (17–60)	0.017±0.004-17-30 M±1SD 0.021±0.007-31-40 yr M±1SD 0.024±0.008-41-50 yr M±1SD 0.027±0.009-51-60 yr M±1SD 0.026±0.009-61-77 yr M±1SD yr	ND	ND	Tibbling et al.[84]
Quanticlone RID Sanofi Pasteur	UD–0.086	ND	ND	Sanofi Technical Bulletin[69]

Note: UD = undetectable, ND = not done.

according to the study population's age and occasionally sex. The impact of genetic factors (race or ethnic origin, blood groups, and histocompatibility antigens); physiological factors (stage in menstrual cycle or pregnancy and physical condition); or socioeconomic and environmental factors on the level of immunoglobulins in healthy study populations has not been definitively determined.

3.5.2   IMMUNOGLOBULIN REFERENCE INTERVALS

3.5.2.1   Serum IgG, IgA, IgM, and IgD

B cells are reportedly produced by the eighth week of gestation. At full term (38 weeks), the healthy newborn has a complete complement of B cells containing surface immunoglobulin of all isotypes. In 1965, Stiehm and Fudenberg[66] reported the first quantitative study of serum IgG, IgA, and IgM levels in humans as a function of age that were measured in mass per volume units (g/L) rather than in previously used arbitrary units or titers. These quantitative immunoglobulin levels were measured by immunodiffusion using polyclonal antiserum produced in their own facility and sera collected from 296 children and 30 adults who were clinically healthy at the time of their study. In 1982, Jolliff et al.[67] used nephelometric

TABLE 3.9
Human Serum IgE Nonatopic Reference Ranges

Total N	Sex (M/F)	Age Range	Total Human IgE Geometric Mean (kU/L)	Total Human IgE Upper 95% Confidence Limit (kU/L)	Reference and Author
26	M (15), F (11)	Cord blood	0.22	1.28	Saarinen et al.[87]
21	M (7), F (14)	6 weeks	0.69	6.12	Scandinavian
20	M (14), F (6)	3 months	0.82	3.76	children[b]
20	M (10), F (10)	6 months	2.68	16.3	
20	M (14), F (6)	9 months	2.36	7.3	
18	M (14), F (4)	1 year	3.49	15.2	
20	M (13), F (7)	2 years	3.03	29.5	
11	M (6), F (5)	3 years	1.80	16.9	
9	M (6), F (3)	4 years	8.58	68.9	
19	M (10), F (9)	7 years	12.9	161	
20	M (11), F (9)	10 years	23.7	570	
22	M (15), F (7)	14 years	20.1	195	
175	Not specified	17–85 years	13.2	114	Swedish adults
72	M	6–14 years	42.7	527	Barbee et al.[88]
73	F	6–14 years	43.3	344	White adults in
109	M	15–24 years	33.6	447	the United
121	F	15–24 years	18.6[b]	262	States
108	M	25–34 years	16.8	275	
89	F	25–34 years	16.6	216	
62	M	35–44 years	21.7	242	
67	F	35–44 years	19.3	206	
88	M	45–54 years	19.2	254	
97	F	45–54 years	13.3	177	
105	M	55–64 years	21.3	354	
172	F	55–64 years	11.7[b]	148	
145	M	65–74 years	21.2	248	
199	F	65–74 years	11.5[b]	122	
69	M	75+ years	18.4	219	
87	F	75+ years	9.2[b]	124	
758	M	6–75 years	22.9	317	
905	F	6–75 years	14.7[b]	189	

Note: M = male, F = female: all total serum IgE levels reported in this table were measured with a noncompetitive paper disk radioimmunosorbent test marked by Kabi-Pharmacia Diagnostics (now known as Phadia).

[a] Study performed with sera from children with no history of atopic disease or first-degree atopic relatives.

[b] Mean serum IgE for females is significantly lower than for males.

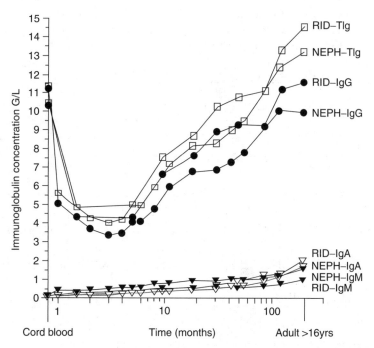

FIGURE 3.1 Mean total serum IgG (closed circles), IgA (open inverted triangles), IgM (closed inverted triangles), and immunoglobulin (open boxes) concentrations as measured by immunodiffusion[66] or nephelometry[67] from birth (cord blood) to adult levels (>192 months). IgE and IgD are not presented as they comprise a <1% of the total immunoglobulin in serum. Levels of almost exclusively maternal IgG in cord serum decrease to a minimum by 3–5 months and then progressively increase to adult levels by age 16 for IgG and IgM or early adulthood for IgA. Table 3.5 presents the associated 95 percentile intervals associated with these mean estimates. (Data extracted with permission from References 66 and 67.)

methods to measure IgG, IgA, and IgM in the serum of 25 boys and 25 girls who were bled at defined intervals from birth to 10 years of age. The mean serum IgG, IgA, and IgM concentrations for both of these studies are depicted in Figure 3.1 and presented together with their 95 percentile reference intervals in Table 3.5. Both studies reported that the mean umbilical cord serum IgG level was 10.3–11.2 g/L, which constituted approximately 90% of the total immunoglobulins measurable in the cord blood. Moreover, cord serum contained trace quantities of IgM (0.11–0.13 g/L) and IgA (0.02 g/L), which could be elevated by *in utero* infections. Transplacentally transferred IgG in the serum of healthy newborns decreased to a minimum by 3–5 months as neonatal IgM production began to increase (Figure 3.1). No difference in immunoglobulin levels was detected between male and female infants during the first year of life in the Stiehm study and thus they combined these values for clinical use. Adult IgG and IgM levels are achieved by the age of 16, whereas levels of

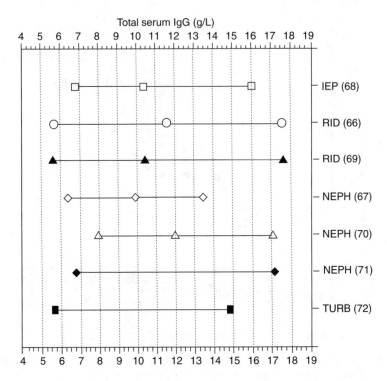

FIGURE 3.2   Mean and 95 percentile reference intervals (mean $\pm$ 2 SD) for total serum IgG in adults as reported by eight groups using the available commercial assays. See Table 3.6 for actual data. Despite the use of different assay procedures and unique sets of immunochemical reagents, the mean IgG level determined in adult sera was 11.0 g/L with an acceptable variance of 8.7% CV. The group's total serum IgG maximum was 17.65 g/L and minimum was 5.60 g/L.

IgA increased into early adulthood.[66] Table 3.6 and Figures 3.2 through 3.4 present the mean and 95 percentile reference intervals for IgG, IgA, and IgM that have been reported for healthy adults in these and five additional studies.[68–72] The demographics of the study populations (number, sex, age range, race, clinical testing, and environment), assay type, and statistical methods employed are summarized where available. These reports cover the principal commercial assays that are employed clinically for human IgG, IgA, and IgM measurements. Although several immunoassays for total human serum IgG are available, none are presently used in clinical testing, possibly because they are more technically complex and labor-intensive and they have narrower working ranges than nephelometric and turbidimetric assays.

3.5.2.2   Serum IgG Subclasses

Of the five human immunoglobulin isotypes, IgG has achieved special importance because it is the principal immunoglobulin that is transported across the placenta, and thus confers humoral immunity on the neonate. Concentrations of the individual

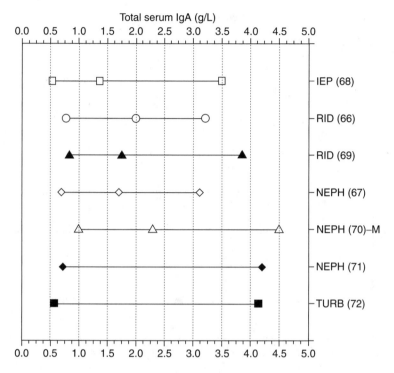

FIGURE 3.3 Mean and 95 percentile reference intervals (mean $\pm$ 2 SD) for total serum IgA in adults as reported by eight groups using the available commercial assays. See Table 3.6 for actual data. Despite the use of different assay procedures and unique sets of immunochemical reagents, the mean IgA level determined in adult sera was 1.95 g/L with an acceptable variance of 13.8% CV. The group's total serum IgA maximum was 4.90 g/L and minimum was 0.57 g/L.

IgG subclasses in human serum have been extensively studied as a function of age using a variety of assays. Figure 3.5 presents a composite of the age-dependent profiles of human and the mean IgG1, IgG2, IgG3, and IgG4 levels in the serum of children and adults as measured by eight groups[66,73–79] using a variety of poly- and monoclonal antibody reagents. Trends in the mean levels of IgG1, IgG2, IgG3, and IgG4 are in general agreement, with a characteristic valley in all four subclasses occurring at 3–6 months as maternally derived IgG is replaced by immunoglobulin synthesized by the infant. IgG1 and IgG3 synthesis occurs earlier and is more rapid in early childhood than IgG2 and IgG4. IgG1 concentrations increase to adult levels at 5–7 years, in contrast to IgG3 at 7–9, IgG2 at 8–10, and IgG4 at 9–11 years.[80,81] Differential exposure to environmental antigens is thought to combine with natural biological variation in the rate of achieving immunological maturity and inherited genetic factors to produce the variation observed in adult IgG subclass concentrations in serum. The 95 percentile reference intervals for human IgG as defined using an MAb-based immunoassay are presented in Table 3.7 as an illustration of representative target ranges.

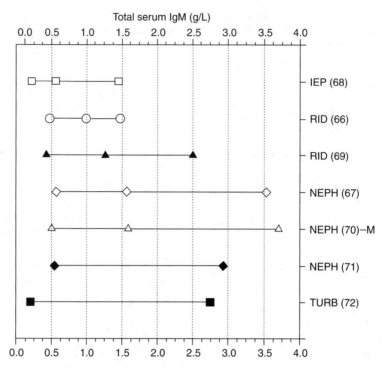

FIGURE 3.4   Mean and 95 percentile reference intervals (mean $\pm$ 2 SD) for total serum IgM in adults as reported by eight groups using the available commercial assays. See Table 3.6 for actual data. Despite the use of different assay procedures and unique sets of immunochemical reagents, the mean IgM level determined in adult sera was 1.30 g/L with a variance of 18.6% CV. The group total serum IgM maximum was 3.70 g/L and minimum was 0.20 g/L.

3.5.2.3   Urinary IgG, IgA, and IgM

The glomeruli of the kidney function as ultrafilters for plasma proteins and normally exclude high-MW proteins such as IgM from reaching the glomerular filtrate, except in trace amounts. The passage of high-MW proteins into the urine (proteinuria) can occur as a result of (a) increased glomerular permeability, (b) defective tubular reabsorption, (c) overload of a particular serum protein (e.g., Bence Jones light chains), or (d) postrenal protein secretion.[82] Urine can be clinically evaluated by electrophoresis–immunofixation for light chains when a plasma-cell dyscrasia is suspected as a result of condition (c). Both immunoglobulins (IgG and IgA) and light chains may also be quantitatively measured as evidence of kidney damage. IgG and IgA are normally present in urine at total levels from 1.2 to 6.5 mg (IgG) and 1.3–5.0 mg (IgA) in a 24-h urine specimen (Table 3.8).[82,83] Levels above this range are considered evidence of kidney dysfunction.

3.5.2.4   Cerebral Spinal Fluid IgG, IgA, and IgM

CSF is secreted by the choroid plexuses around the cerebral vessels and along the walls of the ventricles of the brain. It fills the ventricles and bathes the spinal cord.

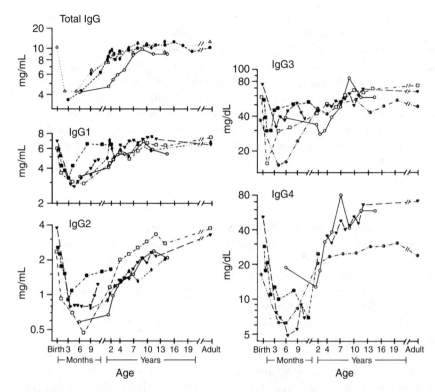

FIGURE 3.5 Mean total IgG, IgG1, IgG2, IgG3, and IgG4 concentrations as measured by immunodiffusion or immunoassay in serum collected from children at birth (cord blood) through adulthood (>16). Table 3.7 presents the associated 95 percentile intervals associated with one of these mean estimates from Ref. 73. Data for figure extracted from Refs. 73 (Schur et al., open circles [----]); 74 (Allansmith et al., closed circles [- - - -]); 66 (Stiehm and Fudenberg, open triangles [......], total IgG only); 75,76 (Zegers et al. and Van der Giessen et al.; closed inverted triangles [-- -- --]); 77 (Shakelford et al., closed diamonds [...---...---]); 78 (Oxelius and Svenningsen, open boxes [.-.-.-.]); 79 (Morell et al., closed boxes [.--..--..--]); and 80 (Lee et al., closed circles [..--..--..]). All studies demonstrate a decrease in all four subclasses of IgG during the 3–5 month period of life. (Reproduced from Reference 80 with permission.)

Eventually, CSF is reabsorbed into the blood through the arachnoid villi. Because CSF is principally an ultrafiltrate of plasma, its protein (typically 0.15–0.45 g/L in lumbar fluid) is primarily composed of low-MW prealbumin, albumin, and transferrin. CSF from healthy adults also normally contains low IgG concentrations that range from undetectable to 0.086 g/L depending on the report, method of measurement, and study group's age (Table 3.7). In one study, a gradual age-dependent increase was observed by Tibbling et al.[84] up to 40 years of age. IgA concentrations in the CSF of healthy adults range from undetectable to 0.006 g/L. IgG and IgA can be detected in the CSF presumably as a result of immunoglobulin synthesis by B cells that infiltrate demyelinated lesions within the central nervous system. An elevation in CSF IgG concentrations and the presence of oligoclonal immunoglobulin bands by electrophoresis–immunofixation are clinically used markers of increased perme-ability of the blood–brain barrier (capillary endothelium of vessels of the central nervous system) that may occur in patients with active multiple sclerosis, subacute sclerosing panencephalitis, and acute aseptic meningitis. The CSF IgG/albumin ratio or immunoglobulin index is used to determine whether there is an increased per-meability or increased local IgG production, or both. Decreased CSF IgG levels have been seen in individuals with active systemic lupus erythematosus with central nervous system involvement. In the few studies where IgM was measured in CSF from healthy adults, it was undetectable (<0.002 g/L). Thus, IgM is not routinely measured in CSF.

3.5.2.5 Serum IgE

As with the other immunoglobulin isotypes, the concentration in IgE in the serum is highly age-dependent. The concentration of IgE in cord serum is low, usually <2 kU/L (<4.8 µg/L), because it does not cross the placental barrier in significant amounts.[85–88] Mean serum IgE levels progressively increase in healthy children up to 10–15 years of age. The rise in serum IgE toward adult levels is slower than that of IgG but comparable to that of IgA. Atopic infants have an earlier and steeper rise in serum IgE levels during their early years of life as compared with nonatopic controls.[88] An age-dependent decline in total serum IgE may occur from the second through eighth decades of life. In contrast to IgG, IgA, IgM, and IgD levels that are routinely compared against 95 percentile reference intervals obtained with serum from healthy individuals, serum IgE levels must be judged against intervals estab-lished with serum from age-adjusted healthy, *nonatopic* individuals.

Representative serum IgE concentrations as measured in the serum of nonatopic children and adults are presented in Table 3.9. After 14 years of age, serum IgE levels >333 kU/L (800 µg/L) are considered abnormally elevated and strongly associated with atopic disorders such as allergic rhinitis, extrinsic asthma, and atopic derma-titis. The overlap between IgE levels in atopic and nonatopic populations, however, is considerable.[85–89] One study of adults with allergic asthma demonstrated a mean serum IgE level of 1589 µg/L (range 55–12,750 µg/L), with only about one-half of them having IgE concentrations above the 800-µg/L upper limit for nonatopic individuals. In a different study, very high levels of serum IgE (mean: 978 kU/L, range 1.3–65,208 kU/L) were observed in approximately 90% of patients with atopic dermatitis.

3.5.2.6 Immunoglobulins in Other Fluids

Of the gut-associated lymphoid tissue, bronchus-associated lymphoid tissue, and human small intestinal lamina propria-associated B cells, approximately 85% contain surface IgA, whereas only 5 and 10%, respectively, contain surface IgG and IgM. Because the surface immunoglobulin reflects the isotype of plasma cells derived from B-cell precursors, secretory IgA is the predominant immunoglobu-lin produced in these tissues and associated secretions. Moreover, secretory IgA is the predominant immunoglobulin in secretions emanating from mucosal tissues in middle ear, urogenital tract, mammary gland, conjunctiva, and salivary glands. IgA in saliva, for instance, appears to reach adult levels (approximately 0.11 g/L) by about 6 weeks of development.[90,91] Reports of immunoglobulins other than IgA are rare in body fluids such as tears, sweat and peritoneal fluid, colostrum, saliva, bronchial secretions, nasal mucosa, prostatic fluid, vaginal secretions, and mucous secretions of the small intestine.

Total IgE has been measured in tears by ophthalmologists. An MAb-based solid-phase immunoassay has been designed to analyze IgE levels in tears that have been collected from the inferior marginal tear duct using a 2-μL capillary tube. Tear IgE levels are reportedly increased in individuals with giant papillary conjunctivitis, secondary to wearing contact lenses.[92] The mean total IgE levels in the tears of allergic symptomatic patients has also been shown to be statistically elevated as compared to tears of nonallergic symptomatic and asymptomatic subjects.[93]

3.5.3   HUMAN LIGHT-CHAIN REFERENCE INTERVALS

As indicated earlier, an alternation in the κ/λ light-chain ratio in serum has been proposed as a marker of monoclonal or M-component-related immunoglobulin abnormalities. The κ chain containing immunoglobulins normally predominate in healthy individuals with an average reported κ/λ light-chain ratio of 1.8–1.9.[30,70,83,94–96] Table 3.10 presents the patient demographics, methods, and published κ and λ levels in serum for selected studies that compute the serum κ/λ ratio. The mean immunoglobulin κ/λ ratios among these studies of healthy subjects varied from 1.8 to 1.9: 1.84,[29] 1.83 ± 0.3,[30] 1.86,[70] 1.87 + 0.23,[94] 1.9 ± 0.3,[83] and 1.83.[96] Ford et al.[94] examined 10 sequential sera from 12 healthy subjects to study intraindividual biological variation as distinct from methodologic analytical variation. They concluded that high interindividual variability in the light-chain ratios makes their measurement more well suited for monitoring changes rather than as a replacement for serum protein electrophoresis in the early detection of monoclonal gammopathies.

3.5.4   TOTAL SERUM IGG LEVELS DURING GAMMAGLOBULIN THERAPY

As indicated earlier, a total IgG <2.5 g/L with nondetectable IgA and IgM in serum can aid in the identification of a patient with primary hypo- or agammaglobulinemia (Table 3.4). Once identified, the patient, especially a child, may be put on exogenous immunoglobulin replacement therapy. The total serum IgG can be useful to determine the optimal dose required for administration and to document the biological half-life and total achieved level of IgG in the particular patient. Figure 3.6 illustrates one such evaluation in which total IgG was measured in sequential sera collected from a 2-year-old male with Bruton's disease, who had a total IgG of 0.7 g/L and nondetectable IgA and IgM before therapy. Following administration of 200 mg/kg of IV immunoglobulin (IVIg), the total IgG increased to 5 g/L but rapidly declined with a biological half-life of 10–11 days. The decision was made based on these results to increase the dose to 400 mg/kg, which in four subsequent administrations was able to maintain the total serum IgG level between the 2.5 and 97.5 percentile

TABLE 3.10
Human Serum Light-Chain Reference Intervals

N	Sex M/F	Age Range	Statistics	Instrument and Reagents	Assay	κ Light-Chain Range (g/L)	λ Light-Chain Range (g/L)	κ/λ Ratio Reference Interval	Reference and Author
30	F	19–57	Normal[a] logarithmic	Behring Diagnostics	Neph	1.86–4.92 Mean = 3.12	0.86–2.80 Mean = 1.68	1.40–2.45 Mean = 1.84	Finland Seppo et al.[29]
70	M	20–65						Median = 1.82	
50	ND	Adults	ND	Behring Diagnostics	Neph	Mean = 2.72	Mean = 1.47	1.35–2.65	Belgium
								1.83 ± 0.3 (Mean ± SD)	–Lievens[30]
1453	773 M 680 F	Adults	2.5–97.5 Percentile	Behring Diagnostics	Neph	2.0–4.40 Mean = 3.00	1.10–2.40 Mean = 1.60	Mean = 1.86	Dati et al.[70]
12	6 M 6F	23–48	Mean ± 2 SD	Cobas Fara Hoffman LaRoche	Turb	Mean = 7.9	Mean = 4.23	(Mean ± SD) 1.87 ± 0.23	Ford et al.[93]

a Serum κ/λ ratios kept closely to a logarithmic normal distribution with a positive skewness.

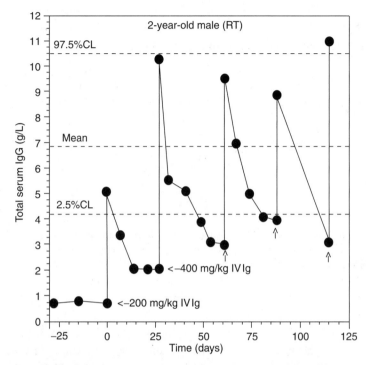

FIGURE 3.6 Changes in total serum IgG levels measured by immunoassay in sequential sera collected from a single 2-year-old male with primary immunodeficiency (Bruton's agammaglobulinemia). Specimens were collected before and during a period when monthly injections of IVIg were administered. Low IgG levels (<1 g/L) were detected before administration of 200 mg/kg of IV gammaglobulin. IgG levels increased to 5 g/L and then diminished with a biological half-life of 10–11 days. Higher doses of IVIg were (400 mg/kg) then administered in the second through fourth doses depicted to achieve an average IgG level around the mean for 2-year-old infants. The biological half-life appeared to decrease to 5–6 days on these subsequent doses.

interval for a 2-year-old child. The biological half-life decreased to 5–6 days with the increased level of administered IgG. This study serves to illustrate the utility of repetitive total serum IgG measurements for monitoring IVIg therapy and for making adjustments in dose to optimize circulating levels of IgG into the range appropriate for the age of the patient.

3.6   EXTERNAL PROFICIENCY SURVEYS

In the United States, the most widely subscribed external proficiency survey for human immunoglobulins and light chains is conducted by the CAP. The design (e.g., number of participating laboratories, cycles per year, sera per cycle, and reporting specimen) for the three surveys of relevance to the diagnostic immunology laboratory are summarized in Table 3.11. The Diagnostic Immunology Series 1 (SM; code S1) and Series 2 (S; code S2) surveys provide 15 specimens per year for the evaluation of laboratory

TABLE 3.11

External Proficiency Surveys for Immunoglobulins and Light Chains

Variable	IgG	IgG1-4	IgA	IgM	IgE	IgD	κ/λ Ratio
Survey code	IG, LN7, M, S2, S4, SPE	S2, S4	IG, LN7, SPE	IG, LN7, SPE	IG, K, SE	S2, S4	IG, S2, S4
Provider agency	CAP[a]	CAP, UKNEQAS	CAP	CAP	CAP	CAP	CAP
No. of cycles per year	3, 3	3, 6	3	3	3, 3	3	3
No. of sera per cycle	5, 1	1, 2	5	5	5, 5	1	1

Note: IG = immunology general, LN7 = immunology calibration verification/linearity, S2 = immunology special, S4 = immunology limited, SPE = serum protein electrophoresis, K = endocrinology (ligand assay general), M = CSF chemistry.

[a] CAP = College of American Pathologists.

methods that are used to measure serum IgG, IgG subclasses 1–4, IgA, IgM, IgE, IgD, and human κ/λ light-chain ratios. There is no available survey for human IgA subclasses. In addition to the immunoglobulins, these surveys also provide external proficiency testing for assays that measure other immunological analytes such as complement components 3 and 4, rheumatoid factor, C-reactive protein, serum hCG, alpha-1 antitrypsin, ceruloplasmin, haptoglobin, prealbumin, and a panel of antibodies against bacterial, viral, and autoantigens. Discussion of these nonimmunoglobulin analytes is beyond the scope of this chapter. The Diagnostic Allergy CAP Survey called "SE" provides 15 sera per year to laboratories performing total serum IgE and allergen-specific IgE antibody assays. The SE Survey has been favorably contrasted to the proficiency program conducted by the U.K. External Quality Assessment Scheme for Autoimmune Serology (UKEQAS).[97] The CAP surveys are generally designed to include three cycles per year and five specimens per cycle, which fulfills the requirements for regulated analytes that are measured by highly complex tests as specified by U.S. Public Law 100-578 and the Clinical Laboratory Improvement Amendments of 1988 (CLIA'88). For the CAP surveys, a 1–2 week turnaround time from receipt of the specimens to submission of results is expected for all participants. Once compiled, a summary of the mean, standard deviation, CV, median, low and high values for each assay method, and a pass/fail designation for regulated analytes is returned to the participating laboratory and state licensing agencies. Laboratories that produce measurements outside the 95 percentile (mean ± 2 SD) of the group data can be considered out of control, provided the number of laboratories is sufficiently large to make statistical inferences. Owing to the availability of excellent, stable primary reference sera, the interlaboratory agreement for total serum IgG, IgA, IgM, and IgE measurements between laboratories using the same assay method and reagents is generally within acceptable levels (<10% CV). Differences in immunoglobulin assignments that occur among laboratories that use different methods can generally be traced to the use of an inappropriate calibration serum by the manufacturer or laboratory.

3.7 INTERFERENCES

Despite the excellent interlaboratory and interassay agreement of immunoglobulin measurements performed in clinical laboratories using FDA-cleared commercial products, there are occasionally outlier results that cannot be traced to hemolysis, lipemia, or high levels of bilirubin. An overview of published work linking human rheumatoid factors to spuriously high or low immunoglobulin levels as a result of immunological method interference is presented elsewhere.[98] This problem can only become more commonplace as murine and engineered chimeric (human–mouse) antibodies are increasingly used to diagnose and treat human diseases.[99,100] In addition to exogenously administered heterologous, chimeric, or humanized antibodies that may elicit human rheumatoid factors, the presence of naturally occurring anti-immunoglobulin autoantibodies are also known to exist.[101] These autoantibodies are thought to play a role in regulating the level of total serum immunoglobulins. They can also be considered as potential interfering factors in immunological assays that are employed by clinical laboratories to quantify human immunoglobulins. Difficulty in documenting the role of these autoantibodies in assay interference in part stems from the fact that human immunoglobulin-specific autoantibodies are heterogeneous and they are difficult to quantify when circulating in a complexed form, bound to their respective immunoglobulin. The use of purified human immunoglobulins and engineered chimeric autoantibodies (e.g., human IgG1 antihuman IgE Fc) are being used as research tools to construct families of dose–response curves by varying either the autoantibody (IgG anti-IgE) or ligand (IgE) concentration in a serum protein matrix.[102] By doing so, reference calibration curves may be established from which the concentration of human immunoglobulin-specific autoantibodies can be estimated. The use of these calibration curves should allow more definitive identification of their presence and role in altering the accuracy of immunological assays that measure total immunoglobulin concentrations in human serum.

In 2003, humanized IgG1 antihuman IgE Fc MAb (omalizumab [also called xolair]) was licensed for use in the management of asthma.[103] Sera from patients who have received omalizumab contain small anti-IgE:IgE complexes and a percentage that is "free IgE" or unbound with omalizumab.[104] Complex formation of anti-IgE with IgE was suspected of causing sufficient interference in IgE assay reagent binding that might reduce the accuracy of results generated by FDA-cleared IgE assays. To examine this potential interference, serum from four atopic adults containing omalizumab at 50 or 200 molar excess to IgE or buffer (sham control) were analyzed by 159 clinical immunology laboratories, using all available IgE antibody assays.[21] The study results showed that two total serum IgE from one manufacturer were minimally impacted (2.4–9.0% reduction in measured IgE) by the presence of omalizumab, whereas five other total IgE assays showed marked reductions ($p < 0.001$) from 13 to 67%. This study confirms that therapeutically administered humanized antibody can interfere sufficiently to markedly reduce the accuracy of some quantitative immunoglobulin assays.

3.8 SUMMARY

The immunoglobulins are among the most important human serum proteins as they confer humoral immunity, promote cellular immunity, and facilitate complement activation. By these actions, immunoglobulins play a key role in the immunological defense of the human against infectious agents. This chapter has examined structural details of immunoglobulins that serve as determinants for immunological reagents that are used in clinical assays for their measurement in human body fluids. The trend in the clinical immunology laboratory has been toward the use of more automated assay platforms that employ increasingly stable and better-characterized reference sera as calibrators and ultraspecific immunological reagents. These improved assay methods will continue to allow accurate analytical measurements of immunoglobulin levels in serum, urine, CSF, and other body fluids by hospital and reference laboratories as well as in research environments throughout the world.

ACKNOWLEDGMENTS

The author extends his appreciation to Deborah Fawcett and Stephen Buescher of the University of Texas School of Medicine in Houston, Texas, who obtained the sequential sera from the 2-year-old immunodeficiency patient that were subsequently used to examine the utility of total serum IgG measurements in monitoring IV-gammaglobulin therapy for dose optimization.

ABBREVIATIONS

ADCC   Antibody-dependent cell-mediated cytotoxicity
CAP   College of American Pathologists
CH   Immunoglobulin heavy-chain constant region domains (e.g., CH1, hinge, CH2, CH3, and CH4)
CSF   Cerebral spinal fluid
Fc   Immunoglobulin fragment that binds complement
Fd   Immunoglobulin heavy chain of the $F(ab')_2$ fragment
IEF   Isoelectric focusing
IgA   Immunoglobulin A
IgD   Immunoglobulin D
IgE   Immunoglobulin E
IgG   Immunoglobulin G
IgM   Immunoglobulin M
κ/λ Kappa to lambda light-chain ratio
MAb   Monoclonal antibody
PAN   An antibody that binds to *all* allotypic forms and subclasses of a particular isotype of human immunoglobulin (e.g., IgG PAN-reactive MA,b = antibody that binds to all allotypic forms of human IgG1, IgG2, IgG3, and IgG4 molecules).

REFERENCES

1. Putnam, F.W., *The Plasma Proteins*, Academic Press, New York, 1977, pp. 1–153.
2. Kabat, E.A., The structural basis of antibody complementarity. *Adv. Protein Chem.* 32, 1, 1978.
3. Goodman, J.W., Immunoglobulins I: structure and function, in *Basic and Clinical Immunology*. Stites, D.P., Stobo, J.D., Wells, J.V., Eds., Appleton & Lange, Norwalk, CT, 1987, chapter 4.
4. Kabat, E.A., Heterogeneity and structure of immunoglobulins and antibodies, in *Structural Concepts in Immunology and Immunochemistry*. Kabat, E.A., Ed., Holt, Rinehart and Winston, New York, 1976, chapter 9.
5. Jefferis, R., Structure–function relationships of IgG subclasses, in *The Human IgG Subclasses: Molecular Analysis of Structure, Function and Regulation*. Shakib, F., Ed., Pergamon Press, New York, 1990, pp. 93–108.
6. Aalberse, R.C., Schuurman, J., van Ree, R., The apparent monovalency of human IgG4 is due to bispecificity. *Inter. Arch. Allergy Immunol.* 118, 187, 1999.
7. Aalberse, R.C., Schuurman, J., IgG4 breaking the rules. *Immunology* 105, 9, 2002.
8. Heremans, J.F., Biochemical features and biological significance of immunoglobulin A, in *Immunoglobulins, Biological Aspects and Clinical Uses*. Merler, E., Ed., National Academy of Sciences, Washington DC, 1970, pp. 52–73.
9. Tomasi, T.B., Jr., The gamma A globulins: first line of defense, in *Immunology, Current Knowledge of Basic Concepts in Immunology and their Clinical Application*. Good, R.A., Fisher, D.W., Eds., Sinauer Assoc., Stanford, CN, 1971, pp. 76–83.
10. Brandtzaeg, P., Human secretory immunoglobulins: quantitation of free secretory piece. *Acta Path Microbiol. Scand.* 79, 189, 1971.
11. Jones, C., Mermelstein, N., Kincaid-Smith, P., Powell, H., Roberton, D., Quantification of human serum polymeric IgA, IgA1 and IgA2 immunoglobulin by enzyme immunoassay. *Clin. Exp. Immunol.* 72, 344, 1988.
12. Mestecky, J., Hamilton, R.G., Magnusson, C.G.M., Jefferis, R., Vaerman, J.P., Goodall, M., de Lange, G.G., Moro, I., Aucouturier, P., Radl, J., Cambiaso, C., Silvain, C., Preud'homme, J.L., Kusama, K., Carlone, G.M., Biewenga, J., Kobayashi, K., Reimer, C.B., Evaluation of monoclonal antibodies with specificity for human IgA, IgA subclasses and allotypes and secretory component. Results of an IUIS/WHO collaborative study. *J. Immunol. Meth.* 193, 103, 1996.
13. Mestecky, J., Moldoveanu, Z., Julian, B.A., Prichal, J.T., J chain disease: a novel form of plasma cell dyscrasia. *Am. J. Med.* 88, 411, 1990.
14. Mesteky, J., Russell, M.W., IgA subclasses. *Monogr. Allergy* 19, 277, 1986.
15. Morell, A., Skvaril, F., Noseda, G., Bbarandun, S., Metabolic properties of human IgA subclasses. *Clin. Exp. Immunol.* 13, 521, 1973.
16. Drenth, J.P., Cuisset, L., Grateau, G., Vasseur, C., van de Velde-Viser, S.D., de Jong, J.G., Beckmann, J.S., van der Meer, J.W., Delpech, M., Mutations in the gene encoding mevalonate kinase cause hyper-IgD and periodic fever syndrome. International Hyper-IgD Study Group. *Nat. Genet.* 22, 178, 1999.
17. Bennich, H.H., Ishizaka, K., Johansson, S.G., Rowe, D.S., Stanworth, D.R., Terry, W.D., Immunoglobulin E: a new class of human immunologlobulin. *Immunochemistry* 5, 327, 1968.
18. Hamilton, R.G., The science behind the discovery of IgE (invited, Allergy Archives). *J. Allergy Clin. Immunol.* 115, 648, 2005.
19. Ishizaka, K., Ishizaka, T., Lee, E.H., Biologic functions of the Fc fragment of E myeloma protein. *Immunochemistry* 7, 687, 1970.
20. Hamilton, R.G., Adkinson, N.F., Jr., *In vitro* assays for IgE mediated sensitivities. *J. Allergy Clin. Immunol.* 114, 213, 2004.

21. Hamilton, R.G., Accuracy of Food and Drug Administration-cleared IgE antibody assays in the presence of anti-IgE (omalizumab). *J. Allergy Clin. Immunol.* 117, 759, 2006.
22. Spiegelberg, H.L., Biological activities of immunoglobulins of difference classes and subclasses. *Adv. Immunol.* 19, 259, 1974.
23. Burton, D.R., Gregory, L., Jefferis, R., Aspects of molecular structure of IgG subclasses. *Monogr. Allergy* 19, 7, 1986.
24. Turner, M.W., Structure and function of immunoglobulins, in *Immunochermistry: An Advanced Textbook*. Glynn, L.E., Steward, M.W., Eds., Wiley, Chichester, 1977.
25. Whicher, J.T., The role of immunoglobulin assays in clinical medicine. *Ann. Clin. Biochem.* 21, 461–466, 1984.
26. Paul, M.E., Shearer, W.T., Approach to the evaluation of the immunodeficient patient, in *Clinical Immunology, Principles and Practice*. Rich, R.R., Fleisher, T.A., Shearer, W.T., Kotzin, B.L., Schroeder, H.W., Eds., Mosby, Philadelphia, PA, 2001, chapter 33.
27. Schroeder, H.W., Jr., Primary antibody deficiencies, in *Clinical Immunology, Principles and Practice*. Rich, R.R., Fleisher, T.A., Shearer, W.T., Kotzin, B.L., Schroeder, H.W., Eds., Mosby, Philadelphia, PA, 2001, chapter 34.
28. Kyle, R.A., Dispenzieri, A., Monoclonal gammopathies, in *Clinical Immunology, Principles and Practice*. Rich, R.R. Fleisher, T.A., Shearer, W.T., Kotzin, B.L., Schroeder, H.W., Eds., Mosby, Philadelphia, PA, 2001, chapter 97.
29. Seppo, T.L., Soppi, E.T., Morsky, P.J., Critical evaluation of the serum kappa/lambda light chain ratio in the detection of M proteins. *Clin. Chim. Acta* 207, 143, 1992.
30. Lievens, M.M., Medical and technical usefulness of measurement of kappa and lambda immunoglobulin light chains in serum with an M-component. *J. Clin. Chem. Clin. Biochem.* 27, 519, 1989.
31. Peebles, S., Liu, M., Lichtenstein, L.M., Hamilton, R.G., IgA, IgG and IgM quantification in bronchoalveolar lavage fluids from allergic rhinitics, allergic asthmatics and normal subjects by monoclonal antibody-based immunoenzymetric assays. *J. Immunol. Meth.* 179, 77, 1995.
32. Leonardi, A., *In vivo* diagnostic measurements of ocular inflammation. *Curr. Opin. Allergy Clin. Immunol.* 5, 464, 2005.
33. Michishiqe, F., Kanno, K., Yoshinaga, S., Hinode, D., Takehisa, Y., Yasuoka, S., Effect of saliva collection method on the concentration of protein components in saliva. *J. Med. Invest.* 53, 140, 2006.
34. Matsson, P., Hamilton, R.G. and Diagnostic Allergy Techniques Working Group. *Analytical Performance Characteristics and Clinical Utility of Immunological Assays for Human IgE Antibody of Defined Allergen Specificities: Guideline*, 2nd Edition, Clinical and Laboratory Standards Institute (formally the National Committee on Clinical Laboratory Standards), Wayne, PA, 1/LA20-A, 2007.
35. Jefferis, R., Reimer, C.B., Skvaril, F., De Lange, G., Ling, N.R., Lowe, J., Walker, M.R., Phillips, D.J., Aloiso, C.H., Wells, T.W., Vaerman, J.P., Magnusson, C.G., Kubagawa, H., Cooper, M., Vartdal, F., Vandvik, B., Haaijnan, J.J., Makela, O., Sarnesto, A., Lando, Z., Gergely, J., Rajnavolgyi, E., Laszlo, G., Radl, J., Molinaro, G., Evaluation of monoclonal antibodies having specificity for human IgG subclasses: result of an IUIS/WHO collaborative study. *Immunol. Lett.* 10, 223, 1985.
36. Hamilton, R.G., Production and epitope location of monoclonal antibodies to the human IgG subclasses, in *The Human IgG Subclasses: Molecular Analysis of Structure, Function and Regulation*. Shakib, F., Ed., Pergamon Press, New York, 1990, pp. 79–91.
37. Reimer, C.B., Phillips, D.J., Aloisio, C.H., Moore, D.D., Galland, G.G., Wells, T.W., Black, C.M., McDougal, J.S., Evaluation of thirty-one mouse monoclonal antibodies to human IgG epitopes. *Hybridoma* 3, 263, 1984.

38. Phillips, D.J., Wells, T.W., Reimer, C.B., Estimation of association constants of mono-clonal antibodies to human IgG epitopes using fluorescent sequential saturation assays. *Immunol. Lett.* 17, 159, 1987.
39. Reimer, C.B., Phillips, D.J., Aloisio, C.H., Black, C.M., Wells, T.M., Specificity and association constants of 33 monoclonal antibodies to human IgA epitopes. *Immunol. Lett.* 21, 209, 1989.
40. Reimer, C.B., Five hybridomas secreting monoclonal antibodies against human IgE. *Monoclonal Antibody News* 4, 2, 1986.
41. Hamilton, R.G., Roebber, M., Reimer, C.B., Rodkey, S.L., Isoelectric focusing patterns of mouse monoclonal antibodies to the four human IgG subclasses. *Electrophoresis* 9, 127, 1987.
42. Neuberger, M.S., Williams, G.T., Mitchell, E.B., Joubal, S.S., Flanagan, J.G., Rabbitts, T.H., A hapten-specific chimeric IgE antibody with human physiological effector func-tion. *Nature* 314, 268, 1985.
43. Morrison, S.L., Johnson, J.J., Herzenberg, L.A., Oi, V.T., Chimeric human antibody molecules: mouse antigen-binding domains with human constant region domains. *Proc. Natl. Acad. Sci. USA* 81, 6851, 1984.
44. Hamilton, R.G., Morrison, S.L., Epitope mapping of human immunoglobulin specific murine monoclonal antibodies with domain switched, deleted, and point-mutated chi-meric antibodies. *J. Immunol. Meth.* 158, 107, 1993.
45. Hamilton, R.G., Application of engineered chimeric antibodies to the calibration of human antibody standards. *Ann. Biol. Clin.* 49, 242, 1991.
46. Rowe, D.S., Grab, B., Anderson, S.G., An international reference preparation for human serum immunoglobulins G, A and M: content of immunoglobulins by weight. *Bull. WHO* 46, 67, 1972.
47. Reimer, C.B., Smith, S.J., Wells, T.W., Nakamura, R.M., Keitges, P.W., Ritchie, R.F., Williams, G.W., Hanson, D.J., Dorsey, D.B., Collaborative calibration of the U.S. National and the College of American Pathologists reference preparations for specific serum proteins. *Am. J. Clin. Pathol.* 77, 12, 1982.
48. Whicher, J.T., Hunt, J., Perry, D.E., Method specific variations in the calibration of a new immunoglobulin standard suitable for use in nephelometric techniques. *Clin. Chem.* 24, 531, 1978.
49. Rowe, D.S., Anderson, S.G., Tackett, L., A research standard for human serum immu-noglobulin D. *Bull. World Health Org.* 43, 607, 1970.
50. *WHO ECBS Technical Report,* Series No. 658, 23, 21, 1981.
51. Baudner, S., Haupt, H., Hubner, R., Manufacture and characterization of a new reference preparation for 14 plasma proteins-CRM 470 = RPPHS lot 5. *J. Clin. Lab. Anal.* 8, 177, 1994.
52. Dati, F., Schumann, G., Thomas, L., Aguzzi, F., Baudner, S., Bienvenu, J., Blaabjerg, O., Blirup-Jensen, S., Carlstrom, A., Petersen, P.H., Johnson, A.M., Milford-Ward, A., Ritchie, R.F., Svendsen, P.J., Whicher, J., Consensus of a group of professional societies and diagnostic companies on guidelines for interim reference ranges for 14 proteins in serum based on the standardization against the IFCC/BCR/CAP Reference Material (CRI 470). International Federation of Clinical Chemistry, Community Bureau of Reference of the Commission of the European Communities, College of American Pathologists. *Eur. J. Clin. Chem. Clin. Biochem.* 34, 517, 1996.
53. Ledue, T.B., Johnson, A.M., Commutability of serum protein values: persisting bias among manufacturers using values assigned from the certified reference material 470 (CRM 470) in the United States. *Clin. Chem. Lab. Med.* 39, 1129, 2001.
54. Johnson, A.M., Whicher, J.T., Effect of certified reference material 470 (CRM-470) on national quality assurance programs for serum proteins in Europe. *Clin. Chem. Lab. Med.* 39, 1123, 2001.

55. Schauer, U., Stemberg, F., Rieger, C.H., Borte, M., Schubert, S., Riedel, F., Herz, U., Renz, H., Wick, M., Carr-Smith, H.D., Bradwell, A.R., Herzog, W., IgG subclass concentrations in certified reference material 470 and reference values for children and adults determined with the binding site reagents. *Clin. Chem.* 49, 1924, 2003.
56. Whicher, J.T., Warren, C., Chambers, R.E., Immunochemical assays for immunoglobulins. *Ann. Clin. Biochem.* 21, 78, 1984.
57. Mancini, G., Carbonara, A.O., Heremans, J.F., Immunochemical quantitation of antigens by single radial immunodiffusion. *Immunochemistry* 2, 235, 1965.
58. McKelvey, E.M., Fahey, J.F., Immunoglobulin changes in disease: quantitation on the basis of heavy polypeptide chains, IgG, IgA, and IgM and of light polypeptide chains. *J. Clin. Invest.* 44, 1778, 1965.
59. College of American Pathologists, *Diagnostic Immunology Survey*, Northfield IL (www.cap.org), 2006.
60. Sternberg, J., A rate nephelometer for measuring specific proteins by immunoprecipitin reactions. *Clin. Chem.* 23, 1456, 1977.
61. Hills, L.P., Tiffany, T.O., Comparison of turbimetric and light scattering measurements of immunoglobulins by use of a centrifugal analyzer with absorbance and fluorescence/light scattering optics. *Clin. Chem.* 26, 1459, 1980.
62. Normansell, D.E., Quantitation of serum immunoglobulins. *CRC Crit. Rev. Clin. Lab. Sci.* 17, 103, 1982.
63. Hamilton, R.G., Wilson, R., Spillman, T., Roebber, M., Monoclonal antibody based immunoenzymetric assays for quantification of human IgG and its four subclasses, *J. Immunoassay* 9, 275, 1988.
64. Papadea, C., Check, I.J., Reimer, C.B., Monoclonal antibody based solid phase immunoenzymetric assays for quantifying human immunoglobulin G and its subclasses in serum. *Clin. Chem.* 31, 1940, 1985.
65. Hamilton, R.G., Immunological methods in the diagnostic allergy laboratory, in *Manual of Molecular and Clinical Laboratory Immunology*. Detrick, B., Hamilton, R.G., Folds, J., Eds., 7th Edition, American Society for Microbiology, Washington DC, 2006, chapter 107, 995pp.
66. Stiehm, E.R., Fudenberg, H.H., Serum levels of immune globulins in health and disease. *Pediatrics* 37, 717, 1966.
67. Jolliff, C.R., Cost, K.M., Stivrins, P.C., Grossman, P.P., Nolte, C.R., Franco, S.M., Fijan, K.J., Fletcher, L.L., Shriner, H.C., Reference intervals for serum IgG, IgA, IgM, C3 and C4 as determined by rate nephelometry. *Clin. Chem.* 28, 126, 1982.
68. Weeke, B., Krasilnikoff, P.A., The concentrations of 21 serum proteins in normal children and adults. *Acta Med. Scand.* 192, 149, 1972.
69. *Technical Bulletin*, Sanofi Diagnostics Pasteur, 0630891.
70. Dati, F., Lammers, M., Adam, A., Sontag, D., Stienen, L., Reference values for 18 plasma proteins on the Behring Nephelometer System. *Sonderdruck Aus. Lab. Med.* 13, 87–90, 1989.
71. TD×FL×TM assay manual for immunoglobulin G, immunoglobulin A and immunoglobulin M. Abbott Laboratories, USA, *Abbott Technical Bulletin* R-107.
72. Hutson, D.K., The performance characteristics of the immunoglobulin method for the DuPont ACA discrete clinical analyzer, *DuPont Company Technical Bulletin E-61050*.
73. Schur, P.H., Rosen, F., Norman, M.E., Immunoglobulin G subclasses in normal children. *Pediatr. Res.* 13, 181, 1979.
74. Allansmith, M., McClellan, B.H., Butterworth, M., Maloney, J.R., The development of immunoglobulin levels in man. *J. Pediatr.* 72, 276, 1968.
75. Zegers, B.J., van der Giessen, M., Reerink-Brongers, E.E., Stoop, J.W., The serum IgG subclass levels in healthy infants of 13–62 weeks of age. *Clin. Chim. Acta* 101, 265, 1980.

76. Van der Giessen, M., Roussow, E., Algra-van Veen, T., Van Loghem, E., Zegers, B.J.M., Sander, P.C., Quantitation of IgG subclasses in sea of normal adults and healthy children between 4 and 12 years of age. *Clin. Exp. Immunol.* 21, 501, 1975.

77. Shakelford, P.G., Granoff, D.M., Hahm, M.G., Relation of age, race and allotype to immunoglobulin subclass concentrations. *Pediatr. Res.* 19, 846, 1985.

78. Oxelius, V.A., Svenningsen, N.W., IgG subclass concentrations in preterm neonates. *Acta Pediatr. Scand.* 73, 626, 1984.

79. Morell, A., Skvaril, F., Hitzig, W.G., Barandun, S., IgG subclasses: development of the serum concentrations in normal infants and children. *J. Pediatr.* 80, 960, 1972.

80. Lee, S.I., Heiner, D.C., Wara, D., Development of serum IgG subclass levels in children. *Monogr. Allergy* 19, 108, 1986.

81. Black, C.M., Plikaytis, B.D., Wells, T.W., Ramirez, R.M., Carlone, G.M., Chilmoncyk, B.A., Reimer, C.B., Two site immunoenzymetric assays for serum IgG subclass infant/maternal ratios at full term. *J. Immunol. Meth.* 106, 71, 1988.

82. Ritzmann, S.E., Immunoglobulin abnormalities, in *Serum Proteins Abnormalities, Diagnostic and Clinical Aspects.* Ritzmann, S.E., Daniels, J.C., Eds., Alan R. Liss Inc., New York, 1982, pp. 351–485.

83. Skvaril, F., Barandum, S., Morell, A., Kuffer, F., Probst, M., Imbalances of kappa/lambda immunoglobulin light chain ratios in normal individuals and in immunodeficient patients, in *Protides of Biological Fluids.* Peeters, H., Ed., vol. 23, Pergamon Press, Oxford, 1975, pp. 415–420.

84. Tibbling, G., Link, H., Ohman, S., Principles of albumin and IgG analysis in neurological disorders. I. Establishment of reference values. *Scand. J. Clin. Lab. Invest.* 37, 385, 1977.

85. Dati, F., Ringel, K.P., Reference values for serum IgE in healthy non-atopic children and adults. *Clin. Chem.* 28, 1556, 1982.

86. Ringel, K.P., Dati, F., Buchhqiz, E., IgE-Normalwerte bei Kindem. *Laboratorumblatter* 32, 25, 1982.

87. Saarinen, U.M., Juntunen, K., Kajosarri, J., Bjorksten, F., Serum immunoglobulin E in atopic and non-atopic children aged 6 months to 5 years. *Acta Paediatr. Scand.* 71, 489, 1982.

88. Barbee, R.A., Halomen, M., Lebowitz, M., Burrows, B., Distribution of IgE in a community population sample: correlations with age, sex and allergen skin test reactivity. *J. Allergy Clin. Immunol.* 68, 106, 1981.

89. Wittig, H.J., Belloit, J., DeFillippi, I., Royal, G., Age-related serum IgE levels in healthy subjects and in patients with allergic disease. *J. Allergy Clin. Immunol.* 66, 305, 1980.

90. Herbeth, B., Henny, J., Siest, G., Biological variations and reference values of transferrin, immunoglobulin A and orosmucoid. *Ann. Biol. Clin.* 41, 23, 1983.

91. Tietz, N.W., Ed., Clinical Guide to Laboratory Tests, Saunders, Philadelphia, PA, 1983.

92. Insler, M.S., Lim, J.M., Queng, J.T., Wanissorn, C., McGovern, J.P., Tear and serum IgE concentrations by tandem R immunoradiometric assay in allergic patients. *Opthalmology* 94, 945, 1987.

93. McClellan, B.H., Whitney, C.R., Newman, L.P., Allansmith, M.R., Immunoglobulins in tears. *Am. J. Opthalmol.* 76, 89, 1973.

94. Ford, R.P., Mitchell, P.E.G., Fraser, C.G., Desirable performance characteristics and clinical utility of immunglobulin and light chain assays derived from data on biological variation. *Clin. Chem.* 34, 1733–1736, 1988.

95. Lammers, M., Gressner, A.M., Immunoglobulin light chain determination in serum and urine by use of a fully mechanized immunonephelometric method. *J. Clin. Chem. Clin. Biochem.* 24, 786, 1986.

96. Sun, T., de Szalay, H., Lien, Y.Y., Chang, V., Quantitation of kappa and lambda light chains for the detection of monoclonal gammopathy. *J. Clin. Lab. Anal.* 2, 84–90, 1988.

97. Fifield, R., Hamilton, R.G., Inter-laboratory "external" quality assessment programs for the diagnostic allergy laboratory. *J. Clin. Immunoassay* 16, 144, 1993.

98. Hamilton, R.G., Autoantibodies to immunoglobulins: rheumatoid factor interference in immunological assays. *Monogr. Allergy* 26, 27, 1989.

99. Hamilton, R.G., Monoclonal antibodies in the diagnosis and therapy of human diseases. *Ann. Biol. Clin.* 47, 575, 1989.

100. Chang, T.W., Davis, F.M., Sun, N.C., Sun, C.R.Y., MacGlashan, D.W., Hamilton, R.G., Monoclonal antibodies specific for human IgE producing B-cells: a potential therapeutic for IgE mediated allergic diseases. *Biotechnology* 9, 122, 1990.

101. Lichenstein, L.M., Kagey-Sobotka, A., White, J.M., Hamilton, R.G., Anti-human IgG causes basophil histamine release by acting on IgG–IgE complexes bound to IgE receptors. *J. Immunol.* 148, 3929, 1992.

102. Hamilton, R.G., Molecular engineering: applications to the clinical laboratory. *Clin. Chem.* 39, 1988, 1993.

103. Busse, W., Corren, J., Lanier, B.Q., McAlary, M., Fowler-Taylor, A., Chioppa, G.D. Van As, A., Gupta, N., Omalizumab, anti-IgE recombinant humanized monoclonal antibody, for the treatment of severe allergic asthma. *J. Allergy Clin. Immunol.* 108, 184, 2001.

104. Hamilton, R.G., Marcotte, G.V., Saini, S.S., Immunological methods for quantifying free and total serum IgE in asthma patients receiving omalizumab (xolair) therapy. *J. Immunol. Meth.* 303, 81, 2005.

4 The Complement System

Patricia C. Giclas

CONTENTS

4.1 INTRODUCTION

Complement is a complex system of enzymes, regulatory proteins, and cell surface receptors that are involved in host defense, inflammation, and modulation of immune responses. The system provides a fast-acting mechanism for the identification and removal of foreign substances, providing protection before the adaptive immune system can come into play. It is also involved in a wide variety of homeostatic processes including the clearance of immune complexes, effete cells, and cellular debris from damaged tissues. Complement system contributes to inflammation by inducing local changes in blood flow and the influx of inflammatory cells into the affected area. The fragments of complement also induce the release of additional mediators, cytokines, and enzymes from cells near the site of activation. It is a tightly regulated system designed to produce minimum "collateral damage" that might result in tissue damage and loss of function. The pathology that accompanies uncontrolled activation or incomplete performance of complement's functions is often the result of a deficiency or impairment of one of the components.

Complement system comprises major initiation pathways: classical, alternative, and lectin, as well as a terminal pathway (TP) common to all. Each initiating pathway is triggered by a different type of activator, usually a cell, microbe, or molecular aggregate that presents charge patterns that are "recognized" by components of the individual initiating pathway, making complement one of the innate immune system's primary pattern-recognition mechanisms for detecting nonself.[1] The initiating pathways have several things in common. They are triggered by (1) the binding of one of their components to the activator, (2) a cascade of enzyme activation, and (3) generation of biological effects. Although complement is often referred to as a constituent of plasma, the components are also found in other body fluids and tissues. Not only does diffusion of the molecules takes place between the intravascular and extravascular compartments, but local synthesis also occurs. There are physical and chemical mechanisms as well as specific regulators that prevent uncontrolled activation and damage to local cells and tissues. A network of fluid-phase and cell-associated regulatory proteins and specific receptors interact with the components of complement and their split products, and are involved in controlling complement activation at the cell surface, as well as a wide variety of cell-signaling events. This chapter will address the structure of the complement components and its pathways, their activation and control processes, and the diseases that are associated with deficiencies or improper regulation of complement system.

4.2 THE PATHWAYS OF COMPLEMENT ACTIVATION

Each complement pathway has unique proteins for the initiating step, but shares the same or related proteins for the intermediate steps, and uses the same components in the last step, culminating in the same activities. Complement activation represents the dynamic interplay among the different pathways, the control processes, and other protein systems and cells in the local environment. The outcome of an activation event is modulated by all these participants.

Figure 4.1 depicts the major pathways of complement activation. It can be seen that the classical and lectin pathways (LPs) follow parallel routes and share several

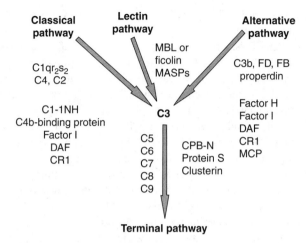

FIGURE 4.1 Schematic of the complement activation pathways. The three initiation pathways (CP, LP, and AP) generate C3 convertases that cleave C3 and allow activation of the terminal components.

components. The alternative pathway (AP) is a separate branch with its own set of proteins and all three pathways converge at the step where C3 is cleaved and the TP begins. The individual proteins involved in activation are depicted in black under each pathway heading, whereas the control proteins for each pathway are listed in gray. The proteins of the TP are listed to the left-hand side of the arrow and its control proteins are to the right-hand side.

4.2.1 CLASSICAL PATHWAY

The first pathway discovered, the classical pathway (CP), was described in the late 1890s and early 1900s as an "activity" of fresh serum that was capable of killing bacteria. By fractionating the serum using increasingly complex methods, investigators found multiple activities that could be separated and then recombined to restore activity. One of the first assays of complement's activity was the lysis of sheep red blood cells (SRBC). Serum from animals that had been immunized with SRBC membranes lysed SRBC efficiently, but if the serum was first heat-inactivated (56°C, 30 min), it had no activity. Likewise, fresh serum from a nonimmunized animal had little or no activity, but if the two sera were combined, the lytic activity was restored. The fresh serum was said to "complement" the activity of the heat-inactivated immune serum. Both the immune factor (antibody) and the heat labile factor (complement) were required for maximum activity. This assay, the CH50, formed the basis for much of the testing in complement field, and is still in use today.

The original components of the CP were given names that included C′ and a number, the latter corresponding to the order in which they were discovered and characterized. The "prime" symbol from C′ was dropped in the 1970s but the numbering system is still used. The first component, C1, is really a macromolecular complex: one C1q, two C1r, and two C1s molecules. Rearranging the components of the pathway in the order in which they act gives the sequence: C1q, C1r, C1s, C4, C2,

C3, C5, C6, C7, C8, and C9. Technically, only C1, C4, and C2 constitute the CP and the final six make up the TP.

Initiation of the CP occurs through the binding of C1q to various compounds, cells, and protein complexes. Small circulating immune complexes (CIC) are common activators of the CP. Other substances that may bind and activate C1q include the lipid A portion of bacterial lipopolysaccharide,[2] subcellular membranes such as mitochondria released from damaged cells,[3] and some of the enveloped viruses.[4] C-reactive protein complexes are also good activators of the CP through binding C1q.[5] Many of these antibody-independent activators play a role in the pathology of certain diseases, but the best-characterized activators of C1 *in vivo* are antigen–antibody complexes, attached either to cells or particulate antigen surfaces, or as parts of an immune complex.

C1q is a complex macromolecule made up of 18 protein strands: six A, six B, and six C, arranged to make up six three-chain (one each of A, B, and C) subunits. Disulfide bonds link the A and B chains within each subunit, and the C chains between the subunits.[6] In the intact C1q molecule, the six subunits are aligned so that their collagenlike regions (CLR; approximately one-third of the length of each subunit) are linked together, forming a "tail" on the molecule. The next portion of each of the subunits is flexible and branches away from the tail. At the end opposite to CLR, globular heads on the ends of flexible stalks are present. From electron micrographs, the resulting molecule resembles a bunch of six tulips branching from the more compact CLR stalk.[7] The six globular regions recognize and bind to antigen-bound immunoglobulins and other activators. The CLR region interacts with C1q receptors, and is the target for autoantibodies in some patients to be discussed later.

Free C1q can bind to activator substances, but without calcium and $C1r_2s_2$, no activation occurs. The proenzymes, C1r and C1s, are linked in tandem: C1s–C1r–C1r–C1s, and bound ionically to the C1q molecule in the presence of calcium. C1r and C1s are similar proteins, with serine protease domains that are active only after the cleavage of a single bond in the proenzyme. C1r has an intrinsic property to rapidly autoactivate when the $C1qr_2s_2$ complex binds to an activating substance.[8,9] Activated C1r cleaves C1s, the active form of which is the enzyme described in the early literature as C1-esterase. Active C1s is a serine protease that cleaves two substrate molecules, C4 and C2, to generate the fragments C4a, C4b, C2a, and C2b.

C4 is synthesized as a single-polypeptide chain, but is posttranslationally converted to three chains in its plasma form: alpha, beta, and gamma. A conformational change in C4 occurs due to its cleavage by C1s or other enzymes and results in the exposure of a highly reactive thioester group that is present in the alpha chain. This thioester gives the resulting C4b fragment, a highly reactive ester link that can bind covalently to amino or hydroxyl groups on nearby molecules or surfaces.[10,11] Although cleavage of the alpha chain, releasing C4a peptide, is the usual method of activating the thioester bond, other forces such as heat or high pH can also cause the thioester to react. The resulting covalently bound C4b plays several important roles in the progression of the activation process:

1. It forms covalent complexes with the activating antigen, antibody, particle, or other proteins or carbohydrates near it.
2. The ester-linked portion of the alpha chain remains attached to the original protein or surface although most of the rest of C4b molecule is removed.

3. It provides a binding site for C3b receptors on phagocytes and other cells to promote removal of microbes or antigen–antibody complexes to which the C4b is bound.
4. It provides a binding site for C2a to form the CP C3 convertase (C4b2a) in which the enzyme site is in the C2a fragment and the C4b acts as a cofactor.

The description of the thioester bond provided also applies to C3 and its activation. The CP C3 convertase, C4b2a, is formed when C2 binds to C4b. C1s can then cleave the C2, releasing C2b and leaving C2a bound to the C4b. C2a is a serine protease like other proenzymes of complement system, and contains the enzyme site that cleaves C3. The half-life of the C4b2a enzyme is about 5 min under physiological conditions.[12] Control of this pathway will be discussed later.

Note that the term C3 convertase, applied to the enzymes that cleave C3, is derived from the observation that cleavage of C3 results in "conversion" of the C3 to a form that migrates faster on immunoelectrophoresis. This was a commonly used assay to study complement activation in the early days of complementology.

4.2.2 LECTIN PATHWAY

The LP, a primitive innate immune mechanism, is similar to the CP.[13] It utilizes several proteins of the collectin family of carbohydrate-binding proteins, including the C1q-like prototype, mannan-binding lectin (MBL), as well as several forms of ficolin. The latter are lectins with CLR and fibrinogenlike domains that recognize sugars on bacterial and other surfaces. The lectins contain carbohydrate recognition domains that bind with high affinity to sugar residues primarily hexoses, commonly found on microbial surfaces.[14] There is evidence that these proteins and other collectins (pulmonary surfactant proteins A and D) also bind free DNA and DNA on apoptotic cells, possibly through the repeating arrays of pentoses that are part of the helix.

The four (MBL associated serine proteases [MASPs]) MASP proteins, MASP-1, MASP-2, MASP-3, and MAp19, are encoded by two genes[15]. They bear homology to C1r and C1s, having a serine protease domain in the B chain. MASP-1 and MASP-3 are alternative splice products of a single gene, MASP1/3. They have the same A chain but different B chains. Another gene encodes MASP-2 and MAp19, the latter of which has only the first two domains of MASP-2 plus an additional four amino acids. MASP-2 is responsible for the cleavage of C4 and C2, generating C4b2a, making it analogous to C1s in function. Like C1r and C1s, the serine proteases associated with the LP become active when binding of the lectin occurs.[16] Selander et al.[17] have recently described a mechanism whereby the LP directly activates C3, leading to AP activation in the absence of C2. This mechanism could explain the relative lack of infections in many C2-deficient patients compared with other component deficiencies.[18] In addition to activating complement, MBL may also be able to act as an opsonin by interacting directly with cell surface receptors. The latter property is also shared with the surfactant proteins A and D.[1]

4.2.3 ALTERNATIVE PATHWAY

Discovered after the CP, the AP, as its name suggests, was proposed as an "alternative" way for complement activation to occur. Because one of the first components of the AP to be identified was properdin (P), the AP was also known as the properdin pathway. The AP proteins include factor D (FD), factor B, C3b or C3•H$_2$O, and P. Like the other pathways, the AP includes the late components, C3 and C5 through C9.

4.2.3.1 Initiation

Unlike the CP or the LP, the AP-initiating step depends on preexisting C3b that has been covalently linked to a surface, a protein, or other molecule. Lutz and Jelezarova[19,20] recently showed that another source of this bound C3b was provided by C3b2–IgG complexes formed when C3 activation occurs in the serum. This bound C3b provides a binding site for factor B so that it can be cleaved by FD, a unique serine protease, in that it does not require enzymatic cleavage for activation, nor inhibitors to control its activity. D is able to change conformation from an inactive state to an active one when it encounters its substrate C3bB and then return to the inactive form after B has been cleaved.[21] D is, in a sense, a renewable resource, and although its low concentration in plasma may be rate limiting, it is adequate for its participation in initiating the AP-activation process.[22]

4.2.3.2 Amplification Loop

Another unique feature of the AP is the amplification loop, or C3 feedback loop, in which P plays the role of a stabilizer to enhance activity of the AP C3 convertase.[23] P not only stabilizes the preformed C3bBb, but also acts as a coordinator of the binding of B to C3b on a surface.[24] Normal control of C3bBb is through dissociation of the enzyme complex, but with the binding of P, the resulting C3bBbP is resistant to dissociation and its half-life as an active enzyme increases from 1.5 to 18 min at 37°C. This makes the AP more efficient at producing large amounts of C3b that deposit on cell or antigen surfaces. Each new C3 that is cleaved provides another C3b to feed back into the amplification loop or deposit on the surface of the activating particle or other nearby residue.

4.2.4 NONCOMPLEMENT ACTIVATION OF C3

Complement activation seldom involves only one pathway. If C3b is deposited on the activating surface or other surfaces or molecules in the area by the CP or LP, the AP can also become involved. The binding of one of the pattern-recognition lectins, MBL or ficolin, to the appropriate sugar residues on microbial surfaces leads to the activation of the LP. Complex polysaccharides such as yeast cell walls (zymosan), high-molecular-weight inulin, sephadex, cellulose acetate, and bacterial cell walls provide appropriate surfaces for AP activation.

Many noncomplement enzymes can cleave C3, C4, C5, and factor B. For example, enzymes of the contact coagulation system (XIIa and Kallikrein), proteases from inflammatory cells (elastase and cathepsin G), or from bacteria can produce active C3a and C5a, and presumably deposit C3b, which could initiate the amplification loop of the AP.[25–28]

A special word about C3—*critical*! Because C3 occupies the central position where the three initiating pathways merge (Figure 4.1), it is critical in the functioning of the system as a whole (Figure 4.2). Cleavage of C3 is accomplished by

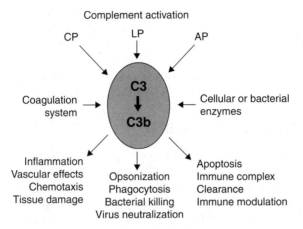

FIGURE 4.2 Cleavage of C3 is the central act of complement activation. Although the complement pathways are predominant in this reaction, enzymes from other sources no doubt play a role in creating C3b that can initiate the amplification (feedback) loop of the AP.

complement and noncomplement enzymes, and the C3 tickover mechanism generates C3b-like C3 (C3•H_2O), assuring that some C3b will be available to the AP at all times. C3 activation is necessary for the production of biological activities associated with complement: anaphylatoxin, chemotaxis, enhanced phagocytosis, clearance of CICs, killing of bacteria and other target cells, removal of apoptotic cells and debris from damaged tissues, modulation of the immune response and memory cells, and so on.

4.2.5 THE TERMINAL PATHWAY

The late components of complement include C3 as well as C5 through C9. C3, as seen from Figure 4.1, is the central component of complement activation mechanisms. All of the activation processes are aimed at producing more C3 cleavage. The biological activities associated with this protein include anaphylatoxin activity, chemotactic activity for eosinophils and induction of mediator release (C3a), opsonization of particles for enhanced uptake by phagocytes and clearance of immune complexes (C3b), and modulation of the immune response (C3d). In addition, C3b is the cofactor for factor Bb enzyme cleavage of C3 (C3bBb) and C5 (C3bBbC3b).

Initiation of the TP leads to two main events: cleavage of C5 to release C5a, the most potent of the anaphylatoxins and a major chemotactic factor for inflammatory cells, and C5b that leads to the formation of the membrane attack complex (MAC). The former activity is accomplished when an additional C3b is cleaved and binds to one of the C3 convertases, C3bBb or C4b2a, to form the C5 convertases, C3bBbC3b, or C4b2a3b. The addition of C3b to the convertase provides a binding site for C5b so that it can efficiently be cleaved by the Bb enzyme. C5 is similar in structure to C3 and C4, but does not contain a thioester bond so it cannot form covalent complexes with other molecules or surfaces.

Cleavage of C5 is the final enzymatic step in complement pathway. Once C5b has been formed, it forms a complex with C6 and C7 (C5b-7) that expresses a transient binding site for membrane surfaces. If a nearby membrane is available, the C5b-7 binds to it. Next, C8, which has a C5b binding site in its beta chain, binds to the C5b-7 complex and the C8 alpha chain provides a site for the binding and polymerization of C9 to form the lytic MAC lesion on the membrane. There have been many arguments in the past about the exact nature of this lesion, but there is no question as to its lethality on an unprotected membrane such as a red blood cell. Lysis occurs when ions are allowed to leak out of the cell and water leaks in, rupturing the cell.

4.3 CONTROL OF COMPLEMENT ACTIVATION

Because activation of complement is potentially harmful to host cells and disrupts homeostasis, complement activation must be very tightly controlled. Passive control of complement activation occurs because the half-life of the thioester sites on C4b and C3b is very short and limits the distance that the molecules can diffuse away from the activation site before becoming inactive (C4bi, C3bi). A positive result of this is that the C3b or C4b is deposited in clusters around the enzyme activating enzyme site, giving the binding of phagocytes and other C3b receptor-bearing cells with higher avidity. Another control is that the multimolecular complexes that form complement enzymes (C4b2a, C3bBb) are inherently unstable. Once dissociation occurs, the enzyme portions of the complex (C2a, Bb) become inactive, but the cofactor parts (C4b, C3b) remain bound to the activator surface and can accept another C2 or B to repeat the enzyme formation.

Active control occurs through specific complement fluid-phase and cell-bound inhibitors and inactivators whose functions include blocking the initiation of the cascade, preventing formation and stabilization of the C3 and C5 convertases, and preventing the formation of the biologically active split products such as the anaphylatoxins and the MAC. In particular, the regulators of complement activation (RCA) are a group of proteins that share genetic and structural motifs. Derived from a gene cluster on human chromosome 1q32, these control proteins are made from 2 to 30 short consensus repeats (SCRs) that are similar but not identical. They share a common ability to bind to C4b and C3b as well, and are active in many phases of controlling of complement activity.

Another control protein that is very important for the CP is the serpin, C1-inhibitor, described in Section 4.3.1 that acts quickly to stop C1 activity. Finally, the *coup de grace* is delivered to the C3 and C5 convertases by a protease present in the circulation in an active state, but requiring a cofactor for the expression of its activity. Factor I has been known by several other names over the years, including conglutinogen activating factor (KAF) and C3b inactivator. Its primary purpose is to degrade C4b and C3b so that they cannot participate in the formation of a new enzyme complex.

4.3.1 CONTROL OF THE CLASSICAL PATHWAY

Figure 4.3 depicts the control of the CP, beginning with C1 inhibitor (C1-Inh) acting as soon as C1r and C1s are activated.[12] C1-Inh is a member of the serpin family of serine protease inhibitors. C1-Inh is the only inhibitor that controls C1r and C1s, but it contributes to the control of other enzymes including the MASP enzymes of the LP, and coagulation factors XIa, XIIa, plasmin, and plasma kallikrein. When an enzyme and an inhibitor interact, the enzyme cleaves a "bait-sequence" on the

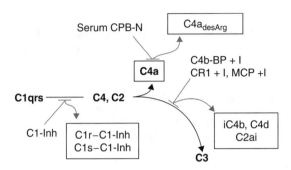

FIGURE 4.3 Control of the CP initiation is effected through the inactivation of C1r and C1s by C1-lnh. Once C4 has been cleaved, the removal of the C-terminal arginine from C4a by serum carboxypeptidase-*N* occurs, and the inactivation of C4b to iC4b, C4d, and C4c prevents its further use as a cofactor for C3 convertase.

inhibitor, getting trapped and thus inactivated by the inhibitor. The inhibitor–enzyme complexes are internalized and degraded by cells bearing the low-density lipoprotein receptor-related protein (LRP).[29]

C4a is released when C1s cleaves C4. Serum carboxypeptidase-*N* rapidly cleaves the C-terminal arginine from the molecule. It used to be thought that C4a had mild anaphylatoxin activity, however more recent experiments with highly purified or recombinant C4a failed to verified this activity, nor has a receptor for C4a been identified. C4a can be used as a good plasma marker for complement activation, provided that precautions are taken to prevent *in vitro* complement activation during sample acquisition, processing, and storage.

The next step in the control of the CP is the C3 convertase. C4b2a is inherently unstable, but there are additional proteins available to accelerate the process. In the fluid phase, C4b binding protein (C4bp), one of the RCA, binds to C4b and displaces the C2a, which becomes inactive C2i.[30] C4bp, a member of the RCA family of control proteins, has seven identical alpha chains (each with eight SCRs and binding sites for C4b), and one beta chain (with three SCRs and a binding site for the anticoagulant protein S). The intact C4bp molecule resembles a spider and can serve as a cofactor for factor I. Once the C2 is displaced, C4bp facilitates factor I cleavage of C4b in several sites on the alpha chain on either side of the thioester binding site. The result is that the fragment (C4d) that was bound to the activator remains bound and the remainder of the molecule (C4c) is shed from the activator surface. The activator carries the C4d fragment and can interact with CR2 receptors that recognize C4b as well as C3b. It is possible that these bound C4d fragments on various cell types can serve as useful markers of complement activation.[31,32] In addition to C4bp, other RCA proteins are active in the control of the CP C3 convertase. Decay-accelerating factor (DAF; CD55) is a small RCA protein (four SCRs) that is linked to almost all cell surfaces by a glycophosphoinositol anchor. As its name implies, it accelerates the decay of the C4b2a enzyme complexes, but it lacks cofactor activity. Membrane cofactor protein (MCP; CD46) is another RCA protein similar in structure to DAF, but with cofactor activity. Complement receptor 1 (CR1; CD35) is another membrane-linked RCA protein with 30 SCRs. It has both decay-accelerating and cofactor activities.

The net result of all of the controls for the CP is that unless the AP-amplification loop is triggered by the C3b generated in the early stages of CP activation, the reaction generally burns out before it reaches the TP, and C3 depletion by this pathway alone is rare.[33]

4.3.2 CONTROL OF THE LECTIN PATHWAY

Specific LP controls are still unknown, but C1-Inh does inactivate the MASPs, and the same controls that take care of the C4b2a regulation also apply to any C3 convertase derived from LP activation.

4.3.3 CONTROL OF THE ALTERNATIVE PATHWAY

The AP differs from the other pathways in that it contains both positive and negative regulators. The negative side of the control pathway (Figure 4.4) is similar to the CP mechanisms that regulate C4b2a. C3bBb can be dissociated through the action of factor H in the fluid phase.[34] H is an RCA protein with 20 tandem SCRs. There are multiple binding sites for C3b as well as sites for binding numerous other polyanion molecules such as heparin. In terms of controlling AP activation, H acts like C4bp does in the CP: it binds to C3bBb, displaces the Bb, and allows factor I to cleave the C3b alpha chain in two places on either side of the thioester site. The resulting fragments of C3 include iC3b from the first cleavage by I, then C3d (bound by the thioester site to the activator surface) and C3c in the fluid phase after the second cleavage occurs. As with the CP control of the convertase, the same molecules control the AP. DAF, MCP, and CR1 also dissociate the C3bBb, and except for DAF, have cofactor activity that enables factor I to degrade the C3b.[35,36]

The positive control of the AP discussed earlier, relies on the property of P to circumvent the negative mechanisms provided by factor I and the RCA proteins, and enhance the stability of the C3bBb complex so that more C3b can be generated.

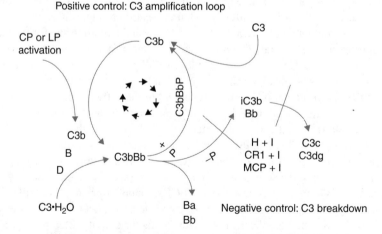

FIGURE 4.4 Control of the alternative pathway includes both the positive action of properdin, allowing the amplification loop to proceed, and the negative effects of C3b inactivation by factor I and its various cofactors.

4.3.4 CONTROL OF THE TERMINAL PATHWAY

Figure 4.5 shows the control mechanisms associated with the TP. The C5 convertases are under the same controls as the C3 convertase of the AP. Dissociation of the enzyme complex and degradation of the C3b occurs as described earlier. Inactivation of the potent anaphylatoxins, C3a and C5a, is accomplished by cleavage of the C-terminal arginine from the peptides. This prevents interaction with their receptors (C3aR, C5aR [CD88]) and expression of activity. C5a$_{desArg}$ does retain a percentage of its chemotactic activity and this may be enhanced by interaction with Gc-globulin as a cochemotaxin.[37,38]

The MAC is controlled through two mechanisms. First, its completion on the membrane surface is prevented by cell surface receptor CD59 binding to C8, which interferes with the binding and polymerization of C9. Second, the newly formed C5b-7 and other intermediates that form in the fluid phase are prevented from binding to cell surfaces by two inhibitors. Clusterin (SP40,40) and S Protein (vitronectin) can bind to the C5b-7 complexes and keep them soluble although C8 and C9 can continue to bind to the complex. This is important in controlling the phenomenon known as reactive lysis or bystander lysis that would occur to cells in the vicinity of complement activation. The resulting soluble C5b-9 (SC5b-9) complexes are known as terminal complement complexes (TCC). These complexes can be found in the circulation for hours following the activation event, and can serve as markers of complement activation.

4.4 COMPLEMENT DEFICIENCIES AND THEIR DIAGNOSIS

Deficiencies or mutations in complement components, whether inherited or acquired, predispose the individual to infections, autoimmune diseases, impaired immune responses, and an increasingly long list of diverse conditions including fetal loss,

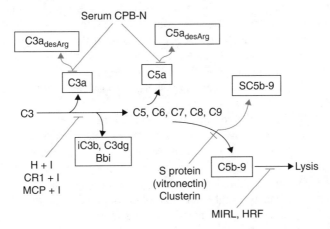

FIGURE 4.5 Control of the terminal pathway includes inactivation of the anaphylatoxins, C3a and C5a, by carboxypeptidase-*N*. The C3 and C5 convertases are inactivated by factor I and its cofactors, generating the fragments iC3b, C3d, and C3c. The membrane attack complex is prevented from forming in the fluid phase by S Protein or clusterin, and on the cell surface by CD46.

renal disease, vasculitis, angioedema, and macular degeneration. Deficiency of any one of the regulator proteins results in excessive complement consumption leading to an inappropriate inflammatory response, destruction of self-tissue, and depletion of C3 or other components downstream of the missing control protein.

Identifying a complement deficiency and determining whether it is due to a genetic defect or an acquired process is one of the problems the practicing physicians face. Several screening tests are available that make it easier to find the answers. It is important to know as much as possible about the reason(s) for low or absent complement so that decisions regarding appropriate treatment can be made, including when to use antibiotics and immunizations as well as genetic counseling for inherited deficiencies.

Indications for possible complement abnormalities include recurrent mild or serious bacterial infections, angioedema, vasculitis, certain renal conditions, and autoimmune disease. A family history of members having the same presentation should increase the suspicion about a complement deficiency. The initial tests done to evaluate a patient's complement system are critical because they can rule out a defect or indicate what further testing must be done if there is a problem. The aim of the evaluation process is to rule out or clearly define complement component deficiency with as few tests as possible.

4.4.1 Deficiencies of Classical Pathway Components

4.4.1.1 C1q, C1r, C1s, C2, C4

Deficiencies of the CP components are associated with immune complex diseases such as systemic lupus erythematosus (SLE) and rheumatoid arthritis (RA) as well as recurrent infections. The incidence of SLE in patients with known C1q, C4, or C2 deficiency is roughly 90, 75, and 15%, respectively.[39] These patients present with several characteristics that distinguish them from other patients: early age of onset, prominent photosensitivity, and fewer renal symptoms.[40] The ANA titers in these patients are variable and often normal. The role of these complement components in the clearance of apoptotic cells that contain many of the antigens against which the autoimmune response is directed, and the impairment of this process in the absence of C1q or C4 has been suggested as a possible mechanism leading to autoimmunity.[41,42]

C4A and C4B, the two haplotypes of human C4, are coded by two different genes that are present in the major histocompatability complex on human chromosome 6. The two forms were named A and B because one was slightly more acidic and the other more basic so they migrated differently on immunoelectrophoresis gels. Variation in C4 gene number in the population is not uncommon, and the variability of mutations within the genes is high. Only 75% of the population has two C4A and two C4B genes. The total sum of C4A and C4B genes in an individual can range from none to eight or more copies, giving this protein a wide range of concentrations and even wider range of functions in the general population.[43,44] Most of the partial C4 deficiencies are without consequence, but some individuals have a total deficiency of C4A or C4B or both. The differences between the two forms of C4 are minor—a few amino acids near the thioester group. This difference only affects the ability of the C4b fragment to bind preferentially to protein (C4A) versus carbohydrate residues on cell surfaces (C4B). Generally, the C4A-derived C4b molecules are more important in immune complex disease; C4b bound to antigen and antibody in CICs is required for efficient clearance. Deficiency of C4A is associated with a 15% incidence of SLE.[44] C4B-derived C4b molecules have higher hemolytic activity in the CH50 assay because the red cell surfaces are largely unposed of carbohydrate residues. There is some evidence that C4B deficiency is associated with a higher incidence of infection or with Henoch–Schönlein purpura.[45,46]

Standard assays for C4 do not distinguish between these forms since the minor differences between C4A and C4B do not affect the reactivity of the C4 proteins with the antibodies used in the measurement of C4. Close to 100 different polymorphisms in the genes make C4 one of the most variable proteins in the circulation. C4 is also an acute-phase reactant, with levels that can double within 24 h after trauma or acute inflammation. Another issue confounding the measurement of C4 in serum is that the antibodies used by most diagnostic laboratories react with the C4 breakdown product C4c as well as native C4. The final report represents the mixture of C4 and C4c that is present at the time of the measurement and is not an accurate measure of the amount of native C4 that would be available for complement activation in the patient.

Recurrent infections with *Streptococci*, *Staphylococci*, and other common bacteria, but not usually with *Neisseria* are another signal for possible CP deficiencies. Complement analysis in children with this presentation should not be overlooked in the work-up for immune dysfunction. The best screens for these deficiencies are the CH50 and AH50 assays. By doing both tests before ordering levels or functions, it is possible to identify the pathway that is blocked and avoid tests and expense that are not necessary. The absence of any one of the CP proteins (C1q, C1r, C1s, C4, or C2) will block CP activity (CH50), the absence of an AP protein (D, B, P) will block AP activity (AH50), and the absence of a terminal protein (C3, C5, C6, C7, C8, or C9) will block both assays.[47] Deficiencies of C1r and C1s are rare and may be partial or combined, and predispose the patient to recurrent pyogenic infections as well as immune complex disease.[48]

C2-deficient patients have increased susceptibility to infections, often presenting with streptococcal pneumonia. The link to autoimmunity in this group is not as strong as it is with C1q or C4 deficiency, but Caucasian patients with rheumatologic disorders have a higher incidence of homozygous C2 deficiency than the general population, where incidence of homozygous deficiency is estimated to be 1/10,000.[49,50] There are two forms of C2 deficiency. Type I C2 deficiency, comprising over 90% of patients, results from complete lack of synthesis, whereas type II C2 deficiency results from a secretion defect.[51] The former deficiency is characterized by little or no detectable protein in the circulation, and the latter by up to 10% of normal levels.[52,53]

4.4.1.2   C1-Inhibitor and C4bp

For the patient with recurrent angioedema and a family history consistent with C1-Inh deficiency, low C4 levels (or elevated C4a) and absent or dysfunctional C1-Ihn protein confirms the diagnosis of hereditary angioedema (HAE). CH50 and

C2 may also be low, but C3, C1q, and other complement proteins should be normal in HAE unless an unrelated complement-activating process such as an infection or injury occurs. In the latter case, the proteins should normalize when the activating event has stopped.

HAE is predominantly a heterozygous deficiency state that is inherited in an autosomal dominant fashion. There are two forms of HAE. Type I includes patients who have one gene that produces a functional C1-Inh protein and one gene that produces no protein. These patients represent 75–80% of HAE patients. Type II patients have one normal gene and one gene with a defect that is due to minor change such as a single nucleotide polymorphism (SNP). This defective gene produces a C1-Inh protein with little or no activity in the functional screening assays, but with enough similarity to the normal protein to be recognized by the antibodies used to measure the C1-Inh levels. The amount of C1-Inh protein in the circulation in these patients is within the normal range or even elevated in some cases. There are many case reports in the literature where an HAE patient had no family history and represented a spontaneous mutation.[54,55] The first case on record of homozygous deficiency of C1-Inh was recently reported.[56] Two members of eight were studied in the family of the propositus, who had coding area mutations in both genes for C1-Inh. The other six family members were heterozygous for the deficiency.

When the family history is negative, then acquired angioedema (AAE) must be ruled out.[57] The same tests, C4 and C1-Inh level and function, should be done, but C1q levels should be added to the list. As in HAE, type I AAE is characterized by low levels of C1-Inh and low C4, but unlike HAE, C1q is also low in the AAE patients. The mechanism in this form of AAE is hypercatabolism of the inhibitor by an underlying disease process, often a lymphoproliferative disorder. Type II AAE patients may have normal levels of C1-Inh, but it is inactive due to the presence of an autoantibody against the inhibitor molecule.[54] The presence of the autoantibody can be detected by enzyme-linked immunosorbent assay (ELISA) methods[58] or inferred from a lower-molecular-weight form of the inhibitor when it is observed on polyacrylamide gel electrophoresis. The onset of the acquired forms of AE is usually later in life.

Deficiencies of C4bp are very rare and would logically be associated with excessive activation of C3 through the CP. The only documented case was a patient with an atypical form of Behçet's disease and angioedema.[59] Circulating concentrations of C4bp vary according to genetic determinants as well as fluctuations due to consumption during disease states.[60–62]

4.4.2 DEFICIENCIES OF THE LECTIN PATHWAY

MBL serum concentrations vary widely, in part due to the presence of gene variants that produce different amounts of the protein. Approximately 10% of individuals studied to date are MBL deficient.[63] LP activation of complement is antibody-independent so it is thought to play an important role in innate immunity during infancy at the time that maternal antibodies are clearing but before the infant's immune responses are mature.[64] Deficiency of MBL is associated with increased frequency of pyogenic infections and sepsis, particularly in neonates and young children.[65] MBL deficiency is increased in patients with autoimmune disease, often identified as those who have more frequent and severe infections and worse outcomes.[66,67] Like Clq, MBL seems to be a target for autoantibodies in lupus patients, but the significance of these antibodies was not evident from the disease characteristics of the patients.[68]

The ficolins are a group of proteins containing a CLR, but they also have a C-terminal fibrinogen domain that is thought to be the ligand-binding site for N-acetylglucosamine.[16] Ficolin-L and ficolin-H are found in the circulation, whereas ficolin-M is a secretory protein of neutrophils, monocytes, and type II alveolar epithelial cells. All forms have similar opsonizing capacity for pathogens, and are associated with the MASP enzymes. Although the ficolins were first described as opsonins in 1996, little is known of the prevalence of deficiencies.[69]

Mutations have been located in the genes for the MASP proteins, but mutations in MBL that prevent the assembly of the complete complex are the major known cause of functional deficiency of these proteins.[70]

4.4.3 Deficiencies of the Alternative Pathway

4.4.3.1 Factors D, B, and Properdin

Two families with FD deficiency have been described to date. In the first, the propositus was a 23-year-old Dutch woman who presented with meningococcemia from *Neisseria meningitidis*. Three relatives had FD deficiency with no health problems, and a deceased family member had a history of meningitis in his twenties and died from pneumonia and meningitis caused by *Streptococcus pneumoniae* at the age of 81.[71] The second family had two children with septic shock, one case was fatal. Complement studies in the first family included normal C3, C4, and CH50 in all members, but AH50 was approximately 10% of normal in deficient members, normal in all others including those heterozygous for the deficiency. When FD was analyzed, it was slightly elevated in those members who were homozygous for the wild-type FD gene, about half the normal level in those heterozygous for the mutation, and undetectable in the four patients homozygous for the mutation. In the second family, MBL, Clq, C4, C3, B, P, H, and I were all normal but D was not measurable.[72]

Factor B is an acute-phase protein and increases during inflammation. There are two polymorphic forms: BF*F and BF*S, based on electrophoretic mobility. The amount of B in the circulation is higher in BF*F homozygous individuals and lower in the BF*S homozygous individuals, with heterozygous individuals being intermediate.[73]

There has only been one unconfirmed report of a B deficiency in humans. No naturally occurring deficiencies have been identified in animals, but Bf gene knock-out mice survive in pathogen-free conditions.[74]

P is the only complement protein that is X-linked, with its gene at Xp11.3–Xp11.23. The protein is synthesized by monocytes, granulocytic cells, and T cells. Several mutant forms of the protein have been identified that result in decreased AP function. P deficiency increases the susceptibility to bacterial infections including *Neisseria*.[75]

4.4.3.2 AP Control Proteins

Deficiencies of factor H are linked with a wide variety of symptoms. Complete deficiency of H leads to uncontrolled ac tivation of AP, and depletion of C3 occurs.

The first patient with factor H deficiency was initially thought to be C3-deficient. This form of factor H deficiency is similar in presentation to the late component deficiencies due to the low or absent levels of C3. Recent data has been published that demonstrates how critical the role for this complement control protein is in maintaining health in a number of tissues. In addition to pyogenic infections, deficiency or dysfunction of factor H is associated with type II membranoproliferative glomerulonephritis (MPGN-II),[76,77] atypical hemolytic uremic syndrome (aHUS),[78,79] and age-related macular degeneration (AMD).[80–82] It has become clear that these diseases are examples of control processes that have become impaired on the surfaces within the organs affected, and that not only factor H but also other control proteins such as factor I and MCP (CD46) are involved.[83] Many of the defects in these proteins have been shown to be due, not to gross deficiency of the protein, but to a subtle alteration in the amino acid sequence that interferes with the normal function of the inhibitor. New studies continue to identify additional SNPs in the sequences of these control proteins.

4.4.4 DEFICIENCIES OF THE TERMINAL PATHWAY

4.4.4.1 C3, C4, C5, C6, C7, C8, C9

C3 deficiency is rare and generally associated with recurrent fulminant infections. Deficiency of C3, the major opsonin, results in severe, recurrent pyogenic infections that begin shortly after birth, a clinical presentation and course similar to that observed in hypogammaglobulinemia. Acquired C3 deficiency that occurs with factor H deficiency, factor I deficiency, or in the presence of C3 nephritic factor (C3NeF) predisposes the patient to the same risks as congenital C3 deficiency. Deficiencies of C3, C5, C6, C7, and C8 are strongly linked with susceptibility to pyogenic infections, particularly with *Neisseria*.[84] These components are required for the formation of the MAC and lysis of the bacteria.

C8 is made up of three chains that are encoded by different genes.[85] Because C8 requires all three chains to be functional in the MAC, assays that measure only the protein can be misleading, whereas the functional assay is diagnostic. Deficiency of C9 is not strongly linked with infections.[84]

4.5 LABORATORY EVALUATION OF COMPLEMENT

4.5.1 SPECIMEN PREPARATION AND HANDLING

Because complement is a very labile system, the way that specimens are collected and processed can influence the result of testing. Tubes drawn for serum collection should be allowed to clot for a minimal amount of time (either at room temperature or at 37°C if possible). Once the clot has formed it should be removed by centrifugation in the cold, if possible. The use of tubes with gel separators does not affect the quality of the specimen. Functional assays for complement (CH50, AH50, and individual component functions) should be done with serum that was collected as described and frozen at −70°C within an hour after it was separated from the clot. Ideally, the tube containing the serum should be kept on ice once it has been separated, and transferred to a −70°C freezer or placed on dry ice as quickly as possible.

Many of complement tests require EDTA plasma as the specimen. The presence of EDTA prevents activation of complement enzymes and reduces the likelihood that additional activation will occur in the tube after it has been drawn. EDTA specimens should be drawn, centrifuged, and the plasma separated and frozen at $-70°C$ within 30 min.

If specimens are to be used to measure complement split products, an additional precaution can be taken to prevent noncomplement activation of the proteins. Futhan, a broad spectrum serine protease inhibitor added to EDTA blood-collection tubes, has been used successfully to prevent *ex vivo* complement activation.[86,87]

4.5.2 TESTS FOR FUNCTIONAL ACTIVITY

Functional tests measure the ability of complement to perform activities associated with the system. These tests are designed to measure the activity of entire pathways (e.g., CH50 and AH50), individual components (e.g., C1-Inh function, factor B hemolytic activity), complement fragments (e.g., C5a chemotactic activity), or auto-antibodies (e.g., C3NeF activity). Functional tests are often more sensitive than measures of protein concentration, and have the advantage of measuring only the native (active) components or pathways.

The CH50 was one of the earliest complement tests developed and is still in use today.[88] It is based on a hemolytic assay in which an immune complex is formed on the surface of SRBC by antibodies that react with a cell surface antigen. When complement is activated by the antigen-fixed antibodies on the cell surface, the cell is lysed and hemoglobin is released. The formation of the MAC on the cell requires the sequential action of all nine components of the classical (C1, C4, and C2) and terminal (C3, C5, C6, C7, C8, and C9) pathways. By titration of complement source (serum in most cases), such that only a portion of the cells present are lysed, the amount of active complement can be calculated. The results are expressed as the reciprocal of the dilution of serum that causes lysis of 50% of the cells in the assay. Although this assay served for years as the "gold standard" for complement analysis and was the basis for many other tests developed later, many modern labs prefer new methods that are not so labor intensive. Only a few clinical labs in the United States still perform the hemolytic titration method of the CH50.

Two alternative methods analogous to the CH50 assay for measuring CP function are currently in use in clinical laboratories. In one assay, the activation of the MAC by the CP induces the lysis of liposomes used as a substitute for SRBCs. The end point is the release of an enzyme that can be read on an automated chemistry analyzer of the sort used for other clinical lab tests (WAKO Clinical Diagnostic Reagents, Richmond, Virginia). Several solid-phase assays (ELISAs) based on the detection of the final C9 neoantigen that is formed when the complete pathway has been activated are also available (Quidel, San Diego, California; Diasorin, Stillwater, Minnesota). All these methods are adequate screens for CP complement deficiency and make the CH50 the best single screen for complement abnormalities, since the absence or decrease of activity in the CH50 implies that at least one of the necessary complement components is missing, low, or functionally aberrant.

The analogous assay for AP activity, the AH50, is not as widely available as the CH50, but is useful as a screen for complement deficiency especially when used in conjunction with the CH50. The AH50 for human serum depends on the unique properties of rabbit erythrocytes to activate the AP, with sequential activation of factors D, B, P, C3, C5, C6, C7, C8, and C9.[89] P is required for the stabilization of the C3 convertase (C3bBb) and inefficient activation of the AH50 occurs if P is low or absent. The buffer used for the AH50 contains ethylene glycol tetraacetic acid (EGTA) to block activation of the calcium-dependent CP or LP, and magnesium that is required for the formation of the C3 convertase. The AH50 measures the percentage of erythrocytes lysed by a given amount of serum. The result is expressed in units that represent the reciprocal of the dilution that lyses 50% of the cells used in the assay.

If both CH50 and AH50 are used to screen for complement deficiency, the number of additional tests required to pinpoint the defect can be minimized. Because both assays include the same six terminal components (C3, C5, C6, C7, C8, and C9), the results will be low or absent in both tests if one or more of these components are missing. If a CP component is missing, the CH50 will be low or absent, but the AH50 will be normal, whereas if an AP component is low or missing, the reverse will be true. A missing control protein such as H or I, will lead to a decrease in both pathways due to uncontrolled activation of C3. C3NeF can also have this effect. To determine whether the defect is in the TP or a control protein, it is necessary to look at multiple components. A true deficiency of one of the terminal proteins will affect only the level or function of that particular protein, whereas the lack of a control protein such as H or I will cause the decrease predominantly of C3 and variably decrease the late components.

Several assays for C1-Inhibitor function have been developed. One that is commonly used is an ELISA that measures complexes formed between biotinylated C1s (in its active form) and C1-Inh that are captured on an avidin-coated plate (Quidel, San Diego, California). Another test for C1-Inh function relies on the observation that binding of the inhibitor to C1r masks the site on C1r that is recognized by some polyclonal antibodies.[90] After activation of the test serum with aggregated immunoglobulins, the amount of detectable C1r decreases in proportion to the amount of active C1-Inh present.

Functional tests can be done for each of the individual components by utilizing variations on the CH50 or AH50 assays in which an excess of all components except the one being evaluated is added to the appropriate cells and then the patient's serum provides the only source of the particular component in question. Component-depleted sera are available from several commercial sources. Alternatively, purified components can be added to the patient's serum to determine which one(s) restores activit.

The function of the MBL pathway can be determined by using an ELISA in which the patient's serum is placed into wells coated with mannan. After MBL binds to the mannan-coated surface, the MASP enzymes cleave C4 and the resulting C4b and C4d are deposited on the plate and can be measured using enzyme-conjugated monoclonal antibodies.

4.5.2.1   Quantitative Tests for Component Concentrations

Like most other circulating proteins, complement components can be measured by immunochemical methods common in most laboratories. These include technologies such as nephelometry, radial immunodiffusion (RID), radioimmunoassay (RIA), and ELISA techniques. The critical points are the specificity of the antibodies used, and the reliability of the standard and controls. There are as yet very few complement assays that have been standardized and validated for FDA approval, and most laboratories must rely on in-house methods and proven research technology to perform diagnostic procedures for complement analysis.

The most definitive method for evaluating complement activation is by quantification of the fragments formed during the enzymatic cleavage steps.[76] Because many of complement components are acute-phase reactants, decreases due to activation may be masked by increases in the synthesis rates during an inflammatory episode. The split products can be used to determine if activation has occurred because their increase occurs only when complement enzymes are formed and active. An added bonus is that the pathway of activation can be determined: C4a and C4d are markers for CP or LP activation; Bb is a marker for AP activation; and C3a, iC3b, C5a, and soluble C5b-9 can be used to determine TP activation.

4.5.2.2 Tests for Complement Autoantibodies

C1q and C1-Inh autoantibodies can be determined using ELISA methods. The antibodies to C1-Inh bind to the inhibitor molecule and prevent it from attaching to the enzyme, but do not prevent the enzyme from cleaving the inhibitor. The resulting lower molecular-weight inhibitor fragment can be detected by polyacrylamide gel electrophoresis.[91,92] There are several assays available for C3 nephritic factor that rely on its function and measure either lysis due to generation of C3b on a red cell surface, or look directly at the amount of C3 that is cleaved when the patient's serum is mixed with normal serum.[93,94]

4.6 CONCLUDING REMARKS

Complement deficiencies have been documented for almost all the known components of complement system, including the cell-associated receptors and control proteins. These deficiencies manifest themselves in the form of recurrent infections, edema, increased incidence of autoimmune disorders, urticaria, vasculitic syndromes, renal disease, and macular degeneration. At present, there is little one can do for these patients, but it is hoped that in the near future recombinant proteins or targeted gene therapy will be available to help patients with deficiencies.

Pharmaceutical companies are studying complement to discover how activation can be blocked by specific compounds including monoclonal antibodies and new inhibitors. These new compounds may be used to prevent autoimmune or trauma sequellae. Complement is a powerful tool for the destruction of target cells and, if controlled activation could be accomplished, could be directed against tumors. There are many exciting possibilities on the horizon for the study, intervention, and use of complement in treating a wide range of diseases (Table 4.1).

TABLE 4.1

Complement Proteins

Component Name (Other Names)	Function(s)	Chain Structure of Circulating Protein (M_r)	Chromosomal Location of Gene	Primary Site of Synthesis	Serum Concentration
C1q	CP activation; Binding; Immune complex clearance; Clearance of apoptotic cells	18 Chains: 6 each of A, B, and C chains; A chain (27.5 K); B chain (25.2 K); C chain (23.8 K)	1p34–1p36.3	Macrophages and monocytes	83–125 µg/mL
C1r	CP serine protease; Cleavage of C1s	Single-polypeptide chain (173 K)	12p13	Hepatocytes	34 µg/mL (mean)
C1s (C1-esterase)	CP serine protease; Cleavage of C4 and C2	Single-polypeptide chain (79.8 K)	12p13	Hepatocytes	31 µg/mL (mean)
C1	CP activation; Complex of $C1qC1r_2C1s_2$; Requires Ca^{2+} for function	Sum of the subunits (964 K)			
C2	CP and LP serine protease; C2a fragment has enzyme sites for C3 and C5 cleavage	Single-polypeptide chain (100 K)	6p21.3 within MHC class III	Liver; Monocytes and macrophages	11–35 µg/mL
C4	CP and LP cofactor for C2 function; Requires Mg^{2+} to form C4bC2a; Provides the binding site for C3; C4b fragment is opsonin; C4a fragment is an activation marker	Three chains: Alpha (93 K); Beta (75 K); Gamma (33 K); Internal thioester provides covalent bond to other molecules	6p21.3 within MHC class III; Note: C4A and C4B isotypes are both present	Liver (hepatocytes); Acute-phase reactant	150–450 µg/mL

C3	CP, AP, LP, and TP Central component in activation Cleaved by C4bC2a or C3bBb C3bBbC3b cleaves C5 (2nd C3b provides the binding site for 5) C3b fragment is opsonin C3a fragment is anaphylatoxin, and is chemotactic for eosinophils C3d fragment is the ligand for CR2	19p13.3–p13.2	Two chains: Alpha (115 K) Beta (75 K) Internal thioester provides covalent bond to other molecules	Liver (hepatocytes) Acute-phase reactant	1–1.5 µg/mL
C5	TP Cleaved by C4bC2aC3b or C3bBbC3b (2nd C3b provides the binding site for C5) C5b initiates the formation of MAC C5a is a potent chemoattractant and anaphylatoxin for inflammatory cells	9q33	Two chains Alpha (115 K) Beta (75 K)	Liver (hepatocytes)	55–113 µg/mL
C6	TP Forms a complex with C5b Participates in the formation of MAC	5p12–14	Single-polypeptide chain (100 K)	Liver (hepatocytes)	28–60 µg/mL
C7	TP Forms a complex with C5b6 Participates in the formation of MAC	5p12–14	Single-polypeptide chain (95 K)	Granulocytes Liver (hepatocytes) Weak acute-phase reactant	27–74 µg/mL

(continued)

TABLE 4.1 (continued)
Complement Proteins

Component Name (Other Names)	Function(s)	Chain Structure of Circulating Protein (M_r)	Chromosomal Location of Gene	Primary Site of Synthesis	Serum Concentration
C8	TP Forms complex with C5b67 Participates in formation of MAC	Oligomer with three subunits Alpha (64 K) Beta (64 K) Gamma (22 K)	C8 alpha: 1p32 C8 beta: 1p32 C8 gamma: 9q34.3	Liver (hepatocytes) Acute-phase reactant	49–106 μg/mL
C9	TP Final component of MAC with C5b-8 Forms tubular polymers of 12–18 C9 molecules	Single-polypeptide chain (71 K)	5p13	Liver Acute-phase reactant	33–95 μg/mL
FD (CFD) (C3 convertase activator, adipsin)	AP serine protease Cleaves FB bound to C3b to make C3bBb	Single-polypeptide chain (25 K)	19p13.3	Adipose tissue	1–5 μg/mL
FB	AP serine protease Catalytic subunit of C3/C5 convertases	Single-polypeptide chain (83 K)	6p21.1–6p21.3 within MHC class III	Liver Acute-phase protein	74–286 μg/mL
Properdin	AP Stabilizes C3bBb Inhibits cleavage by FI	Diverse polymers, single-polypeptide chain (55 K)	Xp 11.3–Xp 11.23	Monocytes, T cells, and granulocytes	10–36 μg/mL
MBL (MBP) (Mannose-binding lectin) (Mannose-binding protein)	LP activation by carbohydrates opsonin	2–6 Oligomers of trimers, single-polypeptide chain (32 K)	10q11.2–q21	Liver Acute-phase protein	0–5 μg/mL

	Function	Structure	Chromosome	Synthesis	Concentration
MASP-1 (MBL-associated serine protease 1) P100	LP serine protease Cleaves C4 and C2 to form C4bC2a, also cleaves C3 directly	Proenzyme circulates as single-polypeptide chain (93 K) Two chains (when activated) Heavy (66 K) Light (31 K) Interchain disulfide provides covalent bond	3q27–28	Liver	6 µg/mL
MASP-2 (MBL-associated serine protease 2)	LP serine protease Cleaves C4 and C2 to form C4bC2a	Proenzyme single-polypeptide chain (76 K) Two chains (when activated) A (52 K) B (31 K) Interchain disulfide	1p36.3–p36.2		Not established
FH (CFH)	AP inhibitor Displaces Bb from C3bBb Cofactor with FI to inactivate C3b Chemoattractant, adhesion protein	Single-polypeptide chain (155 K)	1q32	Liver	550 µg/mL
FI (C3b/C4b inactivator) (KAF)	AP control protein Serine protease for C3b or C4b with cofactors FH or C4bp, respectively	Two chains: Heavy (50 K) Light (38 K) Interchain disulfide provides covalent bond	4q25	Liver	35 µg/mL in plasma
C4bp (C4b binding protein)	CP, TP Accelerates decay of C3/C5 convertases Cofactor with FI to inactivate C4b Binds serum amyloid P	Heterogeneous oligomers of two chains Major isoform is $\alpha 7\beta 1$ Alpha (70 K) Beta (45 K)	1q32	Liver hepatocytes	150–300 µg/mL

(continued)

TABLE 4.1 (continued)
Complement Proteins

Component Name (Other Names)	Function(s)	Chain Structure of Circulating Protein (M_r)	Chromosomal Location of Gene	Primary Site of Synthesis	Serum Concentration
CI-INH	CP Serine protease inhibitor (serpin) for CIr and CIs	(104 K)	11q11–q13.1	Liver Acute-phase reactant	150 µg/mL
CR1, CD35 Complement receptor type 1, C3b/C4b receptor	Receptor for C3b, C4b, and CIq	Membrane glycoprotein (205–250 K)	Iq32	Found on many hematopoietic cells and others	Varies with cell type
CR2 Complement receptor type 2, CD21, C3dR, EBVR2		Membrane glycoprotein (146 or 148 K)	Iq32		NA
CR3 Complement receptor type 3, CD1 1b/CD 18, MAC-1, $\alpha_M\beta_2$ integrin	Protects host cells from complement Accelerates decay of C3/C5 convertases Has cofactor activity for FH and FI	Membrane glycoprotein heterodimer Alpha (CDIIb) (160 K) Beta (CD 18) (95 K)	Alpha 16pl 1–pl3.1 Beta 21q22.3		NA

CR4 Complement receptor type 4, CD11c/CD18, $A_x\beta_2$ integrin		Membrane glycoprotein heterodimer Alpha CD11c (150 K) Beta CD18(95 K) same beta as in CR3	Alpha 16p11.2 Beta 21q22.3		NA
DAF Decay-accelerating factor, CD55	Protects host cells from complement, Accelerates decay of C3/C5 convertases	Single-chain membrane glycoprotein (75 K)	1q32	Hematopoietic cells and many others, especially, endothelial cells	NA
MCP Membrane cofactor protein, CD46	Protects host cells from complement Accelerates decay of C3/C5 convertases Has cofactor activity for FH and FI		1q32		NA
C1qRp, CD93 C1q receptor 1 MXRA4		Single-polypeptide chain (100 K)	20p11.21		NA
C3a receptor		Membrane glycoprotein (54 K)	12p13		NA
C5a receptor		Membrane glycoprotein 50–55 K on eosinophils	19q13.3–q13.4		NA

REFERENCES

1. Sim RB, Clark H, Hajela K, Mayilyan KB, Collectins and host defense, *Novartis Found Symp* 279, 170–181, 2006.
2. Cooper NR, Morrison DC, Binding and activation of the first component of complement by the lipid A region of lipopolysaccharides, *J Immunol* 120, 1862–1868, 1978.
3. Giclas PC, Pinckard RN, Olson MS, *In vitro* activation of human complement by isolated human heart subcellular membranes, *J Immunol* 122, 146–151, 1979.
4. Perrin LH, Joseph BS, Cooper NR, Oldstone MB, Mechanism of injury of virus-infected cells by antiviral antibody and complement: participation of IgG, F(ab')2, and the alternative complement pathway, *J Exp Med* 143, 1027–1041, 1976.
5. McGrath FD, Morgan BP, Arlaud GJ, Daha MB, Hack CE, Roos A, Evidence that complement protein C1q interacts with C-reactive protein through its globular head region, *J Immunol* 176, 2950–2957, 2006.
6. Reid KB, Porter RR, Subunit composition and structure of subcomponent C1q of the first component of human complement, *Biochem J* 155, 19–23, 1976.
7. Tschopp J, Villiger W, Fuchs H, Kilchherr E, Engel J, Assembly of subcomponents C1r and C1s of first component of complement: electron microscopic and ultracentrifugal studies, *Proc Natl Acad Sci USA* 77, 7014–7018, 1980.
8. Thielens NM, Illy C, Bally IM, Arlaud GJ, Activation of human complement serine proteases C1r is down-regulated by a Ca^{2+} dependent intramolecular control that is released in the C1 complex through a signal transmitted by C1q, *Biochem J* 301, 509–516, 1994.
9. Villiers CL, Arlaud JG, Colomb MG, Autoactivation of human complement subcomponent C1r involves structural changes reflected in modifications of intrinsic fluorescence, circular dichroism and reactivity with monoclonal antibodies, *Biochem J* 215, 369–375, 1983.
10. Law SK, Lichtenberg NA, Levine RP, Covalent binding and hemolytic activity of complement proteins, *Proc Natl Acad Sci USA* 77, 7194–7198, 1980.
11. Law SKA, Dodds AW, The internal thioester and the covalent binding properties of complement proteins C3 and C4, *Protein Sci* 6, 263–274, 1997.
12. Ziccardi R, The first component of human complement (C1): activation and control, *Springer Semin Immunopathol* 6(2–3), 213–230, 1983.
13. Matsushita M, The lectin pathway of complement system, *Microbiol Immunol* 40(8), 887–893, 1996.
14. Turner M, Mannose-binding lectin: the pluripotent molecule of the innate immune system, *Immunol Today* 17, 532–540, 1996.
15. Schwaeble W, Dahl MR, Thiel S, Stover C, Jensenius JC, The mannan-binding lectin-associated serine proteases (MASPs) and MAp19: four components of the lectin pathway activation complex encoded by two genes, *Immunobiology* 205(4–5), 455–466, 2002.
16. Matsushita M, Endo Y, Fujita T, Cutting edge: complement-activating complex of ficolin and mannose-binding lectin-associated serine protease, *J Immunol* 164(5), 2281–2284, 2000.
17. Selander B, Mårtensson U, Weintraub A, Holmström E, Matsushita M, Thiel S, Jensenius JC, Truedsson L, Sjöholm AG, Mannan-binding lectin activates C3 and the alternative complement pathway without involvement of C2, *J Clin Invest* 116, 1425–1434, 2006.
18. Atkinson JP, Frank MM, Bypassing complement: evolutionary lessons and future implications, *J Clin Invest* 116(5), 1215–1218, 2006.
19. Lutz HU, Jelezarova E, Complement activation revisited, *Mol Immunol* 43, 1–12, 2006.
20. Jelezarova E, Luginbuehl A, Lutz HU, C3b$_2$-IgG complexes retain dimeric C3 fragments at all levels of inactivation, *J Biol Chem* 278, 51806–51812, 2003.
21. Volanakis JE, Narayansa SV, Complement factor D, a novel serine protease, *Protein Sci* 5, 553–564, 1996.

22. Harboe M, Ulvund G, Vien L, Fung M, Mollnes TE, The quantitative role of alternative pathway amplification in classical pathway induced terminal complement activation, *Clin Exp Immunol* 138, 439–446, 2004.
23. Smith CA, Pangburn MK, Vogel W, Müller-Eberhard HJ, Molecular architecture of human Properdin, a positive regulator of the alternative pathway of complement, *J Biol Chem* 259, 4582–4588, 1984.
24. Hourcade D, The role of properdin in the assembly of the alternative pathway C3 convertases of complement, *J Biol Chem* 281, 2128–2132, 2006.
25. Ghebrehiwet B, Silverberg M, Kaplan AP, Activation of the classical pathway of complement by Hageman factor fragment, *J Exp Med* 153, 665–676, 1981.
26. DiScipio R, The activation of the alternative pathway C3 convertase by human plasma kallikrein, *Immunology* 45, 587–595, 1982.
27. Hiemstra PS, Daha MR, Bouma BN, Activation of factor B of complement system by kallikrein and its light chain, *Thromb Res* 38, 491–503, 1985.
28. Kirschfink M, Borsos T, Binding and activation of C4 and C3 on the red cell surface by non-complement enzymes, *Mol Immunol* 25, 505–512, 1988.
29. Storm D, Herz J, Trinder P, Loos M, C1 inhibitor-C1s complexes are internalized and degraded by the low density lipoprotein receptor-related protein, *J Biol Chem* 272, 31043–31050, 1997.
30. Gigli I, Sorvillo J, Halbwachs-Mecarelli L, Regulation and deregulation of the fluid-phase classical pathway C3 convertase, *J Immunol* 135, 440–444, 1985.
31. Liu CC, Manzi S, Ahearn JM, Biomarkers for systemic lupus erythematosus: a review and perspective, *Curr Opin Rheumatol* 17, 543–549, 2005.
32. Navratil JS, Manzi S, Kao AH, Krishnaswami S, Liu CC, Ruffing MJ, Shaw PS, Nilson AC, Dryden ER, Johnson JJ, Ahearn JM, Platelet C4d is highly specific for systemic lupus erythematosus, *Arthritis Rheum* 54, 670–674, 2006.
33. Gigli I, Fujita T, Nussenzweig V, Modulation of the classical pathway C3 convertase by plasma proteins C4 binding protein and C3b inactivator, *Proc Natl Acad Sci USA* 76, 6596–6600, 1979.
34. Wyatt RJ, Forristal J, Davis CA, Coleman TH, West CD, Control of serum C3 levels by beta 1H and C3b inactivator, control of serum C3 levels by beta 1H and C3b inactivator, *J Lab Clin Med* 95, 905–917, 1980.
35. Liszewski MK, Farries TC, Lublin DM, Control of complement system, *Adv Immunol* 61, 201–283, 1996.
36. Medof ME, Walter EI, Rutgers JL, Knowles DM, Nussenzweig V, Identification of complement decay-accelerating factor (DAF) on epithelium and glandular cells and in body fluids, *J Exp Med* 165(3), 848–864, 1987.
37. Haddad J, Plasma vitamin D-binding protein (Gc-globulin): multiple tasks, *J Steroid Biochem Mol Biol* 53, 579–582, 1995.
38. Kew RR, Fisher JA, Webster RO, Co-chemotactic effect of Gc-globulin (vitamin D binding protein) for C5a. Transient conversion into an active co-chemotaxin by neutrophils, *J Immunol* 155, 5369–5374, 1995.
39. Pickering M, Walport MJ, Links between complement abnormalities and systemic lupus erythematosus, *Rheumatology* 39, 133–141, 2000.
40. Bala Subramanian V, Liszewski MK, Atkinson JP, complement system and autoimmunity, In: *Textbook of Autoimmune Diseases*, Lahita RG, Chiorazzi N, Reeves W (Eds), Lippincott-Raven, Philadelphia, PA, 2000.
41. Botto M, Walport MJ, C1q, Autoimmunity and apoptosis, *Immunobiology* 205, 395–406, 2002.
42. Taylor PR, Carugati A, Fadok VA, Cook HT, Andrews M, Carroll MC, Savill JS, Henson PM, Botto M, Walport MJ, A hierarchical role for classical pathway complement proteins in the clearance of apoptotic cells *in vivo*, *J Exp Med* 192, 359–366, 2000.

43. Yang Y, Chung EK, Zhou B, Blanchong CA, Yu CY, Fust G, Kovacs M, Vatay A, Szalai C, Karadi I, Varga L, Diversity in intrinsic strengths of the human complement system: serum C4 protein concentrations correlate with C4 gene size and polygenic variations, hemolytic activities, and body mass index, *J Immunol* 171, 2731–2745, 2003.

44. Yu CY, Blanchong CA, Chung EK, Rupert KL, Yang Y, Yang Z, et al., Molecular genetic analysis of human complement components C4A and C4B, In: *Manual of Clinical Laboratory Immunology, 6th edition*, NR Rose, RG Hamilton and B Detrick (Eds), ASM press, Washington, 2002, pp. 117–131.

45. Stefansson TV, Kolka R, Sigurdardottir SL, Edvardsson VO, Arason G, Haraldsson A, Increased frequency of C4B*Q0 alleles in patients with Henoch-Schönlein purpura, *Scand J Immunol* 61, 274–278, 2005.

46. Yang Y, Lhotta K, Chung EK, Eder P, Neumair F, Yu CY, Complete complement components C4A and C4B deficiencies in human kidney diseases and systemic lupus erythematosus, *J Immunol* 173, 2803–2814, 2004.

47. Giclas P, Choosing complement tests: differentiating between hereditary and acquired deficiency, In: *Manual of Clinical Laboratory Immunology, 6th edition*, Rose NR, Hamilton RG, Detrick B (Eds), ASM publications, Washington, 2002, pp. 111–116.

48. Morgan BP, Walport MJ, Complement deficiency and disease, *Immunol Today* 12, 301, 1991.

49. Atkinson J, Complement deficiency: predisposing factor to autoimmune syndromes, *Clin Exp Rheumatol* 7, 95–101, 1989.

50. Figueroa JE, Densen P, Infectious diseases associated with complement deficiencies, *Clin Microbiol Rev* 4, 359–395, 1991.

51. Johnson CA, Densen P, Wetsel RA, Molecular heterogeneity of C2 deficiency, *N Engl J Med* 326, 871–874, 1992.

52. Johnson CA, Densen P, Hurford RK, Colten HR, Wetsel RA, Type I human complement C2 deficiency. A 28-base pair gene deletion causes skipping of exon 6 during RNA splicing, *J Biol Chem* 267, 9347–9353, 1993.

53. Wetsel RA, Kulics J, Lokki ML, Kiepiela P, Akama H, Johnson CA, Densen P, Colten HR, Type II human complement C2 deficiency. Allele-specific amino acid substitutions (Ser189→Phe; Gly444→Arg) cause impaired C2 secretion, *J Biol Chem* 271, 5824–5831, 1996.

54. Davis A, C1 inhibitor gene and hereditary angioedema, In: *The Human Complement System in Health and Disease*, Volanakis JE, Frank MM, (Eds), Marcel Dekker, New York, 1998, pp. 229–283.

55. Donaldson VH, Bissler JJ, C1 inhibitors and their genes: an update, *J Lab Clin Med* 119, 330–333, 1992.

56. Blanch A, Roche O, Urrutia I, Gamboa P, Fontan G, Lopez-Trascasa M, First case of homozygous C1 inhibitor deficiency, *J Allergy Clin Immunol* 118, 1330–1335, 2006.

57. Fremeaux-Bacchi V, Guinnepain MT, Cacoub P, Prevalence of monoclonal gammopathy in patients presenting with acquired angioedema type 2, *Am J Med* 113, 194–199, 2002.

58. Giclas P, Complement tests, In: *Manual of Clinical Laboratory Immunology, 5th edition*, Rose NR, Conway de Macario E, JD Folds, HC Lane, RM Nakamura (Eds), ASM publications, Washington, 1997, pp. 181–186.

59. Trapp RG, Fletcher M, Forrestal J, West CD, C4 binding protein deficiency in a patient with atypical Behcet's disease, *J Rheumatol* 14, 135–138, 1987.

60. Bergamaschini L, Miedico A, Cicardi M, Coppola R, Faioni EN, Agostoni A, Consumption of C4b-binding protein (C4bp) during *in vivo* activation of the classical complement pathway, *Clin Exp Immunol*, 116(2), 220–224, 1999.

61. Esparza-Gordillo J, Soria JM, Buil A, Souto JC, Almasy L, Blangero J, Fontcuberta J, de Cordoba SR, Genetic determinants of variation in the plasma levels of the C4b-binding protein (C4bp) in Spanish families. *Immunogenetics* 54, 862–866, 2003.

62. Hesselvik JF, Malm J, Dahlback B, Blomback M, Protein C, protein S and C4b-binding protein in severe infection and septic shock, *Thromb Haemost* 65, 126–129, 1991.

63. Casanova JL, Abel L, Human mannose-binding lectin in immunity: friend, foe, or both? *J Exp Med* 17, 1295–1299, 2004.

64. Turner, M., The role of mannose-binding lectin in health and disease, *Mol Immunol* 40, 423–429, 2003.

65. Garred P, Pressler T, Madsen HO, Association of mannose-binding lectin gene heterogeneity with severity of disease and survival in cystic fibrosis, *J Clin Invest* 104, 431–437, 1999.

66. Garred P, Madsen HO, Marquart H, Two edged role of mannose binding lectin in rheumatoid arthritis: a cross sectional study, *Rheumatology* 27, 2000.

67. Kilpatrick D, Mannan-binding lectin: clinical significance and applications, *Biochim Biophys Acta* 15772, 401–413, 2002.

68. Takahashi R, Tsutsumi A, Ohtani K, Goto D, Matsumoto I, Ito S, Wakamiya N, Sumida T, Anti-mannose binding lectin antibodies in sera of Japanese patients with systemic lupus erythematosus, *Clin Exp Immunol* 136, 585–590, 2004.

69. Matsushita M, Endo Y, Taira S, Sato Y, Fugita T, Ichikawa N, Nakata M, Mizuochi T, A novel human serum lectin with collagen- and fibrinogen-like domains that functions as an opsonin, *J Biol Chem* 271, 2448–2454, 1996.

70. Sorensen R, Thiel S, Jensenius JC, Mannan-binding-lectin-associated serine proteases, characteristics and disease associations, *Springer Semin Immunopathol* 28, 1–21, 2005.

71. Biesma DH, Hannema AJ, van Velzen-Blad H, Mulder L, van Zwieten R, Kluijt I, Roos D, A family with complement factor D deficiency, *J Clin Invest* 108(2), 233–240, 2001.

72. Sprong T, Roos D, Weemaes C, Neeleman C, Geesing CL, Mollnes TE, van Deuren M, Deficient alternative complement pathway activation due to factor D deficiency by 2 novel mutations in complement factor D gene in a family with meningococcal infections, *Blood* 107, 4865–4870, 2006.

73. Mortensen JP, Lamm LU, Quantitative differences between complement factor-B phenotypes, *Immunology* 42, 505–511, 1981.

74. Matsumoto M, Fukuda W, Circolo A, Goellner J, Strauss-Schoenberger J, Wang X, Fujita S, Hidvegi T, Chaplin DD, Colten HR, Abrogation of the alternative complement pathway by targeted deletion of murine factor B, *Proc Natl Acad Sci USA* 94, 8720–8725, 1997.

75. Fijen CAP, van den Bogaard R, Schipper M, Properdin deficiency: molecular basis and disease association, *Mol Immunol* 36, 863–868, 1999.

76. Thompson RA, Winterborn MH, Hypocomplementaemia due to a genetic deficiency of beta 1H globulin, *Clin Exp Immunol* 46, 110–119, 1981.

77. Varade WS, Forristal J, West CD, Patterns of complement activation in idiopathic membranoproliferative glomerulonephritis types I, II, and III, *Am J Kidney Dis* 16, 196, 1990.

78. Richards A, Goodship JA, Goodship TH, The genetics and pathogenesis of haemolytic uremic syndrome and thrombotic thrombocytopenic purpura, *Curr Opin Nephrol Hypertens* 11, 431–435, 2002.

79. Warwicker P, Goodship TH, Donne RL, Pirson Y, Nicholls A, Ward RM, Turnpenny P, Goodship JA, Genetic studies into inherited and sporadic hemolytic uremic syndrome, *Kidney Int* 53, 836–844, 1998.

80. Edwards AO, Ritter, R III, Abel KJ, Manning A, Panhuysen C, Farrer LA, Complement factor H polymorphism and age-related macular degeneration, *Science* 308, 421–424, 2005.

81. Haines JL, Hauser MA, Schmidt S, Scott WK, Olson LM, Gallins P, Spencer KL, Kwan SY, Noureddine M, Gilbert JR, Schnetz-Boutaud N, Agarwal A, Postel EA, Pericak-Vance MA, Complement factor H variant increases the risk of age-related macular degeneration, *Science* 308, 419–421, 2005.

82. Klein RJ, Zeiss C, Chew EY, Tsai JY, Sackler RS, Haynes C, Henning AK, SanGiovanni JP, Mane SM, Mayne ST, Bracken MB, Ferris FL, Ott J, Barnstable C, Hoh J, Complement factor H polymorphism in age-related macular degeneration, *Science* 308, 385–389, 2005.

83. Liszewski MK, Leung MK, Schraml B, Goodship TH, Atkinson JP, Modeling how CD46 deficiency predisposes to atypical hemolytic uremic syndrome, *Mol Immunol* 44, 1559–1568, 2007.

84. Ross SC, Densen P, Complement deficiency states and infection: epidemiology, pathogenesis and consequences of Neisserial and other infections in an immune deficiency, *Medicine* 63, 243–273, 1984.

85. Kaufman KM, Snider, JV, Spurr NK, Schwartz CE, Sodetz JM, Chromosomal assignment of genes encoding the alpha, beta and gamma subunits of human complement protein C8. Identification of a close physical linkage between the alpha and beta loci, *Genomics* 5, 475–480, 1989.

86. Pfeifer PH, Kawahara MS, Hugli TE, Possible mechanism for *in vitro* complement activation in blood and plasma samples: futhan/EDTA controls *in vitro* complement activation, *Clin Chem* 45, 739–743, 1999.

87. Watkins J, Wild G, Smith S, Nafamostat to stabilise plasma samples taken for complement measurements, *Lancet* 1, 896–897, 1989.

88. Mayer M, Complement and complement fixation, In: *Experimental Immunochemistry. 2nd edition*, Kabat EA, Mayer MM (Eds), Charles Thomas Publisher, Springfield, IL, 1961, pp. 133–239.

89. Platts-Mills TA, Ishizaka K, Activation of the alternate pathway of human complements by rabbit cells, *J Immunol* 113, 348–358, 1974.

90. Ziccardi RJ, Cooper NR, Development of an immunochemical test to assess C1 inactivator function in human serum and its use for the diagnosis of hereditary angioedema, *Clin Immunol Immunopathol* 15, 465–471, 1979.

91. Wisnieski JJ, Baer AN, Christiansen J, Cupps TR, Flagg DN, Jones JV, et al., Hypocomplementemic urticarial vasculitis syndrome. Clinical and serologic findings in 18 patients, *Medicine* 74, 24–41, 1995.

92. Giclas PC, Wisnieski JJ, Autoantibodies to complement components, In: *Manual of Clinical Laboratory Immunology*, Rose NR, Conway de Macario E, Folds JD, Lane HC, Nakamura RM 5th edition (Eds), ASM Publications, Washington, 1997, pp. 960–967.

93. Giclas P, C3 Nephritic Factor (C3-NeF) analysis, In: *Diagnostic Immunology Laboratory Manual*, Harbeck RJ, Giclas PC, Raven Press, New York, 1991, pp. 99–107.

94. Daha MR, Fearon DT, Austen KF, C3 nephritic factor (C3NeF): stabilization of fluid phase and cell-bound alternative pathway convertase, *J Immunol* 116, 1–7, 1976.

5 Cellular Immunology: Monitoring of Immune Therapies

Theresa L. Whiteside

CONTENTS

5.1 INTRODUCTION

Cellular immunity, as opposed to humoral immunity, encompasses a broad spectrum of immune phenomena mediated by several well-characterized cell populations. The latter comprises various subsets of lymphocytes: T, B, and natural killer (NK) cells, which are derived from a common lymphoid progenitor cell in the fetal liver and bone marrow. Monocytes, dendritic cells (DCs), and granulocytes represent hematopoietic cells with lineages distinct from that of lymphocytes, but which also participate in cellular immune reactions. They have the capability to mediate non-major histocompatibility complex (MHC)-restricted cytotoxicity and to release a variety of enzymes and cytokines. Lymphocytes interact with these other hematopoietic cells in the peripheral blood as well as in tissues, and the immune response represents a network of carefully balanced interactions responsible for maintaining homeostasis. A diagram of cells mediating innate and adaptive immunity is presented in Figure 5.1. Any perturbation of the immune network leads to a response—a series of events involving the immune effector cells—which is transient and which culminates in restoration of the baseline level of immune activity.

Cellular immunity includes antigen-driven responses induced by the MHC-restricted interactions of the antigen with the specific T-cell receptor (TCR) expressed on a subset of presensitized T lymphocytes (specific immunity) as well as responses that do not require previous sensitization and are largely not restricted by the MHC (nonspecific or innate immunity). These two arms of the immune system are reviewed in Chapter 1. Although T cells are required for the initiation and are the main mediators of specific cellular immunity, all cells of the immune system may participate in both types of responses. Soluble products of immune cells, lymphokines, and chemokines serve as hormones of the immune system. Immune effector cells express receptors for and are responsive to cytokines or chemokines (see more on cytokines or chemokines in Chapter 16) released by cells found in the microenvironment. Immune cells are able to extravasate and migrate from blood to tissues. Consequently, they are found not only in the main lymphoid organs but are

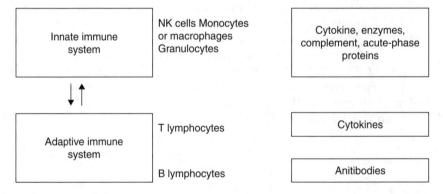

FIGURE 5.1 A diagram of components of the innate and adaptive immune systems. The arrows indicate that considerable interactions exist between the two systems. In general, cells of the innate immune system are nonspecific, whereas those of the adaptive immune system are antigen-specific.

also distributed throughout the body. Cellular immune responses can be either local or systemic depending on the route and dose of the antigenic challenge.

Interactions among immune cells and those of immune cells with surrounding tissue cells are highly complex and not yet completely understood. They are necessary for health and when disturbed may be followed by disease. For this reason, a great deal of effort has been invested in studying and monitoring these interactions as well as phenotypic and functional characteristics of immune cells in blood and tissues. Today, changes in the number or functions of immune cells can be adequately monitored in health and disease. Furthermore, when immune therapies are administered to patients who have immunologic abnormalities, it is possible to accurately monitor effects of a drug or a biologic agent on various cellular components of the immune system. This chapter, reviews the principles and guidelines of immune monitoring during clinical trials with biologics and the usefulness of some of the immunologic assays in such monitoring. The aspects of immune monitoring that are discussed appear to be particularly appropriate in today's context. With increasing frequency, biologic agents such as vaccines, cytokines, or adoptively transferred immune cells are being used for immunotherapy, and various diseases are being treated with agents that either stimulate or suppress immune functions. Thus, it has become necessary to assess the impact of these therapeutic interventions on various subsets of immune cells and to define the mechanisms that are involved in clinical responses. Furthermore, when immune effector cells are used for therapy in conjunction with adjuvants or cytokines, adoptive transfer of these cells to immunocompromised patients is dependent on successful *in vitro* generation and activity of the effector cells. Cellular products released for human therapy have to be generated under current good manufacturing practice (cGMP) conditions (21 CFR 211) and meet criteria establishing their safety.

5.2   CLINICAL TRIALS WITH BIOLOGIC AGENTS

Several categories of biologic agents available for therapies are (a) monoclonal antibodies, (b) cytokines, (c) growth factors, (d) activated cells, (e) cellular by-products, (f) immunotoxins, and (g) other targeting agents. The basic premise behind the selection of a biologic agent for therapy is that it can enhance the ability of the innate or adaptive immune system to control the disease process. In general, such a selection is preceded by extensive *in vitro* studies and *in vivo* experiments in animal models of the disease to define the presumed mechanism of action of a biologic agent and assess its potential toxicity.

Following *in vitro* and *in vivo* (in animals) preclinical evaluations, a new biologic agent is introduced to the clinic as a phase I clinical trial protocol. The design of a phase I study is based on its mechanism of action tentatively identified in preclinical experiments. The developmental strategy involves determination of the optimal biologic dose (OBD), that is, the dose that maximally activates the postulated mechanism of action with tolerable toxicity. A clinical trial designed to determine the OBD is referred to as phase Ib, in contrast to a phase Ia trial, in which only clinical toxicity of the new agent is defined as a maximal tolerated dose (MTD).[1] It is important to remember that biologic agents may exert optimal effects at lower rather than higher doses, resulting in a bell-shaped dose–response curve, and that the OBD may be different from the MTD of the agent. Often the OBD is considerably smaller than the MTD and, in general, adverse events are infrequent with biologic therapies. In addition, the same biologic agents might have quite disparate OBDs for two different immunologic characteristics. A systemically administered cytokine, interleukin-2 (IL-2), for example, which is currently approved for therapy of metastatic melanoma and renal cell carcinoma (RCC), has a much lower OBD for the activation of circulating T and NK cells than that for supporting the generation of lymphokine-activated killer (LAK) activity.[2]

Experience suggests that in phase Ib trials with biologic agents, various surrogate immunologic endpoints, such as increased numbers or activity of immune effector cells, are achieved more easily than clinical responses. In the case of IL-2 therapy in RCC, for example, nearly all patients achieved high levels of endogenous LAK activity during therapy, whereas only a proportion of these patients experienced objective antitumor responses.[2] The same trend is observed in the case of therapeutic antitumor vaccines. This might reflect the fact that the choice of surrogate endpoints, which is usually based on limited preclinical data, may not reflect the entire spectrum of physiologic mechanisms that are mediated by the biologic agent used for therapy, including those responsible for its therapeutic effects.

It could be argued that in phase Ib trials, direct determinations of the dose of a biologic agent that has the best therapeutic effects, for example, antitumor effects in a patient with cancer, would be preferable to the use of surrogate measurements. However, the use of surrogate measures remains the only reasonable alternative because practical and ethical considerations rule out the possibility of directly determining the best therapeutic dose. Antitumor response rates for single agents generally range between 10% and 20% and, therefore, large-scale dose-finding studies would require excessively large number of patients, while exposing many of them to ineffective doses of the agent tested in an effort to establish the optimal therapeutic dose. In the subsequent clinical trials, it may be possible to expand the earlier clinical observations and begin to ask questions about the biology of the agent or its mechanisms of antitumor response by gradually incorporating additional studies designed to answer more mechanistic questions.

5.3 SELECTION OF IMMUNE ASSAYS FOR MONITORING

"Immunologic monitoring" is defined as serial measurements of selected immune parameters over a period of time, generally weeks or months. Such serial measurements are usually taken before, during, and after therapy, as part of a clinical trial. However, they can also be taken during disease, during recovery from disease, or simply to establish immune responses of an individual over time. A baseline measurement (or better, two or three baseline assays) may be necessary because one of the goals of monitoring is to define changes from the baseline that occurs within a designated time period. Biologic fluctuations that are likely to occur in immunologic responses during this period of time, for example, those associated with hormonal changes, infections, stress, and exercise situations, have to be distinguished from changes due to therapeutic interventions. This aspect of monitoring is most difficult,

TABLE 5.1

Immunologic Assays Used in Monitoring of Early-Phase Clinical Trials with Biologic Agents

Categories of Assays	Sample Type
Soluble cellular products	
Immunoglobulin levels	Serum, plasma, body fluids
Cytokine or chemokine levels	Serum, plasma, body fluids
Cytokine receptors and antagonists (soluble)	Serum, plasma, body fluids
Ligands and growth factors	Serum, plasma, body fluids
Enzymes (e.g., 2′,5′-adenylate synthetase, arginase, and metaloproteinases)	Serum, plasma, body fluids
Neopterin	Serum, plasma, body fluids
β2 Microglobulin, soluble MHC molecules	Serum, plasma, body fluids
Phenotypic assays	
Proportions of cells	Whole blood, tissue biopsy, body fluids
Absolute numbers of cells	Whole blood
Cellular subpopulations (e.g., Th1 or Th2 or memory or naïve T cells)	Whole blood, body fluids, isolated lymphocytes
Treg quantification	Isolated lymphocytes
Functional assays	
In vivo DTH skin test	Visual inspection of the site or biopsy
Cytotoxicity: ADCC, LAK, NK, T-cell specific	MNC or subpopulations of MNC
Cytokine production	MNC or subpopulations of MNC
Proliferation	MNC or subpopulations of MNC
Chemotaxis	MNC or subpopulations of MNC
Signal transduction	MNC or subpopulations of MNC
Superoxide generation	MNC or subpopulations of MNC
Apoptosis or necrosis	MNC or subpopulations of MNC

and it requires selection of assays that can reliably measure immunologic changes as well as the ability to accurately correlate physiologic or clinical observations with immunologic results.

A list of assays more frequently used for monitoring are provided in Table 5.1. Selection of assays for monitoring of immune parameters is best guided by a hypothesis. The hypothesis to be tested in a clinical trial is formulated on the basis of preliminary evidence that is considered by the investigators to be the best indication of mechanisms that could lead to therapeutic effects. Obviously, because most biologic agents have multiple biologic effects, more than a single hypothesis of action can be postulated, and the choice of the hypothesis to test depends on individual insights of the investigator. Testing of several hypotheses at once is discouraged, however, because it complicates the design of a clinical trial and makes the accompanying monitoring too extensive. There is no way to guarantee that a hypothesis selected for testing in a clinical trial is a correct one and, thus, its selection represents a risk that an investigator is obliged to undertake.

Although the selection of one or a panel of immunologic assays for monitoring of a clinical trial is primarily guided by the scientific hypothesis that is being tested, a number of other factors might influence the decision to choose a particular assay in preference of another. The frequency of immunologic monitoring is an important component of the trial design. The time points selected for monitoring clearly need to include the baseline (pretherapy) and final (posttherapy) measurements. Beyond that, the schema for monitoring is usually determined by extrapolation from animal models or from previous clinical experience with the same or similar biologic agents, and it may have to be modified depending on the frequency and dose of the administered agent.

A decision to select cellular versus humoral immunologic assays for monitoring (Table 5.1) is directly related to the proposed mechanism of action for a biologic agent. Although it may be difficult to prioritize based on the current understanding of such mechanisms, consideration of practical aspects of monitoring can help in reaching a reasonable decision.[3] Limitations in the frequency or volume of samples obtained from the peripheral blood as well as variability of cell yields restrict the use of cellular assays. Cellular assays are more difficult and more costly to perform than serum assays. Therefore, it is always advisable to consider serum or plasma assays first. A variety of serum assays are available, which can substitute for cellular assays, including those measuring levels of released cytokines or chemokines; enzymes known to be induced by a particular biologic agent (e.g., interferon [IFN]-induced $2',5'$-adenylate synthetase, arginase, or indoleamine 2,3 dioxygenase [IDO]); inhibitors or antagonists of biologic agents or immune cells (e.g., IL-10, transforming growth factor β [TGF-β], and IL-1 receptor antagonist [IL-1ra]); or products of activated cell subsets (e.g., soluble cytokine receptors, β2 microglobulin, soluble MHC molecules, and neopterin). From a practical point of view, it is preferable to use serum or plasma instead of cells for monitoring whenever possible.

In many instances, monitoring of numbers or functions of mononuclear cells (MNCs) in the peripheral blood during therapy does not adequately reflect immunologic events that take place at the site of disease. Approximately 2% of total lymphocytes are in the peripheral circulation at any given time point, and monitoring of these cells is not likely to reflect events occurring in tissue. Systemic effects of cytokines and certain other soluble factors are often distinct from their locoregional effects. Therefore, immunologic monitoring of lesions, tissues, or organs involved in a disease is likely to yield more informative data than the same assays performed with MNC obtained from the peripheral blood. The obvious difficulties with this strategy are that blood is more readily available than tissues and that only superficial lesions or sites accessible to repeated biopsies can be monitored. Nevertheless, during a clinical trial, it might be feasible to obtain serial biopsies, and this opportunity for *in situ* studies should be taken advantage of as often as possible. An alternative possibility is to obtain body fluids (pleural fluids or ascites), in addition to the peripheral blood to be able to detect changes in the organ-associated immune cells or their products relative to those in peripheral circulation.

Whether immune cells, serum, or plasma are used for immunologic monitoring, the samples have to be harvested from peripheral blood, body fluids, or tissues at time points specified in a protocol schema. For studies of pharmacokinetics, a special effort has to be made to collect the specimens at precisely designated time intervals and to process them in accordance with experimental protocol to avoid degradation, inactivation, or loss of activity. While planning immunologic monitoring, not only the timing of specimen collections but also the nature of anticoagulants used needs to be considered. For example, to measure levels of cytokines in body fluids, plasma rather than serum is preferable because cytokines tend to be trapped in the clot, with subsequently low levels of cytokines measured in the serum.[4] Separation of serum or plasma or MNC from the peripheral blood are routine laboratory procedures, but recovery and fractionation of cells from tissues or body fluids containing tumor or tissue cells is time-consuming, costly, and requires special expertise and considerable effort.[5] The use of individual separated cell subsets of tissue infiltrating or blood MNC is increasingly being adopted due to technical advances allowing a rapid and effective recovery of desired subsets.[6] It should be remembered, however, that large blood volumes are needed for cell separations. Also, unpredictable yields of cells recoverable from tissues might hamper the use of cell-separation techniques for monitoring of tissue samples. Nevertheless, the advantage of studies performed with highly purified subsets of immune cells obtained from the site of disease is obvious, and when it seems feasible to obtain a biopsy, even if it is only pre- and posttherapy, investigators are encouraged to incorporate such procedures in their protocols. Two approaches to immunologic monitoring are outlined in Figure 5.2.

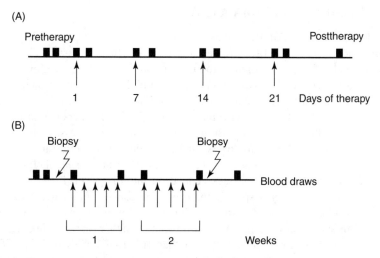

FIGURE 5.2 Examples of protocol schemas that might be used for immunologic monitoring. In (A), immunotherapy is administered systemically at weekly intervals for three successive weeks. Blood draws for monitoring are obtained each week just before therapy to monitor long-term effects on the immune cells and 24 h after therapy to monitor short-term effects. Three pretherapy (baseline) and one posttherapy blood draws are obtained. In (B), immunotherapy is administered locoregionally (e.g., around the tumor) on every working day of the week (five successive daily treatments) for two successive weeks. Tumor biopsy is obtained pre- and posttherapy for *in situ* monitoring, in addition to blood samples. Arrows indicate days on which therapy is given.

It is often unclear which immunologic assays can be reliably performed with fresh versus cryopreserved cells. For longitudinal immunologic monitoring, it is always more practical to use serially harvested frozen aliquots of serum or plasma, or batched cryopreserved immune cells instead of freshly harvested specimens. The ability to reliably cryopreserve immune effector cells for future use in retrospective studies or for batching of serially collected specimens to eliminate the interassay variability is highly desirable. However, some immunologic measurements are best performed with fresh samples because the use of cryopreserved effector cells may introduce artifacts in certain functional assays, specifically those measuring cytotoxicity. Many other functional assays, including proliferation or cytokine production, can be reliably performed with frozen cells, and it may be preferable to select these assays for monitoring rather than those requiring the use of fresh cells. In any case, each monitoring laboratory is obliged to compare fresh versus frozen cells to ascertain that a test can be reliably performed with either of them.[3]

In other words, the choice of a strategy for immune monitoring of clinical trials with biologic agents is neither simple nor straightforward. No single assay or experimental approach is appropriate for all biologic therapeutic interventions. The choice requires familiarity with principles of immunologic assays; a great deal of judgment; and considerable understanding of biologic, immunologic, and therapeutic effects induced by biologics.

5.4   SERUM AND PLASMA ASSAYS

In Table 5.1, several serum and plasma assays are listed that are widely used in immune monitoring. The list is representative rather than inclusive. These assays measure levels of various cellular products in body fluids. Immunoglobulin quantification, for example, has become a routine and highly accurate estimate of quantitative and qualitative changes in B-cell function. Recently, assays for levels of cytokines, growth factors, and enzymes in body fluids have become an important part of immunologic monitoring. Both bioassays and immunoassays are available for assessments of soluble factor levels[7,8] in body fluids, but only the latter (mainly in the form of commercial immunoassays) are practical for monitoring of clinical trials. Cytokine or growth factor bioassays depend on the use of cell lines, are difficult to perform reliably, are less specific than immunoassays, require neutralization to confirm cytokine identities, and are labor-intensive. For these reasons, bioassays are only used in special circumstances as confirmatory rather than monitoring assays. In contrast, immunoassays are less sensitive than bioassays, but are more specific and easily performed using commercially available reagents. High-throughput, immunobead multiplex assays are available allowing for detection of numerous cytokines or other growth factors in 50 μL of body fluid.[7] Levels of cytokines or soluble factors in body fluids of patients treated with biologic agents may change very rapidly during therapy. Figure 5.3 illustrates the usefulness of an immunoassay in monitoring plasma levels of sIL-2R (CD25) in a phase Ib trial, showing rapid and dramatic changes during therapy with IL-2.

A number of problems exist with assessments of cytokines in sera, including the ability of cytokines to bind to and form complexes with serum proteins or the

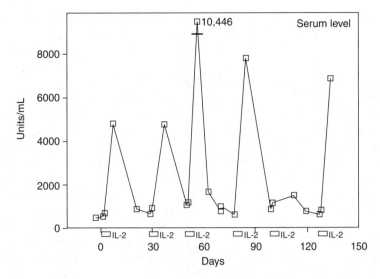

FIGURE 5.3 An example of dramatic changes that may occur in plasma levels of sIL-2R (CD25) during administration of IL-2. In phase Ib trial, a patient with metastatic melanoma was treated with adoptively transferred autologous-activated natural killer (A-NK) cells and with cycles of IL-2 (10^6 U/m^2/day $\times$ 5) administered systemically as continuous infusions. Note that plasma levels of sIL-2R peaked during the periods of IL-2 therapy and decreased to a near-baseline level when IL-2 therapy was stopped.

presence of soluble cytokine receptors or antagonists, natural inhibitors, or anticytokine antibodies.[8] Also, cytokines have a short half-life (minutes) in serum, and their levels are generally low or undetectable by conventional immunoassays even when substantial cytokine production occurs at a disease site. All of these factors may contribute to difficulties in the interpretation of cytokine or growth factor assays. Knowledge of cytokine biology is also important for accurate assessments and interpretation of cytokine levels in biologic fluids.[8] The latter can be frozen and batched for immunoassays. Profiling for soluble factors in body fluids based on high-throughput platforms such as Luminex or by flow cytometry (discussed further in Chapter 17) greatly facilitate monitoring, allowing simultaneous quantification of many proteins and the establishment of functional profiles that might better correlate with clinical endpoints than single cytokine levels.

5.5 PHENOTYPIC ASSAYS

In many trials using biologic agents for therapy, high priority is given to cell-surface phenotype assays generally performed with fresh whole blood specimens and, less frequently, with MNC separated from the peripheral blood, body fluids, or tissues. As described in this volume, multiparameter flow cytometry assays are now easily accessible and highly accurate. (See Chapter 6 for more details on flow cytometry.) A broad array of labeled monoclonal antibodies suitable for multiparameter flow analyses are commercially available, allowing the identification, sorting, and enumeration of various cell subsets. Changes in these subsets (proportions or absolute numbers) as well as changes in activation levels (based on surface marker expression) can be readily assessed during immunotherapy. Unfortunately, flow cytometry assays are often selected without adequate justification and used in combinations that are not optimally meaningful. For example, the popular use of the CD4/CD8 ratio is often not informative, whereas the determination of the ratio of memory to naive lymphocytes or that of activated to nonactivated cells might be. Large panels of immunologic markers are used indiscriminately to follow changes induced by biotherapy without considering the hypothesis advanced as part of a clinical trial. Through proper selection of marker panels, it is possible to focus on a single population of effector cells, on several phenotypically distinct subsets, on cells expressing specific activation markers, or on those that express a particular constellation of adhesion molecules. A recent recognition of regulatory T cells (Treg), for example, and the ability to phenotypically discriminate Treg (FOXP3[+]) from CD25[+] activated CD4[+] T cells has opened a way to testing the hypothesis that Treg accumulations are associated with a poor prognosis in cancer,[9] but with an improved clinical outcome in patients with autoimmune diseases.[10] Attempts at correlating phenotypic changes with functional assays may be particularly informative. Various receptors on the cell surface for growth factors and cytokines can further be used to subset the cells and to relate their phenotypes to their functional potential. Thus, phenotypic analysis is a powerful tool for monitoring the effects of therapy on immune cells, given that the selection of markers to be monitored on T, B, NK cells, monocytes, or granulocytes is based on the predicted mechanism of action of the agent being used for therapy.

Four-color flow cytometry is most widely used for monitoring immune therapies today, although multicolor flow (e.g., nine-color) can provide a more specific analysis of cell subsets. The usefulness of multicolor flow cytometry in monitoring is currently limited due to the complexities associated with the analysis of results. From a practical point of view, it is difficult for an average laboratory to analyze a large volume of multicolor flow data in a timely fashion. The alternatives are to limit phenotyping to four-color analysis of a single cell type considered the most important, or to send the specimens for phenotyping to a specialized laboratory equipped to handle multicolor flow data. This alternative is recommended for monitoring of multicenter or cooperative group trials, as whole blood specimens can be shipped by overnight carriers for flow analysis without compromising their quality.

In addition to surface phenotyping, flow cytometry has been extensively used to measure intracytoplasmic markers using permeabilized cells as well as to perform certain functional assays. Recent technologic advances allow these approaches to be applied in clinical trials. Because of the substantial advantages they offer, such as the possibility of correlating expression of a receptor on the cell surface with signaling events downstream, these methods are being incorporated in the roster of useful monitoring assays.[11] Intracytoplasmic proteins (including enzymes, cytokines, signal transduction molecules, or transcription factors) measured in phenotypically defined cells by flow cytometry can provide useful information about the activation state and functional attributes of individual cells or a subset of cells. Proliferation using carboxyfluorescein diacetate succinimidyl ester (CFSE)-labeled cells[12] or flow-based cytotoxicity assays[13] can be performed simply and more efficiently by flow than by the familiar radiolabel assays, providing the identity of effector cells at the population or even the single-cell level. Disadvantages of these methods include the requirement for a flow cytometer, relatively complex preparation steps, including cell permeabilization with combinations of surface and intracellular staining, and the current paucity of comparative data confirming the validity of these techniques for monitoring.

Interpretation of flow cytometry data in serial monitoring requires stringent controls for interassay variability. Some of the lymphocyte subsets (e.g., peptide- or epitope-specific T cells detectable with the use of tetramers) encompass only a small proportion of cells in the gate, and it is necessary to acquire sufficient number of these T cells to reliably distinguish shifts induced by therapy (see Chapter 2 for laboratory statistics). To enhance the quality of monitoring, changes in the phenotype of effector cells during therapy should be interpreted in conjunction with changes observed in cellular function. Specifically, it is important to consider whether changes in function induced during therapy correlate with changes in the number of effector cells or are due to the augmentation of effector cell function. It is also possible that therapy alters both the number and function of effector cells, as indicated in Table 5.2. During therapy, these changes can be dramatic or subtle, and they can simultaneously occur in several subsets of cells. Analysis of these changes and the correlation between their magnitude and frequency requires special statistical

TABLE 5.2

Changes in the Proportion of NK-Cell Subsets and NK and LAK Activities in Response to Locoregional Administration of IL-2 in the Peripheral Blood of Patients with Head and Neck Cancer

	Pretherapy	Posttherapy	*p*Value
NK-cell subsets (%)			
CD3$^-$CD56$^+$	16 ± 2	22 ± 2	0.003
CD16$^+$CD56$^+$	9 ± 2	16 ± 2	0.013
CD16$^+$HLA-DR$^+$	1.3 ± 0.2	2.7 ± 0.6	0.003
Cytotoxicity (log LU)[a]			
NK (versus K562)	2.0 ± 0.07	2.2 ± 0.07	0.005
LAK (versus Daudi)	2.9 ± 0.08	3.3 ± 0.12	0.006

Note: The data are mean values ± SEM. Paired analysis (the signed rank test) was performed for pre- and posttherapy measurements obtained in 26 patients with inoperable head and neck cancer who were treated with perilesional IL-2.

[a] Cytotoxicity data were obtained in 4-h ^{51}Cr-release assays using the indicated targets. LAK cells were generated by incubation of MNC in the presence of 6000 IU of IL-2 for 3 days. The data are log 10 LU$_{20}$/10^7 effector cells.

Source: Adapted from Whiteside, T. L., Letessier, E., Hiraayashi, H., Vitolo, D., Bryant, J., Barnes, L., Snyderman, C., Johnson, J. T., Myers, E., Herberman, R. B., Rubin, J., Kirkwood, J. M., Vlock, D. R. *Cancer Res.*, 53, 5654, 1993. With permission.

approaches, as discussed elsewhere. Because entirely different mechanisms may cause alterations in the proportion or absolute number of effector cells versus functional changes (e.g., increased blood vessel permeability and up-regulation of NK activity during therapy with IL-2, respectively), it should not be expected that phenotypic and functional data will necessarily correlate with one another. Furthermore, it is possible that shifts in one, but not the other parameter, will correlate with clinical response or, more likely, that no significant correlations will be detected, although profound changes in immune cell numbers or function are registered during therapy. For example, perilesional therapy with escalating doses of IL-2 in 36 patients with advanced head and neck cancer resulted in highly significant increases in the number of CD3$^-$CD56$^+$ cells and NK activity in the peripheral blood of most patients, but no complete clinical response in a phase I trial at our institution.[14] This reflects the complexity of mechanisms that are involved in mediating therapeutic effects of biologic agents administered to patients with cancer.

5.6   FUNCTIONAL ASSAYS

Functional assays that can be used for monitoring include a variety of technologies, many of which are highly sensitive and adaptable to samples containing a small number of cells or even single cells. Many of these new technologies including genomics and proteomics employ high-throughput platforms that are particularly suitable for performance of serial monitoring. Their advantages are simplicity, substantially lower cost, and elimination of radioactive labels. This section briefly describes the functional assays that are currently available for monitoring. This overview is followed by more detailed descriptions of functional monitoring of T and NK cells.

5.6.1   *IN VIVO* DELAYED TYPE HYPERSENSITIVITY SKIN TEST

The skin test for delayed type hypersensitivity (DTH) is the only *in vivo* assay available for measurements of cellular immunity in man. This test is underutilized in monitoring, largely because of rigorous requirements for recording its results. The DTH assay is often misinterpreted because the results are not measured correctly (induration not erythema should be measured) or at the proper time (at >24 h and not more than 48 h after test application). Yet, when performed and read according to the guidelines (Figure 5.4),[15] the DTH assay remains the best available measure of specific cell-mediated immunity in patients with immunodeficiencies, particularly, when it is combined with a biopsy to confirm the nature of cells infiltrating the skin Although reagents for Merieux DTH testing devices are no longer available, it is possible to obtain tetanus, candida, and purified protein derivative (PPD) of tuberculin from commercial sources and construct the individual DTH panel for measurements of recall responses. Other antigens, for example, tumor-derived purified or synthetic peptides, which are not yet commercially available from vendors can also be used, provided they have passed safety requirements. A change in the DTH skin test from nonreactive to reactive as a result of biotherapy is a significant result. In some cases, it may be advisable to biopsy the DTH sites to identify lymphocytes accumulating *in situ* in response to an *in vivo* challenge with the epitope of interest.

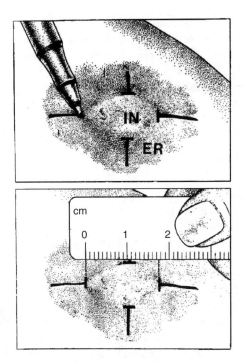

FIGURE 5.4 DTH skin test is the only available *in vivo* measure of immunocompetence. The skin test is read 24–48 h after intradermal antigen application. Generally, swelling and induration (IN) develop with accompanying redness (erythema [ER]). The figure shows how to correctly read a DTH skin test: the area of IN has to be measured and recorded as the cross product or the sum of two greatest perpendicular diameters. In general, IN >5 mm is considered as a positive DTH reaction.

5.6.2 LOCAL VERSUS SYSTEMIC ASSAYS

Because measurements of phenotypic characteristics do not adequately convey the functional capabilities of a cell or a population of cells, cellular assays for monitoring the functions of immune effector cells are necessary. These assays are usually performed with peripheral blood MNC and only rarely with effector cells obtained from the site of disease. Thus, functions that are being monitored are not those of the effector cells that may be directly involved in disease, but those of substitute cells that are easily accessible for monitoring. This is by far the greatest limitation of monitoring because systemic effects of biologic agents are likely to be different from local or locoregional effects on the immune cells. It should not be expected that the "surrogate" results obtained with peripheral blood lymphocytes will closely reflect biologic activity of the drug on functions of immune cells *in situ*. To overcome this limitation, investigators are attempting to study serial tissue biopsies utilizing immunostaining or *in situ* hybridization (ISH). Another approach utilizes the reverse transcriptase polymerase chain reaction (RT-PCR) for messenger ribonucleic acid (mRNA) coding for proteins involved in cytotoxicity (perforin, granzymes, and tumor necrosis factor [TNF]), proliferation (growth factors and cytokines) or signal transduction (changes in the TCR zeta-chain expression and phosphorylation of a variety of substrates mediated by protein tyrosine kinases). In some cases, it may be possible to recover a limited number of cells from serial biopsies to perform functional studies. Obviously, these studies are very difficult to organize in humans and are not applicable when the biopsy cannot be obtained. Nevertheless, they are extremely important because they allow for a comparison between local and systemic effects of a biologic agent on immune effector cells and eventually justify or discredit the common practice of monitoring alterations in cellular functions in the peripheral blood alone.

5.6.3 CYTOTOXICITY

Among functional assays for lymphocytes, cytotoxicity has occupied a special place, especially, in the monitoring of immune therapies. There is a good rationale for monitoring cytotoxicity in clinical trials with biological agents, which tend to augment this effector function in many different cell types. In oncology protocols, antitumor cytotoxicity, whether class I MHC-restricted (mediated by the tumor antigen-specific T cells), nonspecific (mediated by NK cells or monocytes), or antibody-mediated (depending on the presence of FcγR on effector cells), has been extensively monitored because of the evidence derived from numerous animal models that tumor growth and elimination of metastases are mediated, at least in part, by cytotoxic effector cells. In human studies, it had been reported early on that autotumor cytotoxicity was the only significant *in vitro* correlate of clinical responses in phase I trials of adoptive therapy with *in vitro*–activated human immune cells and IL-2.[16] However, autologous tumor targets are seldom available for cytotoxicity measurements and surrogate targets; generally tumor cell lines are used.

Recent data indicate that tumor regression or metastasis elimination are complex tissue events, generally involving cell necrosis or apoptosis that may be mediated by perforin or granzyme release from effector cells, cytokine secretion by effector cells, or receptor-mediated or mitochondrial apoptosis induced by effector cell interactions with its target.[17] There is considerable evidence suggesting that the nature of cell death (e.g., necrosis versus apoptosis) determines the response of immune cells to the offending agent. Thus, silent death by apoptosis induces less response from immune effector cells, whereas necrosis results in signaling and inflammatory cell infiltration to sites of cytotoxicity.[18]

In view of the documented existence of several distinct cellular mechanisms involved in cytotoxic death of target cells, the question then arises as to which type of cytotoxicity assays should be selected for monitoring (Figure 5.5). Once again, the choice depends on the preliminary data available, the hypothesis tested, and practical considerations of investigators' ability to perform the assay reliably. The ^{51}Cr-release assay (Figure 5.5) that was for a long time considered the "gold standard" for cytotoxicity measurements is being slowly replaced by nonradioactive, flow cytometry–based assays (Figure 5.6).

5.6.4 PROLIFERATION ASSAYS

Proliferation of immune cells in response to mitogens or antigens has been a part of the monitoring repertoire for a long time.[19] It is a good measure of immunocompetence

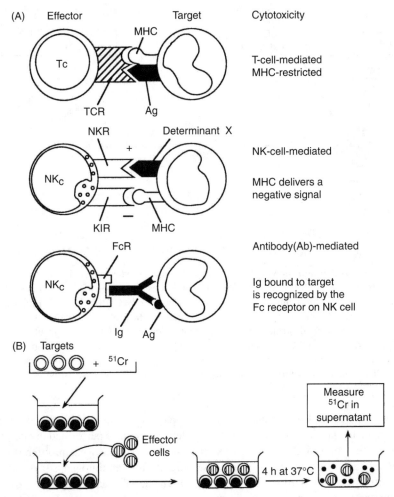

FIGURE 5.5 Various types of cytotoxicity mediated by immune effector cells. In (A), interactions between effector and target cells are presented. Cytotoxic T cells (Tc) bind to the target, which presents antigen and class I MHC determinants recognized by the TCR. NK cells recognize a triggering determinant X on the susceptible target. Class I MHC molecules on the target are also recognized by the killer inhibitory receptors (KIRs) on NK cells. Negative signals received by KIRs prevent NK activity, whereas triggering receptors induce NK-mediated cytotoxicity. Lysis or lack of lysis depends on the balance of negative versus positive signals received by the NK cell. In the presence of target-specific antibodies, NK cells recognize the Fc portion of IgG bound to antigens on the target cell surface. Cross-linking of FcγRIII (CD16) on NK cells by the Fc fragment of the target-specific antibodies (Abs) leads to lysis of the target cell (ADCC). In (B), a ^{51}Cr-release assay, considered the "gold standard" of cytotoxicity measurements for a long time is shown. Targets are labeled with ^{51}Cr and coincubated with MNCs as effector cells. Lysis of targets results in release of ^{51}Cr into the medium. The assay must be performed at a minimum of four different effector (E):target (T) ratios to generate a lytic curve defining the relationship between the percent specific lysis and the number of effector cells used in the assay.

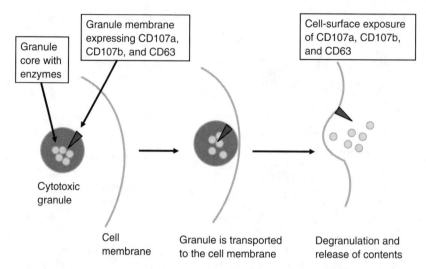

FIGURE 5.6 A diagram showing the principle of a flow cytometry-based assay for measuring cytotoxicity. The CD107 degranulation assay measures expression of CD107a molecule on the target cell surface following coincubation with effector cells. CD107a (LAMP-1) originates from lytic granules containing perforin and granzymes (serine protases). When these granules are transported to the cell surface and their contents released, the CD107a molecule becomes transiently expressed on the T- or NK-cell surface and is detected with a fluorochrome-labeled antibody by flow cytometry. Because degranulation in T or NK cells can be correlated to target cell killing, this assay is being used as an alternative to ^{51}Cr-release assays.

and can be performed with banked, cryopreserved cells, or as a whole blood assay.[19] For monitoring, proliferation assays with very few cells can be performed in microtest plates holding only 5–20 μL (Terasaki plates). Assays measuring proliferation in response to a specific antigen are, perhaps, more useful than mitogen assays, but are more difficult to standardize and perform. Often the stimulating antigens are not available as purified or recombinant reagents, making the interpretation of results difficult. It is important to remember that antigen-presenting cells (APCs), which are generally present in the peripheral blood, are necessary for optimal proliferation, and in assays with purified lymphocyte subsets, APCs have to be added (Figure 5.7A). Titrations of antigens or mitogens to select the optimal stimulatory concentrations are obligatory. Proliferation assays have to be performed with at least three different doses of the antigen because lymphocytes obtained from various individuals respond optimally to quite different concentrations of the same stimulatory agent. A stimulation index, calculated as the ratio of experimental to control values, is used to measure proliferative responses. Recent nonradioactive assays utilize labeling of responding cells with CFSE and flow cytometry for assessment of the label redistribution to daughter cells as the cells divide (Figure 5.7B). The phenotype of proliferating cells can be determined by concomitant surface labeling and flow cytometry.[20]

Proliferation assays seem to be replaced by measurements of cytokines and cytotoxicity in recent clinical trials, yet they are informative, practical, and easily adjustable to fit the specific trial design or circumstances in a monitoring laboratory.[19]

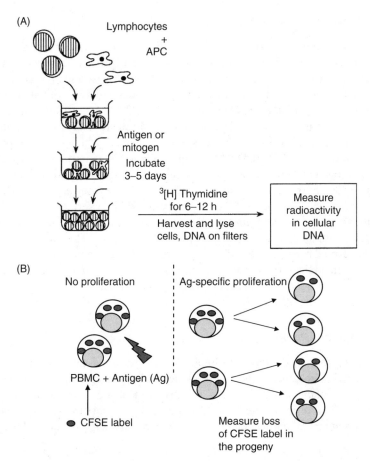

FIGURE 5.7 Proliferation assays. In (A), a classic proliferation assay is illustrated, in which lymphocytes and APC are incubated in the presence of a mitogen or antigen. APCs express class II MHC molecules and present antigens to T cells. The initiation of proliferation involves recognition and signaling events. Cytokines, generated as a result of T-cell activation and interaction with APC, are necessary for proliferation. The amount of radioactivity (counts per minute [cpm]) generated by incorporation of radiolabeled nucleotides into DNA of proliferating cells is measured in experimental and control wells. In (B), a schema illustrating the principle of a proliferation assay based on CFSE labeling of responding cells. R cells are incubated with a dye (CFSE), washed, and cocultured with mitogens or antigens for 3–5 days in a growth medium. As R cells divide, CFSE is equally redistributed between daughter cells (effectively halving the amount of dye and the level of fluorescence). Irradiated R cells + antigen (left-hand side) do not divide and conserve the dye. Flow cytometry is used to measure the dye dilution (read as decrease in fluorescence) as each successive generation of cells contains less label or fluorescence (right-hand side). A ModFit software program is used to determine the number of divisions, proliferation index, and the frequency of proliferating cell precursors present in the population.

5.6.5 Suppressor T-Cell Assays

Regulatory or suppressor (S) cells (Treg) inhibit activities, including proliferation, of other immune cells. The presence and activity of Treg is of interest in monitoring because these cells are known to accumulate at sites of disease and interfere with the generation, function, or survival of effector T-cell populations.[21] Although intracytoplasmic or surface markers (FOXP3 and CD4+CD25high expression, respectively) that identify Treg are now available, functional S assays are likely to be more meaningful. To this end, a coculture assay has been developed, which utilizes separated or not separated Treg (CD4+CD25high T cells). These cells are titrated into CFSE-labeled autologous responder (R) cells activated with beads coated with anti-CD3 and anti-CD28 Abs or with a relevant epitope on APC. The assay is performed at different R/S ratios, and following flow cytometry, a Modfit® (Verity Software House, Topsham, ME) software program is used to compare proliferation in control cultures (i.e., R cells with no S cells) to that in R plus S cocultures.[22] This assay is likely to be widely adapted because as elimination of Treg with, for example, anti-CD25 antibodies (Abs) in clinical trials aiming at vaccination of cancer patients is increasingly common[23] as is a trend for transfers of Treg to suppress autoimmune sequellae[10] or graft versus host disease (GVHD).[24]

5.6.6 Cytokine Production

The ability of immune effector cells to produce and release cytokines in response to stimulation with biologic agents is a useful monitoring strategy. In contrast to assays of serum levels of cytokines, this strategy calls for incubation of cells alone (spontaneous production) or in the presence of an activating agent (stimulated production) and the subsequent quantification of cytokines in cellular supernatants.[25] Figure 5.8

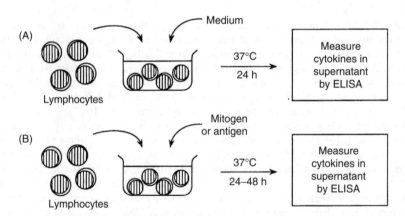

FIGURE 5.8 Cytokine production by immune cells. A diagram of spontaneous or stimulated cytokine production *in vitro* is shown. In (A), peripheral blood MNCs are incubated in medium alone for 24–48 h, and the supernatants are tested for the presence of spontaneously produced cytokines. *In vivo*–activated MNC will tend to spontaneously produce cytokines. Alternatively, MNCs are incubated in the presence of stimulators (mitogens or antigens) to determine their immunocompetence, that is, the ability to produce a panel of cytokines in (B).

shows the approach for measuring spontaneous and stimulated cytokine production. The assay can be used to measure immune competence of cells by including stimulators such as PHA or LPS, or the ability to produce cytokines in response to the stimulator of choice. The assay also allows for assessment of *in vivo* activation of cells by measuring spontaneous cytokine release. Unlike cytotoxicity assays, which require fresh and not cryopreserved effector cells, cytokine production assays can be reproducibly performed with either fresh or cryopreserved cells,[25] allowing for batching of cells or cellular supernatants. Moreover, multiple cytokines can now be quantified in small volumes (50 µL) of peripheral blood MC (PBMC) supernatants or body fluids using a Luminex platform. Multiplex cytokine assays use antibody-coated beads and can simultaneously detect several cytokines or growth factors providing a reasonable assessment of an individual's immunocompetence and allowing for the definition of cytokine profiles characteristic of T-lymphocyte subsets (Th0, Th1, or Th2).[26,27] Serial monitoring of individuals over a period of several months has demonstrated that stimulated cytokine production is a stable trait.[28] In addition, spontaneous cytokine release by PBMC indicates that they have been activated *in vivo*.[25] It is recognized that cytokines are produced by a variety of cells, whereas PBMC serves as a convenient vehicle for cytokine assays, primarily because of its accessibility. The ability of PBMC to spontaneously secrete cytokines *in vitro* has been shown to be modulated by immunotherapies administered *in vivo*.[25] It is possible that this ability may prove to be a more useful correlate of response to therapy than the determination of cytokine levels in plasma. There may be several problems associated with testing of plasma, as reviewed elsewhere and these are largely eliminated in the PBMC system. However, the PBMC assay requires the use of cells and, as with all indirect procedures, is more time-consuming than direct serum-based assays.

Although cytokine measurements in plasma are easily accomplished, those in tissues are more difficult but desirable. Cytokines are local mediators, and their *in vivo* production is probably compartmentalized to achieve optimal physiologic effects. Thus, assays of local cytokine production at the site of inflammation or organ injury and repair are more likely to be biologically relevant than those in the peripheral blood. To this end, we have described and are using a microdialysis procedure to obtain interstitial fluids from human tissues for cytokine detection by Luminex.[29] These assays require special equipment and are technically demanding. Nevertheless, for *in situ* cytokine detection they represent the best approach. Immunocytohistology (ICH) for cytokines with cytokine-specific poly- or monoclonal antibodies can be used to detect these proteins in tissue samples (Figure 5.9A), with the caveat that it may be difficult to microscopically distinguish cytokines that occupy cellular receptors, having been produced elsewhere in the body, from those that are present in the cytoplasm of the cell.

Single-cell cytokine assays are described later in this chapter and also in Chapter 17.

5.6.7   *IN SITU* HYBRIDIZATION

Another approach to evaluation of cytokines in tissue is to measure their gene expression.[30] ISH for cytokine mRNA may be preferable to protein-based assays,

(A)   (B)

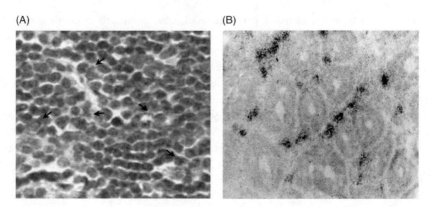

FIGURE 5.9 Cytokine detection *in situ*. In (A), ICH for IL-10 protein in tumor tissue is performed using paraffin-embedded tissue sections and fluorescein-labeled Ab specific for IL-10. Magnification × 400. In (B), the result of ISH is shown, in which an antisense probe for TNF-α was used for hybridization to show the presence of cells positive for TNF-α mRNA in the intestinal tissue. Magnification × 120.

although it is by no means simpler or more reliable. Figure 5.9B illustrates the use of ISH for detection of mRNA encoding TNF-α in tissue. Assays for cytokine gene expression are generally difficult to perform and interpret because mRNAs for cytokines are tightly regulated and rapidly processed. Therefore, very sensitive assays performed at the time when the message is likely to be expressed are necessary. Thus, ISH for cytokine mRNA is usually performed on frozen sections.[30,31] ISH provides information about the number, localization, and distribution of cells expressing cytokine genes in tissue, but not about their identity. A combination of ISH with ICH is necessary to accomplish the latter. Although cryosections have been successfully used for ISH to detect cytokine mRNA in freshly cryopreserved biopsy tissues, the possibility of using archival paraffin-embedded tissues has also been explored with variable and, so far, not truly convincing results[32] because expression of mRNA for cytokines is frequently decreased or absent in archival paraffin-embedded specimens. Like all other RNA determinations, ISH for cytokine mRNA depends on the presence of the undergraded RNA in tissue. To this end, tissues must be snap-frozen immediately after surgery and never allowed to warm up for fear of activating endogenous ribonucleases (RNases). The quality of tissue submitted for ISH or other mRNA studies remains the most critical aspect of this technology. For this reason, recent successes in improving RNA preservation go a long way in facilitating cytokine gene detection in tissue. This is especially crucial for microarrays that are extensively used for *in situ* studies. Although not yet validated for monitoring, the microarray platform provides useful information in terms of cytokine profiles expressed by different subsets of immune cells. The newer quantitative techniques for cytokine mRNA often used to validate microarray results involve the competitive RT-PCR, which combines the exquisite sensitivity of detection with specificity.[33]

5.6.8   MOLECULAR ASSAYS FOR SIGNALING MOLECULES

Molecular assays have been emerging as a novel way of immune cell monitoring. In cancer and certain immunodeficiency diseases, abnormalities in signal transduction have been identified, and they may be responsible for defective functions of immune cells in these diseases.[34] Because it might be possible to reverse these abnormalities and repair immune functions with immunotherapy, monitoring for the presence and extent of signaling defects is indicated before and after such therapy. The main concerns about signal transduction studies for monitoring are the large number of cells needed for preparation of cell lysates (atleast $5-10 \times 10^6$) to be used in Western blots, the semiquantitative nature of Western blots, and the possibility of degradation of signaling molecules during the process of lysis. These concerns have been largely addressed by adapting flow cytometry to measure quantitative changes in the levels of expression of signaling molecules or changes in their phosphorylation status (Figure 5.10) or changes in the phosphorylation status of their substrate(s). Flow-based assays require smaller numbers of cells ($2.5-5 \times 10^6$) and depend on the availability of antibodies specific for signaling molecules and their specific phosphorylated versus nonphosphorylated substrates. As an example, decreased expression or absence of the zeta chain, associated with TCR in T cells and FcγRIII in NK cells, has been observed by flow cytometry in tumor-infiltrating lymphocytes and in circulating lymphocytes of patients with advanced cancers.[34] Decreased

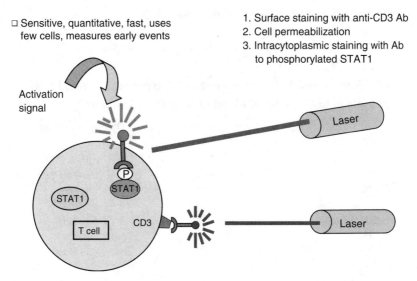

FIGURE 5.10   Multicolor flow cytometry can be used to measure levels of phosphorylated STAT1 in activated immune cells. Two different signals, one delivered by a fluorochrome-labeled Ab binding to phosphorylated STAT and the other by differently labeled Ab to CD3 on the cell surface, are simultaneously processed by the flow cytometer and provide a quantitative measure of phosphorylated STAT1 expression in these T cells.

expression of signaling molecules in immune cells can be confirmed in tissue biopsies by immunostaining with antibodies specific for these molecules. Both quantitative flow cytometry and immunostaining are applicable to monitoring, and should contribute to the improved understanding of the mechanism through which immunologic functions are modulated during disease or therapy.

5.7 MONITORING OF T LYMPHOCYTES

T lymphocytes represent 66–86% of peripheral blood lymphocytes, with a normal absolute number in adults ranging from 1133 to 2125/mm^3 (see Table 5.3). Changes in the total lymphocyte count generally reflect those in T lymphocytes. However, flow cytometry combined with the differential and white blood cell (WBC) counts allows for very precise quantification of T cells and their subsets in whole blood, other body fluids, or in cells isolated from tissue sites. Monitoring of the *percentages* of T-cell subsets in the peripheral blood is not adequate because the total blood lymphocyte number may be altered during therapy with a biologic agent, or in the course of disease without a change in proportions of various lymphocyte subsets. By determining only the percentages of T-cell subsets, changes in their absolute number could be missed if the WBC or differential counts fluctuate during therapy. Drug-induced changes in the T-cell number could be due to altered production, increased or decreased margination, extravasation with movement into tissue, or destruction, for example, apoptosis of these lymphocytes. It is generally not possible to discern the mechanisms responsible for these changes on the basis of phenotypic analyses, unless more extensive studies are undertaken.

TABLE 5.3
Percentages and Absolute Numbers of T Lymphocytes and T-Cell Subsets in the Peripheral Blood of Normal Individuals

	Positive Cells (%)	Positive Cells/mm^3
T cells		
CD3$^+$	66–86	1133–2125
CD4$^+$	38–61	662–1481
CD8$^+$	21–41	385–897
Activated T cells		
CD3$^+$HLA-DR$^+$	3–12	48–239
CD4$^+$HLA-DR$^+$	2–5	39–139
CD8$^+$HLA-DR$^+$	1–8	26–175
Naive T cells		
CD3$^+$CD45RA$^+$	24–48	493–1237
Memory T cells		
CD3$^+$CD45RO$^+$	26–45	495–1237

Note: The data are mid-80% normal ranges obtained by testing peripheral blood MNC of 100 normal individuals. The percentages of positive cells were determined by two-color flow cytometry.

5.7.1 DETERMINATION OF ABSOLUTE NUMBERS AND PERCENTAGES OF T CELLS AND T-CELL SUBSETS

Subsets of T lymphocytes associated with distinct functions cannot be always quantified by phenotypic analysis. This is because phenotypic markers have not been identified for many specific functional subpopulations of lymphocytes. For example, two distinct subsets of human CD4+ T helper cells Th1 and Th2 are best defined by mutually exclusive patterns of cytokine secretion (Figure 5.11). Th1 but not Th2 cells produce IL-2, IFN-γ, and TNF-α, whereas Th2 but not Th1 produce IL-4, IL-5, and IL-10. Different cytokine profiles imply distinct effector functions. Thus, phagocyte-dependent defense is mediated by Th1 cells, which trigger both cellular responses and production of opsonizing and complement-fixing antibodies. In contrast, Th2 cells mediate phagocyte-independent responses, which include IgE and IgG antibody production as well as differentiation and activation of mast cells and eosinophils.[35] Many CD4+ T cells exhibit aspects of both Th1 and Th2 subsets, and some of these cells may represent precursors (Th0) of Th1 and Th2 cells. The Th2 subset of CD4+ T cells in humans appears to be associated with expression of a membrane marker, CD30, a member of the TNF/nerve growth factor (NGF) receptor superfamily.[36] Originally described as a surface marker of Hodgkin and Reed–Steinberg (H–RS) cells, CD30 is strongly and selectively expressed on Th2+ clones of T cells, which are either CD4+ or CD8+. Th2 clones or lines of T cells also release CD30 into their supernatants, and CD30 expression may be a marker of cells predominantly secreting Th2-type cytokines.[36] A combination of phenotypic (e.g., CD4 or CD8) and functional (e.g., cytokine production) assays can provide an adequate profile of T-cell subsets in normal individuals or patients with various disease.[37] Subsets of naive or memory T lymphocytes can be distinguished by flow analysis using antibodies specific for isoforms of CD45, that is, CD45RA or CD45RO, respectively, which are differentially expressed on the surface of these cells. Although a certain degree

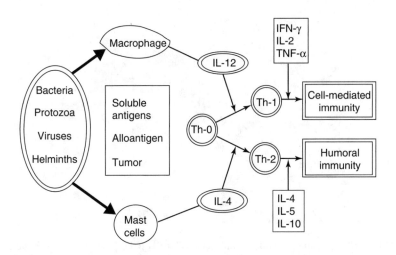

FIGURE 5.11 Interactions between Th0, Th1, and Th2 subsets of human CD4+ T lymphocytes and distinct cytokines produced by these subsets.

of overlap may exist between these markers, they are useful in subsetting CD4+ or CD8+ T cells. Expression of CCR7, a chemokine receptor on the cells marked as CD45RA or CD45RO further characterized them as naïve, memory, or terminally differentiated subsets.[38] The use of multi-color flow cytometry allows distinctions to be drawn between central memory, naive, effector, and effector memory populations (Section 6.16.2).

To monitor the phenotype or function of tissue-infiltrating T lymphocytes, it is preferable to isolate them from tissue, although *in situ* monitoring can provide a certain amount of useful information about their local interactions. *In situ* monitoring is limited by the availability and size of tissue biopsies that can be obtained from patients, and by the quality of antibodies or molecular probes available for the definition of phenotypic and functional molecules in tissue. Some of the antibodies, for example, do not work well in paraffin-embedded tissues, and mRNA analyses cannot be performed in tissues that are not fresh or handled in a special way to ensure the integrity of cellular RNA. Despite these limitations, *in situ* monitoring is encouraged in conjunction with T-cell studies in peripheral blood. T lymphocytes (CD3+) as well as CD4+ and CD8+ subsets, including Treg, are readily identifiable in cryosections of tissues (Figure 5.12), and the creative use of immunostaining with different antibodies specific for, for example, activation markers, adhesion molecules, or triggering receptors in combination with ISH can provide crucially important assessments of the ability of effector cells to produce cytokines, enzymes, or signaling molecules at the site of disease.

5.7.2 FUNCTIONAL ASSAYS FOR T LYMPHOCYTES

Functional assays for T lymphocytes that are used for monitoring include proliferation, cytotoxicity, and cytokine production. The assays for these functions are described in

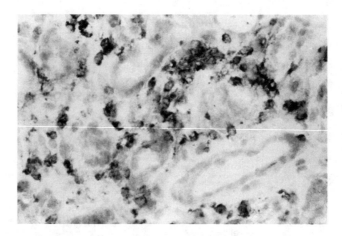

FIGURE 5.12 Immunoperoxidase staining for T cells (CD3+) in a cryostat tissue section of kidney. The biopsy was obtained from a renal transplant patient developing rejection. Magnification × 360.

detail elsewhere.[39] From the monitoring point of view, it is important to distinguish antigen-specific, MHC-restricted T-cell responses from those mediated by activated, but not MHC-restricted T lymphocytes. The latter can be easily assessed by a combination of one of the functional assays, for example, cytotoxicity against a tumor target, with flow cytometry to confirm that activated CD3[+] T cells whose cytotoxicity is not inhibited by class I MHC antibodies are the effector cells. Antigen-specific, MHC-restricted T-cell responses are more difficult to measure because such T cells are rare in the peripheral blood, and their *in vitro* expansion from precursor T-cells may be necessary before functional assays. The main difficulty with T-cell-specific assays is that the antigen or antigen plus target cells have to be available for *in vitro* sensitization as well as specificity assessments. In bacterial- or viral-specific responses, where the nature of antigenic molecules is known, T-cell-specific responses can, of course, be monitored, but even in these situations such monitoring is difficult to implement and perform on a large scale. T-cell-specific assays performed with *in vitro* expanded effector cells require a panel of blocking antibodies (anti-TCR, anti-MHC-class I, anti-CD3, etc.) to confirm the specificity of T cells and a panel of target cells expressing the relevant antigen or pulsed with the antigen for testing the effector cell function.

Frequencies of proliferating or cytolytic T-cell precursors (CTL-p) can be determined directly in the peripheral blood by limiting dilution and clonal analyses,[40] but this approach does not lend itself easily to serial monitoring of clinical trials because it is both labor-intensive and expensive. In vaccination trials, where monitoring of CTL-p or proliferating T-cell precursors (PTL-p) appears to be necessary, the skillful use of limiting dilution technology and subsequent functional assays with the immunizing antigen or peptide allow the determination of the frequency of precursor cells.[41] An alternative to the limiting dilution analysis in the fresh peripheral blood is *in vitro* expansion of a population of antigen-specific T cells, followed by specificity testing to confirm the presence of such T cells among lymphocytes that proliferate in cultures routinely supplemented with IL-2, IL-7, and IL-15. Such *in vitro* expansion requires repeated sensitization with the immunizing antigen and is both time-consuming as well as highly subjective.

Because the frequency of antigen-specific precursors in the peripheral blood (or in culture of T cells) obtained from tumor-bearing patients or human immunodeficiency virus (HIV)-1 positive individuals is very low, special assays have been developed to measure the number of cells able to secrete cytokines, for example, IFN-γ, in response to the cognate antigen instead of measuring cytotoxicity after limiting dilution.

Among the variety of procedures for determination of cytokine production, single-cell assays are particularly useful for defining the cytokine profile of T-cell subsets. Among these, cytokine flow cytometry (CFC), ELISPOT, and tetramer binding measure the frequency of epitope-specific effector T lymphocytes in the mononuclear cell specimens (Figure 5.13). All the three assays are based on TCR recognition of cognate peptides presented by MHC class I or II molecules on the surface of APC to the R T cells. However, no consensus exists as to which of these three assays should be selected to best monitor vaccination results. The common perception that all these assays provide the same information may not be correct.

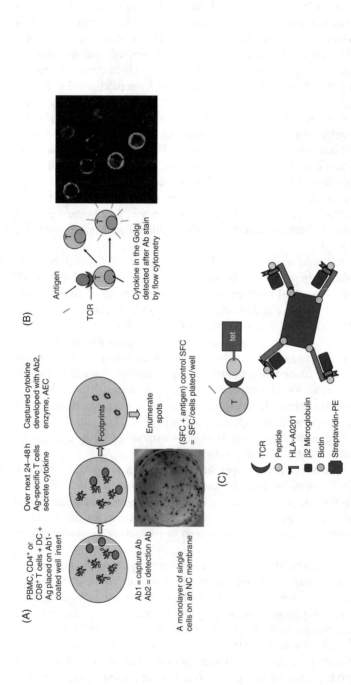

FIGURE 5.13 Single-cell assays for enumeration of cytokine-secreting or antigen-reactive T cells. In (A), ELISPOT assay is depicted including the photograph of real spots in a positive well. (B) gives a diagrammatic representation of cytokine flow cytometry (CFC), in which T cells are incubated for 4–6 h with an antigen in the presence of monensin or brefeldin, stained with an Ab to CD3 or other surface markers, permeabilized and then stained with a labeled, cytokine-specific Ab. T cells responding to activation by intracytoplasmic expression of a cytokine (white rim) can also be visualized and enumerated by immunofluorescence or confocal microscopy (right-hand side). In (C), the principle of tetramer analysis is presented. In this rare-event analysis, T cells expressing receptors for a cognate peptide bind tetramers and are thus identified and counted by flow cytometry. The recognition is MHC class I or II restricted. Tetramers must be custom-made for every epitope. They detect rare T cells that are epitope specific, but do not inform about functionality of such T cells.

The ELISPOT assay measures the production of cytokines (most commonly either IFN-γ or IL-5) by individual T cells in the plated population with a theoretical detection sensitivity of 1/100,000 cells.[42] CFC identifies single responding T cells (1/50,000) with expression of a cytokine in the Golgi zone.[43] Tetramer binding detects peptide-specific T cells expressing the relevant TCR with a theoretical detection sensitivity of 1/10,000 cells.[44] The assays not only have different sensitivities of detection, but also differ in specificity. ELISPOT and CFC are antibody-based and, by definition, are highly specific. In contrast, tetramers, which are complexes of peptides sitting in grooves of four MHC molecules (i.e., human leukocyte antigen [HLA] restricted to appropriate class I genes of each individual) held together by a streptavidin–biotin scaffold[44] bind to T lymphocytes expressing the relevant TCR with variable affinity (Figure 5.13). Tetramers must be designed with the appropriate peptide in the context of the appropriate MHC molecules for each individuals HLA type. Tetramers are not available for all MHC gene products and peptides. *In vitro*, tetramers might easily dissociate or bind nonspecifically to B cells, monocytes, or apoptotic cells, so that tetramer specificity is carefully controlled. Further, T cells that bind tetramers may not be functional because TCR signaling could be compromised, as often happens in cancer.

CFC offers a possibility of the identification of lymphocytes simultaneously stained for T-cell surface marker (e.g., CD4 or CD8) and for a cytokine (e.g., IFN-γ), which is detectable in the Golgi of those cells in the population that specifically respond to the antigen (Figure 5.13B). By flow cytometry, it is feasible to rapidly enumerate tens of thousands of cells; thus, even rare cytokine-producing T cells in the population can be identified and quantitated.[43] However, permeabilization of cells before antibody staining requires a careful selection of a permeabilizing agent (usually saponin that should be freshly prepared) and of controls to correct for background or nonspecific staining. Since the possibility exists that a cytokine may be released from T cells before permeabilization and staining, pretreatments of antigen-sensitized populations with monensin or brefeldin A (agents which block the secretory pathway) are recommended.[43] Also, CFC measures cytokine expression in a cell and not its secretion (although it is commonly assumed that the expressed cytokine would be secreted). The CFC is presented in detail in Chapter 17.

ELISPOT is based on the similar principle as CFC, it only measures cytokine *secretion* from stimulated R cells that are plated as a monolayer of individual cells on a nitrocellulose membrane to avoid cell-to-cell contact and allow for adequate spot detection. Only ELISPOT measures the function of individual R cells by identifying those that produce and secrete the measured cytokine. ELISPOT also does not require cell permeabilization or the use of a flow cytometer for cytokine detection.[42] Because it is always preferable to measure function rather than phenotype, ELISPOT would be an assay of choice for monitoring. However, CFC and tetramer binding are flow cytometry–based assays, and thus allow for surface labeling of R cells and their identification. It is possible to select CD8$^+$ or CD4$^+$ T-cell subsets on antibody-charged columns before ELISPOT, and the two-color ELISPOT now available offers a possibility of identifying T cells simultaneously producing two cytokines. In addition, supernatants from ELISPOT wells can be collected and tested for cytokine levels in multiplex assays. However, tetramer binding can be combined both with surface staining to determine cellular phenotype and intracytoplasmic staining for the detection of cytokine production.[44] Although most informative, especially in situations when some of tetramer-binding cells do not express cytokines, this technology is time-consuming and labor-intensive: clearly, not the best choice for serial monitoring. The recommended solution would be to monitored by ELISPOT or CFC (but not both), depending on considerations such as sample numbers, time, labor, cost, and access to a flow cytometer, and to use tetramer binding (if available for a given patient or vaccine) as a confirmatory assay in situations where it is important to demonstrate a functional deficiency of tetramer-binding T cells. ELISPOT performed under strictly controlled, preferably good laboratory practice (GLP), conditions can provide accurate estimates of the frequency of functionally competent effector T cells in batched, serial samples obtained from subjects enrolled in clinical studies. Compared to CFC and tetramer binding, the cost of ELISPOT is reasonable enough to permit its use in high-volume testing. However, the ELISPOT assay is not easy to standardize, and responder–stimulator interactions might result in unacceptably high background spot counts, in which case the assay becomes uninterpretable.

ELISPOT, tetramer analysis, and CFC for the quantification of T cells able to secrete a cytokine protein have all been used for clinical monitoring. These methods also allow for the definition of the Th0, Th1, or Th2 subsets of T lymphocytes. In combination with phenotypic markers, cytokine production assays and especially single-cell frequency assays are applicable to monitoring of abnormalities in T-cell subsets in various body compartments, including those at the site of the disease. The most common T-cell abnormalities are deficiencies induced by infections and those associated with systemic diseases, including autoimmune disease or cancer. Monitoring of T-cell numbers and functions in disease or in response to various therapies has recently become more firmly established in most clinical centers. The possibility of dissecting complex patterns of changes in T-cell subsets and correlating them to clinical responses in patients treated with biologic therapies is an exciting and promising aspect of clinical immunology.

5.8 MONITORING OF NATURAL KILLER CELLS

NK cells are a subset of effector lymphocytes that, in contrast to T cells, do not require prior sensitization with the antigen for their functionality. The lytic functions of NK cells are regulated by a set of inhibitory receptors (killer inhibitory receptors [KIRs]) recognizing MHC class I determinants on target cells (see Figure 5.14 and Table 5.4).[44] NK cells also express triggering or activating receptors responsible for recognition and mediating positive signaling, and recognize ligands that are upregulated in tumor cells, stressed cells, or virally infected targets. It is believed that NK-cell function is controlled through a fine balance of signals from the functionally opposing activating and inhibitory receptors. These receptors act as a fail-safe mechanism, so that the engagement of activating receptors by their ligands identifies the potential targets, but the engagement of inhibitory receptors blocks the killing of normal cells (Figure 5.14). Table 5.4 lists receptors that are known to be expressed on human NK cells and ligands that bind to these receptors.

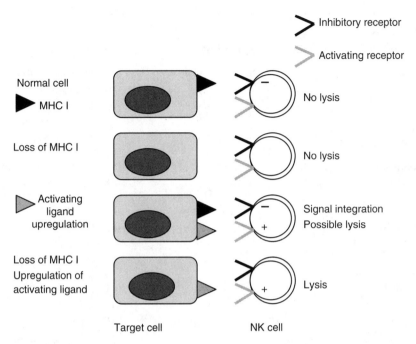

FIGURE 5.14 NK-cell interactions with targets are controlled by inhibitory and activating receptors expressed on all NK cells. A balance between positive and negative signals delivered to the NK cell from the target determines its fate. In case of upregulation of activating ligands by the target, the NK cell will proceed to lyse if the strength of activating signal is greater than that of inhibitory signal.

5.8.1 DETERMINATION OF ABSOLUTE NUMBERS AND PROPORTIONS OF NK CELLS AND NK SUBSETS

For the purpose of monitoring, human NK cells are phenotypically identified as $CD3^-CD56^+CD16^+$ lymphocytes, which represent 5–21% of lymphocytes in the peripheral blood (103–425 positive cells/mm³ in normal adult volunteers [see Table 5.5]). Although $CD3^-CD56^+ CD16^+$ cells are the largest subset of peripheral blood NK cells (90%), two smaller subsets, $CD56^-CD16^+$ and $CD56^+CD16^-$, are also recognized. The $CD3^-CD56^+$ NK cells may outnumber $CD3^-CD16^+$ cells in the peripheral blood lymphocytes of normal individuals, but the proportions of these subsets fluctuate and a large degree of individual variability exists. CD16, a receptor that binds IgG complexes with low affinity, has been designated as FcγRIII and is responsible for antibody-dependent cellular cytotoxicity (ADCC) mediated by NK cells. CD56, an isoform of the neural adhesion molecule that serves as a pan NK-cell marker, is expressed on nearly all NK cells. CD56 expression distinguishes functional subsets of NK cells.[45] The majority of NK cells express low levels of CD56 ($CD56^{dim}$) and high levels of FcγRIII ($CD16^{high}$), whereas about 10% of all NK cells are $CD56^{bright} CD16^{dim/-}$. These two NK-cell subsets differ in receptor expression, cytokine production, and the ability to mediate cytotoxicity. $CD56^{dim}$ NK cells are more cytotoxic and mediate high ADCC, but produce low levels of cytokines.

TABLE 5.4

NK-Cell Receptors and Ligand Specificity

	Activating	Inhibitory	Ligand Specificity
KIRs	+	+	MHC class I alleles including HLA-A, -B, and -C specificities
NCRs			
NK p30 (CD337)	+		Unknown
NK p44	+		Influenza hemagglutinin (HA)
NK p46 (CD335)	+		Influenza HA
C-type lectin receptors			
CD94/NKG2A/B		+	HLA-E
CD94/NKD2C	+		HLA-E
NKG2D (CD314)	+		MICA, MICB, ULBP-1, -2, -3
Other receptors			
CD16 (FcγRIII)	+		Fc of IgG
CD2	+		CD58
2B4	+		CD48
CD 40 Ligand	+		CD40
CD27	+		CD70
CD28	+		CD80, CD86

Note: KIR = killer Ig like receptor, NCR = natural cytotoxicity receptor, MICA = MHC class I chain-related peptide A, MICB = MHC class I chain-related peptide B, ULBP = UL-16 binding protein.

TABLE 5.5

Percentages and Absolute Numbers of NK Cells and NK Activity in the Peripheral Blood of Normal Individuals

	Positive Cells (%)	Positive Cells/mm³
NK activity		
CD3⁻CD56⁺	6–21	96–425
CD3⁻CD16⁺	3–20	62–383
DR⁺CD56⁺	1–4	15–90

NK-cell activity 66–341 $LU_{20}/10^7$ effector cells

Note: The data are mid-80% normal ranges obtained by testing peripheral blood MNC of 50 normal individuals. The percentages of positive cells were determined by two-color flow cytometry. NK activity was measured in 4-h ^{51}Cr-release assays.

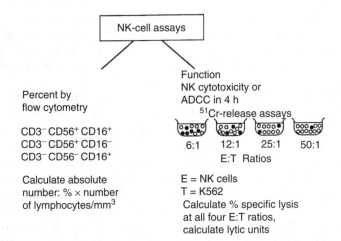

FIGURE 5.15   A schematic outline for monitoring human NK-cell numbers and function in peripheral blood specimens.

In contrast, CD56[bright] NK cells can produce abundant cytokines, but exert low natural cytotoxicity and ADCC.[45] Figure 5.15 indicates that assessments of absolute NK-cell numbers should accompany phenotyping by flow cytometry.

5.8.2   FUNCTIONAL ASSAYS FOR NATURAL KILLER CELLS

In addition to eliminating or inhibiting growth of transformed cells, intracellular pathogens, and certain immature normal cells, NK cells perform additional functions including a broad range of activities influencing hematopoiesis, immunoregulation, and fetal development. They are also involved in interactions with nonimmune tissue cells.[46] The most commonly used functional assay for NK cells is a 4-h ^{51}Cr-release assay,[47] which can be performed either with isolated MNC or with whole blood.[48] NK-cell assays are performed at several different effector to target cell ratios to be able to construct a lytic curve and calculate lytic units (LU) of activity.[46] The target cell most commonly used for monitoring of NK activity is the K562 human leukemia line maintained in culture and passaged frequently to assure that the targets used for the assay are in the log phase of growth (Figure 5.15). Using this line, it has been possible to measure NK activity in the peripheral blood of healthy individuals and to confirm that it is a stable trait.[47] NK activity is generally higher in males (201 LU/10[7] effector cells, with $n = 49$) than females (115 LU/10[7] effector cells, with $n = 103$). In disease, NK activity changes and the NK cell is a sensitive monitor of physiologic alterations occurring during stress, infection, exercise, or other events. NK activity is rapidly upregulated by cytokines including IL-2, IL-12, IL-15, IL-18, or by direct interaction with another cell such as dendiritic cell.[49]

Activated NK cells use a variety of mechanisms to eliminate tumor cells or virally infected targets: (a) perphorin- or gramzyme-mediated natural cytotoxicity, (b) TRAIL/FasLigand-mediated apoptosis, (c) ADCC, and (d) cytokine secretion including not only IFN-γ but also many other cytokines and chemokines. Activated NK cells are a major component of innate immunity. In oncology protocols, activity of NK cells primed by cytokines (the so-called LAK cells) is monitored by serial assessments of killing of the Daudi cell line (an NK-resistant target) by peripheral blood MNC.

In general, it has been difficult to correlate NK activity with responses to therapy in clinical trials with biologic agents. There could be many reasons for this lack of correlation between NK function and clinical responses in cancer patients treated with biologic agents, but the most likely possibility is that NK cells mediate antitumor effects by more than one mechanism, as indicated earlier. Thus, commonly used 4-h ^{51}Cr-release assays, measuring the perforin or granzyme secretory pathway, only reflect one mechanism of antitumor activity used by NK cells *in vivo*.[50] It is highly likely that nonsecretory pathways involving cytokine-mediated deoxyribonucleic acid (DNA) fragmentation and apoptosis contribute to antitumor activity of NK cells. NK-cell-mediated apoptosis can be measured *in situ* by a terminal deoxynucleotidyl transferase (Tdt)-mediated d-UTP nick-end labeling (TUNEL) assay or *in vitro* by apoptosis assays that test for DNA fragmentation, mitochondrial permeability changes, or activation of various caspases.[50] Recent evidence indicates that NK activity is strongly influenced by DC. This is referred to as "NK–DC cross talk."[51] These NK–DC interactions appear to be crucial for the development of effective adaptive immune responses. To achieve a better understanding of NK-cell-mediated events associated with clinical response in cancer and other diseases it may be necessary to monitor a broader spectrum of NK activities including cytokine secretion. Recently, a new genre of NK-cell assays that are flow cytometry–based has been introduced, replacing the 4-h ^{51}Cr-release assay. These new assays can simultaneously measure granule movement and release, perforin or granzyme cell content, cytokine profile, conjugate formation, and phenotypic markers, thus identifying the NK cell and the target with which it interacts.[52] None of these assays have been exhaustively compared to the ^{51}Cr-release procedure, which remains as the "gold standard." Nevertheless, the flow-based cytotoxicity assays offer an opportunity to monitor several NK-cell functions that result in death of NK-sensitive targets and give a more realistic picture of the killing potential that characterizes the NK cell.

5.9   IMMUNOTHERAPY WITH HUMAN T, DC, AND NK CELLS

In patients with cellular immunodeficiencies, a good rationale exists for adoptive transfer of *in vitro*–activated immune effector cells. When the patients' own immune effector cells are decreased, absent, or nonfunctional, adoptive transfer of competent effector cells along with cytokines supporting their activities seems to be indicated. This therapeutic strategy is particularly attractive when other therapies, including attempts at *in vivo* activation of endogenous effector cells, are ineffective. Therapeutic transfer of both T and NK cells, generally in conjunction with high-dose IL-2 therapy, have been performed in patients with advanced cancer or AIDS.[53,54] Recently, autologous T cells cultured *in vitro* have been adoptively transferred to patients with cancer following lymphodepletion to "provide room" for the transferred dividing T cells and to provide them with sufficient endogenous cytokines.[55]

Transfers of the allogeneic irradiated NK-92 cell line *in lieu* of activated NK cells have been performed.[56] But by far the most widely used has been immunotherapy with DC used as adjuvants in antitumor therapeutic vaccines.[57] A general scheme for the adoptive transfer of immune cells is shown in Figure 5.16.

The rationale for adoptive immunotherapy with T cells is to increase the number of specific immunocompetent effector cells and to generate long-term memory T cells. The transfer of antigen-specific T cells is feasible in only some situations because of practical difficulties in expanding a sufficiently large number of such T cells in humans. Transfers of human T cells bearing the TCR that recognizes specific viral antigens, for example, CMV proteins, have been performed in recipients of allogeneic bone marrow at high risk of developing CMV disease.[58] Safe and effective restoration of CMV-specific T-cell responses has been achieved in such patients by the adoptive transfer of *in vitro* expanded, CD8+ CMV-specific T-cell clones derived from MHC-identical allogeneic bone marrow donors.[58] These clones appear to

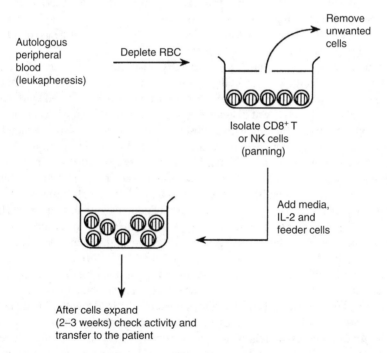

FIGURE 5.16 Preparation of human effector cells for adoptive immunotherapy. To enrich autologous MNC in CD8+ T cells, NK cells, or DC, panning on plastic surfaces, which are coated with monoclonal antibodies for positive selection of effector cells, can be performed. Alternatively, autoMACS™ Ab-charged columns can be used for positive or negative cell selection using magnets. Elutriation is another way of obtaining enriched subsets of MNC without using Abs for cell separation. The captured effector cells are expanded by culturing them in the presence of cytokines and irradiated feeder cells for 1–3 weeks. Before reinfusion of effector cells into the patient, testing of sterility, endotoxin, and mycoplasma is obligatory. Phenotypic and functional assays are also performed to characterize the product used for adoptive therapy.

recognize structural viral proteins. Adoptive immunotherapy with EBV-reactive T cells (un-irradiated MNC obtained from the allogeneic EBV seropositive bone marrow donors) has been used in patients who developed a lymphoproliferative syndrome following transplantation with T-cell-depleted allogeneic bone marrow.[59] Clinical trials with CD8$^+$ T lymphocytes for the treatment of adenovirus infections in transplanted patients on immunosuppression are in progress.[60] The goal of these therapies is to provide an antiviral effect by increasing the number of competent virus-reactive T cells present in the host. It is yet unclear whether virus-specific immuno-logic memory can be induced as a result of adoptive immunotherapy or not.

In addition to these successful transfers of virus-specific T cells, adoptive immunotherapy of cancer using tumor-specific or -reactive T-cells has been remarkably successful when used in combination with nonmyeloablative che-motherapy to lymphodeplete the host before transfer of autologous T cells.[55,61] Therapy with IL-2-activated autologous tumor-infiltrating (TIL) T cells expanded in culture resulted in objective tumor regression in more than 50% of treated melanoma patients.[61,62] Therapeutic transfers of activated MHC class I-restricted tumor-reactive T cells originating from autologous TIL or genetically modified T cells expressing a TCR of choice[63] have also been performed without evidence of adverse effects and with substantial clinical benefits.[64]

An alternative strategy to therapy with MHC-restricted antigen-specific T cells is to transfer nonspecific effector cells such as activated NK cells or subsets of acti-vated NK cells. NK cells obtained from cancer patients show impaired cytotoxicity and low levels of NK activity at diagnosis and appear to predict subsequent metasta-ses.[46] Adoptive immunotherapy with a subset of activated NK cells (generated from the MNC of cancer patients by culture of MNC with high concentrations [22 nM] of IL-2) has been used in phase I trials.[65,66] These effector cells appear to be able to extravasate and actively migrate to target tissues. They exert antitumor activities leading to reduction of established metastases in animal models showing metastasis[67] and in some patients who have received this form of adoptive immunotherapy.[65] HLA-matched allogeneic hematopoietic cell transplantation is used in clinical practice to treat leukemia. A report by Ruggeri et al.[68] showed that haploidentical trans-plant with MHC class I mismatches (i.e., mismatched for KIR ligands) generates an alloreactive NK-cell response that eradicates leukemia, improves engraftment, and protects from T-cell mediated GVHD. These effects were attributed to killing of host APC and leukemia cells by activated alloreactive NK cells. Recently, Miller et al.[69] reported antitumor effects by adoptive transfer of haploidentical NK cells after an intense immunosuppressive regiment in poor prognosis acute myelogenous leukemia patients. Here also the use of KIR ligand-mismatched donors significantly improved therapeutic effects leading to complete remission in some patients.[69]

The NK-92 cell line established some years ago from a patient with leukemia has been used for immunotherapy of malignant melanoma in a severe combined immunodeficiency (SCID) mouse model with promising results.[70] Therapeutic use of NK cells is likely to grow in the future concomitant with a better understand-ing of their receptor systems and signaling pathways. NK-cell functions can be enhanced with soluble cytokines such as IL-15, monoclonal antibodies blocking inhibitory receptors, and manipulating the NKG2D receptor pathway that plays a key role in NK-cell activation. Adoptive transfers of *ex vivo*–activated NK cells to tumor sites, where they can kill tumor targets and directly interact with DC may be a particularly attractive strategy to be pursued.

Autologous DCs have been extensively used as adjuvants in antiviral and anticancer vaccines.[71] DCs are the professional APC of the immune system with the potential to either stimulate or inhibit immune responses. Because of this potential, therapeutic utilization of DC offers a great promise for treatment of autoimmune diseases, cancer, or prevention of transplant rejection. Therapeutic DCs are generated from monocytes or CD34$^+$ precursors by culture in the presence of cytokines, IL-4 and GM-CSF. Following *ex vivo* exposure to antigens, DCs are matured and delivered as therapeutic vaccines through intradermal, intralymphatic, or intravascular routes. Early clinical trials with DC showed that transfers of antigen-pulsed autologous DC were safe and free of adverse events. Immune responses to vaccines have been detected in patients with cancer or HIV-1, but clinical responses are few.[71] It is likely that numerous logistical issues associated with preparation and standardization of DC for therapy hamper their use as standard therapy. For example, several methods for activating immature (i)DC exist. Conventional cytokines (IL-1β, IL-6, and TNF-α) are used to produce iDC, whereas the mixes of cytokines including IFNs α and γ will mature iDC into αDC1, which are highly activated and very efficient in antigen presentation *in vitro* (Figure 5.17). Expression of surface markers such as costimulatory

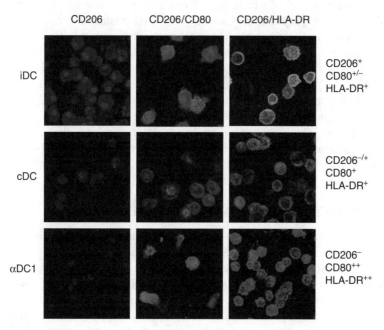

FIGURE 5.17   DC maturation *ex vivo* can be monitored by following expression of surface molecules by immunostaining or flow cytometry. iDCs express mannose receptor and are phagocytic. Few iDCs are CD206$^+$, and they express CD80. αDC1 are strongly positive for CD80 and HLA-DR but negative for CD206. DC maturation is critical for their *in vivo* performance as adjuvant to the vaccines.

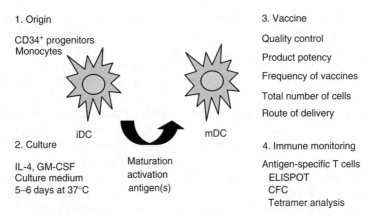

1. Origin

CD34⁺ progenitors
Monocytes

iDC

2. Culture

IL-4, GM-CSF
Culture medium
5–6 days at 37°C

Maturation
activation
antigen(s)

mDC

3. Vaccine

Quality control

Product potency

Frequency of vaccines

Total number of cells

Route of delivery

4. Immune monitoring

Antigen-specific T cells
ELISPOT
CFC
Tetramer analysis

FIGURE 5.18 *Ex vivo* production of human autologous DC for therapy involves many steps and requires extensive laboratory support as well as product standardization under cGMP or cGTP conditions.

molecules (HLA-DR), and mannose receptors (CD206) can be used to monitor the progress of DC maturation *ex vivo* and define the product destined for therapy (Figure 5.17). Mapping the optimal way for therapeutic preparation of DC requires numerous strategies, all of which may be critically important for achieving successful vaccinations. Further, the quality of the final product and its potency has to be measured *ex vivo* to be able to correlate the latter with the clinical efficacy of the administered vaccine (Figure 5.18). Of utmost importance is full activation of DC, which can be achieved by various approaches (combinations of cytokines, toll-like receptor [TLR] ligands, and CD40 ligand) and which should be sustained *in vivo* after DC transfer.[72] Monitoring for DC-induced T-cell responses after vaccinations is a complex process, which generally depends on ELISPOT, CFC, and tetramer analysis. The early clinical trials support a promising role for DC vaccination in human diseases. However, future therapy must address the DC product standardization, mode of DC preparation, maturation, and *in situ* activation of DC as well as timing, frequency, and routes of their delivery (Figure 5.18).

5.10 QUALITY CONTROL AND DATA ANALYSIS

Monitoring of immunologic assays, particularly cellular assays, is difficult and requires a well-designed and rigorously maintained quality control (QC) program be in place in a monitoring laboratory. Such a QC program contains several components: (a) availability of standard operating procedures (SOP), (b) training of personnel, (c) implementation of GLP guidelines, (d) instrument maintenance program, (e) review of the quality of performance, and (f) proficiency testing. Laboratories monitoring immunologic assays for clinical trials currently have to implement their own QC programs to ensure that acceptable results are generated. No model QC programs exist, although various QC software packages are becoming available. Monitoring laboratories are encouraged to follow the GLP defined by professional groups such as the College of American Pathologists or departments of health in some states.

A central and most crucial aspect of immune monitoring is the ability to document changes from baseline over time. This is only possible when measurements taken and accumulated overtime are consistent. It is essential to control variability and ensure reproducibility of each assay selected for monitoring. The objective is to assure reproducibility of test results from individual patients tested at different time points, often over a period of weeks, months, or even years. Laboratories involved in preparation of effector cells for immunotherapy are obliged to follow Food and Drug Administration (FDA) standards for operation of cGMP facilities and product generation (21 CFR 211) as well as regulations designed to prevent transmission of infectious diseases (21 CFR 1271) and known as good tissue practices (cGTP). The quality of cellular products released for therapy has to be rigorously controlled and documented. Regular checks of reagent quality and equipment performance have to be in place, and all products must be checked for sterility, viability, purity, stability, and potency before release.

Implementation of the QC program requires considerable effort. The process of QC begins with sample collection and processing, which have to be organized to meet the protocol schema and occur at specified times of the day and, preferably, before the next cycle of therapy. Blood for immunologic monitoring needs to be routinely harvested in the morning to avoid diurnal variability. The flow of specimens and recording the collection and arrival times of samples is a major effort in a monitoring laboratory. Although immunologic monitoring assays should be scheduled in advance, sample collection and arrival times tend to vary. The laboratory must establish strict rules for sample acceptance and handling. Samples that are obviously outdated or mislabeled are not tested and a precise history of each untested sample is maintained. Sample processing should be uniform and follow the SOP. The SOP must be available in writing and be reviewed and updated regularly. The decision to cryopreserve cells or use fresh cells should be made before the clinical trial, and has to be based on preliminary comparative studies using fresh and cryopreserved lots of the same normal MNC. These comparisons need to be performed and documented for every assay. This is a crucial process because if assays can be batched (i.e., testing all samples in the trial at the same time), day-to-day variability can be avoided and the cost of monitoring can be decreased considerably.

The importance of the reproducibility of assays used for longitudinal monitoring cannot be overemphasized. Regardless of whether cryopreserved or fresh samples are used, assay standardization has to be performed before a clinical trial begins. The standardization data are obtained by repeatedly performing the assay with cells obtained from normal individuals under the invariant and previously optimized experimental conditions to establish the mean, median, 80% normal range, and coefficient of variation. Intra-assay variability is also determined. When the assay is standardized, a set of appropriate controls has to be selected and these depend on whether fresh or cryopreserved cells are used. With fresh cells collected at different time points, repeated testing of preserved control samples (e.g., cryopreserved lots of normal MNC with a predetermined range of reactivity) is necessary to control day-to-day variability. With fresh cells, it may also be advisable to include fresh control cells obtained from a healthy volunteer. A pool of volunteers repeatedly tested over time can be established and used for this purpose. The data obtained from control samples, evaluated in parallel with each patient sample, helps ensure the validity of the results for a particular day's assay. With cryopreserved cells or frozen serum or plasma samples, it is best to batch and test all serial samples of a patient in one assay. Even in this case, however, it is necessary to control day-to-day variability to ensure that the assay performs equally well for all patients on a protocol. If universally accepted standards are available (e.g., the World Health Organization [WHO] standards for cytokines), these should be regularly included in monitoring assays. Alternatively, internal controls, initially compared to the standards (which are often available only in finite quantities), can replace the latter for routine use. A properly designed QC program will ensure that the values obtained for control samples remain within an acceptable range over time and will specify when patient results are abnormal or altered from baseline values.

Analysis of serial immunologic data should be performed by a qualified biostatistician working closely with the clinical immunologist. In preparation for final analysis, immunologic data have to be "cleaned," that is, purged of errors that occur during data entry into computer. All outlier values are checked against the laboratory records and collection timing of all specimens is verified. The analysis selected depends on the hypothesis tested and the trial design, but it generally seeks to determine if the changes from baseline that occur during various points of therapy or, overall, as a result of therapy, are significant. The monitoring data are generally presented as a series of time plots, which are adjusted by subtraction of the estimated contribution of each patient's baseline level, so that the plots depict the relationship that would exist if baseline values for all patients receiving therapy or a particular dose of therapy were equal to the overall pretreatment average.

5.11 SUMMARY

A wide range of immunotherapies for correcting deficiencies or restoring the balance of the immune system altered by disease has become available in recent years. Administration of biologic agents to patients with immunologic abnormalities is based on newly obtained insights into functional attributes of immune effector cells, and it is constantly evolving. It is now possible to accurately evaluate the numbers and many functions of effector cells in humans longitudinally in health and in disease as well as during therapy. Cellular products used for immunotherapy have to be precisely characterized for key phenotypic and functional properties that might determine their *in vivo* effectiveness. Such potency assays for biologic agents need to be developed and validated before phase III clinical trials. New technologies are available that utilize fewer cells, permit *in situ* studies (i.e., in diseased tissues as well as peripheral blood), and allow for more comprehensive monitoring of cellular functions in disease. Molecular technologies in combination with highly specific immunomethods offer a possibility for defining functional defects in unique subsets of immune cells and for documenting a reversal of these defects during therapy. High-throughput technologies utilizing small samples are being gradually introduced to clinical monitoring, expanding the number of assays performed and the available data sets. It is reasonable to expect that in the near future it will be possible to obtain detailed immunologic profiles of patients and to correlate changes in functions of immune effector cells induced by therapy to clinical endpoints. To achieve this goal, however, monitoring of immune therapies has to be performed under strictly defined and controlled conditions, ensuring that changes occurring in effector cells in response to therapeutic interventions are accurately and precisely measured. The single most important requirement of immunologic monitoring, regardless of the type of assay used, is the day-to-day reproducibility of the assay. For this reason, it is recommended that immunologic monitoring of cellular functions be performed in clinical immunology laboratories with established QC programs and the capabilities to handle serial specimens and analyze serially acquired results.

REFERENCES

1. Herberman, R. B. Design of clinical trials with biological response modifiers. *Cancer Treat. Rep.*, 69, 1161, 1985.
2. Gambacorti-Passerini, C., Hank, J. A., Albertini, M. R., Borchert, A. A., Moore, K. H., Schiller, J. H., Bechhofer, R., Borden E. C., Storer, B., Sondel, P. M. A pilot phase II trial of continuous-infusion interleukin-2 followed by lymphokine-activated killer cell therapy and bolus-infusion interleukin-2 in renal cancer. *J. Immunother.*, 13, 43, 1993.
3. Whiteside, T. L. Monitoring of immunologic therapies. In: *Manual of Molecular and Clinical Laboratory Immunology*, eds. B. Detrick, R. G. Hamilton, J. D. Folds, 7th Edition, ASM Press, Washington, pp. 1171–1182, 2006.
4. Whiteside, T. L. Introduction to cytokines as targets for immunomodulation. In: *Cytokines in Human Health: Immunotoxicology, Pathology and Therapeutic Applications*, eds. R. V. House, J. Descotes, Humana Press Inc., Totowa, NJ, 2006.
5. Elder, E. M., Whiteside, T. L. Processing of tumors for vaccine and/or tumor infiltrating lymphocytes. In: *Manual of Clinical Laboratory Immunology*, eds. H. Friedman, N. R. Rose, E. C. deMacario, J. L. Fahey, H. Friedman, G. M. Penn, 4th Edition, American Society For Microbiology, Washington, pp. 123, 817, 1992.
6. Sallusto, F., Geginat, J., Lanzavecchia, A. Central memory and effector memory T cell subsets: function, generation and maintenance. *Annu. Rev. Immunol.*, 22, 745, 2004.
7. Vignali, D. A. Multiplex particle-based flow cytometric assays. *J. Immunol. Meth.*, 243, 243, 2000.
8. Whiteside, T. L. Cytokine assays. *Biotechniques*, 33, 4, 2002.
9. Curiel, T. J., Coukos, G., Zou, L., Alvarez, X., Cheng, P., Mottram, P., Evdemon-Hogan, M., Conejo-Garcia, J. R., Zhang, L., Burow, M., Zhu, Y., Wei, S., Kryczek, I., Daniel, B., Gordon, A., Myers, L., Lackner, A., Disis, M. L., Knutson, K. L., Chen, L., Zou, W. Specific recruitment of regulatory T cells in ovarian carcinoma fosters immune privilege and predicts reduced survival. *Nat. Med.*, 10, 942, 2004.
10. Bacchetta, R., Gregori, S., Roncarolo, M. G. CD4+ regulatory T cells: mechanisms of induction and effector function. *Autoimmun. Rev.*, 4, 491, 2005.
11. Suni, M. A., Maino, V. C., Maecker, H. T. *Ex vivo* analysis of T-cell function. *Curr. Opin. Immunol.*, 17, 434, 2005.
12. Givan, A. L., Fisher, J. L., Waugh, M. G., Bercovici, N., Wallace, P. K. Use of cell-tracking dyes to determine proliferation precursor frequencies of antigen-specific T cells. *Meth. Mol. Biol.*, 263, 109, 2004.
13. Burkett, M. W., Shafer-Weaver, K. A., Strobl, S., Baseler, M., Malyguine, A. A novel flow cytometric assay for evaluating cell-mediated cytotoxicity. *J. Immunother.*, 28, 396, 2005.
14. Whiteside, T. L., Letessier, E., Hiraayashi, H., Vitolo, D., Bryant, J., Barnes, L., Snyderman, C., Johnson, J. T., Myers, E., Herberman, R. B., Rubin, J., Kirkwood, J. M.,

Vlock, D. R. Evidence for local and systemic activation of immune cells by peritumoral injections of interleukin 2 in patients with advanced squamous cell carcinoma of the head and neck. *Cancer Res.*, 53, 5654, 1993.

15. McCormick, T., Shearer, W. T. Delayed hypersensitivity skin testing. In: *Manual of Clinical Laboratory Immunology*, eds. B. Detrick, R. G. Hamilton, J. D. Folds, 7th Edition, ASM Press, Washington, p. 234, 2006.

16. Aebersold, P. M., Hyatt, C., Johnson, S., Hines, K., Korcak, L., Sanders, M., Lotze, M., Topalian, S., Yang, J., Rosenberg, S. A. Lysis of autologous melanoma cells by tumor-infiltrating lymphocytes: association with clinical response. *JNCI*, 8, 932, 1991.

17. Screpanti, V., Wallin, R. P., Grandien, A., Ljunggren, H. G. Impact of FASL-induced apoptosis in the elimination of tumor cells by NK cells. *Mol. Immunol.*, 42, 495, 2005.

18. Zeh, H. J., III, Lotze, M. T. Addicted to death: invasive cancer and the immune response to unscheduled cell death. *J. Immunother.*, 28, 1, 2005.

19. Whiteside, T. L. Antigen/mitogen-stimulated lymphocyte proliferation. In: *Measuring Immunity: The Immunologic Surogates Handbook*, eds. M. T. Lotze, A. W. Thomson, 1st Edition, Academic Press, London, p. 361, 2005.

20. Bercovici, N., Givan, A. L., Waugh, M. G., Fisher, J. L., Vernel-Pauillac, F., Ernstoff, M. A., Abastado, J. P., Wallace, P. K. Multiparameter precursor analysis of T-cell responses to antigen. *J. Immunol. Meth.*, 276, 5, 2003.

21. Shevach, E. M. CD4+CD25+ suppressor T cells: more questions than answers. *Nat. Rev. Immunol.*, 2, 389, 2002.

22. Strauss, L., Whiteside, T. L., Knights, A., Bergmann, C., Knuth, A., Zippelius, A. Selective survival of naturally occurring human CD4+CD25+FOXP3+ regulatory T cells cultured with rapamycin. *J. Immunol.*, 178, 320, 2007.

23. Kubler, H., Vieweg, J. Vaccines in renal cell carcinoma. *Semin. Oncol.*, 33, 614, 2006.

24. Zou, W. Regulatory T cells, tumor immunity and immunotherapy. *Nat. Rev. Immunol.*, 6, 295, 2006.

25. Friberg, D., Bryant, J., Shannon, W., Whiteside, T. L. *In vitro* cytokine production by normal human peripheral blood mononuclear cells as a measure of immunocompetence or the state of activation. *Clin. Diag. Lab. Immunol.*, 1, 261, 1994.

26. Whiteside, T. L. Introduction to cytokines as targets for immunomodulation. In: *Cytokines in Human Health*, ed. House, Descotes, in press, 2006.

27. Romagnani, P., Annunziato, F., Piccinni, M. P., Maggi, E., Romagnani, S. Th1/Th2 cells, their associated molecules and role in pathophysicology. *Eur. Cytokine Netw.*, 11, 510, 2000.

28. Rossi, M. I., Bonino, P., Whiteside, T. L., Flynn, W. B., Kuller, L. H. Determination of cytokine production variability in normal individuals over time. *Am. Geriatr. Soc. Suppl.*, 41, SA3, 1993.

29. Rosenbloom, A. J., Ferris, R., Sipe, D., Riddler, S., Connolly, N., Abe, K., Whiteside, T. L. *In vitro* and *in vivo* protein sampling by combined microdialysis and ultrafiltration. *J. Immunol. Meth.*, 309, 55, 2006.

30. Vitolo, D., Zerbe, T., Kanbour, A., Dahl, C., Herberman, R. B., Whiteside, T. L. Expression of mRNA for cytokines in tumor-infiltrating mononuclear cells in ovarian adenocarcinoma and invasive breast cancer. *Int. J. Cancer*, 51, 573, 1992.

31. Vitolo, D., Kanbour, A., Johnson, J. T., Herberman, R. B., Whiteside, T. L. *In situ* hybridization for cytokine gene transcripts in the solid tumor microenvironment. *Eur. J. Cancer*, 3, 371, 1993.

32. Bromley, L., McCarthy, S., Stickland, J. E., Lewis, C. E., McGee, J. O'D. Non-isotopic *in situ* detection of mRNA for interleukin 4 in archival human tissue. *J. Immunol. Meth.*, 167, 47, 1994.

33. O'Garra, A., Vieira, P. Polymerase chain reaction for detection of cytokine gene expression. *Curr. Opin. Immunol.*, 4, 211, 1992.

34. Whiteside, T. L. Down-regulation of ζ chain expression in T cells: a biomarker of prognosis in cancer? *Cancer Immunol. Immunother.*, 53, 865, 2004.

35. Romagnani, S. Lymphokine production by human T cells in disease states. *Annu. Rev. Immunol.*, 12, 227, 1994.

36. Romagnani, S., Del Prete, G., Maggi, E., Chilosi, M., Caligaris-Cappio, F., Pizzolo, G. CD30 and type 2 helper (Th2) responses. *J. Leuk. Biol.*, 57, 726, 1995.

37. Romagnani, S. Biology of human Th1 and Th2 cells. *J. Clin. Immunol.*, 15, 121, 1995.

38. Lanzavecchia, A., Sallusto, F. Understanding the generation and function of memory T cell subsets. *Curr. Opin. Immunol.*, 17, 326, 2005.

39. Dietrick, B., Hamilton, R. G., Folds, J. D., eds., *Manual of Clinical Laboratory Immunology*, 7th Edition, ASM Press, Washington, 2006.

40. Bonnefoix, T., Bonnefoix. P., Mi, J. Q., Lawrence, J. J., Sotto, J. J., Leroux, D. Detection of suppressor T lymphocytes and estimation of their frequency in limiting dilution assays by generalized linear regression modeling. *J. Immunol.*, 170, 2884, 2003.

41. Coulie, P. G., Connerotte, T. Human tumor-specific T lymphocytes: does function matter more than number? *Curr. Opin. Immunol.*, 17, 320, 2005.

42. Whiteside, T. L. ELISPOT assays. In: *Analyzing T Cell Responses*, eds. D. Nagorsen, F. M. Marincola, Dordrecht, The Netherlands, p. 143, 2005.

43. Jung, T., Schauer, V., Heusser, C., Meumann, C., Reiger, C. Detection of intracellular cytokines by flow cytometry. *J. Immunol. Meth.*, 159, 197, 1993.

44. Altman, J. Flow cytometry application of MHC tetramers. *Meth. Cell Biol.*, 75, 433, 2004.

45. Lewis, L. L. NK cell recognition. *Annu. Rev. Immunol.*, 23, 225, 2005.

46. Whiteside, T. L., Herberman, R. B. Role of human natural killer cells in health and disease. *Clin. Diag. Lab. Immunol.*, 1, 125, 1994.

47. Whiteside, T. L., Bryant, J., Day, R., Herberman, R. B. Natural killer cytotoxicity in the diagnosis of immune dysfunction: criteria for a reproducible assay. *J. Clin. Lab. Anal.*, 4, 102, 1990.

48. Whiteside, T. L., Herberman, R. B. Measurements of natural killer cell numbers and function in humans. In: *Neuroimmunology: Vol. 24 Methods in Neuroscience*, eds. M. J. Phillips, E. E. Evans, Academic Press, Orlando, FL, p. 10, 1994.

49. Di Santo, J. P. Natural killer cell developmental pathways: a question of balance. *Annu. Rev. Immunol.*, 24, 257, 2006.

50. Vujanovic, N. L., Nagashima, S., Herberman, R. B., Whiteside, T. L. The nonsecretory killing pathway of natural killer cells. *Nat. Immun.*, 14, 73, 1995.

51. Degli-Esposti, M. A., Smyth, M. J. Close encounters of different kinds: dendritic cells and NK cells take center stage. *Nat. Rev. Immunol.*, 5, 112, 2005.

52. Kim, G. G., Donnenberg, V. S., Donnenberg, A. D., Gooding, W., Whiteside, T. L. A novel multiparametric assay to immunophenotype effector natural killer cells and simultaneously measure tumor cell death. *Proc. Abst. 1228*, 95th AACR Annual Meeting, Orlando, FL, 2004.

53. Cooley, S., June, C. H., Schoenberger, S. P., Miller, J. S. Adoptive therapy with T cells/NK cells. *Biol. Blood Marrow Transpl.*, 13, 33, 2007.

54. Dudley, M. E., Wunderlich, J. R., Shelton, T. E., Even, J., Rosenberg, S. A. Generation of tumor-infiltrating lymphocyte cultures for use in adoptive transfer therapy for melanoma patients. *J. Immunother.*, 26, 332, 2003.

55. Dudley, M. E., Wunderlich, J. R., Robbins, P. F., Yang, J. C., Hwu, P., Schwartzentruber, D. J., Topalian, S. L., Sherry, R., Restifo, N. P., Hubicki, A. M., Robinson, M. R., Raffeld, M., Duray, P., Seipp, C. A., Rogers-Freezer, L., Morton, K. E., Mavroukakis, S. A., White, D. E., Rosenberg, S. A. Cancer regression and autoimmunity in patients after clonal repopulation with anti-tumor lymphocytes. *Science*, 298, 850, 2002.

56. Yan, Y., Steinherz, P., Klingemann, H. G., Dennig, D., Childs, B. H., McGuirk, J., O'Reilly, R. J. Antileukemia activity of a natural killer cell line against human leukemias. *Clin. Cancer Res.*, 4, 2859, 1998.

57. Whiteside, T. L., Odoux, C. Dendritic cell biology and cancer therapy. *Cancer Immunol. Immunother.*, 53, 240, 2004.

58. Riddell, S. R., Greenberg, P. D. Principles for adoptive T cell therapy of human viral diseases. *Annu. Rev. Immunol.*, 13, 545, 1995.

59. Koehne, G., Smith, K. M., Ferguson, T. L., Williams, R. Y., Heller, G., Pamer, E. G., Dupont, B., O'Reilly, R. J. Quantification, selection, and functional characterization of Epstein-Barr virus-specific and alloreactive T cells detected by intracellular interferon-gamma production and growth of cytotoxic precursors. *Blood*, 99, 1730, 2002.

60. Leen, A. M., Myers, G. D., Bollard, C. M., Huls, M. H., Sili, U., Gee, A. P., Heslop, H. E., Rooney, C. M. T-cell immunotherapy for adenoviral infections of stem-cell transplant recipients. *Ann. NY Acad. Sci.*, 1062, 104, 2005.

61. Dudley, M. E., Wunderlich, J. R., Yang, J. C., Sherry, R. M., Topalian, S. L., Restifo, N. P., Royal, R. E., Kammula, U., White, D. E., Mavroukakis, S. A., Rogers, L. J., Gracia, G. J., Jones, S. A., Mangiameli, D. P., Pelletier, M. M., Gea-Banacloche, J., Robinson, M. R., Berman, D. M., Filie, A. C., Abati, A., Rosenberg, S. A. Adoptive cell transfer therapy following non-myeloablative but lymphodepleting chemotherapy for the treatment of patients with refractory metastatic melanoma. *J. Clin. Oncol.*, 23, 2346, 2005.

62. Gattinoni, L., Powell, D. J., Jr., Rosenberg, S. A., Restifo, N. P. Adoptive immuno-therapy for cancer: building on success. *Nat. Rev. Immunol.*, 6, 383, 2006.

63. Kershaw, M. H., Westwood, J. A., Parker, L. L., Wang, G., Eshhar, Z., Mavroukakis, S. A., White, D. E., Wunderlich, J. R., Canevari, S., Rogers-Freezer, L., Chen, C. C., Yang, J. C., Rosenberg, S. A., Hwu, P. A phase I study on adoptive immunotherapy using gene-modified T cells for ovarian cancer. *Clin. Cancer Res.*, 12, 6106, 2006.

64. Morgan, R. A., Dudley, M. E., Wunderlich, J. R., Hughes, M. S., Yang, J. C., Sherry, R. M., Royal, R. E., Topalian, S. L., Kammula, U. S., Restifo, N. P., Zheng, Z., Nahvi, A., de Vries, C. R., Rogers-Freezer, L. J., Mavroukakis, S. A., Rosenberg, S. A. Cancer regression in patients after transfer of genetically engineered lymphocytes. *Science*, 314, 126, 2006.

65. Lister, J., Rybka, W. B., Donnenberg, A. D., deMagalhaes-Silverman, M., Pincus, S. M., Bloom, E. J., Elder, E. M., Ball, E. D., Whiteside, T. L. Autologous peripheral blood stem cell transplantation and adoptive immunotherapy with A-NK cells in the immediate post-transplant period. *Clin. Cancer Res.*, 1, 607, 1995.

66. Kirkwood, J. M., Ernstoff, M. S., Vlock, D. R., Herberman, R. B., Whiteside, T. L. New approaches to the use of IL2 in melanoma: adoptive cellular therapy utilizing purified populations of A-LAK effectors. In: *Biological Agents in the Treatment of Cancer: Proceedings of the International Conference held in Newcastle, September 4–7, 1990*, ed. P. Hersey, Government Printing, Newcastle, NSW, 1990.

67. Yasumura, S., Lin, W.-C., Hirabayashi, H., Vujanovic, N. L., Herberman, R. B., Whiteside, T. L. Immunotherapy of liver metastases of human gastric carcinoma with interleukin 2-activated natural killer cells. *Cancer Res.*, 54, 3808, 1994.

68. Ruggieri, L., Capanni, M., Urbani, E., Perruccio, K., Shlomchik, W. D., Tosti, A., Posati, S., Rogaia, D., Frassoni, F., Aversa, F., Martelli, M. F., Velardi, A. Effectiveness of donor natural killer cell alloreactivity in mismatched hematopoietic transplants. *Science*, 295, 2097, 2002.

69. Miller, J. S., Soignier, Y., Panoskaltsis-Mortari, A., McNearney, S. A., Yun, G. H., Fautsch, S. K., McKenna, D., Le, C., Defor, T. E., Burns, L. J., Orchard, P. J., Blazar, B. R., Wagner, J. E., Slungaard, A., Weisdorf, D. J., Okazaki, I. J., McGlave, P. B. Successful

adoptive transfer and *in vivo* expansion of human haploidentical NK cells in patients with cancer. *Blood*, 105, 3051, 2005.

70. Tam, Y. K., Miyagawa, B., Ho, V. C., Lingemann, H. G. Immunotherapy of malignant melamona in a SCID mouse model using the highly cytotoxic natural killer cell line NK-92. *J. Hematother.*, 8, 281, 1999.

71. Rosenberg, S. A., Yang, J. C., Restifo, N. P. Cancer immunotherapy: moving beyond current vaccines. *Nat. Med.*, 10, 909, 2004.

72. Figdor, C. G., de Vries, I. J. M., Lesterhuis, W. J., Melief, C. J. M. Dendritic cell immunotherapy: mapping the way. *Nat. Med.*, 10, 475, 2004.

6 Understanding Clinical Flow Cytometry

Albert D. Donnenberg and Vera S. Donnenberg

CONTENTS

6.1 INTRODUCTION

One of the major advances in laboratory medicine has been the ability to immuno-logically identify specific subsets of leukocytes and other cells by their expression of proteins and other markers. This process, termed "immunophenotyping," employs monoclonal antibodies reactive against cell-associated determinants to distinguish specific subsets within a heterogeneous mixture of cells. Quantification of a subset of interest can be readily accomplished by the use of a *flow cytometer*, an instrument that electronically measures the intensity of scattered or emitted light from individual cells as they pass before a laser.

In clinical laboratories, immunophenotyping by flow cytometry has been invaluable for defining the cell of origin of specific neoplasms, particularly in patients with acute leukemia or non-Hodgkin's lymphoma (see Chapter 7). A second major application is to provide information critical for the staging and management of human immunodeficiency virus (HIV)-1 infected patients and patients with other immune deficiency syndromes (see Chapters 8, 12 respectively); the data derived from analysis of T-cell subsets play an important role in guiding therapeutic decisions. A host of additional applications, including quantification of hematopoietic stem cells [1] and reticulocytes [2], diagnosis of acquired and congenital immune deficiency syndromes (see Chapter 9), paroxysmal nocturnal hemoglobinuria [3], detection of minimal residual leukemia after therapy [4], and many others are either approved as *in vitro* diagnostic (IVD) tests, or serve as diagnostic adjuncts to licensed tests. In addition to clinical considerations, these analytic techniques provide essential information concerning the structure and function of the cells comprising normal hematopoietic and lymphoid cells. In addition to the analysis of cells that occur naturally as single-cell suspensions (blood, lymph, bone marrow, cerebrospinal fluid, ascites, and effusions), flow cytometry can also be used to analyze marker expression and function in disaggregated tissue specimens. This review will focus on the factors that make cytometry as robust as it is, and also on avoiding the pitfalls inherent in such a flexible technique.

Flow cytometry is a mature clinical technique at present. When it was first introduced as a research tool a little more than 30 years ago, both instrumentation and analytical reagents were of limited usefulness [5]. The first instruments measured fluorescence on a linear scale at a single wavelength and displayed the values on an oscilloscope. The only means of recording and analyzing data was to make Polaroid photographs of the oscilloscope screen. Major advances in hardware and software spurred by the digital revolution, which together with the development of monoclonal antibodies and new dye technologies, have resulted in the emergence of instruments and procedures that are readily adaptable to clinical laboratories. Several publications are available that detail the principles of cytometry and the operation of flow cytometers [6–12].

6.2   MONOCLONAL ANTIBODIES

The invention of monoclonal antibodies by Milstein and colleagues in 1977 [13] changed the field of flow cytometry, providing an inexhaustible supply of reagents of exquisite specificity. By definition, a monoclonal antibody is an immunoglobulin produced by a clonal population of immortalized hybridoma cells. Accordingly, the antigen-binding sites are completely homogeneous, a situation very different from naturally occurring (polyclonal) antibodies. Because of their homogeneity, hybridomas can be selected which produce high-affinity antibodies with great specificity for particular antigenic determinants. Within a year of Milstein's invention, monoclonal antibodies specific for rat [14] and murine [15] helper T cells and murine MHC antigens [16] were described. Shortly thereafter, the description of three murine monoclonal antibodies specific for human T-cell surface determinants, designated OKT1,

OKT3, and OKT4 [17] paved the way for human studies. Today, these hybridomas are recognized as producing antibodies against the CD5, CD3, and CD4 determinants, respectively. Most of the monoclonal antibodies currently in use for the detection of determinants on human cells are of murine origin. The antibodies are generated to proteins, glycoproteins, glycolipids, and carbohydrates. The biological functions of these molecules are varied and include serving as membrane receptors or ligands for soluble and contact-dependent signaling, acting as enzymes, membrane pumps, or adhesion molecules. Antibodies that recognize intracellular components such as histones, lysosomes, signal transducing molecules, and cytokines, to name but a few are available. In particular, the detection of protein activation states with antibodies directed against specific phosphoepitopes, has opened up a new area in which flow cytometry can be used to probe cellular function [18,19].

Antibodies recognize and react against small peptide sequences of complex antigenic molecules. Thus, each antigenic protein contains numerous small peptide sequences, termed epitopes, to which monoclonal antibodies may be raised. Antibodies to any epitope of the parent molecule can be used to identify the entire molecule. Initially, this led to a nomenclature dilemma; each laboratory and commercial source had unique designations for their monoclonal antibodies. To facilitate communication, an international workshop established a consensus nomenclature (see historical review of consensus nomenclature in Chapter 11) [20]. Cell surface antigens were assigned a *cluster of differentiation* (CD) number based on their recognition by monoclonal antibodies. All antibodies to the same antigen, regardless of their epitope specificity, were grouped together. Thus, two or more antibodies, each reactive with a different epitope on the same antigen are considered as directed against the same CD. In general, the antigens themselves are identified by a CD number (e.g., CD3 on T cells) and the monoclonal antibodies recognizing them are grouped as *anti-CD number* (e.g., anti-CD3).

At a workshop held in Vienna in 2006, 350 CD clusters were recognized. These are listed and annotated as Tables 11.2 and 11.3 in Chapter 11. It is important to note that few of the CD antigens are truly restricted to a single cellular lineage. With rare exceptions (e.g., CD3 on T cells), lineage assignments cannot be made on the basis of a single CD antigen. Rather, they are determined by the presence and absence of a combination of markers.

6.3 FLUOROCHROMES AND FLUORESCENCE DETECTION

To detect the presence of a cell-bound monoclonal antibody by flow cytometry, the antibody must be coupled either directly or indirectly to a fluorescent dye (fluorochrome). These are the dyes that, when excited by light of a specific wavelength, emit light at a longer wavelength (this is known as the Stokes shift). Because different dyes have different Stokes shifts, they can be distinguished by passing emitted light through a series of optical filters that permit only specific bandwidths of light to reach the photomultiplier tubes (PMTs) (Figure 6.1). The PMTs generate a signal proportional to the intensity of the incident light. Through spectral compensation, which will be discussed in detail later, the processed signals can be quantitatively related to the amount of dye bound to a particular cell. With appropriate standards, this in turn, can be used to estimate the number of molecules targeted by the fluorescence-tagged antibody.

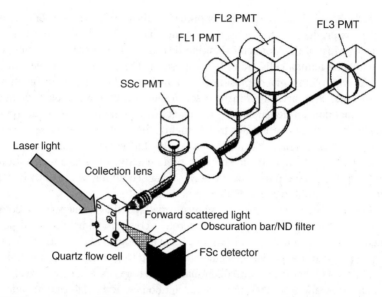

FIGURE 6.1   Schematic operation of a three-color single-laser flow cytometer. The fluoro-chrome-labeled cells are pumped through the flow cell, where they enter the light path of a 488-nm argon laser (arrow). A photodiode aligned with the laser path captures light scattered between 1.5° and 19° (FSc). Before encountering this FSc detector, the scattered laser light can be attenuated by a neutral density (ND) filter if needed; the laser beam itself is blocked by an obscuration bar. A collection lens is positioned 90° from the incident laser light. This light (which consists of both laser light scattered by the cell and light emitted by the fluorochromes associated with the cell) passes through a series of dichroic mirrors and filters that separates the light sources on the basis of their wavelengths. In the instrument depicted here, light from the argon laser (488 nm) is deflected to the side scatter photomultiplier tube (SSc PMT) by a 488 dichroic long-pass (DL) mirror. Before reaching the SSc PMT, this light passes through a 488-nm band-pass (BP) filter, which blocks all other wavelengths. The use of these filters ensures that the SSc PMT measures only light emitted by the argon laser and scattered at 90° by the cell. The light transmitted through the 488 DL mirror then passes through blocking filters that attenuate any remaining scattered laser light, leaving only the light emitted by the cell-bound fluorochromes. This emitted light is then deflected to the FL1 PMT, using 550-nm DL and 525-nm BP filters. Thus, the FL1 PMT captures a narrow bandwidth of light centering on 525 nm (green fluorescence emitted, e.g., by the dye FITC). In the three-color instrument shown here, this process is repeated two more times, with the FL2 and FL3 PMTs capturing light of progressively longer wavelengths, emitted by additional fluorochromes. Commercially available cytometers with multiple lasers and photodetectors have been used to measure as many as 14 fluorochromes simultaneously, using the same principles (Modified from Beckman Coulter Manual. COULTER EPICS XL/XL-MCL FLOW CYTOMETER, PN 4237340A1-1, 6 April 1998, Elite XL Cytometer.).

In the simplest flow cytometers, light is supplied by a single air-cooled argon or diode laser that emits blue light at a wavelength of 488 nm. Ultraviolet, violet, green, yellow, and red lasers are becoming increasingly common as the list of available fluorochromes grows and multiparameter flow cytometry is more widely adopted. The most commonly used fluorochromes include the old standbys fluorescein isothiocyanate (FITC), phycoerythrin (PE) and the PE-tandem dyes (PE-Texas red, PE-Cy5, PE-Cy5.5, and PE-Cy7), peridinin chlorophyll protein (PerCP), allophyco-cyanin (APC), and the APC tandems: APC-Cy5.5 and APC-Cy7. The tandem dyes are of particular importance because they allow a single laser to excite as many as six separate fluorochromes (e.g., FITC, PE, and the PE-tandems can all be excited by a 488-nm laser). PE is a large fluorescent protein that can be excited at 488 nm (but is more efficiently excited by a green or yellow laser), and it emits at 578 nm. It can be decorated with small dye molecules such as Texas red, Cy5, Cy5.5, or Cy7, which are not excited by themselves at 488 nm. Known as *free energy transfer* in physics, the excited PE, instead of emitting a photon at a higher wavelength, directly excites the coupled dye, which then emits light at its characteristic wavelength. Several dif-ferent PE-tandem dyes with increasingly longer Stokes shifts can be used simultane-ously with a single blue or green laser capable of exciting PE. The cyanine and Alexa dye families represent a new generation of *designer dyes*, which are specifically engineered for their excitation and emission characteristics.

Increasingly, antibodies conjugated to a variety of fluorochromes are commercially available. For the direct staining procedure, a cell suspension is initially incubated with one or more dye-coupled monoclonal antibodies directed against the determinants of interest. In the indirect assay, an unlabeled antibody against one target molecule is used to stain the cells. The presence of this antibody is detected by a second labeled antibody. For example, T cells can be identified in human peripheral blood by stain-ing with an unlabeled murine antibody directed against CD3. After washing away the unbound antibody, a fluorescence-labeled goat antimouse immunoglobulin antibody can be used to detect the presence of anti-CD3 on the human T cells. Indirect staining can be used in concert with direct staining, provided the indirect staining is performed first, and all free antimouse immunoglobulin-binding sites are blocked (with mouse serum or purified mouse IgG) before adding the directly conjugated murine antibod-ies. A relatively new development in which small quantities of monoclonal antibody can be labeled with the antigen-specific portion (FAB) of an antiglobulin conjugated to a dye of choice, provides additional flexibility [21]. Finally, antibodies that are directly labeled with biotin can be purchased. Biotin is not fluorescent by itself, but binds to the tetrameric molecule avidin or streptavidin with an unusually high affinity (hence it's name), which in turn can be conjugated to the dye of choice. This provides versatility, since a biotin-conjugated antibody can be used with your choice of avidin-conjugated fluorochromes, and conversely, the same avidin-conjugated fluorochrome can be used with an array of different biotinylated antibodies.

In addition to the fluorochromes used for labeling monoclonal antibodies, a wide variety of dyes with useful biological properties can be used to identify cell popula-tions or measure functions in living cells (Table 6.1).

6.4   HOW THE CYTOMETER WORKS

After cells have been labeled with dye-conjugated antibodies or other fluorescent molecules, they are resuspended at an appropriate concentration and introduced "single file" into the analytical cytometer. The cell sample is injected into a stream of sheath fluid (saline or distilled water), such that the sample stream forms a stable

TABLE 6.1
Some Commonly Used Dyes and Fluorescent Molecules

Dye	Maximal Excitation/ Emission (nm)	Properties
DAPI [22–25]	358/461 nm, bound to DNA	Binds to AT-rich regions of DNA. Viability staining (viable cells exclude dye); Nuclear (DNA) staining in permeabilized cells
7-AAD [26]	546/647	Viability staining (viable cells exclude dye); Nuclear (DNA) staining in permeabilized cells
Acridine orange [27–29]	495/519	Measurement of DNA and (ribonucleic acid) RNA
Aldefluor [30]	505/513	Nonfluorescent molecule (BODIPY-aminoacetaldehyde) that fluoresces when converted to BODIPY-aminoacetate by aldehyde dehydrogenase (active in stem cells)
BODIPY [31,32]	505/513	Labeling drugs for uptake and efflux studies
CFSE [33]	494/514	Labeling cytoplasmic proteins of viable cells (cell tracking and proliferation assays)
Draq5 [34]	647/670 (huge tail allows excitation with 488 line)	Nuclear (DNA) staining of living (nonpermeabilized) cells
Green fluorescent protein (GFP) and its variants [35,36]	395/509	Identification and sorting of transfected cells. Variants provide alternative excitation/ emissions maxima
Hoechst 33342 [23,37,38]	350/461	Nuclear (DNA) staining of living (nonpermeabilized) cells; Measurement of DNA content (cell cycle); Detection of cells with constitutive MDR transporter activity (side population)
Indo-1 [39,40]	330/401, 475	Intracellular calcium concentration (ratiometric)
JC-1 [41,42]	514/529	Mitochondrial membrane potential
Monochlorobimane [43]	394/490	Measurement of intracellular glutathione
Phi Phi Lux [44]	505/530	Quenched fluorescent molecule linked to the caspase-3 substrate amino acid sequence (DEVDGI). Fluoresces when cleaved by caspase-3 (detection of apoptosis)
Propidium iodide (PI) [45–48]	535/617 nm, bound to nucleic acids	Viability staining (viable cells exclude dye); Nuclear (DNA) staining in permeabilized cells
Rhodamine 123 [37,49]	507/529	MDR transporter-mediated dye efflux, mitochondrial mass

Note: DAPI—4,6-diamidinozphenylindole; CFSE—carboxy fluorescein succinimydal ester; 7AAD—
7-amino actinomycin-D; BODIPY—DIPYrromethene BOron Difluoride.

core within the sheath stream. Because laminar flow is established between the two streams, sample and sheath fluid do not normally mix, and the cells are concentrated near the center of the sample stream by a process called hydrodynamic focusing. In an analytical cytometer, which does not have the additional task of sorting the cells, all this takes place within a quartz flow cell in which the cells are illuminated by one or more lasers. Light scattered by the cells is captured at a small angle (forward light scatter [FSc]) and at 90° (side scatter [SSc]) from the incident light as they cross the laser path (Figure 6.1). FSc correlates well with cell size, whereas SSc measures the cell's internal complexity (nuclear convolutions, vacuoles, etc.). Forward scattered light, because of its intensity, is usually detected by a relatively insensitive and inexpensive photodiode, whereas side scattered light, such as light emitted by the fluorochromes with which the cells are labeled, is measured using very sensitive PMTs. In peripheral blood, small cellular debris, lymphocytes, granulocytes, and monocytes can be distinguished on the basis of their intrinsic light-scattering properties alone (Figure 6.2). The photons emitted by fluorescent dyes associated with the cell are detected by PMTs, which with the aid of a series of filters, limit the window of detection to the major emission region of particular fluorochromes (e.g., with FITC, the intensity of green light, between 515 and 535 nm). The analog output from the PMT, a voltage pulse lasting for the duration of time that it takes the cell to transit the laser line, is converted into one or more numbers that quantify the pulse (area, peak, or duration of the pulse). Precisely, how this is accomplished differs from cytometer to cytometer. The digitized data, often corrected for spectral overlap (also called color compensation) are stored in a data file called a *list-mode* file [50]. Conventionally, these are given the extension *.fcs* or *.lmd*. In some cytometers, both compensated and uncompensated data can be stored in the same file. A typical *list-mode* file can be likened to a large data table in which the columns are the parameters (e.g., FSc, SSc, green, orange, and red fluorescence intensity), and each row represents a distinct *event*. In the parlance of flow cytometrists, *events* are signals that may represent cells but can also be debris or other sources of noise. The process of data collection

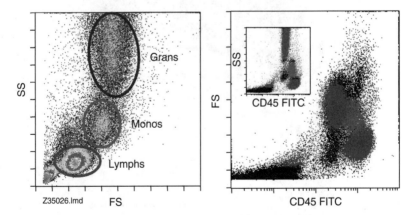

FIGURE 6.2 Discrimination of lymphocyte, monocyte, and granulocyte populations by light scatter. This fixed control cell preparation (CD Chex CD34, Streck Laboratories) was also stained with anti-CD45, a marker expressed to varying degrees on all leukocytes. Color eventing was used to mark the lymphocyte (green), monocyte (red), and granulocyte (blue) populations identified by light scatter alone. Subcellular debris can be seen on the left-hand side of the lymphocyte population. The populations visualized by light scatter also form discrete populations based on CD45 and light scatter (right panel and inset). Note that debris and red cells can be identified by light scatter but are better resolved with anti-CD45.

in which cells are run through the cytometer and parameters describing individual events are stored in data files, is known as *acquisition*. Depending on the complexity of the data, they can either be analyzed in real time during sample acquisition or *offline* by recalling the stored *list-mode* data files.

6.5 SAMPLES

Flow cytometric assays can be performed using anticoagulated blood or bone marrow, isolated cell suspensions, cell cultures, or disaggregated tissue. In peripheral blood, where erythrocytes outnumber white cells by 1000 to 1, it is usually necessary to remove them from the sample. It is not necessary to physically remove platelets and small debris, because these can be *thresholded* out during acquisition (in this case, the cytometer does not count them), *live gated* during acquisition (events are counted but not saved in the list-mode file), or *gated* out on the basis of their low light scatter during analysis. Similarly, cell doublets and clumps can be recognized by their light-scatter properties, and gated from the analysis (Figure 6.3).

As mentioned earlier, normal leukocytes can be segregated into three distinct groups, corresponding to lymphocytes, monocytes, and granulocytes, on the basis of light scatter. A typical three-part differential was illustrated in Figure 6.2. Lymphocytes comprise the leukocyte population with the smallest cell diameter (FSc) and the least internal complexity (SSc). Monocytes are larger cells and have greater internal complexity; neutrophils are cells of approximately the same size as monocytes but

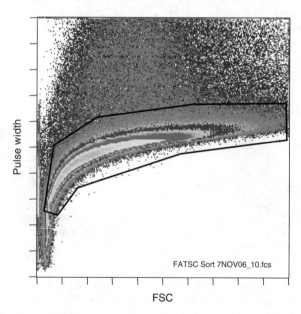

FIGURE 6.3 Doublet discrimination can be performed by comparing pulse analysis parameters. The pulse parameters used for doublet discrimination differ between cytometer manufacturers. The data shown here were collected using a Dako MoFlo sorter, where FSc pulse height (*x*-axis) is plotted against forward scatter pulse width. Doublets and cell clusters have widths too large for their peaks. Singlets are shown within the polygonal region.

with still greater internal complexity. Because the scatter and fluorescence characteristics of each event are recorded in the *list-mode* file, analysis of collected data can
be performed on all events or can be restricted to gated events (i.e., cells that meet
predetermined scatter and fluorescence criteria).

Measurement of the percentage of peripheral blood T cells illustrates the principles of flow cytometry analysis. A unique characteristic of mature T cells is the
presence of the membrane determinant CD3, a polypeptide that is part of the T-cell
antigen receptor complex [51]. At a minimum, T cells can be detected by incubating a
cell suspension with a FITC-labeled monoclonal antibody specific for an epitope present on the CD3 determinant. The characteristics of each cell in the suspension are then
analyzed and recorded. This includes correlates of two inherent physical characteristics, size (FSc) and internal complexity (SSc), and the intensity of green fluorescence
(anti-CD3 FITC) emitted by each cell. Even for a simple three-parameter problem,
such as this, there are several ways to approach data analysis. The conventional way
is to first establish a *gate* based on forward and side scatter alone to delineate the
lymphocyte population. The subsequent analysis of fluorescence is then limited to
events with scatter parameters falling within this *lymphocyte gate*. T cells are quantified as a fraction of cells falling within the lymphocyte gate. Although this approach
benefits from its simplicity, it does not exploit the fact that T cells and debris are very
well separated in the three-parameter space of this assay. Plotting CD3 versus SSc
(Figure 6.4) yields a distinct population (raw CD3 gate), which can then be cleaned of
early apoptotic T cells or other artifacts by subsequently measuring forward and side
scatter on this population. This true T-cell gate then provides guidance for where to
make the lower and upper cuts on the lymphocyte gate. The former is often obscured
by debris, and the latter by monocytes if we try to draw a lymphocyte gate on ungated
events. Once the gates are established, they can be validated against an isotype control and adjusted if necessary. The upper limit of staining with the isotype control can
be used to set a fluorescence cutoff; higher fluorescence is considered to indicate specific antibody binding, and therefore the presence of the target antigen. In the present
example, the isotype control consists of an FITC-labeled mouse IgG without specific
reactivity to human leukocytes and of the same isotype as the anti-CD3 antibody.
Although the use of isotype controls is routine in many laboratories, interpretation of
the isotype control is subject to several pitfalls. A few of the commonly encountered
problems are: control antibodies should be matched with respect to concentration and
fluorescence to protein ratio as well as isotype; specific binding can compete with
nonspecific binding, such that the negative peak in the control antibody is brighter
than the negative peak in the test sample; and positive and negative populations are
not always discretely bimodal, but may fuse to form a single skewed distribution that
overlaps with the negative control. In this example, CD3+ and negative events are so
well separated that an isotope control is not strictly necessary to establish a cutpoint
between fluorescence positive and negative cells. In addition to measuring the percent
of T cells within the lymphocyte gate, the absolute circulating T-cell count can be calculated by multiplying the percentage of FITC+ cells in the lymphocyte scatter gate
by the total number of circulating lymphocytes (cells/μL) determined independently
with a hematology analyzer. Inherent in this procedure is the assumption that the
flow cytometer and the hematology analyzer detect the same events as lymphocytes.

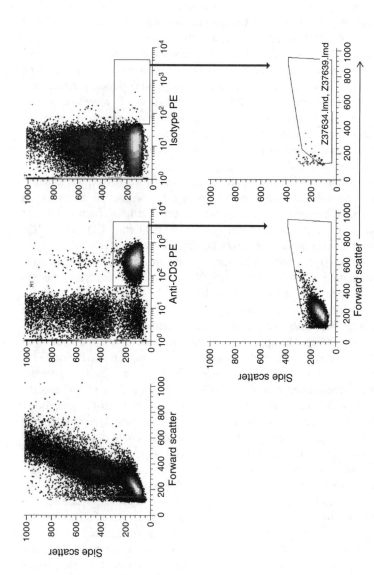

FIGURE 6.4   Gating on CD3 versus SSc prior to creating a lymphocyte gate. Many of us were taught to begin analysis by creating gates on forward and side light scatter. It is our goal to discriminate between cells (the subject of our analysis) and debris, but it is often difficult to see a clear demarcation between these populations on the basis of light scatter alone (top left panel), especially if many events have been acquired. It is often much easier to visualize the population of interest (T cells, in this example) versus SSc. Here (top center panel), a rectangular gate has been created with reference to the isotype control and the expected low SSc characteristic of peripheral T cells. This population can be further cleaned with a conventional "lymphocyte gate" (bottom panels). Note that events nonspecifically binding the isotype control antibody (bottom right) have low FSc and relatively high SSc characteristic of apoptotic lymphocytes.

This assumption is usually correct for fresh blood samples, but not bone marrow or other samples for which the analyzer has not been validated. A more robust *single-platform* method of determining absolute counts by comparison to reference beads is illustrated in Section 6.16.

A major advantage of flow cytometers is their capacity to analyze many different properties simultaneously. The recent invention of the imaging flow cytometer allows a peek at what the cytometer is actually measuring, when cells pass single file before the laser. The principal difference between an imaging flow cytometer (commercially manufactured by Amnis, Seattle, Washington) and a conventional analytical cytometer is that the fluorescent light emitted by the cells, after passing through optical filters, is captured by a sensitive charge-coupled device camera in such a way that separate images of each cell are captured at the wavelength bands corresponding to fluorochrome emissions. As mentioned earlier, a conventional flow cytometer quantifies emitted light using PMTs, and therefore is not capable of assembling this information into images of individual cells. Each individual image captured by an imaging flow cytometer can be analyzed for the intensity of light scatter or fluorescent signals and displayed in much the same way as conventional flow cytometry data are. Additionally, fluorescent signals can be localized to cellular features (plasma membrane, nucleus, cytoplasm, Golgi, lysosomes, etc.). By registering several images of the same cell obtained at different wavelengths, it can be determined whether different fluorescent signals colocalize. This is very important for functional measurements, since signaling molecules often translocate from the cytoplasm to the nucleus when they are activated.

The example given in Figure 6.5 gives a glimpse of what an imaging flow cytometer sees when it looks at a sample of peripheral blood leukocytes stained with anti-CD45 (a pan leukocyte marker), anti-CD14 (a monocyte marker), anti-CD3, and a nuclear dye—Draq5.

6.6 QUALITY CONTROL AND QUALITY ASSESSMENT

If there is a dark side to flow cytometry, it is the perception that artful dial tweaking can make any sample "show whatever the investigator wants." Make a positive sample negative or a negative sample positive? No problem. Even without intentional malfeasance, potential variability in samples, reagents, staining procedures, instrumentation performance and settings, compensation of spectral overlap and data analysis, conspire such that the default condition is often irreproducibility. As the flow cytometry field has matured, so have quality control (QC) and quality assessment procedures. Religious application of these standard measures is a necessary step toward minimizing the most important sources of variation and assuring that the results are interpretable (although not necessarily what the investigator would like). Performance and documentation of QC and assessment measures are routine in clinical laboratories (and are required for certification), but are no less important in research or assay development settings. Rote performance of quality control measures, without understanding their rationale and without knowing how to respond when deviations are detected, may be sufficient to satisfy an inspector, but does

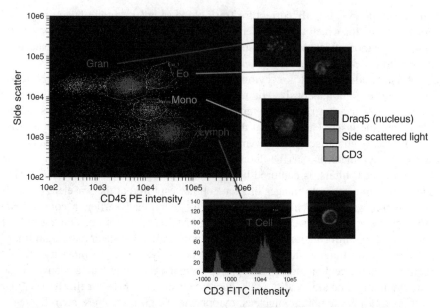

FIGURE 6.5   Imaging flow cytometry provides a glimpse into what a flow cytometer actually "sees" when stained cells are interrogated by a laser, and scattered and emitted light are picked off at 90°. In this example, we chose something familiar: CD45, CD3, and CD14 staining. The DNA intercalating dye Draq5 was used as a nuclear stain. The upper histogram shows patterns of CD45 intensity versus SSc that we would see by conventional flow cytometry. Populations of granulocytes, eosinophils, monocytes, and lymphocytes are identified based on their characteristic patterns. The bottom histogram shows CD3 intensity on cells falling within the lymphocyte gate. All this is very familiar to the flow cytometrist. What is remarkable about imaging flow cytometry is that each event, shown as a dot in the histogram, corresponds to an actual image captured by the cytometer. These images can be recalled simply by clicking on a dot. Alternatively, all the images within a gate can be displayed as a gallery (not shown).

improve assay consistency. This section will consider the most important aspects of quality assessment and control.

QC in flow cytometry includes determining that

- Sample processing is performed correctly
- The instrument is functioning properly
- The instrument settings are consistent
- The regents are behaving as expected
- Data analysis is performed consistently

6.7   SAMPLE ACCESSION AND PROCESSING

As the applications for flow cytometry have grown so have the variety of samples and methods of sample processing. Clinical labs have sample accession criteria, which state the attributes that a sample must have before it is accepted into the laboratory.

For peripheral blood, these may include how the sample is labeled, the elapsed time since sample collection, sample volume, the presence of a particular anticoagulant, and the adequacy of anticoagulation. Failure to meet these criteria results either in sample rejection, or sample acceptance after noting the particular deviation and its probable significance. Prior to performing cytometry, peripheral blood samples are commonly subjected to some form of erythrocyte lysis or removal. Common methods of red cell lysis include a very brief exposure to a hypotonic solution or to formic acid, or a more protracted incubation with ammonium chloride solution. Red cells, mature granulocytes, and platelets may be removed or greatly reduced with a Ficoll–Hypaque density gradient (see details in Section 6.7.2) [52].

6.7.1 ANTICOAGULATION

The common anticoagulants used for peripheral blood and bone marrow samples are heparin, ethylene diamine tetraacetic acid (EDTA), and acid citrate dextrose (ACD). Heparin is a biological product isolated from bovine lung or porcine gut. It is a long negatively charged highly sulfated glycosoaminoglycan that interferes with the clotting cascade by binding to and potentiating antithrombin III. Sodium heparin is nontoxic at a dose effective to prevent clotting (10 Units/mL). It is the preferred anticoagulant when cells are to be used for assays of immune function. Despite its highly charged nature, exposure to sodium heparin does not appear to adversely affect the function of lymphocytes or monocytes. Heparin has an *in vitro* half-life, so it may be exhausted in samples held overnight at room temperature or warmer. EDTA and ACD both work by chelating divalent cations (calcium and magnesium), which are necessary for clotting. EDTA is often the preferred anticoagulant for immunophenotyping samples, whereas ACD is the major anticoagulant used in blood banking and leukapheresis. Neither loses anticoagulating activity with time. Exposure to both agents inhibits calcium and magnesium-dependent cellular processes, so they should be avoided when functional tests are to be performed.

6.7.2 RED BLOOD CELL DEPLETION

Ficoll 400 is a sucrose polymer that is perceived as hypotonic by mature granulocytes, but not lymphocytes, monocytes, or immature myeloid cells. Exposure to Ficoll causes granulocytes to dehydrate, which increases their density. Hypaque, a dense radiological contrast medium, is used to adjust the specific gravity of the Ficoll–Hypaque solution to 1.077. Dilute anticoagulated blood is layered carefully on a cushion of Ficoll–Hypaque requiring the same skills needed to make a clean "black and tan" by layering Guinness stout on a cushion of Bass ale. After centrifugation (of the sample, not the drink), the mononuclear cells and platelets are concentrated in a layer at the diluents' Ficoll–Hypaque interface, and the red cells, granulocytes, and dead or dying mononuclear cells form a pellet at the bottom of the tube. The buffy coat is harvested from the gradient interface and washed, removing platelets and residual Ficoll–Hypaque.

After the cells have been processed, cell viability is the major criterion for cell quality. This can be measured independently using Trypan blue dye and a hemocytometer, but it is often convenient to use an exclusion dye directly in the flow assay, provided the cells are not fixed. Appropriate dyes excluded by living cells include deoxyribonucleic acid (DNA) intercalating agents such as propidium iodide (PI), 7-amino actinomycin D (7-AAD), or 4,6-diamidino2phenylindole (DAPI). The first two are excitable with a blue (488 nm) laser, whereas DAPI requires an ultraviolet (UV) or near-UV violet laser. Because loss of membrane integrity is a late event in the process of cell death, a cell cannot be presumed to be healthy solely on the basis of dye exclusion. However, PI, 7-AAD, or DAPI positive cells have never been known to come back to life. Mario Roederer is fond of saying that you should not try to measure anything on a dead cell. Freshly isolated peripheral blood cells uniformly display excellent viability by dye exclusion, therefore, less than 95% viability may indicate sample mishandling. In contrast, cultured cell samples or single-cell suspensions isolated from disaggregated solid tissue normally contain a significant proportion of dead or dying cells and subcellular debris. These are bad actors in flow cytometric assays, often displaying bright autofluorescent streaks and binding antibodies nonspecifically. Assays on these types of samples often require a viability dye or some other means of identifying and removing these spurious events. If cells are to be fixed after staining, they can be gently permeabilized with saponin and incubated with an intercalating dye shortly before acquisition on the cytometer. Diploid cells with intact DNA will appear as a uniform bright peak, whereas subcellular debris and cells with fragmented DNA will have lower fluorescence (Figure 6.6).

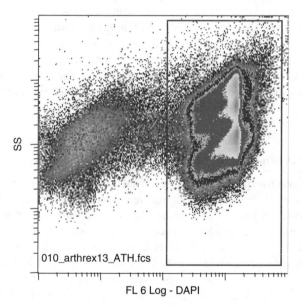

FIGURE 6.6   Eliminating debris with a nuclear stain. This plot represents 2 million Ficoll–Hypaque-separated bone marrow mononuclear cells that were stained, fixed, gently permeabilized, and stained with DAPI. Despite the Ficoll–Hypaque separation, which removes most red cells, mature granulocytes, and platelets, about 6% of the events have less than a complete complement of DNA, indicating that they are red cells, apoptotic cells, or subcellular fragments. Dead cells are often a source of interference in rare event analysis and should be routinely removed by gating.

Cell clumping is sometimes a problem in bone marrow aspirates or single-cell suspension obtained from solid tissue by digestion with collagenase. This clumping is often due to the release of DNA by short-lived granulocytes or other dead cells within the preparation. DNA clumps can be identified by their stringy glistening appearance. DNA slimeballs entrap viable as well as dead cells and should be broken up with DNAase as soon as they are detected.

Cells are often fixed after staining, both to stabilize them and render them non-infectious. Formaldehyde, either freshly prepared from paraformaldehyde, or commercially supplied as methanol-free electron microscopy grade is commonly used at final concentrations ranging from 0.25 to 2%. Most, but not all bound antibodies are unaffected by formaldehyde fixation.

6.8 ASSURING PROPER INSTRUMENT FUNCTION

Flow cytometers are complex pieces of equipment, but the types of failures that result in data corruption are easily identified. The major instrument vendors specify daily QC regimens that differ slightly from one another, but the principles apply to all instruments. The minimum recommended daily QC includes tests of instrument-related variability and validation of PMT settings (voltage and gain). The latter assumes that you already have arrived at settings for each individual fluorochrome (as well as light scatter), and now you want to make certain that a signal of a given intensity gives the same reading (mean fluorescence channel) today as it did yesterday. Although it is desirable to standardize on PMT target channels for a given instrument, the use of very bright or dim fluorochromes or very autofluorescent cells may necessitate assay-specific settings. Arriving at appropriate settings for a given test will be discussed in detail later.

In an improperly functioning instrument, additional variability can be introduced in a variety of ways. Debris in the flow cell can disrupt the laminar flow and hydrodynamic focusing, resulting in turbulence or causing the sample stream to go off center. A misaligned laser or dichroic mirror may decrease sensitivity and add variability. When the sample stream is hydrodynamically focused, the optics are well aligned and the electronics are functioning properly; reference beads manufactured to a very close tolerance limit will show little bead-to-bead variability in the cytometer, as manifested by a very sharp fluorescence intensity peak. A single-bead preparation with broad fluorescence emissions properties such as FlowCheck® (Beckman Coulter) can be used to test all fluorescence parameters and light scatter. Such beads give a single sharp peak in each fluorescence parameter. Typically, signals are acquired with linear amplification, and the mean and standard deviation (SD) of fluorescence intensity are acquired for each fluorescence parameter. The coefficient of variation (CV), computed by dividing the SD by the mean, is determined for each fluorescence parameter and compared to a target CV (typically ≤2%). When this test fails, check that the beads are gated in forward and SSc on singlets (beads can clump), clean the instrument with water or bleach followed by water, and try again. The persistence of broad peaks is an indication that service is required. The instrument may still appear to function, but the data will be compromised by the addition of unwanted noise.

Next comes the validation of PMT target channels. For a given cell type and combination of reagents, you will have previously determined target channels for light scatter and fluorescence parameters. Precisely how this is accomplished will be discussed in detail later, but for now assume that you have an assay with which you are satisfied, and you wish to be able to replicate it from day to day. Previously, during assay development you determined the appropriate gain for FSc, and voltages and gains for SSc and fluorescence parameters; you next ran a standard fluorescent bead (e.g., FlowSet®, Beckman Coulter) and determined its geometric mean fluorescence intensity (MFI) in each parameter. These bead intensities were recorded and now comprise the target channels to be hit each time this assay is run. You start with the exact settings (voltages and gains) used last time the assay was run, but adjust these settings for each fluorescence parameter, until the geometric MFI of the standard fluorescent bead is placed exactly in the target channel. The fact that very little adjustment is usually required does not minimize the critical importance of this element of QC. Here, it is important to stress two things: (1) it is the target channel and not the PMT voltage that must be held constant from day to day and (2) if experiments are to be run that require different PMT settings, each experimental setup will have its own set of target channels, which must be validated. In our laboratory, we assign each set of PMT target channels a nickname. When a data file is acquired using these settings, an abbreviation for this nickname is included in the file name. As discussed later, keeping PMT target settings straight is very important when offline compensation is to be used, since the single-stained compensation files must be acquired with the same settings as the experimental data.

The optimal standard bead for validation of PMT settings depends on the instrument configuration. In single-laser systems, a single bead with a broad fluorescence emission spectrum such as FlowSet® may suffice for all fluorescence parameters. In multilaser configurations, you may wish to use beads optimized for each light source. We find that a single-bead preparation (Spherotech Inc., 8-peak Rainbow Calibration Particles®) is sufficiently bright when excited by our violet (405 nm), blue (488 nm), and red (635 nm) lasers. We use the 7th or 8th peak to define our target channels, the lowest detectable peak to validate sensitivity, and all of the peaks to validate linearity.

Sensitivity is a measure of the dimmest signal that an instrument can distinguish from a negative (nonfluorescent) signal. It depends both on the measured difference in fluorescence intensity between dim and negative peaks, and their respective spreads (usually expressed as their CV). Sensitivity can be quantified as mean channel separation of the negative and dimmest peak, which is calculated by taking the difference of fluorescence intensities of the negative and dim peaks divided by a pooled SD of both populations. For a given set of PMT targets, an instrument will have characteristic sensitivities in each fluorescence parameter, which can be validated using multipeak beads (Figure 6.7).

Linearity refers to the relationship between the detected signal and the actual (known) fluorescence intensity over a particular dynamic range. Linearity is assessed using a cocktail of carefully calibrated beads of known fluorescence intensity. The measured fluorescence (geometric MFI) of each peak is plotted versus the reference values provided by the manufacturer and a linear regression analysis is performed (Figure 6.8). The coefficient of correlation (r^2), gives the proportion of

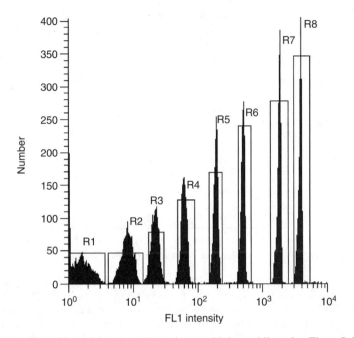

FIGURE 6.7   Use of 8-peak beads to determine sensitivity and linearity. These Spherotech rainbow calibration particles are a mixture of negative beads (R1) and seven bead populations (R2-R7), each containing a known amount of a broadly fluorescent dye (calibrated in equivalents of a given fluorochrome). In the present example, the dimmest peak is easily distinguished from the negative peak, and the brightest peak is on scale.

variation in the measured parameter that is explained by variation in the known values (as reviewed in Chapter 2). A perfect correlation ($r^2 = 1$) means that all of the observed and expected pairs lie along a straight line. A slope of unity means that the measured fluorescence increases in direct proportion to the expected fluorescence. The degree and range of linearity depends chiefly on the instrument design and changes little from day to day. Although it is not necessary to check sensitivity and linearity daily, the data are available if multipeak beads are used for daily validation of PMT settings.

6.9   REAGENT AND PROCESS QUALITY CONTROL

Laboratories that are accredited by the College of American Pathologists (CAP) must satisfy a checklist that includes many aspects of reagent and process quality assessment and control [53]. According to CAP, "The laboratory has the responsibility for ensuring that all reagents, calibrators, and controls, whether purchased or prepared by the laboratory, are appropriately reactive." Further, "verification of reagent performance is required and must be documented." Any of the several methods for reagent validation may be appropriate, such as direct analysis with reference materials, parallel testing of old versus new reagents (known as overlap testing), and comparison with routine controls. When individually packaged reagents or kits are

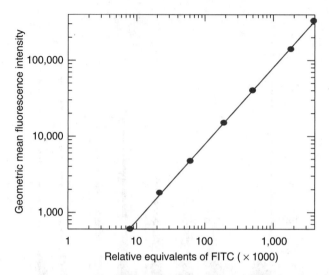

FIGURE 6.8   The use of multipeak beads to determine linearity. The data shown here are the geometric mean fluorescent intensities of the peaks shown in Figure 6.7 (*y*-axis), plotted versus the known FITC equivalents per particle. Note that the unstained peak is not used because 0 is undefined on a log scale. The line shown is the least squares line of best fit of the log-transformed values. The coefficient of correlation ($r^2 = 1.000$) indicates that virtually all variations in *y*-axis measurements (measured fluorescence) are explained by variations in *x*-axis measurements (fluorescent molecules per particle). The slope of the line is also very close to unity (0.989, standard error = 0.007), indicating a one-to-one relationship between measured and actual fluorescence intensity. We were very pleased to learn that our new cytometer was so linear.

used, criteria must be established to monitor reagent quality and stability. Processing of periodic *wet controls* to validate reagent quality and operator technique is also a typical component of such a system. Records must be kept of each reagent used for each test, its lot number, and the expiration date. Clinical laboratories in the United States that manufacture cellular products for therapy are required by the Food and Drug Administration to follow current Good Tissue Practices (cGTP, as specified by 21 CFR 1270), as well as elements of current Good Manufacturing Practices (cGMP). According to cGMP, newly acquired reagents must be stored separately (quarantined) from reagents currently in use until they have been overlap tested.

    In research laboratories, where reagent management is more relaxed and conservation of expensive materials is a priority, it is very important to understand which reagents are stable and which require special care. Monoclonal antibodies themselves have exceptional long-term stability when stored in a sterile environment at 4°C in the presence of a carrier protein such as bovine serum albumin. This is how they are usually supplied, and may also include sodium azide to inhibit microbiological overgrowth. Antibodies conjugated directly with single dyes such as FITC, PE, and APC can also be used for years beyond their nominal expiration dates, providing that they are well protected from exposure to light. In contrast, the tandem dyes, particularly PE-Cy7 and APC-Cy7, pose a QC challenge to clinical and research laboratories alike. Tandem dyes rely on energy transfer from the donor dye (e.g., PE) to the acceptor dye (e.g., Cy7). When all are working properly, virtually all of the energy emitted by the donor is captured by the acceptor, and there is little or no emission at the wavelength of the donor. However, photobleaching of the acceptor results not only in a decreased intensity at the wavelength of the acceptor, but also in significant emission at the wavelength of the donor. A vial of PE-Cy7-conjugated antibody, if left unprotected in ambient light, will degrade in this manner in a matter of hours. Quality assessment of tandem dye–conjugated antibodies can be incorporated into the daily spectral compensation routine (see Section 6.11).

In our laboratory, we use directly labeled beads (Calibrite Beads, Becton Dickinson) as compensation standards for antibodies conjugated with single dyes (e.g., FITC, PE, and APC). For these dyes, the emission spectrum is a matter of physics, and the emission of a bright FITC-conjugated plastic bead can be used as a standard for any FITC-conjugated antibody. Because the emission spectra of tandem dyes is not identical between antibody preparations, and can change with time for a given conjugated antibody, we run separate single stained controls for each tandem dye–conjugated antibody used on a given day. When the epitope detected by the antibody is both prevalent and bright, readily available cells, such as peripheral blood mononuclear cells from a healthy donor are often used for single-stained controls. We find that the routine use of antimouse Ig capture beads (BD CompBead®, Becton Dickinson) provides a far more reliable compensation standard. These beads are stained with tandem dye–conjugated antibody as one would stain cells, but binding occurs because the antimurine Ig antibodies on the bead's surface recognize a constant region on the murine monoclonal antibody. The result is that virtually any murine antibody (even those that recognize epitopes expressed at low density on rare cell populations) stains the beads with a sharp and bright peak.

The health of a tandem dye can be gauged by examining the fluorescence intensity at the acceptor wavelength, or the amount of compensation needed to correct the *spillover* from the acceptor to the donor wavelength. For example, capture beads single-stained with a bad PE-Cy7 conjugate will have significant spillover into the PE channel. It is important to remember when working with tandem dyes that the stained sample must also be protected from ambient light (e.g., with aluminum foil) and acquired as soon as possible after staining.

The advent of commercial cellular control reagents consisting of gently fixed or lyophilized human cells with published reference values has greatly facilitated process assessment for flow cytometry. These control samples are stained in parallel with test samples. Because acceptable ranges (percent positive and absolute number) are published by the manufacturer, the assay provides an overall metric for quality assessment of the instrument, reagents, staining, and data analysis procedures. An example using CD-Chex Plus® (Streck Laboratories) as a control for a T-cell absolute count assay is shown in Figure 6.9.

Summarizing the total quality assessment process:

- Determine the CV of a homogeneous test bead in each channel. Make certain that the target CV is not exceeded
- Adjust PMT voltage and gain such that standard beads are placed in their target channel

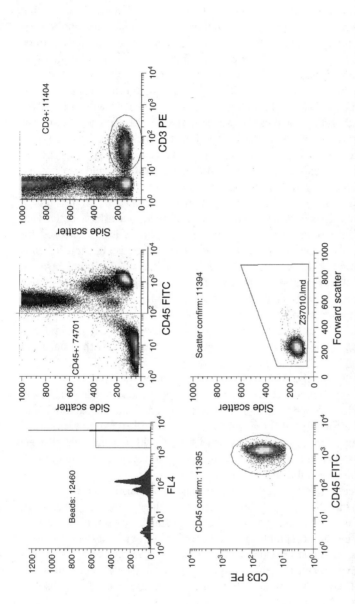

FIGURE 6.9 CD Chex Plus reference standards for the single-platform determination of CD3 absolute count. CD Chex Plus reference cells were stained with anti-CD45 and anti-CD3 monoclonal antibodies and processed according to a *lyse, no wash* protocol. Calibration beads (StemCount, Beckman Coulter) were detected as a sharp bright peak (*top left*) and the number of events within the peak was determined (12,460). Beads were removed from subsequent plots with a *not* gate. Next CD45+ cells (leukocytes) were identified (*top center*). CD3+ cells with low SSc were identified within the CD45+ population (*top right*) and this population was cleaned up on the basis of CD45 expression (*bottom left*) and light scatter (*bottom right*). The absolute leukocyte count was calculated as the number of CD45+ events (74,701) divided by the number of bead events (12,460), multiplied by the known concentration of the beads (1,022 beads/μL). The T-cell count was calculated similarly. In this example, the measured leukocyte count was 6,127 cells/μL and the measured T-cell count was 935 cells/μL. These values were within the published reference ranges (6,100 ± 500 for leukocytes and 1,179 ± 264 for T cells).

- Determine linearity and sensitivity of the instrument in each channel using multipeak beads
- Stain Ig capture beads with all tandem dye–conjugated reagents to be used for the day
- Determine and evaluate PE spillover for all antibodies conjugated to tandem dyes by staining Ig capture beads
- Create compensation matrices using Ig capture beads for all tandem dye antibodies and integrally stained beads in place of all single-dye conjugated antibodies
- Stain a verify tube using control cells (e.g., CD-Chex Plus®) (acquire data and compare observed and expected values)
- Graph and monitor PMT voltages required to meet target channels as a function of time (Levey-Jennings plots, see Section 6.14)

6.10   CHOOSING PHOTOMULTIPLIER TUBE GAIN SETTINGS

In the early days of flow cytometry, choosing proper PMT settings was a simple task. Unstained cells, usually human peripheral blood mononuclear cells, were run and the PMT voltage and gain settings were adjusted until the peak of this negative sample was placed within the lower portion of the first decade of log fluorescence intensity plots. Care was taken not to let more than half of the cells fall *off axis* (i.e., in the first channel). This process was repeated for each fluorescence parameter. Two things have changed that make this process a bit more challenging. First, the use of multiple laser lines and far red and infrared emitting dyes means that the intrinsic fluorescence (autofluorescence) of unstained cells is not equivalent across the fluorescence parameters. Second, in the days of entirely analog signal processing, a good part of the signal detected in the first decade was due to instrument noise, and thus independent of sample. Modern cytometers are still imperfect at the lower end of the first decade, but improved signal processing requires one to define where in first decade artifact ends and the signal begins. For newer instrumentation with digital signal processing, Holden Maecker of BD Bioscience recommends plotting the CV of an unstained sample as a function of PMT voltage, gradually moving the peak across the first decade. As the voltage is increased, the CV decreases sharply initially and then levels off (Figure 6.10). This is due both to raising a greater proportion of events off axis (increasing the mean and therefore lowering the CV) and to an actual sharpening of the peak as the contribution of instrument noise diminishes. After the voltage setting at which the CV levels off has been determined, standard beads are run at that setting to establish the target bead channel (channel in which the bead peak falls). This target channel must be reproduced during daily QC. Beyond this, there are two more factors to be considered in selecting PMT target channels: (1) The brightest positive signals to be encountered in stained samples must be on scale (data in the last channel cannot be compensated for spectral overlap) and (2) The PMT settings for the different fluorescence parameters should be sufficiently balanced to permit compensation for spectral overlap. Why this is important will be explained in Section 6.11.

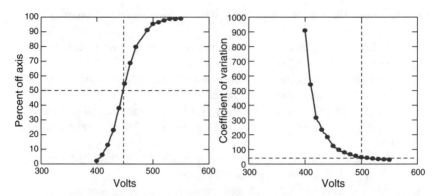

FIGURE 6.10 Determination of the optimal PMT voltage on unstained peripheral blood lymphocytes. Unstained peripheral blood cells were acquired within a lymphocyte light scatter gate. For each fluorescence parameter, the PMT voltage was gradually increased and the percent of events falling on scale and the CV of the peak were determined at each voltage setting. The results were plotted, and negative exponentially weighted curve smoothes were applied to the points. The example shown here is for the FL1 PMT, which was set up to measure green fluorescence. At the lowest voltage tested, most of the cells were off scale (i.e., piled up in the first channel) and the CV was very high. At 450 V, about half of the cells were on scale (left panel, dashed lines), but the CV was still relatively high. This would be the lowest recommended setting, and would be used if very bright events were to be acquired in FL1. The CV continued to decrease with increasing voltage and was close to its lower asymptote at 500 V (right panel, dashed lines). This would be the preferred setting for the majority of cases. It is important to note that this setting is optimized for peripheral blood lymphocytes, which have relatively little autofluorescence. Different settings may be required for cultured cells or disaggregated solid tissues.

6.11   COLOR COMPENSATION MADE SIMPLE

Spectral or color compensation has been treated extensively elsewhere [54–57]. Yet, it remains the most common source of error and vexation for cytometrists everywhere. This section is meant to be a practical guide to understanding and avoiding the pitfalls of color compensation, rather than a complete exploration of the physical principles and mathematics. Color compensation can be thought of as correcting for the spillover of a given fluorochrome from the fluorescence channel in which it is intended to be measured to all of the other fluorescence channels. The spillover coefficient from channel A into channel B ($SC_{A \to B}$) (e.g., from the FITC channel into the PE channel) is measured by running a single-stained sample meant to be detected in channel A and dividing the geometric MFI of the peak detected in channel B (FI_B) by that of channel A (MFI_A) (Figure 6.11):

$$(SC_{A \to B}) = \frac{MFI_B}{MFI_A}$$

This spillover coefficient, $SC_{A \to B}$, gives the proportion of the signal detected in channel A that will be spuriously detected in channel B. It is dependent not only on the emission spectrum of the particular fluorochrome, but also on the relative gains

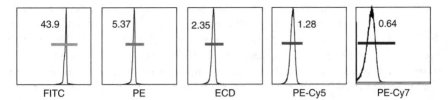

FIGURE 6.11 Determining spillover coefficients. FITC-labeled Calibrite beads were acquired and the geometric MFI was determined for each fluorescence parameter. The x-axes are labeled corresponding to the fluorochromes to be measured in the experimental setup. The spillover of FITC into the PE channel can be expressed as the MFI of FITC beads in the PE channel divided by the MFI of FITC beads in the FITC channel. In this example, it is 12.2%. The greater the difference in emission wavelength between FITC and the other fluorochromes, the lower the spillover. This is not the case for tandem dyes, which often have significant spillover at the emission wavelength of the donor dye.

of the two PMTs. Spillover coefficients can also be conceived of and calculated as slopes, since $SC_{A \to B}$ is the change in MFI_B as a function of MFI_A.

In a two-color instrument, where spillover and compensation coefficients are almost identical, the fluorescence intensity of an event detected in channel B can be corrected for spectral overlap (FI_{BComp}) by subtracting the fluorescence intensity detected in channel A multiplied by the spillover coefficient of A into B:

$$FI_{BComp} = FI_B - (FI_A \times SC_{A \to B})$$

The two-way spillover coefficients (A into B and B into A) are minimized, and thus compensation is minimized, when both PMTs are balanced. PMT balance cannot be achieved by applying equal voltages to the two PMTs, because PMT sensitivity is not standardized and often very different voltages are required to achieve equal results from different PMTs. Further, perfectly balanced PMTs are not always desirable. If you wish to detect a dim signal in channel B, the gain of PMT_B can be increased relative to that of A. This will result in greater spillover of A into B (and hence greater compensation) and less spillover of B into A (less compensation). This may seem counterintuitive, but consider how we defined the spillover coefficient. Similarly, it may be necessary to lower the sensitivity of a PMT if it is used to detect an exceptionally bright fluorochrome. It is for these reasons that PMT target channels cannot be chosen solely by evaluation of unstained controls, and it is often necessary to have different PMT target channel sets for different applications.

One more item remains to be addressed before we leave the theoretical, and that is the difference between spillover and compensation coefficients. As we mentioned earlier, they are very close, but not identical in a two-color problem. As the number of fluorescence parameters increases they further diverge. For example, when three fluorochromes are used, the total signal detected in channel A results from the true signal from fluorochrome A, plus the spillover of fluorochrome B into A, plus the spillover of C into A. We wish to correct (compensate), such that the effects of spillover are eliminated and only the true signal from fluorochrome A is measured. But the proportion of the signal in B to be subtracted from A is not correctly described by the spillover coefficient of B into A, because some of the signal detected in channel B is due to spillover of A into B and C into B. Fortunately, this type of a problem is addressed by the field of linear algebra and the true compensation coefficients can be described by a series of simultaneous equations and solved with matrix algebra. All modern automated compensation software use this algorithm, sometimes referred to as the *matrix inversion* method [58].

Now that we have seen why color compensation is necessary, we can examine practical ways in which it can be implemented and assessed in simple and complex applications. The CD34 test used in our clinical laboratory (StemKit®, Beckman Coulter) is nominally a four-color assay, but one fluorescence channel is used to identify viable cells by 7-AAD exclusion, and one channel is used to detect StemCount® beads as a reference for absolute cell counting. Thus, we have a two-color compensation problem (anti-CD45 FITC versus anti-CD34 PE). Because these are single dyes for which the emission spectra are constant, the amount of color compensation required should never change, provided the PMT gains are consistent from day to day (this is ensured by validating the PMT target channels). However, accrediting bodies such as the CAP require color compensation to be validated on a regular basis. For the CD34 assay, we validate our compensation settings weekly using unstained, FITC and PE Calibrite® beads. We use the same fluorescence PMT settings to be used in our assay (but different forward and side scatter, adjusted for bead detection). These settings are adjusted daily against FlowSet® beads. In the clinical laboratory, we are still using a very *mature* software package (System II®, Beckman Coulter) that includes a semiautomated compensation algorithm, which mimics old-fashioned manual compensation by matching the fluorescence intensity of the negative population with that of the stained beads (e.g., compensation is increased until the intensity of the FITC standard in the PE channel matches that of the unstained beads). We have not found it necessary to change the compensation values over a period of years, providing an additional independent validation of the PMT settings. Real-time data analysis is performed during data acquisition using the validated compensation values; compensated data are saved to list-mode files containing the original analysis protocol, in case reanalysis is necessary.

For complex multicolor analyses, real-time color compensation and data analysis may not always be feasible or desirable. Further, the classical technique of matching median fluorescence intensities described earlier is not the most efficient or the most accurate. All modern flow cytometry analysis software packages support offline compensation. Compensation is computed from measurements of spillover or slopes of single-stained samples, using the sweet range of the cytometer rather than the troublesome first decade. The advisability of performing a fully automated compensation on multicolor data cannot be over emphasized.

As detailed previously, PMT setting validation, linearity, and sensitivity testing are done daily, and single-stained beads (directly labeled or Ig capture) are acquired for each run. It is critical that the single-stained control beads be acquired at the same fluorescence PMT settings as the actual samples, and for tandem dyes, that the single-stained controls be performed with the actual labeled antibodies used to stain the sample.

Compensation matrices derived from mix-and-match single-stained bead list-mode files are computed and applied during data analysis. When tandem dyes are used, it is not unusual to create four separate compensation matrices for a panel consisting of four tubes. Data-presentation aids that allow on-axis events to be visualized (e.g., Baseline Offset from Beckman Coulter and Log Bias from Verity Software House) must never be used during compensation, because even grossly overcompensated events appear to be perfectly compensated.

With the exception of DNA, ploidy measurements, and a few other specific applications, most fluorescence data are conventionally displayed on a logarithmic scale. Compared to a linear scale, a log scale emphasizes (spreads out) lower values and compresses higher ones. It also normalizes the distribution of fluorescence intensity within a given cell population, and is a very useful way of visualizing data that is distributed over a wide dynamic range. Zero and negative numbers are undefined on a log scale, and this often creates a problem causing data to *pile up* on the axes. A new family of data display scales (biexponential, hyperbolic sine function, HyperLog [59]) offers the best of linear and log displays. These scales are approximately linear on the low end of the scale and become logarithmic at higher values. Like a linear scale, 0 and negative values are defined. This is very useful for evaluating compensation, since the mean FI_{BComp} for a sample single-stained with A should be 0. Because measurements have variance (spread), a symmetrical distribution of the data around the mean of 0 means that there will be both positive and negative values. This is easily visualized in a HyperLog plot (Figure 6.12), but missed entirely in a conventional log display where 0 and negative values are undefined, and events with fluorescence less than or equal to the lowest value defined in the scale are crowded together on the axis. It will take time before this type of display is universally accepted and it is not without problems (determining how much graphic space to devote to negative values

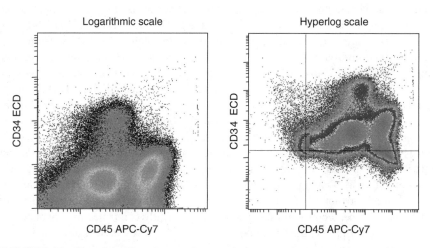

FIGURE 6.12 Use of the HyperLog scale to evaluate spectral compensation. In this example, the same data are displayed on a conventional logarithmic scale (left panel) and on a HyperLog scale (right panel). The origin of the quadrants on the right panel indicates fluorescence intensities of zero. Zero and negative values are undefined on the logarithmic scale, and all such values pile up on the axes. When data are properly compensated, negative populations are symmetrically distributed about the origin.

is difficult because the range of negative values can be quite variable). However, it solves far more problems than it creates, and is likely to become the standard.

One more word on compensation standards: When using tandem dyes we strongly recommend against using cells stained with a bright antibody (e.g., CD45 PE-Cy7, CD8 APC-Cy7) as a compensation standard for other antibodies conjugated with the same tandem dye. Although this is recommended by some cytometer manufacturers, it makes the assumption that both antibodies require the same amount of compensation. This is asking for trouble.

6.12   QUALITY CONTROL FOR ABSOLUTE COUNTS

The single-platform absolute count method using the *lyse no wash* method is widely used for determining absolute CD34 or T-cell counts (i.e., quantifying the concentration in cells/microliters). The principle is simple. According to one popular method, a known and carefully measured volume of sample is deposited in a tube and stained by the addition of directly labeled monoclonal antibodies. After staining, the red blood cells are lysed by the addition of a large excess of ammonium chloride–based lysing solution. Calibration beads of known concentration, of volume equal to the sample volume and measured with equal care, is then added. After gentle mixing, a viability dye is added and the sample is acquired on the flow cytometer. Because the sample was never washed, there is no opportunity for cell loss. It is of little consequence that the unbound antibody is still present in the sample; pulses caused when labeled cells are interrogated by the laser are measured relative to the background fluorescence of the sample stream (baseline) and thus are seen above the noise caused by unbound fluorochrome. During analysis, the internal standard beads are identified by their unique scatter and fluorescence profile. The number of bead events is counted and compared to the number of events of interest (e.g., CD45+ CD34+ cells). The absolute count is determined as the number of events of interest divided by the number of beads and multiplied by the known bead concentration. Given proper instrument performance and accurate gating of the beads and the events of interest, there are only two key steps in the process: The beads must be handled properly (brought to temperature and gently mixed) to assure that the bead concentration added is the same as that stated by the manufacturer (usually about 1×10^6/mL). The same pipetting device should be used for beads and sample, because the precision of volume measurement is critical (see Chapter 2, Statistics, for a discussion of precision versus accuracy). A positive displacement pipette is best for this purpose. Note that the volumes of antibody and lysing solution do not enter into the absolute count equation and are therefore not critical. Beyond relying on the bead concentration stated by the manufacturer, bead concentration can be independently confirmed on a hematology counter. An example of the single-platform method is given in Section 6.16.

6.13   QUALITY CONTROL FOR QUANTITATIVE FLUORESCENCE

Flow cytometers are so linear and perform well over such a large dynamic range that it is natural that they should be used for making quantitative measurements of proteins and other molecules detected on single cells. The problem has always been one of calibration. How many molecules do you have on a cell when they have a particular MFI? The fluorescence of CD4+ T-cells can be used as a *poor man's* quantitative standard, providing that (1) normal peripheral blood mononuclear cells, which are known to express on average 50,000 CD4 molecules per T-cell [60], are used as the standard; (2) antibody conjugation is well controlled, such that antibodies being compared have identical fluorescence to protein ratios (PE is easiest, since a well-conjugated antibody has one PE molecule per antibody molecule); and (3) the antibodies are used at epitope saturating concentrations. The number of molecules of interest can then be calculated by taking the ratio of the geometric MFI of the molecule of interest divided by that of the CD4+ population and multiplying by 50,000 [61]. A more elegant and better-accepted method involves the use of a series of multipeak beads calibrated in terms of molecules of equivalent soluble fluorescence (MESF) units (Quantum Beads®, Bangs Laboratories). The beads are run daily and a linear regression is performed on geometric MFI of the beads versus known MESF. The MESF of the experimental sample is then determined by plugging the MFI of the unknown into the regression equation. The measurement of ZAP 70 expression in chronic lymphocytic leukemia provides a clinical example of the utility of quantitative fluorescence as a prognostic indicator. An excellent paper by Chen et al. [62] compares several methodologies, including quantitative fluorescence determined as molecules of equivalent soluble fluorochrome, in the context of ZAP 70 measurement.

6.14   USE OF LEVEY–JENNINGS PLOTS

Levey–Jennings plots provide a handy way to examine QC data for trends over time and a suitable means for the periodic review of QC data (Figure 6.13). Levey–Jennings plots can be used to show results (percent positive of a process control standard) or settings (such as the voltage required to hit a PMT target channel) as a function of time. Rather than show raw results on the *y*-axis, Levey–Jennings plots transform the data by subtracting the mean of a series of historical observations, and then scaling the results in terms of SDs (calculated from the same historical values). This transformation provides an enormous advantage over merely plotting the raw data. Assuming that normally occurring random fluctuations in performance have a Gaussian distribution,

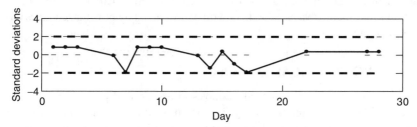

FIGURE 6.13   Levey–Jennings plot of changes in FL1 voltage with time. The *y*-axis is scaled in SD units. Dashed lines show a range 2 SD below and above the mean. Over the plotted time period, the average voltage required to meet the target channel, as defined with FlowSet beads, was 760 V ± 2.2 (SD). Here, the SD (2.2 V) is so small that we are not concerned that the voltage dropped almost 2 SD below the mean on days 7 and 17.

the Levey–Jennings scale facilitates a probabilistic interpretation of daily QC results. All positive data are above the mean historical value, and all negative data are below it. A value one SD above or below the mean has a 32% probability of occurring as a result of random fluctuation. Results 2 and 3 SDs beyond the mean would be expected only 4.6 and 0.3% of the time, respectively. When it is improbable that a fluctuation of a given magnitude has occurred by chance, it implies that something has changed that has caused your results to fall outside your historical norms. Even for samples falling well within the expected range, the probability that two consecutive values fall above the mean (0 on the Levey–Jennings plot) is high (1 in 2), but the probability of six consecutive values lying above the mean is only 1 in 32, and therefore indicative of drift.

6.15 DON'T QC LIKE A MONKEY

We know of laboratories in which QC is performed religiously but seldom if ever acted on, and laboratories in which thorough QC is performed for a standard setting, but not the predominant settings currently used by the laboratory. Regrettably, such missteps have even cropped up from time to time in laboratories under the author's direction. The key is understanding deeply that QC is not *just* a thankless chore (which it undoubtedly is); but, additionally, it is the only thing that separates the reproducible from the irreproducible. How to avoid QC'ing like a monkey? Look at the results with a practiced eye:

- Are the CVs of your standard beads getting wider?
- Has there been a sudden increase or an increasing trend in the PMT voltages needed to meet your target channels (the Levey–Jennings plot is a great way to spot this)?
- Has the dimmest peak that you can resolve with multipeak beads changed?
- Is the PE or APC spillover of individual tandem dye conjugated antibodies increasing?
- Do you have to make unexplained adjustments in scatter or fluorescence analytical regions?

Such changes may indicate an instrumentation problem, such as a failing component, misalignment, a vacuum leak, or a dirty flow cell, any of which could compromise data integrity.

6.16 TWO PRACTICAL EXAMPLES

This section provides two practical examples of assays used in the clinic that came directly out of the research laboratory. The first provides a means of measuring absolute T-cell counts in samples where the frequency of T cells is low. We use this test in the hematopoietic stem cell laboratory to quantify T cells in T-depleted stem cell graft products. It provides an example of a bead-calibrated single-platform assay, and provides an opportunity to review the principles of rare event detection. The second example examines the detection of cutaneous T-cell lymphoma in the blood by T-cell receptor (TCR) v-beta analysis. It provides an example of the classifier/outcome gating strategy recommended for multiparameter flow cytometry problems, as well as a clever use of fluorochromes to measure six specificities using a five-color instrument.

6.16.1 SINGLE-PLATFORM DETECTION OF ABSOLUTE T-CELL COUNTS AS RARE EVENTS

Beckman Coulter pioneered the clinical use of the single-platform *lyse no wash* assay when it introduced the StemKit, a test for quantifying CD34+ cells. StemKit is labeled for IVD use in the United States and is also CE-marked (the equivalent in the European Union). Single platform means that absolute counts (cell/µL) can be obtained using only a conventional flow cytometer (in contrast, a dual platform assay would relate CD34 percent in a sample to the white blood cell [WBC] determined on a hematology analyzer to determine CD34+ cells/µL). *Lyse no wash* means that a whole blood sample is stained and erythrocytes are lysed without a wash step. This eliminates the inevitable loss of cells when centrifuging and decanting. We have used the principles of the StemKit assay to develop an in-house test to quantify T cells in samples in which they are very rare, for example, in the blood of patients immediately after myeloablative or immunoablative therapy, or in T-cell depleted graft products. Such *home brew* assays are classified as *analyte specific reagent* (ASR) tests by the Food and Drug Administration (FDA), and can be reported for clinical use providing that the laboratory has proper Clinical Laboratory Improvement Amendments (CLIA) certification, and the report is accompanied by an ASR disclosure. In our CD3 rare event assay, the number of replicates to be run depends on the WBC, with six replicates run when the WBC count is 250 cells/µL or less. In this case, heparinized blood (200 µL) is added to seven tubes very precisely with a positive displacement pipettor. Six tubes are stained with anti-CD45-FITC/anti-CD3-PE and one is stained with anti-CD45-FITC/isotype-PE. After staining, cells are lysed by the addition of 2 mL of ammonium chloride lysing solution. 100 µL of StemCount Fluorospheres® (at a known concentration stated by the manufacturer) are added with the same pipettor used for the sample. After the addition of the viability dye, 7AAD, the cells are held on ice and exhaustively acquired on the flow cytometer. The six replicate stained samples are acquired first. Because the entire volume of each of the 6 samples is acquired, this is the equivalent of counting every cell in 1.2 mL of whole blood. A blank tube is acquired next to minimize sample carryover, followed by the isotype control tube. Figure 6.14 shows typical results for a patient sample assayed 3 days after receiving an immunoablative regimen consisting of cyclophosphamide, fludarabine, and antithymocyte globulin. Figure 6.15 shows data from the same patient immediately after having completed the therapy. The data are shown in reverse chronological order to explain the gating strategy better. Only one of six replicate determinations is shown, but the replicates were used to calculate confidence intervals about the determinations. The patient's T-cell count nadired at 0.84 ± 0.11 CD3/µL at day 0 (mean $\pm$ SEM), and had recovered to 7.83 ± 1.40 CD3/µL 3 days later.

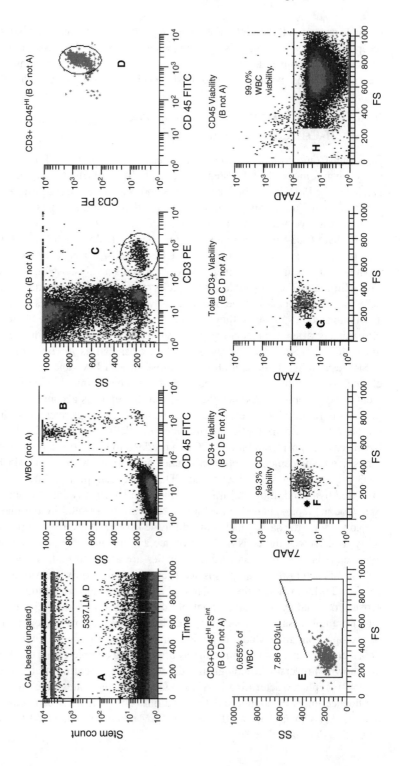

FIGURE 6.14   Detection of CD3+ T cells as rare events in the peripheral blood of a patient three days after completing immunoablative therapy. To achieve maximum sensitivity, whole heparinized blood (200 μL) was stained with anti-CD45-FITC and anti-CD3-PE in six-plicate and cells were acquired to exhaustion. StemCount beads are identified by their high fluorescence in FL4 versus time (gate A). These are removed from subsequent analyses using a *not gate*. Collecting time as a parameter facilitates identification of fluidic and other instrument problems during sample acquisition. WBCs are identified by CD45 expression (gate B). Note the extraordinary amount of debris. The WBC gate, compounded with the not bead gate is passed to a histogram of anti-CD3 versus SSc. CD3+ events with low SSc are identified (gate C). The compound gate of WBC, not bead, and CD3+ is passed to a histogram of CD45 versus SSc, where T cells are distinguished from noise by their bright expression of CD45 (gate D). No events fitting these criteria were seen in the isotype control (not shown). The compound gate of WBC, not bead, CD3+, and bright CD45 expression is passed to a histogram of forward scatter versus SSc (gate E), which is used to eliminate events with low forward scatter. The number of CD3+ cells obtained in gate E, taken as a percent of WBCs identified in gate B, represents the percent of CD3+ cells. For determination of the viability of CD3+ cells, the compound gate of WBC, not bead, CD3+, bright CD45 expression, and intermediate to high forward scatter is passed to a histogram of forward scatter versus intracellular 7AAD concentration (FL3). Because 7AAD is excluded by viable cells, viable CD3+ cells are detected in region F. Even in samples with low overall viability, the proportion of dead cells within the F region is invariably low. This is because the majority of dead and dying cells have low forward scatter and are eliminated by gate E. It is for this reason that we measure and report CD3 viability using a compound gate of WBC, not bead, CD3+, and bright CD45 expression, where viable CD3+ cells are detected in region G and reported as a percent of cells in gate D. Similarly, viable WBCs are detected in region H and reported as a percent of cells in gate B. Absolute WBC and CD3 counts are obtained by dividing the events in gates B and E, respectively, by 2 (the ratio of sample volume to bead volume); dividing by the number of beads in gate A; and then multiplying by the known StemCount bead concentration (1×10^6/mL). In this sample, CD3+ cells represented 0.655% of CD45+ events. The viability of total WBCs (CD45+) and CD3+ cells exceeded 99%.

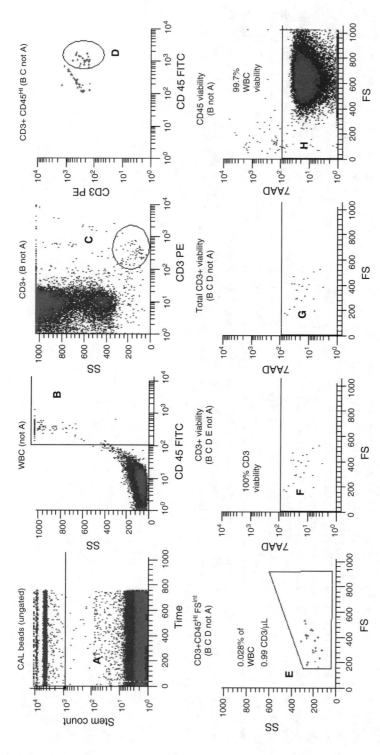

FIGURE 6.15 Detection of CD3+ T cells as rare events in the peripheral blood of a patient on day 0 after completing immunoablative therapy. The same analysis shown in the preceding figure was performed on blood collected immediately after completing immunoablative therapy. Even after acquiring the entire sample, few WBCs were seen in gate B), and the CD3 population did not form a discrete population versus SSc (gate C). The reason for this is apparent in gate D, where T cells with bright CD45 expression are separated from a diagonal of "noise." Following this gating procedure, identified T cells have credible light scatter (gate E) and high viability. The peripheral CD3 count was less than 1 cell/µL. For gates A, G, and H are applied as described in Figure 6.14.

6.16.2   DETECTION OF CIRCULATING TUMOR IN CUTANEOUS T-CELL
LYMPHOMA BY THE ANALYSIS OF T-CELL RECEPTOR V-BETA USAGE

Cutaneous T-cell lymphoma is a general term that covers several types of T-cell lymphoma of the skin. In a variant known as Sézary syndrome, the presenting features often include widespread redness of the skin (erythroderma) with severe itching. Lymph nodes are enlarged and the malignant T cells found in the skin are also found in the circulation. Because the malignant cells are clonal, they have identical TCR specificities, and therefore, identical v-beta usage. If the v-beta specificity of the skin lesion is known, a circulating tumor can be identified by the excess proportion that they comprise in the peripheral blood. T-cell subsets, roughly corresponding to naïve, memory, effector memory, and late effector cells can be defined by their expression of CD45 isoform [63,64] and CD27 [65–67], CD28 [68,69], or CD62L [70,71]. If the malignant clone is uniquely or predominantly confined to a particular T-cell subset, then examining v-beta usage by the subset provides greater sensitivity to detect the circulating tumor.

In the present example, we examine peripheral blood from a Sézary patient. After separation on a Ficoll–Hypaque gradient, the mononuclear cells were stained in bulk with a cocktail of anti-CD4-PECy7, anti-CD45RA-ECD, and anti-CD27-PC5. The stained cells were aliquoted into 10 microtiter wells (~1 million cells/well), centrifuged, and the supernatant decanted. Eight wells were used to measure 24 v-beta specificities, one well for TCR alpha–beta/gamma–delta and one for the FITC and PE isotype controls. The Beta Mark anti-TCR v-beta staining kit, commercialized by Beckman Coulter, was used for v-beta detection. The v-beta reagents occupied the FITC and PE channels, but covered three v-beta specificities per well. This is possible because only one v-beta specificity is expressed on any given T cell. Thus for a mixture of antibodies to the v-beta families X, Y and Z, X was FITC conjugated, Y was PE conjugated, and Z was a mixture of FITC and PE conjugated antibodies.

We used a *classifier/outcome* gating strategy to examine TCR v-beta usage in T-cell subsets. Primary classifiers are sequential, whereas secondary classifiers branch. In this case, the *primary classifiers* used to identify CD4+ T cells and clean them up were: forward versus log SSc, CD4 versus log SSc, and forward versus log SSc again (Figure 6.16). The secondary classifiers, CD45RA versus CD27 were used to define four subsets within CD4+ cells. The outcomes, namely v-beta specificities, were measured on total CD4+ lymphocytes, and each CD4 subset.

Figure 6.16 details the analysis of 1 of the 10 wells in the panel (v-betas 13.1, 13.6, and 8). The v-beta specificities are given as a percentage of total CD4+ lymphocytes, and of each CD4 subset. The values would be meaningless without reference to expected values derived from a population of healthy control subjects (Figure 6.17). The reason that these ranges are computed and displayed in a logarithmic scale is critical to their interpretation and is explained in detail in Chapter 2, Statistics. From this figure, it is clear that our patient has a marked expansion of v-beta 13.1 CD4+ T cells (12.5% versus the 95th percentile reference range of 2.2–6.6%). Further, the expansion is confined to the CD45RA+ CD27 negative *effector* population, in which they represent 68.6% of all gated cells. This is an important finding, since this patient's

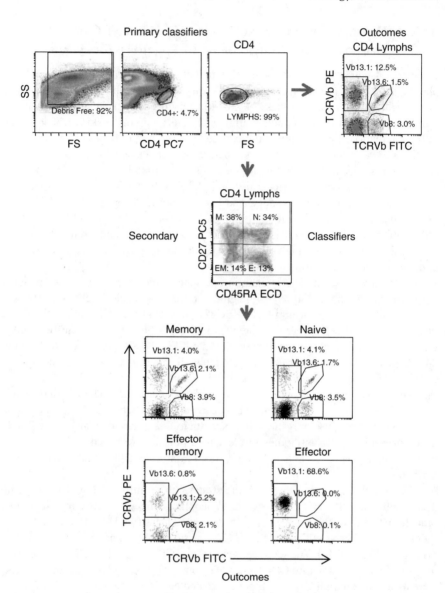

FIGURE 6.16 Use of the classifier/outcome strategy to determine TCR v-beta distribution among subsets of CD4+ lymphocytes. The primary classifiers (light scatter, CD4) are used sequentially to identify a population of CD4+ lymphocytes free of dead cells, debris, and monocytes. The secondary classifiers, CD45RA and CD27 are used to define the CD4 subsets. The outcome parameters (v-beta families) are measured on all the populations defined by the classifiers (total CD4, CD4 subsets). This particular subject, a patient with cutaneous T-cell lymphoma, has a marked expansion in v-beta 13.1, as determined by comparison to expected values defined in Figure 6.17. Virtually, all excess v-beta 13.1+ cells are members of the CD45RA+, CD27− effector subset.

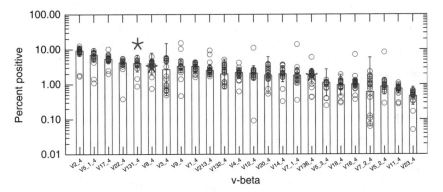

FIGURE 6.17 TCR v-beta usage in healthy adult subjects. The geometric mean values (bars), 95th percentiles (brackets), as well as individual control subject values (open circles) are shown. The proportion of v-beta 13.1, 13.6, and 8 positive CD4 T cells, determined in a peripheral blood sample from a patient with cutaneous T-cell lymphoma (Figure 6.16) are shown with stars. The patient has a significantly higher than expected proportion of v-beta 13.1+ cells, which represents an expansion of the malignant clone. This expansion was even more pronounced in the CD45RA+, CD27− effector subset.

WBC was within normal limits, and circulating Sézary cells are morphologically difficult to differentiate from normal activated T cells. (For more details on Sézary syndrome see Chapter 7.)

This analysis illustrates several principles: (1) Cells were stained in bulk for markers common to all wells; (2) After making a rough cut on forward versus log SSc to eliminate unambiguous debris, CD4+ lymphocytes were distinguished from CD4dim monocytes using CD4 versus log SSc; (3) These CD4+ cells were further cleaned up with an additional forward versus log SSc gate. The rare blast population (high forward scatter) could have been analyzed separately; (4) The analysis was performed according to a *classifier/outcome* strategy. Isotype controls were run on the outcome parameters only, allowing the regions for v-beta populations to be set with confidence; and (5) Time versus forward scatter was examined to detect any irregularities occurring during acquisition (not shown).

6.17 CLOSING REMARKS

This chapter attempts to explain the rudiments of clinical flow cytometry with particular emphasis on avoiding some of the most common pitfalls in instrument setup, experimental design, and data analysis. The most important point that I would like to communicate is that it is the very flexibility of the flow cytometer that makes it dangerous when placed into inattentive hands. Attention to detail, which means proper calibration, proper experimental controls, and a basic knowledge of the workings of the cytometer, all help ensure that the results will be interpretable. Even if you will never operate a flow cytometer, you will almost certainly need to interpret results created by someone else, or even analyze primary data yourself. If you are a laboratorian, it is your responsibility to ask your flow operator how she or he QCs the instrument (and to understand the answer), as well as to learn the controls and standards appropriate to your experiment. Flow cytometers are almost uniquely robust, such that new tests can be devised as new needs arise. However, the further we move from well-validated clinical assays into the realm of the experimental, the more vigilant we should be. In the words of Albert Einstein, "If we knew what it was we were doing, it would not be called research, would it?"

REFERENCES

1. Sutherland DR, Anderson L, Keeney M, Nayar R, Chin-Yee I. The ISHAGE guidelines for CD34+ cell determination by flow cytometry. International Society of Hematotherapy and Graft Engineering. *Journal of Hematotherapy*, 5(3):23–226, 1996.
2. Lee LG, Chen CH, Chiu LA. Thiazole orange: a new dye for reticulocyte analysis. *Cytometry*, 7(6):508–217, 1986.
3. Hall SE, Rosse WF. The use of monoclonal antibodies and flow cytometry in the diagnosis of paroxysmal nocturnal hemoglobinuria. *Blood*, 87(12):5332–5340, 1996.
4. Voskova D, Schnittger S, Schoch C, Haferlach T, Kern W. Use of five-color staining improves the sensitivity of multiparameter flow cytomeric assessment of minimal residual disease in patients with acute myeloid leukemia. *Leukemia & Lymphoma*, 48(1): 80–88, 2007.
5. Herzenberg LA, Sweet RG, Herzenberg LA. Fluorescence activated cell sorting. *Scientific American*, 234:108–117, 1976.
6. Andreeff M. *Clinical Cytometry*. New York: The New York Academy of Sciences, 1986.
7. Shapiro HM. *Practical Flow Cytometry* (4th edition). Hoboken, NJ: WileySons, 2003.
8. Melamed M, Lindmo T, Mendelsohn M (Editors). *Flow Cytometry and Sorting* (2nd edition). New York: Wiley-Liss, 1990.
9. Givan A. *Flow Cytometry. First Principles*. New York: Wiley-Liss, 1993.
10. Robinson J. *Handbook of Flow Cytometry Methods*. New York: Wiley-Liss, 1993.
11. Snow C. Flow cytometer electronics. *Cytometry Part A: The Journal of the International Society for Analytical Cytology*, 57(2):63–69, 2004.
12. Shapiro HM. The evolution of cytometers. *Cytometry Part A: The Journal of the International Society for Analytical Cytology*, 58(1):13–20, 2004.
13. Pearson T, Galfre G, Ziegler A, Milstein C. A myeloma hybrid producing antibody specific for an allotypic determinant on "IgD-like" molecules of the mouse. *European Journal of Immunology*, 7(10):684–690, 1977.
14. White RA, Mason DW, Williams AF, Galfre G, Milstein C. T-lymphocyte heterogeneity in the rat: separation of functional subpopulations using a monoclonal antibody. *Journal of Experimental Medicine*, 148(3):664–673, 1978.
15. Trowbridge IS. Interspecies spleen-myeloma hybrid producing monoclonal antibodies against mouse lymphocyte surface glycoprotein, T200. *Journal of Experimental Medicine*, 148(1):313–323, 1978.
16. Hammerling GJ, Lemke H, Hammerling U, Hohmann C, Wallich R, Rajewsky K. Monoclonal antibodies against murine cell surface antigens: anti-H-2, anti-Ia and anti-T cell antibodies. *Current Topics in Microbiology and Immunology*, 81:100–106, 1978.
17. Kung P, Goldstein G, Reinherz EL, Schlossman SF. Monoclonal antibodies defining distinctive human T cell surface antigens. *Science*, 206(4416):347–349, 1979.
18. Chow S, Minden MD, Hedley DW. Constitutive phosphorylation of the S6 ribosomal protein via mTOR and ERK signaling in the peripheral blasts of acute leukemia patients. *Experimental Hematology*, 34(9):1183–1191, 2006.

19. Perez OD, Nolan GP. Phospho-proteomic immune analysis by flow cytometry: from mechanism to translational medicine at the single-cell level. *Immunological Reviews*, 210:208–228, 2006.
20. IUIS-WHO Nomenclature Subcommittee. Nomenclature for clusters of differentiation (CD) of antigens defined on human leukocyte populations. *Bulletin of the World Health Organization*, 62:809–811, 1984.
21. Raab L. Simple Complexes Molecular Probes' Zenon technology uses immune complexes to label primary antibodies. *The Scientist*, 16(12):48, 2002.
22. Hamada S, Fujita S. DAPI staining improved for quantitative cytofluorometry. *Histochemistry*, 79(2):219–226, 1983.
23. Park CH, Kimler BF, Smith TK. Comparison of the supravital DNA dyes Hoechst 33342 and DAPI for flow cytometry and clonogenicity studies of human leukemic marrow cells. *Experimental Hematology*, 13(10):1039–1043, 1985.
24. Takahama M, Kagaya A. Hematoporphyrin/DAPI staining: simplified simultaneous one-step staining of DNA and cell protein and trial application in automated cytological screening by flow cytometry. *Journal of Histochemistry and Cytochemistry*, 36(8):1061–1067, 1988.
25. Otto F. DAPI staining of fixed cells for high-resolution flow cytometry of nuclear DNA. *Methods in Cell Biology*, 33:105–110, 1990.
26. Gaforio JJ, Serrano MJ, Algarra I, Ortega E. Alvarez de Cienfuegos G. Phagocytosis of apoptotic cells assessed by flow cytometry using 7-Aminoactinomycin D. *Cytometry*, 49(1):8–11, 2002.
27. Kapuscinski J, Darzynkiewicz Z, Melamed MR. Interactions of acridine orange with nucleic acids. Properties of complexes of acridine orange with single stranded ribonucleic acid. *Biochemical Pharmacology*, 32(24):3679–3694, 1983.
28. Traganos F, Darzynkiewicz Z, Sharpless TK, Melamed MR. Erythroid differentiation of friend leukemia cells as studied by acridine orange staining and flow cytometry. *Journal of Histochemistry and Cytochemistry*, 27(1):382–389, 1979.
29. Bauer KD, Clevenger CV, Williams TJ, Epstein AL. Assessment of cell cycle-associated antigen expression using multiparameter flow cytometry and antibody-acridine orange sequential staining. *Journal of Histochemistry and Cytochemistry*, 34(2):245–250, 1986.
30. Armstrong L, Stojkovic M, Dimmick I, Ahmad S, Stojkovic P, Hole N, Lako M. Phenotypic characterization of murine primitive hematopoietic progenitor cells isolated on basis of aldehyde dehydrogenase activity. *Stem Cells*, 22(7):1142–1151, 2004.
31. Makrigiorgos GM. Detection of lipid peroxidation on erythrocytes using the excimer-forming property of a lipophilic BODIPY fluorescent dye. *Journal of Biochemical and Biophysical Methods*, 35(1):23–35, 1997.
32. Rosati A, Candussio L, Crivellato E, Klugmann FB, Giraldi T, Damiani D, Michelutti A, Decorti G. BODIPY-FL-verapamil: a fluorescent probe for the study of multidrug resistance proteins. *Cellular Oncology*, 26(1–2):3–11, 2004.
33. Lyons AB. Analysing cell division *in vivo* and *in vitro* using flow cytometric measurement of CFSE dye dilution. *Journal of Immunological Methods*, 243(1–2):147–154, 2000.
34. Smith PJ, Wiltshire M, Davies S, Patterson LH, Hoy T. A novel cell permeant and far red-fluorescing DNA probe, DRAQ5, for blood cell discrimination by flow cytometry. *Journal of Immunological Methods*, 229(1–2):131–139, 1999.
35. Coffin RS, Thomas SK, Thomas NS, Lilley CE, Pizzey AR, Griffiths CH, Gibb BJ, Wagstaff MJ, Inges SJ, Binks MH, Chain BM, Thrasher AJ, Rutault K, Latchman DS. Pure populations of transduced primary human cells can be produced using GFP expressing herpes virus vectors and flow cytometry. *Gene Therapy*, 5(5):718–722, 1998.

36. Sorensen SJ, Sorensen AH, Hansen LH, Oregaard G, Veal D. Direct detection and quantification of horizontal gene transfer by using flow cytometry and gfp as a reporter gene. *Current Microbiology*, 47(2):129–133, 2003.
37. Lizard G, Roignot P, Maynadie M, Lizard-Nacol S, Poupon MF, Bordes M. Flow cytometry evaluation of the multidrug-resistant phenotype with functional tests involving uptake of daunorubicin, Hoechst 33342, or rhodamine 123: a comparative study. *Cancer Detection and Prevention*, 19(6):527–534, 1995.
38. Goodell MA, Brose K, Paradis G, Conner AS, Mulligan RC. Isolation and functional properties of murine hematopoietic stem cells that are replicating *in vivo. Journal of Experimental Medicine*, 183(4):1797–1806, 1996.
39. Rabinovitch PS, June CH, Grossmann A, Ledbetter JA. Heterogeneity among T cells in intracellular free calcium responses after mitogen stimulation with PHA or anti-CD3. Simultaneous use of indo-1 and immunofluorescence with flow cytometry. *Journal of Immunology*, 137(3):952–961, 1986.
40. Lopez M, Olive D, Mannoni P. Analysis of cytosolic ionized calcium variation in polymorphonuclear leukocytes using flow cytometry and indo-1 AM. *Cytometry*, 10(2): 165–173, 1989.
41. Reers M, Smiley ST, Mottola-Hartshorn C, Chen A, Lin M, Chen LB. Mitochondrial membrane potential monitored by JC-1 dye. *Methods in Enzymology*, 260:406–417, 1995.
42. Bernas T, Dobrucki J. Mitochondrial and nonmitochondrial reduction of MTT: interaction of MTT with TMRE, JC-1, and NAO mitochondrial fluorescent probes. *Cytometry*, 47(4):236–242, 2002.
43. Rice GC, Bump EA, Shrieve DC, Lee W, Kovacs M. Quantitative analysis of cellular glutathione by flow cytometry utilizing monochlorobimane: some applications to radiation and drug resistance *in vitro* and *in vivo. Cancer Research*, 46(12, Part 1):6105–6110, 1986.
44. Belloc F, Belaud-Rotureau MA, Lavignolle V, Bascans E, Braz-Pereira E, Durrieu F, Lacombe F. Flow cytometry detection of caspase 3 activation in preapoptotic leukemic cells. *Cytometry*, 40(2):151–160, 2000.
45. Deitch AD, Law H, deVere White R. A stable propidium iodide staining procedure for flow cytometry. *Journal of Histochemistry and Cytochemistry*, 30(9):967–972, 1982.
46. Jacobs DB, Pipho C. Use of propidium iodide staining and flow cytometry to measure anti-mediated cytotoxicity: resolution of complement-sensitive and resistant target cells. *Journal of Immunological Methods*, 62(1):101–108, 1983.
47. Prosperi E, Giangare MC, Bottiroli G. Nuclease-induced DNA structural changes assessed by flow cytometry with the intercalating dye propidium iodide. *Cytometry*, 12(4):323–329, 1991.
48. Ormerod MG, Kubbies M. Cell cycle analysis of asynchronous cell populations by flow cytometry using bromodeoxyuridine label and Hoechst-propidium iodide stain. *Cytometry*, 13(7):678–685, 1992.
49. Darzynkiewicz Z, Traganos F, Staiano-Coico L, Kapuscinski J, Melamed MR. Interaction of rhodamine 123 with living cells studied by flow cytometry. *Cancer Research*, 42(3):799–806, 1982.
50. Murphy RF, Chused TM. A proposal for a flow cytometric data file standard. *Cytometry*, 5(5):553–555, 1984.
51. Fowlkes B, Pardoll D. Molecular and cellular events of T cell development. *Advances in Immunology*, 44:207–264, 1989.
52. Boyum A. Isolation of leucocytes from human blood. Further observations. Methylcellulose, dextran, and ficoll as erythrocyte aggregating agents. *Scandinavian Journal of Clinical and Laboratory Investigation Supplement*, 97:31–50, 1968.
53. College of American Pathologists. College of American Pathologists (CAP) Flow Cytometry—Section II Checklist, edition October 6, 2005.

54. Roederer M. Spectral compensation for flow cytometry: visualization artifacts, limitations, and caveats. *Cytometry*, 45(3):194–205, 2001.
55. Bayer J, Grunwald D, Lambert C, Mayol JF, Maynadie M. Thematic workshop on fluorescence compensation settings in multicolor flow cytometry. *Cytometry Part B, Clinical Cytometry*, 72(1):8–13, 2007.
56. Schwartz A, Marti GE, Poon R, Gratama JW, Fernandez-Repollet E. Standardizing flow cytometry: a classification system of fluorescence standards used for flow cytometry. *Cytometry*, 33(2):106–114, 1998.
57. Mario R. Compensation is not dependent on signal intensity or on number of parameters. *Cytometry*, 46(6):357, 2001.
58. Bagwell CB, Adams EG. Fluorescence spectral overlap compensation for any number of flow cytometry parameters. *Annals of the New York Academy of Sciences*, 677:167–184, 1993.
59. Bagwell CB. HyperLog—A flexible log-like transform for negative, zero, and positive valued data. *Cytometry Part A*, 64A:34–42, 2005.
60. Hultin LE, Matud JL, Giorgi JV. Quantitation of CD38 activation antigen expression on CD8+ T cells in HIV-1 infection using CD4 expression on CD4+ T lymphocytes as a biological calibrator. *Cytometry*, 33:123, 1998.
61. Donnenberg VS, O'Connell PJ, Logar AJ, Zeevi A, Thomson AW, Donnenberg AD. Rare-event analysis of circulating human dendritic cell subsets and their presumptive mouse counterparts. *Transplantation*, 72(12):1946–1951, 2001.
62. Chen YH, Peterson LC, Dittmann D, Evens A, Rosen S, Khoong A, Shankey TV, Forman M, Gupta R, Goolsby CL. Comparative analysis of flow cytometric techniques in assessment of ZAP-70 expression in relation to IgVH mutational status in chronic lymphocytic leukemia. *American Journal of Clinical Pathology*, 127(2):182–191, 2007.
63. Schraven B, Roux M, Hutmacher B, Meuer SC. Triggering of the alternative pathway of human T cell activation involves members of the T 200 family of glycoproteins. *European Journal of Immunology*, 19(2):397–403, 1989.
64. Mason D, Powrie F. Memory CD4+ T cells in man form two distinct subpopulations, defined by their expression of isoforms of the leucocyte common antigen, CD45. *Immunology*, 70(4):427–433, 1990.
65. de Jong R, Brouwer M, Kuiper HM, Hooibrink B, Miedema F, van Lier RA. Maturation- and differentiation-dependent responsiveness of human CD4+ T helper subsets. *Journal of Immunology*, 149(8):2795–2802, 1992.
66. Baars PA, Maurice MM, Rep M, Hooibrink B, van Lier RA. Heterogeneity of the circulating human CD4+ T cell population. Further evidence that the CD4+CD45RA-CD27- T cell subset contains specialized primed T cells. *Journal of Immunology*, 154(1):17–25, 1995.
67. Kuss I, Donnenberg AD, Gooding W, Whiteside TL. Effector CD8+CD45RO-CD27- T cells have signalling defects in patients with squamous cell carcinoma of the head and neck. *British Journal of Cancer*, 88(2):223–230, 2003.
68. Rotteveel FT, Kokkelink I, van Lier RA, Kuenen B, Meager A, Miedema F, Lucas CJ. Clonal analysis of functionally distinct human CD4+ T cell subsets. *Journal of Experimental Medicine*, 168(5):1659–1673, 1988.
69. Monteiro J, Batliwalla F, Ostrer H, Gregersen PK. Shortened telomeres in clonally expanded CD28-CD8+ T cells imply a replicative history that is distinct from their CD28+CD8+ counterparts. *Journal of Immunology*, 156(10):3587–3590, 1996.
70. Currier JR, Stevenson KS, Kehn PJ, Zheng K, Hirsch VM, Robinson MA. Contributions of CD4+, CD8+, and CD4+CD8+ T cells to skewing within the peripheral T cell receptor beta chain repertoire of healthy macaques. *Human Immunology*, 60(3):209–222, 1999.
71. Hamann D, Baars PA, Rep MH, Hooibrink B, Kerkhof-Garde SR, Klein MR, van Lier RA. Phenotypic and functional separation of memory and effector human CD8+ T cells. *Journal of Experimental Medicine*, 186(9):1407–1418, 1997.

7 Leukemia and Lymphoma Immunophenotyping and Cytogenetics

Maria A. Proytcheva

CONTENTS

7.1 INTRODUCTION

Leukemia and lymphoma are hematologic malignancies that originate from hematopoietic stem cells. They are stratified primarily according to the lineage of the neoplastic cells into myeloid, lymphoid, histiocytic or dendritic cell, and mast cell neoplasms. Within each category, a combination of morphologic features, immuno-phenotype, genetics, and clinical syndromes define further distinct diseases.[1] The cell lineage is determined by identifying the antigenic determinants (epitopes) of the cells with specific antibodies, a process called immunophenotyping. Furthermore, the neoplastic cells have consistent and reproducible patterns of antigen expression. Thus, these patterns are used in disease identification and are a valuable tool in diagnostic hematopathology. In some instances, the immunophenotype has prognostic significance; however, other factors such as the presence or absence of specific genetic alterations appears to be more informative than immunophenotyping alone in predicting disease progression and response to therapy.[2]

Following an overview of the methods used for immunophenotyping and normal hematopoietic cell differentiation, a discussion of an array of hematopoietic disorders with distinctive immunophenotypic features are presented. Where appropriate, the immunophenotype is correlated with cytogenetic abnormalities; however, readers interested in a comprehensive review of cytogenetic abnormalities can refer to other sources (http://atlasgeneticsoncology.org).

7.2 IMMUNOPHENOTYPING: ANTIBODIES AND TECHNIQUES

7.2.1 ANTIBODIES

Mono- and polyclonal antibodies along with flow cytometry (FCM) form the primary tools for immunophenotyping. Markers on the surface of leukocytes that are characterized by specific monoclonal antibodies are systemically categorized into the so-called cluster of differentiation (CD). A total of 350 CDs have been characterized to date and are reviewed and presented in Chapter 11. The success of immunophenotyping is primarily dependent on the specificity of the antibodies. Although Table 7.1 presents a list of the most common antibodies used in immunophenotyping of leukemia and lymphoma; the readers can access a full list of the CD molecules through the 8th HLDA Workshop website (www.hlda8.org).[3]

Two major immunophenotyping techniques—FCM and immunohistochemistry (IHC)—are used in clinical practice and each of them has its own advantages and disadvantages. Surface or cytoplasmic antigens are detected by specific antibodies conjugated with either fluorochromes (FCM) or specific enzymes and appropriate substrates (IHC). FCM is more useful for the immunophenotypic analysis of peripheral blood and bone marrow specimens, whereas IHC is more suitable for antigen detection on formalin-fixed tissue.

7.2.2 FLOW CYTOMETRY

FCM measures physical and chemical properties of cells or other biological particles as they pass single file in a fluid through a highly focused light source, most commonly lasers. As one or more lasers interrogate each particle, the laser light is scattered and cells labeled with fluorescent molecules adsorb the laser light and reemit the light at characteristic wavelengths. All of the light scattered or emitted from the particle of interest is carefully directed along a series of mirrors and filter

TABLE 7.1
Most Common Antibodies Used in Immunophenotyping of Leukemia and Lymphoma

Marker	Description and Function	Normal Cell Expression	Disease Association	Method of Detection
CD1a	49 kDA GP, β chain noncovalently binds $\beta2$ microglobulin	Thymic T cells, dendritic and Langerhans cells, cytoplasm of activated T cells	T-cell ALL and lymphoma, Langerhans cell histiocytosis	FCM, IHC
CD2	50 kDa transmembrane GP, LFA-1, LFA-3 (CD58) ligand, sheep E-rosette receptor, alternative T-cell activation, T- and NK-cell cytolysis	Thymic and mature T cells, NK cells, thymic B cells	T-cell ALL and lymphoma, subsets of M3 and M4 AML, some MDS-related AML	FCM, IHC
CD3	20–50 kDa complex of six polypeptides (γ/ε, δ/ε, and ζ/ζ dimers) bind TCRα/β and γ/δ, signal tranducer after antigen recognition by TCR	Thymic and mature T cells	T-cell leukemia and lymphoma	FCM, IHC
CD3ζ	Zeta chains are expressed independently of other CD3 dimers, signal transducer after antigen recognition by TCR, regulation of cell surface expressions of CD3–TCR complex, NK-cell activation	Thymic and mature T cells, NK cells	T- and NK-cell leukemia and lymphoma	FCM
CD4	55 kDa transmembrane GP, coreceptor with TCR for MHC class II antigen-induced activation, thymic T-cell differentiation, receptor for HIV retrovirus	Thymic and mature T cells recognizing MHC class II antigens, monocytes, histiocytes, rare NK subset	T-cell ALL and lymphoma, M4 and M5 AML, some NK-cell blastic malignancies	FCM, IHC
CD5	67 kDa transmembrane GP, signal transduction	Thymic and mature T cells, B-cell subset, higher intensity on Th than Tc	T-cell leukemia and lymphoma, CLL	FCM, IHC
CD7	40 kDa transmembrane protein, costimulatory molecule	Stem cells, thymic and mature T cells, NK cells	T-cell ALL and lymphoma, 15% of Ly+ AML	FCM, IHC

(continued)

TABLE 7.1 (continued)
Most Common Antibodies Used in Immunophenotyping of Leukemia and Lymphoma

Marker	Description and Function	Normal Cell Expression	Disease Association	Method of Detection
CD10	100 kDa protein; common ALL antigen (CALLA); neutral endopeptidase, regulator of B-cell growth and proliferation	Precursor B- and T cells, germinal center B cells, mature neutrophils	Precursor B- and T-ALL and LBL, Burkitt and FCC lymphoma	FCM, IHC
CD11b	170 kDa GP, phagocytosis, chemotaxis, apoptosis	Granulocytes, monocytes, NK cells, subsets of B- and T cells	AML, precursor B- and T-ALL, NK-cell leukemia, absent in patients with leukocyte adhesion deficiency	FCM
CD11c	150 kDa GP, functions similar to CD11b	Granulocytes, monocytes, NK cells, subsets of B- and T cells	AML, precursor B- and T-ALL, NK-cell leukemia, absent in patients with leukocyte adhesion deficiency	FCM
CD13	150 kDa GP, aminopeptidase catalyses, removal of amino acids or small peptides, receptor for coronavirus, CMV uses CD13 to interact with target cells	Early committed progenitors through late stages of granulocytic and monocytic maturation, large granular lymphocytes, bone marrow stromal cells, osteoclasts	Most AML, 20–30% of My$^+$ ALL	FCM
CD14	55 kDa phosphoinositol-linked GP, LPS receptor for endotoxin to release cytokines	Mature monocytes, macrophages, weak expression by neutrophils	M4 and M5 AML with maturing monocytic cells	FCM
CD15	Carbohydrate (x-hapten; Lewis-x), adhesion or phagocytosis	Monocytic and myelocytic cells, Langerhans cell	AML, granulocytic sarcoma, 10% of My$^+$ ALL, R–S cells	FCM, IHC
CD16	Fc receptor of IgG (Fcγ RIII), receptor for antibody-dependent cellular cytotoxicity	Granulocytes, monocytes, NK cell	NK-cell LGLL and lymphoma	FCM
CD19	95 kDa GP, signal transduction for B-cell developments, activation, and differentiation	Precursor and mature B cells, absent on plasma cells, follicular dendritic cells	Precursor and mature B-cell leukemia and lymphoma, t(8;21)$^+$ AML	FCM
CD20	35–37 kDa phosphoprotein, cellular activation and proliferation	Precursor and mature B cells, absent on differentiated plasma cell	Precursor and mature B-cell leukemia and lymphoma	FCM, IHC[a]

CD21	145 kDa GP; receptor for C3d, C3dg, iC3b, and EBV; signal transduction	Surface Ig-positive B cells (lost with activation), thymic T-cell subset, follicular dendritic cell	B-cell leukemia and lymphoma, T-ALL	FCM
CD22	135 kDa GP; adhesion and activation	Precursor and mature B cells	Precursor and mature B-cell leukemia and lymphoma, rare cases of AML	FCM
CD24	35–45 kDa GP; early B-cell development and apoptosis	Throughout B-cell maturation but not plasma cells, neutrophils	Most precursor B-ALL and LBL cases	FCM
CD30	105 kDa GP; member of the TNF receptor family	Activated T-, B-, NK cells, monocytes	R–S cells; immunoblastic and anaplastic large cell lymphoma; lymphocytes infected with EBV, HTLV-1, or HIV; ATLL	FH, IHC
CD33	67 kDa transmembrane GP; sialoadhesin (lectin activity for sugar chains containing sialic acid)	CRU-GEMM, CFU-GM, CFU-G, BFU-E, myeloblast → neutrophil maturation, monocytic cells, megakaryoblasts, early erythroblasts	80% of AML, 30% of My$^+$ALL	FCM
CD34	40 and 116 kDa transmembrane GP, sialomucin, two forms with one having a truncated cytoplasmic domain, stromal cell adhesion	Lympho and hematopoietic stem cells and progenitors, small-vessel endothelium	ALL, AML, LBL, vascular tumors	FCM, IHC[b]
CD36	Platelet gpIV, collagen, *Plasmodium falciparum* receptor recognition and phagocytosis of apoptotic cells, platelet adhesion and aggregation	Platelets, megakaryocytes, erythroblasts and RBC's monocytes, macrophages	M4, M5, M6, and M7 AML	FCM
CD38	ADP-ribosyl cyclase cell adhesion and proliferation	Progenitor and activated leukocytes and plasma cells	Lymphoblastic leukemia, multiple myeloma	FCM, IHC
CD41	gpIIb (α subunit) of CD41–CD61 (gpIIb–gpIIIa) complex, composed of gpIIb-α and gpIIb-β subunits, platelet aggregation, receptor for fibronectin, fibrinogen, and von Willebrand factor	Megakaryocytes, platelets	M7 AML, Glanzmann's thromboasthenia	FCM
CD42b	Platelet gpIb α forms heterodimer with beta chain (CD42c) and complexes with gpIX/CD42a, von Willebrand factor-ristocetin receptor	Megakaryocytes, platelets	M7 AML, Bernard–Soulier syndrome	FCM

(continued)

TABLE 7.1 (continued)
Most Common Antibodies Used in Immunophenotyping of Leukemia and Lymphoma

Marker	Description and Function	Normal Cell Expression	Disease Association	Method of Detection
CD45	180–220 kDa GP, leukocyte common antigen (LCA), tyrosine phosphatase, different isoforms characteristic of different subsets of hematopoietic cells, critical for lymphocyte activation	All leukocytes, weakly expressed by very early erythroblasts	ALL, AML, lymphoma	FCM, IHC
CD45RO	180 kDa GP, CD45 isoform	Thymic and mature T cells, monocytes, neutrophils	T-cell leukemia and lymphoma, M5 AML	FCM, IHC
CD56	175–220 kDa transmembrane GP, neural adhesion molecule (N-CAM), many isoforms	140 kDa isoform on NK cells, Tc, subset of CD4+ T cells, neural tissues	T- and NK-cell leukemia and lymphoma, some t(8;21)+ and t(15;17)+ AMLs, M4 and M5 AML, neuroendocrine tumors	FCM, IHC
CD61	gpIIIa (β subunit) of CD41–CD61 (gpIIb–gpIIIa) and CD51–CD61 complexes, adhesion to diverse matrix proteins	Megakaryocytes and platelets with CD41 and CD61, monocytes[c]	M7 AML, Glanzmann's thromboasthenia	FCM, IHC
CD64	72 kDa GP, FcγRI, endocytosis of IgG antigen complexes, phagocytosis	Monocytes, histiocytes, neutrophils, early myeloid and monocytic precursors, dendritic cells	Myeloid and monocytic AML, neutrophil expression increases during infection	FCM
CD65	Carbohydrate carried by lipid and maybe protein, poly-N-acetyl-lactosamine, unknown function	Myeloid and monocytic cells	Myeloid and monocytic AML, some My+ ALL	FCM
CD66c	90 kDa GPI-linked GP, member of CEA antigen family, regulator of adhesion activity	Granulocytes, epithelial cells	Subset of precursor –B-ALL	FCM
CD68	110 kDa transmembrane GP, primarily in cytoplasmic granules, endocytosis, lysosomal trafficking	Monocytes, macrophages, mast cells, neutrophils, basophils, dendritic cells, subset of myeloid progenitors, activated T cells, subset of B cells, osteoclasts	M4 and M5 AML, subset of precursor T-ALL and LBL	FCM, IHC

CD79a	*Mn-1* gene product Igα, associates with CD79b, signal transduction for surface Ig	Precursor and mature B cells	Precursor and mature B-cell leukemia and lymphoma, some T-ALL and AML	FCM, IHC
CD79b	*B29* gene product Igβ, associates with CD79a, signal transduction for surface Ig	Precursor and mature B cells	Mature B-cell leukemia and lymphoma	FCM
CD83	43 kDa transmembrane GP	Nonfollicular dendritic cells, Langerhans cells	Dendritic leukemia, Langerhans histiocytosis	FCM
CD117	143 kDa transmembrane GP, tyrosine kinase c-kit, stem cell factor receptor, a growth factor receptor; cell adhesion	Hematopoietic stem and progenitor cells, mast cells	AML, rare precursor T-ALL, mastocytosis	FCM, IHC
CD179a	16–18 kDa polypeptide, V_{preB} surrogate light-chain component disulfide-linked to Igμ to form pre-BCR, transduces signals for B-cell differentiation and proliferation	Early pre-B and pre-B cells	Precursor B-ALL, rare precursor T-ALL and AML	FCM
CD179b	Lambda 5/14.1 surrogate light-chain component disulfide-linked to Igμ to form pre-BCR, transduces signals for B-cell differentiation and proliferation	Early pre-B and pre-B cells	Precursor B-ALL, rare precursor T-ALL and AML	FCM
CD235a	Sialoglycoprotein glycophorin A, proposed functions include minimizing RBC aggregation and inhibition of lysis	Erythroid cells	M6 AML, some M7 AML	FCM, IHC

Note: GP, glycoprotein; FCM, flow cytometry; IHC, immunohistochemistry; NK, natural killer; TCR, T-cell receptor; MHC, major histocompatibility complex; Th, helper T cells; Tc, cytotoxic T cells; CLL, chronic lymphocytic leukemia; LBL, lymphoblastic lymphoma; FCC, follicular center cell; BCR, B-cell receptor; MDS, myelodysplastic syndrome; LPS, lipopolysaccharide; My$^+$ ALL, myeloid antigen-positive ALL; R–S, Reed–Sternberg cell; LGLL, large granular lymphocytic leukemia; CFU, colony forming unit; GEMM, granulocyte/erythroid/myeloid/monocytic; GM, granulocyte/monocyte; G, granulocytic; BFU-E, burst forming unit-erythroid; EBV, Epstein-Barr virus; HIV, human immunodeficiency virus; HTLV-1, human T-cell lymphotropic virus; ATLL, adult T-cell leukemia and lymphoma; Ig, immunoglobulin; GPI, glycosylphosphatidylinositol; AML, acute myeloid leukemia; LFA - lymphocyte function-associated antigen; CMV - Cytomegalovirus; TNF - Tumor necrosis factor; ADP Adenosine diphosphate; CEA - carcinoembryonic antigen; RBC - Red blood cell

[a] CD20 antibodies differ for FCM and IHC.

[b] Different antibodies optimal for blast cells and endothelium.

[c] Nonspecific binding to monocytes.

to a light-collecting apparatus, usually a photomultiplier tube, where the signal is amplified, converted into a digital signal, and then collected by appropriate software for analysis. Scattered light provides information about the innate physical characteristics of the cells such as their size and cytoplasmic or nuclear complexity. Fluorochrome-labeled monoclonal antibodies, fluorescent dyes specific for nucleic acid, and a variety of fluorescent probes specific for a variety of cell-associated physiological measurements are assessed accurately and reproducibly to permit the clustering of very specific cell populations.[4,5]

The advantages of FCM over conventional microscopy are numerous and include the ability to analyze large number of cells (10,000 or more) at extremely rapid rates (thousands per second) and to identify several parameters simultaneously on single cells, which allows for the detection of multiple populations in a single sample. Three main disadvantages of the technique are the need for fresh tissue, lack of morphologic confirmation of the findings, and the call for special instrumentation that is expensive and requires trained personnel to perform and analyze the data.

Flow cytometric immunophenotyping (FCI) of hematologic neoplasms includes several interrelated stages, from the initial medical decision that the suspected hematologic condition is appropriate for FCM analysis to the final step of making a diagnosis that includes the assessment and correlation of the FCM data with other relevant clinical and laboratory findings. FCI involves three major steps: (1) Preanalytical (specimen collection, processing, and antibody staining), (2) Analytical (acquiring data on the flow cytometer), and (3) Postanalytical (data analysis and interpretation). This brief introduction is only meant to orient the reader for the remainder of the information presented in this chapter. Details regarding the technical aspects of FCM (instrumentation, reagents, quality control, etc.) are presented in Chapter 6 and other references.[6–8] This chapter focuses on specimen collection, data analysis, and the correlation of phenotypic profiles with specific hematologic malignancies.

Successful immunophenotyping involves proper specimen collection, sample preparation, accurate data interpretation, and appropriate reporting of the results. Peripheral blood, bone marrow, or body fluids should be collected with heparin, ethylenediaminetetraacetic acid (EDTA), or acid citrate dextrose (ACD) anticoagulant. To preserve tissue integrity and prevent bacterial overgrowth, solid tissue (lymph nodes, spleen, or other tissues) should be placed in a culture medium, such as RPMI-1640, containing fetal bovine serum and antibiotic. Storing the cells in a refrigerator may prolong their viability, whereas low temperatures can downregulate the expression of some antigens, which can lead to erroneous results. Thus, it is recommended that the cells are stored at room temperature.

It is essential to use a comprehensive approach to FCM analysis and interpretation of the data.[9] The optimal method is for the laboratory medical staff to apply a visual approach to FCM data analysis rather than relying on percentages of positive cells. List mode data should be collected unselected (i.e., it includes all cells in the sample). After the data are collected ungated, certain gating procedures can be applied during the analysis step (Figure 7.1). Some of the most useful gating strategies include gating of blasts in a forward or side scatter versus CD45 (Figure 7.2) dot plot or gating on specific clusters (characterized by appropriate surface markers) of B cells to determine clonality and the coexpression of other critical antigens.[10]

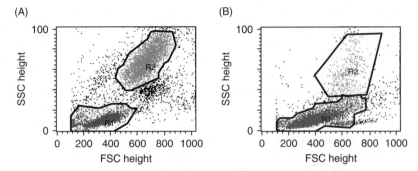

FIGURE 7.1 A two-parameter dot plot histogram illustrating the light scattered character-istics of normal peripheral blood (A) and acute leukemia (B). (A) Three distinct populations are present. The small lymphocytes (gate R1) have a low side scatter (SSC) and low forward scatter (FSC), whereas the mature granulocytes (R2) have a high SSC and FSC. The mono-cytes are larger cells with abundant, but agranular cytoplasm and can be noticed as a small population in the area between the granulocytes and lymphocytes. This population is small and not always detectable because the number of monocytes is relatively low (<10%). (B) In acute leukemia, however, there is no maturation and only a few granulocytes are detected in gate R2. The blasts have agranular cytoplasm and low SSC and are seen in R1. The size of the blasts is variable, which is demonstrated by the higher than normal lymphocyte FSC.

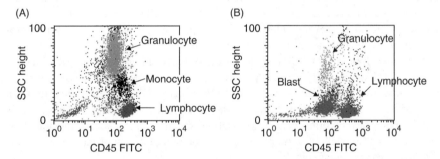

FIGURE 7.2 A two-parameter dot plot histogram illustrating CD45 versus SSC on normal peripheral blood (A) and acute leukemia (B). (A) The degree of CD45 expression in a com-bination with side scatter separates the populations seen in (A) into more precisely defined populations of small lymphocytes, monocytes, and granulocytes. This approach is useful in resolving low SSC and FSC populations on heterogeneous specimens into mature lym-phocytes (CD45 bright) and blasts (CD45 dim) (B). Back gating on the CD45 dim population is a useful approach in identifying the antigenic profile of these blasts.

To achieve this, the United States–Canadian Consensus on the immunophenotypic analysis of hematologic neoplasia recommends the use of a multiparameter approach to FCM analysis, employing multiple fluorochromes in addition to forward and side-light scatters.[11] The goals of data analysis in immunophenotyping of leukemia and lymphoma are to properly distinguish populations of abnormal cells from normal cells and to characterize the antigenic profile of those abnormal cells. Abnormalities expressed by leukemic cells include the expression of markers not normally present, the absence of markers that are normally expressed, and the aberrant level of expression (either increased or decreased). The interpretation of flow cytometric data is not based on rigid rules for diagnosis and it is best done in conjunction with morphology. For some diseases, it is possible to make a definitive diagnosis by FCM. For example, precursor B- or T-lymphoblastic leukemia and lymphoma, B-cell chronic lymphocytic leukemia (CLL), hairy cell leukemia (HCL), and acute myeloid leukemia (AML) have characteristic immunophenotypic signatures that are crucial for the proper diagnosis. However, for other diseases (such as low-grade B-cell lymphoproliferative disorders, large B-cell lymphoma, Burkitt lymphoma and leukemia (BL), myelodysplasia, or plasma-cell neoplasms) FCM can contribute to defining or resolving a differential diagnosis based on the morphologic examination.[12] Thus, the FCM technique is widely used and is a well-established tool in leukemia and lymphoma diagnosis.

7.2.3   IMMUNOHISTOCHEMISTRY

Another methodological approach for the detection of cellular antigens is the immunohistochemical staining of routinely fixed and paraffin-embedded tissue sections. IHC is designed to allow maximum sensitivity consistent with precise localization of antigens. A marker enzyme, most often horseradish peroxidase conjugated to mono- or polyclonal antibodies, is used to produce a visible, insoluble product at the site of antigen–antibody reaction.[13] There are numerous ICH methods that are used to localize antigens; however, readers interested in a comprehensive review of immunohistochemical techniques can refer to other sources (http://www.ihcworld.com).

The development of sensitive detection systems, along with antigen retrieval techniques, has made IHC a valuable tool for the diagnosis, prognosis, and treatment of hematopoietic and nonhematopoietic diseases. IHC has traditionally been used to detect tissue proteins; however, the development of new antibodies enables the detection of mutations, gene amplification, specific chromosomal translocations associated with novel chimeric proteins, and the identification of molecular targets for novel therapeutic agents.[14,15]

A major advantage of the IHC over FCM is that the morphology and antigen expression can be assessed simultaneously. With the advance of antigen retrieval techniques IHC has been successful on paraffin sections with almost all antibodies used in clinical practice. Major limitations of the technique include lack of standardization, absence of ideal tissue controls, a limited ability to assess more than one antigen on a single cell, and a relatively slow turnaround time given that the IHC is performed after the initial morphologic assessment.

Standardization of IHC is difficult to achieve. Numerous preanalytical variables such as the duration of fixation, delay in fixation, or state of preservation of the tissue sample are often beyond the control of the histology laboratory. In contrast, analytical variables such as tissue processing, antigen retrieval, staining, and detection systems can be controlled in the laboratory and hence standardized. Postanalytical variables such as interpretation and reporting, are also readily amenable to standardization. Regardless of these limitations, IHC is an integral part of the routine diagnostic histopathology and remains the immunophenotyping method of choice for tissue biopsies.

7.3   OVERVIEW OF LINEAGE-ASSOCIATED MARKERS

The cellular antigens commonly used in the immunophenotyping of leukemia and lymphoma are broadly categorized into B cell, T cell, natural killer (NK) cell, myelomonocytic, erythroid, and nonlineage-associated markers (including activation markers such as CD38 and HLA-DR). Information about these markers is derived from studies on the differentiation and maturation of hematopoietic stem cells. The totipotent hematopoietic stem cell has the capability of self renewal and can be identified by the earliest differentiation antigen, CD34. There is a considerable heterogeneity within the CD34 positive population and the more mature hematopoietic stem cells also express major histocompatibility complex (MHC) class II HLA-DR molecules and CD38.[16] Both CD34 and human leukocyte antigen (HLA) class II molecules are important in the interaction of the stem cells with the stromal cells in the bone marrow. Another marker of immaturity is terminal deoxynucleotidyl transferase (TdT), which is most often associated with the lymphoid lineage.

7.3.1   MYELOMONOCYTIC LINEAGE

Cells from the *myeloid* lineage are most commonly identified by the expression of CD13, CD33, CD117, and myeloperoxidase. During maturation from myeloblasts to promyelocytes, cells lose CD34 and HLA-DR expression and acquire CD15, CD11b, and CD16 at the myelocyte and the metamyelocyte stage. CD33 is a much more sensitive marker for the *myelomonocytic* lineage than CD13, and the expression levels of CD33 differs between mature granulocytes and mature monocytes (the latter expressing more CD33). Bright CD14 is the characteristic of mature monocytes. In addition, the monocytes that express CD64 are myeloperoxidase negative and nonspecific esterase positive.

Erythroid progenitors are characterized by downregulation of CD45 and high levels of surface CD71 (transferrin receptor). Hemoglobin and glycophorin antibodies have only limited utility because they are expressed at late stages of erythroid maturation, when the cells can be morphologically recognized as being erythroid progenitors.

Megakaryocyte-associated antigens include CD41, CD42b, CD61, and CD56. The latter antigen is nonspecific and can be expressed on NK cells and other hematopoietic and nonhematopoietic cells.

7.3.2   B-CELL LINEAGE DIFFERENTIATION AND MATURATION SEQUENCE

The earliest B-cell precursors can be identified by cytoplasmic CD22, which is present before any detectable rearrangement of the immunoglobulin (Ig) genes (Figure 7.3). The B-cell receptor (BCR) complex is the most B-cell-specific antigen and is composed of an antigen-specific surface Ig component, which is associated with two intracytoplasmic heterodimers of CD79a and CD79b. Thus, cytoplasmic CD22 and CD79a are considered to be the hallmark of B-cell lineage. The *PAX-5* gene encoding the B-cell-specific activator protein (BSAP) appears to mediate B-lineage commitment by repressing the transcription of non-B-lymphoid genes and by simultaneously activating the expression of B-lineage-specific genes.[17] Early B cells express surface CD10, CD19, CD34, HLA-DR, and TdT. The subsequent maturation and differentiation of B-cell progenitors in the bone marrow is characterized

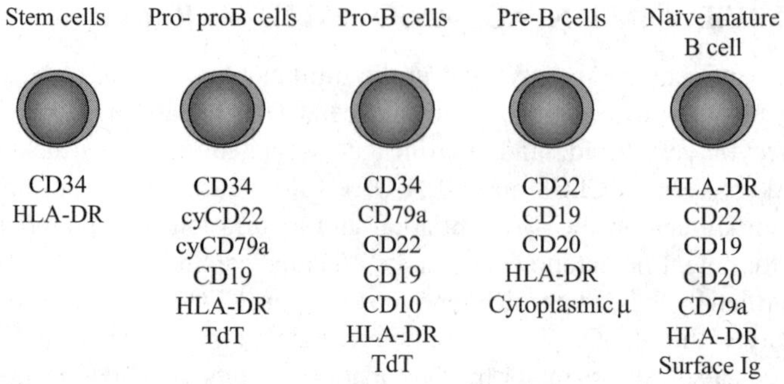

Stem cells	Pro- proB cells	Pro-B cells	Pre-B cells	Naïve mature B cell

CD34	CD34	CD34	CD22	HLA-DR
HLA-DR	cyCD22	CD79a	CD19	CD22
	cyCD79a	CD22	CD20	CD19
	CD19	CD19	HLA-DR	CD20
	HLA-DR	CD10	Cytoplasmic μ	CD79a
	TdT	HLA-DR		HLA-DR
		TdT		Surface Ig

FIGURE 7.3 B-cell maturation sequel and antigen expression. The expression of relevant B-cell antigens is indicated for each differentiation stage. The first indication of commitment toward B-cell lineage is the intracytoplasmic expression of CD22 and CD79a. CD10 expression is transient during B-cell maturation. It appears at the pro-B-cell stage and disappears during the pre-B stage of maturation. This differentiation stage is characterized by the presence of the cytoplasmic μ chain. The naïve B cells express surface Igs.

by a gradual decrease in CD10 along with a gradual gain of CD20. The appearance of surface Igs and disappearance of CD34 and TdT define a mature B cell.

B-cell maturation takes place in the bone marrow and proceeds through stages that can be identified by the rearrangement and pattern of expression of the Ig genes. The earliest bone marrow cells committed to become B cells have the diversity and joining (DJ) regions of the Ig heavy-chain gene rearranged, but do not synthesize Igs. The functional rearrangement of the Ig heavy-chain gene results in the appearance of cytoplasmic μ heavy chains and indicates further maturation of pro-B cells to a pre-B-cell stage. During this stage, the cells do not express Ig light chains, kappa (κ) or lambda (λ). Rearrangement and transcription of the Ig light-chain genes lead to the formation of complete Ig molecules and to the transition from pre-B- to mature B lymphocytes. These mature B cells are referred to as naïve or virgin B cells and they express surface IgM and later may also express IgD. The naïve B cells leave the bone marrow and migrate to colonize the follicles of the lymphoid tissue—lymph nodes, spleen, Peyer's patches, tonsils, and other extranodal lymphoid tissue. In response to antigen stimulation, mature B cells proliferate and form a germinal center, where the BCR undergoes somatic hypermutation of the Ig genes, which results in further diversification of the mature B lymphocytes. In the secondary lymphoid tissue, the Ig heavy-chain genes undergo an isotype switch to produce IgD, IgG, or IgA B cells.[18,19] B cells that have been exposed to specific antigen are also characterized by the surface expression of CD27, and can be considered memory B cells.

The germinal center B lymphocytes share features similar to the pre-B lymphocytes and they express CD19, CD20, and CD10. They are BCL2 negative and tend to undergo apoptosis if BCR is not rearranged properly. These cells, however, are TdT negative. B cells that successfully rearranged their Ig genes leave the germinal center and terminally differentiate into plasma cells. Plasma cells downregulate CD19, CD20, and CD45 and do not express surface Ig, but contain high levels of cytoplasmic Ig. They express CD79a as well as CD38 and CD138 on their cell surface.

7.3.3   T-Cell Lineage Differentiation Antigens and Maturation Sequence

Similar to B cells, T-cell maturation involves the expression of specific antigens associated with specific stages of maturation (Figure 7.4). T-cell maturation begins in the bone marrow and is completed in the thymus, where mature but naïve T cells (not previously exposed to antigen) are generated and released to the peripheral blood. During thymic maturation, T-cell precursors undergo rearrangement of the T-cell receptor (TCR). The earliest and most specific T-cell lineage antigen is cytoplasmic CD3. Other T-cell antigens expressed on early T cells include CD2 and CD7 as well as TdT and CD1a. In the majority of T-cells, the resulting functional TCR molecule is composed of $\alpha\beta$ heterodimers. A minor population (3–5%) of T cells expresses $\gamma\delta$ TCR. The $\gamma\delta$ T cells migrate from the fetal thymus to reside in the cutaneous and mucosal tissue, particularly gastrointestinal mucosa. Thymic maturation is accompanied by the appearance of other T-cell-associated markers (CD5, CD4, and CD8). CD2, CD5, and CD7 are considered as pan-T-cell antigens, although they can be expressed on other cell types. For example, CD2 and CD5 can be expressed on B cells and CD7 on myeloid cells. The stages of thymic maturation and antigen expression of T cells are depicted in Figure 7.4. Mature T cells are TdT and CD1a negative and express either CD4 or CD8, although a small proportion are CD4 and CD8 double negative. Finally, cytotoxic CD8 positive T cells contain several

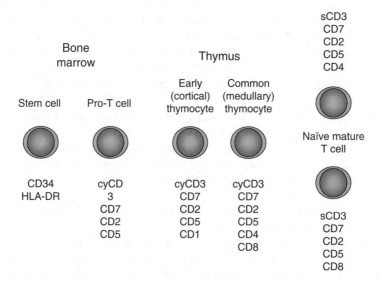

FIGURE 7.4   T-cell maturation sequel and antigen expression. The expression of relevant T-cell antigens is indicated for each differentiation stage. The most immature T cells express CD34, but no CD4, CD8, or surface CD3. As the cells mature, they lose CD34 expression while gaining CD4 and CD8, thus becoming double positives. Early double-positive cells initially express CD1 and CD10 (early cortical thymocytes). The cells that successfully rearrange their TCR gain CD3 expression and undergo both positive and negative selection. Cells surviving this process proceed through a final step of differentiation (in which either CD4 or CD8 is expressed) to become mature single-positive T cells.

cytotoxic proteins, including perforin, granzyme B, and T-cell-restricted intracellular antigen (TIA)-1. Following antigen stimulation, both B- and T cells express a variety of *activation antigens*, including HLA-DR, CD25, CD30, and CD69.

7.3.4 NATURAL KILLER-CELL MARKERS

NK cells express pan-T-cell antigens CD7 and often CD2, but do not express the TCR–CD3 complex. Most of the NK cells are positive for CD16 or CD56 antigens and some express CD57. CD8 is expressed on 5–15% of the NK cells. Recently, several new receptors that trigger the process of natural cytotoxicity, natural cytotoxic receptors (NCRs), have been discovered.[20,21] NK cells also differentially express various inhibitory receptors—NK-cell immunoglobulin-like receptors (KIRs)—that interact specifically with HLA class I molecules. KIRs act to prevent cytotoxicity against autologous HLA class-I positive cells.[22] Although NCRs are exclusively expressed on NK cells, CD16, CD56, CD57, and KIRs are also expressed on cytotoxic CD8 positive T lymphocytes. In addition, the NK cells contain cytotoxic molecules such as perforin, granzymes B, and TIA-1.

7.4 IMMUNOPHENOTYPING OF LEUKEMIA AND LYMPHOMA

Leukemia usually present with widespread involvement of the bone marrow accompanied by the presence of neoplastic cells in the peripheral blood. In contrast, lymphoma generally present as discrete tissue masses that most often originate in a lymph node and occasionally in extranodal tissue (e.g., skin, gastrointestinal tract, and brain). This distinction between leukemia and lymphoma, however, is not absolute because some lymphoma involve the bone marrow and can present in a leukemic phase (i.e., low-grade B-cell lymphoproliferative disorders). Similarly, some leukemia can present with tissue involvement (i.e., acute monoblastic leukemia infiltrating skin).

With a few exceptions, leukemia are divided into acute and chronic based on whether the neoplastic cells are immature blasts or mature cells respectively. The diagnosis of acute leukemia requires that the bone marrow or peripheral blood have more than 20% blasts according to the World Health Organization (WHO) classification of leukemia and lymphoma, or have more than 30% blasts according to the French-American-British (FAB) classification.[23–25] Acute leukemia are aggressive diseases and if untreated can cause rapid death. Although the chronic leukemia usually have slower progression, they can undergo blastic transformation to acute leukemia.

The antigen expression of leukemia and lymphoma often parallels the normal stages of myeloid, B-, and T-cell differentiation and maturation and as a result provides a framework for the classification of hematopoietic and lymphoid malignancies. However, unlike normal cells, neoplastic cells may express more than one lineage-specific antigen. For example, CD15, a myeloid marker, is often expressed in precursor B- or T-cell lymphoblastic leukemia and CD7, a T-cell marker, is often seen on myeloid leukemia. Knowledge of the antigen expression patterns of hematologic malignancies facilitates their specific lineage identification and proper classification, which ultimately affects the choice of treatment and their predicted response to therapy.

7.4.1 ACUTE LEUKEMIA

Acute leukemia are a heterogeneous group of disorders involving the bone marrow and peripheral blood with characteristic clinical presentation, genetic alterations, with variable patterns of disease progression and response to therapy. The WHO classification of acute leukemia and antigen expression of the neoplastic blasts are shown schematically in Figure 7.5. Based on the cell of origin they are divided into two broad categories—AMLs and acute lymphoblastic leukemia or lymphoma (ALLs). Furthermore, each category is subclassified based on the degree of neoplastic cell differentiation and lineage stratification.

Morphologically the blasts seen in AML and ALL are very similar. They have a high nuclear cytoplasmic ratio, immature chromatin, nucleoli, and a variable amount of cytoplasm. Although in a minority of the cases the neoplastic cells may have a specific morphologic landmark (e.g., Auer rods indicating myeloid origin), most of the blasts cannot be appropriately classified by morphologic examination alone.

The cell lineage is identified to be pivotal because different leukemia have different natural histories and require different treatments. For this reason, immunophenotyping is an indispensable tool in the diagnostic evaluation. However, many studies have established that there is a heterogeneity based on the molecular or cytogenetic alterations even in the same category of acute leukemia.[2,26] These abnormalities are more predictive of the behavior of the neoplastic process than immunophenotyping alone. Furthermore, with a few exceptions, there is no straightforward correlation between the immunophenotype and the molecular alterations; thus genetic studies should be performed for all acute leukemia.

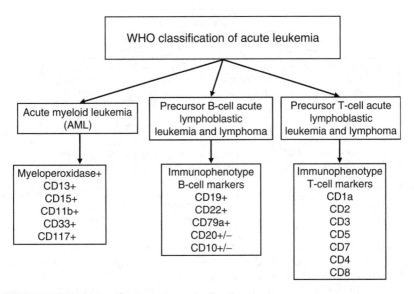

FIGURE 7.5   WHO classification of acute leukemia.

7.4.1.1   Acute Myeloid Leukemia

AML is a clonal expansion of myeloid blasts in bone marrow, peripheral blood, or other tissues.[1] AMLs are immunologically and genetically heterogeneous group of disorders, as shown in Figure 7.6, and are more common in adults than in children. It is thought that AML often originates from a previously unrecognized myelodysplastic syndrome. The blasts in AML fail to mature further to neutrophils, monocytes, erythrocytes, or megakaryocytes. The degree of immaturity varies from undifferentiated cases where the cells express only a few myeloid markers to more mature types. Immunophenotypic analysis in the diagnosis of AML distinguishes between minimally differentiated AML and ALL. Immunophenotyping is also useful in the differentiation of acute promyelocytic leukemia (APL) from other types of AML, and in the recognition of acute megakaryoblastic leukemia (M7).

Almost all AMLs express CD13 and CD33, myeloid markers in combination with CD34, CD117, and HLA-DR.[27] The use of CD13 and CD33 is sufficient in the initial leukemia-screening panel. The addition of CD11c and CD15 may help to confirm the myeloid lineage of cases that are either CD13 or CD33 negative. Overall, CD33 is a much more sensitive marker than CD13 for the myeloid lineage, but it is also less specific.

Neoplastic myeloblasts may show "lineage infidelity" and express B- or T-cell antigens, that is, CD7, CD2, CD4, CD19, or CD56. The frequency of expression of lymphoid markers in AML is estimated to be between 10% and 30%, but the significance of this expression is still controversial. Although CD7 is the most commonly expressed lymphoid antigen in AML and has been associated with a poor prognosis

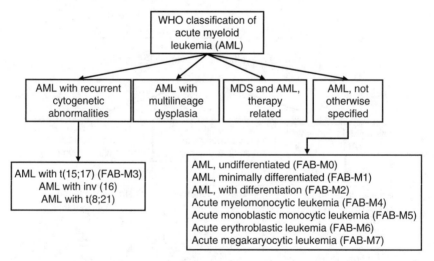

FIGURE 7.6 WHO classification of AML with the appropriate correlation with the FAB classification of AML. WHO classifies AML into several categories. AML not otherwise categorized encompasses those cases that do not fulfill the criteria for inclusion in one of the other groups. The primary bases for subclassification within this category is the morphologic and cytochemical features of the leukemic cells and the degree of maturation. When appropriate, WHO's categories are correlated with the FAB classification of AML.

in AML,[28] CD2 and CD19 expression on AML blasts have been shown to portend a favorable prognosis in some studies,[29] whereas others have reported that their expression signifies a poor prognosis.[28,30]

Immunophenotyping is particularly helpful in lineage determination of minimally differentiated AML—classified as M0 according to FAB criteria because myeloperoxidase may not be expressed in some cases. The absence of both cytoplasmic CD3 and CD79a indicates the diagnosis of AML-M0 more often than the expression of myeloperoxidase. The expression of CD2, CD4, CD7, CD10, CD19, or CD20 is not considered sufficient for a lymphoid lineage designation because many AMLs express these antigens. Figure 7.7 shows an example of antigen expression profile of AML with an aberrant expression of T-cell markers, CD2 and CD7. Of the lymphoid markers, CD5 is the most helpful in distinguishing AML from precursor T-cell ALL because CD5 is not expressed in AML-M0, but is expressed in precursor T-cell ALL.[5]

APL with t(15;17)(q22;q21) resulting in the fusion of the *PML* and *RARAα* genes is characteristically HLA-DR and CD34 negative and frequently expresses CD2, but lacks CD4, CD11c, CD36, and CD117. The combination of HLA-DR and CD34 is much more helpful in distinguishing APL from non-APLs and non-AMLs than any of these antigens alone. However, approximately 10% of the non-APL and non-AMLs express one of these antigens.[5] The lack of HLA-DR and CD34 expression in myeloperoxidase-positive AML should warrant a search for the t(15;17) or *PML–RARAα* variants. These leukemia dictate a particular line of therapy and it is essential that this rearrangement be sought out by molecular or cytogenetic methods. A rapid, specific, and accurate immunofluorescent assay for the detection of PML protein in clinical specimens has recently been developed but is not yet widely used.[31]

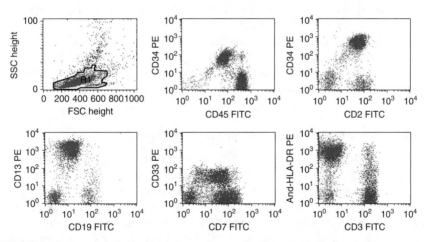

FIGURE 7.7   An example of antigen expression profile of AML. The leukemic blasts express CD45 (dim), CD34, HLA-DR, and myeloid markers—CD13 and CD33. They show an aberrant expression of CD2 and CD7 T-cell markers. Cytochemical stain with myeloperoxidase is strongly positive (not shown). In addition, a small population of mature T lymphocytes is present and can easily be separated from the leukemic blasts by the bright expression of CD45.

AMLs with *monocytic* differentiation show a gradient of CD45 expression from dim in monoblastic leukemia to bright in acute monocytic leukemia. CD13 and CD33 are the most commonly expressed pan-myeloid markers in acute monocytic leukemia, whereas the CD34 and CD117 markers of immaturity are usually absent. The monocytic blasts typically express brighter CD33 than do other myeloid leukemia. It has been suggested that the combination of bright CD33 with negative CD13 and negative CD34 is highly predictive of AML with monocytic features.[32] Furthermore, these leukemia express other monocytic antigens: CD64, CD36, CD14, or CD4.[33] Although the last two markers are highly sensitive, they are not specific to monocytic differentiation because nonspecific Ig binding may also occur due to the presence of avid Fc receptors on the leukemic cells.

Acute megakaryocytic leukemia are usually best recognized by the expression of platelet glycoproteins, CD61, and CD41. Mature platelet antigens such as CD42b tend not to be expressed on megakaryoblasts, but show a strong expression in platelets. It is important to exclude false-positive platelet glycoprotein expression due to adherence of platelets to other blast types.[34] The combination of CD61 and CD42b can be helpful in distinguishing between megakaryoblasts and mature platelets. Myeloid antigens, CD33 and CD13, are variably expressed on megakaryoblasts, as are CD34 and CD45. Their expression can vary from completely negative to bright positive.

Except for APL, immunophenotyping is not very useful for the determination of specific cytogenetic translocations in other types of AML, that is, t(8;21)(q22;q22) and inv(16)(p13;q22)/t(16;16)(p13;q22). Associations of expression of CD19 and CD34 with the t(8;21) and the expression of CD2 in myelomonocytic leukemia with the inv(16) have been reported but are not confirmed.[27] Furthermore, many AMLs with the t(8;21) do not express CD19 and CD34 and many CD19 and CD34 positive AMLs do not have a t(8;21). Although the expression of these antigens is not helpful in detecting specific cytogenetic abnormalities, it can be used as a tumor marker in detecting minimal residual disease.

Finally, an indicator of poor response to therapy in AML is the expression of multidrug-resistant glycoprotein (MDR1) gene. High expression of MDR1 is seen in elderly patients and is associated with an unfavorable cytogenetic profile (-5/5q-, -7/7q-, inv(3), or abnormalities at 17p) and a poor prognosis.[35,36]

7.4.1.2   Acute Lymphoblastic Leukemia or Lymphoma

ALL and lymphoblastic lymphoma (LBL) share a common lymphoid origin and arise from progenitor lymphoid cells or lymphoblasts of B- and T-cell lineage. B-cell ALL is more common than the T-cell ALL and comprises approximately 85–90% of the pediatric and 75% of the adult ALLs.[1] However, precursor T-cell LBLs are more common than B-cell LBLs comprising 85% of the cases.[1] Immunophenotyping in ALL and LBL can be used to establish the B- or T-cell lineage of the lymphoblasts to determine the stage of differentiation of the blasts, to identify markers with prognostic significance that can be used to monitor residual disease after chemotherapy, or to assess the expression of molecules that may be targets for immunotherapy. FCM of peripheral blood, bone marrow, or lymphoid tissue is the most useful technique in determining the B- or T-cell lineage of the leukemic blasts. However, IHC can also be used to detect many of the antigens required to assess these disorders. With FCM, the lymphoblasts are typically small and agranular and have a low SSC and FSC. They are distinguished from the mature, small lymphocytes by the expression of CD45, which is dim or negative on lymphoblasts and bright on mature lymphocytes.

The two most important factors in predicting the outcome in pediatric ALL are patients' age and white blood cell count at diagnosis (50,000/μL or higher). The immunophenotype of the blasts has also been shown to have prognostic significance, although the impact of this variable has been lessened with the improvement of therapy. The T-cell immunophenotype has been associated with inferior event-free survival, yet these cases usually present in patients of older age as a mediastinal mass or massive lymphadenopathy with high white blood cell count, which alone can account for such inferior survival.[37] The significance of coexpression of myeloid antigens on lymphoid blasts remains controversial. Some studies have reported that ALLs expressing myeloid markers have a poor prognosis, whereas other studies have shown no significant difference in the outcome of myeloid-positive ALLs compared to the typical precursor B-cell ALLs.[38]

Numerous studies employ microarrays to characterize gene expression signatures of tumors, and use the information for classification or determination of clinically important prognostic factors.[39,40] Moos et al.[41] described gene expression signatures that discriminate between AML versus ALL, T versus B-lineage ALL, and *ETV6-RUNX1 (TEL-AML1)*+ ALL. Furthermore, studies of more than 327 pediatric ALLs have identified seven distinct ALL subtypes: T-ALL, *E2A-PBX1, BCR-ABL, ETV6-RUNX1, MLL* rearranged, hyperdiploid (>50 chromosomes), and a new novel subtype. Although microarrays have the potential to be used in clinical practice, additional validation in independent experiments and cost-effective analyses are needed before they are widely adopted.

7.4.1.2.1 Precursor B-Cell Acute Lymphoblastic Leukemia

The precursor B-cell ALL accounts for approximately 85–90% of pediatric ALLs and approximately 75% of the adult ALLs. The blasts are usually at least partially positive for CD34, a stem cell marker, and the immature lymphoid markers TdT and HLA-DR. Because intense HLA-DR expression is seen in a majority of AMLs and a small number of precursor T-cell ALLs, negative HLA-DR expression is more informative than positive expression in the diagnosis of precursor B-cell ALL. Blasts express the B-cell-associated antigens, CD19, CD22, CD79a; and to a variable degree, CD20. Furthermore, CD9 and CD24—markers that typically present in the intermediate and late stage of normal B-cell development—are frequently expressed in precursor B-cell ALLs.

The degree of differentiation of precursor B-cell lymphoblasts has clinical and genetic correlates. Based on the expression of CD10, CD20, CD34, and cytoplasmic IgM heavy chains (μ), the precursor B-cell ALLs are divided into several groups that follow the normal B-cell maturation patterns (see Figure 7.3). In the earliest pro-pro-B-cell stage, the blasts express CD19, cytoplasmic CD22 and CD79a, and nuclear TdT. More mature pro-B cells are seen in a vast majority of the precursor B-cell ALLs. These blasts express CD10 and CD34, have a variable expression of CD20, and are cytoplasmic μ negative. Because these leukemia have a high frequency, they are referred to as *common B-cell ALLs*. An immunophenotypic example

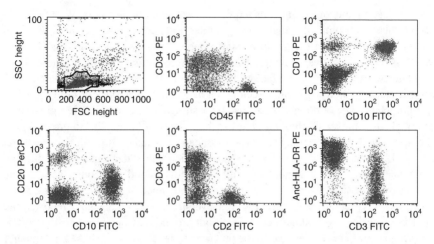

FIGURE 7.8 An example of antigen expression profile of precursor B-cell ALL. The blasts express CD10 (CALLA), CD19, CD34, and HLA-DR. They are CD20 negative.

of precursor B-cell ALL is presented as Figure 7.8. In the more mature pre-B cell stage, the blasts are CD34 negative, express CD20 and cytoplasmic μ, and they have a variable expression of CD10. Surface Igs are usually absent, but their presence does not exclude a diagnosis of precursor B-cell ALL.

Chromosomal abnormalities play a major role in risk assessment and stratification of the pediatric B-cell ALLs. Several well-defined prognostic groups are known to influence the outcome and are used to modify treatment in childhood leukemia. Approximately one-third of the ALLs have an increase in the modal chromosome number. They are the so-called hyperdiploid leukemia. Another one-third show chromosomal translocations along with or without change in the chromosome number. This group includes major translocations—t(12;21)(p12;q22), t(1;19)(q23;p13), t(9;22)(q34;q11.2), and rearrangements of the *MLL* gene on chromosome band11q23 with over 40 partners—each defining a unique biologic subset of patients. Finally, the near hypodiploid group of B-cell ALLs is infrequent and includes leukemia with less than 45 chromosomes.

With the current treatment protocols, more than 90% of the pediatric ALLs are curable. The favorable prognostic group includes the hyperdiploid ALLs and ALLs with t(12;21). These ALLs are more often seen in patients between 1 and 10 years of age. The increased number of chromosomes in the hyperdiploid ALLs is not random, and the favorable outcome is attributed to the positive prognostic impact of trisomies of chromosomes 4, 10, and 17 (triple trisomies).[42] In the second group, ALLs with t(12;21), a chimeric product, *ETV6-RUNX1* is formed as a result of fusion of the *ETV6* gene at 12q13 with the transcription factor-encoding *RUNX1* on chromosome 21. t(12;21) is not detectable by standard cytogenetics. As a result, fluorescent *in situ* hybridization or other molecular techniques are required for its detection. Immunophenotypically, ALLs with t(12;21) show bright expression of CD19, CD10, and HLA-DR, frequently a bimodal pattern of CD34, partial expression or loss of CD9, and partial or dim expression of CD20.[43,44]

Increased resistance to therapy and an unfavorable prognosis are characteristic of lymphoblastic leukemia with a t(1;19), t(9;22), and abnormalities involving chromosome 11q23 resulting in the rearrangement of the *MLL* gene. The t(1;19) generates *E2A–PBX1* fusion product[45] and is present in about one-third of the precursor B-ALLs with a pre-B immunophenotype. The blasts are brightly positive for CD19, CD10, and CD9 and express cytoplasmic μ. The blasts are negative for CD34, and CD20 is often absent or dimly expressed.[46] Immunophenotypic characteristics of ALLs associated with abnormalities at 11q23 are the expression of CD15 and the absence of expression of CD10 and CD24. These leukemia also have strong CD19 expression, variable expression of CD34, and are more often seen in infants below 1 year of age.

Another B-cell leukemia with a poor response to therapy is ALL associated with t(9;22). This translocation results in the formation of the *BCR–ABL* fusion gene and overexpression of tyrosine kinase *ABL*. These leukemia are much more common in adults (20% of adult B-cell ALLs) and are rarely seen in children (4% pediatric B-cell ALLs). Although aberrant expression of myeloid markers (CD13 or CD33) is often seen in these leukemia, such an occurrence is not limited to this genetic background and can be seen in other types of precursor B-cell ALL.

Mature B-cell ALL accounts for 2–5% of all cases of ALL and represents the leukemic manifestation of BL. The leukemic blasts have abundant basophilic cytoplasm, express bright CD45, and tend to merge with the more mature lymphocytes and monocytes on a CD45 versus side scatter histogram. The neoplastic cells express high levels of surface Igs, restricted to either κ or λ, and are notable for the bright expression of CD19, CD20, CD22, and CD24, whereas CD34 and TdT are negative. BL rarely presents with a leukemic phase as is discussed in more detail in Section 7.4.2.2 covering diffuse large B-cell lymphoma (DLBCL).

It is important to be able to distinguish leukemic precursor B lymphoblasts from *normal B-cell progenitors (hematogones)* that are usually present in the bone marrow of infants and young children. An increased number of hematogones are also found in patients with congenital cytopenias, acquired immune deficiency, solid tumors metastatic to the bone marrow, and postmyeloablative chemotherapy.[47,48] Hematogones may persist for more than a year after hematopoietic stem cell transplantation.[49] Distinct from the bone marrow, hematogones are rarely found in the peripheral blood. Although both leukemic blasts and hematogones are CD10 and TdT positive, hematogones lack aberrant antigen expression and show a reproducible pattern of coexpression of markers associated with normal B-cell development— CD10, CD19, CD20, CD34, and CD45. Antigen expression of normal B-cell progenitors (hematogones) is presented in Figure 7.9. Hematogones are a mixture of predominantly intermediate B cells (CD10 and CD19 positive and TdT and surface Ig negative) and more mature B lymphocytes (CD19, CD20, and surface Ig positive). In contrast, the neoplastic blasts differ from normal with a predominance of immature cells and a paucity of mature forms. The maturation pattern of hematogones is easily depicted by multiparameter FCM as shown in Figure 7.9. The population of hematogones is markedly heterogeneous in terms of CD20 intensity (from negative to bright), and an inverted "J" from the CD10/CD20 antigen expression profile is easily identified. If present, the expression of CD20 on the leukemic blasts is less

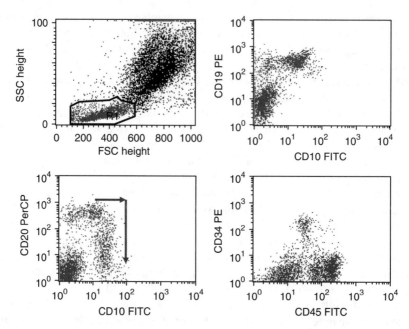

FIGURE 7.9 Antigen expression of normal B-cell progenitors (hematogones). A two-parameter dot plot histogram of bone marrow shows multiple cell populations. A small population of normal, immature B cells (hematogones) expressing CD10, CD19, CD20, and CD34 (partial) is of interest. This population is markedly heterogeneous in terms of CD20 intensity and produces a characteristic inverted J-shaped trail pattern on the CD10–CD20 dot plot. Although the precursor B-cell ALLs can be CD20 positive, the CD20 intensity is more homogeneous when compared to the CD20 intensity in hematogones.

heterogeneous, than on hematogones. Furthermore, asynchronous expression of the earliest and latest antigens (i.e., concurrent CD34 and CD20), aberrant expression of myeloid markers (such as CD13 or CD33), and aberrant under- or overexpression of antigens are absent in hematogones and often present in leukemic blasts.

7.4.1.2.2 Precursor T-Cell Acute Lymphoblastic Leukemia

T-cell immunophenotype is present in 25% of adult and 15% of pediatric ALL cases.[32] In contrast to precursor B-cell ALLs, the T-cell lesions are present more often with tissue involvement as lymphoma and less often as leukemia. Precursor T-cell ALL has an immunophenotype that resembles normal maturing thymocytes. They are characterized by the expression of CD7, CD5, and CD2 with negative surface CD3, but positive cytoplasmic CD3. Although CD4 and CD8 can be expressed in all combinations, they are often either both negative or both positive (Figure 7.10). CD4/CD8 double-positive cells, when present, can be considered diagnostic of T-cell ALL as can the expression of CD1a. TdT and CD34 also are often positive. CD10 is expressed in approximately 25% of the precursor T-cell ALLs.[5]

Based on the expression of CD1a and surface CD3, the T-cell ALLs can be classified according to the stages of intrathymic T-cell differentiation (see Figure 7.3). The immature thymocytes do not express CD1a or surface CD3. They are CD7 and

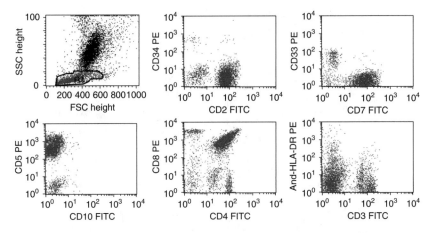

FIGURE 7.10   An example of antigen expression profile of precursor T-cell ALL. The neoplastic cells express CD2, CD5, and CD7 and coexpress CD4 and CD8, but are surface CD3, CD34, and HLA-DR negative.

cytoplasmic CD3 positive and may coexpress CD34 and myeloid-associated antigens (e.g., CD33). The intermediate or cortical T-ALL is defined by the expression of CD1a and coexpression of CD4 and CD8. Most of these ALLs are surface CD3 negative, but in some cases they may be positive. The most mature T-ALLs are CD1a negative, express either CD4 or CD8, and are surface CD3 positive.[50]

The immunophenotypic classification of T-cell ALL is important and has been shown to correlate with survival. The highest level of disease-free survival is seen in those cases of T-ALL that recapitulate the immature or "common thymocyte" stage of T-cell maturation (see Figure 7.4). These leukemia coexpress CD4 and CD8 in addition to CD7, CD1a, CD2, and CD5. Recent microarray gene expression studies in precursor T cell have provided further understanding of the biologic heterogeneity of the disease and have also revealed clinically significant subtypes. Using oligonucleotide microarrays, Ferrando et al.[51,52] identified several gene signatures that are indicative of leukemic arrest at specific stages of normal T-cell development: *LYL1+* is seen in pro-T-cells, *HOX11+* in early cortical thymocytes, and *TAL1+* in late cortical thymocytes (Figure 7.11). Overexpression of *HOX11* gene occurs in approximately 5–10% of the childhood T-cell ALLs, 30% of the adult T-cell ALLs, and is associated with cortical thymocyte immunophenotype and a favorable prognosis. However, T-ALLs with overexpression of *TAL1, LYL1,* or, surprisingly, *HOX11L2* exhibit a much poorer response to treatment.[51,52] The prognostic significance of these genomic features is yet to be determined. At present, phenotyping for the gene products is not routinely performed.

7.4.2 NON-HODGKIN LYMPHOMA

Non-Hodgkin lymphoma (NHLs), a heterogeneous group of disorders, can be divided into mature B-cell and mature T-cell neoplasms based on the cells of origin. The usefulness of immunophenotyping in making the correct diagnosis for NHLs depends

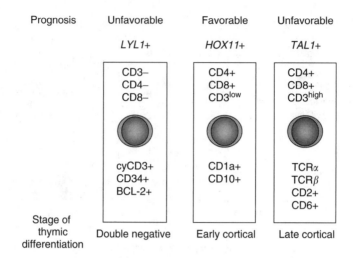

Prognosis	Unfavorable	Favorable	Unfavorable
	LYL1+	*HOX11+*	*TAL1+*

CD3– CD4– CD8–	CD4+ CD8+ CD3low	CD4+ CD8+ CD3high
cyCD3+ CD34+ BCL-2+	CD1a+ CD10+	TCRα TCRβ CD2+ CD6+

Stage of thymic differentiation   Double negative   Early cortical   Late cortical

FIGURE 7.11 Correlation of gene expression profiles in LYL1+, HOX11+, and TAL1+ with precursor T-cell ALL with recognized stages of thymic differentiation. Cell surface markers normally associated with each developmental stage are shown with corresponding microarray findings. The leukemia with an immunophenotype of early cortical thymocytes show constitutive expression of *HOX11* and are associated with a favorable prognosis. The *LYL1* and *TAL1* expressions seen in leukemia with double negative or late cortical thymocyte are associated with a poor response to therapy and a shorter 5-year survival. (Adapted from Ferrando, A. A., et al., *Lancet*, 363(9408), 535–536, 2004; Ferrando, A. A., et al., *Cancer Cell*, 1(1), 75–87, 2002. With permission.)

on the specific disease. For some lymphoma, such as follicular, marginal zone B-cell lymphoma of mucosa-associated lymphoid tissue (MALT) type, or small lymphocytic lymphoma, information on the immunophenotype does not increase the diagnostic accuracy significantly. However, for other types, such as mantle cell, diffuse large B-cell, or T-cell lymphoma, immunophenotyping is useful in many cases in reaching the correct diagnosis and improving the diagnostic accuracy by 10–45%.[53] The remainder of the chapter focuses on the most common NHLs with characteristic immunophenotypes.

7.4.2.1 Mature B-Cell Non-Hodgkin Lymphoma

The mature B-cell neoplasms are clonal proliferations of B cells at various stages of differentiation, from naïve B cells to mature plasma cells, and they comprise over 85% of the NHLs worldwide.[1] Some of them present as disseminated leukemia and lymphoma with major involvement of the bone marrow and peripheral blood, whereas others essentially involve lymph nodes or extranodal tissue. Mature B-cell lymphoma typically have a distinctive morphology and immunophenotype that allow them to be readily classified according to their postulated cell of origin. Several mature B-cell neoplasms have distinctive cytogenetic abnormalities, usually involving translocations that place a potential cellular oncogene under the influence of the Ig gene promoter.

Immunophenotyping is used to distinguish the lymphoma of small cell types—CLL and chronic lymphocytic lymphoma, follicular lymphoma (FL), MALT lymphoma, and plasma-cell neoplasms—from a benign reactive process and to distinguish the DLBCL and BL from nonlymphoid tumors. In general, no one antigen is specific for any of these neoplasms. As a result, a combination of morphologic features and a panel of antigens are necessary for the correct diagnosis. Table 7.2 presents the immunophenotype and genetic features of common small B-cell neoplasms.

FCM of fresh cells is particularly helpful in demonstrating the expression of pan-B-cell markers (CD19, CD20, CD22, and CD79a) of the neoplastic cells and in detecting their light-chain restriction. The expression of only one Ig light chain, either κ or λ, or a skewed $\kappa:\lambda$ ratio (i.e., $\kappa:\lambda$ of more than 6:1 or less than 1:1), or lack of expression of surface Ig is immunophenotypic evidence of presence of monotypic B-cell population.[12,54] When the clonal nature of the B cells cannot be determined by immunophenotyping, however, molecular techniques, such as Ig heavy gene rearrangement study, can be employed to determine clonality of B cells.

7.4.2.2 Chronic Lymphocytic Leukemia and Small Lymphocytic Lymphoma

CLL and SLL are generally regarded as blood and tissue phases of the same disease (respectively) and have the same immunophenotype. The neoplastic cells express

---

TABLE 7.2
Immunophenotype and Genetic Features of Small B-Cell Lymphoma and Leukemia

B-Cell NHL	SIg	CD5	CD10	CD23	CD11c	CD103	Cyclin D1	Genetic Abnormalities
CLL/SLL[a]	Dim	+	–	++/+	++/– wk	–	–	+12; 13q del
MCL	+	+	–	–	–	–	+	t(11;14), 13q del, +12
FL	+	–	+	–	–	–	–	t(14;18), +7, +18, 3q27–28, 6q23–26, 17p
MALT lymphoma	+	–	–	–	++/– wk	++/– wk	–	+3, t(11;18), t(1;14), t(14;18)
HCL	+	–	–	–	++/– wk	+	–	Not known

Note: NHL, non-Hodgkin lymphoma; SIg, surface immunoglobulin; CLL/SLL, chronic lymphocytic leukemia and small lymphocytic lymphoma; MCL, mantle cell lymphoma; FL, follicular lymphoma; MALT, mucosa-associated lymphoid tissue; HCL, hairy cell leukemia; +/– wk, weak positive.

[a] CLL/SLL can be differentiated from MCL by expression of CD79b and FMC-7 (both absent in CLL/SLL and bright in MCL).

Source: Adapted from Jaffe, E. S., Harris, N. L., Stein, H., Vardiman, J. (Eds.). *World Health Organization Classification of Tumours: Pathology and Genetics of Tumours of Haematopoietic and Lymphoid Tissues*. Lyon: IARC Press, 2001.

most of the pan-B-cell antigens and are CD5 and CD23 positive. They have dim expression of surface IgM and are FMC-7 negative. Neither CD5 nor CD23 alone is sufficient for the diagnosis of CLL and SLL because other low-grade B-cell lymphoproliferative disorders can express these antigens.

Although most of the CLLs progress slowly and have relatively benign course, a small group of them have aggressive behavior and show a lack of response to therapy. This variability in disease progression has been correlated with the mutational status of the heavy-chain Ig-variable region V(H) gene. Surrogate markers for V(H) status are CD38 and ZAP-70 expression of the neoplastic cells, but their validity has yielded controversial results. Current data indicate that CLL cells with ZAP-70 expression in more than 20% of the cells, high CD38 antigen density, and lack of somatic hypermutation of VH genes, have more aggressive behavior and therefore a poorer prognosis.[55–58] Patients with somatic VH hypermutations, or few CD38 and ZAP-70 positive cells, appear to have a better prognosis.

Genomic aberrations are the other genetic parameters shown to be of prognostic relevance in CLL. Deletion at 17p and 11q is associated with shorter survival time, whereas trisomy 12q and deletion 13q are associated with longer survival. The significance of trisomy 12 seen in 15–20% of CLLs, however, is still unresolved.[59,60]

Mantle cell lymphoma (MCL) is another CD5+ B-cell lymphoma. CD23 is usually dimly expressed or is negative and FMC-7 is bright positive, which is helpful in distinguishing MCL from CLL and SLL. Virtually all cases of the MCLs have a t(11;14)(q13;q32) between CCND1 (Cyclin D1/BCL1) and Ig heavy-chain genes. This translocation results in the overexpression of Cyclin D1, which can be detected by IHC in virtually all cases of MCL. The t(11;14)(q13;q32) can be detected using fluorescence *in situ* hybridization.[61]

The immunophenotype of FL reflects normal germinal center B cells. The tumor cells characteristically express CD10 (in approximately 75% of the cases) and BCL6. The cells are CD5 and CD23 negative with bright expression of FMC-7. Virtually all cases of FL have cytogenetic abnormalities. Of the FLs, 70–80% carry t(14;18)(q32;q21) as a sole cytogenetic abnormality, involving rearrangement and overexpression of the BCL2 gene and the Ig heavy-chain locus. Rare cases with t(2;18)(p11;q21) and t(18;22)(q21;q11) have been reported which place the BCL2 gene with the light-chain gene on chromosome 2 or 22. Overexpression of BCL2 confers a survival advantage of B cells *in vitro*, by preventing apoptosis under conditions of growth factor deprivation. Although immunophenotypic detection of BCL2 in germinal center B lymphocytes is useful to distinguish between FL and reactive lymphoid hyperplasia, its utility in differentiating among different B-cell lymphoma is limited because most of them are BCL2 positive. Furthermore, other mature B-cell neoplasms, such as BL and a subset of DLBCLs, express CD10 and should be distinguished from FL. Approximately 15% of the FLs show abnormalities at chromosome 3q27 involving the *BCL6* gene.

HCL is a neoplasm of small B lymphoid cells with oval nuclei and abundant cytoplasm with "hairy" projections in peripheral blood, diffusely infiltrating the bone marrow and splenic red pulp. The tumor cells are surface Ig positive, and express B-cell-associated antigens except for CD79b. They are typically CD5, CD10, and CD23 negative, and strongly express CD11c and CD25 as well as FMC7 and CD103.

Tartrate-resistant acid phosphatase is present in most of the cases, but is neither specific nor required for the diagnosis.[62]

Splenic marginal-zone lymphoma (SMZL) is another mature B-cell lymphoproliferative disorder. The immunophenotype shares similarities with HCL. The neoplastic cells are usually DBA-44 positive and may also be CD11c positive. However, unlike HCL, these cells are CD25 and CD103 negative.

Extranodal marginal-zone lymphoma of MALT are slowly progressing B-cell lymphoma involving the gastrointestinal tract, lung, head and neck, ocular adnexae, skin, thyroid, and breast. Antigenic stimulation by infectious agents contributes to lymphoma pathogenesis. For example, in more than two-thirds of the cases, infection with *Helicobacter pylori* is associated with gastric MALT lymphoma and eradication of the infection results in clinical regression of the disease. The MALT lymphoma typically express pan-B-cell markers and IgM and show a light-chain restriction. They are typically CD5, CD10, CD23, and Cyclin D1 negative. The cells express the marginal zone cell associated antigens—CD21 and CD35—and in approximately 50% of the cases are CD43, a T-cell marker, positive. Trisomy 3 is a common cytogenetic abnormality found in 60% of the MALT lymphoma, and t(11;18)(q21;q21) is seen in 25–50% of the cases. Other translocations identified in MALT lymphoma include t(1;14)(p22;q32) and t(14;18)(q32;q21).[63]

DLBCL constitutes 30–40% of the adult NHLs and patients require aggressive chemotherapy for survival. The immunophenotype of DLBCL is variable and reflects the heterogeneity of this group of tumors. The neoplastic cells have strikingly a high forward scatter on FCM and variably express pan-B-cell markers (CD19, CD20, CD79a, and PAX-5). Despite their B-cell origin, a subset of these lymphoma does not express surface or cytoplasmic Ig.

Response to therapy and gene expression profiling has confirmed that the DLBCLs are a heterogeneous group of disorders and can be divided into several prognostically distinct groups[2,64,65] DLBCL with a gene expression profile similar to germinal center B cells (CD10 and BCL6 positive and activation markers, MUM1/IRF4 and CD138 negative) have a much better prognosis than DLBCL cases with gene expression profiles resembling activated B cells (CD10 and BCL6 negative and MUM1/IRF4 and CD138 positive).[66]

Another subtype of DLBCL, primary mediastinal B-cell lymphoma (PMBCL), which more often affects young women, has a characteristic expression profile.[67] The neoplastic cells have a B-cell phenotype and express CD30 and BCL2. CD23 is positive in most of the cases, in contrast to the other DLBCLs. The cells usually express BCL6, but are CD10 and CD21 negative. MUM1 is variably positive. Although the expression profile of PMBCL shares some similarities with Hodgkin lymphoma (HL), the overall pattern of B-cell signaling molecules in this disease is closer to that of DLBCL.

BL is a highly aggressive B-cell lymphoma characterized by translocations involving *cMYC* proto-oncogene. The tumor cells express pan-B-cell-associated antigens (e.g., CD19, CD20, and CD22), CD10, and BCL6, which are surface IgM positive and show Ig light-chain restriction. The cells are negative for CD5, CD23, BCL2, and TdT. The blasts in BL presenting as leukemia have a mature B-cell phenotype, that is, a brighter expression of CD45 when compared to the precursor

B lymphoblasts, CD34 and TdT negative. BL is a tumor with a very high growth fraction in which almost 100% of the cells are proliferating and Ki-67 (MIB1) positive. All cases of BL have a translocation of *cMYC* at 8q24 with the Ig heavy-chain gene on 14q32 (t(8;14)(q24;q32)), or less commonly with one of the Ig light-chain genes, 2q11 (t(2;8)(p12;q24)), or 22q11 (t(8;22)(q24;q11)). In all cases, *cMYC* is constitutively expressed, which drives the tumor cells into the cell cycle.

Plasma-cell dyscrasias (PCDs) are always bright CD38 and CD138 positive and express cytoplasmic but not surface Igs. The cytoplasmic Ig expression can be detected by FCM using an additional permeabilization step or by IHC. Expression of other B-cell antigens by the plasma cells is variable. Almost always at least one marker, most commonly CD20, is lost. CD45 is also poorly expressed. Aberrant expression of CD56 is frequent.

7.4.2.3   Mature T-Cell Lymphoproliferative Disorders

The mature T-cell neoplasms are relatively uncommon and represent approximately 12% of all NHLs.[53] They are derived from the mature or postthymic T cells and they share some immunophenotypic and functional properties with NK cells.

Establishing B-cell clonality by immunophenotyping and demonstration of Ig light-chain restriction is straightforward, whereas establishing T-cell clonality by immunophenotyping is less clear-cut. Immunophenotypic criteria helpful in the diagnosis of mature T-cell lymphoma include (1) an aberrant loss of one or more pan T-cell markers relative to the normal T-cell population, (2) T-cell subset antigen restriction (i.e., CD4 or CD8) or predominance of either CD4 or CD8, (3) dual-positive or -negative CD4 and CD8 expression, (4) loss or abnormal expression of CD45, (5) restricted $\gamma\delta$ TCR rather than $\alpha\beta$ TCR expression, and (6) expression of additional or aberrant antigens not normally expressed on T cells. Although the presence of aberrant immunophenotype is suggestive of T-cell malignancy, it is not sufficient for such determination because some reactive T-cell proliferations may also present with a loss of a pan-T-cell marker. Thus, the immunophenotypic results should be interpreted in conjunction with morphologic findings.

In the last decade, new antibodies against the protein product of the variable gene segment of the TCRβ, -γ, and -δ (Vβ, Vγ, and Vδ) have been developed, are commercially available, and can be used for the determination of T-cell clonality.[68,69] (For further discussion on TCR $\lambda\beta$ immunophenotyping see Chapters 9 and 10.) However, these are expensive and, therefore, rarely used for diagnosis. Currently, in the diagnostic workup of T-cell malignancy, the test of choice for determining the T-cell clonality is polymerase chain reaction (PCR)–based *TCR* gene rearrangement studies.

Presently, specific genetic abnormalities have not been identified for many T- and NK-cell neoplasms. One of the few exceptions is anaplastic large cell lymphoma (ALCL), which is characterized with translocations involving *ALK1* locus on chromosome 2. In all other cases, therefore, the clinical features and immunophenotype play a major role in the subclassification of the T- and NK-cell neoplasms.

T-cell prolymphocytic leukemia (T-PLL) is an aggressive T-cell neoplasm characterized by proliferation of small to medium-sized T prolymphocytes with a mature postthymic phenotype. These cells are CD1a and TdT negative and are characteristized by the strong expression of CD7. The T cells express CD4 in most of the cases, but dual CD4/CD8 and rarely CD8 expression are seen. T-PLLs are HLA-DR and CD25 negative, which distinguishes them from human T-cell leukemia virus type 1 (HTLV)-1-associated adult T-cell leukemia and lymphoma (ATLL). Most of the T-cell PLL express $\alpha\beta$ TCR and only rare $\gamma\delta$ T-cell PLL cases have been reported.

ATLL is a peripheral T-cell neoplasm often composed of highly pleomorphic lymphoid cells. The disease is usually widely disseminated in the body, and is caused by the HTLV-1. The tumor cells have an activated T-helper phenotype expressing T-cell antigens (CD2, CD3, and CD5), CD4, CD25, CD95, and HLA-DR. They are usually negative for CD7, CD8, and CD56, and do not express cytotoxic-associated molecules, granzyme B and TIA-1.

T-cell large granular lymphocytic leukemia (T-LGL) generally exhibits a persistent (more than 6 months) increase in the number of large granular lymphocytes without other clearly identified causes. Approximately 80% of the cases are CD3, TCR$\alpha\beta$, CD8 positive, and CD4 negative. Rare variants may be CD4 positive, express both CD4 and CD8, or be of TCR$\gamma\delta$ type. They commonly express CD16 and CD57, and NK-associated antigen; however, they are CD56 negative. T-LGL expresses cytotoxic granule-associated proteins, TIA-1 and granzyme B.

Mycosis fungoides (MF) and Sezary syndrome (SS) share similarities to the point that SS is traditionally regarded as a variant of MF. MF presents with skin patches or plaques, whereas SS is generalized and usually more aggressive. The neoplastic cells are usually of a helper type, CD4 positive, and express pan-T-cell antigens (CD2, CD3, CD5, and CD45RO). Aberrant expression of T-cell antigens (i.e., dim expression of CD3 and CD4) is common. A majority of the cases are negative for CD7, CD8, and CD25.

The cells in Anaplastic Large Cell Lymphoma (ALCL) are usually large with pleomorphic, often horseshoe-shaped nuclei, and abundant cytoplasm. The vast majority of the ALCLs express anaplastic large cell lymphoma kinase (ALK) protein due to genetic alteration of the *ALK* locus on chromosome 2.[70] Although t(2;5)(p23;35) between *ALK1* gene and the *nucleophosmin* (*NPM1*) gene on chromosome 5 is most common, variant translocations involving *ALK* and other partner genes on chromosomes 1, 2, 3, 17, and 22 have been reported. The tumor cells have a characteristic CD30 expression (membrane and Golgi region pattern) and most of them express at least one pan-T-cell marker. CD3 is negative in more than 75% of the cases. In contrast, CD2 and CD4 are positive in most of the cases and are the most helpful immunophenotypic markers. The neoplastic cells express cytotoxic molecules, TIA-1 and granzyme B, whereas CD8 is usually negative. In some cases several pan-T-cell antigens are lost and the neoplastic cells have an apparent "null-cell" phenotype. These cases, by definition, express CD30 and ALK1 and molecular studies can confirm the T-cell lineage of the neoplastic process.[71]

Hepatosplenic T-cell lymphoma is a rare clonal T-cell proliferation derived from cytotoxic T cells that are usually of $\gamma\delta$ TCR type. The neoplastic cells are CD3 positive and express cytotoxic-associated protein, TIA-1, but are usually negative for perforin and lack both CD4 and CD8. In rare cases, CD8 is positive. A minority of the cases can be of the $\alpha\beta$ TCR type.[72]

A large number of T-cell lymphoma have no recognizable T-cell subtype and do not correspond to a particular clinicopathologic entity. These lymphoma are grouped together as peripheral T-cell lymphoma, unspecified. Although peripheral T-cell lymphoma, unspecified, express T-cell-associated antigens, aberrant T-cell phenotypes are common. Most nodal cases are CD4 positive and expression of cytotoxic granule-associated proteins is rare. TCR genes are clonally rearranged and a complex karyotype is common.

7.4.3 HODGKIN LYMPHOMA

Unlike the NHLs where a vast majority of the cells are neoplastic and there is a minor population of reactive cells, in HLs, the large neoplastic cells represent a minor population in an abundant heterogeneous admixture of nonneoplastic inflammatory and accessory cells. The large cells are known as Hodgkin and Reed–Sternberg cells and they are usually ringed by T lymphocytes in a rosette-like manner. The Hodgkin and Reed–Sternberg cells are derived from germinal center B cells. Biological and clinical studies in the last 20 years have shown that HLs are composed of two disease entities—nodular lymphocyte predominant HL (NLPHL) and classical HL. These lymphoma have a characteristic morphologic appearance and immunophenotype. Characteristic of the classical HL immunophenotype is the expression of CD15 and CD30 and a lack of expression of CD45. Approximately 20% of the cases express CD20. In the NLPHL, the neoplastic large cells express CD20, CD79a, BCL6, and CD45 in nearly all cases, and are negative for CD15 and CD30. IHC is the technique of choice because the small number of neoplastic cells has not been reliably detected by FCM.

Immunohistochemical staining with OCT-2 is helpful in the differential diagnosis of classical HL and NLPHL. OCT-2 is a transcription factor that induces Ig synthesis by activating the promoter of Ig genes in conjunction with its coactivator BOB.1. Both molecules are consistently expressed in NLPHL, whereas both are absent in 80% of the classical HL and one of them can be expressed in the remaining 20%. Classical HLs in which both OCT-2 and BOB.1 are expressed have not been observed. Table 7.3 summarizes the immunophenotypic profiles of HLs and other large cell lymphoma.

7.5 CONCLUDING REMARKS

Immunophenotyping of leukemia and lymphoma is an essential tool in diagnostic hematopathology. It allows proper lineage stratification and identification of diseases. Apart from its value in diagnosing a particular disease, immunophenotyping also permits the development of the differential diagnosis, which can be resolved in conjunction with morphologic features and other ancillary studies. With respect to predicting disease progression and response to therapy, however, the role of immunophenotyping alone is relatively limited. Other factors, such as the presence or absence of specific genetic alterations, appear to be more informative in this area and must be sought out.

ACKNOWLEDGMENTS

The author extends her great appreciation to John E. Schwarz from the University of Arizona for editorial assistance; Elizabeth Hyjek from the University of Chicago and Katrin Carlson and Lawrence Jennings from Children's Memorial Hospital

TABLE 7.3
Immunophenotypic Profiles of Large Cell Lymphoma

Marker	NLPHL	Classical HL	DLBCL	ALCL
CD30	−	+	−/+	+
CD15	−	−/+	−	−
CD45	+	−	+	+−/
CD20	+	−/+	+	−
CD79a	+	−/+	+	−
PAX-5	+	+	+	+
OCT-2	+	−/+	+	n.a.
BOB.1	+	−	+	n.a.
CD3	−	−	−	−/+
CD2	−	−	−	+−/
Perforin/Granzyme B	−	−	−	+
CD43	−	−	−/+	+−/
ALK-1	−	−	−	+−/

Note: NLPHL, nodular lymphocyte predominant Hodgkin lymphoma; HL, Hodgkin lymphoma; DLBCL, diffuse large B-cell lymphoma; ALCL, anaplastic large cell lymphoma +, positive in all cases; +/− majority of cases positive; −/+ minority of the cases positive; − all cases negative.

Source: Adapted from Jaffe, E. S., Harris, N. L., Stein, H., Vardiman, J. (Eds.). *World Health Organization Classification of Tumours: Pathology and Genetics of Tumours of Haematopoietic and Lymphoid Tissues*. Lyon: IARC Press, 2001.

for their critical reading of the manuscript and valuable suggestions; N. Bensen, L. Wolfsen, L. Lott, J. Little, C. McMorran, and E. Fullmer from the Immunology Laboratory of Children's Memorial Hospital in Chicago for their excellent work and providing flow cytometric data for illustration; and Elyse DeVries from Children's Memorial Hospital in Chicago for her outstanding technical support.

REFERENCES

1. Jaffe, E. S., Harris, N. L., Stein, H., Vardiman, J. (Eds.), *Pathology and Genetics of Tumours of Haematopoietic and Lymphoid Tissues*. Lyon: IARC Press, 2001, p. 351.
2. Staudt, L. M., Dave, S., The biology of human lymphoid malignancies revealed by gene expression profiling. *Advances in Immunology* 2005, 87, 163–208.
3. Zola, H., Swart, B., Nicholson, I., Aasted, B., Bensussan, A., Boumsell, L., Buckley, C., Clark, G., Drbal, K., Engel, P., Hart, D., Horejsi, V., Isacke, C., Macardle, P., Malavasi, F., Mason, D., Olive, D., Saalmueller, A., Schlossman, S. F., Schwartz-Albiez, R., Simmons, P., Tedder, T. F., Uguccioni, M., Warren, H., CD molecules 2005: human cell differentiation molecules. *Blood* 2005, 106(9), 3123–3126.
4. Jennings, C. D., Foon, K. A., Recent advances in flow cytometry: application to the diagnosis of hematologic malignancy. *Blood* 1997, 90(8), 2863–2892.
5. Kaleem, Z., Crawford, E., Pathan, M. H., Jasper, L., Covinsky, M. A., Johnson, L. R., White, G., Flow cytometric analysis of acute leukemia. Diagnostic utility and critical analysis of data. *Archives of Pathology & Laboratory Medicine* 2003, 127(1), 42–48.

6. Kaleem, Z., Flow cytometric analysis of lymphoma: current status and usefulness. *Archives of Pathology & Laboratory Medicine* 2006, 130(12), 1850–1858.

7. Stewart, C. C., Behm, F. G., Carey, J. L., Cornbleet, J., Duque, R. E., Hudnall, S. D., Hurtubise, P. E., Loken, M., Tubbs, R. R., Wormsley, S., U.S.–Canadian Consensus recommendations on the immunophenotypic analysis of hematologic neoplasia by flow cytometry: selection of antibody combinations. *Cytometry* 1997, 30(5), 231–235.

8. Stelzer, G. T., Marti, G., Hurley, A., McCoy, P., Jr., Lovett, E. J., Schwartz, A., U.S.–Canadian Consensus recommendations on the immunophenotypic analysis of hematologic neoplasia by flow cytometry: standardization and validation of laboratory procedures. *Cytometry* 1997, 30(5), 214–230.

9. Braylan, R. C., Atwater, S. K., Diamond, L., Hassett, J. M., Johnson, M., Kidd, P. G., Leith, C., Nguyen, D., U.S.–Canadian Consensus recommendations on the immunophenotypic analysis of hematologic neoplasia by flow cytometry: data reporting. *Cytometry* 1997, 30(5), 245–248.

10. Borowitz, M. J., Guenther, K. L., Shults, K. E., Stelzer, G. T., Immunophenotyping of acute leukemia by flow cytometric analysis. Use of CD45 and right-angle light scatter to gate on leukemic blasts in three-color analysis. *American Journal of Clinical Pathology* 1993, 100(5), 534–540.

11. Borowitz, M. J., Bray, R., Gascoyne, R., Melnick, S., Parker, J. W., Picker, L., Stetler-Stevenson, M., U.S.–Canadian Consensus recommendations on the immunophenotypic analysis of hematologic neoplasia by flow cytometry: data analysis and interpretation. *Cytometry* 1997, 30(5), 236–244.

12. Braylan, R. C., Impact of flow cytometry on the diagnosis and characterization of lymphomas, chronic lymphoproliferative disorders and plasma cell neoplasias. *Cytometry Part A: The Journal of the International Society for Analytical Cytology* 2004, 58(1), 57–61.

13. Polak, J. M., Van Noorden, S., *Introduction to Immunocytochemistry.* J. M. Polak, S. Van Noorden (Eds.), 3rd ed. Oxford: BIOS Scientific Publishers, 2003, p. 176.

14. Leong, A. S. Y., Leong, T. Y. M., Newer developments in immunohistology. *Journal of Clinical Pathology* 2006, 59(11), 1117–1126.

15. Gown, A. M., Genogenic immunohistochemistry: a new era in diagnostic immunohistochemistry. *Current Diagnostic Pathology* 2002, 8(3), 193–200.

16. Bender, J. G., Unverzagt, K., Walker, D. E., Lee, W., Smith, S., Williams, S., Van Epps, D. E., Phenotypic analysis and characterization of CD34+ cells from normal human bone marrow, cord blood, peripheral blood, and mobilized peripheral blood from patients undergoing autologous stem cell transplantation. *Clinical Immunology and Immunopathology* 1994, 70(1), 10–18.

17. Nutt, S. L., Eberhard, D., Horcher, M., Rolink, A. G., Busslinger, M., Pax5 determines the identity of B cells from the beginning to the end of B-lymphopoiesis. *International Reviews of Immunology* 2001, 20(1), 65–82.

18. Selsing, E., Ig class switching: targeting the recombinational mechanism. *Current Opinion in Immunology* 2006, 18(3), 249–254.

19. Dudley, D. D., Chaudhuri, J., Bassing, C. H., Alt, F. W., Mechanism and control of V(D)J recombination versus class switch recombination: similarities and differences. *Advances in Immunology* 2005, 86, 43–112.

20. Moretta, L., Moretta, A., Unravelling natural killer cell function: triggering and inhibitory human NK receptors. *EMBO Journal* 2004, 23(2), 255–259.

21. Moretta, A., Biassoni, R., Bottino, C., Mingari, M. C., Moretta, L., Natural cytotoxicity receptors that trigger human NK-cell-mediated cytolysis. *Immunology Today* 2000, 21(5), 228–234.

22. Moretta, L., Moretta, A., Killer immunoglobulin-like receptors. *Current Opinion in Immunology* 2004, 16(5), 626–633.

23. Bennett, J. M., Catovsky, D., Daniel, M. T., Flandrin, G., Galton, D. A., Gralnick, H. R., Sultan, C., Proposed revised criteria for the classification of acute myeloid leukemia. A report of the French-American-British Cooperative Group. *Annals of Internal Medicine* 1985, 103(4), 620–625.

24. Harris, N. L., Jaffe, E. S., Diebold, J., Flandrin, G., Muller-Hermelink, H. K., Vardiman, J., Lister, T. A., Bloomfield, C. D., The World Health Organization classification of neoplastic diseases of the haematopoietic and lymphoid tissues: Report of the Clinical Advisory Committee Meeting, Airlie House, Virginia, November 1997. *Histopathology* 2000, 36(1), 69–86.

25. Harris, N. L., Jaffe, E. S., Stein, H., Banks, P. M., Chan, J. K., Cleary, M. L., Delsol, G., De Wolf-Peeters, C., Falini, B., Gatter, K. C., Grogan, T. M., Isaacson, P.G., Knowles, D.M., Mason, D.Y., Muller-Hermelink, H.K., Pireli, S.A., Piris, M.A., Ralfkiaer, E., Warnke, R.A., A revised European–American classification of lymphoid neoplasms: a proposal from the International Lymphoma Study Group. *Blood* 1994, 84(5), 1361–1392.

26. Heerema, N. A., Bernheim, A., Lim, M. S., Look, A. T., Pasqualucci, L., Raetz, E., Sanger, W. G., Cairo, M. S., State of the art and future needs in cytogenetic/molecular genetics/arrays in childhood lymphoma: summary report of workshop at the First International Symposium on childhood and adolescent non-Hodgkin lymphoma, April 9, 2003, New York City, NY. *Pediatric Blood and Cancer* 2005, 45(5), 616–622.

27. Khalidi, H. S., Medeiros, L. J., Chang, K. L., Brynes, R. K., Slovak, M. L., Arber, D. A., The immunophenotype of adult acute myeloid leukemia: high frequency of lymphoid antigen expression and comparison of immunophenotype, French-American-British classification, and karyotypic abnormalities. *American Journal of Clinical Pathology* 1998, 109(2), 211–220.

28. Cross, A. H., Goorha, R. M., Nuss, R., Behm, F. G., Murphy, S. B., Kalwinsky, D. K., Raimondi, S., Kitchingman, G. R., Mirro, J., Jr., Acute myeloid leukemia with T-lymphoid features: a distinct biologic and clinical entity. *Blood* 1988, 72(2), 579–587.

29. Ball, E. D., Davis, R. B., Griffin, J. D., Mayer, R. J., Davey, F. R., Arthur, D. C., Wurster-Hill, D., Noll, W., Elghetany, M. T., Allen, S. L., Rai, K., Lee, E. J., Schiffer, C. A., Bloomfield, C.D., Prognostic value of lymphocyte surface markers in acute myeloid leukemia. *Blood* 1991, 77(10), 2242–2250.

30. Solary, E., Casasnovas, R. O., Campos, L., Bene, M. C., Faure, G., Maingon, P., Falkenrodt, A., Lenormand, B., Genetet, N., Surface markers in adult acute myeloblastic leukemia: correlation of CD19+, CD34+ and CD14+/DR–phenotypes with shorter survival. Groupe d'Etude Immunologique des Leucemies (GEIL). *Leukemia* 1992, 6(5), 393–399.

31. Samoszuk, M. K., Tynan, W., Sallash, G., Nasr, S., Monczak, Y., Miller, W. H., Jr., An immunofluorescent assay for acute promyelocytic leukemia cells. *American Journal of Clinical Pathology* 1998, 109(2), 205–210.

32. Weir, E. G., Borowitz, M. J., Flow cytometry in the diagnosis of acute leukemia. *Seminars in Hematology* 2001, 38(2), 124–138.

33. Krasinskas, A. M., Wasik, M. A., Kamoun, M., Schretzenmair, R., Moore, J., Salhany, K. E., The usefulness of CD64, other monocyte-associated antigens, and CD45 gating in the subclassification of acute myeloid leukemia with monocytic differentiation. *American Journal of Clinical Pathology* 1998, 110(6), 797–805.

34. Betz, S. A., Foucar, K., Head, D. R., Chen, I. M., Willman, C. L., False-positive flow cytometric platelet glycoprotein IIb/IIIa expression in myeloid leukemia secondary to platelet adherence to blasts. *Blood* 1992, 79(9), 2399–403.

35. Leith, C. P., Kopecky, K. J., Chen, I. M., Eijdems, L., Slovak, M. L., McConnell, T. S., Head, D. R., Weick, J., Grever, M. R., Appelbaum, F. R., Willman, C. L., Frequency and clinical significance of the expression of the multidrug resistance proteins MDR1/P-glycoprotein, MRP1, and LRP in acute myeloid leukemia: a Southwest Oncology Group Study. *Blood* 1999, 94(3), 1086–1099.

36. Leith, C. P., Kopecky, K. J., Godwin, J., McConnell, T., Slovak, M. L., Chen, I. M., Head, D. R., Appelbaum, F. R., Willman, C. L., Acute myeloid leukemia in the elderly: assessment of multidrug resistance (MDR1) and cytogenetics distinguishes biologic subgroups with remarkably distinct responses to standard chemotherapy. A Southwest Oncology Group study. *Blood* 1997, 89(9), 3323–3329.

37. Uckun, F. M., Gaynon, P. S., Sensel, M. G., Nachman, J., Trigg, M. E., Steinherz, P. G., Hutchinson, R., Bostrom, B. C., Sather, H. N., Reaman, G. H., Clinical features and treatment outcome of childhood T-lineage acute lymphoblastic leukemia according to the apparent maturational stage of T-lineage leukemic blasts: a Children's Cancer Group study. *Journal of Clinical Oncology* 1997, 15(6), 2214–2221.

38. Uckun, F. M., Sather, H. N., Gaynon, P. S., Arthur, D. C., Trigg, M. E., Tubergen, D. G., Nachman, J., Steinherz, P. G., Sensel, M. G., Reaman, G. H., Clinical features and treatment outcome of children with myeloid antigen positive acute lymphoblastic leukemia: a report from the Children's Cancer Group. *Blood* 1997, 90(1), 28–35.

39. Raetz, E. A., Moos, P. J., Impact of microarray technology in clinical oncology. *Cancer Investigation* 2004, 22(2), 312–320.

40. Raetz, E. A., Moos, P. J., Szabo, A., Carroll, W. L., Gene expression profiling. Methods and clinical applications in oncology. *Hematology—Oncology Clinics of North America* 2001, 15(5), 911–930.

41. Moos, P. J., Raetz, E. A., Carlson, M. A., Szabo, A., Smith, F. E., Willman, C., Wei, Q., Hunger, S. P., Carroll, W. L., Identification of gene expression profiles that segregate patients with childhood leukemia. *Clinical Cancer Research* 2002, 8(10), 3118–3130.

42. Sutcliffe, M. J., Shuster, J. J., Sather, H. N., Camitta, B. M., Pullen, J., Schultz, K. R., Borowitz, M. J., Gaynon, P. S., Carroll, A. J., Heerema, N. A., High concordance from independent studies by the Children's Cancer Group (CCG) and Pediatric Oncology Group (POG) associating favorable prognosis with combined trisomies 4, 10, and 17 in children with NCI Standard-Risk B-precursor Acute Lymphoblastic Leukemia: a Children's Oncology Group (COG) initiative. *Leukemia* 2005, 19(5), 734–40.

43. Borowitz, M. J., Rubnitz, J., Nash, M., Pullen, D. J., Camitta, B., Surface antigen phenotype can predict TEL-AML1 rearrangement in childhood B-precursor ALL: a Pediatric Oncology Group study. *Leukemia* 1998, 12(11), 1764–1770.

44. De Zen, L., Orfao, A., Cazzaniga, G., Masiero, L., Cocito, M. G., Spinelli, M., Rivolta, A., Biondi, A., Zanesco, L., Basso, G., Quantitative multiparametric immunophenotyping in acute lymphoblastic leukemia: correlation with specific genotype. I. ETV6/AML1 ALLs identification. *Leukemia* 2000, 14(7), 1225–1231.

45. Crist, W. M., Carroll, A. J., Shuster, J. J., Behm, F. G., Whitehead, M., Vietti, T. J., Look, A. T., Mahoney, D., Ragab, A., Pullen, D. J., Land, V.J., Poor prognosis of children with pre-B acute lymphoblastic leukemia is associated with the t(1,19)(q23,p13): a Pediatric Oncology Group study. *Blood* 1990, 76(1), 117–122.

46. Borowitz, M. J., Hunger, S. P., Carroll, A. J., Shuster, J. J., Pullen, D. J., Steuber, C. P., Cleary, M. L., Predictability of the t(1,19)(q23,p13) from surface antigen phenotype: implications for screening cases of childhood acute lymphoblastic leukemia for molecular analysis: a Pediatric Oncology Group study. *Blood* 1993, 82(4), 1086–1091.

47. McKenna, R. W., Asplund, S. L., Kroft, S. H., Immunophenotypic analysis of hematogones (B-lymphocyte precursors) and neoplastic lymphoblasts by 4-color flow cytometry. *Leukemia and Lymphoma* 2004, 45(2), 277–285.

48. McKenna, R. W., Washington, L. T., Aquino, D. B., Picker, L. J., Kroft, S. H., Immunophenotypic analysis of hematogones (B-lymphocyte precursors) in 662 consecutive bone marrow specimens by 4-color flow cytometry. *Blood* 2001, 98(8), 2498–2507.

49. Leitenberg, D., Rappeport, J. M., Smith, B. R., B-cell precursor bone marrow reconstitution after bone marrow transplantation. *American Journal of Clinical Pathology* 1994, 102(2), 231–236.

50. Szczepanski, T., van der Velden, V. H., van Dongen, J. J., Flow-cytometric immuno-phenotyping of normal and malignant lymphocytes. *Clinical Chemistry and Laboratory Medicine* 2006, 44(7), 775–796.
51. Ferrando, A. A., Neuberg, D. S., Dodge, R. K., Paietta, E., Larson, R. A., Wiernik, P. H., Rowe, J. M., Caligiuri, M. A., Bloomfield, C. D., Look, A. T., Prognostic importance of TLX1 (HOX11) oncogene expression in adults with T-cell acute lymphoblastic leukae-mia. *Lancet* 2004, 363(9408), 535–536.
52. Ferrando, A. A., Neuberg, D. S., Staunton, J., Loh, M. L., Huard, C., Raimondi, S. C., Behm, F. G., Pui, C. H., Downing, J. R., Gilliland, D. G., Lander, E. S., Golub, T. R., Look, A. T., Gene expression signatures define novel oncogenic pathways in T cell acute lymphoblastic leukemia. *Cancer Cell* 2002, 1(1), 75–87.
53. Anonymous, A clinical evaluation of the International Lymphoma Study Group classification of non-Hodgkin's lymphoma. The Non-Hodgkin's Lymphoma Classifica-tion Project. *Blood* 1997, 89(11), 3909–3918.
54. Gudgin, E. J., Erber, W. N., Immunophenotyping of lymphoproliferative disorders: state of the art. *Pathology* 2005, 37(6), 457–478.
55. Kampalath, B., Barcos, M. P., Stewart, C., Phenotypic heterogeneity of B cells in patients with chronic lymphocytic leukemia/small lymphocytic lymphoma. *American Journal of Clinical Pathology* 2003, 119(6), 824–832.
56. Chen, L., Apgar, J., Huynh, L., Dicker, F., Giago-McGahan, T., Rassenti, L., Weiss, A., Kipps, T. J., ZAP-70 directly enhances IgM signaling in chronic lymphocytic leukemia. *Blood* 2005, 105(5), 2036–2041.
57. Chen, L., Widhopf, G., Huynh, L., Rassenti, L., Rai, K. R., Weiss, A., Kipps, T. J., Expression of ZAP-70 is associated with increased B-cell receptor signaling in chronic lymphocytic leukemia. *Blood* 2002, 100(13), 4609–4614.
58. Mainou-Fowler, T., Dignum, H. M., Proctor, S. J., Summerfield, G. P., The prognostic value of CD38 expression and its quantification in B cell chronic lymphocytic leukemia (B-CLL). *Leukemia & Lymphoma* 2004, 45(3), 455–462.
59. Athanasiadou, A., Stamatopoulos, K., Tsompanakou, A., Gaitatzi, M., Kalogiannidis, P., Anagnostopoulos, A., Fassas, A., Tsezou, A., Clinical, immunophenotypic, and molecu-lar profiling of trisomy 12 in chronic lymphocytic leukemia and comparison with other karyotypic subgroups defined by cytogenetic analysis. *Cancer Genetics and Cytogenet-ics* 2006, 168(2), 109–119.
60. Crossen, P. E., Genes and chromosomes in chronic B-cell leukemia. *Cancer Genetics and Cytogenetics* 1997, 94(1), 44–51.
61. Li, J. Y., Gaillard, F., Moreau, A., Harousseau, J. L., Laboisse, C., Milpied, N., Bataille, R., Avet-Loiseau, H., Detection of translocation t(11,14)(q13,q32) in mantle cell lymphoma by fluorescence *in situ* hybridization. *American Journal of Pathology* 1999, 154(5), 1449–1452.
62. Foucar, K., Chronic lymphoid leukemia and lymphoproliferative disorders. *Modern Pathology* 1999, 12(2), 141–150.
63. Isaacson, P. G., Update on MALT lymphoma. *Bailliere's Best Practice in Clinical Haematology* 2005, 18(1), 57–68.
64. Alizadeh, A. A., Eisen, M. B., Davis, R. E., Ma, C., Lossos, I. S., Rosenwald, A., Boldrick, J. C., Sabet, H., Tran, T., Yu, X., Powell, J. I., Yang, L., Marti, G. E., Moore, T., Hudson, J., Jr., Lu, L., Lewis, D. B., Tibshirani, R., Sherlock, G., Chan, W. C., Greiner, T. C., Weisenburger, D. D., Armitage, J. O., Warnke, R., Levy, R., Wilson, W., Grever, M. R., Byrd, J. C., Botstein, D., Brown, P. O., Staudt, L. M., Distinct types of diffuse large B-cell lymphoma identified by gene expression profiling. *Nature* 2000, 403(6769), 503–511.
65. Rosenwald, A., Staudt, L. M., Gene expression profiling of diffuse large B-cell lymphoma. *Leukemia & Lymphoma* 2003, 44(Suppl 3), S41–S47.

66. Chang, C. C., McClintock, S., Cleveland, R. P., Trzpuc, T., Vesole, D. H., Logan, B., Kajdacsy-Balla, A., Perkins, S. L., Immunohistochemical expression patterns of germinal center and activation B-cell markers correlate with prognosis in diffuse large B-cell lymphoma. *American Journal of Surgical Pathology* 2004, 28(4), 464–470.

67. Marafioti, T., Pozzobon, M., Hansmann, M. L., Gaulard, P., Barth, T. F., Copie-Bergman, C., Roberton, H., Ventura, R., Martin-Subero, J. I., Gascoyne, R. D., Pileri, S. A., Siebert, R., Hsi, E. D., Natkunam, Y., Moller, P., Mason, D. Y., Expression pattern of intracellular leukocyte-associated proteins in primary mediastinal B cell lymphoma. *Leukemia* 2005, 19(5), 856–861.

68. van den Beemd, R., Boor, P. P., van Lochem, E. G., Hop, W. C., Langerak, A. W., Wolvers-Tettero, I. L., Hooijkaas, H., van Dongen, J. J., Flow cytometric analysis of the Vbeta repertoire in healthy controls. *Cytometry* 2000, 40(4), 336–345.

69. Langerak, A. W., van Den Beemd, R., Wolvers-Tettero, I. L., Boor, P. P., van Lochem, E. G., Hooijkaas, H., van Dongen, J. J., Molecular and flow cytometric analysis of the Vbeta repertoire for clonality assessment in mature TCRalphabeta T-cell proliferations. *Blood* 2001, 98(1), 165–173.

70. Stein, H., Foss, H. D., Durkop, H., Marafioti, T., Delsol, G., Pulford, K., Pileri, S., Falini, B., CD30(+) anaplastic large cell lymphoma: a review of its histopathologic, genetic, and clinical features. *Blood* 2000, 96(12), 3681–3695.

71. Foss, H. D., Anagnostopoulos, I., Araujo, I., Assaf, C., Demel, G., Kummer, J. A., Hummel, M., Stein, H., Anaplastic large-cell lymphoma of T-cell and null-cell phenotype express cytotoxic molecules. *Blood* 1996, 88(10), 4005–4011.

72. Cooke, C. B., Krenacs, L., Stetler-Stevenson, M., Greiner, T. C., Raffeld, M., Kingma, D. W., Abruzzo, L., Frantz, C., Kaviani, M., Jaffe, E. S., Hepatosplenic T-cell lymphoma: a distinct clinicopathologic entity of cytotoxic gamma delta T-cell origin. *Blood* 1996, 88(11), 4265–4274.

8 Guidelines for the Use of Flow Cytometry in the Management of Patients Infected with the Human Immunodeficiency Virus Type-1 and the Acquired Immunodeficiency Syndrome

Maurice R.G. O'Gorman

CONTENTS

8.1 INTRODUCTION

Human immunodeficiency virus type-1 (HIV-1) infection is the most common cause of acquired immune deficiency. The virus infects CD4+ leukocytes and ultimately leads to a profound CD4+ T-cell lymphopenia. The progressive decline in CD4+ T cells is associated with the increased risk of HIV-associated diseases including opportunistic infections, wasting, and death.[1-3] Flow cytometry plays a dominant role in the characterization and monitoring of patients with HIV-1 infection.

Additionally, flow cytometry as a technology has played and continues to play a key role in the investigation of HIV-1 pathogenesis, vaccine efficacy, response to therapy, and prognosis, with over 2500 peer-reviewed publications on this topic since the mid-1980s. This chapter focuses on the role of flow cytometry in the management of immune deficiency in HIV-1-infected individuals.

Although several informative and complex procedures have been developed and evaluated for assessing and defining the severity of HIV-1-related immunodeficiency, it is the simple measurement of peripheral blood CD4+ T cells that remains the single most important parameter in the management of HIV-1-infected patients, and is regarded as the standard for the immunological assessment of HIV-infected patients.[4–6]

Several major societies, including the Centers for Disease Control and Prevention in the United States, Infectious Diseases Society of America, National Institutes of Health, the World Health Organization, International AIDS Society, and the Pediatric European Network for the Treatment of AIDS (and others) have and continue to provide guidelines for the surveillance and management of HIV-infected individuals. All the guidelines include a requirement for CD4 T-cell measurements in the care and monitoring of HIV-infected individuals with respect to the classification of the level of immunodeficiency, the prognosis as related to the development of acquired immunodeficiency syndrome (AIDS) and death, time point for determining when to start prophylactic/antiretroviral therapy (ART), and when to change the therapy due to failure. The level of CD4 T cells (especially at levels below 50 cells/mm^3) at the initiation of highly active ART is also a recognized risk for the development of immune reconstitution inflammatory syndrome (IRIS).[7]

8.2 CD4 T-CELL MEASUREMENTS: GENERAL CONSIDERATIONS

Several technologies for the enumeration of absolute CD4 counts have been and continue to be developed. The combination of the critical information provided by peripheral blood CD4 T-cell counts (and percentage), and the urgent necessity to provide ART on a global scale is driving the development of cost-effective CD4 enumeration technologies. Several nonflow cytometry–based technologies have come and gone;[8] however, flow cytometry remains the platform of choice for measuring peripheral blood CD4 T-cell counts and percentages. There are several new platforms in the testing and validation stages that will be appropriate for adoption in resource-rich as well as resource-limited settings. Simplification of the procedures and a decreased reliance on electricity must be considered, where in addition to cost considerations, the availability of experienced technologists and appropriate resources (e.g., power and water) may be limiting.

In the United States, the recommendations published in 2002[9] suggest that three- or four-color monoclonal antibody combinations with CD45 gating on a "single platform" (i.e., not flow and hematology dual-platform testing) should be adopted for the most accurate measurement of CD4 peripheral blood absolute counts and percentages. Trials evaluating the adoption of single-platforms systems have shown improved precision over dual-platform systems,[10–13] leading to their recommendation[10,11] and widespread adoption in Europe and Canada (Frank Mandy and David Barnett, personal communication). The adoption of single-platform methods in the United States has been slower than in Europe and Canada. In the United States, there are 82 laboratories approved under the Immunology Quality Assessment Program of the NIH-NIAID to perform CD4 T-cell enumeration studies, and only 19 (23%) of these laboratories have so far completed the validation studies required for approval to switch to the single-platform CD4 counting methodology (Raul Louzao, Program Manager, IQA, 2007, personal communication).

When measuring CD4 T-cell counts and percentages in HIV-1-infected persons, it is very important to consider changes in the trend of CD4 levels as opposed to individual values. As individuals approach new levels of immune deficiency, it is prudent to increase the frequency of CD4 measurements. Owing to the high variability of CD4 T-cell levels in children, it is strongly recommended that two CD4 T-cell measurements below the appropriate threshold be obtained[4] before initiating treatment.

All laboratories performing CD4 testing should enroll in an external proficiency program. It is also noteworthy that (a) intercurrent illness may affect CD4 counts and (b) values vary during the day (the so-called diurnal variation). Therefore, it is recommended that CD4 testing be interpreted in the context of infections, the test be repeated following any significant changes from previous results, and the blood be drawn around the same time of the day for each visit.

8.3   CLASSIFYING THE LEVEL OF IMMUNE DEFICIENCY

There are several "guidelines" published that describe (a) different cutoffs for age categories and (b) different CD4 T-cell levels used to characterize the level of immune deficiency in HIV-1-infected individuals. Three recent guidelines published by the World Health Organization (WHO) (and available online): (1) "Antiretroviral therapy of HIV infection in infants and children in resource-limited settings: toward universal access. Recommendations for a public health approach"; (2) "Antiretroviral therapy for HIV infection in adults and adolescents: Recommendations for a public health approach, 2006 revision"; and (3) "WHO case definitions of HIV for surveillance and revised clinical staging and immunological classification of HIV-related disease in adults and children" provide a consolidated approach to characterizing HIV-1-infected adults and infants. These documents are designed to provide standardized guidelines for large centers caring for HIV-infected persons anywhere in the world. The last reference serves to "harmonize" the staging of HIV-1-related diseases in infants and children with the classification and staging of HIV disease in adults and adolescents into one document.

Although several recommendations and guidelines exist, the immune deficiency categories as published by the WHO consolidate all the recommendations into the most appropriate age categories (three for children and one for adults) and levels of immune deficiency for widespread adoption. Table 8.1 indicates the new classification schema for HIV-1-infected adults and children in specific age categories.[4] As described in Table 8.1, these levels of immune deficiency (when available) provide key information in the management of HIV-infected individuals.

TABLE 8.1

Classification of HIV-Associated Immunodeficiency

Classification of HIV-Associated Immunodeficiency	Age-Related CD4 Values			
	<11 Months (% CD4)	12–35 Months (% CD4)	36–59 Months (% CD4)	>5 Years (Absolute Number per mm³ or % CD4)
None or not significant	>35	>30	>25	>500
Mild	30–35	25–30	20–25	350–499
Advanced	25–29	20–24	15–19	200–349
Severe	<25	<20	<15	<200 or <15%
Severe[a]	<1500 cells/mm³	<750 cells/mm³	<350 cells/mm³	<200 cells/mm³

[a] It is highly recommended that the percentage of CD4+ T cells be utilized to establish the level of immunodeficiency in all HIV-1-infected children below 5 years of age.

For adult and adolescent guidelines please refer to the following document accessible via (http://www.who.int/hiv/pub/guidelines/artadultguidelines.pdf).

Source: Reprinted from www.who.int/hiv/pub/guidelines/paediatric020907.pdf (Reference# 4). With permission.

8.4   PROGNOSIS

CD4 T-cell percentages (in children below 5 years of age) and absolute CD4 T-cell counts provide the single most effective prognostic factor for estimating the risk of AIDS and death in HIV-1-infected children and adults. The CD4 cell count remains the strongest predictor of HIV-related complications, even in patients on antiretroviral medication.[14–18] The prognostic information gained from the CD4 level has resulted in the CD4 level being included in all guidelines on when to initiate ART and when to change therapy due to failure. In children, the CD4 levels change significantly between birth and 5 years of age,[19,20] and are considerably higher than the levels observed in adults. It is also well known that in children, the absolute CD4 T-cell count varies considerably more than the percentage of CD4 T-cell measurements. For these reasons, age is a key determinant in the decision to start ART, and it is strongly recommended that the percentage of CD4 values (as opposed to absolute counts) be used in management decisions for all HIV-1-infected children below 5 years of age. A comprehensive meta-analysis of the short-term risk of disease progression (to AIDS) or death in HIV-infected children was performed by the HIV Pediatric Prognostic Markers Collaborative Study Group.[17,18] In this analysis, the percentages of CD4 and HIV-1 viral load (VL) were independent predictors of AIDS and death, although the percentage of CD4 was the stronger predictor.[16,17] This information has formed the basis of new recommendations on when to start ART in children.[4,5,21] Studies investigating the surrogates of risk of progression to AIDS and death have also been evaluated in women and adults in South Africa.[16,22] Progression to AIDS and death in women was predicted by CD4 T cells of less than 200 cells/mm³, whereas those with values between 200 and 350 cells/mm³ did not differ from the group with values greater than 350 cells/mm³, supporting the recommendations for delaying ART until CD4 T-cell values are between 200 and 350 cells/mm^3. In the South African cohort, the CD4 count was also a strong predictor of the 6-month risk of AIDS or death. In this cohort, ranges of CD4 counts between 200 and 350 cells/mm^3 were also strong predictors of AIDS and death, and the rates of clinical events were higher than those reported in high-income countries. This information is relevant in establishing guidelines for the initiation of ART in resource-limited settings. The latter data support all initiatives to reduce the cost and expand the use of CD4 measurements in resource-limited settings.

8.5 CD4 MEASUREMENTS AND WHEN TO START OPPORTUNISTIC INFECTION PROPHYLAXIS AND ANTIRETROVIRAL TREATMENT (ART)

The strong association of CD4 T-cell counts with risk of infection provides the ideal measure of when to initiate ART. The baseline count is not only important in the decision on when to start therapy but is also a key in the monitoring of the response to treatment. Decisions on when to start ART are now tempered with the knowledge of the toxicities and the risk of resistance from long-term ART therapy. It is for this reason that ART is delayed until the risk of progression is "significant."[21] This risk is defined primarily by the age-associated CD4 measurements (age groups differ in the recommendations published by the different societies [i.e., NIH[5] versus WHO[4] versus PENTA[21]] and the level of clinical disease [see Tables 8.2 and 8.3 for pediatrics and Tables 8.4 and 8.5 for adults]).

TABLE 8.2
CD4 Levels and the Recommendations for Initiating ART in HIV-Infected Infants

WHO Pediatric Stage	Availability of CD4	Age-Specific Treatment Recommendations	
		<11 Months	>12 Months
4	CD4	Treat all	
	No CD4		
3	CD4	Treat all	Treat all, CD4 guided in those children with tuberculosis (TB), lymphocytic interstitial pneumonitis (LIP), oral hairy leukoplakia (OHL), and thrombocytopenia
	No CD4		Treat all
2	CD4	CD4 guided	
	No CD4	Total lymphocyte count (TLC) guided	
1	CD4	CD4 guided	
	No CD4	Do not treat	

Note: Stabilize any opportunistic infection before the initiation of ART; baseline CD4 is useful for monitoring ART even if it is not required to initiate ART; in children with pulmonary or lymph node TB, the CD4 level and clinical status should be used to determine the need for and timing of ART in relation to the treatment of TB.

Source: Reprinted from www.who.int/hiv/pub/guidelines/paediatric020907.pdf. With permission.

TABLE 8.3
Recommendations for Initiating ART in HIV-Infected Children

Immunological Marker	Age-Specific Recommendation to Initiate ART			
	<11 Months	12–35 Months	36–59 Months	>5 Years
CD4 (%)	25	<20	<15	<15
CD4 counts (cells/mm³)	<1500	<750	<350	<200

Note: Immunological markers supplement clinical assessment and should therefore be used in combination with clinical staging. CD4 is preferably measured after stabilization of acute presenting conditions. ART should be initiated by these cutoff levels, regardless of the clinical stage: a drop of CD4 below these levels significantly increases the risk of disease progression and mortality. The percentage of CD4 T cell (% CD4) is preferred for children below 5 years of age.

Source: Reprinted from www.who.int/hiv/pub/guidelines/paediatric020907.pdf. With permission.

TABLE 8.4
CD4 Criteria for the Initiation of ART in Adults and Adolescents

CD4 (Cells/mm³)	Treatment Recommendations
<200	Treat irrespective of clinical stage
200–350	Consider treatment and initiate before CD4 count drops below 200 cells/mm³
>350	Do not initiate treatment

Note: CD4 cell count should be measured after stabilization of any intercurrent illness; CD4 count supplements clinical assessment and should therefore be used in combination with clinical staging in decision making; a drop in the CD4 cell count below 200 cells/mm³ is associated with a significant increase in opportunistic infections and death; the initiation of ART is recommended for all patients with any WHO clinical stage 4 disease and some WHO clinical stage 3 conditions, notable pulmonary TB, and severe bacterial infections; the initiation of ART is recommended in all HIV-infected pregnant women with WHO clinical stage 3 disease and CD4 <350 cells/mm³.

Source: Reprinted from www.who.int/hiv/pub/guidelines/adultguidelines.pdf. With permission.

Table 8.4 summarizes the immunological criteria on when to initiate ART in adults. These recommendations are now very similar for most societies.

Highly active ART is indicated for all patients (children, pregnant women, and adults) with an AIDS-defining illness (WHO,[4] NIH,[5] and PENTA[21]), regardless of CD4 T-cell measurements or VL.[4,23] In patients with non-AIDS-defining illness, the decision to initiate highly active antiretroviral therapy (HAART) is based on immune compromise as determined by percentage of CD4 in children below 5 years of age and absolute CD4 T-cell counts in adolescents and adults. CD4 measurements are also utilized for recommendations on when to initiate prophylaxis for opportunistic infections.[23]

TABLE 8.5
Recommendations for Initiation of ART in Adults and Adolescents in Accordance with Clinical Stages and the Availability of Immunological Markers

WHO Clinical Staging	CD4 Testing Not Available	CD4 Testing Available
1—Asymptomatic	Do not treat	Treat if CD4 count is below 200 cells/mm^3
2—Mild	Do not treat	
3—Advanced	Treat	Consider treatment if CD4 count is below 350 cells/mm^3 and initiate ART before CD4 count drops below 200 cell/mm^3
4—Severe	Treat	Treat irrespective of CD4 cell count

Note: Details of the clinical indications, which define each of the clinical stages, can be obtained in the Annexes 1 and 2 of Ref. 6. CD4 cell count advisable to assist with determining the need for immediate therapy for situations such as pulmonary TB and severe bacterial infections that may occur at any CD4 level. A total lymphocyte count of 1200 mm^3 or less can be substituted for the CD4 count when the latter is unavailable and mild HIV disease exists. It is not useful in asymptomatic patients. Thus, in the absence of CD4 cell counts and TLCs, patients with WHO adult clinical stage 2 should not be treated. The initiation of ART is recommended in all HIV-infected pregnant women with WHO clinical stage 3 disease and CD4 counts below 350 cells/mm^3. The initiation of ART is recommended for all HIV-infected patients with CD4 counts below 350/mm^3 and pulmonary TB or severe bacterial infection. The precise CD4 cell level above 200/mm^3 at which ART treatment should be started has not been established.

Source: Reprinted from www.who.int/hiv/pub/guidelines/artadultguidelines.pdf. With permission.

ABBREVIATIONS

AIDS   Acquired immunodeficiency syndrome
HIV   Human immunodeficiency virus
IRIS   Immune reconstitution inflammatory syndrome
LIP   Lymphocytic interstitial pneumonitis
OHL   Oral hairy leukoplakia
TB   Tuberculosis
TLC   Total lymphocyte count

REFERENCES

1. Vlahov D, Graham N, Hoover D, Flynn C, Bartlett JG, Margolick JB, Lyles CM, Nelson KE, Smith D, Holmberg S, Farzadegan H. Prognostic indicators for AIDS and infectious disease death in HIV-infected injection drug users: plasma viral load and CD4+ cell count. *JAMA*, 1998;279:35–40.
2. MacDonell KB, Chmiel JS, Poggensee L, Wu S, Phair JP. Predicting progression to AIDS: combined usefulness of CD4 lymphocyte counts and p24 antigenemia. *Am J Med*, 1990;89:706–712.
3. Nishanian P, Taylor JM, Manna B, Aziz N, Grosser S, Giorgi JV, Detels R, Fahey JL. Accelerated changes (inflection points) in levels of serum immune activation markers

and CD4+ and CD8+ T cells prior to AIDS onset. *J Acquir Immune Defic Syndr Hum Retrovirol*, 1998;18:162–170.

4. World Health Organization Antiretroviral Therapy for HIV Infection in Infants and Children: Towards Universal Access. Recommendations for a public health approach, 2006. http://www.who.int/hiv/pub/guidelines/paediatric020907.pdf.

5. *Guidelines for the Use of Antiretroviral Agents in HIV-1 Infected Adults and Adolescents.* Developed by the DHHS Panel on Antiretroviral Guidelines for Adults and Adolescents—A Working Group of the Office of AIDS Research Advisory Council (OARAC) October 10th, 2006. http://aidsinfo.nih.gov/contentfiles/adultandadolescentGL.pdf.

6. Antiretroviral Therapy for HIV Infection in Adults and Adolescents. Recommendations for a Public Health Approach, 2006. http://www.who.int/hiv/pub/guidelines/artadultguidelines.pdf

7. French MA, Lenzo N, John M, Mallal SA, McKinnon EJ, James IR, Price P, Flexman JP, Tay-Kearney ML. Immune restoration disease after the treatment of immuno-deficient HIV-infected patients with highly active antiretroviral therapy. *HIV Med*, 2000;1:107–115.

8. O'Gorman M, Paul R. Scholl role of flow cytometry in the diagnostic evaluation of primary immunodeficiency disease. *Clin Appl Immunol Rev*, 2002;2:321–335.

9. Schnizlein-Bick CT, Mandy FF, O'Gorman MR, Paxton H, Nicholson JK, Hultin LE, Gelman RS, Wilkening CL, Livnat D. Use of CD45 gating in three and four-color flow cytometric immunophenotyping: guideline from the National Institute of Allergy and Infectious Diseases, Division of AIDS. *Cytometry*, 2002;15;50(2):46–52.

10. Barnett D, Granger V, Whitby L, Storie I, Reilly JT. Absolute CD4+ T-lymphocyte and CD34+ stem cell counts by single-platform flow cytometry: the way forward. *Br J Haematol*, 1999;106:1059–1062.

11. O'Gorman MRG, Nicholson JKA. Adoption of single-platform technologies for the enumeration of absolute T-lymphocyte subsets in peripheral blood. *Clin Diag Lab Immunol*, 2000;7:333–335.

12. Schnizlein-Bick CT, Spritzler J, Wilkening CL, Nicholson JKA, O'Gorman MRG, Site Investigators, The NIAID DAIDS New Technologies Evaluation Group. Evaluation of TruCount absolute-count tubes for determining CD4 and CD8 cell numbers in human immunodeficiency virus-positive adults. *Clin Diag Lab Immunol*, 2000;7(3):336–343.

13. Reimann KA, O'Gorman MRG, Spritzler J, Wilkening CL, Sabath DE, Helm K, Campbell DE, The NIAID DAIDS New Technologies Evaluation Group. Multisite comparison of CD4 and CD8 T-lymphocyte counting by single-versus multiple-platform methodologies: evaluation of Beckman Coulter flow count fluorospheres and the tetraONE system. *Clin Diag Immunol*, 2000;7(3):344–351.

14. Gadelha A, Accacio N. Morbidity and survival in advanced AIDS in Rio de Janeiro, Brazil. *Rev Inst Trop S Paulo*, 2002;449:179–186.

15. Mellors J, Munoz A, et al. Plasma viral load and CD4+ lymphocytes as prognostic markers of HIV-1 infection. *Ann Intern Med*, 1997;126(12):946–954.

16. Anastos K, Barrón Y, Miotti P, Weiser B, Young M, Hessol N, Greenblatt RM, Cohen M, Augenbraun M, Levine A, Munos A. Risk of progression to AIDS and death in women infected with HIV-1 initiating highly active antiretroviral tratment at different stages of disease. *Arch Intern Med*, 2002;162:1973–1980.

17. Dunn D, HIV Paediatric Prognostic Markers Collaborative Study Group. Short term risk of disease progression in HIV-1 infected children receiving no antiretroviral therapy or zidovudine monotherapy: a meta-analysis. *The Lancet*, 2003;363:1605–1611.

18. Dunn D, HIV Paediatric Prognostic Markers Collaborative Study. Predictive value of absolute CD4 cell count for disease progression in untreated HIV-1-infected children. *AIDS*, 2006;20(9):1289–1294.

19. Shearer W. Lymphocyte subsets in healthy children from birth through 18 years of age. The pediatric AIDS clinical trials group P1009 study. *J Allergy Clin Immunol*, 2003;112(5):973–980.

20. O'Gorman MRG, Millard DD, Lowder JN, Yogev R. Lymphocyte subpopulations in healthy one to three day old infants. *Cytometry*, 1998;34:235–241.

21. Sharland M, Blanche S, Castelli G, Ramos J, Gibb DM on behalf of the PENTA Steering Committee. PENTA guidelines for the use of antiretroviral therapy, 2004. *HIV Med*, 2004;5(Suppl. 2):61–86.

22. Badri M, Lawn SD, Wood R. Short-term risk of AIDS or death in people infected with HIV-1 before antiretroviral therapy in South Africa: a longitudinal study. *Lancet*, 2006;368(9543):1254–1259.

23. Gupta SB, Pujari SN, Patel AK. API consensus guidelines for use of antiretroviral therapy in adults. Endorsed by the AIDS Society of India. *J Assoc Physicians India*, 2006;54:57–74.

9 Role of Flow Cytometry in the Diagnosis and Monitoring of Primary Immunodeficiency Disease

Maurice R.G. O'Gorman

CONTENTS

9.1 INTRODUCTION

Accurate diagnosis and classification of primary immunodeficiency disease (PID) is necessary to decide on appropriate clinical management, enable informed genetic counseling, and permit the systematic collection of data on PID through registries that will facilitate future investigation, diagnoses, and treatments of these rare diseases. There are several components involved in making the correct diagnosis, including traditional history, physical examinations, routine laboratory evaluations, and finally specific diagnostic tests. Early screening laboratory tests include a complete blood count (CBC), quantitative immunoglobulins for suspected humoral immunodeficiency, complement levels for suspected complement deficiencies, and a "routine" immunophenotypic assessment of the relative proportion and absolute number of the major peripheral blood lymphocyte subsets. These early screening tests are extremely valuable and can provide important information to aid in the accurate diagnosis of the suspected PID. More specific tests may, however, be required to identify, classify, and ultimately diagnose the specific disorder. The most accurate test would be the identification of the specific mutation in the specific gene associated with the suspected disease. Although it is conceivable that deoxyribonucleic acid (DNA) sequencing for more than 120 known genes involved in PID will one day be feasible and widely available, such testing is currently not widely available, neither it is cost effective nor timely. *In lieu* of DNA sequencing, several methods have been developed that allow the assessment of defects associated with specific mutated genes. Such defects include failure of development and maturation of a specific cell subset(s) and absence of a specific protein or a specific function. Flow cytometry, with the ability to identify up to 20 parameters per cell at thousands of cells per second is ideally suited to detect such cellular abnormalities and has been exploited over the past 20 years for the development of numerous methods used to evaluate patients suspected of primary immunodeficiency disease. In many cases, the results obtained by flow cytometry serve to hone in on the specific underlying genetic abnormality or at least limit the possibilities to only a few potential suspected abnormalities. In other words, the abnormality detected by flow cytometry narrows the possible etiology and allows for a much quicker intervention in the disease process before the actual genetic abnormality can be specifically ascertained. Although extremely valuable, flow cytometric data need to be considered in the light of the overall picture that includes clinical information, other laboratory data, and where appropriate, confirmatory genetic testing. In this regard, the reader is referred to the formal diagnostic criteria for PID that were proposed recently [1].

Flow cytometry has emerged as an invaluable technology in the clinical laboratory and has contributed significantly to both the understanding and the evaluation (monitoring) of the immune system. The unparalleled ability to simultaneously identify numerous parameters per cell at rates of thousands of cells per second has resulted in the development of a large repertoire of diagnostic, prognostic, and monitoring assays. The "Immunodeficiency Resource" (http://bioinf.uta.fi/idr/index.shtml) is a knowledge base that integrates clinical, biochemical, genetic, proteomic, and other specific information on the primary immunodeficiencies. The site currently lists over 158 diseases of which more than 120 have known genetic etiologies [2]. The PIDs are formally classified by a group of worldwide experts. A recent meeting held in Budapest, Hungary in late 2005, resulted in the formal classification/reclassification of over 120 specific diseases [3]. The detection of many of the abnormalities by flow cytometry in peripheral blood leukocytes that are caused by these genetic mutations can broadly be grouped as (1) relative or absolute decrease in a specific subset or subsets, (2) loss or abnormal expression of a specific cell-associated marker or markers, and (3) loss or abnormal function.

9.2 BRIEF HISTORY OF PRIMARY IMMUNODEFICIENCY CLASSIFICATION

In 1970, the World Health Organization (WHO) convened a group of experts to classify (unified nomenclature) and define the PIDs [4,5]. This process continued under the aegis of WHO until recently when the sponsorship of the committee was assumed by the International Union of Immunological Societies, the Jeffrey Modell Foundation, and the National Institute of Allergy and Infectious Diseases of the NIH [3,4,6,7]. The completion of the human genome project and the development of robust molecular techniques have resulted in the progression from a clinical description and quantification of cells and proteins in blood for the diagnosis of primary immunodeficiencies to a molecular understanding based on the identification of the responsible gene(s). At a recent meeting, new diseases were included, and the nomenclature of some PIDs were amended to describe the nature of the underlying genetic defect more closely [3]. Although the genetic defects are known for more than 120 of the primary immunodeficiencies, flow cytometry has and continues to be an invaluable surrogate technology for the detection of the associated cellular and molecular abnormalities caused by the specific genetic defects. In this chapter, a brief clinical description of those diseases amenable to flow cytometric analysis within each of the eight PID groups (as summarized in Table 9.1) is provided.

9.3 THE PRIMARY IMMUNODEFICIENCY DISEASES

Groups I through VIII are summarized from the corresponding tables in Ref. 3.

9.3.1 GROUP I: COMBINED T- AND B-CELL IMMUNODEFICIENCY [1]

1. T−B+(NK−) (severe combined immunodeficiency [SCID])
   a. Common gamma chain (γC, CD132, IL-2Rγ) deficiency
   b. JAK3 deficiency
   c. IL-7R alpha deficiency
   d. CD45 deficiency
   e. CD3 delta/CD3 epsilon deficiency

TABLE 9.1
Primary Immunodeficiency Classification by Category

Group	Category	Flow Application
Group 1	Combined T- and B-cell immunodeficiencies	Several PIDs contained in this group are directly characterized by flow cytometry
Group II	Predominantly antibody deficiencies	Absence of B cells or of B-cell markers readily assessed
Group III	Other well-defined immunodeficiency syndromes	Only a few PIDs in this group can be characterized
Group IV	Diseases of immune dysregulation	A few very specific abnormalities can be characterized
Group V	Congenital defects of phagocyte number function or both	Several specific procedures virtually diagnostic of the specific PID
Group VI	Defects in innate immunity	Limited applications involving flow cytometry
Group VII	Autoinflammatory disorders	Limited applications involving flow cytometry[a]
Group VIII	Complement deficiencies	Limited applications involving flow cytometry[a]

[a] Will not be discussed further.

Source: Adapted from Notorangelo, L., Casanova, J-L., Conley, M.E. et al., *J. Allergy Clin. Immunol.*, 117, 883–896, 2006. With permission.

2. T−B−(NK+), (SCID)
   a. RAG 1/2 deficiency
   b. DCLRE1C (Artemis deficiency)
   c. Adenosine deaminase deficiency (ADA)
   d. Reticular dysgenesis
3. Omenn syndrome
4. DNA ligase IV
5. CD40 ligand deficiency (X-linked hyper-IgM [XHIGM] syndrome)
6. CD40 deficiency (autosomal Hyper IgM syndrome)
7. Purine nucleoside phosphorylase deficiency (PNP)
8. MHC class II deficiency (bare lymphocyte syndrome)
9. CD3 gamma deficiency
10. CD8 deficiency
11. ZAP-70 deficiency
12. TAP-1/2 deficiency
13. Winged helix deficiency (nude)

Each of the combined immunodeficiencies listed in group I is associated with peripheral blood lymphocyte subset and/or specific protein abnormalities that can be detected by flow cytometry. To detect many of the latter abnormalities, we have adopted a basic panel of monoclonal antibodies designed to measure the major circulating peripheral blood lymphocyte subsets (referred to as the "routine panel" throughout the chapter).

9.3.1.1 Routine Immunophenotyping Panel for the Identification of Subset Abnormalities in Primary Immunodeficiency Disease

Routine immunophenotyping is designed to measure qualitative and quantitative abnormalities in the major peripheral blood lymphocyte subsets, and is an extremely valuable procedure for the screening of several PIDs. My laboratory currently utilizes the following panel to measure the relative and absolute number of B cells (CD19+), T cells (CD3+), T helper cells (Th, CD3+, and CD4+), T cytotoxic cells (Tc, CD3+, and CD8+), natural killer (NK) cells (CD3− and CD16+ or CD56+), and activated T cells (CD3+ and HLA-DR+). The CD3, HLA-DR tube assesses activated T cells, but also serves as a control for the measurement of B cells (CD3− and bright HLA-DR+) and HLA-DR+ non-B, non-T cells. Currently, each tube contains up to four fluorochrome-conjugated antibodies; however, the same information can and has successfully been obtained using 2, 3, 5, or 6 color combinations (see current panel in Table 9.2). Several commercially available monoclonal antibody products are available for the detection of the subsets described earlier; however, it is the laboratory's responsibility to ensure that both the equipment and the reagents are properly maintained and quality controlled before their implementation.

Historically, absolute lymphocyte subset counts have been derived as the product of the absolute lymphocyte count generated by an automated hematology instrument and the individual lymphocyte subset percentage obtained by flow cytometry (dual-platform method). It is rapidly becoming apparent that absolute counts derived directly from the flow cytometer (single-platform assays) combined with the use of CD45 versus right-angle light scatter lymphocyte gating is more precise and more robust than the traditional dual-platform methods [8,9]. Although the majority of these developments were validated in populations of HIV-1-infected patients, the recommendations are very appropriate for immunophenotyping patients suspected of PID. The adoption of single-platform technologies has become the norm in Europe and Canada, but has been slower to take hold in the United States. In a quick evaluation of the flow cytometry laboratories that have been accredited by the NIAID to perform CD4 enumerations for clinical

TABLE 9.2
Monoclonal Antibody Panel and Fluorochromes Used in Four-Color Routine Immunophenotyping

Tube	Fluorochrome			
	FITC	PE	PerCP	APC
#1	CD3	CD8	CD45	CD4
#2	CD3	CD16 & 56	CD45	CD19
#3	CD3	HLA-DR	CD45	

Note: FITC = fluorescein isothiocyanate; PE = pycoerytherin; PerCP = peridinin chlorophyll protein; and APC = allophycocyanine.

TABLE 9.3
Mean Absolute Counts of Routine Immunophenotyping

	CD3 Abs	CD4 Abs	CD8 Abs	CD19 Abs	NK Cells	HLA-DR+ T Cells
Dual platform	2186.6	1386.8	721.2	596.5	217.4	143.4
Single platform	2020.1	1268.1	670.4	544.1	196.0	130.2
10th ‰ (new–old)	−428.3	−303	−138	−144.1	−58	−36
Median (new–old)	−166.5	−117.5	−50.8	−51.9	−21.1	−13.0
90th ‰ (new–old)	73.7	102	54	19.1	20	16

Note: Dual-platform versus single-platform methods. Results were generated on 100 consecutive samples received in the Diagnostic Immunology Laboratory at the Children's Memorial Hospital. The absolute T-cell count for this group ranged from 566 to 4732 cells/mm³.

trials, only 23% had switched to the single-platform methods (Raul Louazo, 2007, personal communication).

Table 9.3 summarizes an example of the results obtained in a "switch study" performed in my laboratory (unpublished). Briefly, dual platform–derived absolute lymphocyte subset counts (FACSCalibur, MultiTest-flow cytometry and a Bayer Advia 2120-automated hematology analyzer) were compared to the absolute lymphocyte subset counts derived from the single-platform TrueCount® method on a FACSCalibur four-color instrument (Becton Dickinson) from 100 consecutive patients. For the single-platform method, a known volume of blood is added to TrueCount tubes that contain a known concentration of beads. The ratio of the number of cells of interest (e.g., CD3+CD4+) counted in the sample to the number of beads counted in the sample multiplied by the known bead concentration (corrected for the volume of whole blood added) provides the absolute number of the cells of interest in whole blood (often described as the ratiometric method of single-platform absolute counting). As seen in Table 9.2, the TrueCount single-platform method resulted in a consistently lower (bias) absolute count than the dual-platform method. The mean absolute counts derived from both methods on the 100 samples, as well as the median, the tenth percentile, and the ninetieth percentile of the differences between the old (dual-platform) and new methods (TrueCount) are presented in Table 9.2.

The recommendation for the use of CD45 expression levels versus right-angle light scatter properties for lymphocyte gating [9] versus gating on lymphocytes by their inherent light scatter properties only is particularly appropriate for the analysis of PIDs. This is because many of the underlying defects in several PIDs affect the resolution of the lymphocyte cluster based on light scatter properties alone (see improvement in the resolution of the lymphocyte cluster by use of CD45 and right-angle light scatter in Figure 9.1).

Our laboratory and others have been involved in the establishment of age-associated lymphocyte subset reference ranges [10,11]. This information is particularly helpful in the evaluation of very young infants suspected of a PID. Tables 9.4 and 9.5 provide the age-associated lymphocyte subset reference ranges for percentages and

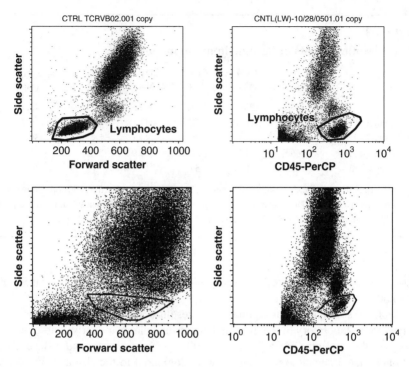

FIGURE 9.1 Flow cytometric methods for lymphocyte gating in a healthy nondisease control (top row) and a SCID patient (bottom row). The dot plots on the left-hand side column illustrate traditional lymphocyte gating using only the integral light scatter properties of forward-angle light scatter (FALS) versus right-angle light scatter or "side scatter." The dot plots in the right-hand side column illustrate lymphocyte gating on dot plots featuring CD45 surface expression (on the x-axis) versus side scatter on the y-axis as an alternative (and recommended) method of gating on the lymphocyte population. The top left and top right dot plots illustrate lymphocyte gating in a normal healthy control where the lymphocyte cluster is clearly demarcated and easily gated using either method. The bottom left and right dot plots clearly illustrate the differences between the two lymphocyte gating methods. Note the improved resolution of the lymphocyte cluster when CD45 is combined with side scatter (lower right) as compared to light scatter gating only (lower left). CD45 versus side scatter is the recommended method for gating on the lymphocyte population for flow cytometric analysis in most applications.

absolute counts (respectively) that were generated in our laboratory with informed consent from a group of healthy children enrolled in a new vaccine protocol at the Children's Memorial Hospital in Chicago, Illinois. A manuscript describing the statistical methods used to create the specific age categories and validate these reference ranges is in preparation and the reference values for young infants were published previously [10]. None of the primary immunodeficiencies are more closely aligned with flow cytometry than the SCIDs as listed in group I, since they are actually categorized according to the lymphocyte subset abnormalities that were traditionally detected by flow cytometry.

TABLE 9.4
Absolute Cell Counts (Cells/μL) Reference Ranges within Age Ranges (Days)

Age Range (Days)	N	33 <52.5	40 <124.5	45 <336.5	40 <691	35 <1115	37 <1528.5	27 >1528.5
CD4	5th	1168.9	1392.4	1785.9	1088.7	769.4	517.9	547.7
	50th	2695.8	3160.5	3355.8	2379.7	1551.0	1358.6	1020.0
	95th	5623.4	5209.8	5140.7	4552.2	2441.8	2355.6	1720.2
CD8	5th	267.2	651.9	945.8	749.5	428.7	269.5	331.8
	50th	1058.4	1329.3	1442.8	1370.1	969.8	793.0	684.0
	95th	1859.9	2449.0	2790.6	3748.5	1743.6	1530.2	1306.9
CD3	5th	1375.0	2532.8	2888.7	2206.5	1346.9	894.6	1051.2
	50th	3742.9	4427.8	4896.0	3812.2	2459.7	2331.4	1792.0
	95th	7128.5	6777.6	7994.6	8192.1	4085.0	3654.4	3031.4
CD19	5th	104.3	745.4	858.4	703.9	522.7	396.5	203.4
	50th	607.6	1441.1	1719.9	1372.0	900.0	699.2	528.0
	95th	1447.6	3498.7	3774.0	2710.8	1779.3	1539.1	1138.8
CD16 + 56	5th	60.0	194.2	174.1	181.9	164.3	121.0	138.3
	50th	204.1	470.6	430.1	440.0	342.2	291.9	315.0
	95th	434.3	993.6	1102.1	1581.4	1171.1	641.8	1027.0

TABLE 9.5
Subset Percentages Reference Ranges within Age Ranges (Days)

Age Range (Days)	N	45 <52.5	40 <124.5	45 <336.5	40 <691	35 <1115	37 <1528.5	27 >1528.5
CD4	5th	48.0	34.9	38.1	25.8	25.9	25.9	28.0
	50th	62.0	49.0	49.0	41.0	38.0	40.0	36.0
	95th	74.8	62.0	60.8	53.1	50.3	53.3	49.1
CD8	5th	13.2	14.0	14.1	15.9	10.9	12.8	16.7
	50th	23.0	21.0	21.0	23.0	25.0	24.0	25.0
	95th	32.9	30.1	33.7	39.1	34.5	37.1	32.4
CD3	5th	69.0	54.7	58.8	50.2	51.5	43.7	54.2
	50th	82.8	68.0	69.0	65.1	65.8	68.8	67.0
	95th	92.6	79.0	80.2	77.8	77.5	79.3	79.4
CD19	5th	4.1	14.0	16.1	15.9	16.9	16.9	9.0
	50th	14.0	23.5	25.0	26.0	24.0	22.0	22.0
	95th	25.9	39.0	36.0	33.1	34.2	37.4	31.9
CD16 + 56	5th	2.0	3.0	3.0	4.0	4.9	4.0	5.7
	50th	4.0	7.0	6.0	9.5	10.0	8.0	11.0
	95th	13.6	14.1	12.9	20.1	18.8	22.3	25.4

9.3.1.2 Th−Tc−NK−B+ X-Linked Severe Combined Immunodeficiency

Common gamma chain (γc/CD132): X-linked SCID is the most common form of the SCID syndromes [12]. It is caused by mutations in the gene encoding the IL-2 receptor gamma chain (IL-2RG, also referred to as the common gamma chain (γc), or CD132). A functional IL-2RG protein is required for the maturation of NK and T cells, both of which are abnormally low or absent in the peripheral blood of patients with mutation in this gene [13,14]. An example of a patient diagnosed with X-linked SCID is presented in Figure 9.2, with the characteristic finding of very low NK- and T-cell subsets detected in the peripheral blood. The common gamma chain (γc/CD132) is also the signaling component of the IL-4, IL-7, IL-9, IL-15, and IL-21 receptors [15] that are

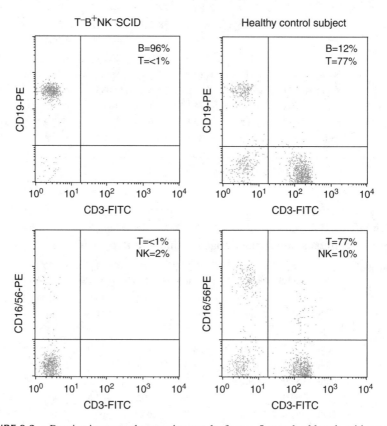

FIGURE 9.2 Routine immunophenotyping results from a 5-month-old male with an absolute lymphocyte count of 400 cells/mm^3 and a clinical history consistent with SCID. The lymphocyte subset analysis is remarkable for the paucity of CD3+ T cells (<15 of all lymphocytes) with 96% B cells (upper left) and a low 2% NK cells (lower left). This pattern is consistent with both X-linked SCID (common gamma chain deficiency) as well as autosomal SCID (mutations in Janus kinase 3). Molecular analysis confirmed the diagnosis of X-linked SCID due to a nonsense mutation in the gene encoding the common gamma chain (CD132). Routine immunotyping results on a healthy control run simultaneously are presented in the right-hand side column for comparison.

involved in the maturation and function of all peripheral blood lymphocytes. Although present in normal numbers, B cells from X-SCID patients are abnormal, do not mature, and fail to undergo class-switch recombination [16]. Anti-γc monoclonal antibodies are available directly fluorescinated from several companies and have been successfully employed to specifically detect absent or abnormal protein expression [17,18].

9.3.1.3   Th−Tc−NK−B+, Autosomal Severe Combined Immunodeficiency (JAK3)

This autosomal form of SCID that is clinically and immunophenotypically indistinguishable from the X-linked form is associated with mutations in gene encoding the Janus 3 (JAK3) kinase [19]. JAK3 kinase is directly associated with a common gamma chain and is downstream in the signaling pathway of the cytokine receptors, which explains why the phenotypic abnormalities are identical to patients with mutations in the IL-2RG gene. Autosomal JAK3 represents the fourth most common genetic defect in SCID [16].

Patients with either X-linked or autosomal JAK-3 SCID, who undergo bone marrow transplantation but have not received pretransplant myeloablative therapy can retain their B- and NK cells that fail to function normally. This appears to be due to the fact that B- and NK cells of the donor do not engraft, and the abnormalities related to IL-4 and IL-21 cytokine receptors in the B cells and abnormal IL-15R in NK cells prevent normal functioning even in the presence of normal T cell help by donor-derived T cells [16].

9.3.1.4   Th−Tc−B+NK− (CD45)

Patients with mutations in the gene encoding CD45 are also reported to have significantly reduced percentages and absolute numbers of T- and NK cells [20]. In addition to the abnormal subset representation described, the level of CD45 expression on the surface of the majority of circulation leukocytes is absent or severely reduced. Detection of the latter clearly distinguishes this group of patients from the either the X-linked or JAK3-associated SCID.

9.3.1.5   Th−Tc−B+NK+ (IL-7R Alpha, CD3 Delta, and CD3 Zeta)

IL-7 receptor alpha mutations represent the third most common cause of SCID [16,21]. These patients have a preponderance of B cells with relatively normal levels of NK cells suggesting that a functional IL-7 receptor is not required for the maturation of NK cells. Mutations in the delta chain of the CD3 receptor have also been described to cause clinical SCID, with a profound deficiency of CD3+, CD4+, and CD8+ circulating T cells, and a severe reduction of gamma/delta T-cell receptor (TCR) positive T cells [22]. Recently, the description of a mutant CD3 zeta chain leading to SCID was provided by Buckley's laboratory; however, this specific deficiency has not yet been officially classified [23].

9.3.1.6   Th−Tc−B−NK+ (Rag1, Rag2, Lig IV, Artemis)

The absence of circulating B and T cells with the presence of circulating NK cells (Figure 9.3) forms another distinct group of PID. This abnormal phenotype is the

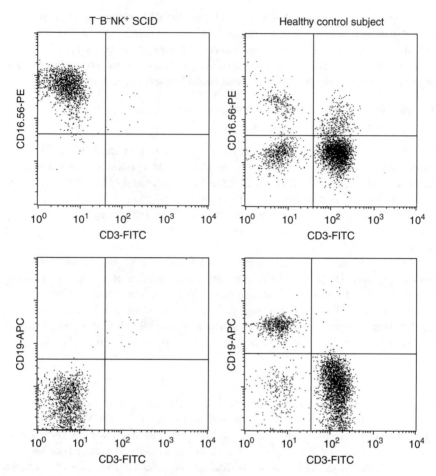

FIGURE 9.3 The flow cytometry results from a 7-month-old female patient with an absolute lymphocyte count of 350 cells/mm^3 and a clinical and family history consistent with autosomally inherited SCID are presented in the left-hand side column. The absence of B and T cells (both less than 1%) is suggestive of a mutation in one of the genes involved in the formation of mature antigen receptor genes (RAG 1, RAG 2, Lig IV, or Artemis).

result of mutations in genes involved in the process of specific antigen receptor formation on B and T cells. The large variability in the antigen-recognition motifs of B- and T-cell receptors is generated by the process of somatic V(D)J recombination in the latter in the T- and B-cell receptor genes. This highly regulated molecular event is essential for the maturation of circulating peripheral blood B and T cells. Recombination activating genes 1 and 2 (RAG1 and RAG2) interact with the appropriate sequences in the V, J (and sometimes D) regions of the T-cell receptor and immunoglobulin receptor genes to induce a double-strand break. The recombination (repair) process is then effected by components of the nonhomologous end joining (NHEJ) machinery. This complex process involves at least six proteins, Ku70,

Ku80, DNA-PKcs, Artemis, DNA ligase IV (Lig4), and Xrcc4 [24]. Mutations in genes inducing the double-strand breaks, that is, RAG1 and RAG2 [25], as well as mutations in the genes encoding the proteins responsible for the repair of the double-strand breaks (Artemis [26] and Lig4 [27,28]) lead to SCID disease associated with no or very low levels of T and B cells. V(D)J recombination defects are responsible for approximately 20% of the SCID cases in Europe [29], whereas in the United States, the V(D)J recombination defects are less common, representing less than 3% of all SCID patients' genetic abnormalities [16].

9.3.1.7 Th−Tc−B−NK− (Adenosine Deaminase Gene)

Patients with severe lymphopenia (<500 cells/μL) and a virtual absence of all lymphocyte subsets should be suspected of harboring mutations in the adenosine deaminase gene (ADA), which accounts for approximately 15–20% of all SCID patients in Europe and the United States [29–31]. In addition to more severe lymphopenia than other forms of SCID, ADA patients exhibit other distinguishing features including multiple skeletal abnormalities on an x-ray [16].

9.3.1.8 Th+Tc+B+NK− (IL-2R/IL-15Rβ)

The complete absence of NK cells in the peripheral circulation, decreased T cells, and normal levels of B cells have been described in a 17-month-old male [18]. The IL-2R/IL-15Rβ expression levels were less than 10% of normal and the messenger ribonucleic acid (mRNA) levels were detected at less than 10% of the control, however, no mutations were detected. Whether or not there is a mutation in the gene encoding IL-2R/IL-15Rβ remains to be seen; however, it was noted by the authors that the sensitivities of the mutation detection techniques utilized were not 100%.

9.3.1.9 Omenn Syndrome

This syndrome represents a unique clinical picture but the syndrome is actually associated with the same mutations that cause many of the SCID diseases described earlier. Omenn syndrome appears to occur primarily in patients with hypomorphic mutations (leading to some expression of an abnormal gene product) in Artemis [26], RAG1 or RAG2 [32], or IL-7R [33] genes. Clinically, these patients present with erythematous rash (98%), hepatosplenomegaly (88%), lymphadenopathy (80%), and recurrent infections as reviewed by Aleman et al. [34]. In addition to frequent infections, patients often have high serum IgE, and an elevated white blood cell count due to eosinophilia and lymphocytosis. Omenn syndrome patients can have normal to elevated levels of circulating T-cell numbers, but they are abnormally activated (very high HLA-DR expression; Figure 9.4) (activated T-cell phenotype) and of limited diversity, i.e., can have a skewed, mono, or oligoclonal TCR repertoire (see Section 9.3.1.10 and Figure 9.5).

9.3.1.10 T-Cell Receptor V-Beta Repertoire Analysis in CD4 and CD8 T-Cell Subsets

The assessment of the T-cell repertoire can be performed by a variety of methods as reviewed by Dr. Uzel in Chapter 10. In our laboratory, abnormalities in the TCR

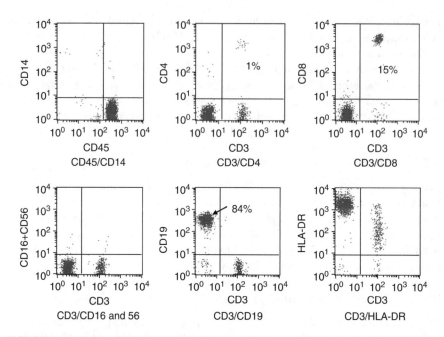

FIGURE 9.4 Increased HLA-DR (MHC class II) expression levels on CD3+ T cells from a patient diagnosed with Omenn syndrome due to a mutation in Janus kinase 3 (JAK3). This is an example of our two color–six tube panel for routine immunophenotyping using light scatter gating on lymphocytes. Note that the patient was lymphopenic (abs. lymphocyte count = 600/mm³), and had significantly reduced T cells (16%), inverted CD4:CD8 ratio (0.07), reduced NK cells (<2%), and an increased proportion of B cells (84%). The T cells were notable for the significantly increased percentage expressing HLA-DR on the cell surface (90%), which is significantly greater than the upper limit of normal (10%) for this age. Patient's with Omenn syndrome can have normal absolute counts but their T-cell receptor repertoire are often abnormally skewed with significant over and underrepresentation of several V-beta families as illustrated in Figure 9.5.

V-beta repertoire are assessed within both CD4 and CD8 T-cell subsets using a modified commercial kit (Beckman Coulter Corporation). Briefly, each tube contains three different V-beta monoclonal antibodies each with a unique combination of fluorescein isothiocyanate (FITC) only, pycoerytherin (PE) only, or FITC+PE, as well as CD3-PerCP (BD Biosciences) and CD8-APC (BD Biosciences), that is, a four-color panel. This combination of monoclonal antibodies and fluorochromes allows us to measure the relative representation of three individual TCR V-beta families within both the CD3+CD8+ and CD3+CD8− T-cell (primarily CD4+) subsets in each tube for a total assessment of 24 different TCR V-beta families (as illustrated in Figure 9.6) per patient. This panel is useful in supporting the diagnosis of Omenn syndrome as described earlier and represented in Figure 9.5, and has been reported to be useful in assessing the degree of immunological compromise in patients with the Di George syndrome as described in Section 9.3.3.3 [35].

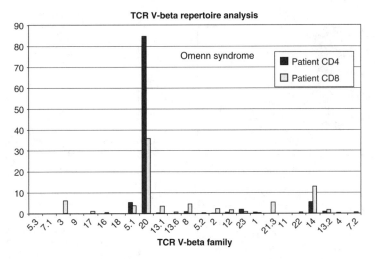

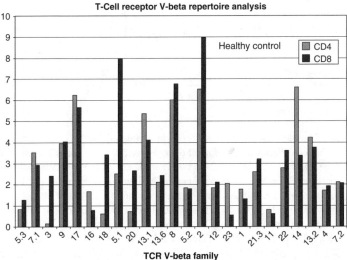

FIGURE 9.5   A basic assessment of the T-cell receptor repertoire can be performed quickly and relatively easily by flow cytometry (see Section 9.3.1.10 as well as Chapter 10 by Dr. G. Uzel) by measuring the relative proportion of 24 different TCR V-beta families expressed within the T helper (CD3+CD8−) and T cytotoxic (CD3+CD8+) cells. This figure illustrates the distribution of each of the 24 V-beta TCR families in a normal healthy control individual as well as in a patient suspected of Omenn syndrome (the actual method of measuring the V-beta families is illustrated in Figure 9.6). The CD8+ T-cell subpopulation contained over 80% TCR V-beta 20 families and a complete absence of several other TCR V-beta families (top graph). The CD4 T-cell subset also contained an abnormally high proportion of V-beta 20 as well as V-beta 14, with a similar absence of several other V-beta families. All 24 TCR V-beta families were normally represented in the T-helper and T-cytotoxic cells of the healthy control subject (lower graph). Note that the "y"-axis on each of the graphs is different. Similar figure was published previously by O'Gorman, M.R.G., *Clin. Lab. Med.* 27, 591, 2007.

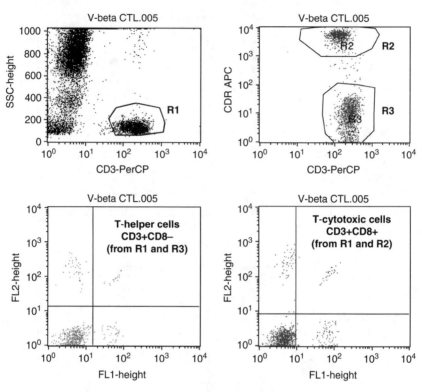

FIGURE 9.6 The basic flow cytometry procedure adopted for the measurement of 24 different TCR V-beta families with T helper and T cytotoxic T-cell subsets. Briefly, whole blood is labeled with CD3-PerCP, CD8-APC, and three different TCR V-beta monoclonal antibodies (one of the three TCR V-beta monoclonal antibodies is conjugated with FITC, the other with PE, and the third is conjugated with equal amounts of PE and FITC) in each of eight tubes. The individual TCR V-beta families within T helper and T cytotoxic cells are analyzed by first drawing a region around the CD3+ cluster (R1) on a dot plot of CD3-PerCP versus RALS (upper left-hand side dot plot). The events in this gate are then displayed on a dot plot of CD3-PerCP versus CD8-APC (upper right-hand side dot plot), and two regions are drawn to define CD3+CD8+ (T cytotoxic cells = R2) and CD3+CD8− (T-helper cells = R3). The events that satisfy both R1 and R2 (T-cytotoxic cells) are then assessed for the expression of each of the three different TCR V-beta families (lower right-hand side dot plot), and the events that satisfy R1 and R3 (T helper) are then assessed for the expression of each of the same V-beta families (lower left-hand side dot plot) by measuring the percentage of events in each of the appropriate quadrants. Figure was published previously by O'Gorman, M.R.G., *Clin. Lab. Med.* 27, 591, 2007.

9.3.1.11   Summary of Severe Combined Immunodeficiency

Currently, there are 11 genes (*ADA, CD132, RAG1, RAG2, LIG4, Artemis, JAK3, CD3 delta, CD3 zeta, CD45,* and the *IL-7* receptor) known to be associated with SCID disease. Many of the mutations lead to characteristic abnormalities in circulating peripheral blood lymphocyte subsets. A routine flow cytometric assay for the measurement of relevant proportions and absolute numbers of the major lymphocyte subsets can provide valuable clues to the underlying genetic abnormality in a very fast and relatively inexpensive manner. It must be noted that mutations in different genes lead to identical phenotypic abnormalities, indicating that the ultimate diagnosis and determination of the molecular etiology may have to be determined with the appropriate genetic sequencing technology.

9.3.1.12   Hyper IgM Syndromes: CD40 Ligand, CD40 Receptor-Group I

The nomenclature for the HIGM syndromes with immunoglobulin class-switch defects were reclassified at the last Classification Committee meeting [3] to more accurately reflect their genetic defect and are now represented both in group I (combined immunodeficiencies [CD40 ligand, CD40 receptor]) and group II (primarily antibody deficiency groups, NEMO, uracil-DNA glycolusase [UNG], activation-induced cytidine deaminase [AICDA]). Mutations in any of the latter genes result in low levels of serum IgG and IgA with normal to elevated IgM, defective Ig class-switch recombination, absence of Ig gene somatic hypermutation, and a variable level of cellular immunodeficiency. Patients with mutations in the genes encoding the CD40 receptor or the CD40 ligand tend to have increased cellular immunodeficiency and experience more severe clinical disease (hence the classification as group I). The most frequently observed cause of the HIGM syndromes is due to mutations in the X-linked gene encoding the CD40 ligand. In a recent review of 130 patients with HIGM syndrome, 75% were found to have a mutation in CD40 ligand gene [36]. Flow cytometry–based procedures have been utilized in assessing all forms of the HIGM syndromes but have been particularly useful (and more specific) in the detection of patients with either CD40 ligand– or CD40 receptor–related abnormalities. The CD40 ligand (CD154), expressed primarily on the surface of activated CD4+ T cells, binds to the CD40 receptor of B cells and along with the appropriate cytokines induces their proliferation, differentiation, and immunoglobulin class switching of the immunoglobulin genes. CD154 can only be reliably measured following a potent *in vitro* activation step that has led to the development of a function-based flow cytometry assay [37] (see details of the procedure in Section 9.3.1.13). Following the ligation of the CD40 ligand with its receptor, the B cells signal back to the T cell to continue its maturation and differentiation. Disruption of this costimulatory interaction is believed to be associated with the abnormal cellular immunity (in addition to the humoral immune abnormalities) observed in these patients. Patients with mutations encoding genes downstream of the CD40 receptor in B cells have primarily humoral immune defects (group II) and less-severe clinical disease.

9.3.1.13   Measurement of CD40 Ligand on *in Vitro*–Activated
T-Helper Cells and CD40 Receptor on B Cells

Following the discovery of the cause of XHIM and the development of a monoclonal antibody specific for CD40 ligand, my laboratory designed, developed, and validated a clinical flow cytometry procedure for the screening diagnosis of XHIM patients and carriers [37]. Briefly, whole blood is stimulated with phorbol ester and a calcium ionophore for 4 h, followed by the staining of lymphocytes with a panel of monoclonal antibodies including CD3, CD8, CD40 ligand (TRAP-10 clone conjugated with phycoerytherin (Pharmingen, San Diego, California), and CD69. Following the labeling procedure, the red blood cells are lysed, the cells are fixed, and the samples are acquired and analyzed on the flow cytometer using a novel gating strategy (see Figure 9.7). Although the procedure is now routine, in the development process we discovered the following: phorbol 12-myristate 13-acetate (PMA) induces the rapid modulation of the CD4 molecule off of the cell surface (so you must use a negative gating strategy, i.e., CD3+CD8− cells); the stimulation and upregulation of CD40 ligand is strongly calcium dependent, therefore, you cannot use EDTA anticoagulated blood.

Abnormal CD40 ligand expression has been detected in all of our X-linked HIGM patients (to date), their carrier mothers (when available), and in one of three patients with common variable immunodeficiency (see Figure 9.7, right-hand side column, for an example of an XHIM in patient and carrier). Most laboratories with a moderate amount of flow cytometry experience could perform this assay [38]; however, it must be cautioned that others have reported the possibility that some XHIM associated mutations could be missed if their detection relied exclusively on monoclonal antibodies and flow cytometry [36].

Mutations in the gene encoding the CD40 receptor (formerly classified as HIGM3) also (not surprisingly) lead to a clinical phenotype virtually identical to the X-linked form of the HIGM syndrome. Abnormal expression of the CD40 receptor is detected by gating on B cells (CD19 or CD20) and measuring the percentage of CD40 positive events (normal range is greater than 90% of B cells, that express CD40; unpublished result observed in our laboratory). The latter form of autosomal recessive HIGM syndrome is very rare [39].

9.3.1.14   Abnormal HLA-DR Expression (Omenn Syndrome—Abnormally Elevated and MHC Class II Deficiency—Abnormally Decreased)

The percentage of T cells expressing HLA-DR+ T can be abnormally elevated in patients with PID in two different circumstances: (1) SCID patient with maternally engrafted T cells and (2) Patients with hypomorphic mutations in RAG1, RAG2, and Artemis or IL-7RA genes presenting with Omenn syndrome (as discussed earlier with abnormal oligoclonal T cells circulating in the periphery). Maternal engraftment can cause significant problems pre- and post-bone marrow transplantation in SCID patients and can be ruled out with the appropriate molecular testing [40].

Support for a diagnosis of Omenn syndrome, however, can be obtained by the detection of abnormal HLA-DR expression on T cells combined with T-cell repertoire abnormalities as presented in Figures 9.4 and 9.5.

Interestingly, MHC class II deficiency is not due to mutations in the class II genes themselves, rather the abnormal MHC class II expression is due to mutations in genes encoding transcription factors binding to the promoter sites of the class II genes. Mutations in genes encoding the class II transcription activation factor (CIITA) and the regulatory factor X proteins (RFX5, RFXAP, and RFXANK) lead to abnormal MHC class II expression on the surface of B cells, NK cells, and

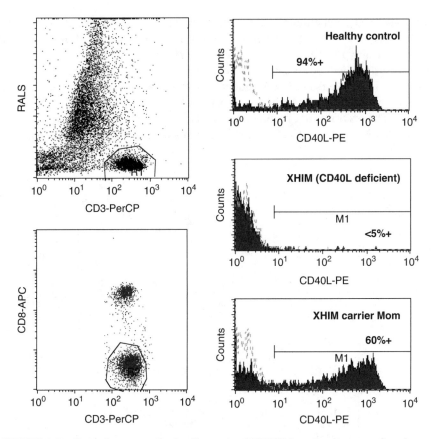

FIGURE 9.7 Screening assay for the diagnosis of XHIGM syndrome now referred to as CD40 ligand deficiency [1]. Briefly, whole, heparinized peripheral blood is activated with optimal concentrations of phorbol myristate acetate and the calcium ionophore ionomycin (Sigma Chemical Co., St. Louis, Missouri) or diluent alone. After 4 h at 37°C, samples are washed and labeled with CD3-PerCP, CD40 ligand-PE, and CD8-APC (or FITC). This monoclonal antibody combination allows for the "negative gating of CD4+ T cells by drawing a region around the CD3+CD8− T cells (bottom left-hand side dot). The histogram in the right-hand side column illustrates the level of CD40 ligand expression levels on CD4+ T cells (filled curves), whereas as the stippled curves represents the level of fluorescence generated on a nonhuman cell–antigen-specific isotype matched control antibody. The results of this assay on a patient diagnosed with XHIM indicate that the CD40 ligand was not upregulated (<5%, middle histogram, right-hand side column), whereas his mother's cells expressed below normal levels of CD40 ligand (60%) as seen in the bottom histogram of the right-hand side column. Results obtained from a healthy nondiseased control run concurrently with all patients tested indicates a robust and normal (i.e., >80%) level of CD40 ligand upregulation on T-helper cells. Not shown in the figure are the results of induced CD69 expression on the *in vitro* activated CD3+ T cells on samples from the patient, the mom, and the control that were all consistent with normal (i.e., >90%), eliminating the possibility of a nonspecific *in vitro* activation abnormality in the patient. Figure was published previously by O'Gorman, M.R.G., *Clin. Lab. Med.* 27, 591, 2007.

activated T cells, and ultimately cause the "bare lymphocyte syndrome"—a form of SCID (reviewed in Ref. 41). Patients with this disorder are detected in the routine immunophenotyping panel. B cells, which normally express high levels of HLA-DR on their surface, will be absent in a dot plot that has been gated on lymphocytes and displays HLA-DR versus CD3. Additionally, CD4+ T-cell percentages and absolute counts of the patients will be below normal values due to abnormal selection in the thymus. The results of our routine immunophenotyping analysis on a patient diagnosed with MHC class II deficiency (bare lymphocyte syndrome) is presented in Figure 9.8.

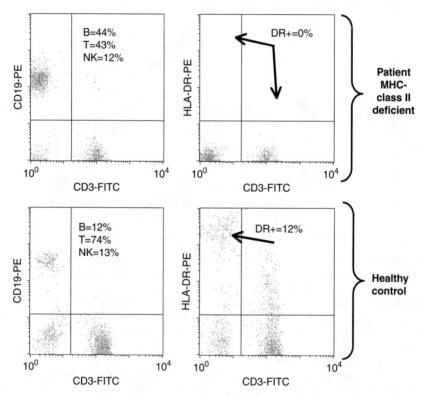

FIGURE 9.8 Detection of MHC class II deficiency in the routine immunophenotyping panel. This 7-month-old male patient with a presumed diagnosis of SCID had an absolute lymphocyte count of 1900 cells/mm³, %CD3+CD4+ (T helper) cells below normal and a CD4:CD8 ratio of 0.2, which is significantly below normal (not shown). Other abnormalities included an elevated proportion of B cells (44%) with a correspondingly low proportion of T cells (CD3 = 43%). The most significant observation, however, was the complete absence of bright HLA-DR+ CD3− events (corresponding to B cells) that should have been 44% of the lymphocytes and the lack of any detectable HLA-DR on the T cells (see arrows in upper right-hand side dot plot). Compare the results of the patient (upper row) to those obtained in the healthy control (bottom row), who as the same proportion of B cells (CD19+) as the proportion of bright HLA-DR+CD3− cells (12%) in the healthy control. Figure was published previously by O'Gorman, M.R.G., *Clin. Lab. Med.* 27, 591, 2007.

9.3.1.15 Decreased CD8+ T Cells (TAP-1/2 Deficiency, ZAP-70 Deficiency)

Class I deficiencies have also been reported and like the class II defects are not due to mutation in the class I genes themselves. Mutations in the gene encoding the TAP2 subunit of the peptide transporter associated with antigen processing (TAP) leads to failure of MHC class I to be loaded with peptide [42], which in turn leads to the failure of class I expression on the surface of nucleated cells. Lack of class I expression leads to abnormal selection of CD8+ T cells in the thymus. Patients with TAP deficiency have been reported to have decreased numbers of alpha/beta TCR positive CD8 T cells, a higher proportion of gamma/delta positive CD8+ T cells, and normal levels of NK cells (both CD3-CD56 and CD16+ and CD8+ NK cells); however, NK function is abnormal [42]. This disease has also been referred to as Bare Lymphocyte syndrome.

Patients with mutations in ZAP70 exhibit an absence of circulating CD8+ T cells, and their CD4+ T cells fail to respond to TCR signaling *in vitro* [43]. Patients with ZAP-70 mutations have abnormal TCR-mediated activation [44], resulting in severely depressed T-cell function and a clinical presentation very similar to other SCID patients [43,45].

9.3.1.16 Decreased CD4+ T Cells

The idiopathic CD4+ T-lymphopenia syndrome is characterized by low CD4 counts (<300 cell/µL) or low CD4% (<20%) on at least two occasions with evidence of opportunistic infections and rigorous exclusion of all other known primary and acquired immunodeficiency states (e.g., HIV infection) [46]. This syndrome has been reported in children and adults as well as in patients with Down syndrome [47].

There is a report of immunodeficiency associated with homozygous caspase 8 mutations in two children born to consanguineous parents [48]. The percentage of CD4+ T cells in both patients was significantly below normal. Other heterozygous family members did not have a history of immunodeficiency.

9.3.2 GROUP II: PREDOMINANTLY ANTIBODY DEFICIENCIES

1. Severe reduction in all serum Ig isotypes with absent B cells
   a. Bruton's tyrosine kinase (Btk) deficiency
   b. Mu heavy-chain deficiency
   c. I 5 deficiency
   d. Ig alpha deficiency
   e. B cell linker protein (BLNK) deficiency
   f. Thymoma with immunodeficiency
2. Severe reduction in at least two serum Ig isotypes with normal or low numbers of B cells
   a. Common Variable Immunodeficiency Disorders (CVID)
   b. Inducible Costimulator (ICOS) deficiency
   c. CD19 deficiency
   d. transmembrane activator and calcium-modulator and cyclophillin ligand interactor (TACI) deficiency
   e. B cell-activating factor (BAFF) receptor deficiency

3. Severe reduction in serum IgG and IgA with increased IgM and normal numbers of B cells (previously classified as autosomal HIGM syndromes)
   a. activation-induced cytidine deaminase (AID) deficiency
   b. urasil DNA glycosylase (UNG) deficiency
4. Isotype or light-chain deficiencies with normal numbers of B cells
   a. Ig heavy-chain deletions
   b. Kappa chain deficiency
   c. Isolated IgG subclass deficiency
   d. IgA with IgG subclass deficiency
   e. Selective IgA deficiency
5. Specific antibody deficiency with normal Ig concentrations and normal numbers of B cells
6. Transient hypogammaglobulinemia of infancy.

9.3.2.1 Absent B Cells: Th+Tc+B−NK+

The absence of B cells can be readily detected in the routine immunophenoping panel (as described previously) by two separate and supporting results: very few events will be CD3− and CD19+ and 2. CD3$^-$ HLA-DR$^+$ bright events will be absent (see Figure 9.9). Mutations in genes that result in arrested B-cell maturation can be broadly separated into (1) X-linked agammaglobulinemia (XLA) and (2) autosomal recessive (rarely autosomal dominant) inheritance forms. The majority of patients with extremely low or absent B cells will be XLA, that is, males with a mutation in the gene encoding Btk [49,50], and are easily identified in the routine immunophenotyping panel. The abnormally expressed Btk protein can however also be detected by flow cytometry (see Section 9.3.2.2).

9.3.2.2 Measuring Btk by Flow Cytometry in X-Linked Agammaglobulinemia

Abnormal expression of Btk in platelets and monocytes can be assessed by flow cytometry. Peripheral blood labeled with CD14 PE followed by the fixation, permeabilization, and intracellular labeling of Btk correctly identified abnormal Btk protein expression in the monocytes of 40 of 41 XLA patients and cellular mosaicism in 35 of 41 obligate XLA carriers [51]. This test provides a very fast assessment of Btk expression, however, it must be noted that patients with missence mutations in the Btk gene can express normal levels of Btk by flow cytometry. While measuring Btk in platelets, Futatani observed that Btk was abnormally expressed in 37 out of 45 unrelated families with XLA and that obligate XLA carriers had both normal and abnormal expressing platelets [52]. In eight of the families, detection of an abnormality by flow cytometry was not possible due to the normal levels of the mutated Btk protein. Those patients with detectable Btk protein (even at very low levels) in their monocytes, presented with significantly higher percentages of B cells (0.7%) compared to XLA patients with no detectable Btk protein (0.1% B cells) [53]. However, no relationship was observed between the level of expression of Btk in platelets and the clinical phenotype as reported by Futatani [52]. Although flow cytometry is very

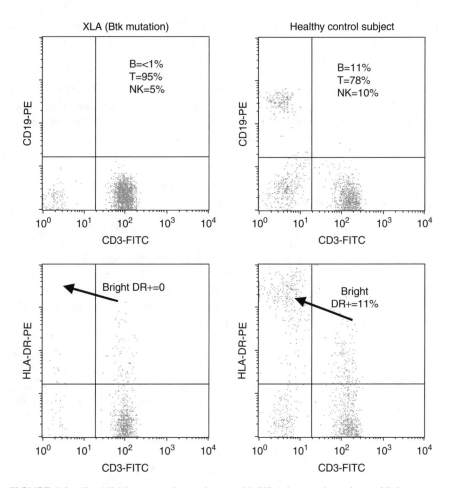

FIGURE 9.9 Establishing a result consistent with XLA in a male patient with hypogammaglobulinemia. The results indicated a (left-hand side column) virtual absence of CD19+ cells (less than 1%), as well the absence of bright HLA-DR+ CD3− cells, which is consistent with XLA. The absolute lymphocyte count in the 11-month-old was 3300 cells/mm³ (within age-associated reference range) and the serum immunoglobulin levels (IgG, IgA, and IgM) were all greater than two standard deviations below normal. Patients with autosomal forms of agammaglobulinemia can express the identical immunophenotypic abnormalities and without a consistent family history must have sequencing performed to identify the underlying molecular abnormality. Figure was published previously by O'Gorman, M.R.G., *Clin. Lab. Med.* 27, 591, 2007.

useful in detecting patients with abnormal Btk expression, the absence of the abnormality does not rule out a mutation in the Btk gene.

9.3.2.3 Autosomally Inherited Agammaglobulinemia

In a smaller proportion of agammaglobulinemia patients, failure of B-cell maturation is associated with mutations in the genes encoding the B-cells receptor gene,

IgM, the surrogate light chain, lambda 5/14.1, the B-cell signaling receptor, Ig alpha, and the B-cell linker adapter protein (BLNK) [12]. These patients experience clinical findings similar to patients with XLA. A novel form of agammaglobulinemia is inherited in an autosomal dominant fashion (rare for immunodeficiency disease) [54] and is caused by mutations in the leucine-rich repeat-containing 8 gene (*LRRC8*) [55]. Patient's B cells were completely absent in the peripheral blood and the patient's cells in the bone marrow expressed both the normal and the mutant protein. Surface marker analysis of the B-cell lineage indicated that arrest had occurred at the pro-B-cell stage (slightly earlier than in XLA) [54].

9.3.2.4 Memory B Cells in Common Variable Immunodeficiency and X-Linked Hyper IgM Syndromes

Peripheral blood memory B-cells can be immunophenotypically identified by cell surface expression of CD27 [56] as well as switched surface Ig (i.e., not IgM and IgD positive). Human peripheral blood contains approximately 35% memory B cells (although this is age related), and the majority of the memory B cells are of nonisotype-switched variety (i.e., do not secrete and express immunoglobulins other than IgM and IgD) [57]. A rapid flow cytometry procedure used to identify CD27+ memory B cells is illustrated in Figure 9.10.

CVID represents a heterogeneous group of diseases with a variety of underlying genetic causes. The identification of the absence of memory B cells in some patients with CVID [58] has resulted in attempts to classify CVID patients based on the level of this subset [59,60]. Cunningham–Rundles' laboratory adopted the classification system developed by Warnatz to evaluate a group of 53 CVID patients [61].

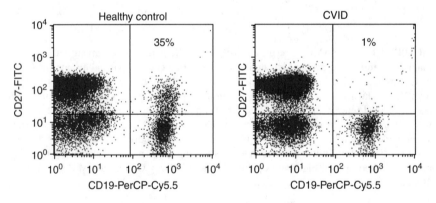

FIGURE 9.10 Measurement of memory B cell levels by flow cytometry. Memory B cells can be detected by the surface expression of CD27 on CD19+ (or CD20+) lymphocytes. Memory B cells are expressed as the percentage of B lymphocytes expressing CD27. In the figure, B-lymphocytes are first gated using CD19 versus right-angle light scatter and the expression of CD27 is displayed on CD19+ cells. The percentage illustrated in the figure represents only the percentage of B cells expressing CD27. The abnormally low percentages of memory B cells illustrated are consistent with a diagnosis of common variable immunodeficiency and in some forms of the HIGM syndrome).

Patients with less than 0.4% switched memory B cells (CD19+CD27+IgM-IgD-) were categorized as CVID-group I and patients with greater than 0.4% were categorized as CVID-group II. Group I patients had the lowest levels of serum IgG, poor responses to pneumococcal vaccine, and increased rates of autoimmune and granulomatous disease, that is, more severe disease [61]. There are only a few laboratories in the United States currently performing routine memory B-cell analysis in CVID patients (and other patients with associated immunoglobulin abnormalities), but studies such as that reported earlier suggest that this procedure may have a role in characterizing higher risk CVID patients and may eventually allow for the improved identification of additional genetic abnormalities. Patients with some forms of the HIGM syndrome have also been reported to express reduced levels of CD27+ memory B cells [62], indicating that a low percentage of memory B cells is not specific for CVID.

A recently described flow cytometry procedure very similar to the CD40 ligand upregulation protocol described previously has been developed for the detection of ICOS deficiency in patients with CVID [63]. Briefly, whole blood is incubated with the appropriate concentration of PMA and ionomycin for 20 h and then the cells are labeled with CD3, CD8, and anti-ICOS (eBioscience, Germany). After gating on the CD3+CD8− T cells, the level of expression of ICOS is assessed. A patient with a genetically defined homozygous ICOS mutation expressed ICOS on less than 1% of the CD4 T cell compared with a range of $86\pm14\%$ (2 SD) in a group of nondiseased healthy control subjects. This test requires confirmation on a larger group of patients but appears to have potential in identifying CVID and characterizing the underlying genetic abnormality.

Autosomal HIGM (due to defects in genes other than CD40 receptor and CD40 ligand) is associated with mutations in the genes encoding activation-induced cytidine deaminase (AICDA) (formerly classified as HIGM2) and uracil DNA glycosylase (UNG) (now classified with group II, predominantly antibody deficiencies [3]). Unlike patients with the common variable immunodeficiency syndrome, B cells from most patients with the autosomal form of the HIGM syndrome express CD27, implying that the generation of memory B cells in these patients is not necessarily impaired even with an inability to switch Ig classes [64,65]. *In vitro* induction of class-switch recombination (i.e., from IgM+IgD+ to other immunoglobulin classes) with soluble CD40 ligand and appropriate cytokines is impaired in the autosomally inherited forms of HIGM, but occurs normally in the X-linked form of HIGM [65].

Detection of kappa-chain deficiency is easily observed following surface staining with the appropriate antilight-chain antisera, and a B-cell marker.

9.3.3   GROUP III: OTHER WELL-DEFINED IMMUNODEFICIENCY SYNDROMES

1. Wiscott–Aldrich syndrome
2. DNA repair defects (other than those in Table 9.1)
   a. Ataxia telangiectasia
   b. Ataxia-like syndrome
   c. Nijmegen breakage syndrome
   d. Bloom syndrome

3. Thymic defects Di George anomaly
4. Immuno-osseous dysplasias
   a. Cartilage hair hypoplasia
   b. Schimke syndrome
5. Hermansky–Pudlak syndrome type 2
6. Hyper-IgE syndrome
7. Chronic mucocutaneous candidiasis

9.3.3.1 Wiscott–Aldrich Syndrome

Wiscott–Aldrich syndrome (WAS) is an X-linked recessive disease characterized by thrombocytopenia (with small platelets), eczema, and variable levels of both humoral and cellular immunodeficiency [66]. WAS is associated with mutations in the gene encoding the WAS protein (WASP). Monoclonal antibodies specific for WASP [67,68] have been combined with intracellular flow cytometry staining for the diagnosis of WAS and WAS carriers [68–71], to evaluate mixed chimera status in WAS patients post-bone marrow transplant [72] and to separate patients with WAS from thrombocytopenias of unknown etiology [71]. There are only a few laboratories in the United States performing flow-based studies for the detection of abnormal WASP expression.

9.3.3.2 Ataxia Telangiectasia and the Ataxia Telangiectasia Mutated Gene

Immunodeficiency in the context of telangiectasias and cerebellar ataxia portends a diagnosis of ataxia-telangiectasia (AT). The disease is inherited in an autosomal recessive fashion and is caused by mutations encoding the gene called ataxia telangiectasia mutated (ATM). Diagnostic confirmation includes testing serum alpha fetal protein (AFP) levels, abnormalities in the routine immunphenotyping assay, and assessment of radiosensitivity [73–75]. Routine flow cytometry confirms the presence of abnormalities in AT; however, the abnormalities are not specific. To date there are no reports of successfully detecting abnormal expression of the ATM protein by flow cytometry. Heinrich et al. published a summary of flow cytometry–based cell-cycle testing assay on a group of 330 patients referred for "exclusion of AT" and were able to ascertain AT-negative versus AT-positive in 94.2% of the cases tested [76]. Although the authors concluded that "cell cycle testing complemented AFP measurements and fulfills the criteria as a rapid and economical screening procedure for the differential diagnosis of juvenile ataxias" [76], this particular functional flow-based assay is complex and time-consuming, suggesting it is not ready for adoption in a routine clinical setting.

9.3.3.3 Di George Syndrome

Di George syndrome or Di George anomaly is a congenitally acquired disorder characterized by classical facial features, and defects of the heart, parathyroid glands, and thymus [77]. Although Di George syndrome is often grouped with the 22q11.2 deletion syndrome, only about 90% of the patients have this anomaly [78]. Depending on the degree of thymic involvement, the immune system will be more or less compromised. Those with below normal T-cell function (*in vitro* mitogen-induced proliferation responses) or low T-cell numbers are referred to as partial Di George, whereas those with no T-cell responses and no peripheral T cells are referred to as complete Di George. Although rare at less than 1% of all Di George patients, complete Di George is a medical emergency and unless patients are successfully treated they will die usually before 1 year of age. Patients with complete Di George have been cured with the experimental treatment of thymic transplantation to the quadriceps muscle [79]. Flow cytometry is useful for assessing the level of immunodeficiency (complete versus partial), which is important since the level of deficiency may be associated with increased risk of infection [35]. Significant abnormalities in the TCR V-beta repertoire have also been observed at baseline [35]. Longitudinal studies of patients with partial Di George usually show improvement in the immunological defects over time [35,80], which may be due to a disease-associated reduction in the normal rate of T-cell decline observed with aging [81].

9.3.4   GROUP IV: DISEASES OF IMMUNE DYSREGULATION

1. Immunodeficiency with hypopigmentation
   a. Chediak–Higashi syndrome
   b. Griscelli syndrome, type 2
2. Familial hemophagocytic lymphohistiocytosis (FHL) syndromes
   a. Perforin deficiency
   b. Munc 13-D deficiency
   c. Syntaxin 11 deficiency
3. X-linked lymphoproliferative (XLP) syndrome
4. Syndromes with autoimmunity
   a. Autoimmune lymphoproliferative syndromes (ALPS)
      i. CD95 (Fas) defects, ALPS type 1a
      ii. CD95L (Fas ligand) defects, ALPS type 1b
      iii. Caspase 10 defects, ALPS type 2a
      iv. Caspase 8 defects, ALPS type 2b
   b. Autoimmune polyendocrinopathy with cadidiasis and ectodermal dystrophy (APECED)
   c. Immune dysregulation, polyendocrinopathy, enteropathy (X-linked) (IPEX)

9.3.4.1   Familial Hemophagocytic Lymphohistocytosis

Hemophagocytic lymphohistiocytosis (HLH) is a disorder of early childhood characterized by excessive, uncontrolled T-lymphocyte and macrophage activation. Clinical symptoms include fever, hepatosplenomegaly, cytopenias, hyperlipidemia, and increased hemophagocytes in the reticuloendothelial system (especially bone marrow and liver) [82]. HLH represents two different groups of diseases categorized as either familial or secondary. Familial forms are usually fatal in infancy unless patients undergo bone marrow transplant, whereas most secondary forms respond to appropriate treatment [83]. Three genes (encoding perforin, Munc 13-D, and Syntaxin 11) are known to be associated with the familial form of the disease [84–88]. Perforin abnormalities have been detected in 20–30% of all HLH patients [85]. Abnormal perforin levels can be detected by intracellular staining and flow cytometric analysis of NK cells and cytotoxic T cells [85]. Perforin protein is reported to be "uniformly deficient" in FHL patients with perforin gene mutations [85]; however, one must be cautious of the possibility that a functional mutation could be present although expression levels as assessed by flow cytometry are normal.

9.3.4.2 The Autoimmune Lymphoproliferative Syndrome (ALPS)

The molecular basis of this chronic nonmalignant lymphoproliferative disorders was first characterized in 1995 to be caused by FAS, a gene product involved in apoptosis [89,90]. Although the majority of patients with ALPS have mutations in the gene encoding FAS, it is now known that other genes including FAS-ligand, caspase 8, and caspase 10 (all involved in FAS-mediated apoptosis) also lead to ALPS [91]. Impaired programmed cell death results in the accumulation of lymphocyte and causes the clinical phenotype of lymphadenopathy, hypersplenism, autoimmune cytopenias, and an increased risk of lymphoma. ALPS diagnosis can be substantiated by the flow cytometric assessment of specific subsets (see below) and of FAS-induced apoptosis. Several flow cytometry methods have been developed to assess induced apoptosis including annexin cell surface labeling, terminal deoxynucleotidyl transferase biotin-dUTP nick end labeling (the so-called TUNEL assay), hypodiploid nuclear DNA content, and propidium dye exclusion [91]. Flow cytometry has also been used to screen patients suspected of ALPS by measuring for increases in alpha/beta TCR positive, CD4 and CD8 double negative T cells [92]. We have developed a panel of antibodies (TCR alpha/beta-FITC, CD4-PE, CD3-PerCP, CD8-APC, BD Bioscience, San Jose, California) to measure this subset in patients suspected of ALPS. The procedure for identifying and measuring this subset is illustrated in Figure 9.11. Normal ranges for this subset have been reported from less than 1% of T cells [92] up to less than 2.6% [93]. The flow cytometric assessment of double negative T cells is an easy screening test. Patients with increases in the percentage of double negative T cells bearing the alpha/beta form of the TCR and the appropriate clinical signs warrant significant consideration for the diagnosis of ALPS.

9.3.4.3 Autoimmune Polyendocrinopathy-Candidiasis-Ectodermal Dystrophy (APECED)

APECED, also known as autoimmune polyglandular syndrome type 1 (APS-1), is an autosomal recessive monogenetic disease characterized (as the name suggests) by endocrine organ-specific autoimmunity and candidiasis [94]. The gene identified to be associated with APECED is known as "autoimmune regulator" (AIRE) [95,96]. Measuring the expression levels of AIRE in cells from suspected APECED patients by flow cytometry has not been reported; however, the recent description of positive AIRE expression in CD14+ monocytes and differentiated dendritic cells (but not CD4+ T cells and polymorphonuclear leukocytes [PMNs]) by immunocytochemistry [97] suggests that it may be possible to assess AIRE expression by flow cytometry as an aid in the diagnosis of APECED in the near future.

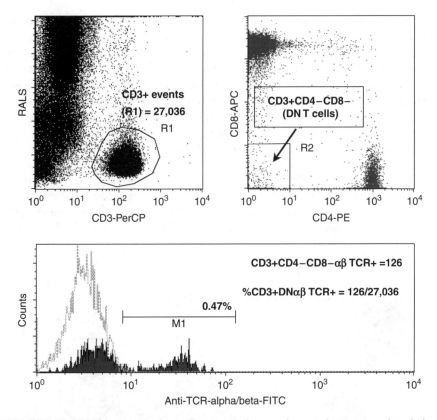

FIGURE 9.11 Illustrated in the figure is our analysis procedure used to measure the relative proportion of CD4 and CD8 double negative, alpha/beta TCR positive, and CD3+ lymphocytes in patients suspected of a diagnosis of ALPS. Briefly, whole blood is labeled anti-TCR alpha/beta-FITC, CD4-PE, CD3-PerCP, and CD8-APC (BD Bioscience), and the flow cytometric analysis involves first gating on T (R1 in the upper left-hand side dot plot), and then noting the number of T-cell events. The CD3+ events are then displayed in a dot plot of CD4 versus CD8 (upper left-hand side dot plot), and a region (R2) is drawn to include the CD4− and CD8− events. Events included in both regions 1 and 2 (i.e., CD3+CD4−CD8−) are then analyzed for the number of events that are positive for the alpha/beta form of the TCR. The number of CD3+CD4−CD8− TCR alpha/beta+ events is divided by the number of T cells (27,036 in the upper left-hand side dot plot of region 1) to obtain the percentage of CD3+CD4−CD8− TCR alpha/beta+ cells. Patients with a diagnosis of ALPS can have any high proportions of this subset as described in the text. Figure was published previously by O'Gorman, M.R.G., *Clin. Lab. Med.* 27, 591, 2007.

9.3.4.4 Immunoregulation-Polyendocrinopathy-Enteropathy-X-Linked (IPEX)

IPEX is a rare X-linked recessive monogenic disorder characterized by overwhelming systemic autoimmunity. The most common clinical features of this X-linked recessive disorder are early onset insulin-dependent diabetes, severe watery diarrhea, and dermatitis often correlated with high level of serum IgE [94]. In addition to the autoimmune manifestations, patients can succumb to severe infections if immune suppressive treatment is not initiated early on [94]. The first complete phenotypic description of IPEX syndrome as a distinct clinical entity was published in 1982 [98]. In 2001, mutations in the Forkhead Box P3 (*FoxP3*), a gene located on the X chromosome were reported as the cause of the IPEX syndrome [99]. FOXP3 protein is the key to the maintenance of immune homeostasis through the development and function of regulatory CD4+CD25+ T cells (Tregs). Regulatory T cells (Tregs) are highly specialized immunosuppressive cells that actively suppress immune responses to both self and pathogens and play a key role in maintaining self-tolerance and immunologic control [100]. Flow cytometry has been used to screen patients suspected of IPEX by the intracellular detection of reduced FOXP3 expression in CD4+CD25+ T cells. In humans, FOXP3 is primarily located in the CD25+CD4+ T regulatory lymphocyte subset. Most IPEX patients lack FoxP3 positive cells, however, we must be cautioned of the possibility that the detection of FoxP3 does not necessarily rule out a mutant *FoxP3* gene. See gating procedure and representative levels of FoxP3+CD25+CD4+ (Tregs) in Figure 9.12.

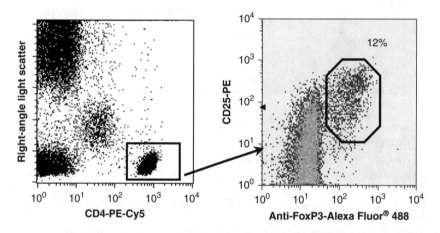

FIGURE 9.12 Measurement of abnormal FoxP3+ CD25+ T cells. Patients suspected of immune dysregulation, polyendocrinopathy, enteropathy, X-linked (IPEX) syndrome due to mutations in the *FOXP3* gene can be screened by flow cytometry. Illustrated is a method we are in the process of validating for the measurement of T-regulatory cells, that is, the percentage of CD4+ T cells that are both CD25+ and FoxP3+. Briefly, cells are labeled for surface expression of CD4-PECy5 and CD25-PE (BioLegend, San Diego, California), followed by permeabilization and fixation, then labeled for the intracellular expression of FoxP3 (conjugated with Alexa Fluor® 488, BioLegend, which emits at a wavelength virtually identical to FITC). Analysis is completed by first gating on CD4+ T cells in a dot plot of CD4 versus right-angle light scatter (left-hand side dot plot), then measuring the percentage of CD4 T cells expressing bright CD25 and FoxP3 (right-hand side dot plot). We are in the process of developing normal ranges. The normal range for the percentage of CD4+CD25 bright positive cells expressing FOXP3 is approximately 80–90% and an absolute number of circulating CD4+CD25+FoxP3+ lymphocytes of 105–259/mm³ (Alexandra Filipovich and Jack J.H. Bleesing, 2007, personal communication; *Note*: Alexa Fluor 488 is a registered trademark of Molecular Probes, Inc.

9.3.5   GROUP V: CONGENITAL DEFECTS OF PHAGOCYTE NUMBER, FUNCTION, OR BOTH

1. Severe congenital neutropenia (*ELA2* gene)
2. Severe congenital neutropenia (*GFI1* gene)
3. Severe congenital neutropenia (granulocyte-colony stimulating factor receptor)
4. Kostmann syndrome
5. Cyclic neutropenia
6. X-linked neutropenia/myelodysplasia
7. Leukocyte adhesion deficiency type-1 (*INTG2* gene, CD18)
8. Leukocyte adhesion deficiency type-2 (FUCT1 GDP-fucose transporter)
9. Leukocyte adhesion deficiency type-3 (defective Rap-1 activation of integrins)
10. Rac 2 deficiency
11. b-Actin deficiency
12. Localized juvenile perodontitis
13. Papillon–Lefevre syndrome
14. Specific granule deficiency
15. Shwachman–Diamond syndrome
16. X-linked chronic granulomatous disease (CGD) (*CYBB* gene, p91phox)
17. Autosomal CGD (*CYBA* gene, p22phox)
18. Autosomal CGD (*NCF1* gene, p47phox)
19. Autosomal CGD (*NCF2* gene, p67phox)
20. Neutrophil G-6PD deficiency
21. IL-12 and IL-23 receptor B1 chain deficiency
22. IL-12p40 deficiency
23. Interferon gamma receptor 1 (IFN-γR1) deficiency
24. IFN-γR2 deficiency
25. STAT1 deficiency (autosomal recessive inheritance)
26. STAT1 deficiency (autosomal dominant inheritance)

9.3.5.1   Neutropenia

Flow cytometry is particularly well suited for the detection of neutropenia; however, this technology is rarely used to assess the relative or absolute numbers of circulating neutrophils due to the ubiquitous availability of reliable and inexpensive automated hematology analyzers.

9.3.5.2   Chronic Granulomatous Disease: gp91phox, p22phox, p47phox, and p67phox

CGD is the most common immunodeficiency of the phagocyte primary immunodeficiencies. The genetic abnormalities in CGD result in a severely reduced capability to generate superoxide and other reactive oxygen species, which in turn results in abnormal microbial killing within phagocytes. Abnormal nicotinamide dinucleotide phosphate (NADPH) oxidase activity is caused by mutations in one of four known NADPH subunit genes, that is, gp91phox, p22phox, p47phox, and p67phox. CGD is relatively rare with a reported incidence of approximately 1 in 200,000 persons [101]. The most common form of CGD is caused by mutations to the X-linked gene, gp91phox (70% of CGD cases), the remainder of the cases occur as a result of mutations in autosomally encoded genes: p22phox (the other component of membrane flavocytochrome b558), p47phox, and p67phox [102].

Monoclonal antibodies specific for the cytochrome b558 complex [103–105] of the NADPH oxidase have been combined with flow cytometry to detect abnormal membrane expression in CGD patients [106]. The use of monoclonal antibodies in the diagnosis of CDG is, however, largely relegated to highly specialized clinical/ research laboratories using Western blot methodology to confirm the mutated gene products. Screening for CGD is much more commonly performed by flow cytometry using quicker and less-expensive functional flow cytometry–based oxidative burst screening assays.

9.3.5.3 The Oxidative Burst Assay

The basic principle of the flow cytometry–based oxidative burst procedure is that cells preloaded with a nonfluorescent oxygen-sensitive dye, will become brightly fluorescent when the dye is exposed to reactive oxygen intermediates generated in the course of an induced oxidative burst. Patients with CGD are unable to generate an oxidative burst and are easily identified since they cannot oxidize the dye (i.e., the dye will remain nonfluorescent). Bass first reported the use of a flow cytometry–based procedure for the diagnosis of CGD with the oxygen-sensitive dye dichlorofluorescein diacetate (DCFH-DA) in 1983 [107]. The discovery of a more sensitive dye led [108] to the development of improved flow cytometry–based assays for the diagnosis of CGD, however, the tests were still performed on samples that had been processed before dye loading and stimulation [109]. By utilizing the uncharged nonfluorescent dye, dihydrorhodamine 123 (DHR-123), which could be added directly to whole blood, we were able to develop a new assay that required both less blood and less time than previously published methods [110]. This assay has become widely adopted. Briefly, appropriately titred DHR-123 is added directly to diluted whole blood and incubated in a shaking water bath at 37°C for 15 min followed by the addition of PMA for an additional 15 min at 37°C. After these two incubations, the red blood cells are lysed with an ammonium chloride solution and the remaining cells are fixed in 1% paraformaldehyde and run immediately on the flow cytometer. Results are expressed as a normal oxidative index (NOI), which is the ratio of the fluorescence in stimulated cells to the fluorescence observed in unstimulated cells. The laboratory's normal range is an NOI of greater than 30 (usually over 100) as compared to most CGD patients who have NOI results of less than 3 (i.e., very little change in fluorescence following *in vitro* stimulation). Occasionally, patients with either the X-linked form or the autosomal form of the disease will produce NOI results approaching the normal range (personal experience). In the absence of confirmed CGD in a family member, such a result requires follow-up genetic testing to confirm mutations in one of the components of the NADPH oxidase system. Patients with the autosomal form of CGD (p47phox) can have significantly higher NOI than the patients with X-linked CGD (gp91phox), and the fluorescence levels can be more variable (increased coefficient of variation (CV)) than both the X-linked CGD patients and healthy nondiseased controls [111]. We and others have observed false positive results (i.e., no increase in fluorescence) in patients with complete myeloperoxidase deficiency and care must be taken to ensure such patients are not misdiagnosed as CGD patients [112].

Biological mothers of patients with an abnormal NOI can be assessed by flow cytometry for their carrier status. Those mom's whose granulocytes generate two different granulocyte populations with clearly separated levels of fluorescence (one normal >30 and one abnormal <30) confirm both (a) diagnosis of X-linked CGD in the child and (b) their status as a carrier of the X-linked mutation. We have successfully diagnosed X-linked CGD carriers with as few as 15% normal granulocytes. Parents of CGD patients with the autosomal forms of the disease, in contrast, generate oxidative burst assay results that are consistent with normal (personal experience). The flow cytometry–based oxidative burst assay is very sensitive. Vowells et al. [113] were able to consistently detect 0.1% normal granulocytes when mixed with the cells of a CGD patient using a procedure very similar to ours. The test can be performed on sodium heparin anticoagulated blood samples that are up to 72 h old [114], although this needs to be validated in each individual laboratory. Recently, a whole blood procedure that did not involve either lysis or wash steps was described. The results of this new procedure generated "activity inde[ces]" (equivalent to the NOI) that were fourfold lower than the lyse wash methods [115], suggesting it will not be widely adopted as a diagnostic test. The gating and the results obtained from a normal control, an X-linked CGD patient, and the mother of the X-linked CGD patient (an X-linked carrier) are presented in Figure 9.13.

9.3.5.4 Leukocyte Adhesion Deficiency Type-1, CD18 Deficiency

Leukocyte adhesion deficiency type-1 (LAD-1) is a rare disorder characterized clinically by defective wound healing, infections by Gram-negative bacteria without pus, delayed umbilical cord separation, gingivitis, and peridontitis. The underlying genetic defect is mutations in the gene encoding the beta chain (CD18) of the β2-class of leukocyte integrins. An abnormal beta chain (CD18) results in its inability to complex with and carry the alpha chain complexes (CD11a, CD11b, and CD11c) to the cell surface. Patients with LAD-1 have reduced cell surface levels of CD11a, CD11b, and CD11c, and CD18 on all resting leukocytes as well as the inability to upregulate levels of the latter molecules in response to activation. A patient with a clinical history and laboratory findings (very high white blood cell count) consistent with LAD-1 presented to our emergency room and proved to have cell surface abnormalities consistent with LAD-1. With informed consent from the patient, our laboratory studied various stimuli and monoclonal antibody combinations in an effort to develop a whole blood flow cytometry–based diagnostic test [116]. An efficient whole blood procedure was developed and remains in use in our laboratory. Briefly, an optimized concentration of phorbol myristate acetate is added to a sample of EDTA anticoagulated whole blood for 15 min (stimulated sample) followed by the staining of the stimulated sample and a nonstimulated sample with anti-CD11b (CD11b-PE, BD

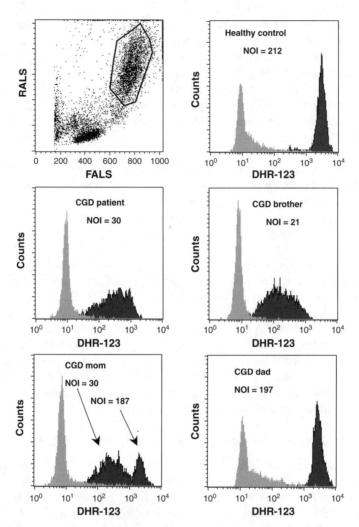

FIGURE 9.13 Flow cytometry screening assay to detect abnormal oxidative burst activity in patients suspected of CGD. Illustrated are the flow cytometry results obtained in a family with an unusual case of X-linked CGD. The basic analysis procedure is illustrated in the top left-hand side dot plot. Briefly, cells with light scatter properties consistent with PMNs are gated on, and the level of fluorescence emitted in the FL1 channel (approximately 535 nm) is measured for unlabeled cells not shown, cells exposed to the oxygen-sensitive dye, DHR-123 but not stimulated (light coloured histograms), and cells exposed to DHR-123 and stimulated with a phorbol ester to undergo an oxidative burst (black histograms). The results of the oxidative burst assay are reported as NOI, which is simply the ratio of median channel of the labeled and activated cells divided by the median fluorescent channel of the labeled but not activated cells. The normal range is an NOI greater than 30. In each of the remaining histograms, the fluorescence generated by the unstimulated (light color) and the stimulated cells (dark peaks) are presented with the calculated NOI. Results from a nondiseased control individual (run with every patient) are consistent with normal (NOI > 30). The middle histograms in the left- and right-hand side columns were generated on samples obtained from brothers with mild symptoms suggestive of CGD. The results were difficult to interpret as they were

Bioscience). A fluorochrome matched nonhuman antigen-specific monoclonal isotype control is used to set the background level of fluorescence of the granulocyte cluster (gating on forward versus right-angle light scatter parameters), which is higher than the level of fluorescence generally observed when gating on the lymphocyte cluster. The level of fluorescence of CD11b+ granulocytes is then measured (as the median fluorescent channel) on both resting (nonactivated) and *in vitro* activated granulocytes. A normal range established on a group of 30 nondiseased control subjects is used to evaluate the results of the patient. Patients with LAD-1 express CD11b significantly below normal on both resting and stimulated granulocytes. Patients with severe LAD-1 do not survive infancy and express no detectable CD18, CD11a, CD11b, or CD11c protein on the cell surface, whereas patients with the moderate form of LAD-1 express 1–10% of the level of normal and can survive into adulthood [117].

9.3.5.5   Interferon-Gamma Receptor-1, Interferon-Gamma Receptor-2, IL-12p40, and IL-12 and IL-23 Receptor-Beta 1 Chain: Interferon-Gamma Pathway Defects: Increased Susceptibility to Poorly Pathogenic Mycobacteria, *Salmonella*, and Other Intracellular Infectious Pathogens

Recurring and unusually severe infections with intracellular organisms including Bacille Calmett—Guerin (BCG), and nontuberculous mycobacteria (*Mycobacterium avium, Mycobacterium fortuitum*, and *Mycobacterium chelonae*) in family members led to the identification of mutations in genes involved in the IFN-γ pathway [118,119]. The IFN-γ pathway involves IFN-γ receptors (IFN-γR1 and IFN-γR2), interleukin-12 (IL-12p40), and the IL-12-receptor-beta 1 chain (IL-12Rβ1). IL12 secreted primarily by monocytes/macrophage stimulates T- and NK cells to synthesize and secrete INF-γ, which in turn activates the macrophage and further stimulates cytolytic CD4+ T cells and effector CD8+ T cells, which then kill the infected macrophage [120]. Decreased IL-12 receptor expression and both increases and decreases in the level of expression of the INF-γ receptors have been reported and can be detected by flow cytometry [118–121]. In the dominant form of inheritance, mutations in the gene encoding the IFN-γ receptor lead to an increase in cell surface expression due a deletion encompassing both the signaling and the recycling domains [118,119,121]. The lack of specificity in associating surface receptor expression levels to specific genetic mutations has led to the development of surrogate functional flow cytometry assays. The latter are based on the ability to detect phoshorylated versus

---

very close to our normal cutoff of 30 (NOI on CDG patient = 30, NOI on brother = 22). Repeats on the same samples generated similar results. A sample from the mom was obtained and clearly showed two peaks, one normal (NOI = 187) and one abnormal (NOI = 30). The results on the mom clarified the probable patient diagnosis and were reported as consistent with X-linked CGD for the two sons and consistent with an X-linked carrier of CGD for the mom (protein analysis confirmed the absence of pg91phox, confirming the flow cytometry diagnosis (results not shown). Figure was published previously by O'Gorman, M.R.G., *Clin. Lab. Med.* 27, 591, 2007.

nonphosphorylated kinase substrate components of the IFN-γ pathway following *in vitro* stimulation with IFN-γ [118]. Lack of STAT-1 phosphorylation as detected by flow cytometry following *in vitro* stimulation with INF-γ indicates either abnormal IFN-γ receptor molecules (IFN-γR1 or IFN-γR2) or abnormalities in STAT-1 itself, whereas normal *in vitro* STAT-1 phosphorylation levels in a patient with repeat atypical mycobacterial infections might suggest abnormalities in either the IL-12 gene or the IL-12 receptor genes, which can then be sequenced. The determination of the specific underlying genetic mutation in these patients is very important since (a) the prognosis differs with the different forms of genetic lesions and (b) the specific defect has implications for the design of optimal therapeutic interventions [119,122]. Detection of intracellular phosphorylated STAT-1 following *in vitro* culture is a relatively complex procedure performed in more specialized flow cytometry laboratories and has been used primarily to help focus on the target defect for gene sequencing.

9.3.6 GROUP VI: DEFECTS IN INNATE IMMUNITY

1. Anhidrotic ectodermal dysplasia with immunodeficiency (EDA-ID) (*NEMO* gene)
2. Anhidrotic ectodermal dysplasia with immunodeficiency (EDA-ID) (*IKBA* gene)
3. IL-1 receptor-associated kinase (IRAK-4) deficiency (*IRAK4* gene)
4. Warts, hypogammaglobulinemia, infections, myelokathexis (WHIM) syndrome (*CXCR4* gene)
5. Epidermodysplasia verruciforis (*EVER1* and *EVER2* genes)

Toll-like receptors (TLR) are involved in the effective elimination of pathogens by the innate immune systems (reviewed in Chapter 1). The TLRs have received considerable attention over the past few years, and recently, abnormalities in the TLR signaling pathways have been associated with specific patterns of infection [123]. Mutations in the gene *IRAK-4* are associated with a predisposition to invasive infections with *Streptococcus pneumoniae* and less often *Staphylococcus aureus* [124]. Recently (and not yet officially classified as a PID), Casrouge's laboratory identified mutations in UNC-93B, another toll-like receptor signaling pathway component as associated with predisposition to meningoencephalitis caused by herpes simplex virus [125]. Abnormalities in the toll-like receptor signaling pathways described earlier have been detected by flow cytometry. Stimulation of whole blood with specific TLR ligands allows for the screening of specific TLR defects. The secretion of cytokines by flow cytometry (and also by enzyme-linked immunosorbent assay [ELISA]) has been reported to "rapidly screen" patients with suspected defects in the toll-like receptor pathway [126,127]. The latter cytokine-based flow assays have been successfully utilized to screen patients with *IRAK-4* mutations. Intracellular cytokine detection assays are, however, relatively laborious, time-consuming, and expensive. A new flow cytometry assay was recently described that serves as a surrogate of specific TLR activation and has been successfully adopted to detect abnormalities in TLR pathways involving both IRAK-4 and UNC-93B [128]. The latter assay exploits the physiological shedding of L-selectin (CD62L) from the surface of granulocytes following *in vitro* activation with a variety of stimuli, including specific TLR agonists. Stimulation of whole blood with specific TLR agonists, lipopolysaccharide (LPS) (TLR4), Pam3CSK4 (TLR1/2), Pam2CSK4 (TLR2/6), or R-848 (TLR7 and TLR8) induced no shedding in five genetically confirmed *IRAK-4* deficient patients as compared with the complete shedding in a group of 38 healthy control subjects [128]. Additionally, stimulation with R-848 resulted in no shedding of CD62L in two patients with confirmed mutations in the UNC-93B gene [128]. This rapid and simple flow cytometry procedure has the potential for broad applications such as a diagnostic screening test for IRAK-4- and UNC-93B-deficient patients, will help to complete the clinical characterization of these defects and will probably aid in the discovery of new TLR signaling defects in the future.

Group VII: Autoinflammatory disorders and Group VIII: Complement deficiencies are not routinely diagnosed and monitored by flow cytometry, however, this does rules out the possibility in the future. For a discussion of the complement and complement deficiencies please refer to Chapter 4 by Dr. Giclas.

9.4 SUMMARY

Flow cytometry has been and continues to be an invaluable tool for the diagnosis and monitoring of patients suspected of PID. As new molecules are discovered and their functions and roles in various physiological pathways are understood, more flow cytometry–based assays will be developed to provide rapid screening for defects associated with primary immunodeficiencies.

REFERENCES

1. Conley, M.E., Notarangelo, L.D., and Etzioni, A., Diagnostic criteria for primary immunodeficiencies, *Clin. Immunol.*, 93, 190, 1999.
2. Samarghitean, C., Valiaho, J., and Vihinen, M., IDR knowledge base for primary immunodeficiencies, *Immunome Res.*, 3, 6, 2007.
3. Notorangelo, L., Casanova, J.-L., Conley, M.E., Fischer, A., Hammarström, L., Nonoyama, S., Ochs, H.D., Puck, J.M., Roifman, C., Seger, R., and Wedgwood, J., Primary immunodeficiency diseases: an update from the International Union of Immunological Societies Primary Immunodeficiency Diseases Classification Committee Meeting in Budapest, 2005, *J. Allergy Clin. Immunol.*, 117, 883, 2006.
4. Fudenberg, H.H., Good, R.A., Hitzig, W., Kunkel, H.G., Roitt, I.M., Rosen, F.S., Rowe, D.S., Seligmann, M., and Soothill, J.R., Classification of the primary immune deficiency deficiencies: WHO recommendation, *N. Engl. J. Med.*, 283, 656, 1970.
5. Fudenberg, H.H., Good, R.A., Goodman, H.C., Hitzig, W., Kunkel, H.G., Roitt, I.M., Rosen, F.S., Rowe, D.S., Seligmann, M., and Soothill, J.R., Primary immunodeficiencies. Report of a World Health Organization Committee, *Pediatrics*, 47, 927, 1971.
6. Notorangelo, L., Casanova, J.L., Fischer, A., Puck, J., Rosen, F., Seger, R., and Geha, R., Primary immunodeficiency diseases: an update, *J. Allergy Clin. Immunol.*, 114, 677, 2004.
7. Shearer, W.T. and Fischer, A., The last 80 years in primary immunodeficiency: how far have we come, how far need we go? *J. Allergy Clin. Immunol.*, 117, 748, 2006.
8. O'Gorman, M.R. and Nicholson, J.K., Adoption of single-platform technologies for enumeration of absolute T-lymphocyte subsets in peripheral blood, *Clin. Diagn. Lab. Immunol.*, 7, 333, 2000.

9. Schnizlein-Bick, C.T., Mandy, F.F., O'Gorman, M.R., Paxton, H., Nicholson, J.K., Hultin, L.E., Gelman, R.S., Wilkening, C.L., and Livnat, D., Use of CD45 gating in three and four-color flow cytometric immunophenotyping: guideline from the National Institute of Allergy and Infectious Diseases, Division of AIDS, *Cytometry*, 50, 46, 2002.

10. O'Gorman, M.R.G., Millard, D.D., Lowder, J.N., and Yogev, R., Lymphocyte subpopulations in healthy one to three day old infants, *Cytometry*, 34, 235, 1998.

11. Shearer, W.T., Rosenblatt, H.M., Gelman, R.S., Oyomopito, R., Plaeger, S., Stiehm, E.R., Wara, D.W., Douglas, S.D., Luzuriaga, K., McFarland, E.J., Yogev, R., Rathore, M.H., Levy, W., Graham, B.L., and Spector, S.A., Lymphocyte subsets in healthy children from birth through 18 years of age: the Pediatric AIDS Clinical Trials Group P1009 study, *J. Allergy Clin. Immunol.*, 112, 973, 2003.

12. Buckley, R.H., Primary immunodeficiency disease due to defects in lymphocytes, *N. Engl. J. Med.*, 343, 1313, 2000.

13. Noguchi, M., Yi, H., Rosenblatt, H.M., Filipovich, A.H., Adelstein, S., Modi, W.S., McBride, O.W., and Leonard, W.J., Interleukin-2 receptor gamma chain mutation results in X-linked severe combined immunodeficiency in humans, *Cell*, 73, 147, 1993.

14. Puck, J.M., Deschênes, S.M., Porter, J.C., Dutra, A.S., Brown, C.J., Willard, H.F., and Henthorn, P.S., The interleukin-2 receptor gamma chain maps to Xq13.1 and is mutated in severe combined immunodeficiency, SCIDX1, *Hum. Mol. Genet.*, 2, 1099, 1993.

15. Asao, H., Okuyama, C., Kumaki, S., Ishii, N., Tsuchiya, S., Foster, D., and Sugamura, K., Cutting edge: the common gamma-chain is an indispensable subunit of the IL-21 receptor complex, *J. Immunol.*, 167, 1, 2001.

16. Buckley, R.H., Molecular defects in human severe combined immunodeficiency and approaches to immune reconstitution, *Annu. Rev. Immunol.*, 22, 625, 2004.

17. Gilmour, K.C., Cranston, T., Loughlin, S., Gwyther, J., Lester, T., Espanol, T., Hernandez, M., Savoldi, G., Davies, E.G., Abinun, M., Kinnon, C., Jones, A., and Gaspar, H.B., Rapid protein-based assays for the diagnosis of T-B+ severe combined immunodeficiency, *Br. J. Haematol.*, 112, 671, 2001.

18. Gilmour, K.C., Fujii, H., Cranston, T., Davies, E.G., Kinnon, C., and Gaspar, H.B., Defective expression of the interleukin-2/interleukin-15 receptor β subunit leads to a natural killer cell-deficient form of severe combined immunodeficiency, *Blood*, 98, 877, 2001.

19. Roberts, J.L., Lengi, A., Brown, S.M., Chen, M., Zhou, Y.J., O'Shea, J.J., and Buckley, R.H., Janus kinase 3 (JAK3) deficiency: clinical, immunologic, and molecular analyses of 10 patients and outcomes of stem cell transplantation, *Blood*, 103, 2009, 2004.

20. Kung, C., Pingel, J.T., Heikinheimo, M., Klemola, T., Varkila, K., Yoo, L.I., Vuopala, K., Poyhonen, M., Uhari, M., Rogers, M., Speck, S.H., Chatila, T., and Thomas, M.L., Mutations in the tyrosine phosphatase CD45 gene in a child with severe combined immunodeficiency disease, *Nat. Med.*, 6, 343, 2000.

21. Puel, A., Ziegler, S.F., Buckley, R.H., and Leonard, W.J., Defective IL7R expression in T(−) B(+)NK(+) severe combined immunodeficiency, *Nat. Genet.*, 20, 394, 1998.

22. Dadi, H.K., Simon, A.J., and Roifman, C.M., Effect of CD3δ deficiency on maturation of α/β and γ/δ T-cell lineages in severe combined immunodeficiency, *N. Engl. J. Med.*, 349, 1821, 2003.

23. Roberts, J.L., Lauritsen, J.P., Cooney, M., Parrott, R.E., Sajaroff, E.O., Win, C.M., Keller, M.D., Carpenter, J.H., Carabana, J., Krangel, M.S., Sarzotti, M., Zhong, X.P., Wiest, D.L., and Buckley, R.H., T−B+NK+ severe combined immunodeficiency caused by complete deficiency of the CD3zeta subunit of the T-cell antigen receptor complex, *Blood*, 109, 3198, 2007.

24. Lieber, M.R., Ma, Y., Pannicke, U., and Schwarz, K., The mechanism of vertebrate non-homologous DNA end joining and its role in V(D)J recombination, *DNA Repair (Amst)*, 3, 817, 2004.

25. Villa, A., Sobacchi, C., Notarangelo, L.D., Bozzi, F., Abinun, M., Abrahamsen, T.G., Ark-wright, P.D., Baniyash, M., Brooks, E.G., Conley, M.E., Cortes, P., Duse, M., Fasth, A., Filipovich, A.M., Infante, A.J., Jones, A., Mazzolari, E., Muller, S.M., Pasic, S., Rechavi, G., Sacco, M.G., Santagata, S., Schroeder, M.L., Seger, R., Strina, D., Ugazio, A., Väliaho, J., Vihinen, M., Vogler, L.B., Ochs, H., Vezzoni, P., Friedrich, W., and Schwarz, K., V(D)J recombination defects in lymphocytes due to RAG mutations: severe immu-nodeficiency with a spectrum of clinical presentations, *Blood*, 97, 81, 2001.

26. Ege, M., Ma, Y., Manfras, B., Kalwak, K., Lu, H., Lieber, M.R., Schwarz, K., and Pannicke, U., Omenn syndrome due to Artemis mutations, *Blood*, 105, 4179, 2005.

27. Van der Burg, M., van Veelen, L.R., Verkaik, N.S., Wiegant, W.W., Hartwig, N.G., Barendregt, B.H., Brugmans, L., Raams, A., Jaspers, N.G., Zdzienicka, M.Z., van Don-gen, J.J., and van Gent, D.C., A new type of radiosensitive T−B−NK+ severe combined immunodeficiency caused by a LIG4 mutation, *J. Clin. Invest.*, 116, 137, 2006.

28. Buck, D., Moshous, D., de Chasseval, R., Ma, Y., Le Deist, F., Cavazzana-Calvo, M., Fischer, A., Casanova, J.L., Lieber, M.R., de Villartay, J.P., Severe combined immuno-deficiency and microcephaly in siblings with hypomorphic mutations in DNA ligase IV, *Eur. J. Immunol.*, 36, 224, 2006.

29. Fischer, A., Have we seen the last variant of severe combined immunodeficiency? *N. Engl. J. Med.*, 349, 1789, 2003.

30. Buckley, R.H., Schiff, S.E., Schiff, R.I., Markert, L., Williams, L.W., Roberts, J.L., Myers, L.A., and Ward, F.E., Hematopoietic stem-cell transplantation for the treatment of severe combined immunodeficiency, *N. Engl. J. Med.*, 340, 508, 1999.

31. Stephan, J.L., Vlekova, V., Le Deist, F., Blanche, S., Donadieu, J., De Saint-Basile, G., Durandy, A., Griscelli, C., and Fischer, A., Severe combined immunodeficiency: a retrospective single-center study of clinical presentation and outcome in 117 patients, *J. Pediatr.*, 123, 564, 1993.

32. Santagata, S., Villa, A., Sobacchi, C., Cortes, P., and Vezzoni, P., The genetic and bio-chemical basis of Omenn syndrome, *Immunol. Rev.*, 178, 64, 2000.

33. Giliani, S., Bonfim, C., de Saint Basile, G., Lanzi, G., Brousse, N., Koliski, A., Malvezzi, M., Fischer, A., Notarangelo, L.D., and Le Deist, F., Omenn syndrome in an infant with IL7RA gene mutation, *J. Pediatr.*, 148, 272, 2006.

34. Aleman, K., Noordzijm, J.G., de Grootm, R., van Dongen, J.J., and Hartwig, N.G., Reviewing Omenn syndrome, *Eur. J. Pediatr.*, 160, 718, 2001.

35. Cancrini, C., Romiti, M.L., Finocchi, A., Di Cesare, S., Ciaffi, P., Capponi, C., Pahwa, S., and Rossi, P., Post-natal ontogenesis of the T cell receptor CD4 and CD8 V beta rep-ertoire and immune function in children with DiGeorge syndrome, *J. Clin. Immunol.*, 25, 265, 2005.

36. Lee, W.I., Torgerson, T.R., Schumacher, M.J., Yel, L., Zhu, Q., and Ochs, HD., Molecu-lar analysis of a large cohort of patients with the hyperimmunoglobulin M (IgM) syn-drome, *Blood*, 105, 1881, 2005.

37. O'Gorman, M.R.G., Zaas, D., Paniagua, M., Corrochano, V., Scholl, P.R., and Pachman, L.M., Development of a rapid whole blood flow cytometry procedure for the diagnosis of x-linked hyper IgM syndrome (XHIM) patients and carriers, *Clin. Immunol. Immu-nopathol.*, 85, 172, 1997.

38. Freyer, D.R., Gowans, L.K., Warzynski, M., and Lee, W.I., Flow cytometric diagnosis of X-linked Hyper IgM syndrome: application of an accurate and convenient procedure, *J. Pediatr. Hematol. Oncol.*, 26, 363, 2004.

39. Ferrari, S., Giliani, S., Insalaco, A., Al-Ghonaium, A., Soresina, A.R., Loubser, M., Avanzini, M.A., Marconi, M., Badolato, R., Ugazio, A.G., Levy, Y., Catalan, N., Durandy, A., Tbakhi, A., Notarangelo, L.D., and Plebani, A., Mutations of CD40 gene cause an autosomal recessive from of immunodeficiency with hyper IgM, *Proc. Natl. Acad. Sci., USA*, 98, 12614, 2001.

40. Palmer, K., Green, T.D., Roberts, J.L., Sajaroff, E., Cooney, M., Parrott, R., Chen, D.F., Reinsmoen, N.L., and Buckley, R.H., Unusual clinical and immunologic manifestations of transplacentally acquired maternal T cells in severe combined immunodeficiency, *J. Allergy Clin. Immunol.*, 2007 (epub ahead of print).

41. Masternak, K., Muhlethaler-Mottet, A., Villard, J., Peretti, M., and Reith, W., Molecular genetics of the bare lymphocyte syndrome, *Rev. Immunogenet.*, 2, 267, 2000.

42. de la Salle, H., Hanau, D., Fricker, D., Urlacher, A., Kelly, A., Salamero, J., Powis, S.H., Donato, L., Bausinger, H., and Laforet, M., Homozygous human TAP peptide transporter mutation in HLA class I deficiency, *Science*, 265, 237, 1994.

43. Elder, M.E., Hope, T.J., Parslow, T.G., Umetsu, D.T., Wara, D.W., and Cowan, M.J., Severe combined immunodeficiency with absence of peripheral blood CD8+ T cells due to ZAP-70 deficiency, *Cell. Immunol.*, 165, 110, 1995.

44. Gelfand, E.W., Weinberg, K., Mazer, B.D., Kadlecek, T.A., and Weiss, A., Absence of ZAP-70 prevents signaling through the antigen receptor on peripheral blood T cells but not on thymocytes, *J. Exp. Med.*, 182, 1057, 1995.

45. Mazer, B., Harbeck, R.J., Franklin, R., Schwinzer, R., Kubo, R., Hayward, A., and Gelfand, EW., Phenotypic features of selective T cell deficiency characterized by absence of CD8+ T lymphocytes and undetectable mRNA for ZAP-70 kinase, *Clin. Immunol. Immunopathol.*, 84, 129, 1997.

46. Smith, D.K., Neal, J.J., and Holmgren, S.D., Centers for Disease Control Idiopathic CD4+ T-lymhopenia Task Force, unexplained opportunistic infections and CD4+ T-lymphopenia without HIV infection: an investigation of cases in the United States, *N. Engl. J. Med.*, 328, 373, 1993.

47. Tanaka, S., Teraguchi, M., Hasui, M., Taniuchi, S., Ikemoto, Y., and Kobayashi, Y., Idiopathic CD4+ T-lymphocytopenia in a boy with Down syndrome. Report of a patient and a review of the literature, *Eur. J. Pediatr.*, 163, 122, 2004.

48. Chun, H.J., Zheng, L., Ahmad, M., Wang, J., Speirs, C.K., Siegel, R.M., Dale, J.K., Puck, J., Davis, J., Hall, C.G., Skoda-Smith, S., Atkinson, T.P., Straus, S.E., and Lenardo, M.J., Pleiotropic defects in lymphocyte activation caused by caspase-8 mutations lead to human immunodeficiency, *Nature*, 419, 395, 2002.

49. Tsukada, S., Saffran, D.C., Rawlings, D.J., Parolini, O., Allen, R.C., Klisak, I., Sparkes, R.S., Kubagawa, H., Mohandas, T., and Quan, S., Deficient expression of a B cell cytoplasmic tyrosine kinase in human x-linked agammaglobulinemia, *Cell*, 72, 279, 1993.

50. Vetrie, D., Vorechovský, I., Sideras, P., Holland, J., Davies, A., Flinter, F., Hammarström, L., Kinnon, C., Levinsky, R., and Bobrow, M., The gene involved in X-linked agammaglobulinaemia is a member of the src family of protein-tyrosine kinases, *Nature*, 361, 226, 1993.

51. Futatani, T., Miyawaki, T., Tsukada, S., Hashimoto, S., Kunikata, T., Arai, S., Kurimoto, M., Niida, Y., Matsuoka, H., Sakiyama, Y., Iwata, T., Tsuchiya, S., Tatsuzawa, O., Yoshizaki, K., and Kishimoto, T., Deficient expression of Bruton's tyrosine kinase in monocytes from x-lined agammaglobulinemia as evaluated by a flow cytometric analysis and its clinical application to carrier detection, *Blood*, 91, 595, 1998.

52. Futatani, T., Watanabe, C., Baba, Y., Tsukada, S., and Ochs, H.D., Bruton's tyrosine kinase is present in normal platelets and its absence identifies patients with X-inked agammaglobulinaemia and carrier females, *Br. J. Haematol.*, 114, 141, 2001.

53. Lopez-Granados, E., Pérez de Diego, R., Ferreira Cerdán, A., Fontán Casariego, G., and García Rodríguez, M.C., A genotype–phenotype correlation study in a group of 54 patients with X-linked agammaglobulinemia, *J. Allergy Clin. Immunol.*, 116, 690, 2005.

54. Meffre, E., Le Deist, F., de Saint-Basile, G., Deville, A., Fougereau, M., Fischer, A., and Schiff, C., A human non-XLA immunodeficiency disease characterized by blockage of B cell development at an early proB cell stage, *J. Clin. Invest.*, 98, 1519, 1996.

55. Sawada, A., Takihara, Y., Kim, J.Y., Matsuda-Hashii, Y., Tokimasa, S., Fujisaki, H., Kubota, K., Endo, H., Onodera, T., Ohta, H., Ozono, K., and Hara, J., A congenital

mutation of the novel gene *LRRC8* causes agammaglobulinemia in humans, *J. Clin. Invest.*, 112, 1707, 2003.

56. Agematsu, K.S., Hokibara, S., Nagumo, H., and Komiyama, A., CD27: a memory B cell marker, *Immunol. Today*, 21, 204, 2000.

57. Bleesing, J.J. and Fleisher, T.A., Human B cells express a CD45 isoform that is similar to murine B220 and is downregulated with acquisition of the memory B-cell marker CD27, *Cytometry, B. Clin. Cytom.*, 51, 1, 2003.

58. Agematsu, K., Futatani, T., Hokibara, S., Kobayashi, N., Takamoto, M., Tsukada, S., Suzuki, H., Koyasu, S., Miyawaki, T., Sugane, K., Komiyama, A., and Ochs, HD., Absence of memory B cells in patients with common variable immunodeficiency, *Clin. Immunol.*, 103, 34–42, 2002.

59. Warnatz, K., Denz, A., Dräger, R., Braun, M., Groth, C., Wolff-Vorbeck, G., Eibel, H., Schlesier, M., and Peter, H.H., Severe deficiency of switched memory B cells (CD27(+)IgM(−)IgD(−)) in subgroups of patients with common variable immunodeficiency: a new approach to classify a heterogeneous disease, *Blood*, 99, 1544, 2002.

60. Piqueras, B., Lavenu-Bombled, C., Galicier, L., Bergeron-van der Cruyssen, F., Mouthon, L., Chevret, S., Debré, P., Schmitt, C., and Oksenhendler, E., Common variable immunodeficiency patient classification based on impaired B cell memory differentiation correlates with clinical aspects, *J. Clin. Immunol.*, 23, 385, 2003.

61. Ko, J., Radigan, L., and Cunningham-Rundles, C., Immune competence and switched memory B cells in common variable immunodeficiency, *Clin. Immunol.*, 116, 37, 2005.

62. Agematsu, K., Nagumo, H., Shinozaki, K., Hokibara, S., Yasui, K., Terada, K., Kawamura, N., Toba, T., Nonoyama, S., Ochs, H.D., and Komiyama, A., Absence of IgD-CD27(+) memory B cell population in X-linked hyper-IgM syndrome, *J. Clin. Invest.*, 102, 853, 1998.

63. Bunk, R., Dittrich, A.M., Schulze, I., Horn, J., Schmolke, K., Volk, H.D., Wahn, V., and Höflich, C., Rapid whole blood flow cytometric test to detect ICOS deficiency in patients with common variable immunodeficiency, *Int. Arch. Allergy Immunol.*, 140, 342, 2006.

64. Imai, K., Zhu, Y., Revy, P., Morio, T., Mizutani, S., Fischer, A., Nonoyama, S., and Durandy, A., Analysis of class switch recombination and somatic hypermutation in patients affected with autosomal dominant hyper-IgM syndrome type 2, *Clin. Immunol.*, 115, 277, 2005.

65. Durandy, A., Revy, P., Imai, K., and Fischer, A., Hyperimmunoglobulin M syndromes caused by intrinsic B-lymphocyte defects, *Immunol. Rev.*, 203, 67, 2005.

66. Sullivan, K.E., Mullen, C.A., and Blaese, R.M., A multiinstitutional survey of the Wiskott–Aldrich syndrome, *J. Pediatr.*, 125, 876, 1994.

67. Stewart, D.M., Treiber-Held, S., Kurman, C.C., Facchetti, F., Notarangelo, L.D., and Nelson, D.L., Studies of the expression of the Wiskott–Alrich syndrome protein, *J. Clin. Invest.*, 97, 2627, 1996.

68. Kawai, S., Minegishi, M., Ohashi, Y., Sasahara, Y., Kumaki, S., Konno, T., Miki, H., Derry, J., Nonoyama, S., Miyawaki, T., Horibe, K., Tachibana, N., Kudoh, E., Yoshimura, Y., Izumikawa, Y., Sako, M., and Tsuchiya, S., Flow cytometric determination of intracytoplasmic Wiskott–Aldrich syndrome protein in peripheral blood lymphocyte subpopulations, *J. Immunol. Methods*, 260, 95, 2002.

69. Yamada, M., Ohtsu, M., Kobayashi, I., Kawamura, N., Kobayashi, K., Ariga, T., Sakiyama, Y., Nelson, D.L., Tsuruta, S., Anakura, M., and Ishikawa, N., Flow cytometric anlysis of Wiskott–Aldrich syndrome (WAS) protein in lymphocytes from WAS patients and their familial carrier, *Blood*, 93, 756, 1999.

70. Yamada, M., Ariga, T., Kawamura, N., Yamaguchi, K., Ohtsu, M., Nelson, D.L., Kondoh, T., Kobayashi, I., Okano, M., Kobayashi, K., and Sakiyama, Y., Determination of carrier status for the Wiskott–Aldrich syndrome by flow cytometric analysis of Wiskott–Aldrich syndrome protein in peripheral blood mononuclear cells, *J. Immunol.*, 166, 1119, 2001.

71. Ariga, T., Nakajima, M., Yoshida, J., Yamato, K., Nagatoshi, Y., Yanai, F., Caviles, A.P., Nelson, D.L., and Sakiyama, Y., Confirming or excluding the diagnosis of Wiskott–Aldrich syndrome in children with thrombocytopenia of an unknown etiology, *J. Pediatr.Hematol./Oncol.*, 26, 435, 2004.

72. Yamaguchi, K., Ariga, T., Yamada, M., Nelson, D.L., Kobayashi, R., Kobayashi, C., Noguchi, Y., Ito, Y., Katamura, K., Nagatoshi, Y., Kondo, S., Katoh, H., and Sakiyama, Y., Mixed chimera status of 12 patients with Wiskott–Aldrich syndrome (WAS) after hematopoietic stem cell transplantation: evaluation by flow cytometric anlysis of intracellular WAS protein expression, *Blood*, 100, 1208, 2002.

73. Stray-Pedersen, A., Jónsson, T., Heiberg, A., Lindman, C.R., Widing, E., Aaberge, I.S., Borresen-Dale, A.L., and Abrahamsen, T.G., The impact of an early truncating founder ATM mutation on immunoglobulins, specific antibodies and lymphocyte populations in ataxia-telangiectasia patients and their parents, *Clin. Exp. Immunol.*, 137, 179, 2004.

74. Chun, H.H. and Gatti, R.A., Ataxia-telangiectasia, an evolving phenotype, *DNA Repair*, 3, 1187, 2004.

75. Perlman, S., Becker-Catania, S., and Gatti, R.A., Ataxia-telangiectasia: diagnosis and treatment, *Semin. Pediatr. Neurol.*, 10, 173, 2003.

76. Heinrich, T., Prowald, C., Friedl, R., Gottwald, B., Kalb, R., Neveling, K., Herterich, S., Hoehn, H., and Schindler. D., Exclusion/confirmation of Ataxia-telangiectasia via cell-cycle testing, *Eur. J. Pediatr.*, 165, 250, 2006.

77. Sullivan, KE., The clinical, immunological and molecular spectrum of chromosome 22q11.2 deletion syndrome and DiGeorge syndrome, *Curr. Opin. Allergy Clin Immunol.*, 4, 505–512, 2004.

78. Corey, A.H., Kelly, D., Halford, S., Wadey, R., Wilson, D., Goodship, J., Burn, J., Paul, T., Sharkey, A., and Dumanski, J., Molecular genetic study of the frequency of monosomy 22q11 in DiGeorge syndrome, *Am. J. Hum. Genet.*, 51, 964, 1992.

79. Rice, H.E., Skinner, M.A., Mahaffey, S.M., Oldham, K.T., Ing, R.J., Hale, L.P., and Markert, M.L., Thymic transplantation for complete DiGeorge syndrome: medical and surgical considerations, *J. Pediatr. Surg.*, 39, 1607, 2004.

80. Muller, W., Peter, H.H., Wilken, M., Jüppner, H., Kallfelz, H.C., Krohn, H.P., Miller, K., and Rieger, C.H., The DiGeorge syndrome. Clinical evaluation and course of partial and complete forms of the syndrome, *Eur. J. Pediatr.*, 147, 496, 1988.

81. Kanaya, Y., Ohga, S., Ikeda, K., Furuno, K., Ohno, T., Takada, H., Kinukawa, N., and Hara, T., Maturational alterations in peripheral T cell subsets and cytokine gene expression in 22q11.2 deletion syndrome, *Clin. Exp. Immunol.*, 144, 85, 2006.

82. Henter, J., Elinder, G., and Ost, A., Diagnostic guidelines for hemophagocytic lymphohistiocytosis, *Sem. Oncol.*, 18, 29, 1991.

83. Arico, M., Janka, G., Fischer, A., Henter, J.I., Blanche, S., Elinder, G., Martinetti, M., and Rusca, M.P., Hemophagocytic lymphohistiocytosis. Report of 122 children from the International Registry, *Leukemia*, 10, 197, 1996.

84. Risma, K.A., Frayer, R.W., Filipovich, A.H., and Sumegi, J., Aberrant maturation of mutant perforin underlies the clinical diversity of hemophagocytic lymphohistiocytosis, *J. Clin. Invest.*, 116, 182, 2006.

85. Ueda, I., Ishii, E., Morimoto, A., Ohga, S., Sako, M., and Imashuku, S., Correlation between phenotypic heterogeneity and gene mutational characteristics in familial hemophagocytic lymphohistiocytosis (FHL), *Pediar. Blood Cancer*, 46, 482, 2006.

86. Santoro, A., Cannella, S., Bossi, G., Gallo, F., Trizzino, A., Pende, D., Dieli, F., Bruno, G., Stinchcombe, J.C., Micalizzi, C., De Fusco, C., Danesino, C., Moretta, L., Notarangelo, L.D., Griffiths, G.M., and Aricò, M., Novel Munc13-4 mutations in children and young adult patients with haemophagocytic lymphohistiocytosis, *J. Med. Genet.*, 43, 953, 2006.

87. zur Stadt, U., Schmidt, S., Kasper, B., Beutel, K., Diler, A.S., Henter, J.I., Kabisch, H., Schneppenheim, R., Nürnberg, P., Janka, G., and Hennies, H.C., Linkage of familial hemophagocytic lymphohistiocytosis (FHL) type-4 to chromosome 6q24 and identification of mutations in syntaxin 11, *Hum. Mol. Genet.*, 14, 827, 2005.
88. Bryceson, Y., Rudd, E., Zheng, C., Edner, J., Ma, D., Wood, S.M., Bechensteen, A.G., Boelens, J.J., Celkan, T., Farah, R.A., Hultenby, K., Winiarski, J., Roche, P.A., Nordenskjöld, M., Henter, J.I., Long, E.O., and Ljunggren, H.G., Defective cytotoxic lymphocyte degranulation in syntaxin-11-deficient familial hemophagocytic lympho-histiocytosis 4 (FHL4) patients, *Blood*, 2007 (epub).
89. Rieux-Laucat, F., Le Deist, F., Hivroz, C., Roberts, I.A., Debatin, K.M., Fischer, A., and de Villartay, J.P., Mutations in Fas associated with human lymphoproliferative syndrome and autoimmunity, *Science*, 268, 1347, 1995.
90. Fisher, G.H., Rosenberg, F.J., Straus, S.E., Dale, J.K., Middleton, L.A., Lin, A.Y., Strober, W., Lenardo, M.J., and Puck, J.M., Dominant interfering Fas gene mutations impair apoptosis in a human autoimmune lymphoproliferative syndrome, *Cell*, 81, 935, 1995.
91. Worth, A., Thrasher, A.J., and Gaspar, H.B., Autoimmune lymphoproliferative syndrome: molecular basis of disease and clinical phenotype, *Br. J. Haematol.*, 133, 124, 2006.
92. Infante, A.J., Britton, H.A., DeNapoli, T., Middelton, L.A., Lenardo, M.J., Jackson, C.E., Wang, J., Fleisher, T., Straus, S.E., and Puck, J.M., The clinical spectrum in a large kindred with autoimmune lymphoproliferative syndrome caused by a Fas mutation that impairs lymphocyte apoptosis, *J. Pediatr.*, 133, 623, 1998.
93. Teachey, D.T., Manno, C.S., Axsom, K.M., Andrews, T., Choi, J.K., Greenbaum, B.H., McMann, J.M., Sullivan, K.E., Travis, S.F., and Grupp, S.A., Unmasking Evans syndrome: T-cell phenotype and apoptotic response reveal autoimmune lymphoproliferative syndrome (ALPS), *Blood*, 105, 2443, 2005.
94. Notorangelo, L.D., Bambineri, E., and Badolato, R. Immunodeficiencies with autoimmune consequences, *Adv. Immunol.*, 89, 321, 2006.
95. Nagamine, K., Peterson, P., Scott, H.S., Kudoh, J., Minoshima, S., Heino, M., Krohn, K.J., Lalioti, M.D., Mullis, P.E., Antonarakis, S.E., Kawasaki, K., Asakawa, S., Ito. F., and Shimizu, N., Positional cloning of the APECED gene, *Nat. Genet.*, 17, 393, 1997.
96. The Finnish–German APECED consortium. An autoimmune disease, APECED, caused by mutations in a novel gene featuring two PHD-type zinc-finger domains. The Finnish–German APECED Consortium. Autoimmune Polyendocrinopathy-Candidiasis-Ectodermal Dystrophy, *Nat. Genet.*, 4, 399, 1997.
97. Kogawa, K., Nagafuchi, S., Katsuta, H., Kudoh, J., Tamiya, S., Sakai, Y., Shimizu, N., and Harada, M., Expression of AIRE gene in peripheral monocyte/dendritic cell lineage, *Immunol. Lett.*, 80, 195, 2002.
98. Powell, B.R., Buist, N.R., and Stenzel, P., An X-linked syndrome of diarrhea, polyendocrinopahty and fatal infections in infancy, *J. Pediatr.*, 100, 731, 1982.
99. Bennett, C.L., Christie, J., Ramsdell, F., Brunkow, M.E., Ferguson, P.J., Whitesell, L., Kelly, T.E., Saulsbury, F.T., Chance, P.F., and Ochs, H.D., The immune dysregulation polyendocrinopathy, enteropathy X-linked syndrome (IPEX) is caused by mutations of FOXP3, *Nat. Genet.* 27, 20, 2001.
100. Sakaguchi, S., Foxp3+ CD25+ CD4+ natural regulatory T cells in dominant self-tolerance and autoimmune disease, *Immunol. Rev.*, 212, 8, 2006.
101. Segal, B.H., Leto, T.L., Gallin, J.I., Malech, H.L., and Holland, S.M., Genetic, biochemical and clinical features of chronic granulomatous disease, *Medicine*, 79, 170, 2000.
102. Winkelstein, J.A., Marino, M.C., Johnston, R.B. Jr, Boyle, J., Curnutte, J., Gallin, J.I., Malech, H.L., Holland, S.M., Ochs, H., Quie, P., Buckley, R.H., Foster, C.B., Chanock, S.J., and Dickler, H., Chronic granulomatous disease. Report on a national registry of 368 patients, *Medicine (Baltimore)*, 79, 155, 2000.

103. Nakamura, M., Murakami, M., Koga, T., Tanaka, Y., and Minakami, S., Monoclonal antibody 7D5 raised to cytochrome b558 of human neutrophils: immunocytochemical detection of the antigen in peripheral phagocytes of normal subjects, patients with chronic granulomatous disease, and their carrier mothers, *Blood*, 69, 1404, 1987.

104. Mizuno, Y., Hara, T., Nakamura, M., Ueda, K., Minakami, S., and Take, H., Classification of chronic granulomatous disease on the basis of monoclonal antibody-defined surface cytochrome b deficiency, *J. Pediatr.*, 113, 458, 1988.

105. Verhoeven, A.J., Bolscher, B.G., Meerhof, L.J., van Zwieten, R., Keijer, J., Weening, R.S., and Roos, D., Characterization of two monoclonal antibodies against cytochrome b558 of human neutrophils, *Blood*, 73, 1686, 1989.

106. Emmendorffer, A., Nakamura, M., Rothe, G., Spiekermann, K., Lohmann-Matthes, M.L., and Roesler, J., Evaluation of flow cytometric methods for diagnosis of chronic granulomatous disease variants under routine laboratory conditions, *Cytometry*, 18, 147, 1994.

107. Bass, D.A., Parce, J.W., Dechatelet, L.R., Szejda, P., Seeds, M.C., and Thomas, M., Flow cytometric studies of oxidative product formation by neutrophils: a graded response to membrane stimulation, *J. Immunol.*, 30, 1910, 1983.

108. Rothe, G., Oser, A., and Valet, G., Dihydrorhodamine 123: a new flow cytometric indicator for respiratory burst activity in neutrophil granulocytes, *Naturwissenschaften*, 75, 354, 1988.

109. Emmendorffer, A., Hecht, M., Lohmann-Matthes, M.L., and Roesler, J., A fast and easy method to determine the production of reactive oxygen intermediates by human and murine phagocytes using dihydrorhodamine 123, *J. Immunol. Methods*, 131, 269, 1990.

110. O'Gorman, M.R.G. and Corrochano, V., Rapid whole-blood flow cytometry assay for diagnosis of chronic granulomatous disease, *Clin. Diag. Lab Immunol.* 2, 227, 1995.

111. Vowells, S.J., Fleisher, T.A., Sekhsaria, S., Alling, D.W., Maguire, T.E., and Malech, H.L., Genotype-dependent variability in flow cytometric evaluation of reduced nicotinamide adenine dinucleotide phosphate oxidase function in patients with chronic granulomatous disease, *J. Pediatr.*, 128, 104, 1996.

112. Mauch, L., Lun, A., O'Gorman, M.R., Harris, J.S., Schulze, I., Zychlinsky, A., Fuchs, T., Oelschlägel, U., Brenner, S., Kutter, D., Rösen-Wolff, A., and Roesler, J., Chronic granulomatous disease (CGD) and complete myeloperoxidase deficiency both yield strongly reduced dihydrorhodamine 123 test signals but can be easily discerned in routine testing for CGD, *Clin. Chem.*, 53, 890, 2007.

113. Vowells, S.J., Sekhsaria, S., Malech, H.L., Shalit, M., and Fleisher, T.A., Flow cytometric analysis of the granulocyte respiratory burst: a comparison study of fluorescent probes, *J. Immunol. Methods*, 178, 89, 1995.

114. Prince, H.E. and Lape-Nixon, M., Influence of specimen age and anticoagulant on flow cytometric evaluation of granulocyte oxidative burst generation, *J. Immunol. Methods*, 188, 129, 1995.

115. Alvarez-Larran, A., Toll, T., Rives, S., and Estella, J., Assessment of neutrophil activation in whole blood by flow cytometry, *Clin. Lab Haematol.*, 27, 41, 2005.

116. O'Gorman, M.R.G., McNally, A.C., Anderson, D.C., and Myones, B.L., A rapid whole blood lysis technique for the diagnosis of moderate or severe leukocyte adhesion deficiency (LAD), *Annals NY Acad. Sci.*, 677, 427, 1993.

117. Dimanche-Boitrel, M.T., Le Deist, F., Quillet, A., Fischer, A., Griscelli, C., and Lisowska-Grospierre, B., Effects of interferon-gamma (IFN-gamma) and tumor necrosis factor-alpha (TNF-alpha) on the expression of LFA-1 in the moderate phenotype of leukocyte adhesion deficiency (LAD), *J. Clin. Immunol.*, 3, 200, 1989.

118. Uzel, G. and Holland, S.M., Phagocyte deficiencies, in *Clinical Immunology Principles and Practice*, 2nd ed., vol. 1, Rich, R.R., Fleisher, T.A., Shearer, W.T., Kotzin, B.L., and Schroeder, Jr., H.W., Eds., Mosby Int. Ltd., London, UK, 2001, pp. 37.1–37.18.

119. Ottenhoff, T.H.M., de Boer, T., Verhagen, C.E., Verreck, F.A., and van Dissel, J.T., Human deficiencies in type 1 cytokine receptors reveal the essential role of type 1 cytokines in immunity to intracellular bacteria, *Microbes Infection*, 2, 1559, 2000.

120. Ottenhoff, T.H.M. and Mutis, T., Role of cytotoxic T cells in the protective immunity against and immunopathology of intracellular infections, *Eur. J. Clin. Invest.*, 25, 371, 1995.

121. Jouanguy, E., Lamhamedi-Cherradi, S., Lammas, D., Dorman, S.E., Fondanèche, M.C., Dupuis, S., Döffinger, R., Altare, F., Girdlestone, J., Emile, J.F., Ducoulombier, H., Edgar, D., Clarke, J., Oxelius, V.A., Brai, M., Novelli, V., Heyne, K., Fischer, A., Holland, S.M., Kumararatne, D.S., Schreiber, R.D., and Casanova, J.L., A human IFNGR1 small deletion hotspot associated with dominant susceptibility to mycobacterial infection, *Nat. Genet.* 21, 370, 1999.

122. Ottenhoff, T.H., De Boer, T., van Dissel, J.T., and Verreck, F.A., Human deficiencies in type 1 cytokine receptors reveal the essential role of type 1 cytokines in immunity to intracellular bacteria, *Adv. Exp. Med. Biol.*, 531, 279, 2003.

123. Conley, M.E., Immunodeficiency: UNC-93B gets a toll call, *Trends Immunol.*, 28, 99, 2007.

124. Picard, C., Puel, A., Bonnet, M., Ku, C.L., Bustamante, J., Yang, K., Soudais, C., Dupuis, S., Feinberg, J., Fieschi, C., Elbim, C., Hitchcock, R., Lammas, D., Davies, G., Al-Ghonaium, A., Al-Rayes, H., Al-Jumaah, S., Al-Hajjar, S., Al-Mohsen, I.Z., Frayha. H.H., Rucker, R., Hawn, T.R., Aderem, A., Tufenkeji, H., Haraguchi, S., Day, N.K., Good. R.A., Gougerot-Pocidalo, M.A., Ozinsky, A., and Casanova, J.L., Pyogenic bacterial infections in humans with IRAK-4 deficiency, *Science*, 299, 2076, 2003.

125. Casrouge, A., Zhang, S.Y., Eidenschenk, C., Jouanguy, E., Puel, A., Yang, K., Alcais, A., Picard, C., Mahfoufi, N., Nicolas, N., Lorenzo, L., Plancoulaine, S., Sénéchal, B., Geissmann, F., Tabeta, K., Hoebe, K., Du, X., Miller, R.L., Héron, B., Mignot, C., de Villemeur, T.B., Lebon, P., Dulac, O., Rozenberg, F., Beutler, B., Tardieu, M., Abel, L., and Casanova, J.L., Herpes simplex virus encephalitis in human UNC-93B deficiency, *Science*, 314, 308, 2006.

126. Takada, H., Yoshikawa, H., Imaizumi, M., Kitamura, T., Takeyama, J., Kumaki, S., Nomura, A., and Hara, T., Delayed separation of the umbilical cord in two siblings with Interleukin-1 receptor-associated kinase 4 deficiency: rapid screening by flow cytometer, *J. Pediatr.*, 148, 546, 2006.

127. Davidson, D.J., Currie, A.J., Bowdish, D.M., Brown, K.L., Rosenberger, C.M., Ma, R.C., Bylund, J., Campsall, P.A., Puel, A., Picard, C., Casanova, J.L., Turvey, S.E., Hancock, R.E., Devon, R.S., and Speert, D.P., IRAK-4 mutation (Q293X): rapid detection and characterization of defective post-transcriptional TLR/IL-1R responses in human myeloid and non-myeloid cells, *J. Immunol.*, 177, 8202, 2006.

128. von Bernuth, H., Ku, C.L., Rodriguez-Gallego, C., Zhang, S., Garty, B.Z., Maródi, L., Chapel, H., Chrabieh, M., Miller, R.L., Picard, C., Puel, A., and Casanova, J.L., A fast procedure for the detection of defects in toll-like receptor signaling, *Pediatrics*, 118, 2498, 2006.

10 Detection and Characterization of the T-Cell Receptor Repertoire

Gulbu Uzel

CONTENTS

10.1 OVERVIEW

Antigen-specific receptors on T cells are similar, but not identical to those on B cells. Interaction of the cell receptor (T-cell receptor [TCR]) with the antigen in the proper context leads to stimulation of the specific T cell to become activated and to proliferate, leading to clonal selection. The TCR complex is made of membrane proteins including the TCR, which recognizes the antigen, and a set of proteins called the CD3 complex, which leads to signal transduction and activation of T cells. Recognition and binding of the antigen to its specific receptor initiates the activation process, that is, downstream signaling pathways such as protein phosphorylation, the release of inositol phosphates, and the elevation of intracellular calcium levels. Because the adaptive immune system recognizes billions of unique antigens using highly variable TCRs, generation and maintenance of an effective repertoire of TCRs for diverse antigen recognition is essential to the immune system. Detection and monitoring of the TCR repertoire is important in diagnosing a T-cell malignancy, autoimmune phenomena, or response to immunotherapy.

10.2   THE T-CELL ANTIGEN RECEPTOR

The antigen receptor of mature T cells is a heterodimer of two highly variable glycoproteins, which is associated with the CD3 complex [1,2]. Clonal distribution of these chains provides the unique expression pattern to the T cells. Two types of TCRs exist either composed of α and β chains or composed of γ and δ chains. The TCR genes are represented as a number of segments comprising V, D, J, and C in germ line deoxyribonucleic acid (DNA). These segments go through a recombination process in the thymus to form the functional TCR [3].

10.2.1   THE $\alpha\beta$ T-CELL RECEPTOR

This is the most common form of the T-cell antigen receptor, expressed on over 95% of circulating T cells. The development and maturation of $\alpha\beta$ T cells mostly takes place in the thymus. Mature $\alpha\beta$ T cells recognize antigen presented in the context of major histocompatibility complex (MHC) class I or II antigens. Both the α and β TCR chains consist of variable (V) and constant (C) regions. The TCR β chain has four regions of hypervariability, the so-called complementary determining regions [CDR] (Figure 10.1). The CDR interacts directly with the antigen–MHC complex. The α and β chains are encoded by discontinuous gene segments that undergo rearrangement and recombination to generate a large repertoire of functional TCRs during T-cell development. The CDR1 and CDR2 of the VJ chain is encoded by at least 63 different Vβ gene segments that can be grouped into 25 families [4]. CDR3, which has the greatest variability, is represented by the V-D-J junctional region. There are two groups of Dβ and Jβ segments, each associated with a Cβ segment. A complete β chain is formed from the recombination of one of the Vβ segments with one of the two Dβ and one of the 13 Jβ segments [5]. A functional α chain is derived

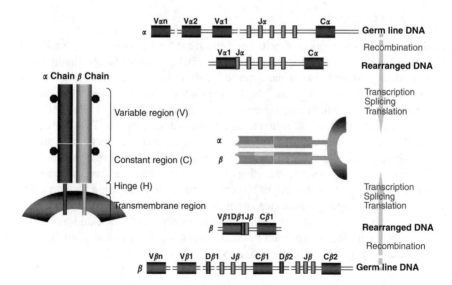

FIGURE 10.1   General scheme of V(D) J recombination for the assembly of T-cell receptor genes.

by the recombination of one of the 50 Vα segments (grouped into 29 families), one of over 70 Jα gene segments, and a single Cα segment [5–7]. The measurement of TCR Vβ distribution by flow cytometry is discussed in Chapters 2 and 6.

10.2.2 THE γδ T-CELL RECEPTOR

A small percentage of T cells express the γδ heterodimeric receptor instead of αβ. Most (approximately 75%) T cells expressing the γδ receptor do not express CD4 or CD8, the so-called double-negative T-cell (DNT) population, with a small percentage expressing CD8 on their cell surface. T cells expressing the γδ receptor have effector functions similar to αβ receptor expressing T cells and may or may not require self-MHC for antigen recognition [8]. There are 14 Vγ segments located upstream of two Jγ segments. The δ-chain gene complex is between the Vα and Jα segment complexes. There are three Vδ, three Dδ, and three Jδ segments [5].

10.3 ASSESSMENT OF T-CELL CLONALITY

It is not possible to assess the clonality of a proliferating T-cell population using standard T-cell surface markers. To establish the clonal character of a suspected T-cell proliferation, molecular analysis of TCR gene rearrangements has been the method of choice. This is especially helpful in malignant T-cell clones, where the determination of identically rearranged TCR genes can help to distinguish between mono- and polyclonal T-cell proliferations. Southern blot (SB) analysis is a highly reliable method for clonality assessment, which can detect every clonal TCR gene rearrangement when optimally positioned probes and the appropriate restriction enzymes are used [9–11]. However, SB is labor-intensive, time-consuming, and requires large amounts of high-quality DNA. In contrast, SB-based detection of clonal TCR rearrangements has been the gold standard to compare and validate other methods of clonality assessment.

Polymerase chain reaction (PCR) analysis of TCRγ genes is currently the most commonly used method. The number of primers required is limited due to the restricted combinatorial repertoire of TCRγ genes. However, this limited repertoire yields to high background amplification of similar rearrangements in normal T cells, which leads to a reduced sensitivity. Analysis of TCRβ genes is mostly done by PCR-based analysis of the Vβ repertoire. Reverse transcription (RT)–PCR amplification of Vβ–Cβ transcripts helps to limit the number of primers used. Flow cytometric analysis of the Vβ repertoire is an alternative approach for molecular clonality studies [12,13]. Combining the TCR Vβ family-specific monoclonal antibodies with surface markers such as CD3, CD8, or CD4, allows for an easy determination of T-cell clonality using multicolor flow cytometry. Using this technology, subpopulations of T cells can be analyzed, quantitated, and even isolated easily. Reference values have been determined in healthy adult controls for the individual Vβ antibodies, and values for percentage of CD3+/CD4+ or CD3+/CD8+ T cells expressing a certain TCR Vβ subset can be compared to the reference values. A listing of the monoclonal antibodies and their specificity is provided in Table 10.1 [14]. Also, an attractive commercial eight-tube kit (IOTest® Beta Mark) is available (Immunotech/Beckman Coulter, Marseille, France); in each of the eight tubes, three distinctly labeled Vβ antibodies are present that not only allow the detection of single Vβ expression,

TABLE 10.1
List of Commercially Available Monoclonal Antibodies Discrete to Vβ Families and Their Specificity

Vβ Antibody Specificity	Company	Vβ Expression Median (%)
Vβ1	Immunotech	4.1 (3.4–5.9)
Vβ2	Immunotech	8.6 (6.2–11.2)
Vβ3	Immunotech	4.5 (1.5–8.5)
Vβ5.1	Immunotech	6.5 (4.0–8.7)
Vβ5.2/5.3	T-Cell Sciences	3.2 (2.1–3.8)
Vβ6.1	Immunotech	1.3 (0.5–2.8)
Vβ6.7	T-Cell Diagnostics	3.0 (0.5–5.5)
Vβ7.1	Immunotech	2.8 (1.6–4.6)
Vβ8.1/8.2	Immunotech	5.0 (3.5–6.5)
Vβ9.2	Immunotech	3.2 (2.1–4.7)
Vβ11.1	Immunotech	0.8 (0.5–1.4)
Vβ12.2	Immunotech	2.4 (1.3–3.6)
Vβ13.1/133	T-Cell Sciences	5.9 (4.3–10.8)
Vβ13.6	Immunotech	2.1 (1.5–3.0)
Vβ14	Immunotech	3.7 (2.1–8.6)
Vβ16	Immunotech	1.3 (1.0–1.7)
Vβ17	Immunotech	5.2 (3.8–6.5)
Vβ18	Immunotech	0.5 (0.5–1.5)
Vβ20	Immunotech	2.4 (1.1–4.1)
Vβ21.3	Immunotech	2.8 (1.8–4.2)
Vβ22	Immunotech	4.0 (0.6–5.3)
Vβ23	Immunotech	1.3 (0.5–2.5)

Source: Adapted from Langerak, A.W., van Den Beemd, R., Wolvers-Tettero, I.L., Boor, P.P., van Lochem, E.G., Hooijkaas, H., and van Dongen, J.J., *Blood*, 98, 165–173, 2001.

but also allow the direct identification of the involved Vβ domain or family. Flow cytometric Vβ repertoire analysis can be used as a (quantitative) screening method for the detection of large, aberrant T-cell populations with single Vβ domain expression. To define single Vβ expression, either the mean or median normal Vβ values can be used, also taking into account the differences in the use of particular Vβ domains between CD4 and CD8 T-cell populations and observations of a more restricted Vβ usage of especially CD8 T lymphocytes in the elderly [12,13]. Specific methodology for determining cutpoints, above which a Vβ expansion is likely, is discussed in 2.7.2. Detection of Vβ expansions is an important criterion for diagnosing T-cell large granular lymphocytic (LGL) leukemia [15,16].

10.3.1 QUANTITATIVE POLYMERASE CHAIN REACTION

By combining PCR with V-segment-specific forward primers and C-region reverse primers and performing the reaction on cDNA derived from T-cell ribonucleic acid

(RNA), it is possible to amplify the expressed TCR repertoire of any one individual T cell or subpopulation of T cells. In addition, by appropriately modifying conditions, PCR can be used to semiquantify the amount of RNA corresponding to a particular sequence or V segment family. This method, known as quantitative PCR (qPCR), has been used successfully to analyze the human TCR repertoire in many different situations; qPCR has been used to assess V segment usage in different antigen responses and in defining the $V\beta$ segments that are recognized by, and respond to, superantigens. qPCR has been used to characterize T cells derived from the peripheral blood and pathological lesions of patients with autoimmune diseases or bacterial or viral infections to define TCR in tumor-infiltrating lymphocytes [17–19].

10.3.2 SPECTROTYPING (IMMUNOSCOPE®)

CDR3 spectratyping is a method that is gaining attention and is increasingly being adopted to analyze the TCR repertoire. This method helps to describe the diversity of a T-cell population repertoire by the analysis of the CDR3 length distribution [20]. In this technique, PCR amplification is performed with V- and C- or J-specific primers. These PCR products are then labeled in runoff experiments with C- or J-specific primers coupled with a fluorophore and loaded on an automated DNA sequencer to separate the different CDR3 lengths in each V–C or V–J combination. GeneScan® (Applied Biosystems, Foster City), Immunoscope (INSERM, Paris), and Genotester® (Amersham, Uppsala) are three popular software packages used to determine nucleotide sizes and areas of the observed CDR3 peaks.

10.4 SUMMARY

Improvements in the standardization of molecular techniques have increased the ease and availability of comprehensive TCR repertoire analysis. Flow cytometry provides a very accessible, more facile general assessment of the relative distribution of individual TCR families. These procedures are however largely relegated to more specialized laboratory facilities.

REFERENCES

1. Blackman, M., Kappler, J., and Marrack, P. 1990. The role of the T cell receptor in positive and negative selection of developing T cells. *Science* 248: 1335–1341.
2. Marrack, P. and Kappler, J. 1987. The T cell receptor. *Science* 238: 1073–1079.
3. Samelson, L.E., Lindsten, T., Fowlkes, B.J., van den Elsen, P., Terhorst, C., Davis, M.M., Germain, R.N., and Schwartz, R.H. 1985. Expression of genes of the T-cell antigen receptor complex in precursor thymocytes. *Nature* 315: 765–768.
4. Robinson, M.A. 1991. The human T cell receptor beta-chain gene complex contains at least 57 variable gene segments. Identification of six V beta genes in four new gene families. *J. Immunol.* 146: 4392–4397.
5. Davis, M.M. and Bjorkman, P.J. 1988. T-cell antigen receptor genes and T-cell recognition. *Nature* 334: 395–402.
6. Griesser, H., Feller, A.C., Mak, T.W., and Lennert, K. 1987. Clonal rearrangements of T-cell receptor and immunoglobulin genes and immunophenotypic antigen expression in different subclasses of Hodgkin's disease. *Int. J. Cancer* 40: 157–160.

7. Mak, T.W., Caccia, N., Reis, M., Ohashi, P., Sangster, R., Kimura, N., and Toyonaga, B. 1987. Genes encoding the alpha, beta, and gamma chains of the human T cell antigen receptor. *J. Infect. Dis.* 155: 418–422.

8. Haas, W., Pereira, P., and Tonegawa, S. 1993. Gamma/delta cells. *Annu. Rev. Immunol.* 11: 637–685.

9. van Dongen, J.J. 1987. Analysis of immunoglobulin genes and T cell receptor genes as a diagnostic tool for the detection of lymphoid malignancies. *Neth. J. Med.* 31: 201–209.

10. van Dongen, J.J. and Wolvers-Tettero, I.L. 1991. Analysis of immunoglobulin and T cell receptor genes. Part II: Possibilities and limitations in the diagnosis and management of lymphoproliferative diseases and related disorders. *Clin. Chim. Acta* 198: 93–174.

11. van Dongen, J.J. and Wolvers-Tettero, I.L. 1991. Analysis of immunoglobulin and T cell receptor genes. Part I: Basic and technical aspects. *Clin. Chim. Acta* 198: 1–91.

12. McCoy, J.P., Jr., Overton, W.R., Schroeder, K., Blumstein, L., and Donaldson, M.H. 1996. Immunophenotypic analysis of the T cell receptor V beta repertoire in CD4+ and CD8+ lymphocytes from normal peripheral blood. *Cytometry* 26: 148–153.

13. Ricalton, N.S., Roberton, C., Norris, J.M., Rewers, M., Hamman, R.F., and Kotzin, B.L. 1998. Prevalence of CD8+ T-cell expansions in relation to age in healthy individuals. *J. Gerontol. A: Biol. Sci. Med. Sci.* 53: B196–B203.

14. Langerak, A.W., van Den Beemd, R., Wolvers-Tettero, I.L., Boor, P.P., van Lochem, E.G., Hooijkaas, H., and van Dongen, J.J. 2001. Molecular and flow cytometric analysis of the Vbeta repertoire for clonality assessment in mature TCRalphabeta T-cell proliferations. *Blood* 98: 165–173.

15. Nash, R., McSweeney, P., Zambello, R., Semenzato, G., and Loughran, T.P., Jr. 1993. Clonal studies of CD3-lymphoproliferative disease of granular lymphocytes. *Blood* 81: 2363–2368.

16. Semenzato, G., Zambello, R., Starkebaum, G., Oshimi, K., and Loughran, T.P., Jr. 1997. The lymphoproliferative disease of granular lymphocytes: updated criteria for diagnosis. *Blood* 89: 256–260.

17. Duchmann, R., Strober, W., and James, S.P. 1993. Quantitative measurement of human T-cell receptor V beta subfamilies by reverse transcription-polymerase chain reaction using synthetic internal mRNA standards. *DNA Cell Biol.* 12: 217–225.

18. Malhotra, U., Spielman, R., and Concannon, P. 1992. Variability in T cell receptor V beta gene usage in human peripheral blood lymphocytes. Studies of identical twins, siblings, and insulin-dependent diabetes mellitus patients. *J. Immunol.* 149: 1802–1808.

19. Panzara, M.A., Oksenberg, J.R., and Steinman, L. 1992. The polymerase chain reaction for detection of T-cell antigen receptor expression. *Curr. Opin. Immunol.* 4: 205–210.

20. Pannetier, C., Cochet, M., Darche, S., Casrouge, A., Zoller, M., and Kourilsky, P. 1993. The sizes of the CDR3 hypervariable regions of the murine T-cell receptor beta chains vary as a function of the recombined germ-line segments. *Proc. Natl. Acad. Sci. USA.* 90: 4319–4323.

11 Human Leukocyte Differentiation Antigens, Cluster of Differentiation: Past, Present, and Future

Maurice R.G. O'Gorman

CONTENTS

11.1   INTRODUCTION

It is probably not fully appreciated the extent to which the invention of monoclonal antibody production revolutionized diagnostic medicine, expanded our ability to investigate the inordinate complexities of the immune system, and has led to a completely new class of biological therapies. Before the invention of monoclonal antibodies, we basically understood that the immune system was bifurcated into the cellular and the humoral components and we were able to identify T and B cells by virtue of their surface binding to sheep rosettes and anti-immunoglobulin reagents, respectively. Today, with literally thousands of well-characterized monoclonal antibodies specific for over 350 equally well-characterized antigens, we are able to identify hundreds of very specific leukocyte cell subsets and function-associated molecules. These remarkable achievements are unlikely to have proceeded at such an incredible rate had it not been for the development of a forum on human leukocyte differentiation antigens (HLDA) that was created to compare the reactivity of monoclonal antibodies specific for distinct leukocyte differentiation antigens [1].

11.2   HISTORY

The first international workshop on human leukocyte differentiation antigens sponsored by Institut National de la Santé et de la recherche médicale (INSERM), World Health Organization (WHO), and the International union of Immunological Societies

(IUIS) was chaired by Jean Dausset, Cesar Milstein, and Stuart F. Schlossman, and included 55 research groups from 14 countries [1]. Monoclonal antibodies were submitted with preliminary data, reviewed, and if appropriate were selected for workshop evaluations. Studies involved the assessment of reactivities against a variety of cell panels. The goal was to identify monoclonal antibodies reacting with similar antigens. This was assessed using a hierarchical clustering algorithm comparing the "distances" between all monoclonal antibody pairs; distance being defined as the "mean absolute difference in reactivity between any two monoclonal antibody pairs tested against target cells included in the study" [1]. All monoclonal antibody pairs are compared and the "distances" between each are added successively until a hierarchical tree is established which links the similarity of all monoclonal antibodies tested in the decreasing order. Once the tree was established, cut points were established to cut the tree into individual monoclonal antibody clusters such that the "distance" within a cluster was minimized and the "distance" between the clusters was maximized [1]. The clusters of monoclonal antibodies that exhibited very similar reactivities to specific cell lines were designated as cluster of differentiation (CDs) [1]. The outcome of the first workshop was 15 distinct CD groups: eight T cells, two B cells, one myeloid group, and four other groups, which were less well-defined and designated as CD Workshop (CDw) to indicate provisional status.

It is important to note that the first workshop generated data (probably for the first time) on the reactivity of defined clusters on normal peripheral blood lymphocytes in an attempt at evaluating the feasibility of establishing standard normal values (15–85th percentile) of lymphocyte subsets [1] (Tables 9.4 and 9.5).

Following the first workshop it was recommended that the CD nomenclature be adopted for common scientific use and was approved by the WHO/IUIS Committee on Standards and Nomenclature [2]. It was also agreed that more workshops should follow. The workshops are based on the submission of antibodies from laboratories around the world that are tested by a variety of methods to assess their reactivity. The ultimate goal of these workshops is to identify all leukocyte cell surface antigens with respect to their biochemical characterization, cell and cell line expression, molecular and cellular functions, and ultimately clinical relevance [3]. Table 11.1 summarizes the next seven HLDA workshops that were held approximately every 2–4 years.

The original CD designations referred to the actual monoclonal antibodies. This was required because in some cases CDs were assigned before the target structures were known and some of the structures were very complex [4]. Around the fourth or fifth workshop, the CD designation evolved to refer to the structures that are recognized by the CD monoclonal antibody [4].

Until the sixth HLDA workshop, the requirements for a new CD cluster were that an antigen had to be recognized by at least two different antibodies submitted to the workshop and that the molecular weight of the antigen had been determined [4,5]. The early workshops used three main techniques to cluster the monoclonal antibodies:

1. Immunofluorescence labeling of the cell surface with analysis by flow cytometry
2. Biochemical analysis of immunoprecipitation or Western blotting to establish the molecular weight
3. Immunohistology of tissue sections.

TABLE 11.1

Summary of the Evolution of Monoclonal Antibody Nomenclature for Specific Leukocyte Antigens through HLDA Workshops in 1980 to HCDM Workshops in 2006

New CD Antigens Identified	Workshop	Location	Year	Lineages Included
CD1–CD15	1st HLDA	Paris, France	1982	T, B, M
CD16–CD26	2nd HLDA	Boston, Massachusett	1984	T, B, M, A
CD27–CD45	3rd HLDA	Oxford, UK	1986	T, B, M, A, NL, P
CD46–CDw78	4th HLDA	Vienna, Austria	1989	T, B, M, A, NL and NK, P
CD79–CDw130	5th HLDA	Boston, Massachusett	1993	T, B, M, A, NK, P, adhesion structures, cytokine receptors, endothelial cell molecules
CD131–CD166	6th HLDA	Kobe, Japan	1996	T, B, M, A, NK, P, adhesion structures, cytokine receptors, endothelial cell molecules
CD167–CD247	7th HLDA	Harrogate, UK	2000	All the preceding plus erythroid cells, dendritic cells, stem and progenitor cells, carbohydrate structures
CD248–CD339	8th HLDA	Adelaide, Australia	2004	All the preceding plus stromal cells
CD340–CD350	HCDM	Quebec, Canada	2006	Added the characterization and validation of monoclonal antibodies against relevant intra-cellular leukocyte antigens (but these will not receive new CDs)

Note: T = T cells, B = B cells, M = myeloid cells, A = activation, P = platelet, NK = natural killer cells, NL = nonlineage associated.

Around the fourth workshop, clustering also included functional studies in addition to the various target cells to cluster the antibodies [5], and it was recognized that cell structures were "now being defined first rather than using the traditional approach of defining functions and then searching for structures" [6]. It was also around the fourth or fifth workshop that molecular methods began to evolve, which would allow for the complementary deoxyribonucleic acids (cDNAs) of specific CDs to be isolated and their chromosomal locations identified [6].

The fifth workshop added cluster assignments to cytokines and growth factor receptors and also to the first nonleukocyte cell type, endothelial cells. The latter cell type was added in recognition that they expressed immunogens that were potentially important in leukocyte biology [7]. The fifth HLDA workshop also introduced quantitative fluorescence in an effort to provide quantitative information on the intensity and heterogeneity of antigen expression levels on various cell types [7]. The seventh HLDA workshop heralded two major advancements. The first was the addition of four new sections, that is, dendritic cells, stem/progenitor cells, erythroid cells, and carbohydrate structures. The second important change was a major revision to the criteria required for establishing a new CD. To accommodate the advancements of the molecular revolution it was established that instead of requiring the existence of at least two independent monoclonal antibodies of the same specificity and molecular weight of the antigen, a new CD could be established for a molecule if its gene had been cloned and at least one specific monoclonal antibody had been submitted in the workshop [8]. This change in criteria led to an explosion of new CDs with over 80 new entities being added as compared to an average of approximately 30 from each of the previous workshops [8].

11.3   PRESENT AND FUTURE

The eighth workshop continued the trend of rapid expansion of new CDs with an additional 95 new CD assignments [9], but it would herald the end of an era and a radical change to traditional workshops. Changes from the eighth workshop are summarized as follows [9] but were not limited to

A name change from HLDA to "Human Cell Differentiation Molecules" (HCDM).

The name change was meant to signify a change in tradition (but kept the "CD" in the acronym), to extend the focus from leukocytes to other cell types because leukocytes do not act alone, and to broaden the scope from cell surface molecules to any molecule whose expression reflects cellular differentiation.

Establish "workshop validation" for submitted antibodies that give a clear-cut positive result and will no longer be restricted to only those molecules that subsequently receive a CD designation. The HCDM office will maintain a database of all validated hybridoma clones and will seek to ensure that these antibodies are available to the research community.

HCDM laboratories will validate antibodies in a yearly cycle and all new antibodies and new CD designations will be posted on the HCDM website at the end of the year.

The HCDM decided that all molecules that are useful in cellular differentiation should be studied by the workshop; however, only those markers that are expressed on the cell surface will be given CDs [9].

The first HCDM workshop was held in Quebec, Canada in May 2006, and it resulted in the changes of some preexisting CDs and eight new CDs using the following criteria [10]:

At least one workshop is characterized by antibody and good molecular data.
The antigen is expressed on the surface of cells involved in immune reactions.
The antibodies react with primary cells, not just transfectants or recombinant proteins. The antibody is available.

Tables 11.2 and 11.3 are derived from a variety of sources including the "Official Poster of the 8th International Workshop on Human Leukocyte Differentiation

TABLE 11.2
CD Table (2007) Alternate Names, Ligands and Molecular Information

CD	Alternative Name	Ligands and Associated Molecules	Molecular Weight (kDa) Unreduced/Reduced	Gene Locus
CD1a	R4,T6/Leu-6, HTA-1	w/β2m	49/—	1q22-q23
CD1b	R1	w/β2m	45	1q22-q23
CD1c	M241, R7	w/β2m	43/—	1q22-q23
CD1d	R3	w/β2m	49	1q22-q23
CD1e	R2		28	1q22-q23
CD2	Sheep red blood cell rosette receptor, T11, LFA-2, Tp50	CD58, CD48, CD59, CD15	50/—	1p13
CD3d	T3, CD3 complex, Leu-4		20–26	11q23
CD3e				
CD3g				
CD4	L3T4, W3/25, T4, OKT4, Leu-3a	MHC class II, gp120, IL-16	55	12pter-p12
CD5	T1, Tp67, Leu-1, Ly-1	CD72, BCR, gp35–37, ZAP-70, TCR, CD21	58/67	11q13
CD6	T12	gp40, gp90, CD166 (ALCAM)	—/105–130	11q13
CD7	gp40, Leu-9, 3A1,T-cell leukemia antigen	PI3-kinase	38/40	17q25.2-q25.3
CD8α	Leu-2, T8, Ly-T2, OKT8	MHC I, Lck	68/30–34	2p12
CD8β	CD8, Leu-2, Ly-T3	MHC I, Lck	32–34	2p12
CD9	p24, DRAP-27, MRP-1, leukocyte antigen MIC3	CD63, CD81, CD82, CD41/CD61, HLA-DR, β1 integrins, PI4-kinase	—/24,26	2p13
CD10	CALLA, NEP, gp100, EC 3.4.24.11, membrane metalloendopeptidase	100/—		3q25.1-q25.2
CD11a	LFA-1α, αL integrin chain, pg180/95	ICAM-1,2,3, 4	170/180	16p11.2

(continued)

TABLE 11.2 (continued)
CD Table (2007) Alternate Names, Ligands and Molecular Information

CD	Alternative Name	Ligands and Associated Molecules	Molecular Weight (kDa) Unreduced/Reduced	Gene Locus
CD11b	αM integrin chain, CR3, alpha chain of C3bi receptor, gp155/95, Mo1, C3niR, mac-1	iC3b, fibrinogen, ICAM-1,2, factor X	165/170	16p11.2
CD11c	αX integrin, gp150/95, AXb2, alpha chain of complement receptor type 4 (CR4)	iC3b, fibrinogen, ICAM-1	145/150	16p11.2
CDw12	p90–120		150–160/120	
CD13	Aminopeptidase N, APN, gp150, EC 3.4.11.2	NGR	150/—	15q25-q26
CD14	LPS receptor	Endotoxin (LPS)	53/55	5q23-31
CD15	X hapten, Lewis X, SSEA-1, 3-FAL	Selectins	CHO	
CD15s	Sialyl Lewis X	E-selectin	CHO	
CD15u	3' Sulfo Lewis X	P-selectin	CHO	
CD15su	6 Sulfo–sialyl Lewis X	L-selectin		
CD16a	FCRIIIA, Fc gamma receptor IIIa	IgG Fc	50–65/—	1q23
CD16b	FCRIIIB, Fc gamma receptor IIIb	IgG Fc	48	1q23
CD17[a]	Lactosylceramide, LacCer		150–160/120	
CD18	β2 Integrin, CD11a,b,c β subunit, macrophage antigen-1, mac-1	CD11a,b,c	90/95	21q22.3
CD19	B4, Bgp95	CD21, CD81, CD225, Leu-13, Lyn, Fyn, VAV, PI3-kinase	>120/95	16p11.2
CD20	B1, Bp35, membrane-spanning four domains, subfamily A, member 1	Lyn, Lck, Fyn, Cell surface protein: 28–30, 180–200, 50–60 kDa	33/35/37	11q12-q13.1

CD	Description	Other	MW	Chromosome
CD21	CR2, EBV receptor, C3d receptor, C3dR	C3d, CD23, CD19, CD81, Leu-13	130–145/110	1q32
CD22	BL-CAM, Lyb-8, Siglec-2, Bgp135	p72sky, p53/56lyn, SHP1, PI3-kinase, CD45	140/130	19q13.1
CD23	Low-affinity IgE receptor, FceRII, B6, BLAST-2, Leu-20, gp50-45	IgE, CD21, CD11b, CD11c	50–45	19p13.3
CD24	BBA-1, heat-stable antigen homologue, HAS	CD62P (P-selectin)	35–45	6q21
CD25	Tac antigen, interleukin-2 receptor alpha chain, IL–2Rα, p55	IL-2	55	10p15-p14
CD26	Dipeptidylpeptidase IV, DPP IV ectoenzyme, gp120, Ta1, ADA binding protein	Adenosine deaminase, collagen, CD45	110	2q24.3
CD27	T14, S152	CD70, TRAF5, TRAF2	110–120	12p13
CD28	Tp44, T44	CD80, CD86, PI3-kinase	90/44	2q33
CD29	Platelet GPIIa, integrin β1 chain, GP, VLA (CD49) beta chain	VCAM-1, MadCAM-1	110–130	10p11.2
CD30	Ber-H2, Ki-1 antigen	CD153, TRAF1,2,3,5	120/105	1p36
CD31	PECAM-1, endocam, platelet GPIIa′	CD38, CD31, GAGs, avb3 integrin	130–140	17q23
CD32	Fc gamma receptor type II, FCgRII, gp40	Phosphatases, IgG	40	1q23
CD33	gp67, My9, p67	Sugar chains containing sialic acid	150/67	19q13.3
CD34	gp105–120, My10	L-selectin, MAdCAM-1	105–120	1q32
CD35	Complement receptor type 1, CR1, C3b/C4b receptor	C3b, C4b, iC3, iC4	160–250	1q32
CD36	GPIIIb, GPIV, PASIV, platelet GPIV, OKM-5 antigen	Thrombospondin, collagen I, IV, V	90	7q11.2
CD37	gp52-40	CD53, CD81, CD82, MHC II	40–52/40–52	19p13-q13.4
CD38	ADP ribosyl cyclase, T10, gp45	CD31, hyaluronic acid	45/45	4p15

(continued)

TABLE 11.2 (continued)
CD Table (2007) Alternate Names, Ligands and Molecular Information

CD	Alternative Name	Ligands and Associated Molecules	Molecular Weight (kDa) Unreduced/Reduced	Gene Locus
CD39	ATP dehydrogenase, NTP dehydrogenase-1, gp80, ectonucleoside triphosphate diphosphohydrolase-1	ADP/ATP	80/80	10q24
CD40	Bp50, TNF receptor 5	CD154, CD40L, TRAP	85/48	20q12-q13.2
CD41	GPIIb, αIIβ integrin, platelet glycoprotein GPIIb	Fibrinogen, fibronectin, vWF	135/120,23	17q21.32
CD42a	Platelet glycoprotein GPIX	vWF, thrombin, CD42b,c,d	22/17–22	3q21
CD42b	Platelet glycoprotein GPIbα	vWF, thrombin, CD42a,c,d	160/145	17pter-p12
CD42c	Platelet glycoprotein GPIbβ	vWF, thrombin, CD42a,b,d	160/24	22q11.21
CD42d	Platelet glycoprotein GPV	vWF, thrombin, CD42a,b,c	82/82	3q29
CD43	Sialophorin, leukosialin, leukocyte sialoglycoprotein, gp95	Hyaluronan	95–135/95–135	16p11.2
CD44	ECMRII, H-CAM, Pgp1, Hermes antigen, Hutch-1, gp80–95	Hyaluronan, ankyrin, fibronectin, MIP1β, osteopontin	85/—	11p13
CD44R	CD44 variant, CD44v9	Hyaluronan, ankyrin, fibronectin, MIP1β, osteopontin	85–200/—	11p13
CD45	LCA, T200, B220, Ly-5, protein tyrosine phosphatase receptor type C	p56Lck, p59Fyn, Src kinases	180–220/—	1q31-q32
CD45RA	Restricted T200, gp220, isoform of leukocyte common antigen	p56Lck, p59Fyn, Src kinases	220	1q31-q32
CD45RB	Restricted T200, gp220, isoform of leukocyte common antigen	p56Lck, p59Fyn, Src kinases	220	1q31-q32

CD	Name	Ligands/Associations	MW	Chromosome
CD45RC	Restricted T200, gp220, isoform of leukocyte common antigen	p56Lck, p59Fyn, Src kinases	220	1q31-q32
CD45RO	UCHL-1, gp180, restricted T200	p56Lck, p59Fyn, Src kinases	180	1q31-q32
CD46	Membrane cofactor protein, MCP	SCR2/3/4, serum factor 1 protease	52–58/64–68	1q32
CD47	gp42, Integrin-associated protein, IAP, ovarian carcinoma antigen, OA3, neurophilin	SIRP, CD61, thrombospondin	45–60/50–55	3q13.1-q13.2
CD47R	MEM-133		120/—	3q13.1-q13.2
CD48	Blast-1, Hulym-3, BCM-1, OX-45	CD2, Lck, Fyn, CD229, CD244	45/45	1q21.3-q22
CD49a	Very late antigen, VLA-1α chain, α1 integrin	Collagen, laminin-1	200/200	5q11.2
CD49b	VLA-2α chain, α2 integrin, platelet GPIa	Collagen, laminin	150/160	5q23-31
CD49c	VLA-3α chain, α3 integrin	Laminin-5, Fn, collagen	145–150/125,30	17q21.31
CD49d	VLA-4α chain, α4 integrin	CD106, MAdCAM, fibronectin	145/150	2q31-q32
CD49e	VLA-5α chain, α5 integrin	Fibronectin, invasin, fibrinogen	160/135,25	12q11-q13
CD49f	VLA-6α chain, α6 integrin, gpI	Laminins, invasin	150/125	2p14-q14.3
CD50	Intracellular adhesion molecule-3, ICAM-3	LFA-1, αd/β2 integrin	110–140/—	19p13.3-p13.2
CD51	αv Integrin, vitronectin receptor alpha chain, VNR-α	Fibrinogen, vitronectin, MMP-2, vWF, TSP	150/124,24	2q31-q32
CD52	CAMPATH-1, HE5		25–29/25–29	1p36
CD53	MRC, OX-44	VLA-4, HLA-DR, other tetraspans	32–42/—	1p31-p12
CD54	Intracellular adhesion molecule-1, ICAM-1	LFA-1, mac-1, rhinovirus	90/95	19p13.3-p13.2
CD55	Decay accelerating factor, DAF	SCR, CD97, echoviruses	55–70/80	1q32

(continued)

TABLE 11.2 (continued)
CD Table (2007) Alternate Names, Ligands and Molecular Information

CD	Alternative Name	Ligands and Associated Molecules	Molecular Weight (kDa) Unreduced/Reduced	Gene Locus
CD56	Leu-19, NKH-1, neural cell adhesion molecule, NCAM	NCAM-1, heparan sulfate	140	11q23-q24
CD57	HNK1, Leu-7	L-selectin, P-selectin, laminin	110–115	11q12-qter
CD58	Lymphocyte-associated antigen-3, LFA-3	CD2, LFA-2	55–70	1p13
CD59	1F5Ag, H19, protectin, MACIF, MIRL, P-18	C8-α, C9, Lck, Fyn	18–25/19–25	11p13
CD60a	GD3			
CD60b	9-O-acetyl GD3		90–94/120	
CD60c	7-O-acetyl GD3			
CD61	Glycoprotein IIIa, GP IIIa, β3 integrin	Fibrinogen	90–110	17q21.32
CD62E	E-selectin, ELAM-1, LECAM-2	Sialyl Lewis X,A, CLA, CD162	115/97	1q22-q25
CD62L	L-selectin, LAM-1, LECAM-1, MEL-14, Leu-8, TQ1	CD34, GlyCAM-1, MAdCAM-1	74	1q23-q25
CD62P	P-selectin, granule membrane protein-140, GMP-140, PADGEM	CD162, CD24	120/140	1q22-q25
CD63	LIMP, MLA1, gp55, neuroglandular antigen, NGA, LAMP-3, granulophysin ME491	VLA-3, VLA-6, CD81, CD9, PI4-kinase	40–60	12-q12-q13
CD64	Fc gamma receptor 1, FcgR1	IgG	72	1q21.2-q21.3
CD65	Ceramide dodecasaccharide 4c, VIM-2	E-selectin		
CD65s	Sialylated-CD65, VIM-2 antigen	Possibly E- or P-selectin		

CD66a	NCA-160, BGP, carcinoembryonic antigen-related adhesion molecule-1	CD62E, CD66c,e, Src kinases	140–180	19q13.2
CD66b	CD67, CGM6, NCA-95	CD66c,e, Src kinases	95–100	19q13.2
CD66c	Nonspecific cross-reaction antigen, NCA, NCA-50/90	CD62E, galectins, CD66a,b,c,e, Src kinases	90	19q13.2
CD66d	CGM1		35	19q13.2
CD66e	CEA	CD66a,c,e	180–200	19q13.2-q13.2
CD66f	SP-1, pregnancy-specific (b1) glycoprotein, PSG		54–72	19q13.2
CD68	gp110, Macrosialin	LDL	110	17p13
CD69	Activation inducer molecule, AIM, EA 1, MLR3, gp34/28, VEA		60	12p13-p12
CD70	Ki-24 antigen, CD27 ligand	CD27	55–170	19p13
CD71	T9, transferrin receptor	Transferrin	190/95	3q26.2-qter
CD72	Ly-19.2, Ly-32.2, Lyb-2	CD5	43/39	9p13.3
CD73	Ecto-5'-nucleotidase	AMP	69–72	6q14-q21
CD74	MHC class-II invariant chain, Ii, invariant chain	HLA-DR, CD44	41	5q32
CD75	Sialo-masked lactosamine	CD22		
CD75S	Alpha 2,6 sialylated lactosamine (formerly CDw75 and CDw76)	CD22 (proposed)		
CD77	Pk blood group antigen, Burkitt's lymphoma–associated antigen, BLA, CTH/Gb3	Shiga toxin, verotoxin 1, CD19	1	
CD79a	Igα, MB1	Ig, CD5, CD19, CD22, CD79b	—/40–45	19q13.2
CD79b	Igβ, B29	Ig, CD5, CD19, CD22, CD79a	—/37	17q23
CD80	B7, B7-1, BB1	CD28, CD152 (CTLA-4)	60/—	3q13.3-q21

(continued)

TABLE 11.2   (continued)
CD Table (2007) Alternate Names, Ligands and Molecular Information

CD	Alternative Name	Ligands and Associated Molecules	Molecular Weight (kDa) Unreduced/Reduced	Gene Locus
CD81	Target of an antiproliferative antibody-1, TAPA-1, M38	Leu-13, CD19, CD21	26/—	11p15
CD82	4F9, C33, IA4, kangai 1, KAI1, R2	MHC I, MHC II, CD4, CD8, β1 integrins	45–90/—	11p11.2
CD83	HB15		—/43	6p23
CD84	p75, GR6		68–80/72–86	1q24
CD85a	ILT5, LIR3, HL9	HLA class I		19q13.4
CD85d[a]	ILT4, LIR2, MIR10	HLA class I	110	19q13.4
CD85j[a]	Immunoglobulin-like transcript 2, ILT2, LIR1, MIR7	HLA class I	110	19q13.4
CD85k[a]	ILT3, LIR5, HM18	HLA class I	60	19q13.4
CD86	B7-2/B70	CD28, CD152 (CTLA-4)	—/80	3q21
CD87	Urokinase plasminogen activator–receptor, uPA-R	uPA, Pro-UPA, vitronectin	35–68/32–66	19q13
CD88	C5aR	C5a/C5a(desArg), anaphylatoxin	43/—	19q13.3-q13.4
CD89	IgA Fc receptor, FcaR	IgA1, IgA2	45–100/45–100	19q13.2-q13.4
CD90	Thy-1	CD45, Lck, Fyn, P100	25–35/25–35	11q22.3-q23
CD91	α2M-R, LRP	RAP, α2M, apoE, lactoferrin, LDLs	600/—	12q13-q14
CD92[a]	p70		70/70	9q31.2
CD93[a]	GR11		110/120	
CD94	Kp43, killer cell lectin-like receptor subfamily D, member 1	HLA class I, NKG2-A, p39	70,30	12q13
CD95	APO-1, FAS, TNFRSF6	CD178 (Fas ligand)	45,90,200/45	10q24.1
CD96	T-cell activation increased late expression, TACTILE		160/—	3q13.13-q13.2

CD	Name	Ligand/partner	MW	Chromosome
CD97	BL-KDD/F12		—/28, 75-85	19p13.2-p13.12
CD98	4F2, FRP-1, RL-388	CD55(DAF)	125/80,45	11q13
CD99	MIC2, E2	Actin	32/32	Xp22.32, Yp11.3
CD99R	CD99 mab restricted		32/32	9q22-q31
CD100	SEMA4D	CD45, serine kinase	300/150	9q22-q31
CD101	IGSF2, P126, V7		240/120	1p13
CD102	ICAM-2	LFA-1, CD11b/CD18, $\alpha L\beta 2$	55-65/—	17q23-q25
CD103	HML-1, integrin αE subunit, ITGAE	E-cadherin, $\beta 7$ integrin	175/150,25	17p13
CD104	Integrin $\beta 4$ subunit, TSP1180	Laminins (I,II,IV,V), CD49F, $\alpha 6$ integrin	205/220	17q11-pter
CD105	Endoglin	TGF-β 1, TGF-β 3	180/90	9q33-q34.1
CD106	Vascular cell adhesion molecule-1, VCAM-1, INCAM-110	$\alpha 4\beta$ 1 Integrin, VLA-4	110/110	1p31-p32
CD107a	Lysosomal-associated membrane protein-1, LAMP-1		100–120/110	13q34
CD107b	Lysosomal-associated membrane protein-2, LAMP-2		100–120/120	8p12-p11
CD108	SEMA7A, John-Milton-Hagen (JMH) blood group antigen, GPI-gp80	CD232, tyrosine kinases	76/80	15q22.3-q23
CD109	Platelet activating factor, 8A3, E123 7D1		170/50	
CD110	Thrombopoietin receptor, TPO-R, MPL, C-MPL	TPO, JAK2	70-95/—	1p34
CD111	PVRL1, HveC1, poliovirus receptor-related 1 protein, PRR1, nectin1, HIgR	Nectin3, afadin gD	—/75	11q23

(continued)

TABLE 11.2 (continued)
CD Table (2007) Alternate Names, Ligands and Molecular Information

CD	Alternative Name	Ligands and Associated Molecules	Molecular Weight (kDa) Unreduced/Reduced	Gene Locus
CD112	HveB, poliovirus receptor-related 2 protein, PRR2, PVRL2, nectin2	PRR3, afadin, CD112	64–72/64–72	19q13.2-13.4
CD113[a]	PVRL3, nectin3, PRR3		83	3q13
CD114	CSF-3R, G-CSFR, HG-CSFR	G-CSF, JAK1, JAK2	—/130	1p35-p34.3
CD115	c-fms, CSF-1R, M-CSFR R alpha subunit	CSF-1, phosphotyrosine binding proteins	150	5q33-q35
CD116	GM-CSF R alpha subunit, GM-CSFRα	GM-CSF, CD131	70–85/—	Xp22.32 or Yp11.3
CD117	c-KIT, stem cell factor receptor, SCFR	SCF, MGF, KL, PI3-kinase	145/145	4q11-q12
CD118	LIFR, gp190	IFNα, IFNβ	190	5p13-p12
CD119[a]	IFNγR, IFNγ R alpha chain	IFNγ	80–95	6q23-q24
CD120a	TNFRI, TNFRp55	TNF, TRADD, TRAF, RiP, LTα	55	12p13.2
CD120b	TNFRII, TNFRp75, TNFR p80	TNF, TRADD, TRAF, RiP, LTα	75	1p36.3-p36.2
CD121a	Type 1 IL-1 receptor	IL-1α and IL-1β, IL1RA	75–85/75–85	2q12
CD121b[a]	Type II IL-1 receptor	IL-1β, IL-1α, IL1RA	60–68/60–68	2q12-q22
CD122	IL-2 receptor beta chain, IL2Rβ	IL-2, IL-15, CD25, CD132, Syk, Lck, JAK1, STAT5	70–75/—	22q13, 22q13.1
CD123	Interleukin-3 receptor alpha chain, IL-3Rα	IL-3, CD131	70/—	Xp22.3 or Yp11.3
CD124	Interleukin-4 receptor alpha chain, IL-4Rα	IL-4, IL-13, CD132, JAK1, Fes, STAT6, IRS-2	140/—	16p11.2-12.1
CD125[a]	Interleukin-5 receptor alpha chain, IL-5Rα	IL-5, CD131	60/—	3p26-p24

CD126	Interleukin-6 receptor alpha chain, IL-6Rα	IL-6, CD130	80/80	1q21
CD127	Interleukin-7 receptor alpha chain, p90, IL-7R, IL-7Rα	IL-7, CD132, Fyn, Lyn, JAK1, PI3-kinase, Lck	65–90/	5p13
CD128a	(Renamed CD181) IL-8Rα, CXCR1	IL-8, GRO	44–59/67–70	2q35
CD128b	(Renamed CD182) IL-8Rβ, CXCR2	IL-8, GRO, NAP-2	44–59/67–70	2q35
CD129	Interleukin-9 receptor alpha chain, IL-9Rα	IL-9		Xq28, Yq12
CD130	gp130	Oncostatin M, LIF, IL-6, IL-11, CNF	130–160/130–160	5q11
CD131[a]	Common β subunit, low affinity (granulocyte macrophage)	CD123, CD125, CD116, JAK2, SHC, Grb2	120–140/—	22q13.1
CD132	Common γ chain, interleukin-2 receptor gamma chain, IL-2Rγ	CD25, CD122, CD124, CD127, IL-9R, JAK3, JAK1, Syk, Lck	65–70/—	Xq13.1
CD133	AC133, PROML1, prominin 1, heamtopoietic stem cell antigen		120/120	4p15.32
CD134	OX40, TNFRSF4	OX40 ligand	35–50/—	1p36
CD135	FLT3, FLK2, STK1	FL (Flt3 ligand)	130/155–160	13q12
CD136[a]	Macrophage-stimulating protein receptor, MSP-R, RON, p158-ron	MSP, HGFl, SHC, PLC-g	180/150,40	3p21.3
CD137[a]	4-1BB, induced by lymphocyte activation, ILA	4-1BB ligand	85/39	1p36
CD138	Syndecan-1, heparan sulfate, B-B4 proteoglycan	Collagen I, III, V, fibronectin, TSP	70–92	2p24.1
CD139	None			
CD140a	Platelet-derived growth factor alpha receptor, PDGF α receptor	PDGF	—/209,228; 160–180/—	4q11-q13
CD140b	PDGF β receptor	PDGF	170–190/—	5q31-q32
CD141	Thrombomodulin, TM, fetomodulin	Thrombin, protein C, TAFI	75/105	20p12-cen

(continued)

TABLE 11.2 (continued)
CD Table (2007) Alternate Names, Ligands and Molecular Information

CD	Alternative Name	Ligands and Associated Molecules	Molecular Weight (kDa) Unreduced/Reduced	Gene Locus
CD142	Tissue factor, thromboplastin, coagulation factor III, F3	Factor VIIa, factor Xa/TFPI	45–47/45–47	1p22-p21
CD143	Angiotensin-converting enzyme, ACE, peptidyl dipeptidase A, kininase II	ANG-1, bradykinin	90,170/90,170	17q23
CD144	VE-cadherin, cadherin-5	β-Catenin, p120 CAS, plakoglobin	135/130	16q22.1
CDw145	None		25,90,110	
CD146	Muc 18, S-endo, MCAM, Mel-CAM		113–118/130	11q23.3
CD147	Basigin, extracellular metalloproteinase inducer, EMMPRIN, M6, OX47		50–60/55–65	19p13.3
CD148	HPTP-η, p260, DEP-1, protein tyrosine phosphatase receptor type J		200–260/200–260	11p11.2
CD150	Signaling lymphocyte activation molecule, SLAM, IPO-3	Tyrosine phosphatase CD45, CD150	65–85/75–95	1q22-q23
CD151	Platelet-endothelial tetra-span antigen-3, PETA-3, SFA-1	α3, α6 Integrins	32/—	11p15.5
CD152	Cytotoxic T lymphocyte antigen-4, CTLA-4	CD80, CD86, PI3-kinase, PTP1D	50/33	2q33
CD153	CD30 ligand, CD30L, TNSF8	CD30	40	9q33
CD154	CD40 ligand, CD40L, gp39, TNF-related activation protein-1, TRAP-1, T-BAM	CD40	33	Xq26

CD	Description / alternate names	Ligand / function	MW	Location
CD155	Poliovirus receptor, PVR		60–90	19q13.2
CD156a	CD156, a disintegrin and metalloproteinase domain 8, ADAM8, MS2	Myeloid	—/69	10q26.3
CD156b	TACE, ADAM17, snake venom-like protease, cSVP	pro-TNF, pro-TGFα, MAD2	100–120/—	2p25
CD156Cª	ADAM10		98	15q2:15q22
CD157	Mo5, BST-1 BP-3/IF7	NAD, cyclic ADP ribose	42–45/42–45	4p15
CD158a	Killer cell Ig-like receptor domain 1, KIR2DL1, p58.1	HLA-Cw4,2,5,6	58/58	19q13.4
CD158b1	KIR2DL2, p58.2	HLA-Cw3,1,7,8	58/58	19q13.4
CD158b2	KIR2DL3, p58.3	HLA-Cw3,1,7,8	58/58	19q13.4
CD158c	KIR2DS6, KIRX			19q13.4
CD158d	KIR2DL4	HLA-Bw4		19q13.4
CD158e1/e2	KIR3DLI/S1, p70	HLA-Bw4	70/70	19q13.4
CD158f	KIR2DL5			19q13.4
CD158g	KIR2DS5			19q13.4
CD158h	KIR2DS1, p50.1	HLA-C		19q13.4
CD158i	KIR2DS4, p50.3	HLA-C	50/50	19q13.4
CD158j	KIR2DS2, p50.2	HLA-C		19q13.4
CD158k	KIR3DL2, p140	HLA-A	140/70	19q13.4
CD158z	KIR3DL7, KIRC1			19q13.4
CD159a	Natural killer cell lectin-like receptor subfamily C, member 1, NKG2A	CD94/CD159a heterodimer binds to HLA-E	70/43	12p12.3–p13.1
CD159c	NKG2C	C type lectin superfamily member	40	12p13
CD160	BY55, NK1, NK28	MHC class I	80/27	1q42.3

(continued)

TABLE 11.2 (continued)
CD Table (2007) Alternate Names, Ligands and Molecular Information

CD	Alternative Name	Ligands and Associated Molecules	Molecular Weight (kDa) Unreduced/Reduced	Gene Locus
CD161	NKR, natural killer cell lectin-like receptor subfamily B, member 1, NKRP1A		80/40	12p13
CD162	P-selectin glycoprotein ligand 1, PSGL-1	Selectins	160–250/110–120	12q24
CD162R	PEN5	CD62L (L-selectin)	240/140	
CD163	M130, GHI/61, RM3/1, GHI/61	Hemoglobin	110	12q13.3
CD164	Multiglycosylated core protein 24, MGC-24, MUC-24		160/80	6q21
CD165	AD2, gp37		37/42	
CD166	Activated leukocyte cell adhesion molecule, ALCAM, KG-CAM, SC-1, BEN, DM-GRASP	CD6, CD166, NgCAM	100–105/100–105	3q13.1
CD167a	Discoidin receptor DDR1 (jCD167a) and DDR2 (CD167b), trkE, cak	SHCA, FRS2, collagens	120	6p21.3
CD168	Receptor for hyaluronan involved in migration and motility, RHAMM, IHABP, HMMR	Ras, Src, Erk, actin, calmodulin, MAPKK, hyaluronic acid	80–88/—	5q33.2
CD169	Sialoadhesin, Siglec-1	CD227, CD206, CD43, $\alpha2$, 3-sialylated ligands	180/200	20p13
CD170	Sialic acid binding Ig-like lectin 5, Siglec-5	Sialylated glycans	140	19q13.3
CD171	L1CAM, NILE, neuronal adhesion molecule	CD171, neurocan, phosphocan, Laminin	200–230	Xq28

CD	Description	Ligand	Size	Location
CD172a	Signal inhibitory regulatory protein family member, SIRP-1α, MyD-1		65–110	20p13
CD172b	SIRPβ		50	20p13
CD172g	SIRPγ		45–50	20p13
CD173	Blood group H type 2		170	20p13
CD174	Lewis Y blood group, LeY, fucosyltransferase 3		170	19p13.3
CD175	Tn antigen (T-antigen novelle)			
CD175s	Sialyl-Tn (s-Tn)			
CD176	Thomsen–Friedenreich antigen, TF antigen		120–198	
CD177	NB1, HNA-2a		49–55/56–64	
CD178	Fas ligand, CD95 ligand	DcR3, CD95 (Fas)	40/40	1q23
CD179a	V pre beta, VpreB	CD179b, Ig μ heavy chain	16–18/—	22q11.22
CD179b	Lambda 5, λ5, 14.1, IGL5	CD179a, Ig μ heavy chain	22/	22q11.23
CD180	RP105, Bgp95, Ly-64	MD-1	95–105/95–105	5q12
CD181	(Formerly CD128A) CXCR1, IL-8Rα	IL-8	39	2q35
CD182	(Formerly CD128B) CXCR2 chemokine receptor, IL-8Rβ		40	2q33-q36
CD183	CXCR3, G protein-coupled receptor 9, GPR9	IP10, Mig, I-TAC	40	Xq13
CD184	CXCR4, NPY3R, fusin	SDF-1, viral MIP-2	45	2q21
CD185	CXCR5		45	11q23.3
CD186[a]	CXCR6		40	3p21
CD191	CCR1 chemokine receptor	MIP-1a, RANTES, MCP-3, MIP-5	39	3p21
CD192	CCR2	MCPs	40	3p21

(continued)

TABLE 11.2   (continued)
CD Table (2007) Alternate Names, Ligands and Molecular Information

CD	Alternative Name	Ligands and Associated Molecules	Molecular Weight (kDa) Unreduced/Reduced	Gene Locus
CD193	CCR3	MIP-1, RANTES, TARC, MCP-1	45	3p21.3
CD194b	CCR4			3p24
CD195	CCR5	MIP-1a,1b,2, RANTES	62, 42	3p21
CD196	CCR6	MIP-3a	45	6q27
CD197	CCR7 (was CDw197)	MIP-3b, SLC (6Ckine)	45	17q12-q21.2
CDw197	CCR7, EBI1, BLR2		90	17q12-q21.2
CDw198	CCR8		50	3p22
CDw199	CCR9		43	3p21.3
CD200	MRC OX2		40–45	3q12-q13
CD201	Endothelial protein C receptor, EPCR	Protein C	49/25	20q11.2
CD202b	TEK/Tie2	Angiopoietin-1,2, and 4	140	9p21
CD203c	PDNP3, B10, PDIβ, E-NPP3	cAMP, NAD, nucleoside phosphates	270/130,150	6q22
CD204	Macrophage scavenger receptor, MSR, SRA	LDL, β-amyloid fibrils	220	8p22
CD205	DEC-205		198	2q24
CD206	Macrophage mannose receptor, MMR	Sialoadhesins and CD45	162–175/—	10p13
CD207	Langerin			2p13
CD208	DC-LAMP	LAMP		3q26.3-q27
CD209	DC-SIGN	ICAM-3, HIV gp120	44	19p13
CDw210	IL-10 receptor, CK	IL-10	90	11q23.3, 21q22.11
CD212	IL-12 receptor beta chain, CK	IL-12	—/110	19p13.1

CD	Other names	Ligands/Associated	MW	Chromosome
CD213a1	CD213a1, IL-13 receptor alpha-1	IL-13		xq24
CD213a2	CD213a2, IL-13 receptor alpha-2			
CD217[a]	IL-17 receptor, CK	IL-17		22q11.1
CD218a[a]	IL-18 receptor alpha, IL18Rα	IL18α	70	
CD218b[a]	IL18Rβ	IL18β	70	
CD220	Insulin receptor	Insulin		19p13.3
CD221	IGF1 receptor, type 1 IGF receptor	IGFI, IGFII		15q25-26
CD222	Mannose-6-phosphate receptor, M6P-R, insulin-like growth factor II receptor, IGFII-R	Plasminogen, M6P, IGFII, LIF, thyroglobulin, cathapsin B,D,L, TGFβ	250/300	6q26
CD223	Lymphocyte activation gene-3, LAG-3	MHC class-II antigen	70	12p13
CD224	Gamma glutamyl transferase, GGT	GSH	27-68	22q11.23
CD225	Leu-13, interferon-induced transmembrane protein	IFNγ	17	11p15.5
CD226	DNAM-1, PTA1	LFA-1	65	18q22.3
CD227	MUC1, PUM, PEM, EMA	CD54, CD169, selectins, Grb2, β-catenin, GSK-3b	300-700/—	1q21
CD228	p97, gp95, Melanotransferrin, MT		97	3q28-29
CD229	Ly-9	SAP protein	100	1q22
CD230	Prion protein, PrPI, PrP(sc) abnormal form	CD56, NCAM	33-37	20pter-p12
CD231	TALLA-1, A15, TM4SF2		150/30-45	Xq11.4
CD232	VESP-R	CD108, viral semaphorin	200	12q23.3
CD233	Band 3, anion exchanger 1, AE1, SLC4A1, Diego blood group antigen	Glycophorin A, ankyrin, hemoglobin	95-100/95-110	17q12-q21

(continued)

TABLE 11.2 (continued)
CD Table (2007) Alternate Names, Ligands and Molecular Information

CD	Alternative Name	Ligands and Associated Molecules	Molecular Weight (kDa) Unreduced/Reduced	Gene Locus
CD234	DARC, Fy-glycoprotein, Duffy blood group antigen	IL-8, MGSA, RANTES, MCP-1	35–43	1q22-23
CD235a	Glycophorin A			4q28-q31
CD235b	Glycophorin B			4q28-q31
CD236	Glycophorin C and D		30–40	2q14-q21
CD236R	Glycophorin C, GYPC		40	2q14-q21
CD238	Kell blood group antigen	Endothelin-3	93	7q33
CD239	Lu/B-CAM, lutheran glycoprotein	Laminin	78–85	19q13.2
CD240CE	Rh blood group system, CD240CE Rh30CE, CD240D		30	1p34.3-p36.1
CD240C	RH30D, CD240DCE Rh30D/CE cross-reactive antigen			
CD240DCE				
CD241	RhAG, Rh50, Rh-associated antigen		50	6p11-p21.1
CD242	ICAM-4, Landsteiner–Wiener blood group antigens, LW blood group	LFA-1, mac-1, VLA-4	37–43	19p13.3
CD243	Multidrug resistance protein-1, MDR-1, P-glycoprotein, pgp170		180	7q21.1
CD244	2B4, P38, NAIL	CD48	70/70	1q23.3
CD245	p220/240, DY12, DY35	Lymphocyte receptor	220–250	
CD246	Anaplastic lymphoma kinase, ALK, p80, Ki-1	Pleiotrophin	200	2p23
CD247	T-cell receptor zeta chain, CD3 zeta chain			2p23

CD	Description	Other names		Chromosome
CD248	TEM1, endosialin	CD164 sialomucin-like 1	175	11q13
CD249	Aminopeptidase A	BP-1, gp160	160	4q25
CD252	OX40L, TNF(ligand) superfamily, member 4	CD134 (OX40)	34	1q25
CD253	TRAIL, TNF (ligand) superfamily, member 10	Apo2	33–34	3q26
CD254	TRANCE, RANKL, TNF (ligand) superfamily, member 11	RANK	35	13q14
CD256	APRIL, TALL2, TNF (ligand) superfamily, member 13	BCMA, TACI	16	17p13.1
CD257	BLYS, TALL1, TNF (ligand) superfamily, member 13b	BCMA, TACI, BAFF-R	45	13q32-34
CD258	LIGHT, TNF (ligand) superfamily, member 14	HVEM	28	19p13.3
CD261	TRAIL-R1, TNF-R superfamily, member 10a	TRAIL	57	8p21
CD262	TRAIL-R2, TNF-R superfamily, member 10b	TRAIL	60	8p22-p21
CD263	TRAIL-R3, TNF-R superfamily, member 10c	TRAIL	65	8p22-p21
CD264	TRAIL-R4, TNF-R superfamily, member 10d	TRAIL	35	8p21
CD265	TRANCE-R, TNF-R superfamily, member 11a	TRANCE	97	18q22.1
CD266	TWEAK-R, TNF-R superfamily, member 12A	TWEAK	14	16p13.3
CD267	TACI, TNF-R superfamily, member 13B	TALL1, BLYS, BAFF	32	17p11.2

(continued)

TABLE 11.2 (continued)
CD Table (2007) Alternate Names, Ligands and Molecular Information

CD	Alternative Name	Ligands and Associated Molecules	Molecular Weight (kDa) Unreduced/Reduced	Gene Locus
CD268	BAFF-R, TNF-R superfamily, member 13C	BAFF	25	22q13.1-q13.3
CD269	BCMA, TNF-R superfamily, member 17	TALL1, BLYS, BAFF	27	16p13.1
CD271	NGFR, p75, TNF-R superfamily, member 16	NGF	75	17q21-q22
CD272	BTLA	B7H4		3q13.2
CD273	B7DC, PDL2	PD2	25	9p24.2
CD274	B7H1, PDL1	PD1	40	9p24.2
CD275	B7H2, ICOSL	ICOS	60	21q22.3
CD276	B7H3		40–45	15q23-q24
CD277	BT3.1, B7 family: butyrophilin 3		56	6p22.1
CD278	ICOS	B7-H2	56	2q33
CD279	PD1	PDL1	55	2q37.3
CD280	TEM22, ENDO180	uPARAP	180	17q23.3
CD281	TLR1, toll-like receptor 1	Bacterial lipoprotein	90	4p14
CD282	TLR2, toll-like receptor 2	Peptidoglycan	85	4q32
CD283	TLR3, toll-like receptor 3	dsRNA	100	4q35
CD284	TLR4, toll-like receptor 4	LPS	85	9q32-q33
CD286[b]	TLR6	TLR2, bacterial lipoproteins		
CD288[b]	TLR8	PAMPs		
CD289[b]	TLR9, toll-like receptor 9	CpG oligonucleotides	115–120	3p21.3
CD290[b]	TLR10	PAMPs		
CD292	BMPR1A		50–58	10q22.3
CDw293	BMPR1B		50–58	4q22-q24

CD	Molecule	Other names	MW (kDa)	Chromosome
CD294	CRTH2, PGRD2, G protein-coupled receptor 44		55–70	11q12-q13.3
CD295	LeptinR, LEPR	Leptin	130–150	1p31
CD296	ART1, ADP-ribosyltransferase 1		37	11p15
CD297	ART4, ADP-ribosyltransferase 4; Dombrock blood group glycoprotein		38	12q13.2-q13.3
CD298	ATP1B3, Na+ K+ ATPase $\beta 3$ subunit		50–60	3q23
CD299	DCSIGN-related, L-SIGN	CD209 antigen-like	45	19p13
CD300a	CMRF-35H		60	17q25.1
CD300c	CMRF-35A			17q25.1
CD300e	CMRF-35L1			
CD301	MGL1, CLECSF14		38	17p13.1
CD302	DCL1		30	2q24.2
CD303	BDCA2		38	12p13.2-p12.3
CD304	BDCA4	Neuropilin 1	140	10p12
CD305	LAIR1		40	19q13.4
CD306	LAIR2			19q13.4
CD307	IRTA2		100	1q21
CD309	VEGFR2, KDR, Flk1	VEGF	230	4q11-q12
CD312	EMR2		90	19p13.1
CD314	NKG2D	MIC A (MHC)	42	12p13.2-p12.3
CD315	CD9P1		135	1p13.1
CD316	EWI2		63	1q23.1
CD317	BST2		29–33	19p13.2
CD318	CDCP1		135	3p21.31

(continued)

TABLE 11.2   (continued)
CD Table (2007) Alternate Names, Ligands and Molecular Information

CD	Alternative Name	Ligands and Associated Molecules	Molecular Weight (kDa) Unreduced/Reduced	Gene Locus
CD319	CRACC, SLAMF7		66	1q23.1-q24.1
CD320	8D6A	8D6 antigen; FDC		19p13.3-p13.2
CD321	JAM1	F11 receptor	32–35	1q21.2-q21.3
CD322	JAM2		45	21q21.2
CD324	E-cadherin		120	16q22.1
CD325[a]	N-cadherin		140	18q11.2
CD326	Ep-CAM		40	2p21
CD327[a]	Siglec6, CD33L	Silylated glycans		19q13.3
CD328[a]	Siglec7	Silylated glycans	75	19q13.3
CD329[a]	Siglec9	Silylated glycans		19q13.4
CD331	FGFR1	FGF	130	8p11.2-p11.1
CD332	FGFR2	FGF	115–135	10q26
CD333	FGFR3	FGF	115–135	4p16.3

CD	Alternative names	Ligand/description	MW	Location
CD334	FGFR4	FGF	110	5q35.1-qter
CD335	NKp46, NCR1, Ly-94	HA, CD3z	46	19q13.42
CD336	NKp44, NCR2, Ly-95	DAP12	44	6p21.1
CD337	NKp30, NCR3	Viral proteins	30	6p21.3
CD338[a]	ABCG2, breast cancer resistance protein, BCRP1		72	4q22
CD339	Jagged-1, JAG1	Notch-1	150	20p12.1-p11.23
CD340[b]	HER-2 (neu)			17q11.2-q12;17q21.1
CD344[b]	Frizzled-4	Wingless type MMTV integration site family of signaling proteins		11q14.2
CD349[b]	Frizzled-9	wnt signaling proteins		7q11.23
CD350[b]	Frizzled-10	Wingless type MMTV integration site family of signaling proteins		12q24.33

[a] Provisional CD(w) not confirmed.

[b] New CD.

Source: Compiled from materials provided by BD Biosciences (San Diego, California), by BioLegend (San Diego, California); the HCDM, www.hcdm.org; HLDA, www.hlda8.org websites; and the recent publication by Zola, H., *J. Immunol. Meth.*, 319, 1, 2007. With permission.

TABLE 11.3
CD Table with Cell-Associated Expression, Function Associations, and Intracellular Interactions

CD Antigen	Cellular Expression	Functions	Intracellular Interactions
CD1a	Cortical thy-c, Langerhans, DC	Nonpeptide Ag presentation	
CD1b	Cortical thy-c, Langerhans, DC	Nonpeptide Ag presentation	
CD1c	Cortical thy-c, Langerhans, DC, B sub	Nonpeptide Ag presentation	
CD1d	Intestinal epi, B sub, DC	Nonpeptide Ag presentation	
CD1e	DC (intracellular)	Nonpeptide Ag presentation	
CD2	T, thy-c, NK, B sub, mono sub	Adh, T activ	Fyn, Lck, SH3KBP1, Sp1, MAD
CD3γ	Mature T, different levels on thy-c	Signaling T activ, regulates TCR expression	
CD3δ	Mature T, different levels on thy-c	Signaling T activ, regulates TCR expression	
CD3ε	Mature T, different levels on thy-c	Signaling T activ, regulates TCR expression	Syk, ZAP-70, Lck, SHC, Grb4, PI3Kα, NCK1
CD4	Thy-c sub, T helper/inducer, Treg, mono/mac	T activ, thymic differen, HIV-R	Lck
CD5	Thy-c, T, B sub, B-CLL	Regulates T–B interaction	Fyn, Lck, ZAP-70, PKCα, PKCβ1, PKCγ
CD6	Thy-c, T, B sub, neuron sub	Thy-c dev, T activ	
CD7	Thy-c, T, NK, myeloid progenitor	T costim	PI3Kα, PI4Kα
CD8a	Thy-c sub, cytotoxic T, NK, DC sub	Coreceptor for MHC class I	Lck, KAT
CD8b	Thy-c sub, cytotoxic T	Coreceptor for MHC class I	
CD9	Pt, pre-B, eosino, baso, act T, endo, epi, stem	Adh and migration, pt activ	PKCα
CD10	B and T precursors, fibro, neutro	Peptidase, regulates B growth	Lyn, SHC, PI3Kα, PI3Kβ
CD11a	All leuko	Adh and costim	RANBP9
CD11b	Gran, mono, NK, T and B sub, DC	Adh, chemotaxis, apoptosis	IRAK1
CD11c	Mono/mac, NK, gran, T and B sub, DC	Adh	
CDw12	Mono, gran, pt, NK	Unknown	
CD13	Gran, mono and their precursors, endo, epi, DC	Aminopeptidase N, adh, coronavirus R	

TABLE 11.3   (continued)
CD Table with Cell-Associated Expression, Function Associations,
and Intracellular Interactions

CD Antigen	Cellular Expression	Functions	Intracellular Interactions
CD14	Mono, mac, Langerhans, gran (low)	R for complex of LPS and LBP, innate immune response	
CD15	Gran, transient in brain	Adh, gran activ	
CD15s	Gran, mono, endo, memory helper T, act T and B, NK, HEV, endo	Adh	
CD15u	Gran, mono, T and B sub, NK, endo	Adh	
CD15su	Gran, mono, T and B sub, NK, endo	Adh	
CD16	Neutro, NK, act mono, mac, DC	Low-affinity Fcγ R, mediates phago-cytosis and ADCC	Lck, ZAP-70, SHC
CD16b	Neutro	Phagocytosis, ADCC	
CD17	Mono, pt, B sub, gran, DC, T	Metabolism, angiogenesis, apoptosis	
CD18	Leuko	Adh	Syk, FAK, ILK, RACK1, RANBP9, PKCα, PKCδ, PKCε, PYK2
CD19	B (not on plasma), FDC	BCR coreceptor, signaling	Fyn, Lyn, Syk, BTK, VAV1, VAV2
CD20	B, T sub	B activ and prolif	Fyn, CK2A1, CK2A2
CD21	Mature B, FDC, T sub	Signaling	p53
CD22	B	Adh, signaling	Lyn, Syk, SHIP1, SLP76, Grb2, PI3Kα, PLCγ1, SHP1
CD23	B (upreg on activ), act mac, eosino, FDC, pt, intestinal epi	Low-affinity R for IgE, allergic response, activ	Fyn
CD24	B, gran, epi, mono, T sub	Cell prolif and differen	Fgr, Lyn
CD25	Act T, B, and mono; DC sub, Treg	IL-2R α chain, w/β and γ chains to form high-affinity IL-2R	STAT3, NFκB1, STAT5B
CD26	Mature thy-c, T (upreg on activ), B sub, NK, mac, epi	Exoprotease, HIV pathogenesis, costim	
CD27	T, medullary thy-c, B sub, NK	Costim	TRAF2, TRAF3, TRAF5, Siva
CD28	Most T, thy-c, plasma, NK	Costim	ITK, Grb2, PI3Kα, PI3Kβ, PI3Kγ, PLCγ1

(continued)

TABLE 11.3   (continued)
CD Table with Cell-Associated Expression, Function Associations,
and Intracellular Interactions

CD Antigen	Cellular Expression	Functions	Intracellular Interactions
CD29	T, B, mono, gran (low), pt, mast, fibro, endo, NK	W/integrin α subunits, adh, activ, embryogenesis	PKCα, PKCε, RACK1, FAK, 14-3-3β, ILK, RhoGAP5, PIK4α
CD30	Act T, act B, mono, act NK, Reed–Sternberg cells	Lympho prolif and death	ALK, TRAF1, TRAF2, TRAF3, TRAF5
CD31	Mono, pt, gran, endo, lympho sub	Adh	Lck, Fyn, c-Src, Hck, c-Yes, Csk
CD32	B, mono, gran, DC, pt, endo	B dev and activ, phagocytosis and mediators release	Fyn, Hck, Lyn, Syk, LAT, BLK, SHC
CD33	Mono, gran, mast, myeloid progenitors	Lectin activity for sugar chains containing sialic acid, adh	c-Src, SHP1, SHP2
CD34	HSC and progenitors, endo	Adh	PKCδ, CRKL
CD35	Eryth, B, mono, neutro, eosino, FDC, T sub	Adh, phagocytosis	
CD36	Pt, mono/mac, endo, erythroid precursors	Scavenger R, adh and phagocytosis	Fyn, Lyn, c-Yes, c-Src
CD37	Mature B, low on T, gran, mono, DC	Adh, signaling	
CD38	Variable levels on majority of hemato and some nonhemato cells, high on plasma	Cell activ, prolif and adh	Lck
CD39	Mac, Langerhans, DC, act B, NK, microglia, endo	ATP and ADP degradation, modulates pt activ, immune response	
CD40	B, mono/mac, FDC, endo, fibro, keratinocytes	Costim, differen and isotype-switching	JAK3, PI3Kα, TRAF1-3, TRAF5-6, Ku80
CD41	Pt, mega	Pt activ and aggregation	
CD42a	Pt, mega	Pt adh and activ	
CD42b	Pt, mega	Pt adh and activ	Grb2
CD42c	Pt, mega	Pt adh and activ	14-3-3 ζ, PKACA
CD42d	Pt, mega	Pt adh and activ	14-3-3 ζ
CD43	Leuko, except resting B, pt (low)	Adh and antiadh	Fyn
CD44	Hemato and nonhemato cells, except pt	Leuko rolling, homing, and aggregation	Csk, Fyn, Lck, Rho, PKN

TABLE 11.3   (continued)
CD Table with Cell-Associated Expression, Function Associations,
and Intracellular Interactions

CD Antigen	Cellular Expression	Functions	Intracellular Interactions
CD44R	Heterogeneous for different isoforms, constitutively expressed on epi, mono, upreg on act leuko	Leuko rolling, homing, and aggregation	
CD45	Hemato cells, except eryth and pt	Activ, signaling	Fyn, Lck, Lyn, RasGAP, SHP1, SLP76, Grb2
CD45RA	B, T sub (naive T), mono, medullary thy-c	Activ, signaling	
CD45RB	T sub, B, mono, mac, gran, DC, NK	Activ, signaling	
CD45RC	B, NK, CD8$^+$ T, sub of CD4$^+$ T, medullary thy-c, mono, DC	Activ, signaling	
CD45RO	Act T and memory T, B sub, act mono, mac, gran, cortical thy-c, DC sub	Activ, signaling	
CD46	Leuko, pt, endo, epi, placental trophoblasts, sperm and variety of tumor cells	Cofactor for factor I, C′ activ, fertilization, R for measles virus	c-Yes, c-Src
CD47	Hemato cells, epi, endo, fibro, brain, mesenchymal cells	Adh	FAK
CD48	Leuko	Adh, costim	Lck
CD49a	Act T, mono, melanoma cells, endo	Adh, embryo dev	
CD49b	Pt, B, mono, act T, epi, endo, NK sub, mega	Adh, pt aggregation, R for echovirus 1	
CD49c	Most adh cell lines, low on B and T	Adh, signaling	
CD49d	T, B, thy-c, mono, eosino, baso, NK, mast, DC, erythroblastic precursors	Adh, cell migration, homing and activ	PRKACA, HIC5
CD49e	Thy-c, T, mono, pt, early and act B, endo, epi	Adh, cell survival and apoptosis	RhoGAP5
CD49f	Memory T, thy-c, mono, pt, mega, epi, endo, cytotrophoblasts	Adh, cell migration, embryogenesis	Fyn, SHC, PKCδ, Grb2
CD50	Leuko, thy-c, Langerhans, endo	Adh and costim	PKCθ
CD51	Pt, act T, endo, osteoblasts, melanoma cells, mega	Adh, signal transduction	Fyn, FAK

(continued)

TABLE 11.3   (continued)
CD Table with Cell-Associated Expression, Function Associations,
and Intracellular Interactions

CD Antigen	Cellular Expression	Functions	Intracellular Interactions
CD52	Thy-c, lympho, mono/mac, epi, sperm cells, mast cells	Costim, antibodies are useful for lysis of target cells	
CD53	Leuko, DC, osteoblasts, osteoclasts	Signal transduction	
CD54	Endo, epi, mono, low on resting lympho (upreg on activ)	Extravasation of leuko from blood vessels, regulates T activ	
CD55	Hemato and nonhemato cells	C' activ, ligand or protective molecule in fertilization, signal transduction	Fyn, Lck
CD56	Neural tissue, NK, T sub, small-cell lung carcinomas	Homophilic and heterophilic adh	Fyn, FAK
CD57	NK and T sub, some B cell lines	Adh	
CD58	Leuko, eryth, epi, endo, fibro	Adh, costim	
CD59	Hemato and nonhemato cells	Prevents C' polymerization, protects cells from C'-mediated lysis	c-Src
CD60a	T sub, thy-c, melanocytes, glial cells, pt, gran	Apoptosis, costim	
CD60b	T sub, act B, melanomas	Costim	
CD60c	T sub	Activ	
CD61	w/CD41 on pt, w/CD51 on mac, endo, pt, fibro, osteoclasts, mast	Mediates cell adh to diverse matrix proteins	c-Src, SHC1, AKT1, talin, FAK, paxillin
CD62E	Act endo	Leuko rolling, tumor cell adh, angiogenesis	PLCγ1, SHP2, FAK, paxillin
CD62L	B, T sub, mono, gran, NK, thy-c	Leuko rolling and homing	Grb2
CD62P	Act pt, endo	Leuko tethering and rolling	
CD63	Act pt, mono, mac, degranulated neutro, fibro, osteoclasts, act baso	Regulates cell motility	PI4Kα
CD64	Mono, mac, DC, IFN-γ, or G-CSF act gran	High-affinity R for IgG, phagocytosis, Ag capture, and ADCC	Hck, Syk, CRKL, LAT

TABLE 11.3   (continued)
CD Table with Cell-Associated Expression, Function Associations,
and Intracellular Interactions

CD Antigen	Cellular Expression	Functions	Intracellular Interactions
CD65	Gran and some mono, myeloid leukemia cells	Unknown	
CD65s	Gran, mono, myeloid leukemia cells	Phagocytosis	
CD66a	Gran, epi, colon, liver, hemato tissues	Homophilic and heterophilic adh, neutro activ	c-Src, SHP2, SHP1, MAP3K10, SHC, paxillin
CD66b	Gran	Adh, neutro activ	
CD66c	Neutro, epi, colon carcinoma	Adh, neutro activ	
CD66d	Neutro	Neutro activ, phagocytosis	c-Src, CKI-alpha-like
CD66e	Adult colon epi, colon cancer	Homophilic and heterophilic adh	
CD66f	Epi, placental syncytiotrophoblasts, fetal liver	Immune regulation and protection of fetus from maternal immune system	
CD68	Mac, neutro, baso, DC, myeloid progenitors, act mono	Phagocytosis	
CD69	Act leuko, NK, thy-c sub, pt, Langerhans	Signaling, costim	
CD70	Act B, act T	Costim	
CD71	Proliferating cells, reticulocytes, erythroid precursors	Transferrin R, iron uptake	Rab5B, TCRζ
CD72	B (except plasma), mac, FDC, T sub	B activ and prolif	Grb2, SHP1, BLNK
CD73	T and B sub, FDC, epi, endo	Dephosphorylation, costim, adh	β-Actin, fibronectin 1, laminin A
CD74	B, act T, mac, Langerhans, DC, act endo and epi	Intracellular sorting of MHC class II, B activ	
CD75	B and T sub, eryth	Adh	
CD75s	Majority B, T sub, endo, epi sub	Adh	
CD77	Germinal center B, high expression on Burkitt's lymphomas	Apoptosis	
CD79a	B	Subunit of BCR complex, signaling	Lck, Fyn, Lyn, Syk, SHP1, BLK

(continued)

TABLE 11.3   (continued)
CD Table with Cell-Associated Expression, Function Associations,
and Intracellular Interactions

CD Antigen	Cellular Expression	Functions	Intracellular Interactions
CD79b	B	Subunit of BCR complex, signaling	Lck, Fyn, Lyn, FAK, Syk, ZAP-70, SHP1, BLK
CD80	Act B, act T, mac, DC	Costim	
CD81	T, B, NK, mono, thy-c, DC, endo, fibro	Activ, costim, differen	SHC
CD82	Leuko, upreg on activ, pt, epi	Costim, adh, tumor metasis	
CD83	Mature DC, act T, act B, Langerhans	Costim	
CD84	Mature B, T sub, mono/mac, pt, thy-c	Adh, activ	EAT2, SAP
CD85a	Mono/mac, gran, DC, T sub	Inhibits NK cytotoxicity	
CD85b	Mono, DC, B, NK, T sub	Activates NK cytotoxicity	
CD85c	Mono, DC, B, NK, T sub	Activates NK cytotoxicity	
CD85d	Mono, DC, B, NK, T sub	Inhibits NK cytotoxicity	SHP1
CD85e	Mono, DC, B, NK, T sub	Activates NK cytotoxicity	
CD85f	Mono, DC, B, NK, T sub	Activates NK cytotoxicity	
CD85g	Mono, DC, B, NK, T sub	Activates NK cytotoxicity	
CD85h	Mono, DC, B, NK, T sub, gran	Activates NK cytotoxicity	
CD85i	Mono, DC, T sub	Activates NK cytotoxicity	
CD85j	Lympho, mono/mac, DC	Inhibits NK cytotoxicity	
CD85k	Gran, mono/mac, DC	Inhibits NK cytotoxicity	SHP1, SHP2
CD85l	NK, T sub, mono/mac, DC, B		
CD85m	T sub, mono/mac, DC, B		
CD86	Mono, act B, act T, DC, endo	Costim of T activ and prolif	
CD87	Gran, mono, NK, T, endo, fibro, hepato	Cell chemotaxis, adh	Fyn, Hck, JAK1, TYK2
CD88	Gran, mono, DC, astrocytes	Gran activ	
CD89	Mono/mac, gran	Phagocytosis, degranulation, respiratory burst	Lyn

TABLE 11.3   (continued)
CD Table with Cell-Associated Expression, Function Associations, and Intracellular Interactions

CD Antigen	Cellular Expression	Functions	Intracellular Interactions
CD90	HSC, neurons, fibro, stromal cells, HEV endo	May inhibit HSC and neuron differen, costim of lympho	Fyn, Lck
CD91	Mono, mac, neurons, fibro	Metabolism, phagocytosis, Ag presentation	c-Src, RAP, JIP, SHC, PRKACA
CD92	Neutro, mono, lympho, endo, epi, fibro, DC	Choline transporter	
CD93	Mono, gran, endo	Phagocytosis, adh	RANBP1, ARHGAP15
CD94	NK, T sub	CD94/NKG2A inhibits NK function, CD94/NKG2C activates NK	DAP12
CD95	Mono, neutro, lympho (upreg on activ), fibro	Induces apoptosis	Fyn, Lck, FADD, Daxx, RIP, FAF1, PKCα, SHP1
CD96	NK and T (upreg on activ)	Adh	
CD97	Gran, mono, low on lympho (upreg on activ), mac, DC	Neutro migration, adh	
CD98	Mono, lympho and NK (upreg on activ), gran	Activ, adh	
CD99	Lympho, NK, mono, gran, endo, epi, some tumor cells	Leuko migration, activ, adh	
CD99R	T, NK, myeloid cells	Isoform of CD99	
CD100	Leuko, oligodendrocytes	Mono migration, T, B activ, T/B and T/DC interaction	
CD101	Mono, gran, DC, act T, Langerhans	T activ and prolif	
CD102	Lympho, mono, pt, endo	Adh, costim, lympho recirculation	Moesin
CD103	IEL, some peripheral blood lympho, act lympho	Lympho retention, activ	
CD104	Epi, endo, Schwann cells, keratinocytes	Cell adh, migration, tumor metastasis	Fyn, c-Yes, FAK, Grb2, PKCα, PKCδ, 14-3-3β, 14-3-3τ
CD105	Endo, mesenchymal stem, erythroid precursors, act mono, mac	Angiogenesis, modulates cellular response to TGF-β1	
CD106	Act endo, FDC, mesenchymal stem	Leuko adh, transmigration and costim	Moesin, ezrin

(continued)

TABLE 11.3   (continued)
CD Table with Cell-Associated Expression, Function Associations,
and Intracellular Interactions

CD Antigen	Cellular Expression	Functions	Intracellular Interactions
CD107a	Act pt, act T, act endo, act gran	Possible role in cell adh	
CD107b	Act pt, act T, act endo	Possible role in cell adh	
CDw108	Act T, eryth	Negative regulation of T function, axon growth	
CD109	Act T, act pt, HSC, mesenchymal stem, endo	Negative regulation	
CD110	HSC and progenitors, mega, pt, endo sub	TPO-R, mega dev, hematopoiesis	IRS2, SHC, SHP2, SOCS1, JAK2
CD111	Stem sub, neurons, endo, epi, fibro	Homophilic and intercellular adh, HSV R	AF6, PARD3
CD112	Mono, neutro, sub of CD34$^+$ cells, endo, epi	Adh	AF6
CD113	Epi, testis, placenta, liver	Adh	AF6
CD114	Myeloid progenitor cells, endo, trophoblastic cells	Myeloid cell differen	Lck, Lyn, Hck, Syk, Grb2, SHIP1
CD115	Mono, mac, monocytic progenitors, neurons, osteoclasts	Monocytic cell differen	Fyn, c-Yes, Lyn, Cbl, Grb2, RasGAP, SHIP1, SHP2
CD116	Mono/mac, gran, DC, endo	Myeloid and DC differen	Lyn, IKKα, IKKβ
CD117	HSC and progenitors, mast	Crucial for HSC, gonadal and pigment stem cell growth and dev	Lck, Fyn, Lyn, c-Src, c-Yes, Hck, Tec, BTK
CD118	Mono, fibro, embryonic stem, liver, placenta	LIF R, cell differen, prolif	PLCγ1, SHP1, SHP2, ERK2
CD119	Lympho, NK, mono/mac, gran, endo, epi, fibro	w/IFNγAF-1, involved in host defense and immunopathological process	JAK1, JAK2, SOCS1, STAT1, SHP2
CD120a	Low level on leuko and most nonhemato cells	Cell differen, apoptosis, necrosis; antibacterial, viral, and parasitic infection	FAK, JAK1, JAK2, SHP1, SHP2, STAT1, TRAF1, TRAF2

TABLE 11.3   (continued)
CD Table with Cell-Associated Expression, Function Associations, and Intracellular Interactions

CD Antigen	Cellular Expression	Functions	Intracellular Interactions
CD120b	Leuko and nonhemato cells	Cell differen, apoptosis, necrosis; antibacterial, viral and parasitic infection	CK1, STAT1, TRAF1-3
CD121a	Low level on fibro, lympho, mono/mac, gran, DC, epi, neural cells	w/IL-1R AcP, mediates IL-1 signaling	
CD121b	B, mono/mac, some T, keratinocytes	Mediates negative signaling	
CD122	NK, T, B, mono	IL-2 and IL-15 R β chain, signaling	Lck, JAK1, JAK3, STAT1, STAT3, STAT5A, STAT5B, SOCS1
CD123	Baso, eosino, hemato progenitors, mac, DC, endo, small sub of lympho	IL-3 R α chain, w/CD131	VAV1, Tec, CISH
CD124	Low level on lympho and their progenitors, mono, endo, epi, fibro	w/CD132 or IL-13Rα chain, R for IL-4 and IL-13	SHC, SHP1, SHP2, RACK1, SHIP, JAK1, IRS1, IRS2
CD125	Eosino, baso, act B, mast	w/CD131, R for IL-5	JAK1, JAK2
CD126	Act B and plasma, T, mono, gran, epi, fibro	Binds IL-6, then w/signaling subunit CD130	c-Src, STAT3, WWP1, WWP2
CD127	B precursors, majority of T, thy-c	IL-7 R α chain, w/CD132	Fyn, Lyn, JAK1, PTK2B
CDw129	Mast, mac, act gran, thy-c, erythroid and myeloid progenitors	w/CD132, IL-9 R	14-3-3 ζ, JAK1
CD130	T, act B, plasma, mono, endo	Transducing biological activities of IL-6, IL-11, LIF, CNF and oncostatin M	TYK2, VAV1, SHP1, SHP2, JAK1, JAK3, SOCS3, STAT3
CD131	Mono, gran, early B, HSC	Signaling for IL-3R, IL-5R and GM-CSFR	Lck, Syk, Lyn, Fyn, JAK1, JAK2, STAT1, STAT3
CD132	T, B, NK, mono/mac, gran, DC	Signaling	JAK1, JAK3, SHB, SHC, STAT1, STAT5A
CD133	HSC sub, epi and endo precursors, neural precursors		

(continued)

TABLE 11.3   (continued)
CD Table with Cell-Associated Expression, Function Associations,
and Intracellular Interactions

CD Antigen	Cellular Expression	Functions	Intracellular Interactions
CD134	Act T, Treg	T activ, prolif, differen, and apoptosis; cell adh	TRAF1-5, Siva
CD135	HSC, myelomonocytic, and primitive B progenitors, thy-c sub	R tyrosine kinase, hemato progenitors growth	SHC, NICK1, Grb2, SHP1, FLT3, SOCS1
CD136	Mac, epi, some hemato and carcinoma cell lines	Induction of migration, morphological change and prolif, antiapoptosis	c-Src, c-Yes, PI3Ka, PLCg1, Grb2, 14-3-3 proteins (β, ε, σ, ζ, θ, η), JAK2
CD137	Act T, FDC, mono, act B, epi	Costim	TRAF1-3
CD138	Plasma, pre-B, epi, neural cells, breast cancer cells	Adh, cell growth	CASK
CD139	B, mono, gran, DC, eryth		
CD140a	Fibro, mesenchymal cells, pt, glial cells and chondrocytes	Cell prolif, differen, and survival	PLCγ1, CRK, Grb2, STAT1, STAT3, STAT5A, STAT5B, JAK1
CD140b	Fibro, mesenchymal cells, pt, glial cells and chondrocytes	Cell prolif, differen, and survival	
CD141	Mono, neutro, pt, endo	Initiation of protein C anticoagulant pathway	
CD142	Mono, epi, astrocytes, Schwann cells, endo, smooth muscle	Initiates blood clotting	
CD143	Endo, epi, DC, neurons, fibro, act mac	Angiotensin converting enzyme, controls blood pressure	
CD144	Endo, stem sub	Adh	c-Src, Csk, SHC, SHP2
CDw145	Endo, some stromal cells		
CD146	Endo, melanoma cells, FDC, act T	Homotypic and heterotypic adh	Fyn
CD147	Leuko, eryth, pt, endo	Adh, T activ, embryonic dev	
CD148	Endo, epi, gran, mono, DC, pt, B, act T	Tyrosine phosphatase, adh, angiogenesis	LAT, PLCγ1

TABLE 11.3 (continued)
CD Table with Cell-Associated Expression, Function Associations, and Intracellular Interactions

CD Antigen	Cellular Expression	Functions	Intracellular Interactions
CD150	T (upreg on activ), Treg, B, DC, endo, HSC	Adh, costim, signaling, measles virus infection	Fyn, Fgr, SHP2, SLAM, EAT2
CD151	Endo, mega, pt, epi	Adh, signaling	
CD152	Act T, act B	Negative regulation of T activ	Fyn, Lck, Lyn, STAT5A, STAT5B
CD153	Act T, act mac, act neutro, act B	Costim of T activ	
CD154	Act T, act pt, act mono	Costim	p53
CD155	Mono, mac, some tumor cells	Cell migration and adh, poliovirus infection	
CD156a	Mono, gran, neuron, oligodendrocytes	Adh, metalloproteases	
CD156b	Lympho, mono, gran, DC, endo, epi	Cleavage of TNF-α, TGF-α, NgR, p75NTR	
CD156c	Articular chondrocytes, leuko, brain, tumor cells	Metalloprotease; cell–cell, cell–matrix interaction	
CD157	Gran, mono, B progenitors, endo, T sub	ADP-ribosyl-cyclic ADP-ribose hydrolase, pre-B growth	
CD158a	Most NK, T sub	Inhibits NK cytotoxicity	
CD158b1	Most NK, T sub	Inhibits NK cytotoxicity	
CD158b2	Most NK, T sub	Inhibits NK cytotoxicity	Lck
CD158c	Most NK, T sub		
CD158d	NK, some T	Activates NK cytotoxicity	
CD158e1	NK, some T	Inhibits NK cytotoxicity	
CD158e2	NK, some T	Activates NK cytotoxicity	CDK3
CD158f	NK, some T	Inhibits NK cytotoxicity	SHP1, SHP2
CD158g	NK, some T	Activates NK cytotoxicity	

(continued)

TABLE 11.3   (continued)
CD Table with Cell-Associated Expression, Function Associations,
and Intracellular Interactions

CD Antigen	Cellular Expression	Functions	Intracellular Interactions
CD158h	NK, some T	Activates NK cytotoxicity	
CD158i	NK, some T	Activates NK cytotoxicity	
CD158j	NK, some T	Activates NK cytotoxicity	
CD158k	NK, some T	Inhibits NK cytotoxicity	
CD158z	NK, some T	Inhibits NK cytotoxicity	
CD159a	NK, some T	Negative regulation of NK activ	SHP1, SHP2
CD159c	NK, CD8$^+$ T sub	Activates NK cytotoxicity	
CD160	NK sub, CTL, IEL	Costim	
CD161	Most NK, NK-T, memory T, thy-c	NK cytotoxicity, induces immature thy-c prolif	SHP1
CD162	Mono, gran, most T, stem	Adh, leuko rolling	Syk
CD162R	NK		
CD163	Mono, mac	Endocytosis	PKCα, CSNK2B
CD164	Epi, mono, lympho, stromal cells, hemato progenitors	Adh, HSC homing	
CD165	Lympho sub, mono, immature thy-c, pt, epi	Adh	
CD166	Act T, mono, epi, fibro, neurons, mesenchymal stem/ progenitor cells	Adh, T activ	
CD167a	Epi, DC, inducible in leuko	Collagen R	Grb4, SHP2, PLCγ1, SHC
CD168	Mono, T and thy-c sub, act lympho	Hyaluronic acid R, cell adh	ERK1
CD169	Tissue mac	Adh	
CD170	Mono, mac, neutro, DC	Adh	
CD171	T and B sub, DC, mono, neurons	Adh	CSNK2A1, RANBP9
CD172a	Mono, DC, gran, stem	Adh	SHP1, SHP2, JAK2
CD172b	Mono, gran, DC, brain, kidney, testis	Phagocytosis, cell activ	DAP12
CD172g	Majority of T, act NK, B sub	Cell adh, costim	
CD173	Eryth, HSC sub, pt		

TABLE 11.3   (continued)
CD Table with Cell-Associated Expression, Function Associations,
and Intracellular Interactions

CD Antigen	Cellular Expression	Functions	Intracellular Interactions
CD174	HSC sub, epi		
CD175	HSC sub, epi		
CD175s	Erythroblasts, endo, epi		
CD176	HSC sub, eryth, endo		
CD177	Neutro sub, baso, NK, T sub, mono, endo		
CD178	Act T, testis, DC, tumor cells	Induces apoptosis	Fyn, Lck, FADD, Daxx, FAF1, c-FLIP
CD179a	Pro- and early pre-B	Early B differen	
CD179b	Pro- and early pre-B	Early B differen	
CD180	B sub, mono, DC	LPS recognition and signaling, B activ	
CD181	Neutro, baso, NK, T sub, mono, endo	Neutro chemotaxis and activ, neoangiogenesis	
CD182	Neutro, baso, NK, T sub, mono, endo	Neutro chemotaxis and activ, neoangiogenesis, hematopoiesis	
CD183	T sub, B, NK, mono/mac, proliferating endo	T chemotaxis	
CD184	T and B sub, DC, mono, endo, HSC	Cell migration, hemato progenitor cell homing, HIV-1 entry	FAK, JAK2, JAK3, STAT1, STAT2, STAT3, STAT5B, SOCS1
CD185	B, T sub, act T, neurons	Cell migration	
CD186	T sub (Th1), B sub, NK sub	T recruitment, HIV-1 coreceptor	
CD191	Mono/mac, lympho, DC, stem	Leuko chemotaxis	JAK1, STAT1, STAT3
CD192	Mono, B, act T, DC	Leuko chemotaxis, HIV-1 coreceptor	JAK2
CD193	Eosino, baso, T sub, DC, microglia	Leuko chemotaxis, HIV-1 coreceptor	Fgr, Hck
CD194	T sub, thy-c, skin T, DC	T chemotaxis, T homing to skin	
CD195	Mono, T sub, DC	Lympho chemotaxis, HIV infection	Lck, FAK, JAK1, JAK2, STAT1, STAT3, STAT5B
CD196	Memory T, B, DC, Langerhans	Cell migration, HIV-1 coreceptor	
CD197	T and B sub, DC	T lympho adh, thy-c migration	

(*continued*)

TABLE 11.3   (continued)
CD Table with Cell-Associated Expression, Function Associations,
and Intracellular Interactions

CD Antigen	Cellular Expression	Functions	Intracellular Interactions
CDw198	Mono, T sub, DC, HUVEC	Cell migration, HIV-1 coreceptor	
CDw199	Thy-c, IEL, melanoma cells	Cell migration, HIV-1 coreceptor	
CD200	Thy-c, B, act T, endo, keratinocyte sub	Inhibits myeloid cell function	
CD201	Endo, HSC	Protein C activ	
CD202b	Endo, stem sub	Angiogenesis, hematopoiesis	Fyn, Grb2, Lck, Lyn, TEK, SOCS1, STAT5A, STAT5B
CD203c	Baso (upreg on act), mast, mega, tumor tissue	Clearance of extracellular nucleotides	
CD204	Mac	LDL uptake, host defense	HSP70
CD205	DC, thymic epi, BM stromal, low on T, B, NK, and mono	Endocytosis, Ag presentation	
CD206	Mac, mono, inflammatory dendritic epidermal cells	Endocytosis	
CD207	Langerhans, DC	Ag recognition and uptake	
CD208	Act DC, type II pneumocytes		
CD209	DC sub	Ag endocytosis and degradation, attachment of HIV and some other viruses	
CDw210a	T, B, NK, mono/mac, thy-c (low), act neutro	R for IL-10	JAK1
CDw210b	T, B, NK, mono, DC, liver, neutro	Signaling	JAK1
CD212	Act T, NK, mac	R for IL-12 and IL-23	STAT4
CD213a1	B, mono, mast, fibro, endo	R for IL-13, signaling	TYK2
CD213a2	B, mono, epi	R for IL-13	
CD217	B, NK, fibro, epi, T, mono/mac, gran	R for IL-17	
CD218a	T sub (Th1), B sub, NK, mono, gran, endo, DC	Binds IL-18, signaling	
CD218b	NK, T sub, mono, endo, DC	Signaling	
CD220	Leuko, fibro, endo, epi	R for insulin, metabolism	c-Src, Csk, Cbl, FAK, JAK1, JAK2

TABLE 11.3   (continued)
CD Table with Cell-Associated Expression, Function Associations, and Intracellular Interactions

CD Antigen	Cellular Expression	Functions	Intracellular Interactions
CD221	Leuko and variety of nonhemato cells	Signaling, cell prolif/differen	c-Src, ASK1, CRK, Csk, JAK1, JAK2
CD222	Lympho, mono, gran, fibro, myocytes, embryonic tissue	Activates latent TGF-β, cell adh, migration, angiogenesis	
CD223	Act T, act NK	Negative regulation of T expansion and homeostasis	
CD224	T and B sub, mac, endo, HSC, renal tubular cells, pancreas	Inhibits apoptosis, cellular detoxification and leukotriene biosynthesis	
CD225	Leuko, endo	Lympho activ and dev	
CD226	T and B sub, NK, mono, pt, thy-c, act HUVEC	Costim, adh	Fyn
CD227	Epi, stem sub, mono, act T, DC	Cell adh and signaling	Lck, c-Src, ZAP-70, Grb2, SOS1
CD228	Melanoma cells, epi, brain, skeletal and heart muscle	Fe transport	
CD229	T, B, thy-c, DC	Adh, costim	SH2D1A, EAT2
CD230	Hemato and nonhemato cells, neurons	Prevents cells from apoptosis, stem cell renewal	BIP, Grb2
CD231	T-ALL and neuroblastoma cells, neurons		
CD232	B, mono, gran, NK, act T, DC	Cell differen and migration	
CD233	Eryth, kidney	Anion exchanger	Lyn, Syk
CD234	Eryth, endo, neurons, epi, cerebellum	Chemokine decoy R, malarial parasite R	
CD235a	Eryth	Parasite R, cell aggregation	
CD235ab	Eryth	Parasite R, cell aggregation	
CD235b	Eryth	Parasite R, cell aggregation	
CD236	Eryth, stem sub	Gerbich antigen	
CD236R	Eryth, stem sub	Gerbich antigen	
CD238	Eryth, hemato progenitors	Endothelin-3 converting enzyme	

(continued)

TABLE 11.3 (continued)
CD Table with Cell-Associated Expression, Function Associations, and Intracellular Interactions

CD Antigen	Cellular Expression	Functions	Intracellular Interactions
CD239	Eryth, fibro, inducible in epi	Eryth differen and trafficking	Laminin A5
CD240CE	Eryth		
CD240D	Eryth		
CD241	Eryth	Rh antigen complex with CD47, LW, glycophorin B	
CD242	Eryth	Adh, LW blood group	
CD243	Stem, multi-drug-resistant tumor cells	Ion pump, regulates drug uptake, distribution, and elimination	
CD244	NK, T sub, baso, mono	Stimulates NK activ, costimulates T cells	LAT, EAT2, SH2D1A
CD245	T, B, NK, mono, gran, pt	Signal transduction, costim	
CD246	Some T lymphomas, endo, some neural cells	R for tyrosine kinase, regulates cell growth and apoptosis	ALK, SHC, JAK3, PLCγ1, IRS1
CD247	T, NK	Ag recognition and signal transduction	Lck, Fyn, ZAP-70, Csk, SHP1, JAK3
CD248	Embryonic endo, tumor endo	Angiogenesis	
CD249	Epi, endo	Converts angiotensin II to angiotensin III	
CD252	DC, act B, endo, mast	Costim	TRAF2
CD253	Act T, NK	Apoptosis	
CD254	Act T, stromal cells, osteoclasts	T–B and T–DC interaction, bone dev	c-Src, AKT1, ERK1, ERK2, TRAF6
CD256	Leuko, pancreas, colon	T and B prolif	
CD257	Act mono, DC	T, B growth and dev	
CD258	Act T, act mono	Costim of T cells, induces apoptosis	TRAF2, TRAF3, SMAC
CD261	Act T, some tumor cells	Induces apoptosis	BCL10, caspase 8, caspase 10, FADD, c-FLIP, BTK
CD262	Leuko, heart, placenta, liver, tumor cells	Induces apoptosis	BCL10, caspase 8, caspase 10, FADD, c-FLIP, BTK
CD263	Low level in most tissues, negative in most tumor tissues	Inhibits TRAIL-induced apoptosis	RAP1α

TABLE 11.3   (continued)
CD Table with Cell-Associated Expression, Function Associations,
and Intracellular Interactions

CD Antigen	Cellular Expression	Functions	Intracellular Interactions
CD264	Low level in most tissues, negative in most tumor tissues	Inhibits TRAIL-induced apoptosis	
CD265	DC, act mono	R for TRANCE	c-Src, Cbl, CblB, Grb2, MAP3K7, TRAF1-3, TRAF5-6, TAB2
CD266	Endo, epi, keratinocytes	Regulates apoptosis, prolif	TRAF1-3
CD267	B, myeloma cells	Inhibits B prolif	TRAF2, TRAF5-6
CD268	B, T sub	B survival and maturation, T activ	
CD269	B, plasma cells	Plasma-cell survival	TRAF1-3
CD271	Neurons, Schwann cells, melanocytes, B, mono, keratinocytes	Low-affinity R for NGF, induces apoptosis, embryo-genesis, hair growth	ERK1, ERK2, Grb2, PRKACB, SHC, TRAF2, TRAF4, TRAF6
CD272	T (upreg on act), B act, NK, mac, DC	Inhib of T-cell function	SHP1, SHP2
CD273	DC, act mono	Costim, inhib	
CD274	T, B, NK, DC, mac, epi	Costim, inhib	
CD275	Act mono, mac, DC	Costim	
CD276	DC, act mono, act T, act B, act NK, epi	May play role in costim or inhib	
CD277	T, B, NK, mono, DC, endo	T activ	
CD278	Act T, thy-c sub	T costim	
CD279	Act T, act B, thy-c sub	T tolerance, negative regulation	
CD280	Fibro, endo, mac, osteoclasts, osteocytes, chondrocytes	Remodeling, uptake and degradation of collagen	
CD281	Mono, mac, DC, keratinocytes	Regulates TLR2 function	
CD282	Mono, gran, mac, DC, keratinocytes, epi	Innate immune response to some bacteria and mycoplasma pathogens	RIP2, TOLLIP, PI3Kα, H-Ras, MYD88
CD283	DC, fibro, epi	Innate immune response to viral pathogens	TRIAD3, MYD88, TRAF6, MAP3K7, TAB2

(continued)

TABLE 11.3   (continued)
CD Table with Cell-Associated Expression, Function Associations, and Intracellular Interactions

CD Antigen	Cellular Expression	Functions	Intracellular Interactions
CD284	Mono, mac, endo, epi	Innate immune response to Gram-negative bacteria	Syk, BTK, IRAK2, MYD88, RIP2, TOLLIP, TRIAD3, MAPK8IP1
CD286	Mono, mac, gran, DC, epi	Innate immune response to some bacteria and mycoplasma pathogens	
CD288	Mono, mac, DC, neurons, axons	Antiviral immune response, brain dev, hematopoiesis	
CD289	pDC, B, mono	Innate immune response to bacteria or virus	BTK, TRIAD3, H-Ras, MAPK8IP3
CD290	B, pDC	Coreceptor with TLR2	
CD292	Mesenchymal cells, epi, bone progenitor, neurons, chondrocytes, skeletal muscle, cardiac myocytes	Kinase, hair morphogenesis, antiapoptosis, embryogenesis	TAB1
CDw293	Mesenchymal cells, bone progenitor, chondrocytes, epi, heart, kidney	Kinase, regulates cartilage formation	PAK1, RhoD, SH3KBP1, SOCS6, SMAD6, SMAD7
CD294	Th2 cells, baso, eosino	Regulates immune and inflammatory response, induces Th2 cells, eosino and baso migration	
CD295	Hemato cells, heart, placenta, liver, kidney, pancreas	Regulates fat metabolism, proliferative/ antiapoptotic T cells, and hemato precursors	
CD296	Neutro, heart, skeletal muscle	Transfers ADP-ribose to target proteins, regulates cellular metabolism	
CD297	Dombrock$^+$ RBC, mono, mac, baso, endo, intestine, ovary	Metabolism, Dombrock blood group antigen	

TABLE 11.3   (continued)
CD Table with Cell-Associated Expression, Function Associations,
and Intracellular Interactions

CD Antigen	Cellular Expression	Functions	Intracellular Interactions
CD298	Broad, highly expressed in CNS, testis	Noncatalytic component of ATPase coupling exchange of Na$^+$, K$^+$	
CD299	Endo of liver and lymph nodes	T-cell trafficking, HCV, EBOV, and HIV infection	
CD300a	Mono/mac, neutro, DC, NK, mast, T and B sub	Inhibitory R	
CD300c	Mono/mac, neutro, DC, NK, T and B sub		
CD300e	Mono/mac, DC sub		
CD301	Mac, immature DC	Cell adh, mac migration, cellular recognition	
CD302	Mono/mac, DC		
CD303	Plasmacytoid DC	Ag capture, inhib of interferon α/β production	
CD304	DC, T, neurons, endo	Angiogenesis, DC–T cell interaction, neuritogenesis	
CD305	T, B, NK, DC, mono/mac	Inhibits cellular activ and inflammation	Csk, SHP2
CD306	T, mono	Inhibits cellular activ and inflammation	
CD307	B, high on germinal center light zone B cells	IgR, may play a role in B activ and neoplasia	
CD309	Endo, primitive stem, some tumor cells	Angiogenesis	Fyn, c-Yes, c-Src, Grb2, Grb10, NCK1
CD312	Mac, act mono, DC, liver, lung	Involved in immune and inflammatory response	
CD314	NK, CD8$^+$ T sub, γ/δ T, mac	NK activ	DAP10, DAP12
CD315	Mega, hepato, epi, endo, weak on B and mono		
CD316	T, B, NK, hepato	Inhibits tumor cell metastasis	
CD317	Plasma, lymphoplasmacytoid cells, stromal cells, fibro, pDC	May play a role in pre-B cell growth	

(continued)

TABLE 11.3   (continued)
CD Table with Cell-Associated Expression, Function Associations,
and Intracellular Interactions

CD Antigen	Cellular Expression	Functions	Intracellular Interactions
CD318	HSC, epi, some tumor cells	Adh	c-Src, PKCδ
CD319	NK, most T, act B, mature DC	Activates NK cytotoxicity, adh	
CD320	FDC	Stimulates B growth	
CD321	Epi, endo, leuko, pt, eryth, lung, placenta, kidney	Cell adh, leuko migration, epithelial barrier maintenance, pt activ	PKCα, CSNK2A1, CASK, PTPB
CD322	Endo, HEV in tonsils, heart	Cell adh, lympho homing	
CD324	Epi, keratinocytes, trophoblasts, pt	Cell–cell, cell–matrix adh; tumor suppression; cell growth and differen	GSK3β, CSNK2A1, RICS, IRS1
CD325	Neurons, skeletal and cardiac myocytes, fibro, epi, pancreas, liver	Cell–cell, cell–matrix adh; cell growth and differen	PI3Kα, RICS
CD326	Epi, low on thy-c and T, tumor cells	Inhibits cellular activ and inflammation	
CD327	B, placenta trophoblasts, gran	Adh	
CD328	NK, T sub, mono, gran	Inhibits NK and T activity	Grb2, SHP2, TRAF4
CD329	Mono; sub of NK, B, T, and neutro; hepato, myeloid leukemia	Inhibits immune response	
CD331	Broad, epi, endo, fibro, mesenchymal, cardiac myocytes, fetal tissue	Cell growth, limb dev	CRK, Grb2, Grb4, Grb14, PI3Kα, PI3Kβ, PLCγ1, SOS1
CD332	Brain, liver, prostate, kidney, lung, spinal cord, fetal tissues	Limb induction, craniofacial dev	Lyn, Fyn, Cbl, PLCγ1, PAK1, PTK2B
CD333	Brain, kidney, testis in adult, small intestine in fetus	Limb induction, craniofacial dev	Grb2, PTK2B, STAT1, STAT3
CD334	Liver, kidney, lung, pancreas, lympho, mac	Bone and muscle dev, cancer progression/metastasis	PLCγ1, STAT1, STAT3
CD335	NK	NK activ	TCRζ
CD336	Act NK	NK activ	DAP12
CD337	NK	NK activ	TCRζ
CD338	Liver, kidney, intestine, lung, endo, melanoma, placenta, side population of stem, certain drug-resistant tumors	Absorption and excretion of certain xenobiotics	

TABLE 11.3 (continued)
CD Table with Cell-Associated Expression, Function Associations, and Intracellular Interactions

CD Antigen	Cellular Expression	Functions	Intracellular Interactions
CD339	BM stromal, thymic epi, endo, Schwann cells, keratinocytes, ovary, prostate, pancreas, placenta, heart	Cell-fate decisions in hematopoiesis, cardiovascular dev	OFUT1
CD340	Epi, endo, keratinocytes, HSC sub, fetal mesodermal and extraembryonic tissues, overexpressed on many malignant cells	Cell growth and differen, tumor cell metastasis	c-Src, JAK2, FAK, Grb2, SOS1, SHC
CD344	Epi, endo, mesenchymal, myeloid progenitors, neuronal progenitors, intestinal neurons	Cell prolif and differen, embryonic dev, retinal angiogenesis	DVL2, β-arrestin 2
CD349	Mesenchymal stem, neural precursor, mammary epi, brain; fetal testis, kidney, eye, and pancreas; some tumors	B dev, hippocampal dev, tissue morphogenesis	
CD350	Brain, embyro, kidney, liver, pancreas, placenta, mammary, and lung epi, some tumors	Dev of limb, nervous system	

Note: Act, activated; ADCC, antibody-dependent cellular cytotoxicity; adh, adhesion; Ag, antigen; B, B cells; baso, basophils; BCR, B-cell receptor; BM, bone marrow; CD, cluster of differentiation; CNS, central nervous system; costim, costimulatory; CTL, cytotoxic T lymphocytes; DC, dendritic cells; endo, endothelium; eosino, eosinophils; epi, epithelium; eryth, erythrocytes; FDC, follicular dendritic cells; fibro, fibroblasts; gran, granulocytes; hemato, hematopoietic; hepato, hepatocytes; HSC, hematopoietic stem cells; HSV, herpes simplex virus; HUVEC, human umbilical vein endothelial cells; IEL, intraepithelial lymphocytes; IL, interleukin; INF, interferon; leuko, leukocytes; mac, macrophage; mega, megakaryocytes; mono, monocytes; NK, natural killer cells; pDC, plasmacytoid dendritic cells; pt, platelets; RBC, red blood cells; stem, stem cells; sub, subset; T, T cells; T-ALL, T-cell acute lymphoblastic leukemia; TCR, T-cell receptor; TGF, transforming growth factor; Th1, T helper type 1 cells; Th2, T helper type 2 cells; Thy, thymus; thy-c, thymocytes; TNF, tumor necrosis factor; TPO-R, thymopoietin receptor; upreg, upregulated; HEV, high endothelial venules; ATP, adenosine triphosphate; ADP, adenosine diphosphate; HIV, human immunodeficiency virus; TRAIL, TNF-related apoptosis induced ligand; TRANCE, TNF-related activation-induced cytokine; RANKL, receptor activator for nuclear factor κ B-ligand; MHC, major histocompatibility complex; NGF, nerve growth factor; HCV, Hepatitis-C Virus; HBOV, human bocavirus; Treg, regulatory T-cells; differen, differentiation; activ, activation; dev, developmental; neutro, neutrophil; prolif, proliferation; mast, mast cell; inhib, inhibitory; lympho, lymphocyte; CLL, chronic lymphocyte leukemia; LPS, lipopolysaccharide; LPB, LPS-binding protein.

Source: Compiled from materials provided by BD Biosciences (San Diego, California), by BioLegend (San Diego, California); the HCDM, www.hcdm.org; HLDA, www.hlda8.org websites; and the recent publication by Zola, H., *J. Immunol. Meth.*, 319, 1, 2007. With permission.

Antigens" reviewed by Heddy Zola (personal communication, 2007) and provided with permission by BD Biosciences (San Diego, California), the HCDM and HLDA websites (www.hcdm.org and www.hlda8.org, respectively, which both go to the same "up-to-date" site); the most recent CD poster, including the first HCDM workshop results as compiled by BioLegend (San Diego, California); and the recent publication by Zola et al. [10]. For other interesting articles on early HLDA meetings see Refs. 11 and 12.

ACKNOWLEDGMENT

I wish to thank Heddy Zola for the helpful suggestions and for encouraging me to add this chapter to this handbook.

REFERENCES

1. Bernard, A. and Boumsell, L., The clusters of differentiation (CD) defined by the First International Workshop on Human Leucocyte Differentiation Antigens, *Hum. Immunol.*, 11, 1, 1984.
2. Nomenclature for clusters of differentiation (CD) of antigens defined on human leukocyte populations. IUIS-WHO Nomenclature Subcommittee. *Bull. World Health Organ.* 62; 809–815, 1984.
3. Chorvath, B. and Sedlak, J., Hematopoietic cell differentiation antigens (CD system 1997). Cancer research relevance, *Neoplasma*, 45, 273, 1998.
4. Springer, T.A., The next cluster of differentiation (CD) workshop, *Nature*, 354, 415, 1991.
5. Erber, W.N., Human leucocyte differentiation antigens: review of the CD nomenclature, *Pathology*, 22, 61–69, 1990.
6. Clark, E.A. and Lanier, L.L., Report from Vienna: in search of all surface molecules expressed on human leukocytes, *J. Clin. Immunol.*, 9, 265, 1989.
7. Pinto, A., Gattei, V., Soligo, D., Parravicini, C., Del Vecchio, L. New molecules burst at the leukocyte surface. *A comprehensive review based on the 15th International Workshop on Leukocyte Differentiation Antigens*. Boston, USA, 3–7 November, 1993. *Leukemia*. 8; 347–358, 1994.
8. Mason, D., André, P., Bensussan, A., Buckley, C., Civin, C., Clark, E., de Haas, M., Goyert, S., Hadam, M., Hart, D., Horejsí, V., Meuer, S., Morrissey, J., Schwartz-Albiez, R., Shaw, S., Simmons, D., Uguccioni, M., van der Schoot, E., Vivier, E., Zola, H. CD antigens 2001: aims and results of HLDA Workshops. *Stem Cells.* 19; 556–562, 2001.
9. Zola, H., CD molecules 2005: human cell differentiation molecules, *Blood*, 106, 3123, 2005.
10. Zola, H., CD molecules 2006: human cell differentiation molecules, *J. Immunol. Meth.*, 319, 1, 2007.
11. Bernard, A. and Boumsell, L., Human leukocyte differentiation antigens, *Presse Med.*, 13, 2311, 1984.
12. Jones, D.B., What's new in the lymphocyte phenotype? *Pathol. Res. Pract.*, 186, 309, 1990.

12 Immunologic Diagnosis of Autoimmunity

Noel R. Rose

CONTENTS

12.1 AUTOIMMUNE RESPONSE VERSUS AUTOIMMUNE DISEASE

12.1.1 DEFINITIONS

Autoimmunity is defined as the immune response to antigens of the host itself. This *autoimmune response* can be demonstrated by the presence of circulating autoantibodies or T lymphocytes reactive with host antigens. A great deal of basic research has been dedicated to unraveling the mechanisms responsible for the body's ability to distinguish its own molecules from foreign molecules (see Chapter 1). Never-theless, exceptions to the rules of governing self-/non-self-discrimination are well known. Most autoimmune responses do not result in disease, but when a harmful response occurs, the pathological consequence of the autoimmune response is called *autoimmune disease*. These disorders can affect virtually any site in the body so that their clinical presentation varies widely. In at least 80 diseases, autoimmunity is now recognized as an important cause or contributor. The immunologic diagnosis of autoimmune disease relies mainly on the demonstration of autoantibodies in the patient's serum. This chapter describes the general approach to diagnosis, emphasiz-ing both the uses and abuses of most widely used test procedures.

12.1.2 Natural Autoantibodies

In undertaking an immunologic diagnosis based on the presence of autoantibod-ies, it is essential to recognize that autoantibodies are normally present. Most of these natural autoantibodies are members of the IgM isotype and have relatively low affinity for their corresponding antigen (see Chapter 1). They are polyreactive and, consequently, highly interconnected. It has even been suggested that naturally occurring autoantibodies have a physiological function. They may be involved in removing effete or damaged cell products that enter the bloodstream. Equally plau-sible is the suggestion that naturally occurring autoantibodies may represent an early mechanism of defense against pathogenic microorganisms. For that reason, their high degree of cross-reactivity may provide a substantial advantage in being able to bind a number of invading pathogens. Although of low affinity, IgM autoantibodies are capable of activating the complement cascade, resulting in lysis or opsonization of the pathogens.

The origin of the naturally occurring autoantibodies is still uncertain. It would appear, however, that a relatively large proportion of the B-cell repertoire is devoted to producing self-reactive antibodies. In support of this view, many myeloma proteins bind self-antigens.[1] In addition, hybridomas made from B cells of normal individuals frequently produce monoclonal autoantibodies. Often, these antibodies are directed to the cytoskeletal matrix or similar large proteins, such as laminin, vimentin, fibronectin, actin, myosin, and collagen.[2] This observation, however, may be biased by the fact that such autoantibodies are relatively easy to demonstrate. More extensive studies show that naturally occurring autoantibodies also react with soluble cell products, such as insulin or thyroglobulin, or even intranuclear con-stituents, such as deoxyribonucleic acid (DNA) or topoisomerase.[3,4] Therefore, the presence of these natural autoantibodies sometimes complicates the demonstration of disease-associated autoimmune responses.

12.1.3 Prevalence of Autoantibodies

Although autoantibodies are common in all human sera, their prevalence is associ-ated with age and sex. The presence and titer of autoantibodies generally increase with age.[5] This dichotomy causes a striking immunological paradox, that is, the increase in autoantibodies in the face of a general decrease in immune responses to exogenous antigens.[6] In addition to age, the prevalence of naturally occurring autoantibodies is associated with sex. Natural autoantibodies are more prevalent in women than men. This observation is intriguing because most autoimmune diseases are more frequent in women. The basis of the sex difference in autoimmunity has not been explained. Although elevated estrogen or progesterone levels are frequently cited as the cause for female bias, this hormonal explanation is inadequate, since many autoantibodies continue to rise in women after menopause.[7]

In most test procedures, it is not the presence or absence of an autoantibody that conveys diagnostic significance, but rather its titer and isotype. Most, but not all, disease-associated autoantibodies are present in high titer and are IgG isotype, whereas most, but not all, naturally occurring autoantibodies are low-titered IgMs. Exceptions to these general guidelines are sufficiently common that the immunologist must always be alert to the complicating presence of naturally occurring autoantibodies in the diagnostic situation. For this reason, a great deal of current research is devoted to delineating the precise specificity of naturally occurring autoantibodies and contrasting them with the specificity of disease-associated autoantibodies. In autoimmune thyroid disease, for example, our group has shown that the natural autoantibodies are generally directed to those conserved epitopes on the thyroglobulin molecule that are shared by many different species. In contrast, disease-associated autoantibodies bind primarily the species-specific epitopes of thyroglobulin.[8] We believe that similar instances of defined specificity at the molecular level will improve the accuracy of laboratory diagnosis of autoimmune disease.

12.2   CRITERIA OF AUTOIMMUNE DISEASE

Autoimmune disease has been defined earlier as the pathological consequence of an autoimmune response. The mere presence of autoantibodies is sufficient to establish a diagnosis of autoimmune disease, which requires clinical and additional laboratory evidence before reaching such a diagnosis. Recently, we reviewed the steps necessary to establish a human disease as autoimmune in etiology.[9] The types of evidence can conveniently be considered as direct, indirect, or circumstantial.

12.2.1   DIRECT EVIDENCE

Direct evidence that a human disease is caused by autoimmunity can be obtained in those instances where the pathological injury is due to an autoantibody. Although it is not ethical to reproduce a disease in humans by deliberate serum transfer, nature has provided us with a number of examples of maternal-to-fetal transfer. In this way, it was possible to show that myasthenia gravis is caused by an antibody to the acetylcholine receptor and Graves' disease produced by an antibody to the thyrotropin receptor.[10,11] In other instances, transfer of the patient's serum to experimental animals successfully reproduces the characteristic pathological changes. Pemphigus vulgaris and bullous pemphigoid have been reproduced by the transfer of serum to newborn mice.[12] Sometimes, autoantibodies can produce characteristic changes *in vitro* that mimic disease process, as seen in some forms of hemolytic anemia.[13] Many autoimmune diseases, however, are not caused by the circulating antibody, but rather cellular immunity. Transfer of such diseases is not yet feasible. Some efforts have been made to utilize severe combined immunodeficient (SCID) mice as *in vivo* test tubes, so that both key lymphocytes and the target organ of a possible autoimmune disease can be placed in juxtaposition to produce the characteristic lesions.[14] This approach, however, is still in its infancy.

12.2.2 INDIRECT EVIDENCE

Because of the logistical and ethical restraints involved in assembling direct evidence for the autoimmune etiology of a human disease, most investigations depend on *indirect evidence* gleaned from experimental animals. Two approaches are widely used. The first requires that the disease be reproduced by experimental immunization, and the second utilizes spontaneously occurring genetic models to replicate the human disease. Each approach has its advantages and shortcomings.

Reproduction of a human disease in an experimental animal requires that the requisite antigen be identified in human patients first. In practical terms, this generally means employing the autoantibody to delineate the antigen. An intrinsic problem arises when the antigen recognized by the antibody is not the one responsible for initiating a pathogenic autoimmune response. Despite this pitfall, the approach has proved to be highly successful in defining the pathogenic antigens involved in such diseases as chronic thyroiditis, uveitis, and myasthenia gravis.[15–17] The strategy usually employed is to isolate the corresponding antigen from an animal source, inject it into a syngeneic recipient (often accompanied by a potent adjuvant, such as complete Freund's adjuvant or bacterial lipopolysaccharide), and demonstrate that such immunization results in the production of autoantibodies and the appearance of characteristic lesions. A major problem of this strategy is to identify a susceptible experimental animal. Generally, autoimmune responses, particularly pathogenic ones, are genetically limited and, therefore, a number of different species and strains may need to be tested before successful reproduction of a disease can be accomplished. Mice are used most frequently because of a large number of genetically diverse strains available. Moreover, the experimental disease in an animal may not exactly replicate its human analog. Because human autoimmune diseases often involve several antigens, there is a special problem in reproducing the disease with a single, purified antigen. This limitation applies to almost all of the induced models described to date. Experimental immunization, therefore, rarely gives a complete picture of a human disease.

The alternative to experimental reproduction of a human disease in animals is to seek out a spontaneous model. Outstanding examples are seen in murine models of lupus, such as (NZB × NZW) F1 hybrids, MRL/lpr-lpr, and BXSB.[18] Although none of these mouse models can be regarded as a full analog of human lupus, each has contributed substantially to the understanding of the pathogenesis and genetics of the human disorder. Insulin-dependent diabetes is another human autoimmune disease that has not been reproduced by experimental immunization. Two spontaneous models, however, are presently available in the NOD mouse and the BB/W rat.[19,20] These models have also contributed to our understanding of this disease. Autoimmune thyroiditis is somewhat unique because both an induced form of disease by experimental immunization with thyroglobulin and spontaneous models present in the OS chicken, the BUF rat, and the NOD H2^{h-4} mouse are available for investigation.[21]

In recent years, great strides have been made in developing models of autoimmunity by manipulating the immune system of the experimental animal. The earliest example, some 4 decades old, is a model of autoimmune thyroiditis produced by irradiation and thymectomy of genetically selected strains of rats.[22] Subsequently, it was found that neonatal thymectomy of mice performed on day 3 may lead spontaneously to autoimmune disease.[23] The particular manifestation of disease depended on the genetic background of the animal. These findings led directly to the isolation of a population of lymphocytes derived from the thymus that is preferentially depleted by the 3-day thymectomy. These cells now referred to as regulatory T cells often play a critical role in suppressing or aborting autoimmune disease.[24] Although it appears that a number of cell populations may serve as suppressors depending on the circumstances and methods of measurement, a T-cell population expressing the surface markers CD4 and CD25 as well as the translational factor FOX P3 are present neonatally and represent an important regulatory T-cell population. Indeed, humans with a genetic deficiency in FOX P3 develop multiple autoimmune manifestations.[25]

Other mediators maintain the normal homeostasis of the adaptive immune system. As examples, IL2 and IL10 are important cytokines in controlling the immune response. Genetic depletion of these cytokines may lead to spontaneous autoimmune inflammatory bowel disease in selected strains of mice. Another important genetic tool to initiate spontaneous autoimmune disease in mice is to substitute the human major histocompatibility complex (MHC) for the mouse MHC. This strategy has been useful in creating models of polychondritis[26] and myocarditis.[27] Finally, a number of investigators have inserted foreign antigens in specific organs under the direction of an organ-specific promoter. An autoimmune disease can then be initiated by adoptively transferring T lymphocytes with the corresponding T-cell receptor (TCR). If the foreign antigen inserted in the organ is also expressed in a virus, infection by the virus can initiate the autoimmune disease process. For example, transgenic mice that produce cardiac myocyte-restricted membrane-bound peptide of ovalbumin (OVA) have been derived; adoptive transfer of OVA-specific CD8+ T cells induces myocarditis, but only after infection with ova-expressing vesicular stomatitis virus.[28] A complimentary strategy is to insert the gene for an autologous antigen in the genome of a virus infection by the virus that can then initiate the autoimmune disease. The possible viral infection etiology of multiple sclerosis has been mimicked in a mouse model by inserting the immunodominant myelin peptide into a nonpathogenic variant of Theiler's virus. Infection of the mice with the virus expressing the myelin peptide developed demyelinating disease.[29] Another instructive model has been created by expressing a glycoprotein or nucleoprotein of lymphocytic choriomeningites virus in beta cells of the mouse pancreas under the control of the rat insulin promoter. Subsequent challenge with the virus produced insulitis and hyperglycemia.[30]

Although they are not exact replicas of human disease, the experimentally induced and spontaneous models have been essential for establishing the autoimmune etiology of the disorder. In animals, it is possible not only to carry out antibody transfer experiments, but also to adoptively immunize recipients with T lymphocytes. In this way, it has been possible to show that the disease is caused by the autoimmune response rather than being the consequence of the disease.

12.2.3 CIRCUMSTANTIAL EVIDENCE

In the case of most human diseases classified as autoimmune, we depend primarily on circumstantial evidence to ascertain the etiology as autoimmune. Many auto-immune diseases tend to cluster, either in the same family or in members of the same family. A patient with autoimmune thyroiditis has heightened probability of demonstrating a second or third autoimmune endocrinopathy, such as adrenalitis or diabetes.[31] Relatives of a patient with lupus have a higher than expected prevalence of lupus or related autoimmune diseases, such as rheumatoid arthritis or thyroiditis.[32] These clinical observations have given rise to the concept of an autoimmune diathe-sis; that is, a broad genetic predisposition to developing autoimmunity.

Many diseases of unknown origin, such as juvenile rheumatoid arthritis, are commonly classified as autoimmune because of their frequent cooccurrence with better-established autoimmune diseases.[33] The presence of an autoimmune diathesis is best explained by the inheritance of a number of diverse genetic traits. In most cases, the strongest genetic signal comes from genes of the MHC, particularly class II MHC. In humans, the associations are usually found with the HLA-DP and -DQ markers.[34] Indeed, the presence of a strong HLA association by itself is sometimes cited as circumstantial evidence for an autoimmune etiology, as in the case of anky-losing spondylitis. Moreover, as MHC typing has become refined, the association of particular HLA alleles with disease has increased greatly. It is not out of line to suggest that HLA may be a major predictor or risk factor for the later development of autoimmune disease. If early intervention is reasonable, the subjects may first be best defined by their HLA genotype. The search for other genes that contribute to the "autoimmune diathesis" will certainly continue, however, and will strengthen the predictability of autoimmune disease.[35]

The association of particular HLA haplotypes with autoimmune disease sug-gests that the diversity of molecular epitopes responsible for initiating the disease is limited. Correspondingly, there are a number of diseases where the use of TCR vari-able genes early in the course of the disease process also appears to be quite limited. Either TCRr V_α or TCRr V_β genes may be involved, although the latter are more often associated with human autoimmune disorders. Indeed, the presence of TCR V_β restriction has, by itself, become an important part of circumstantial evidence of an autoimmune disease.[36]

Often, the clinical identification of a human disease as autoimmune depends mainly on its response to treatment. A number of drugs capable of broadly suppress-ing immunity have been successfully employed to treat autoimmune conditions, although the risks of such drugs are obvious. The effect of many of the drugs should be considered more anti-inflammatory than immunosuppressive, since the dosages given are often insufficient to blunt the entire immune response.

12.3 CLASSIFICATION OF AUTOIMMUNE DISEASES

Autoimmune diseases can affect virtually any organ or tissue of the body and, therefore, the clinical manifestations are highly variable. The site of pathology depends primarily on the distribution and availability of the requisite antigen. Anti-gens that are widely distributed throughout the body are associated with systemic disease, whereas those confined to a particular tissue or organ are involved in organ-localized autoimmune disease. A convenient classification of an autoimmune disease is based on the tissue distribution of pathology. For details of each of these diseases, see Ref. 37.

12.3.1 CONNECTIVE TISSUE AND RHEUMATOLOGIC DISEASES

A most prevalent group of autoimmune diseases attacks the connective tissues or related structures of the body. Systemic lupus erythematosus is the prototype of a multiorgan disease. A number of nuclear and cytoplasmic antigens as well as cell surface molecules are targeted by autoimmune responses in lupus, especially native or denatured DNA.[1] The disease is due to the formation of immune complexes, especially DNA–anti-DNA, and characteristically affects multiple tissues and organs. It may involve the skin, the cardiovascular system, the nervous system, the gastrointestinal system, or the renal glomeruli. Often, kidney damage is the most life-threatening injury in this disease. Although rheumatoid arthritis is thought of primarily as a disease of the joints, it is a systemic condition accompanied by vasculitis, and is caused by immune complexes consisting of rheumatoid factor and immunoglobulin antigen. Scleroderma affects the connective tissue of the skin, but it is often the esophagus that is the site of major pathology. Polymyositis/dermatomyositis involve both the skin and the muscles, whereas polymyalgia rheumatica appears to be an autoimmune response primarily involving muscle. One of the most interesting diseases in the connective tissue group is Sjögren's syndrome (sicca syndrome) with its manifestations of dry eyes and dry mouth. Although the disease clusters with rheumatoid arthritis or lupus, the major pathology involves the lacrimal and salivary glands. Most infrequently, such patients also have manifestations of endocrine autoimmunity.

12.3.2 RENAL DISEASE

Autoimmune diseases of the kidney fall into two major types. Immune complexes deposited in the glomeruli induce an inflammatory response, resulting in glomerular sclerosis and renal failure. Although any immune complex can, in principle, localize in the kidney, frequent candidates are immune complexes involving autoantigens, such as seen in lupus. A second form of glomerulonephritis involves the production of autoantibodies to the glomerular or tubular basement membranes (GBM or TBM) (see Table 12.1). These antibodies may follow infection, for example, by β-hemolytic streptococci or exposure to toxins, such as chlorinated hydrocarbons or heavy metals.

12.3.3 SKIN DISEASE

A number of important skin diseases are associated with autoimmunity. Pemphigus vulgaris is caused by antibodies directed to the intraepithelial desmosomes, whereas bullous pemphigoid antibodies are directed to antigens of the epithelial basement membrane (see Table 12.1). A characteristic finding of dermatitis herpetiformis is the presence of deposits of IgA immunoglobulin in the dermis.

TABLE 12.1
Common Autoantibodies

A. Connective tissue and rheumatic diseases
Systemic lupus erythematosus
   Anti-nuclear
   Anti-dsDNA
   Anti-nucleaosome
   Anti-ribosomal P
Drug-induced lupus
   Anti-histone
Scleroderma
   Anti-nucleolar
   Anti-fibrillarin
   Anti-scl-70 (topoisomerase)
   Anti-RNP III
CREST
   Anti-centromere
Polymyositis/scleroderma
   Anti-Ku
Myositis
   Anti-Jo-1
Rheumatoid arthritis
   Rheumatoid factor
   Anti-CCP
   Anti-RA33
Sjögren's syndrome
   Anti-SS-A (RO)
   Anti-SS-B (La)
   Anti-fodrin
B. Renal disease
Goodpasture's syndrome
   Anti-GBM
C. Skin disease
Pemphigus
   Anti-desmogleins
Bullous pemphigoid
   Anti-hemidesmosomes
Vitiligo
   Anti-tyrosinase
Dermatitis herpetiformis
   Antiepidermal *trans*-glutaminase
D. Nervous system disease
Myasthenia gravis
   Anti-acetylcholine receptor
   Anti-muscle-specific
   kinase
Lambert–Eaton syndrome
   Anti-calcium channels
Peripheral neuropathy
   Anti-gangliosides

Paraneoplastic syndrome
   Anti-Purkinje cell cytoplasm
   Anti-Hu
E. Cardiovascular disease
Myocarditis/dilated cardiomyopathy
   Anti-myosin heavy chain
   Anti-B1 adrenoreceptor
   Anti-M2 cholinergic receptor
Vasculitis/Wegener's granulomatosis
   C-ANCA/anti-proteinase
Crescentic glomerulonephritis
   P-ANCA/anti-myeloperoxidase
F. Gastrointestinal disease
Celiac disease
   Anti-endomysial
   Anti-tissue transglutaminase
   Anti-gliaden
Hepatitis
   Anti-smooth muscle
   Anti-nuclear
   Anti-liver kidney microsomal (LKM)
Primary biliary cirrhosis
   Anti-pyruvic dehydrogenase
G. Endocrine disease
Gastritis
   Anti-parietal cell
   Anti-intrinsic factor
Thyroiditis
   Anti-thyroglobulin
   Anti-thyroperoxidase
Graves disease
   Anti-TSH receptor
Type 1 diabetes
   Anti-islet cell
   Anti-GAD
   Anti-insulin
   Anti-IA-2
Addison's disease (adrenalitis)
   Anti-21 hydroxylase
H. Hemotologic disease
Hemolytic anemia
   Anti-RBC (Coombs' test)
Thrombocytopenia
   Anti-platelet
Antiphospholipid syndrome
   Anti-cardiolipin
   Anti-B_2 glycoprotien 1
   Lupus anticoagulant

12.3.4 NERVOUS SYSTEM DISEASE

Myasthenia gravis is one of the best-characterized autoimmune diseases. Autoantibodies to the acetylcholine receptor are present in almost every case.[1] They block the function of the receptor, interrupting neuromuscular signaling and resulting in progressive weakness. Multiple sclerosis, characterized by plaques of demyelinization in the brain and spinal fluid, is associated with an autoimmune response to myelin antigens. A possible experimental analog of this disease is found in the form of allergic encephalomyelitis, one of the best characterized of the experimentally induced autoimmune diseases. Autoimmunity is also evident in some peripheral neuropathies, including Guillain–Barré syndrome.

12.3.5 CARDIOVASCULAR DISEASE

The classic example of an autoimmune disease associated with microbial infection is rheumatic heart disease, a sequela of infection by *Streptococcus pyogenes*. The mechanism of damage is believed to be molecular mimicry, that is, the presence on the streptococcal membrane of an antigen that resembles a constituent of the heart. Recent experimental evidence suggests that myosin may be one such antigen (see Table 12.1). In the United States, rheumatic heart disease has become rare, although it is still highly prevalent in many of the developing countries. A more common form of myocarditis in the United States follows infection with coxsackievirus. This disease is characterized by the production of multiple antibodies to heart antigens. The disease has been reproduced by experimental immunization of mice with purified cardiac myosin, making myosin the leading candidate as an initiating antigen. Many vasculitides, such as Wegener's granulomatosis and nephrosclerosis, are associated with autoimmune responses, especially to antineutrophil cytoplasmic antigens (ANCA).

Antiphospholipid syndrome is associated with clotting problems, thrombosis, spontaneous abortion, and stroke, and derives its name from the presence of antibodies to anionic phospholipids and a cofactor β_2 glycoprotein-1.

12.3.6 GASTROINTESTINAL DISEASE

Two important diseases of the liver are associated with characteristic autoimmune responses. In some cases of chronic active hepatitis, antibodies are found to smooth muscle cells where the major antigen has been identified as actin. Sometimes, these patients also have antinuclear antibodies. In primary biliary cirrhosis, several antibodies to mitochondria can be demonstrated; the major antigen is the E2 subunit of pyruvic dehydrogenase (see Table 12.1) The inflammatory bowel diseases, Crohn's disease and ulcerative colitis, are often cited as autoimmune disorders, although the antigens responsible have not been identified and no immunological tests are available. Antiendomysial and antitissue transglutaminase antibodies greatly aid the diagnosis of celiac disease.

12.3.7 ENDOCRINE DISEASE

Autoimmune thyroiditis has become the prototypic organ-localized disease. The primary antigen, thyroglobulin, is capable of inducing the disease experimentally (see Table 12.1). Other antigens, such as thyroid peroxidase, however, are useful indicators of clinical activity. In Graves disease, antibodies to the thyrotropin receptor are responsible for the induction of hyperthyroidism. Insulin-dependent diabetes has emerged as one of the most debilitating of the autoimmune diseases. It is characterized by the presence of autoantibodies to glutamic acid decarboxylase, insulin, and several other beta cell constituents, but the initiating antigen has not yet been established with certainty. The result, however, is the immunological destruction of the beta cells of the pancreatic islets. Other endocrine disorders, such as adrenalitis, hypoparathyroidism, or hypophysitis are associated with production of autoantibodies to the particular organ.

Pernicious anemia is caused by antibodies to the gastric secretion, an intrinsic factor, which facilitates absorption of vitamin B12. Antibodies to gastric mucosa are frequently present. This disease is often associated with the autoimmune endocrinopathies, such as chronic thyroiditis.

12.3.8 HEMATOLOGIC DISEASE

Acquired hemolytic anemia represents the classic example of an autoimmune response to circulating red blood cells. IgM antibodies are associated primarily with cold-reactive anemias, in which the blood cells are injured if the temperature falls to subnormal levels. However, warm hemolytic anemias can result in sequestration of the IgG antibody-coated red blood cells in spleen and other reticuloendothelial tissues. Other blood cells are also susceptible to antibody-mediated destruction seen in the form of leukopenias or thrombocytopenias. However, the demonstration of antibodies to these cells is difficult because of spontaneous absorption of immunoglobulin. Finally, one of the most important emerging autoimmune diseases is seen in the form of an autoimmune response to clotting factors. This antiphospholipid syndrome may result in increased clotting as a cause of stroke or spontaneous abortion.

12.4 METHODS FOR DETECTING AUTOANTIBODIES

12.4.1 PRECIPITATION AND AGGLUTINATION

The appropriate method for demonstrating autoantibodies is determined by the position and properties of the antigen and the level of sensitivity desired. Precipitation and agglutination were the first methods employed for demonstrating autoantibodies in human sera. Precipitation of cardiolipin has long served as a method for supporting the diagnosis of syphilis. Immunoprecipitation, using radioisotope-labeled antigen, is a more sensitive method for detecting antibodies, and turbidimetric analyses have replaced precipitation in larger laboratories. Precipitation in agar remains a good method for the precise recognition of the cellular antigens involved in lupus and related connective tissue diseases. The classic Ouchterlony double immunodiffusion test is used for verifying antibodies to Sm or other nuclear antigens.[38] Immunoprecipitation is still regarded as the "gold standard" for several other nuclear antigens such as Ku and Ki. Agglutination reactions are highly sensitive methods for demonstrating antibodies. Indirect, or conditioned, hemagglutination requires that a soluble antigen be attached to a particle, such as a red blood cell or latex. Latex agglutination is a widely used test for the measurement of rheumatoid factor, whereas agglutination of red blood cells treated with chromic chloride or tannic acid is the most sensitive procedure for demonstrating antibodies to thyroglobulin. The direct Coombs' antiglobulin test for immunoglobulin on the surface of red blood cells is the cornerstone for the diagnosis of autoimmune hemolytic anemia.

12.4.2 IMMUNOFLUORESCENCE

A commonly used test for the detection of antibodies to tissue antigens is indirect immunofluorescence.[38,39] The great versatility of this test is due to the fact that the antigen need not be precisely characterized or purified. In fact, multiple tissue substrates can be used and antibodies to a number of tissues recognized. The most common test for antinuclear antibodies utilizes tissue culture cells. Antibodies to smooth muscle or mitochondria as seen in hepatic diseases, such as chronic active hepatitis or primary biliary cirrhosis, are readily measured on composite blocks of several different tissues. ANCA detection utilizes fixed or treated neutrophils as substrate. In some instances, peroxidase is preferable to fluorescein as a tag since the tissue architecture is more readily recognized, and permanent preparations can be obtained. However, endogenous peroxidases can cause false-positive results. Direct immunofluoresence of IgG localization is crucial in the laboratory diagnosis of Goodpasture's syndrome (anti-GBM), pemphigus, and pemphigoid (epidermal desmosomes).

12.4.3 IMMUNOASSAYS

Immunoassays have long been employed to detect autoantibodies in a very sensitive manner.[38,39] The requisite antigen can be attached directly to the plastic vessel or "captured" by a first layer of antibody. Patient serum is then added and immunoglobulin is measured by means of a radioisotope, enzymatic, or fluorescent marker. At this time, the enzyme-linked immunosorbent assay (ELISA) is the most widely used test for measuring autoantibodies to tissue antigens. It is most appropriate when the antigen is available in a purified form.

12.4.4 WESTERN IMMUNOBLOTS

An important addition to the immunological armamentarium has been the western immunoblot. This procedure does not require that the antigen be pure, but it should be present in a relatively high concentration in a tissue preparation. The antigen mixture is first separated by electrophoresis, transferred to nitrocellulose membrane, and reacted with the patient serum. The location of the antibody is identified by secondary antibodies conjugated to an enzyme that will develop a color when a chromogenic substrate substance is added. The method is often used to verify or characterize reactions to complex antigens,[40] and is rapidly finding a place in the clinical immunology laboratory.

12.4.5 FUNCTIONAL TESTS

The final group of methods for detecting autoantibodies is represented by bioassays and receptor assays. They are often employed in demonstrating antibodies to cell receptors; for example, antibody to the thyrotropin receptor present in Graves disease can be demonstrated either by direct binding to the thyrotropin receptor or by measuring the increase in cAMP production by thyroid cells. Other functional tests measure lymphocytotoxic antibodies in lymphopenias or the C3 nephritic factor in membranoproliferative glomerulonephritis.

Detailed descriptions of the many tests available for demonstration of autoantibodies are present in recent publications. The reader is directed to these references for details of the methodology as well as discussions of the appropriate interpretation involved in the procedures.

REFERENCES

1. Yativ, N., Buskila, D., Burek, C.L., Rose, N.R., Blank, M., and Shoenfeld, Y., The detection of antithyroglobulin activity in human serum monoclonal immunoglobulins (monoclonal gammopathies). *Immunnol. Res.*, 12, 330, 1993.
2. Dighiero, G., Guilbert, B., Fernand, J.P., Lymberi, P., Danon, F., and Avrameas, S., Thirty-six human monoclonal immunoglobulins with antibody activity against cytoskeleton proteins, thyroglobulin, and native DNA: immunologic studies and clinical correlations. *Blood*, 62, 1983.
3. Bresler, H.S., Burek, C.L., Hoffman, W.H., and Rose, N.R., Autoantigenic determinants on human thyroglobulin. II. Determinants recognized by autoantibodies from patients with chronic autoimmune thyroiditis compared to autoantibodies from healthy subjects. *Clin. Immunol. Immunopathol.*, 54, 76, 1990.
4. Grabar, P., Autoantibodies and the physiological role of immunoglobulins. *Immunol. Today*, 4, 337, 1983.
5. Bozatto, G.F. and Doniach, D., Autoimmune thyroid disease. *Annu. Rev. Med.*, 37, 217, 1986.
6. Talor, E. and Rose, N.R., Hypothesis: the aging paradox and autoimmune disease. *Autoimmunity*, 8, 245, 1991.
7. Rose, N.R., Thymus function, aging and immunity. *Immunol. Lett.*, 40, 225, 1994.
8. Caturegli, P., Mariotti, S., Kuppers, R.C., Burek, C.L., Pinchera, A., and Rose, N.R., Epitopes on thyroglobulin: a study of patients with thyroid disease. *Autoimmunity*, 18, 41–49, 1994.
9. Rose, N.R. and Bona, C., Defining criteria for autoimmune diseases (Witebsky's postulates revisited). *Immunol. Today*, 14, 426, 1993.
10. Lefvert, A.K., Anti-indiotype antibodies in myasthenia gravis. In *Biological Applications of Antiidiotypes*, Bona, C., Ed., CRC Press, Boca Raton, FL, 1988, p. 22.
11. Davies, T. and DeBernado, E., Thyroid autoantibodies and diseases. in *An Overview in Autoimmune Endocrine Disease*, Davies, T., Ed., Wiley, New York, 1983.
12. Anhalt, G.J., Labib, R.S., Voorhees, J.J., Beals, T.F., and Diaz, L.A., Induction of pemphigus in neonatal mice by passive transfer of IgG from patients with the disease. *N. Engl. J. Med.*, 306, 1189, 1982.
13. Donath, J. and Landsteiner, K., Ueber paroxysmale Hämoglobinurie. *Muench. Med. Wochenschr.*, 51, 1590, 1904.
14. Volpé, R., Kasuga, Y., Akasu, F., Morita, T., Yoshikawa, N., Resetkova, E., and Arreaza, G., The use of the severe combined immunodeficient mouse and the athymic "nude" mouse as

models for the study of human autoimmune thyroid disease. *Clin. Immunol. Immunopathol.*, 67, 93, 1993.

15. Witebsky, E., Rose, N.R., Terplan, K., Paine, J.R., and Egan, R.W., Chronic thyroiditis and autoimmunization. *J. Am. Med. Assoc.*, 164, 1439, 1957.

16. Gery, I. and Streilein, J.W., Autoimmunity in the eye and its regulation. *Curr. Opinion Immunol.*, 6, 938, 1994.

17. Gomez, C.M. and Richman, D.P., Chronic experimental autoimmune myasthenia gravis induced by monoclonal antibody to acetylcholine receptor: biochemical and electrophysiological criteria. *J. Immunol.*, 139, 73, 1987.

18. Theofilopoulos, A.N. and Dixon, F.J., Murine models of systemic lupus erythematosus. *Adv. Immunol.*, 37, 269, 1985.

19. Makino, S., Kunimoto, T., Muraoka, Y., Mizushima, Y., Katagiri, K., and Tochino, Y., Breeding of a non-obese, diabetic strain of mice. *Exp. Anim. (Jikken Dobutsu)*, 29, 1, 1980.

20. Nakhooda, A.F., Like, A.A., Chappel, C.I., Murray, F.T., and Marliss, E.B., The spontaneous diabetic Wistar rat: metabolic and morphologic studies. *Diabetes*, 26, 100, 1977.

21. Kuppers, R.C., Neu, N., and Rose, N.R., Animal models of autoimmune thyroid disease, in *Immunogenetics of Endocrine Disorders*, Farid, N.R., Ed., Alan Liss, New York, 1988, p. 111.

22. Penhale, W. and Ahmed, S., Thyroid transplantation in rats developing autoimmune thyroiditis following thymectomy and irradiation. *Clin. Exp. Immunol.*, 45, 480, 1981.

23. Kojima, A., Tanaka-Kojima, Y., Sakakura, T., and Nishizuka, Y., Spontaneous development of autoimmune thyroiditis in neonatally thymectomized mice. *Lab. Invest.*, 34(6), 550, 1976.

24. Shevach, E.M., DiPaolo, R.A., Andersson, J., Zhao, D.M., Stephens, G.L., and Thornton, A.M., The lifestyle of naturally occurring CD4+CD25+Foxp3+ regulatory T cells. *Immunol. Rev.*, 212, 60, 2006.

25. Gambineri, E., Torgerson, T.R., and Ochs, H.D., Immune dysregulation, polyendocrinopathy, enteropathy, and x-linked inheritance (IPEX), a syndrome of systemic autoimmunity caused by mutations of FOXP3, a critical regulator of T-cell homeostasis. *Curr. Opin. Rheumatol.*, 15(4), 430, 2003.

26. Lamoureux, J.L., Buckner, J.H., David, C.S., and Bradley, D.S., Mice expressing HLA-DQ6alpha8beta transgenes develop polychondritis spontaneously. *Arthritis Res. Ther.*, 8(4), R134, 2006.

27. Elliot, J.F., Liu, J., Yuan, Z.N., Baustista-Lopez, N., Wallbank, S.L., Suzuki, K., Rayner, D., Nation, P., Robertson, M.A., Liu, G., and Kavanagh, K.M., Autoimmune cardiomyopathy and heart block develop spontaneously in HLA-DQ8 transgenic IA (beta) knockout NOD. *Proc. Natl. Acad. Sci., USA*, 2003.

28. Grabie, N., Delfs, M.W., Westrich, J.R., Love, V.A., Stavrakis, G., Ahamd, F., Seidman, C.E., and Lichtman, A.H., IL-12 is required for differentiation of pathogenic CD8+ T cell effectors that cause myocarditis. *J. Clin. Invest.*, 111(5), 671, 2003.

29. Olson, J.K., Ercolini, A.M., and Miller, S.D., A virus-induced molecular mimicry model of multiple sclerosis. *Curr. Top. Microbiol. Immunol.*, 296, 39, 2005.

30. Oldstone, M.B.A., Nerenberg, M., Southern, P., Price, J., and Lewicki, H., Virus infection triggers insulin-dependent diabetes mellitus in a transgenic model: role of anti-self (virus) immune response. *Cell*, 65, 319, 1991.

31. Hall, R., Dingle, P.R., and Roberts, D.F., Thyroid antibodies: a study of first degree relatives. *Clin. Genet.*, 3, 319, 1972.

32. Rose, N.R., The spectrum of autoimmunity: mechanisms of organ specific and nonorgan-specific diseases. *Immunopathol. Immunother. Lett.*, 3, 8, 1988.

33. Rose, N.R. and Burek, C.L., The interaction of basic science and population-based research: autoimmune thyroiditis as a case history. *Am. J. Epidemiol.*, 134, 1073, 1991.

34. Taneja, V. and David, C.S., Genetics and autoimmunity: HLA and MHC genes, in *The Autoimmune Diseases*, 4th Ed., Rose, N.R. and Mackay, I.R., Eds., Elsevier, London, 2006, p. 261, Chap. 20.

35. Tomer, Y., Genetics and autoimmunity: non-MHC genes, in *The Autoimmune Diseases*, 4th Ed., Rose, N.R. and Mackay, I.R., Eds., Elsevier, London, 2006, p. 273, Chap. 21.

36. Maverakis, E., Moudgil, K.D., and Sercarz, E., Generation of T-cell antigenic determinants in autoimmunity and their recognition. In *The Autoimmune Diseases*, 4th Ed., Rose, N.R. and Mackay, I.R., Eds., Elsevier, London, 2006, p. 179, Chap. 14.

37. Rose, N.R. and Mackay, I.R., *The Autoimmune Diseases*, 4th Ed., Elsevier, London, 2006.

38. Shoenfeld, Y., Gershwin, M.E., and Meroni, P.L., *Autoantibodies*, 2nd Ed., Elsevier, London, 2007.

39. Detrick, B., Hamilton, R.G., and Folds, J.D., *Manual of Molecular and Clinical Laboratory Immunology*, 7th Ed., ASM Press, Washington, 2006.

40. Musante, L., Candiano, Bruschi, Santucci, CArnemolla, Orecchia, Giampuzzi, Zennaro, Sanna-Cherchi, Carraro, Oleggini, Camussi, Perfumo, and Ghiggeri. Circulating anit-actin and anti-ATP synthase antibodies identify a subset of patients with idiopathic nephritic syndrome. *Clin. Exp. Immunol.*, 141(3), 491, 2005.

13 Autoimmune and Immune-Mediated Diseases of the Gastrointestinal Tract

Stefano Guandalini, Mala Setty, and Bana Jabri

CONTENTS

13.1 THE INTESTINAL MUCOSAL IMMUNE SYSTEM

13.1.1 INTRODUCTION

The major function of the human gastrointestinal tract (GIT) is to digest and absorb nutrients, electrolytes, and water, while excluding potentially noxious agents such as microorganisms and their toxins.

The GIT has the largest mucosal surface in the human body, with a surface area of over 400 m^2 to provide a sizeable interface between the external environment and the internal milieu. Thus, like the respiratory tract, the GIT remains a major portal of entry that is perpetually exposed to the external environment composed of a myriad of food antigens, allergens, and a mix of organisms of the endogenous enteric flora as well as potential pathogens. Its particular vulnerability to infection has resulted in the development of a highly evolved immune system. This system consists of strategically distributed leukocytes within the lamina propria and intraepithelial compartments involved in the uptake, processing, and presentation of antigens; antibody production and cell-mediated defense.

The primary function of the mucosal immune system is for host defense at the mucosal surfaces. The GIT contains more of macrophages, plasma cells, and T cells than any other lymphoid tissue in the body. Along with other nonimmunologic protective mechanisms, it acts to protect the host. In addition, the mucosal immune system prevents the entry of antigens into circulation, thus protecting the systemic immune system from the inappropriate exposure to antigen.

13.1.2 ANATOMY

Morphologically, the mucosal immune system is divided into organized lymphoid structures such as lymphoid follicles and Peyer's patches (PP), which form the gut-associated lymphoid tissue (GALT) as well as a range of cells scattered throughout the lamina propria.

Like the systemic immune system, the immune system of the gut can be classified into the *adaptive* and the *innate* immune systems (discussed in Chapter 1). The adaptive immune system is further divided into inductive sites, where antigens stimulate naïve T and B lymphocytes, and effector sites, such as the lamina propria and epithelium, where antigen-sensitized cells may be distributed [1].

13.1.3   Mucosal-Associated Lymphoid Tissue

The structures of the mucosal-associated lymphoid tissue (MALT) are the principal sites of induction for priming naïve lymphocytes. Anatomically, the inductive sites occur within PP of the small intestine and appendix, solitary lymphoid follicles of the large intestine and rectum, as well as selected draining lymph nodes, which together make up the MALT [2]. These organized structures, distributed throughout the intestine, are distinct from all other lymphoid structures of the body because they receive antigen directly from the intestinal lumen rather than through the circulation. (See Figure 13.1 for a scheme of the mucosal immune system.)

Their histological architecture resembles lymph nodes with B-cell follicles, intervening T-cell areas, and scattered antigen-presenting cells (APC); however, they do not have afferent lymphatics, thus making them singular structures. PP form dome-like structures extending into the lumen of the gut. They consist of an overlying layer of specialized epithelial cells known as microfold (M) cells. They are specialized cells that express major histocompatibility complex (MHC) class II antigen and lack a thick surface glycocalyx. This structural feature results in direct interaction with molecules within the lumen, allowing the sampling and presentation of antigens to naïve immune cells housed in the subepithelial area. Located strategically within the inductive sites, M cells are believed to guide the broad and sustained priming and expansion of mucosal B and T cells [3].

Several pathogens have been shown to target M cells to gain access to the subepithelial space such as poliovirus, retroviruses, *Salmonella typhi*, and *Salmonella*

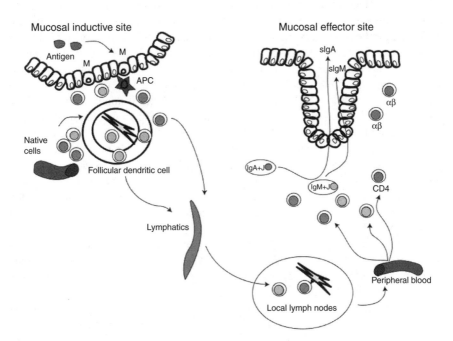

FIGURE 13.1   Scheme of the mucosal immune system.

typhimurium. Owing to this, M cells are currently under investigation as a potential approach for both drug design and orally administered vaccines [4–6].

13.1.3.1 Nonimmune Defenses

Nonspecific intestinal functions play a profound role in modulating the exposure of luminal antigens to the surface epithelium. In the upper small intestine, gastric acidity protects by destroying ingested pathogens and preventing bacterial colonization. Goblet cells that are scattered throughout the intestine, increase in density exponentially from the small to large intestine, and act to provide a physical barrier between the epithelium and the lumen by releasing preformed mucus in response to luminal stimulation. This bars further excessive uptake of luminal antigens and pathogens through the epithelium [7]. The intestinal lumen is replete with digestive enzymes from pancreatic, hepatic, and specialized intestinal epithelial cells to facilitate in the breakdown and absorption of ingested foods. These same enzymes also breakdown antigens to minimize the antigen load on the epithelium. Finally, the intrinsic peristaltic movement of the GIT acts to propulsively eliminate noxious agents [8].

13.1.3.2 Commensals

Normal commensal bacteria in the GIT exist in symbiosis in the colon and terminal ileum, providing competition to pathogenic bacteria for space, nutrients, and synthesis of vitamin K. More than 400 different species of commensal bacteria, plus a variety of infectious agents either directly populate the GIT or are acquired, and hence the innate and adaptive immune mechanisms at the level of the mucosa are important for the protection and survival of humans. A critical function of the mucosal immune system is to preserve a selective and tightly regulated, favorable state of physiological nonresponse due to constant contact to various microflora and antigens. This mechanism of immune restraint is referred to as oral tolerance [9].

The importance of the commensal flora to intestinal immune defense is best illustrated by examples of antibiotic-related adverse effects. In one example, the balance of normal commensals in the colon is disturbed following antibiotic administration, thus *Clostridium difficile* can instigate a severe pseudomembranous colitis. *C. difficile* produces two toxins, which in a perturbed colonic ecosystem can cause severe bloody diarrhea and mucosal injury [10]. In another example, following vascular compromise in the neonatal GIT, nonpathogenic *Escherichia coli* has been observed to cross-damage mucosa to invade the bloodstream resulting in sepsis [11] and enterocolitis.

Under normal conditions, the immune system in conjunction with the nonimmune defenses can selectively enhance and activate protective responses to infectious agents to maintain the integrity of the mucosal barrier. In conditions such as Crohn's disease (CD) and celiac disease, this physiologic balance between nonresponse and active inflammation is believed to be dysregulated, resulting finally in the disruption of tissue integrity [12]. Mechanisms of oral tolerance [12] and manipulations of enteric flora [13] are currently under intense investigation as potential targets for immunotherapy and immunoregulation in autoimmune and inflammatory diseases.

13.1.3.3 Mucosal Barrier

The intestinal epithelial cell provides a physical barrier from the external milieu, and may be central in determining the variety of antigens exposed to the mucosal immune system. Antigens can cross the epithelial barrier by three potential mechanisms: through a cellular route (requiring uptake, processing, and cellular presentation of luminal antigens), a transcellular method (by transcytosis), and paracellular transport through tight junctions and subjacent desmosomes.

Secretory IgA and IgM are released from the submucosal compartment into the gut lumen by transcellular transport. The paracellular route is monitored by tight junctions, which selectively exclude antigens greater than 6–12 Å in diameter. Antigen exposure by this mechanism is modifiable by various humoral factors of the immune system, secreted growth factors, and bacterial toxins. For example, cytokines such as tissue growth factor β (TGF-β) and interleukin (IL)-10 are known to increase paracellular resistance, whereas interferon γ (IFN-γ), IL-6 and tumor necrosis factor α (TNF-α) reduces it.

The epithelium also secretes a variety of protective factors as components of the innate immune response. Such factors include mucin, as mentioned earlier, complement components, alpha defensins (cryptdins), beta defensins, lysozyme, phospholipase A2, lactoferrin, and lactoperoxidase. Alpha defensins, secreted by Paneth cells within the intestinal crypt, exhibit potent and broad microbicidal activity against various pathogens such as *S. typhimurium* and *Listeria monocytogenes*, and act locally to protect intestinal stem cells that are within the crypt epithelium. Table 13.1 summarizes the protective factors secreted by the intestinal epithelium.

Toll-like receptors (TLRs) form a major component of the innate immune system and are involved in mucosal immunity. Epithelial cells express these recognition factors on their cell surface, which act as signaling receptors that stimulate an intracellular cascade of events. The TLRs bind selectively to microbial components such as lipopolysaccharide (LPS) from Gram-negative bacteria, bacterial flagellin,

TABLE 13.1
Protective Factors Secreted by the Epithelium as Part of the Local Innate Immune System

Innate Peptides	Producing Epithelial Cell	Mechanism of Action	Additional Actions
Alpha defensins	Paneth cells	Pore formation	
Beta defensins	Intestinal epithelial cells	Pore formation	Chemotactic for WBC
Lysozyme	Paneth cells	Degrades bacterial peptidoglycan	Modifies inflammatory effects of peptidoglycan
Phospholipase A2	Paneth cells	Degrades bacterial lipids	

microbial DNA, and lipoteichoic acids from Gram-positive bacteria. There are at least 11 recognized members of the TLR family, and they are associated with a complex cascade of events, which culminate in the phosphorylation of nuclear factor kappa B inhibitor (IκBa), the inhibitory molecule of nuclear factor kappa B (NfκB). These pivotal events lead to the translocation of NfκB to the nucleus and further transcriptional event that upregulates the inflammatory cascade.

Intrinsically, epithelial cells secrete numerous inflammatory and regulatory cytokines, which are either inducibly expressed or constitutively expressed. These cytokines can play an important role in promoting intestinal inflammation (TNF-α, IL-1), recruitment of T cells within the epithelium (IL-15, IL-5, IL-7), and B-cell regulation (TGF-β, IL-6, IL-10). The epithelium also expresses numerous surface molecules that are associated with antigen presentation, leukocyte adhesion, and costimulation that interact with adjacent populations of intraepithelial lymphocytes, largely T cells, and dendritic cells.

Paneth cells, first described as granular cells at the base of small intestinal crypts where epithelial stems cells proliferate, have been recognized as an important member of the innate intestinal defense. The granules with Paneth cells contain various local antimicrobial agents such as lysozyme, defensins, and secretory phospholipase A2 (sPLA2), which are released on stimulation.

13.1.3.4 Intraepithelial Lymphocytes

Lymphocytes of the GALT are situated throughout the lamina propria. The effector cells of the mucosal immune system are represented by the scattered foci of lymphocytes and plasma cells of the lamina propria. It is known that these cells migrate to mucosal tissues through the bloodstream and can diffuse through the mucosa via a complex series of receptor homing mechanisms.

13.1.3.4.1 T Cells

T lymphocytes migrate and distribute throughout the mucosa using molecular mechanisms similar to the B cells (addressins). Their primary role is to recognize antigen and distribute within the mucosal immune system to effector sites such as the lamina propria. The major cell types seen predominantly infiltrating the lamina propria are the αβ CD4+ or CD8+ T lymphocytes, which mediate conventional T-cell responses to foreign antigens. The γδ T cells are a more specialized cell type, predominantly CD4/CD8 double negative or CD8+, and are found throughout the intraepithelial compartment. They express a higher proportion of T-cell receptors that bind to the nonclassical class 1B MHC molecules. One subset of γδ T cells carries an activating natural killer (NK) receptor, NKG2D, which binds to MHC class-I-related chains A (MIC-A) and MIC-B expressed by epithelial cells due to cellular stress, and induces killing of these cells. Thus, they survey tissues for injury and destroy cells expressing a stress phenotype. In conditions such as celiac disease, the number of γδ T cells has been shown to be highly increased [14].

T lymphocytes are central to the induction of mucosal tolerance. Regulatory T lymphocytes monitor and adjust immune responses to foreign antigens. T regulatory cells (Tr1), formerly designated as suppressor T cells, exert inhibitory effects primarily through the release of IL-10. Another subset of T regulatory cells, CD4+CD25+ cells or Tregs, suppress activated T cells by direct cell contact. A third type, Th3, is recognized to produce large amounts of TGF-β, which, in turn, modulates and downregulates T-cell activity. Both CD4+CD25+ cells and Tr1 cells have been shown to suppress inflammation in animal models of inflammatory bowel disease (IBD) [15].

13.1.3.4.2   B Cells
B lymphocytes are located throughout the intestinal mucosa, though primarily within PP, lymphoid follicles, and the lamina propria. These cells vary in phenotype depending on their status of activation. The PP is a site containing naïve lymphocytes and large numbers of CD19+ B cells, only a few of which express CD38. The PP CD19+ B cells express surface IgD and secretory IgA. In contrast, lamina propria B cells express CD38 and are predominantly surface and cytoplasmic IgA+. The main functions of intestinal secretory IgA are summarized in Table 13.2.

The migration of B cells from inductive sites to the intestinal lamina propria is also guided by well-defined adhesion molecules and chemokine receptors directing their homing to different segments of the gut.

13.1.3.4.2.1   Lymphocyte Migration
The molecules involved in homing of mucosal lymphocytes to their effector sites include mucosal addressin cell-adhesion molecule (MAdCAM)-1 and selectins and integrins expressed on the surface of lymphocytes. MAdCAM-1 is abundantly expressed by vascular tissue [16] and its glycosylation promotes binding of L-selectin (CD62L) abundantly expressed on naïve lymphocytes. This initial endothelial adherence along with the binding of integrin $\alpha 4\beta 7$ to the N-terminal domains of MAdCAM-1 is crucial for lymphocyte migration into structures such as PP.

13.1.3.4.3   Plasma Cells
Plasma cells are terminally differentiated B cells. They are the most abundant type of lymphoid cell in the GIT, with more than 80% of human plasma cells located in the intestinal mucosa. The majority of plasma cells in the lamina propria are effector cells, programmed to produce dimeric IgA and IgM. Secretory IgA in dimeric form with a binding J chain is predominantly of the IgA2 subclass.

TABLE 13.2
Main Functions of Intestinal Secretory IgA

	IgA Functions
Protective functions	Prevention of antigen absorption
	Mucus trapping of microorganisms
	Virus neutralization
Biological activities	Inhibition of complement activation
	Enhancement of opsonization
	Degranulation of eosinophils
	Antibody-dependent cell-mediated cytotoxicity

13.1.3.4.3.1 IgA

IgA1 is predominantly found within the circulation, whereas IgA2 is primarily secreted into the intestinal lumen. It is released generally as a dimer, which, once secreted, is transported to the lumen through immature epithelial cells at the base of intestinal crypts. The molecular mechanisms of secretion into the intestinal lumen involve the binding of dimeric IgA to a basolateral receptor of the epithelium (called the secretory component [SC]), transcytosis, and the subsequent release from the luminal surface. An adult human secretes approximately 3 g of secretory IgA daily to provide a noninflammatory mechanism to protect the extensive mucosal surfaces from diverse antigens. This dimeric form of IgA is very resistant to proteolysis by the various enzymes present in the intestinal lumen. Secretory IgA and IgM bind to the mucus layer of gut epithelium and neutralize local pathogens and toxins by agglutinating these antigens and forming complexes that prevent uptake.

13.2 DISEASES OF THE GIT WITH IMMUNOLOGICAL INVOLVEMENT

13.2.1 Food Allergy

13.2.1.1 Introduction

Adverse reactions to food are perceived to be extremely common, as up to 20% of adults report some form of food intolerance. However, proven adverse reactions to food, though still very prevalent, are less frequent.

It should be noticed that not all clinically evident food intolerances are due to food allergies. In fact, adverse reactions to ingested foods can be classified as either nonimmune mediated (i.e., due to a variety of other conditions such as disorders of intestinal digestion or absorption and pharmacological reactions to chemicals in food) or immune mediated. Of the latter, only a few represent a true hypersensitivity reaction (i.e., have an IgE-mediated pathogenesis). Nevertheless, the term "food allergy" is used to encompass all the specific reactions to offending food proteins that have an immunological basis, whether IgE- or non-IgE mediated. Table 13.3 is a classification of adverse reactions to food presented in this chapter.

Allergic reactions to food are increasingly common, occurring in approximately 5% of infants and children and in about 3% of adults [17,18]. Several food proteins, both from fluid and solid foods, can act as antigens in humans and cause an immune reaction. Cow's milk proteins are most frequently implicated as a cause of food intolerance during infancy. Soybean proteins rank second as antigens in the first months of life, particularly in infants with primary cow's milk intolerance who are switched to a soy formula. For children of school-age group, egg protein intolerance becomes more prevalent. In childhood, there is evidence of a growing prevalence of allergy to peanuts [19]. Many clinical reactions to food proteins have been reported in children as well as in adults.

Generally, in children GI symptoms predominate [19,20], with a frequency ranging from 50% to 80%, followed by skin lesions (20–40%) and respiratory symptoms (4–25%).

TABLE 13.3
Classification of Adverse Reactions to Food

Type	Pathogenesis	Clinical Entities
Nonimmune mediated	Disorders of digestive–absorptive processes	Glucose–galactose malabsorption
		Lactase deficiency
		Sucrase–isomaltase deficiency
		Enterokinase deficiency
	Pharmacological reactions	Tyramine in aged cheeses
		Histamine in strawberries, caffeine, etc.
	Idiosyncratic reactions	Food additives
		Food colorants
	Inborn errors of metabolism	Phenylketonuria (PKU)
		Hereditary fructose intolerance
		Tyrosinemia
		Galactosemia
		Lysinuric protein intolerance
	Occasionally IgE mediated	Eosinophilic esophagitis
		Eosinophilic gastritis
		Eosinophilic gastroenteritis
Immune mediated (food allergy)	IgE mediated (positive RAST or skin prick tests)	Oral allergy syndrome
		Immediate GI hypersensitivity
	Non-IgE mediated	Food protein–induced Enterocolitis (FPIES)
		Enteropathy
		Proctocolitis
		Chronic constipation
Autoimmune	Innate as well as adaptive immunity	Celiac disease

13.2.1.2 Pathophysiology

The major food allergens are water-soluble glycoproteins (molecular weight [MW] 10,000–60,000) that are resistant to heat, acid, and enzymes. The GIT is permeable to intact antigens, which are taken up through an endocytotic process involving intracellular lysosomes. Some antigens can move through intercellular gaps; however, the penetration of antigens through the mucosal barrier is not usually associated with clinical symptoms. Under normal circumstances, food antigen exposure through the GIT results in a local IgA response and in an activation of suppressor CD8+ lymphocytes that reside in the GALT (oral tolerance).

In some children who are genetically susceptible, or for other as-yet-unknown reasons, oral tolerance does not develop and different immunologic and inflammatory mechanisms can be elicited. Antigen uptake has been found to have increased in children with gastroenteritis and with cow's milk allergy. Local production and systemic distribution of specific reaginic IgE plays a significant role in IgE-mediated reactions to food proteins. In addition, studies have demonstrated the role of GI

TABLE 13.4

Food Proteins Known to Be Responsible for Food Allergies

Food	Specific Protein (When Identified)
Cow's milk	Caseins
	Whey proteins
	β-Lactoglobulin
	α-Lactalbumin
	Bovine serum albumin
Egg	Ovalbumin
Soy	2S-globulin
	Soy trypsin inhibitor
	Soy lectin
Wheat	Gluten
	Glutenin
	Globulin
	Albumin
Corn	50 kDa maize γ-zein
Rice	
Fish	
Shellfish	
Beef	
Pork	
Peanuts, beans, peas	
Tree nuts and seeds, cocoa	

T lymphocytes in the pathogenesis of GI food allergy, so that it is now accepted that both T-cell mediated and delayed hypersensitivity reactions can play a role in food allergies. In spite of the well-known occurrence of IgG antibodies directed against food protein, their actual role in the pathogenesis of clinically relevant symptoms is at best doubtful.

Table 13.3 indicates the clinical conditions manifested by these different underlying pathogenetic mechanisms, whereas Table 13.4 lists the food proteins that are most commonly responsible for food allergies.

13.2.1.3   Clinical Presentation

13.2.1.3.1   History

Food allergy is mainly a problem of infancy and early childhood. Cow's milk allergy typically develops in early infancy. It is thought that about half of the infants who develop cow's milk protein allergy become symptomatic within 7 days of the introduction of milk proteins, and the vast majority within 4 weeks. Although it is usually assumed that food allergy remits by 2 years of age, when the infant's mucosal immune system matures and the child becomes immunologically tolerant, it has more recently become clear that milk protein allergy may actually persist well beyond that time, or even manifest itself initially in children older than 5 years of age.

Occasionally, patients may undergo an apparent remission for years, only to experience a recurrence in their teenage or adulthood.

13.2.1.3.2   Symptoms and Signs

Food allergy can cause a number of different symptoms with GI manifestations being the most common, usually without involvement of other organ systems.

13.2.1.3.2.1   IgE Mediated

  a. *Oral allergy syndrome.* A form of IgE-mediated contact allergy (urticaria-like) that is confined almost exclusively to the oropharynx and is most commonly associated with the ingestion of various fresh fruits and vegetables [21]. Symptoms include itching, burning, and angioedema of the lips, tongue, palate, and throat.
  b. *Immediate GI hypersensitivity.* An IgE-mediated GI reaction that often accompanies allergic manifestations in other organs such as the skin or lungs. The reaction usually occurs within minutes to 2 h of food ingestion. The most commonly involved food antigens are cow's milk or soy proteins, egg, wheat, seafood, and nuts. The patient immediately develops nausea and abdominal pain, followed by vomiting. After 1–2 h, watery diarrhea follows. Like other IgE-dependent allergic disorders, allergy to milk, egg, wheat, and soy generally resolves, whereas allergies to peanuts, tree nuts, and seafood tend to persist [19].

13.2.1.3.2.2   Occasionally IgE Mediated: Eosinophilic Gastroenteropathies

This is a group of several disorders, all characterized by the infiltration of eosinophils into the GI mucosa and various GI symptoms [22]. Peripheral eosinophilia occurs in a variable percentage of these patients, but is rarely massive (i.e., >15–20%). Although there is a definite overlapping among these disorders, the entities are best considered individually as follows:

  a. *Eosinophilic esophagitis.* A recently described entity occurring both in children [23] and in adults [24] is characterized by heavy eosinophilic infiltrates (by definition, >20 eosinophils/high power field [HPF] at pathology) in the esophageal mucosa, with various patterns of morphological alterations described at endoscopy, including furrowing of the mucosa and mucosal rings. Typically, there is no concomitant involvement of the gastric or duodenal mucosa. The condition appears to be rapidly increasing in prevalence. Affected individuals, both children and adults, may present with dysphagia, food impaction, intermittent vomiting, food refusal, epigastric or chest pain, and failure to respond to conventional antireflux medications. Occasionally, esophageal strictures develop. Pediatric patients often show evidence of food hypersensitivity and may respond to elimination diets.
  b. *Eosinophilic gastritis.* Distinct from esophagitis, and typically occurring in the absence of esophageal or duodenal involvement, the rare eosinophilic gastritis is responsive to elimination diets. It mostly occurs in late childhood, adolescence, and adulthood, and is characterized by usual symptoms for gastritis from other etiologies such as postprandial vomiting, abdominal

pain, anorexia, early satiety, and long-term weight loss. Approximately half of these patients have atopic features.

c. *Eosinophilic gastroenteritis.* An ill-defined disease that is characterized pathologically by the infiltration of eosinophils at various sites of the GIT, from the stomach, through the small intestine to the rectum. The syndrome has been reported in children of all ages as well as in adults. The eosinophilic infiltration can be limited to the mucosa, or it can deepen into the other layers of the intestinal wall, including the submucosa, the muscular, and the serosal layers. GI symptoms differ in relation to the area of the GIT that is involved and to the extent of the infiltration. Most commonly they consist of vomiting, diarrhea, and weight loss; however, gastric outlet obstruction, subacute small intestinal obstruction, and appendicitis-like symptoms have all been well documented. Diagnosis requires bioptic samples showing the eosinophilic infiltration. Unfortunately, no clear-cut line can be drawn to distinguish eosinophilic gastroenteritis from other GI diseases or from nonpathologic eosinophilic infiltration of the lower intestine.

13.2.1.3.2.3   Non-IgE Mediated

a. *Food protein–induced enterocolitis syndrome (FPIES).* A symptom complex of profuse vomiting and diarrhea diagnosed in infancy involving both the small and the large intestine [25], food-induced enterocolitis syndrome occurs most frequently in the first months of life. Most of the cases are observed in infants younger than 3 months, who are prone to quickly develop failure to thrive if left untreated. The most common food antigens responsible for this syndrome are cow's milk and soy proteins, although recently it has also been reported that solid food proteins (such as protein from rice, vegetables, and poultry) may cause it in older infants [26,27]. Vomiting generally occurs 1–3 h after feeding, and diarrhea occurs 5–8 h after feeding. Specific descriptions of the histologic findings are less important because the diagnosis can be made clinically. Some small bowel specimens show mild villous injury with edema and inflammatory infiltration, whereas colonic specimens reveal crypt abscesses and a diffuse inflammatory infiltrate. FPIES is non-IgE mediated, but mostly due to a cell-mediated allergy whose intimate mechanisms are poorly understood.

b. *Food protein–induced enteropathy.* A cell mediated form of food allergy involving damage to the absorptive surface of the small intestinal mucosa. Cow's milk and soy proteins are responsible for an uncommon syndrome appearing typically in 3–12 month-old infants and having symptoms of chronic diarrhea, weight loss, and failure to thrive; similar to celiac disease. Vomiting is present in up to two-thirds of patients. Small bowel biopsy shows an enteropathy with variable degrees of villous atrophy. Total mucosal atrophy, histologically indistinguishable from celiac disease, is also a frequent finding. Intestinal protein and blood losses can aggravate the hypoalbuminemia and anemia that are frequently observed in this syndrome. Indeed, protein-losing enteropathy may be a prominent feature of this syndrome, whose frequency and severity have definitely decreased during the past 20 years. Cases more recently described involve patients who tend to present with patchy intestinal lesions.

c. *Food protein–induced proctocolitis.* A common cause of minor rectal bleeding in very young infants, typically 2–8 weeks of age. Again, cow's milk and soy proteins are most often responsible, but interestingly the majority of affected infants are exclusively breast-fed. Symptoms include diarrhea and frequent streaks of fresh blood in the stools. Affected infants generally appear to be healthy and have normal weight gain. The onset of bleeding is gradual and initially erratic over several days. It then progresses to streaks of blood in most stools that can elicit suspicion of an internal anal tear and is generally very alarming to the parents. Endoscopic findings include aphthae, and biopsies show a mild to moderate eosinophilic proctitis. It should be added, however, that endoscopy is generally not needed for diagnosis in such a low-grade, benign condition. In about half of the cases, a strict maternal diet (elimination of all cow's milk–based and soy products from the diet) can resolve the problem.

d. *Chronic constipation.* In the past few years, evidence has accumulated showing that a subset of children with chronic constipation may have allergy to cow's milk as its cause [28]. It appears that these children, who are of the pre-school or school-age groups and often lack clinical landmarks of allergy, have characteristic proctocolonic endoscopic and histologic findings and elevated densities of $\gamma\delta$ T cells [29]. A milk elimination diet has been shown to be helpful in relieving long-standing constipation in the majority of these patients, and a subsequent food challenge typically results in reappearance of the symptom. It is still unclear how common this presentation is and what diagnostic strategy is to be followed to identify it.

13.2.1.3.3 Diagnosis

As in every medical condition, the first step in diagnosis is a sound medical suspicion based on a good history (and in our case family history too, as it is known that the vast majority of patients with food allergy have a positive family history) and physical examination. As a general rule, once a food allergy is suspected, the incriminated food needs to be eliminated, and tests may or may not be requested to confirm the diagnosis. We will address each of the conditions separately.

13.2.1.3.3.1 Oral Allergy Syndrome and Immediate GI Hypersensitivity

As these conditions are always IgE mediated, measuring food-specific IgE antibodies is helpful. This can be done by either prick skin tests or the serum assay for antigen-specific IgE. Different laboratories use different units for allergen-specific IgE, and it is not uncommon for results to be reported in a "class" scheme based on comparison with normal sera that goes from 0 to 5. However, data should be expressed quantitatively using units per liter (U/L), to allow a meaningful comparison among laboratories. Table 13.5 reports current normal values for serum IgE expressed as kilounits per liter (kU/L), as well as the various corresponding "classes" of positive values.

TABLE 13.5
Reference Ranges for Serum Levels of Allergen-Specific IgE Antibodies

IgE (kU/L)	Class	Common Interpretation
<0.35	0	Negative
0.35–0.70	1	Equivocal
0.71–3.5	2	Positive
3.51–17.5	3	Positive
17.6–50.0	4	Strongly positive
50.1–100.0	5	Strongly positive
>100	6	Strongly positive

A word of caution in interpreting results from the reported reference ranges is necessary, as it has been shown that the capacity of a given result to predict the clinical expression of the food allergy varies depending on the specific food. Sampson and Ho [30] showed that allergen-specific serum IgE measurements had excellent sensitivity and negative predictive accuracy, but poor specificity and positive predictive accuracy. In this study, the authors were able to establish diagnostic "decision points" that were at least 95% predictive of clinical reactivity to egg, milk, peanut, soy, wheat, and fish. In a subsequent study [31], they applied such criteria in a population of 100 children also being investigated with a double-blind, placebo-controlled food challenge (DBPCFC). The performance characteristics of such criteria at the levels of 95% and 90% positive predictive value, respectively, are reported in Table 13.6 [31]. It can be seen that the predicted probability for a positive test varies with the food antigen being considered, thus again calling for caution when interpreting the clinical relevance of a "positive" test, especially at lower classes.

Table 13.7 [31] provides a recommended interpretation of food-specific IgE levels in the diagnosis of allergy toward egg, milk, peanut, fish, soy, and wheat in children.

13.2.1.3.3.2 Eosinophilic Gastroenteropathies (Eosinophilic Esophagitis,
Eosinophilic Gastritis, and Eosinophilic Gastroenteritis)

The clinician suspecting any of these conditions may want to measure food-specific IgE antibodies; in approximately 50% of cases, the disorder is mediated by an IgE reaction. The same considerations reported earlier for IgE-mediated conditions apply for the interpretation of allergen-specific serum IgE; with the caveat, however, a negative serum IgE by no means rules out food allergy as a cause of any of the eosinophilic gastroenteropathies. Thus, it is evident that the role of the immunoallergy laboratory here is more limited, with the diagnostic burden resting mostly on the clinician. Therefore, it should also be noted that we still lack strictly defined diagnostic criteria and that the diagnosis requires an experienced gastroenterologist.

The diagnostic workup—among other steps necessary to rule out other conditions that can mimic them—will involve endoscopic procedures (upper or lower GI depending on the presentation). Once a diagnosis of an eosinophilic gastroenteropathy is established, most clinicians would start an elimination diet based on the presumptive

TABLE 13.6
Performance Characteristics of the 95% and 90% Predictive "Decision Points" in 100 Children with Food Allergy

Allergen	Decision Point (kU/L)	Sensitivity	Specificity	Efficiency	PPV	NPV
95%						
Egg	6	64	90	69	96	39
Milk	32	34	100	56	100	44
Peanut	15	57	100	66	100	36
Fish	20	25	100	89	100	36
Soybean	65	24	99	79	86	78
Wheat	100	13	100	77	100	76
90%						
Egg	7	61	95	68	98	38
Milk	15	57	94	69	95	53
Peanut	14	57	100	84	100	36
Fish	3	63	91	87	56	93
Soybean	30	44	94	81	73	82
Wheat	26	61	92	84	74	87

Note: PPV, positive predictive value; NPV, negative predictive value (discussed in Chapter 2).

Source: Based on Sampson, H.A. and Ho, D.G., *J. Allergy Clin. Immunol.*, 100(4), 444–451, 1997. With permission.

TABLE 13.7
Interpretation of Food-Specific IgE Levels (kU/L) in the Diagnosis of IgE-Mediated Food Allergy

	Egg	Milk	Peanut	Fish	Soy	Wheat	
Reactive if ≥ (no challenge necessary)	7	15	14	20	65	80	Probability of reaction
Possibly reactive (physician challenge)					30	26	↓
Unlikely reactive if < (home challenge)	0.35	0.35	0.35	0.35	0.35	0.35	

Source: Adapted from Sampson, H.A., *J. Allergy Clin. Immunol.*, 107(5), 891–896, 2001. With permission.

role of specific food antigens, and then proceed according to the response obtained [22]. In some cases, the use of anti-inflammatory agents (i.e., steroids) may be necessary, especially in cases where no food allergy is detected, something that involves up to 40% of cases [32].

13.2.1.3.3.3   Non-IgE Mediated
Here there is clearly no room for measurement of food-specific IgE antibodies. Presumptive diagnoses are characteristically reached by eliminating the suspected food antigen (most commonly milk protein) and observing the clinical response.

> *Food protein induced enteropathy syndrome (FPIES).* A definitive diagnosis requires reexposure to the offending protein-causing relapse of symptoms; however, widely used standards do not call for such a procedure; rather, most pediatric gastroenterologists (FPIES is seen with great predominance in young infants) would simply delay the reintroduction of the suspected protein to the end of the first year of life or the early part of the second year. Typically, no invasive diagnostic procedure is performed because FPIES is considered as a transient food allergy. Given the similarities with other conditions causing a malabsorption picture (and mostly celiac disease), the diagnostic workup of this condition follows that of any other malabsorptive disorder presenting in infancy and is likely to include an upper GI endoscopy with mucosal biopsies. As mentioned earlier, the pathological findings may be totally indistinguishable from those of celiac disease, including the increased number of intraepithelial lymphocytes.

Several tests on the stools are typically performed during such investigations: semi-quantitative measurement of fecal fat in a random sample (suggesting steathorrhea when positive), measurement of fecal level of calprotectin (a protein whose increased presence in the stools is a nonspecific indication of an inflammatory condition), and measurement of fecal α1-antitrypsin level (an assessment of the presence of protein-losing enteropathy). See Table 13.8 for reference ranges of these fecal measurements. The diagnostic confusion may be aggravated by the fact that specific serology for confirmed celiac disease in this age group (see Section 13.2.3) may be negative. If both gluten and milk proteins are eliminated from the diet of the affected infant

TABLE 13.8

Reference Ranges for Some Fecal Measurements Commonly Utilized in the Diagnostic Workup of Non-IgE-Mediated Food Allergies

Test	Method	Reference Range
α 1 Antitrypsin	Nephelometry	$\leq$54 mg/dL
Calprotectin	ELISA	$\leq$50 μg/g
Fat	NMR	$\leq$20%
Leukocytes	Conventional microscopy	$\leq$5 per HPF

at the time of the diagnostic workup (something that may be occasionally necessary in the presence of a severe disease) it may take a long time and repeated food
challenges before a definitive diagnosis is achieved.

> *Food protein-induced proctocolitis.* The diagnostic workup here is very limited,
> as this benign condition is diagnosed clinically (see Section 13.2.1.3.2.3b)
> and no specific test is available. The presence of increased fecal leukocytes
> can be used as a nonspecific indication of the presence of proctocolitis, with
> a sensitivity and specificity of approximately 80%.
>
> *Chronic constipation.* The current status of this relatively new entity is still ill-
> defined, and no specific recommendation for any test can be made with rea-
> sonable confidence. The immunologic changes described include increased
> peripheral eosinophilic counts, elevated serum levels of milk-specific and
> total IgE, and increased IgG to β-lactoglobulin [29].

13.2.2   INTRACTABLE DIARRHEA WITH PERSISTENT VILLOUS ATROPHY

13.2.2.1   Introduction

The syndrome of intractable diarrhea originally described by Avery et al. [33] comprises several diseases of distinct origin that occur in children within the first 2 years
of life. This rare syndrome is characterized by a diarrhea of noninfectious, nonallergic origin that is associated with villous atrophy and persists under total parenteral nutrition. A number of clinical and histological studies [34–41] have resulted
in an improved appreciation of the complexity of the syndrome and the division of
intractable diarrhea into immune versus nonimmune origin. A better understanding
of the pathology allows for more specific (targeted) therapy (immunosuppressors,
intestinal transplantation, and bone marrow transplantation), which has significantly
improved outcomes of those children suffering from intractable diarrhea.

13.2.2.2   Classification of Intractable Diarrhea

Intractable diarrhea can be separated into immune- and nonimmune-mediated disorders [35] (Figure 13.2). The immune-mediated enteropathies are caused mainly
by a defect in immune regulation, whereas the non-immune-mediated enteropathies
are related to a defect in intestinal epithelial cells. Immunohistochemistry is used to
assess the presence of immune activation by measuring the number of CD3+ T cells
in the mucosa, the expression of the T-cell activation marker CD25 (high-affinity
receptor for IL-2), and the level of expression of MHC class II by crypt epithelial
cells (reflecting the local secretion of IFN-γ). Family history, time of outset, associated diseases, histology, and particular biological features provide additional criteria
used to classify these patients.

13.2.2.2.1   Intractable Diarrhea of Nonimmune Origin
This group is composed of two main disorders: microvillus atrophy [39,42] and tuft
enteropathy [35,43]. There are no signs of immune activation in the intestinal mucosa
of these children, as attested by the absence of an increase in CD3+ T cells or very

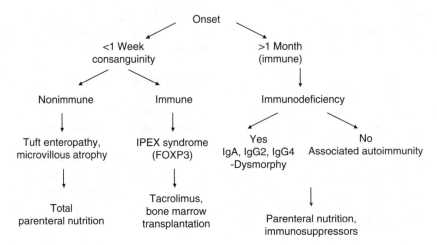

FIGURE 13.2 Classification and treatment options of intractable diarrhea with villous or microvillous atrophy.

low expression levels of MHC class II antigens on crypt epithelial cells. These are inherited (autosomal recessive) disorders whose onset is in the first days or weeks of life and is life long. Outcome has been significantly ameliorated with the improvements in parenteral nutrition management. The only curative treatment for these disorders is intestinal transplantation. As these disorders are not related to derangement of intestinal immunity, they shall not be discussed here.

13.2.2.2.2 Intractable Diarrhea of Immune Origin

This group of intractable diarrhea is characterized by more severe villous atrophy and infiltration of the intestinal mucosa by activated T cells and APC. Classically, the immunohistochemistry analysis of small intestinal biopsies shows a major CD3+ CD25+ activated T-cell infiltrate and expression of MHC class II by crypt epithelial cells [35]. In contrast to celiac disease, the T-cell infiltrate predominates in the lamina propria and there is no crypt hyperplasia. This group includes autoimmune enteropathies and children with immunodeficiencies.

a. *Autoimmune enteropathy (AE).* Unsworth and Walker-Smith first introduced the term "AE" [44]. An important step in the understanding of this entity was the identification of a fatal X-linked syndrome associating diarrhea and polyendocrinopathy [40]. Today, AE can be subdivided into two main subgroups. The first subgroup is called immune dysregulation, polyendocrinopathy, enteropathy, X-linked syndrome (IPEX) [45] because it associates immune dysregulation, polyendocrinopathy, and AE and corresponds to the syndrome initially described by Powell et al. [34,40]. IPEX has a neonatal onset and a dramatic outcome with children dying often within the first months of life. Children with IPEX present with enteropathy, diabetes, eczema, and eventually other endocrinopathies and renal diseases. The only medical treatment with some clear efficacy is tacrolimus, whereas cyclosporine has no effect [46,47]. Overall, the most promising treatment to date is bone marrow transplantation [48]. The majority, but not all, IPEX syndromes are due to a loss of function of the gene encoding FOXP3 located on the X chromosome and hence are X-linked [34,41]. FOXP3 is essential for the development of regulatory T cells, which play fundamental role in immunoregulatory functions by preventing the activation of autoreactive T cells. There are however other undefined genetic mutations not located on the X chromosome that can induce an IPEX [47]. In IPEX patients, antibodies directed against mature enterocytes (antienterocyte antibodies), goblet cells, and a gut- or kidney-specific 75 kDa antigen (AIE75) can be detected [49]. AIE75 shares 99% homology with the NY-Co-38 colon cancer antigen. Importantly, antienterocyte antibodies can also be detected in nonimmune intractable diarrhea and at low titers in IBD and cow's milk allergies. AE comprises a second group of children (girls and boys) [35,38,47] with either no extraintestinal symptoms or with other autoimmune features such as inflammatory arthritis. The onset of diarrhea in these children is classically later (after the first year of life) and the disease might be multigenetic (as in other autoimmune disorders). The treatment is similar to other autoimmune diseases and is based on the usage of immunosuppressants including steroids, azathioprine and cyclosporine. The overall outcome is less severe than in children with IPEX; and in a subset of children, total parenteral nutrition can be discontinued [35,38].

b. *Intractable diarrhea with immunodeficiencies.* These children have moderate to severe villous atrophy and increased lamina propria infiltration [35,38]. Two major types of immunodeficiencies have been associated with intractable diarrhea. Selective immunoglobulin deficiencies (IgA, IgG2, and IgG4) have been described to be associated with intractable diarrhea in a subset of children [35,38,50]. However, the pathogenic mechanism underlying the diarrhea is not understood. Another subgroup of children have a immunodeficiency characterized by defective antibody responses despite normal immunoglobulin levels, and antigen-specific skin tests despite normal proliferative T-cell responses *in vitro* [37]. In addition, these children present with low birth weight, facial dysmorphy, and trichorrhexis. Prognosis is poor and most patients die between the age of 2 and 5 years.

13.2.3 CELIAC DISEASE

13.2.3.1 Introduction

Celiac disease is an inflammatory intestinal disorder with an autoimmune component that occurs in genetically susceptible individuals in response to ingestion of the storage wheat protein gluten, or to related proteins found in barley (hordeins) and rye (secalins) [51–54]. It is a common condition that can be diagnosed at

TABLE 13.9

Main Indications for Serological Screening of Celiac Disease

Family history	A first-degree relative with celiac disease, dermatitis herpetiformis, or any autoimmune disorder
Personal history	Syndromes
	Down
	Williams
	Turner
	Failure to thrive
	Weight loss
	Chronic diarrhea
	Recurrent vomiting
	Idiopathic epilepsy
	Type 1 diabetes
	Hashimoto thyroiditis
Physical examination and	Malnutrition
laboratory data	Osteoporosis
	Iron deficiency anemia
	Peripheral neuropathy
	IgA deficiency
	Unexplained hypertransamina semia

any age. Screening studies suggest that the incidence of celiac disease approaches 1% of the population in Europe and in the United States [36,55]. It occurs almost exclusively in human leukocyte antigen (HLA) DQ2 (DQA1*05/DQB1*02) or DQ8 (DQA1*0301/DQB1*0302) individuals [56]. However, only around 3% of DQ2 or DQ8 patients develop celiac disease. The rates of diagnosis peak after introduction of gluten in the diet of children, and in the fourth and fifth decades in adults. The incidence of celiac disease is highly underestimated because the clinical spectrum is very large. Patients can be asymptomatic, have intestinal and extraintestinal symptoms alone or combined, or may have no clinical symptoms (Table 13.9). Although diarrhea is the most frequent clinical manifestation, celiac patients can have no apparent diarrhea and yet present manifestations linked to malabsorption (e.g., iron deficiency anemia or osteoporosis). Furthermore, celiac disease is associated with several autoimmune conditions such as type 1 diabetes, autoimmune thyroiditis, primary biliary cirrhosis (PBC), and autoimmune myocarditis. Celiac disease occurs with a higher prevalence in neuropathies, Down syndrome, Turner syndrome, and microscopic colitis. The treatment of celiac disease is based on the life-long exclusion of gluten and related proteins.

13.2.3.2 Diagnosis

Diagnosis is based on serological testing, histology of small intestinal biopsies, and the clinical and biological responses to gluten exclusion [37,51,57].

13.2.3.2.1   Serological Testing

Because celiac disease can be silent, it is important to perform screening with serological testing in patients at risk. This includes family members of celiac patients (concordance of 70% in monozygotic twins and prevalence of approximately 10% in first-degree relatives), patients and family members with autoimmune disorders such as type 1 diabetes, as well as other clinical conditions outlined in Table 13.9. Antigliadin antibodies and in particular IgA antigliadin antibodies have been prescribed for many years. However, because of their relatively low specificity their use for diagnosis is now challenged (2004 NIH Consensus Development on Celiac Disease). In the past years, the paramount role of human antitissue transglutaminase antibodies as an effective tool for screening has become consolidated. Before them, antiendomysium antibodies (EMA) had been found to be very useful as well. Indeed, they have a specificity approaching 100% and a sensitivity varying between 80% and 90% [58]. The identification of tissue transglutaminase 2 as the autoantigen against which anti-EMA are in fact directed [59] has led to the development of enzyme-linked immunoassays [60] that are less expensive and less observer-dependent than the EMA fluorescence test. Antitransglutaminase antibodies (based on human recombinant enzymes) are now widely used in the serological screening for celiac disease; they are overall more sensitive, but are somewhat less specific than anti-EMA. See Table 13.10 for their normal values.

IgA antiendomysium or –transglutaminase, because they reveal an immune reaction of intestinal origin, are more specific than IgG antibodies. However, in IgA-deficient patients, who have a relatively high incidence of celiac disease, only IgG-based tests can be used [61]. Recently, a serological test based on *deamidated* gliadin peptides has been developed and has become commercially available. The new test appears to have very good sensitivity and specificity for celiac disease, probably similar to the tissue transglutaminase assay [62].

13.2.3.2.2   Histology

Biopsy of the small intestine remains the gold standard for the diagnosis of celiac disease [63]. The classical features comprise villous atrophy, infiltration of the epithelium by cytotoxic intraepithelial T lymphocytes, and crypt hyperplasia. However, the spectrum can range from intraepithelial lymphocytosis and crypt hyperplasia without villous atrophy to severe villous atrophy with crypt hypoplasia [64].

TABLE 13.10

Normal Values for Human Antitissue Transglutaminase Antibodies Determined by ELISA

Negative	<20.0 U
Equivocal	20.0–30.0 U
Positive	>30.0 U

Note: Some laboratories express their results differently and report the following values: negative, <4.0 U; equivocal, 4.1–7.0 U; positive, >7.0 U.

13.2.3.3   Refractory Sprue and Enteropathy-Associated T-Cell Lymphoma

Refractory sprue and enteropathy-associated T-cell lymphoma (EATL) are two classical complications of celiac disease characterized by persistent villous atrophy despite a gluten-free diet and malignant transformation of intraepithelial T lymphocytes [52,65,66]. In some cases, the diagnosis of celiac disease is made retrospectively in presence of these complications. The diagnosis is evoked in patients with past history of celiac disease, expression of HLA DQ2/DQ8, and the past or present identification of anti-endomysium or -transglutaminase antibodies. The diagnosis of the classical form of refractory sprue is based on the identification of CD3+CD8− intraepithelial lymphocytes by immunohistochemistry [65,67]. The outcome is poor [67]. Several treatments have been attempted. The most classical one comprises steroids and azathioprine. However, because of the poor prognosis a number of alternative treatments are attempted including anti-TNF and bone marrow transplantation. These complications seem to occur exclusively in adults.

13.2.3.4   Treatment

A diet excluding gluten and related proteins remains the treatment for celiac disease today. Oats are normally not harmful in patients with celiac disease [68,69]. Most patients have a very good response to the diet and in general the quality of life improves significantly. Importantly, patients under a gluten-free diet have no increased mortality rate [70]. However, the rapidity and degree of the responses vary [71,72]. Furthermore, the constraints associated with the diet are important and can affect the social life. This is particularly the case of young adolescents and women. Finally, because gluten is ubiquitous, including as a condiment, it is very difficult and stressful for patients to follow a strict diet. All these considerations and the important progress made in understanding the pathogenesis of the disease have led a number of investigators to search for alternative therapies that may ameliorate the life conditions and health of celiac patients in the near future [51,73].

13.2.4   AUTOIMMUNE HEPATITIS

13.2.4.1   Introduction

Autoimmune hepatitis (AIH) is a chronic inflammatory hepatitis of unknown etiology with a variable evolving course occurring both in children and adults. First recognized over 50 years ago, it was initially termed as "lupoid" hepatitis due to association with antinuclear antibodies (ANA). AIH is characterized by a distinctive pattern of clinical, biochemical, and histologic findings, as well as abnormal serum immunoglobulin levels, particularly autoantibodies. A major percentage of patients respond very well to immunosuppressive therapy, and may be effectively controlled with corticosteroids and other immunosuppressive agents.

13.2.4.2   Epidemiology

The worldwide prevalence of AIH is unclear, is more common among women (approximately 4:1), present at any age, and is seen across various ethnic communities. It accounts for 7% of liver transplants every year among adults in the United States.

In Northern Europe, the incidence is 1.9 in 100,000, and causes 10–20% of chronic liver disease among adults. In children, the incidence has yet to be satisfactorily determined [74].

Classically, the presentation is of an acute hepatitis, with chronic malaise, lethargy, epigastric pain, arthralgias, fluctuating jaundice, oligomenorrhea, hirsutism, and acne. Rarely, it may be present as fulminant hepatic failure. Commonly, the disease arises in a setting of coexisting extrahepatic autoimmune diseases such as IBD, diabetes, and thyroiditis.

13.2.4.3 Pathophysiology

Three types of AIH are recognized (see Table 13.11). The presence of smooth-muscle antibody (SMA) or ANA is defined as type 1 AIH. The presence of liver/kidney microsomal type 1 (LKM1) antibody is defined as type 2 AIH. The third form of autoimmune liver disease is described in children and characterized by clinical, histological, and serological similarities with type 1 AIH, but has the discordant radiological features of sclerosing cholangitis. This condition is referred to as AIH/sclerosing cholangitis overlap syndrome or autoimmune sclerosing cholangitis (ASC).

Though the specific provoking agents are unclear, various environmental causes, viruses, drugs, and herbal preparations have been postulated as potential triggers of the T-cell mediated cascade of events directed at specific liver antigens. There have been extensive studies of molecular mimicry between various liver antigens and epitopes of viruses, particularly the hepatitis viruses [75]. There is a clear genetic predisposition, with HLA DRB1*0301(DR3) and DRB1*0401(DR4) as the susceptibility alleles for type I AIH seen in up to 80% of white North Americans and northern

TABLE 13.11
Types of Autoimmune Hepatitis

Findings	Type 1 AIH	Type 2 AIH	Type 3 Overlap
Autoantibodies	ANA <1:40, <1:20 (pediatric)	Antibody to LKM-1 <1:10	Anti-SLA/LP
	SMA <1:40, <1:20 (pediatric) Antiactin antibody Antibody to soluble liver antigen Atypical pANCA	Antibody to LC-1	ANA, SMA, anti-ASGPR
Age at presentation	Any age	Childhood and young adulthood	Any age
Sex	Female 75%	Female 93%	Female 75%
Associated autoimmune disorders	Autoimmune thyroiditis, ulcerative colitis, synovitis, HLA DR3/DR4 association		Same as type 1
Severity	Broad	Severe	
Treatment failure	Infrequent	Frequent	

Europeans. In Japan, Mainland China, and Mexico, other diverse alleles of HLA DR4 have been associated with the disease. DRB1*1301 has been found associated strongly with AIH among South American children [76]. In type II AIH, the presence of immunodominant B-cell epitopes of cytochrome P-450 2D6 (CYP2D6) and evidence of cross-reactivity with homologues of different viruses suggest that the relevant antigens may exist within the mitochondrial oxygen chain [75].

Recent studies suggest that loss of CD4+ regulatory T cells may be involved in the pathogenesis of AIH. Immunoregulatory dysfunction characterized by decreased numbers of CD4+CD25+ regulatory T cells and decreased levels of scurfin, a transcription factor, and the protein product of the FOXP3 gene has been reported [77].

13.2.4.4 Clinical Presentation

The clinical course of AIH is marked with periodic levels of disease activity, during which the patient may complain of fatigue, lethargy, malaise, anorexia, nausea, abdominal pain, pruritus, and arthralgias of the small joints. It should be considered in all who present with systemic or rheumatic features where a background liver condition can be easily overlooked. Acute onset is not uncommon, with patients presenting with fulminant symptoms accompanied by jaundice, prolonged prothrombin time, and profound "transaminitis" [78]. On physical examination, hepatomegaly, splenomegaly, jaundice, and other signs of chronic liver disease may be evident.

13.2.4.5 Extrahepatic Manifestations

13.2.4.5.1 Associated Conditions
AIH has been associated with IgA nephropathy, autoimmune thyroiditis, ulcerative colitis (UC), type 1 diabetes, rheumatoid arthritis, and celiac disease. Coexisting elevations of antitissue transglutaminase antibodies are worth assessing.

13.2.4.5.2 Diagnosis

13.2.4.5.2.1 Autoantibodies
The standard technique used for the detection of autoantibodies is indirect immunofluorescence (IIF). The diagnostic paradigm is developed from a composite substrate of freshly prepared liver, kidney, and stomach from rodents. These three tissues enable the detection of virtually all autoantibodies relevant to the liver autoimmune serology, namely SMA, ANA, anti-LKM1, antimitochondrial antibody (AMA), and anti-liver cytosol type 1 (LC1). See Table 13.12 for the diagnostic methods available.

In adults, the patient serum dilution recommended for autoantibody detection is 1:40, whereas in childhood even titers below 1:20 for ANA and SMA and 1:10 for anti-LKM1 are significant.

Anti-LKM1 is frequently misdiagnosed being commonly confused with AMA, which is a hallmark of PBC. AMA targets an antigen collectively known as M2, which includes enzymes of the 2-oxo-acid dehydrogenase complexes. The confusion occurs because both stain the renal tubules: AMA stains the distal smaller tubules, whereas anti-LKM1 stains the larger proximal ones. AMA also stains the gastric

TABLE 13.12
Diagnostic Methods for Autoimmune Hepatitis

Technique	Autoantibody	Pattern of Staining	Diagnosis
IIF	ANA (<1:20)	Nuclear staining L, K, S HEp2 cell line (pattern definition)	AIH 1—homogeneous, speckled
IIF	SMA (<1:20)	Microfilaments—vascular (V), glomeruli (G), tubules (T) Muscularis mucosae S Arterial walls L, K, S	AIH 1—VGT pattern (80%)
IIF, ELISA	LKM1 (<1:10)	Proximal renal tubules K Hepatocyte cytoplasm L Targets CYP2D6	AIH 2
IIF	ANCA	Neutrophils—atypical pANCA reacts with nuclear membrane components	AIH and ASC—perinuclear
IIF, CIE, DDI	LC1	Liver cytosol—formimino-transferase cyclodeam-inase (FTCD)	
ELISA, RLA	SLA	Soluble liver antigen target—UGA tRNA suppressor-associated antigenic protein (tRNP(Ser)Sec)	

parietal cells of the stomach (which is spared by anti-LKM1 antibodies), whereas anti-LKM1 stains the hepatocytes more strongly. As PBC is rare in the pediatric age group, when AMA is reported in a child with characteristics of AIH, the results should be double-checked. To correctly interpret the diagnosis, the three-tissue substrate of liver, kidney, and stomach is ideal although not typically done in all laboratories. Another way the clinician can verify the diagnosis is with an enzyme-linked immunosorbent assay (ELISA) to detect anti-CYP2D6 and anti-M2. A positive result will confirm the diagnosis of AIH type 2.

In addition to the classical autoantibodies, others are more useful for their diagnostic and prognostic values. These include anti-LC1, originally described in association with AIH type 2 and present in conjunction with anti-LKM1, although it can be detected alone. Anti-soluble liver antigen (SLA) is often associated with a severe clinical course and a poorer outcome. The majority of anti-SLA positive patients are positive for ANA, SMA, or anti-LKM1, although occasionally it is present in isolation. The presence of this autoantibody is of diagnostic importance. Atypical perinuclear antineutrophil cytoplasmic antibody (ANCA) may be present in type 1 AIH as well as in ASC, and virtually absent in type 2 AIH.

13.2.4.5.2.2   Antigen-Based Assays
Identification of the antigenic targets of several autoantibodies, including anti-LKM1, anti-LC1, anti-SLA, and AMA has led to the establishment of much more straightforward commercially based assays. For example, the results of anti-LKM1

identified by IIF and anti-CYP2D6 by ELISA correlate well, thus ELISA is preferable for the diagnosis [79].

13.2.4.5.3 Management
As defined by the American Association for the Study of Liver Diseases, treatment should be instituted in patients with a 10-fold increase in serum aminotransferase levels. Also, those with fivefold elevated levels in association with twofold increase in serum gamma globulins should be treated. Histologic features of bridging necrosis or multiacinar necrosis necessitates therapy as well. In most children, treatment is recommended at the time of diagnosis. The mainstay of therapy for AIH is prednisone alone or in conjunction with azathioprine.

AIH responds to immunosuppressive treatment. It can present insidiously or as an acute hepatitis. The previously accepted requirement of 6-month duration of symptoms before diagnosis has been abandoned and treatment should be instituted as soon as the disease is diagnosed [77].

13.2.5 INFLAMMATORY BOWEL DISEASE

13.2.5.1 Introduction

IBD is a chronic unremitting inflammatory condition of the GIT divided clinically and histologically into two primary types: UC and CD. The hallmark of the disease process is a chronic, unrestrained inflammation of the intestinal mucosa affecting various areas, with UC involving an inflammatory response restricted to the lamina propria of the colon, beginning at the rectum and spreading proximally to varying extents, whereas with CD, the inflammation emerges anywhere in the digestive tract, with a large majority involving the terminal ileum (>70%).

The diagnosis is based on the presence of microscopic changes in the mucosa, principally architectural distortion and infiltration by acute inflammatory cells. Although nonspecific inflammation can be present in normal GITs, in IBD the distinguishing feature is an inability to down modulate the response to contact with pathogens. This persistent state of inflammation is believed to be due to an exaggerated mucosal immune response, the etiology of which remains poorly defined. However, several factors under scrutiny include alterations in the intestinal mucosal barrier [80], changes in the luminal bacterial flora [81], impairment of T-cell activation [82], gut homing [83], and multiple genetic factors [84–86].

13.2.5.2 Epidemiology

A strikingly increased prevalence of IBD has been demonstrated in developed nations, with an estimated 1 million individuals with IBD living currently in the United States, and nearly 30,000 new cases reported annually. The analysis of current prevalence data with respect to racial and ethnic subgroups indicates higher rates among Caucasian and Ashkenazi Jewish origins than among other backgrounds [87]. It is believed to be a disease of clean environments, higher socioeconomic classes, and white-collar workers. The age incidence of UC and CD is evenly divided, with the peak age of onset at 15–30 years of age, and a second smaller peak at 50–70 years [88].

13.2.5.3   Etiology

Although the exact origins of the disease is unknown, IBD occurs in the setting of a multifactorial background, including genetic factors, environmental triggers, mucosal immune defects, and microbial exposures. Current evidence indicates that defective T-cell apoptosis and impairment of the intestinal epithelial barrier function play important roles in the pathogenesis of both the conditions [82].

13.2.5.4   Pathophysiology

13.2.5.4.1   Genetics
A family history of IBD confers a risk to first-degree relatives that is 4–20 times higher than within the general population [87]. A positive family history of CD in a first-degree relative is the major identified risk factor for its development. The clustering of disease among specific populations such as monozygotic twins (44.4–58% rate versus 0–3.8% among dizygotic twins), ethnic groups (Ashkenazi Jews and Caucasians), particular genetic disorders, and within families indicates that multiple susceptibility loci are involved.

Though a simple Mendelian inheritance defect has eluded investigators, in recent years genome-wide screening with microsatellite DNA markers has identified several genetic segments as being potentially associated with UC or CD. One particular site, designated *IBD1*, is an area on chromosome 16 of apparent linkage among CD kindreds, but not those with UC. A gene identified within this region is designated as NOD2 [84] (referred to as caspase activation and recruitment domain [CARD 15]), encodes a cytoplasmic protein, which is expressed in macrophages as a pattern-recognition receptor for bacterial LPS. A high proportion of European and North American CD patients were found to have variants of NOD2 independent of family history [89]. Recent studies have revealed an association between NOD2 mutations and earlier onset of stricturing disease [90]. Thus, this mutation may be used to predict severe stricturing CD, but it does not seem to fully explain the disease process [85].

Immunologically, UC seems to be mediated by Th2 cytokines (IL-4, IL-13) involving specialized cells such as NK T cells (IL-13), and lamina propria T cells that produce significant amounts of IL-13 and are cytotoxic to epithelial cells. Genetically, the MHC class II (IBD3) region on chromosome 6, HLA*DR2, and HLA-DR5 have been associated with UC [91]. Furthermore, MHC class II alleles have been associated more strongly with the pANCA positive UC subpopulation as compared to the UC population as a whole [92].

13.2.5.4.2   Environmental Impact
Although genetic factors can be used to explain some of the compelling familial and ethnic correlations, the prevalence of sporadic disease, which accounts for the majority of cases, argues that a distinct environmental cause could be implicated as well. Interestingly, the incidence rates among particular ethnic groups have been shown to converge based on a change in geography and lifestyles, as demonstrated in migrated second-generation South Asian families. This cohort, now living within

developed countries, have a notable rise in incidence of IBD, comparable to Caucasians and vastly increased relative to their native cousins.

Epidemiological studies have identified a gap in incidence of IBD among various ethnic and socioeconomic classes, and particularly wealthy nations suggesting indirectly that exposure to cleaner environments is part of this environmental trigger. It is thought that the lack of exposure to certain pathogens during the development of mucosal defenses prevents the normal maturation of oral tolerance and T-cell regulation and leads to the adaptive immune defects of IBD.

13.2.5.4.3   Defects in Intestinal Barrier and Innate Immunity
The intestinal barrier may be impaired in IBD. The barrier consists of the biofilm and mucous layer, epithelial cells, and innate immune cells including dendritic cells, Paneth cells, macrophages, and neutrophils. Any defects in these elements can lead to impaired immune regulation. Representative murine models with specific barrier function defects have demonstrated chronic colitis. Of interest is another recently identified susceptibility gene mutation in IBD, *OCTN1* (in the *IBD5* locus), which is primarily expressed in the epithelial layer [86].

13.2.5.5   Bacterial Signal Recognition Defects

The NOD2 gene is involved in cytoplasmic recognition of Gram-negative bacterial LPS. When activated, it leads to NF-κB activation and initiation of the inflammatory response and cytokine expression ultimately leading to tissue destruction. Although NOD2 mutations paradoxically should reduce macrophage activation of NF-κB, this associated genetic mutation suggests that there are multiple innate immune defects in bacterial identification that collectively lead to the unrestrained inflammatory responses seen in IBD [84].

13.2.5.5.1   Paneth Cell and Defensins
Paneth cells are granulated epithelial cells at the base of small intestinal crypts. They occur in abundance in the terminal ileum and are not found in normal colon. However, murine models of colitis demonstrate Paneth cell metaplasia, something that has also been shown in the colon of patients with IBD. In response to bacterial exposure, they release α-defensins, lysozyme, and sPLA2, which are all antibacterial factors. Defensins, in particular, are a family of small peptides that are divided into alpha and beta types.

13.2.5.5.2   Intestinal Flora
The commensal bacteria of the GIT thrive in a symbiotic relationship with gut mucosa in normal individuals. These bacteria provide a competitive environment against other more pathogenic bacteria for space and nutrients, benefiting the host by protecting the microenvironment, providing the synthesis of vitamin K, and breaking down short-chain fatty acids and other waste products. More than 400 different species of commensal bacteria populate the GIT and are believed to be necessary for the normal development of GALT and mucosal immunity. This has been shown by experiments involving germ-free mice [93], and it is exposure to this environment that is thought to propagate the development of selective immune regulation and restraint, referred to as oral tolerance [94]. In IBD, oral tolerance is

poorly governed [95] with the net result being an abnormal innate response to commensal flora leading to activation of unrestrained pathogenic CD4+ T cells, and the subsequent IBD [96].

13.2.5.5.3 Defects in Adaptive Immunity

Normally, activation of CD4 T cells allows differentiation toward the Th1 and Th2 cell types depending on the nature of the stimulatory signal, the background cytokine milieu, and the corresponding costimulatory signals. In CD, the major Th1-inducing cytokine is IL-12, synthesized *en masse* by activated macrophages. The downstream effect of activated Th1 cells includes TNF-α production, which contributes to the recruitment of monocytes, lymphocytes, and neutrophils; and release of mediators causing localized tissue damage. These activated T cells poorly recognize apoptotic signals, thus leading to dysregulated T-cell activity.

13.2.5.6 Clinical Features

GI manifestations of IBD include the presence of diarrhea, grossly bloody, mucoid stools, accompanied by lower abdominal cramping and tenesmus, often with systemic symptoms of fever, anorexia, weight loss, and anemia. In UC, a serious complication is toxic megacolon, leading to acute dilation of the colon, with potential for rupture and intraperitoneal leakage.

13.2.5.6.1 Extraintestinal Manifestations

At least one extraintestinal manifestation is seen in approximately 25–35% of adult patients with IBD. Joint involvement is the most frequent manifestation, with a prevalence of 10–35%. Peripheral arthritis is very common, whereas AS affects 2–8%. Table 13.13 summarizes a few of the findings based on systems.

TABLE 13.13
Main Extraintestinal Manifestations of IBD

Organ	Lesion
Skin	Erythema nodosum (6–15% of CD patients)
	Pyoderma gangrenosum
	Oral–facial granulomatosis
	Inflammatory lymphedema
Joints	Arthralgias, arthritis
	Juvenile ankylosing spondylitis (AS)
	Peripheral arthritis
Pancreas	Acute pancreatitis (UC or CD)
	Secondary to duodenal CD, PSC, or drug therapy
Renal	Oxalate, urate, and phosphate stones
Hepatobiliary	Primary sclerosing cholangitis (UC > CD)
Eye	Acute episcleritis, uveitis, orbital myositis

13.2.5.6.2   Diagnosis

The diagnosis of CD rests on clinical presentation, laboratory studies, radiologic imaging of the small intestine, endoscopy, and pathology of mucosal biopsies, with the exclusion of infectious causes (yersinia, amebiasis, intestinal tuberculosis [TB], and schistosomiasis) of diarrhea.

Routine laboratory studies include comprehensive blood counts with differential, which may show acute on chronic anemia, elevated WBC counts, with a high percentage of neutrophils, and elevated platelet counts indicating inflammation. Additional studies of inflammation are routinely done, including erythrocyte sedimentation rate (ESR) and C-reactive protein (CRP), which in active inflammation are often significantly elevated.

Often, IBD presents in the form of acute infectious diarrhea, therefore, stool studies should be done to reveal the presence of pathogens, especially invasive bacteria such as *C. difficile, Salmonella, Shigella, Yersinia, Campylobacter*, enterohemorrhagic *E. coli* (EHEC), and also rotavirus or adenovirus serotype 40/41. Additional stool studies reveal the presence of WBC and RBC.

13.2.5.6.3   Specific Serological Tests

ANCAs and anti-*Saccharomyces cerevisiae* antibody (ASCA) are screening tools for IBD, and can help differentiate UC from CD [97]. The ANCAs are IgG antibodies with a perinuclear neutrophil-staining pattern. It is more specific to UC, although it has poor sensitivity and is only present in 50–60% of those with UC. Additionally, pANCA is positive in up to 35% of patients with colonic CD. ASCA is the IgG and IgA antibody to cell-wall mannose sequences of *S. cerevisiae*, a species of budding yeast used for baking and brewing. This antibody is detected in approximately 55% of those with CD and 5–10% of controls with other GI issues, thus indicative of a relatively good specificity for CD with poor sensitivity.

13.2.5.6.4   Pathology

Esophago-gastro-duodenoscopy, ileocolonoscopy with histopathological study of biopsies, and x-ray series with small bowel follow-through are necessary for diagnosis of IBD. Pathology rests on finding focal areas of inflammation and granulomas within the mucosal biopsies in the case of CD. In UC, serial biopsies will demonstrate the involvement of the mucosal layer in contiguous areas of the colon.

13.2.5.7   Treatment

Corticosteroids are the mainstay of treatment of acute exacerbations used primarily for their anti-inflammatory effects. Chronic use of steroids is however limited by a long list of possible adverse events. Newer agents, such as the topically active budesonide, have limited systemic bioavailability, and offer an alternative corticosteroid in enteric formulation [98]. In a recent Cochrane database review [99], it was found that budesonide can also be a therapeutic alternative from conventional corticosteroids for disease in the ileum or ascending colon associated with fewer adverse effects, presumably due to poorer absorption. Also, the efficacy of oral 5-ASA agents for the maintenance of remission in IBD has been well documented.

Table 13.14 lists currently available therapeutic options for IBD based on type and location.

TABLE 13.14
Treatment Options for IBD

Course	Distal UC	Extensive UC	Crohn's Disease
Mild	Oral/rectal aminosalyicylates Rectal corticosteroids	Oral aminosalicylates	Oral aminosalicylates
Moderate	Oral/rectal aminosalyicylates Rectal corticosteroids	Oral aminosalicylates	Oral aminosalicylates
Severe	Oral/IV corticosteroids Rectal corticosteroids	Oral/IV corticosteroids IV cyclosporine	Oral/IV corticosteroids SC/IV methotrexate IV Infliximab
Refractory	Oral/IV corticosteroids + oral 6-mercaptopurine (MP)	Oral/IV corticosteroids + oral 6-MP	IV infliximab
Perianal			Oral antibiotic IV infliximab Oral 6-MP
Remission	Oral/rectal aminosalicylates Oral 6-MP	Oral/rectal aminosalicylates Oral 6-MP	Oral 6-MP, mesalamine, metronidazole

Source: Adapted from Jabri, B., Kasarda, D.D., and Green, P.H., *Immunol. Rev.*, 206, 219–231, 2005. With permission.

ABBREVIATIONS

AE	Autoimmune enteropathy
AIH	Autoimmune hepatitis
AMA	Antimitochondrial antibody
ANA	Antinuclear antibody
ANCA	Antinuclear cytoplasmic antibody
APC	Antigen-presenting cells
ASC	Autoimmune sclerosing cholangitis
ASCA	Anti-Saccharomyces antibody
CD	Crohn's disease
CRP	C-reactive protein
DBPCFC	Double-blind, placebo-controlled food challenge
EHEC	Enterohemorrhagic *E. coli*
EMA	Antiendomysium antibodies
ESR	Erythrocyte sedimentation rate
FPIES	Food protein–induced enterocolitis syndrome
GALT	Gut-associated lymphoid tissue
GI	Gastrointestinal
HLA	Human leukocyte antigen
IBD	Inflammatory bowel disease
IFN-γ	Interferon γ
IκBa	Nuclear factor kappa B inhibitor

IL	Interleukin
IPEX	Immune dysregulation, polyendocrinopathy, enteropathy, X-linked syndrome
LKM1	Liver/kidney microsomal type 1
LPS	Lipopolysaccharide
MAdCAM	Mucosal addressin cell-adhesion molecule
MALT	Mucosal-associated lymphoid tissue
MHC	Major histocompatibility complex
MIC-A, MIC-B	MHC class-I-related chains (MIC) A and B
NfκB	Nuclear factor kappa B
NK	Natural killer
PBC	Primary biliary cirrhosis
PP	Peyer's patches
SLA	Antisoluble liver antigen
SMA	Smooth-muscle antibody
sPLA2	Secretory phospholipase A2
TGF-β	Tissue growth factor β
TLR	Toll-like receptor
TNF-α	Tumor necrosis factor α
UC	Ulcerative colitis

REFERENCES

1. Cheroutre, H., IELs: enforcing law and order in the court of the intestinal epithelium. *Immunol Rev*, 2005, 206: 114–131.
2. Elphick, D.A. and Y.R. Mahida, Paneth cells: their role in innate immunity and inflammatory disease. *Gut*, 2005, 54(12): 1802–1809.
3. Brayden, D.J., M.A. Jepson, and A.W. Baird, Keynote review: intestinal Peyer's patch M cells and oral vaccine targeting. *Drug Discov Today*, 2005, 10(17): 1145–1157.
4. Man, A.L., M.E. Prieto-Garcia, and C. Nicoletti, Improving M cell mediated transport across mucosal barriers: do certain bacteria hold the keys? *Immunology*, 2004, 113(1): 15–22.
5. Chehade, M. and L. Mayer, Oral tolerance and its relation to food hypersensitivities. *J Allergy Clin Immunol*, 2005, 115(1): 3–12; quiz 13.
6. Langermann, S., S. Palaszynski, A. Sadziene, C.K. Stover, and S. Koenig, Systemic and mucosal immunity induced by BCG vector expressing outer-surface protein A of *Borrelia burgdorferi*. *Nature*, 1994, 372(6506): 552–555.
7. Mowat, A.M., Anatomical basis of tolerance and immunity to intestinal antigens. *Nat Rev Immunol*, 2003, 3(4): 331–341.
8. Muller, C.A., I.B. Autenrieth, and A. Peschel, Innate defenses of the intestinal epithelial barrier. *Cell Mol Life Sci*, 2005, 62(12): 1297–1307.
9. Mowat, A.M., O.R. Millington, and F.G. Chirdo, Anatomical and cellular basis of immunity and tolerance in the intestine. *J Pediatr Gastroenterol Nutr*, 2004, 39(Suppl 3): S723–S724.
10. Starr, J., *Clostridium difficile* associated diarrhoea: diagnosis and treatment. *BMJ*, 2005, 331(7515): 498–501.
11. Cordero, L., R. Rau, D. Taylor, and L.W. Ayers, Enteric gram-negative bacilli bloodstream infections: 17 years' experience in a neonatal intensive care unit. *Am J Infect Control*, 2004, 32(4): 189–195.

12. Kelsall, B.L. and F. Leon, Involvement of intestinal dendritic cells in oral tolerance, immunity to pathogens, and inflammatory bowel disease. *Immunol Rev*, 2005, 206: 132–148.

13. Rastall, R.A., G.R. Gibson, H.S. Gill, F. Guarner, T.R. Klaenhammer, B. Pot, G. Reid, I.R. Rowland, and M.E. Sanders, Modulation of the microbial ecology of the human colon by probiotics, prebiotics and synbiotics to enhance human health: an overview of enabling science and potential applications. *FEMS Microbiol Ecol*, 2005, 52(2): 145–152.

14. Russell, G.J., H.S. Winter, V.L. Fox, and A.K. Bhan, Lymphocytes bearing the gamma delta T-cell receptor in normal human intestine and celiac disease. *Hum Pathol*, 1991, 22(7): 690–694.

15. Mottet, C., H.H. Uhlig, and F. Powrie, Cutting edge: cure of colitis by CD4+CD25+ regulatory T cells. *J Immunol*, 2003, 170(8): 3939–3943.

16. Brandtzaeg, P. and F.E. Johansen, Mucosal B cells: phenotypic characteristics, transcriptional regulation, and homing properties. *Immunol Rev*, 2005, 206: 32–63.

17. Moneret-Vautrin, D.A. and M. Morisset, Adult food allergy. *Curr Allergy Asthma Rep*, 2005, 5(1): 80–85.

18. Sicherer, S.H., A. Munoz-Furlong, and H.A. Sampson, Prevalence of seafood allergy in the United States determined by a random telephone survey. *J Allergy Clin Immunol*, 2004, 114(1): 159–165.

19. Sicherer, S.H., Clinical aspects of gastrointestinal food allergy in childhood. *Pediatrics*, 2003, 111(6 Pt 3): 1609–1616.

20. Guandalini, S. and A. Nocerino. *Food Protein Intolerance*, e-Medicine, 2003–2005, available at http://www.emedicine.com/ped/topic1908.htm.

21. Mari, A., B.K. Ballmer-Weber, and S. Vieths, The oral allergy syndrome: improved diagnostic and treatment methods. *Curr Opin Allergy Clin Immunol*, 2005, 5(3): 267–273.

22. Rothenberg, M.E., Eosinophilic gastrointestinal disorders (EGID). *J Allergy Clin Immunol*, 2004, 113(1): 11–28; quiz 29.

23. Liacouras, C.A. and E. Ruchelli, Eosinophilic esophagitis. *Curr Opin Pediatr*, 2004, 16(5): 560–566.

24. Sgouros, S.N., C. Bergele, and A. Mantides, Eosinophilic esophagitis in adults: a systematic review. *Eur J Gastroenterol Hepatol*, 2006, 18(2): 211–217.

25. Sicherer, S.H., Food protein-induced enterocolitis syndrome: clinical perspectives. *J Pediatr Gastroenterol Nutr*, 2000, 30(Suppl): S45–S49.

26. Nowak-Wegrzyn, A., H.A. Sampson, R.A. Wood, and S.H. Sicherer, Food protein-induced enterocolitis syndrome caused by solid food proteins. *Pediatrics*, 2003, 111(4 Pt 1): 829–835.

27. Levy, Y. and Y.L. Danon, Food protein-induced enterocolitis syndrome—not only due to cow's milk and soy. *Pediatr Allergy Immunol*, 2003, 14(4): 325–329.

28. Iacono, G., F. Cavataio, G. Montalto, A. Florena, M. Tumminello, M. Soresi, A. Notarbartolo, and A. Carroccio, Intolerance of cow's milk and chronic constipation in children. *N Engl J Med*, 1998, 339(16): 1100–1104.

29. Turunen, S., T.J. Karttunen, and J. Kokkonen, Lymphoid nodular hyperplasia and cow's milk hypersensitivity in children with chronic constipation. *J Pediatr*, 2004, 145(5): 606–611.

30. Sampson, H.A. and D.G. Ho, Relationship between food-specific IgE concentrations and the risk of positive food challenges in children and adolescents. *J Allergy Clin Immunol*, 1997, 100(4): 444–451.

31. Sampson, H.A., Utility of food-specific IgE concentrations in predicting symptomatic food allergy. *J Allergy Clin Immunol*, 2001, 107(5): 891–896.

32. Guajardo, J.R., L.M. Plotnick, J.M. Fende, M.H. Collins, P.E. Putnam, and M.E. Rothenberg, Eosinophil-associated gastrointestinal disorders: a world-wide-web based registry. *J Pediatr*, 2002, 141(4): 576–581.

33. Avery, G.B., O. Villavicencio, J.R. Lilly, and J.G. Randolph, Intractable diarrhea in early infancy. *Pediatrics*, 1968, 41(4): 712–722.

34. Bennett, C.L., J. Christie, F. Ramsdell, M.E. Brunkow, P.J. Ferguson, L. Whitesell, T.E. Kelly, F.T. Saulsbury, P.F. Chance, and H.D. Ochs. The immune dysregulation, polyendocrinopathy, enteropathy, X-linked syndrome (IPEX) is caused by mutations of FOXP3. *Nat Genet*, 2001, 27(1): 20–21.
35. Cuenod, B., N. Brousse, O. Goulet, S. De Potter, J.F. Mougenot, C. Ricour, D. Guy-Grand, and N. Cerf-Bensussan. Classification of intractable diarrhea in infancy using clinical and immunohistological criteria. *Gastroenterology*, 1990, 99(4): 1037–1043.
36. Davidson, G.P., E. Cutz, J.R. Hamilton, and D.G. Gall. Familial enteropathy: a syndrome of protracted diarrhea from birth, failure to thrive, and hypoplastic villus atrophy. *Gastroenterology*, 1978, 75(5): 783–790.
37. Girault, D., O. Goulet, F. Le Deist, N. Brousse, V. Colomb, J.P. Césarini, S. de Potter, D. Canioni, C. Griscelli, and A. Fischer. Intractable infant diarrhea associated with phenotypic abnormalities and immunodeficiency. *J Pediatr*, 1994, 125(1): 36–42.
38. Goulet, O.J., N. Brousse, D. Canioni, J.A. Walker-Smith, J. Schmitz, and A.D. Phillips. Syndrome of intractable diarrhoea with persistent villous atrophy in early childhood: a clinicopathological survey of 47 cases. *J Pediatr Gastroenterol Nutr*, 1998, 26(2): 151–161.
39. Phillips, A.D., P. Jenkins, F. Raafat, and J.A. Walker-Smith. Congenital microvillous atrophy: specific diagnostic features. *Arch Dis Child*, 1985, 60(2): 135–140.
40. Powell, B.R., N.R. Buist, and P. Stenzel, An x-linked syndrome of diarrhea, polyendocrinopathy, and fatal infection in infancy. *J Pediatr*, 1982, 100(5): 731–737.
41. Wildin, R.S., F. Ramsdell, J. Peake, F. Faravelli, J.L. Casanova, N. Buist, E. Levy-Lahad, M. Mazzella, O. Goulet, L. Perroni, F.D. Bricarelli, G. Byrne, M. McEuen, S. Proll, M. Appleby, and M.E. Brunkow. X-linked neonatal diabetes mellitus, enteropathy and endocrinopathy syndrome is the human equivalent of mouse scurfy. *Nat Genet*, 2001, 27(1): 18–20.
42. Pecache, N., S. Patole, R. Hagan, D. Hill, A. Charles, and J.M. Papadimitriou, Neonatal congenital microvillus atrophy. *Postgrad Med J*, 2004, 80(940): 80–83.
43. Goulet, O., M. Kedinger, N. Brousse, B. Cuenod, V. Colomb, N. Patey, S. de Potter, J.F. Mougenot, D. Canioni, and N. Cerf-Bensussan, Intractable diarrhea of infancy with epithelial and basement membrane abnormalities. *J Pediatr*, 1995, 127(2): 212–219.
44. Unsworth, D.J. and J.A. Walker-Smith, Autoimmunity in diarrhoeal disease. *J Pediatr Gastroenterol Nutr*, 1985, 4(3): 375–380.
45. Bennett, C.L. and H.D. Ochs, IPEX is a unique x-linked syndrome characterized by immune dysfunction, polyendocrinopathy, enteropathy, and a variety of autoimmune phenomena. *Curr Opin Pediatr*, 2001, 13(6): 533–538.
46. Bousvaros, A., A.M. Leichtner, L. Book, A. Shigeoka, J. Bilodeau, E. Semeao, E. Ruchelli, and A.E. Mulberg, Treatment of pediatric autoimmune enteropathy with tacrolimus (FK506). *Gastroenterology*, 1996, 111(1): 237–243.
47. Ruemmele, F.M., N. Brousse, and O. Goulet, Autoimmune enteropathy: molecular concepts. *Curr Opin Gastroenterol*, 2004, 20(6): 587–591.
48. Baud, O., O. Goulet, D. Canioni, F. Le Deist, I. Radford, D. Rieu, S. Dupuis-Girod, N. Cerf-Bensussan, M. Cavazzana-Calvo, N. Brousse, A. Fischer, and J.L. Casanova, Treatment of the immune dysregulation, polyendocrinopathy, enteropathy, x-linked syndrome (IPEX) by allogeneic bone marrow transplantation. *N Engl J Med*, 2001, 344(23): 1758–1762.
49. Kobayashi, I., K. Imamura, M. Kubota, S. Ishikawa, M. Yamada, H. Tonoki, M. Okano, W.B. Storch, T. Moriuchi, Y. Sakiyama, and K. Kobayashi, Identification of an autoimmune enteropathy-related 75-kilodalton antigen. *Gastroenterology*, 1999, 117(4): 823–830.
50. McCarthy, D.M., S.I. Katz, L. Gazze, T.A. Waldmann, D.L. Nelson, and W. Strober, Selective IgA deficiency associated with total villous atrophy of the small intestine and an organ-specific anti-epithelial cell antibody. *J Immunol*, 1978, 120(3): 932–938.
51. Green, P.H. and B. Jabri, Coeliac disease. *Lancet*, 2003, 362(9381): 383–391.
52. Jabri, B., D.D. Kasarda, and P.H. Green, Innate and adaptive immunity: the yin and yang of celiac disease. *Immunol Rev*, 2005, 206: 219–231.

53. Sollid, L.M., Coeliac disease: dissecting a complex inflammatory disorder. *Nat Rev Immunol*, 2002, 2: 647–655.
54. Sollid, L.M. and B. Jabri, Is celiac disease an autoimmune disorder? *Curr Opin Immunol*, 2005, 17(6): 595–600.
55. Fasano, A., I. Berti, T. Gerarduzzi, T. Not, R.B. Colletti, S. Drago, Y. Elitsur, P.H. Green, S. Guandalini, I.D. Hill, M. Pietzak, A. Ventura, M. Thorpe, D. Kryszak, F. Fornaroli, S.S. Wasserman, J.A. Murray, and K. Horvath, Prevalence of celiac disease in at-risk and not-at-risk groups in the United States: a large multicenter study. *Arch Intern Med*, 2003, 163(3): 286–292.
56. Louka, A.S. and L.M. Sollid, HLA in coeliac disease: unravelling the complex genetics of a complex disorder. *Tissue Antigens*, 2003, 61(2): 105–117.
57. Hill, I.D., M.H. Dirks, G.S. Liptak, R.B. Colletti, A. Fasano, S. Guandalini, E.J. Hoffenberg, K. Horvath, J.A. Murray, M. Pivor, and E.G. Seidman, Guideline for the diagnosis and treatment of celiac disease in children: recommendations of the North American Society for Pediatric Gastroenterology, Hepatology and Nutrition. *J Pediatr Gastroenterol Nutr*, 2005, 40(1): 1–19.
58. Rostom, A., C. Dube, A. Cranney, N. Saloojee, R. Sy, C. Garritty, M. Sampson, L. Zhang, F. Yazdi, V. Mamaladze, I. Pan, J. MacNeil, D. Mack, D. Patel, and D. Moher, The diagnostic accuracy of serologic tests for celiac disease: a systematic review. *Gastroenterology*, 2005, 128(4 Suppl 1): S38–S46.
59. Dieterich, W., T. Ehnis, M. Bauer, P. Donner, U. Volta, E.O. Riecken, and D. Schuppan, Identification of tissue transglutaminase as the autoantigen of celiac disease. *Nat Med*, 1997, 3(7): 797–801.
60. Dieterich, W., E. Laag, H. Schopper, U. Volta, A. Ferguson, H. Gillett, E.O. Riecken, and D. Schuppan, Autoantibodies to tissue transglutaminase as predictors of celiac disease. *Gastroenterology*, 1998, 115(6): 1317–1321.
61. Korponay-Szabó, I.R., I. Dahlbom, K. Laurila, S. Koskinen, N. Woolley, J. Partanen, J.B. Kovács, M. Mäki, and T. Hansson, Elevation of IgG antibodies against tissue transglutaminase as a diagnostic tool for coeliac disease in selective IgA deficiency. *Gut*, 2003, 52(11): 1567–1571.
62. Sugai, E., H. Vázquez, F. Nachman, M.L. Moreno, R. Mazure, E. Smecuol, S. Niveloni, A. Cabanne, Z. Kogan, J.C. Gómez, E. Mauriño, and J.C. Bai, Accuracy of testing for antibodies to synthetic gliadin-related peptides in celiac disease. *Clin Gastroenterol Hepatol*, 2006, 4(9): 1112–1117.
63. Guandalini, S. and P. Gupta, Do you still need a biopsy to diagnose celiac disease? *Curr Gastroenterol Rep*, 2001, 3(5): 385–391.
64. Marsh, M.N., Gluten, major histocompatibility complex, and the small intestine. A molecular and immunobiologic approach to the spectrum of gluten sensitivity ('celiac sprue'). *Gastroenterology*, 1992, 102(1): 330–354.
65. Cellier, C., E. Delabesse, C. Helmer, N. Patey, C. Matuchansky, B. Jabri, E. Macintyre, N. Cerf-Bensussan, and N. Brousse, Refractory sprue, coeliac disease, and enteropathy-associated T-cell lymphoma. French Coeliac Disease Study Group. *Lancet*, 2000, 356(9225): 203–208.
66. Green, P.H. and B. Jabri, Celiac disease and other precursors to small-bowel malignancy. *Gastroenterol Clin North Am*, 2002, 31(2): 625–639.
67. Cellier, C., N. Patey, L. Mauvieux, B. Jabri, E. Delabesse, J.P. Cervoni, M.L. Burtin, D. Guy-Grand, Y. Bouhnik, R. Modigliani, J.P. Barbier, E. Macintyre, N. Brousse, and N. Cerf-Bensussan, Abnormal intestinal intraepithelial lymphocytes in refractory sprue. *Gastroenterology*, 1998, 114(3): 471–481.
68. Hoffenberg, E.J., J. Haas, A. Drescher, R. Barnhurst, I. Osberg, F. Bao, and G. Eisenbarth, A trial of oats in children with newly diagnosed celiac disease. *J Pediatr*, 2000, 137(3): 361–366.

69. Janatuinen, E.K., T.A. Kemppainen, R.J. Julkunen, V.M. Kosma, M. Maki, M. Heikkinen, and M.I. Uusitupa, No harm from five year ingestion of oats in coeliac disease. *Gut*, 2002, 50(3): 332–335.

70. Collin, P., T. Reunala, E. Pukkala, P. Laippala, O. Keyriläinen, and A. Pasternack, Coeliac disease-associated disorders and survival. *Gut*, 1994, 35(9): 1215–1218.

71. Lee, S.K., W. Lo, L. Memeo, H. Rotterdam, and P.H. Green, Duodenal histology in patients with celiac disease after treatment with a gluten-free diet. *Gastrointest Endosc*, 2003, 57(2): 187–191.

72. Wahab, P.J., J.W. Meijer, and C.J. Mulder, Histologic follow-up of people with celiac disease on a gluten-free diet: slow and incomplete recovery. *Am J Clin Pathol*, 2002, 118(3): 459–463.

73. Sollid, L.M. and C. Khosla, Future therapeutic options for celiac disease. *Nat Clin Pract Gastroenterol Hepatol*, 2005, 2(3): 140–147.

74. Krawitt, E.L., Autoimmune hepatitis. *N Engl J Med*, 2006, 354(1): 54–66.

75. Ichiki, Y., C.A. Aoki, C.L. Bowlus, S. Shimoda, H. Ishibashi, and M.E. Gershwin, T cell immunity in autoimmune hepatitis. *Autoimmun Rev*, 2005, 4: 315–321.

76. Czaja, A.J., Current concepts in autoimmune hepatitis. *Ann Hepatol*, 2005, 4: 6–24.

77. Heneghan, M.A. and I.G. McFarlane, Of mice and women: toward a mouse model of autoimmune hepatitis. *Hepatology*, 2005, 42(1): 17–20.

78. Kanda, T., O. Yokosuka, Y. Hirasawa, F. Imazeki, K. Nagao, Y. Suzuki, and H. Saisho, Acute-onset autoimmune hepatitis resembling acute hepatitis: a case report and review of reported cases. *Hepatogastroenterology*, 2005, 52(64): 1233–1235.

79. Vergani, D., F. Alvarez, F.B. Bianchi, E.L. Cancado, I.R. Mackay, M.P. Manns, M. Nishioka, and E. Penner, Liver autoimmune serology: a consensus statement from the committee for autoimmune serology of the International Autoimmune Hepatitis Group. *J Hepatol*, 2004, 41(4): 677–683.

80. Sanders, D.S., Mucosal integrity and barrier function in the pathogenesis of early lesions in Crohn's disease. *J Clin Pathol*, 2005, 58(6): 568–572.

81. Mueller, T. and D.K. Podolsky, Nucleotide-binding-oligomerization domain proteins and toll-like receptors: sensors of the inflammatory bowel diseases' microbial environment. *Curr Opin Gastroenterol*, 2005, 21(4): 419–425.

82. Dignass, A.U., D.C. Baumgart, and A. Sturm, Review article: the aetiopathogenesis of inflammatory bowel disease—immunology and repair mechanisms. *Aliment Pharmacol Ther*, 2004, 20(Suppl 4): 9–17.

83. Adams, D.H. and B. Eksteen, Aberrant homing of mucosal T cells and extra-intestinal manifestations of inflammatory bowel disease. *Nat Rev Immunol*, 2006, 6(3): 244–251.

84. Ogura, Y., D.K. Bonen, N. Inohara, D.L. Nicolae, F.F. Chen, R. Ramos, H. Britton, T. Moran, R. Karaliuskas, R.H. Duerr, J.P. Achkar, S.R. Brant, T.M. Bayless, B.S. Kirschner, S.B. Hanauer, G. Nunez, and J.H. Cho, A frameshift mutation in NOD2 associated with susceptibility to Crohn's disease. *Nature*, 2001, 411(6837): 603–606.

85. Shaoul, R., A. Karban, B. Weiss, S. Reif, D. Wasserman, A. Pacht, R. Eliakim, J. Wardi, H. Shirin, E. Wine, E. Leshinsky-Silver, and A. Levine, NOD2/CARD15 mutations and presence of granulomas in pediatric and adult Crohn's disease. *Inflamm Bowel Dis*, 2004, 10(6): 709–714.

86. Noble, C.L., E.R. Nimmo, H. Drummond, G.T. Ho, A. Tenesa, L. Smith, N. Anderson, I.D. Arnott, and J. Satsangi, The contribution of OCTN1/2 variants within the IBD5 locus to disease susceptibility and severity in Crohn's disease. *Gastroenterology*, 2005, 129(6): 1854–1864.

87. Hanauer, S.B., Inflammatory bowel disease: epidemiology, pathogenesis, and therapeutic opportunities. *Inflamm Bowel Dis*, 2006, 12(Suppl 1): S3–S9.

88. Podolsky, D.K., Inflammatory bowel disease. *N Engl J Med*, 2002, 347(6): 417–429.

89. Negoro, K., D.P. McGovern, Y. Kinouchi, S. Takahashi, N.J. Lench, T. Shimosegawa, A. Carey, L.R. Cardon, D.P. Jewell, and D.A. van Heel, Analysis of the IBD5 locus and potential gene–gene interactions in Crohn's disease. *Gut*, 2003, 52(4): 541–546.
90. Kugathasan, S., N. Collins, K. Maresso, R.G. Hoffmann, M. Stephens, S.L. Werlin, C. Rudolph, and U. Broeckel, CARD15 gene mutations and risk for early surgery in pediatric-onset Crohn's disease. *Clin Gastroenterol Hepatol*, 2004, 2(11): 1003–1009.
91. Targan, S.R. and L.C. Karp, Defects in mucosal immunity leading to ulcerative colitis. *Immunol Rev*, 2005, 206: 296–305.
92. Yang, H., J.I. Rotter, H. Toyoda, C. Landers, D. Tyran, C.K. McElree, and S.R. Targan, Ulcerative colitis: a genetically heterogeneous disorder defined by genetic (HLA class II) and subclinical (antineutrophil cytoplasmic antibodies) markers. *J Clin Invest*, 1993, 92(2): 1080–1084.
93. Lanning, D.K., K.J. Rhee, and K.L. Knight, Intestinal bacteria and development of the B-lymphocyte repertoire. *Trends Immunol*, 2005, 26(8): 419–425.
94. Mowat, A.M., O.R. Millington, and F.G. Chirdo, Anatomical and cellular basis of immunity and tolerance in the intestine. *J Pediatr Gastroenterol Nutr*, 2004, 39(Suppl 3): S723–S724.
95. Huang, T., B. Wei, P. Velazquez, J. Borneman, and J. Braun, Commensal microbiota alter the abundance and TCR responsiveness of splenic naive CD4+ T lymphocytes. *Clin Immunol*, 2005, 117(3): 221–230.
96. Kelsall, B.L. and F. Leon, Involvement of intestinal dendritic cells in oral tolerance, immunity to pathogens, and inflammatory bowel disease. *Immunol Rev*, 2005, 206: 132–148.
97. Dubinsky, M.C., J.J. Ofman, M. Urman, S.R. Targan, and E.G. Seidman, Clinical utility of serodiagnostic testing in suspected pediatric inflammatory bowel disease. *Am J Gastroenterol*, 2001, 96(3): 758–765.
98. Dilger, K., M. Alberer, A. Busch, A. Enninger, R. Behrens, S. Koletzko, M. Stern, C. Beckmann, and C.H. Gleiter, Pharmacokinetics and pharmacodynamic action of budesonide in children with Crohn's disease. *Aliment Pharmacol Ther*, 2006, 23(3): 387–396.
99. Otley, A. and A.H. Steinhart, Budesonide for induction of remission in Crohn's disease. *Cochrane Database Syst Rev*, 2005, (4): CD000296.

14 Serologic Testing for Infectious Diseases

John L. Schmitz

CONTENTS

14.1   INTRODUCTION

The goal of this chapter is to familiarize the reader with the methods and ratio-nale employed for serologic testing for infectious diseases. Serologic testing entails the detection and quantification of antibodies to infectious agents, the agents them-selves, or a component of the agent. Serology is used when the definitive approach to laboratory diagnosis of infection, isolation, or recently, molecular detection, is not possible, appropriate, or necessary. In addition to diagnostic applications, serologic tests are used to assess immune status, immune competence, and for patient man-agement. Serologic methods can be highly sensitive, specific, and rapid; however, the appropriate application and interpretation of a serologic test requires an appre-ciation of the biology of the humoral immune response, including host factors that influence the response, performance characteristics of the specific method used, and characteristics of the infectious agent being tested for. The results of serologic tests, interpreted in the light of these considerations, provide important information for the diagnosis and management of patients with bacterial, viral, fungal, and parasitic diseases.

14.1.1   THE ANTIBODY RESPONSE

Details of the cellular and molecular aspects of the humoral immune response will not be reviewed here; the reader is referred to other chapters in this edition as well as texts that provide excellent discussions of this topic [1]. It is relevant to note, how-ever, that the characteristics of an antibody response to a pathogenic microorganism in any given individual vary due to the influence of several factors. Among these factors are the nature of the immunoreactive epitopes of the invading microbe (e.g., protein versus carbohydrate), involvement of T cells (T dependent versus T inde-pendent), prior history of antigen exposure (e.g., primary infection, reinfection, and reactivation), the anatomic site of infection, and the genetic background of the host. The interplay of these factors influence features of the humoral immune response. As such, they warrant consideration as they directly relate to the interpretation of the results of serologic tests for infectious diseases. Key features include the kinetics of antibody production, isotype switching, and affinity maturation.

14.1.1.1   Kinetics of Antibody Production

On first exposure to an antigen, there is an initial lag period before the development of detectable antibody. The duration of this seronegative (window) period is quite variable when the antigen is a component of an infectious agent. For some agents, the antibody is detectable within days after infection, whereas for others the antibody may not be detectable for weeks or months. Following the window period, antibod-ies are detected and increased in concentration. After clearance of the pathogen, antibody levels typically decline. On subsequent exposures to the pathogen, antibody levels increase more rapidly and to higher levels. If one measures the concentration (in specific units based on a quantification standard) or titer (the highest dilution of the patient serum that gives a positive test result) of a pathogen-specific antibody at two distinct time points (typically, the first serum being collected during the acute phase of the illness and the second serum 2–3 weeks later, during convalescence), an increase in antibody level suggests that exposure to the agent is recent in contrast to finding a stable antibody level in the two sera. In semiquantitative tests using a two-fold dilution series of the patient serum, a significant increase in the endpoint titer is considered to be a fourfold or greater increase (e.g., 1:16 in the acute serum versus 1:64 in the convalescent serum). Testing of acute and convalescent (paired) sera is the most reliable means to assess the relation of an infectious agent to a disease process when IgG antibody is the primary isotype assessed. In contrast, determination of a single IgG titer is not usually a reliable means of diagnosis due to the persistence of pathogen-specific IgG antibody for long periods of time after resolution of infection or the presence of a chronic (latent) infection.

14.1.1.2   Isotype Switching

IgM antibody is the first isotype of antibody produced following an immunogenic challenge. IgM levels increase and after a period of weeks decline in level, usually to undetectable levels. After isotype switching has occurred, IgG antibodies as well as IgA and IgE are produced. IgG antibodies are typically detected for long periods of time after resolution of infection. Thus, attributing a pathogen-specific antibody response to a current disease process can be made by the determination of IgM-specific responses. The presence of pathogen-specific IgM antibodies is in most cases consistent with the recent exposure to the pathogen, as opposed to the detection of IgG, which indicates infection at some undefined time in the past. Thus, if serum from only a single time point is available for serologic testing to establish a diagnosis, IgM testing is often used. As discussed in Sections 14.2.2.6 and 14.2.2.8, the method used to detect pathogen-specific IgM antibody must be taken into consideration as very sensitive methods may detect IgM antibodies for prolonged periods of time, thus diminishing their role as indicators of recent infection. In addition, reactivation of a chronic infection from its latent state may lead to an increase in pathogen-specific IgM levels.

14.1.1.3   Affinity Maturation

A switch in the isotype of antibody produced in response to infection is usually accompanied by an increase in the affinity of the antibody. Affinity maturation results from two processes, somatic mutation and selection. IgM antibodies, the first isotype produced after infection are typically of low avidity. As the humoral response develops and isotype switching occurs, B-cell clones undergo somatic mutations in regions of immunoglobulin (Ig) genes encoding antigen-binding sites. As the immune response clears the invading microbe, the amount of antigen available to stimulate B cells decreases. As a result, higher-avidity B-cell clones outcompete lower-avidity clones resulting in selection and production of an overall higher-avidity antibody pool over time. Thus, the presence of high-affinity antibodies is characteristic of past exposure to an antigen, whereas a lack of high-affinity antibody suggests that exposure to the antigen is more recent. As discussed in Section 14.2.2.8, the distinction of low- versus high-avidity binding is an approach that has been employed to assess the duration of an infection. This is particularly useful for the diagnosis of infections in pregnant women when it is important to determine if infection occurred before or after conception.

14.2 METHODS TO DETECT ANTIBODIES

A variety of methods, and specimens, are employed for the detection of pathogen-specific antibodies or antigens. They vary in complexity, cost, sensitivity, and specificity. The more common methods used in the clinical laboratory will be discussed. The reader is referred to additional sources [34] that provide detailed procedures, test characteristics, and interpretations.

14.2.1 SPECIMENS

Serum is typically used for the detection of antibodies to infectious agents. It is readily obtained, simple to process, and can be stored for prolonged periods of time. Whole blood collected by venipuncture into tubes containing no anticoagulant is allowed to clot completely and the serum fraction is separated by centrifugation. The serum may be tested immediately or stored at 4°C for several days before testing. If testing will be delayed longer than 7 days, it is advisable to store frozen aliquots (−20 or −70°C) until testing takes place to avoid microbial growth and degradation of Ig. Serum samples with marked hemolysis and lipemia should be avoided if at all possible, as these factors may interfere with some serologic tests. Plasma, obtained from anticoagulated whole blood may also be appropriate for some serologic tests.

Cerebrospinal fluid (CSF) is another specimen type routinely submitted for antibody testing. Testing for pathogen-specific CSF antibodies is performed to determine central nervous system (CNS) infection. Most assays that are used for serum-based testing can be employed for CSF-based testing. However, a critical issue with CSF-based testing is determining whether the antibody detected is actually produced intrathecally or is the result of leakage of serum antibody into the CSF across the blood–brain barrier (BBB). The CSF antibody index [33] is used to confirm intrathecal synthesis of antibody. The ratio of pathogen-specific Ig to total Ig in the CSF is compared to the same ratio in the serum. A significant increase over the serum value is consistent with intrathecal antibody synthesis as opposed to leakage across the BBB.

Although blood and CSF are the most common specimens used for serologic testing, alternative samples such as oral mucosal transudate (OMT), urine, and dried blood spots (DBS) have also been used. These alternative samples share the advantage of less-invasive sample collection. For certain infections, particularly those with public health implications, willingness to be tested is a critical component of control efforts, and the use of less-invasive sample collection methods may help increase the number of individuals tested.

OMT is a recently described specimen for antibody testing [14]. OMT is collected with the use of special fiber pads (OraSure, Inc., Beaverton, Oregon) that are placed between the check and gum after gently rubbing the pad over the gum.

Serum transudate is absorbed into the pad, which is placed into a stabilization fluid. Once in the stabilizer, the sample is stable at room temperature for up to 30 days. Food and Drug Administration (FDA)-approved Enzyme-Linked ImmunoSorbent Assay (ELISA) and other assays specifically designed and validated for use with OMT samples are available. Tests that employ OMT samples may perform nearly as well as blood-based tests [10]. However, OMT sample collection avoids the use of needles (may promote wider acceptance of testing due to the noninvasive nature of the sample collection), and provides a stable matrix for shipment to centralized testing laboratories.

Urine is an additional sample that can be used for serologic testing. Clean catch urine is not necessary. Urine as a sample has similar advantages as OMT including noninvasive collection, stability, and promotion of testing in those not desiring a blood draw, safety for health care work, and the availability of FDA cleared tests specifically for urine samples [6].

DBS are a fourth sample type that can be used for serologic testing [26]. Although not a routine specimen type, DBS are particularly amenable to use in epidemiologic studies in resource-limited settings. The advantages of this sample are stability, simplicity of collection, storage, and transport. DBS can be stored for prolonged periods of time before testing.

14.2.2 Serologic Testing Methods

In 2006, the 100-year anniversary of the report describing the Wasserman test for syphilis was celebrated. Although serologic testing methods have evolved extensively since that report, it is interesting to note that the complement fixation (CF) test (the technology used in the Wasserman test) is still employed today.

Serologic tests can be grouped into those that rely on the functional properties of antibodies for test readout and those that rely on the binding of antibody to its target antigen, which is detected by an indirect indicator system. Serologic tests can be qualitative (positive or negative), semiquantitative (using serial dilutions of patient serum to determine the endpoint titer), or quantitative (if appropriate quantification standards are employed in the test system) [34].

14.2.2.1 Assays Based on Physical Properties of Antibody–Antigen Complexes

Agglutination, precipitation, and flocculation tests are based upon the ability of antibodies to cross-link their target antigens present in solution. Cross-linking results in a readily detectable clumping. Agglutination reactions employ a particulate form of the target antigen such as whole bacterial cells, antigen-coated red cells (indirect hemagglutination), or an inert particle sensitized with the target antigen (e.g., latex or gelatin particles). Precipitation reactions employ a soluble target antigen. These tests have typically been performed in tubes (tube precipitation test) or in semisolid medium (single radial or double immunodiffusion) with precipitation of antibody–antigen complex resulting in visually detectable end product. The speed and sensitivity of precipitation reactions can be enhanced by performing the tests in the presence of an electric current (counter immunoelectrophoresis). Flocculation tests are a third format of test in which the antibody–antigen complexes remain in suspension and are detected by microscopic visualization of the flocculant material (e.g., Venereal Disease Research Laboratory [VDRL] test for syphilis).

14.2.2.2   Complement Fixation

The CF test is of historic interest and although the applications have diminished, CF remains in use today. The CF test is a time-consuming, labor-intensive test accounting in part for its limited use. This two-stage assay begins with the mixing of patient serum with antigen solution. After a suitable incubation period, a source of active complement is added and if target-specific antibody is present in the serum and binds to the antigen, the complement is activated and consumed. The absence of specific antibody in the serum results in no complement activation and it is available for stage 2 of the test. In stage 2, sheep red blood cells (SRBC) sensitized with anti-SRBC antibody (hemolysin) are added to the stage 1 reactants. After a suitable incubation period, the reactions are examined for SRBC lysis. If the patient serum had target-specific antibody, complement was consumed in stage 1 and not available for stage 2 resulting in the presence of a button of red cells in the reaction well (i.e., no complement-induced lysis) indicating a positive test. If no target-specific antibody was present in the serum, complement is available for stage 2 with resulting SRBC lysis indicating a negative test. Serial dilutions of patient serum can be used to determine the endpoint titer of target-specific antibody.

14.2.2.3   Hemagglutination Inhibition

Some infectious agents possess the ability to agglutinate red blood cells (mediated by a hemagglutinin molecule). This feature can be exploited in the development of a read out system for the detection of antibodies that block the agglutination reaction. Hemagglutination inhibition (HI) is carried out by incorporating serial dilutions of the patient serum into viral culture (with hemaglutination activity) in the presence of red blood cells. If the serum contains hemagglutinin-specific antibodies, they will block (neutralize) the hemagglutinin of the pathogen and no hemagglutination will take place. In contrast, a lack of hemagglutinin-specific antibody is indicated by the presence of hemagglutination.

14.2.2.4   Indirect Fluorescent Antibody

Indirect fluorescent antibody (IFA) tests employ target antigen bound to a microscope slide (Figure 14.1). Target antigen may be intact organisms or, for intracellular pathogens, cell lines containing the agent of interest. Antibody from patient serum is added to the slide, and pathogen-specific antibody, if present in the serum, will bind the target antigen. Unbound serum proteins are washed away and an antihuman Ig reagent labeled with fluorescein isothiocyanate (FITC) is added. The FITC-labeled conjugate may be directed to IgG, IgM, IgA, or to all isotypes (polyvalent) of antibody. Serial dilutions of the patient serum are typically used to provide a semiquantitative result.

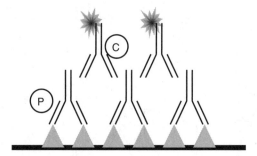

FIGURE 14.1   Indirect fluorescent antibody assay. A glass slide is coated with organisms of interest (e.g., whole bacteria, parasites, or virally infected cell lines). Patient serum is added and if pathogen-specific antibodies (P) are present, they will bind to the slide. Bound patient antibody is detected by the addition of a fluorescently labeled anti-immunoglobulin (C). Slides are examined using a fluorescent microscope for characteristic patterns of fluorescing organisms.

14.2.2.5   Radioimmunoassay

Radioimmunoassay (RIA) is a ligand-binding assay in which the presence of bound antibody is indicated with the use of radiolabeled anti-Ig or in a competitive fashion with the use of radiolabeled antibody to the antigen of interest. This is a sensitive approach that is amenable to higher volume testing but has the disadvantages of using radioisotopes. The latter include disposal of radioisotope (which is of environmental concern and expensive), as well as limitations on reagent storage due to degradation of the isotope. The advent of a similar assay substituting an enzyme for the radioisotope, the ELISA, maintained the advantages of RIA, however, eliminating the disadvantages related to the use of radioisotopes.

14.2.2.6   Enzyme-Linked Immunosorbent Assays

ELISA is the workhorse of the clinical laboratory, because of its capacity to test large numbers of samples, ease of automation, and its sensitivity. There are several variations of ELISA. The more common variants in routing clinical use are the indirect, competitive, and antibody-capture formats (Figure 14.2).

The indirect ELISA format is a common approach for the detection of IgG antibodies to infectious agents. The wells of a microtiter plate or beads are coated with whole organisms or subcellular fractions such as purified or recombinant proteins. Patient serum is added and incubated for a suitable time. Unbound serum components including nonspecific antibodies are washed away. An enzyme-labeled anti-Ig (conjugate) solution, which binds to the antigen-specific antibodies, is then added to the wells. Unbound conjugate is removed by another washing step and a substrate is added that is enzymatically converted to a specific color by the bound conjugate. The color change that occurs after addition of the substrate is detected with a spectrophotometer and read out as optical densities (OD). The amount of color change in a well is reflective of the amount of antigen-specific antibody

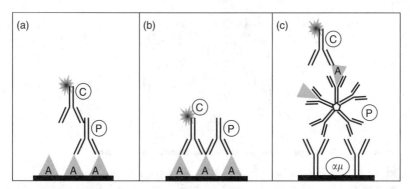

FIGURE 14.2 Representations of the common ELISA variations utilized for serologic testing. (a) indirect ELISA: the solid phase is coated with antigen (A: whole cells or soluble components). Pathogen-specific antibody (P), if present in patient serum, binds to the immobilized antigen. Bound patient antibody is detected by the addition of an enzyme-labeled anti-immunoglobulin (C). Addition of an appropriate substrate will result in a color change in the reaction well. (b) Competitive ELISA: patient serum (P) and enzyme-labeled anti-antigen conjugate (C) are added to the reaction well containing bound antigen (A). Patient antibody and anti-antigen conjugate compete for binding to antigen. Color change is induced by the addition of substrate. In contrast to the indirect ELISA, the color change in a competitive ELISA is inversely related to the amount of patient antibody. (c) IgM-capture ELISA: Solid phase is coated with an anti-IgM antibody ($\alpha\mu$). IgM antibodies (P) in the patient serum are captured. Pathogen-specific IgM antibody is identified by the addition of antigen (A) that binds if specific antibodies are present. Bound antigen is detected by the addition of an antigen-specific enzyme-labeled antibody (C). On addition of substrate, color change is induced indicating the presence of pathogen-specific IgM antibodies.

bound to the antigen in the well (which is directly proportional to the amount of antigen-specific antibody in the patient sample). The OD readings for controls and calibrators are used to set a positive cutoff level to which the patients' sample OD values are compared. The qualitative format can be modified to a quantitative format with the inclusion of appropriate quantification standards. A standard curve is generated of the OD values of the quantification standards containing known concentrations of pathogen-specific antibody. These standards are run in parallel with the patient samples, and the OD values of individual patient sera are compared to the standard curve to determine the concentration of pathogen-specific antibody in the patient serum.

A second version of the ELISA that is in common use is the competitive ELISA. This format obviates the need for the first wash step as the patient serum and anti-antigen conjugate are added simultaneously. The two reactants compete for binding the solid phase, which is the well of a microtiter plate or latex bead coated with antigen. Unbound sample and conjugate are washed away, substrate is added, and the reaction (color) is allowed to develop for a specified time period. Binding of patient antibody is reflected by low OD values (outcompetes the conjugate) and conversely, a lack of patient antibody is indicated by a high OD value (since there is no competition for conjugate binding).

Both indirect ELISA formats reliably detect IgG and other isotypes of pathogen-specific antibody; however, they can be susceptible to false positive and false negative results when used to detect pathogen-specific IgM antibodies. Rheumatoid factor (RF) in sera can cause false positive IgM tests when assessed in a standard indirect ELISA. If pathogen-specific IgG antibodies are bound to the solid phase, IgM RF can bind to the Fc portion of the bound IgG antibodies. An anti-IgM conjugate can then bind to the IgM RF resulting in a positive OD value due to the bound RF and not bound pathogen-specific IgM. High levels of pathogen-specific IgG antibodies can interfere (outcompete) with the detection of pathogen-specific IgM antibodies resulting in false negative IgM results. The IgM-capture ELISA (see the following paragraph) is a commonly used format for the detection of pathogen-specific IgM antibodies that reduces the potential for false positive and false negative results.

IgM-capture ELISA uses a solid phase coated with an anti-IgM antibody preparation. When serum is added, the anti-IgM antibody binds to serum IgM and retains it through a washing step. To determine if there is pathogen-specific IgM among that bound to the solid phase, a preparation of pathogen antigen is added. The antigen will be bound by any antigen-specific IgM antibody and retained through several wash steps. The presence of bound antigen is determined by the addition of an enzyme-labeled antipathogen antibody. On washing followed by addition of substrate, pathogen-specific IgM is indicated by the development of a color change in the supernatant of the test well. By selectively retaining only IgM antibody in the initial steps of the test, this method avoids the problems of RF and competing pathogen-specific IgG antibody in IgM-specific testing.

14.2.2.7   Western Blot

Western blot is a labor-intensive alternative serological pathogen-specific antibody detection system that employs the features of the ELISA, but detects pathogen-specific target antigens that have been separated in a characteristic manner before analysis. The latter step allows detection of antibody reactivity to specific proteins of an intact organism (Figure 14.3). Because this assay is more time consuming and costly, it is typically geared toward use as a confirmatory test following positive screening assays (see Section 14.2.3). The target agent is first solubilized and the constituent proteins are denatured. Treatment with sodium dodecyl sulfate (SDS) is used to allow subsequent separation of protein by size using polyacrylamide gel electrophoresis (PAGE). Separated proteins are then transferred from the SDS-PAGE gel onto a solid matrix such as nitrocellulose. The separated proteins on the nitrocellulose membranes are then used as the target antigens for an indirect ELISA-based detection system similar to that described previously.

14.2.2.8   Avidity-Based Testing

Antibody tests can be modified to detect the relative avidity of binding [16]. ELISA assays are typically used for this purpose. In avidity testing, reactivity of a patient serum is compared between a standard version of the assay and a version in which the well with patient serum is treated with an agent that will disrupt the antigen/antibody binding, such as urea. If a patient serum has only low-avidity antibody,

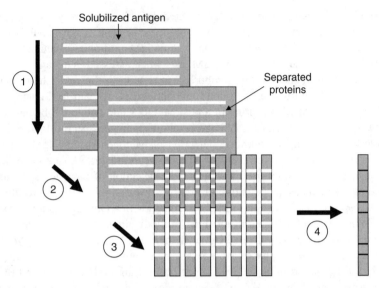

FIGURE 14.3 Western Immunoblot. (1) Solubilized pathogen proteins are separated by size using polyacrylamide gel electrophoresis. (2) Proteins in the gel are transferred to a solid matrix by electrotransfer. (3) The solid matrix with transferred proteins is cut into strips for use in immunoblot detection of pathogen-specific antibody. (4) Patient serum is added to a strip and bound antibody is detected by addition of an enzyme-labeled anti-immunoglobulin. On addition of substrate, a precipitate is deposited at the sites of bound enzyme-labeled anti-immunoglobulin. The Western immunoblot provides enhanced specificity relative to other assays because antibody reactivity can be verified as binding to pathogen-specific proteins.

suggestive of early/acute infection, the urea-treated wells will demonstrate reduced binding. In contrast, serum antibodies from a patient with long-standing or past infection will be urea resistant and demonstrate equivalent binding in the urea-treated well compared to the nontreated well. Using this approach, one can determine an avidity index that reflects the duration of the infection. This approach is particularly applicable to the diagnosis of infections during pregnancy (e.g., Toxoplasmosis) when it is important to determine if infection occurred before or after conception. Avidity testing has potential advantages over IgM testing due to the fact that highly sensitive IgM ELISAs may give positive results for extended periods of time, over 12 months in some cases.

14.2.3 PERFORMANCE CHARACTERISTICS OF SEROLOGIC TESTS

Serologic tests can be very sensitive and very specific; however, like any assay, they are susceptible to false negative and false positive reactions. Technical errors commonly affect assay performance and can be evaluated and corrected with the appropriate controls. Biologic explanations for false negative reactions may be due to testing too early in the disease process, before antibody development or testing an immunosuppressed/deficient patient (defective antibody response). Finally, the use of an assay with suboptimal sensitivity may explain a lack of antibody reactivity in an infected patient.

Likewise, false positive tests may have several explanations, including technical performance problems. Although antibodies are very specific for their target antigen, the target itself or a target with similar structure may be present on organisms other than the infecting agent. When the potential for this cross-reactivity is appreciated, it may be possible to assess certain characteristics of the antibody response to the possible agents to determine which elicited the antibody response. In general, antibody concentration is higher toward the actual infecting agent than toward the cross-reacting agent. Using the appropriate technology, antibody responses to specific components of the infecting agent not present in the cross-reacting agent may be discernable and serve to confirm the true identity of the infecting agent (e.g., Western blot analysis). Finally, some agents, particularly certain viruses, may induce a polyclonal B-cell activation resulting in increases in antibody concentration of many different targets and thus confound the interpretation of serologic tests.

Serologic assays, as with other diagnostic tests, are characterized in terms of sensitivity, specificity, precision (within and between run), positive- and negative-predictive values. Sensitivity is usually characterized in terms of clinical sensitivity (i.e., the proportion of patients with an infection that test positive for antibody). In some instances, however, it is useful to know the analytic sensitivity of a serologic test. This is of particular importance when a known concentration of antibody is correlated with protection from infection and the test is being used to determine the immune status of a patient. The specificity of serologic tests refers to the proportion of individuals without an infection who test negative for pathogen-specific antibody (the lower the specificity, the higher the number of false positive results). Specificity is of critical importance for those tests that are used for confirmation of screening test results. Although the characteristics of sensitivity and specificity define assay-specific performance parameters, the application of these tests in populations with varying prevalence of infection will result in variable test performance in terms of the ability to predict infection or lack thereof. The positive-predictive value (PPV) refers to the proportion of positive test results derived from patients with the infection, whereas the negative-predictive value (NPV) refers to the proportion of negative test results derived from patients without the infection. The predictive value parameters are highly dependent upon the prevalence of infection in the population being tested.

In some situations, combinations of tests are used to maximize diagnostic performance. Common examples of this approach include the use of sensitive, inexpensive screening assays with confirmation of positive results by a second, more specific (confirmatory) test. In this combined testing approach, the screening assay is designed to miss as few infected individuals as possible (few false negative results). However, this increased sensitivity is often associated with a higher likelihood of false positive (lower specificity) results. To detect the false positives, confirmatory assays are used as a second test with characteristics that impart high specificity. A patient sample is considered truly positive only when both the screening and confirmatory tests are positive.

14.2.4   DETECTION OF PATHOGENS OR PATHOGEN-DERIVED ANTIGENS

Many of the same serologic methods used for detection of pathogen-specific antibody can be used to detect pathogens or specific antigenic components of pathogens. Common methods employed include agglutination, IFA, and ELISA. Specific applications are discussed later in this chapter.

14.2.5   NEW DEVELOPMENTS IN SEROLOGIC TESTING

Rapid tests have recently been applied to infectious disease serologic testing [19]. Based on flow-through or wicking technology, rapid tests take a very short time to complete (~15 min). In addition to speed, one of the main advantages of these tests lies in their ability to provide a patient result during the visit thus assuring communication of the result and appropriate follow-up. This has important implications in the public health setting.

Automation is playing an increasing role in the conduct of serologic tests for infectious diseases. As laboratory staffing decreases and workload increases, laboratories must increase their efficiency. A variety of automated systems are available from various vendors of serologic tests. The instrumentation varies extensively by the manufacturer. Systems for automation can be applied to IFA and ELISA. Automation for IFA-based tests involves the processing of slides (i.e., dilution of sample and staining of slides). ELISAs are very amenable to automation. Systems are available that use robotics to handle all aspects of pipetting, plate washing, reading, and data management. In addition, random access systems are available from several vendors that make serologic testing for infectious diseases more akin to clinical chemistry testing.

14.3   APPLICATIONS

14.3.1   INDICATIONS FOR SEROLOGIC TESTING

Indications for serologic testing include the diagnosis of a current infection, determination of immune status (susceptibility to infection), stage of infection (past versus recent versus chronic), screening of blood and tissue donors for blood-borne pathogens, and assessing the response to vaccination (either to verify that protective levels of antibody have been achieved or to assess humoral immunocompetence).

For diagnostic purposes, both IgM and IgG antibodies may be employed to ascertain the cause of a patient's current illness. Additionally, responses to one or more antigens of certain organisms can be used to determine the stage of infection, that is, acute versus chronic versus past infection. Finally, some tests can be used to assess the efficacy of antimicrobial therapy.

14.3.1.1   Assessing Immune Status

Prior infection or vaccination usually imparts immunity to reinfection and the detection of antibody is used to ascertain prior infection and vaccination status. The determination of immune status has important implications in several scenarios.

Examples of infections for which the determination of immune status has practical application include varicella zoster virus (VZV) and rubella virus in pregnant women. Both of these viruses can cause congenital infection if acquired during pregnancy. An immune status test (i.e., determination of the presence of IgG antibodies to the viruses) is performed in pregnant women who have potential exposure to individuals with VZV. Varicella zoster immune globulin is administered as soon as possible in a seronegative woman exposed to a VZV infected individual. Rubella immune status testing is offered to pregnant women and if found to be seronegative they are counseled to avoid exposure to individuals with possible rubella infection and to get vaccinated after delivery.

Transplant patients form a second group for which immune status testing is critical. The use of posttransplant immune suppression for the promotion of allograft survival renders patients susceptible to a variety of infections (and malignancies). Chronic viral infections normally held in check can cause serious disease in an immunosuppressed patient. Transplant candidates and donors are screened for the presence of antibodies to Cytomegalovirus (CMV) to guide donor selection as well as posttransplant management. CMV seronegative individuals are at high risk for CMV disease from donors who are CMV seropositive.

14.3.1.2 Vaccine Response

Assessing the response to vaccination may be used to determine if an individual responds with protective levels of vaccine-specific antibody. For example, it is known that a small proportion of individuals are nonresponders to the Hepatitis B vaccine, suggesting that high-risk patients should be assessed postvaccination to ascertain protection. Measuring vaccine responsiveness is also commonly used to assess immune competence in the investigation of suspected primary immunodeficiency disease.

Individuals with defects in humoral immunity may fail to make antibody upon vaccination with polysaccharide-based antigens (e.g., pneumococcal polysaccharide) and protein-based antigen (e.g., tetanus toxoid). A blood sample is collected before vaccination and a second sample is collected 2–3 weeks postvaccination. Seroconversion (i.e., conversion from negative prevaccination to positive postvaccination), attainment of a fourfold increase in titer, or achievement of a specified level of antibody all indicate a functional humoral immune system.

14.3.1.3 Diagnosis

Diagnosis of an acute infection relies upon patient history, clinical presentation, and classically, isolation of the organism. Recently, molecular methods have been employed to help diagnose infection for which culture or antigen-detection methods are suboptimal (for details see Chapter 15). In cases where the methods mentioned are either inappropriate or not available, detection of pathogen-specific antibody is used for the laboratory diagnosis of infection.

As discussed, the mere presence of antibody does not prove that the current manifestations are attributable to the particular pathogen tested, and stronger evidence for a link to diagnosis may be required such as the presence of IgM pathogen-specific antibody and a fourfold increase in the titer of IgG antibody. An isolated, positive IgG result is difficult to interpret as it may be the result of either a current or a past infection.

14.3.1.4   Screening for Blood/Tissue Borne Infections

Blood and tissue donors are screened for a variety of transmissible infectious agents. Donors are tested for pathogen-specific antibody and ruled out for donation if the test is positive. However, molecular methods may also be employed to identify donors in the window period before seroconversion. Blood and tissue donors are routinely screened using a variety of serological assays for Human Immunodeficiency Virus (HIV), Human T-Lymphotrophic Virus-1/2 (HTLV-1/2), Hepatitis B Virus (HBV), Hepatitis C Virus (HCV), CMV, Epstein-Barr Virus (EBV), and syphilis. As alluded to earlier, it is important to note that a limitation of antibody testing—the seronegative window period—is addressed for certain pathogens by the inclusion of highly sensitive nucleic acid amplification testing (NAAT) in additional to antibody testing. Inclusion of NAAT testing decreases the likelihood that blood from a donor with HIV or other chronic viral infections will be transfused.

14.3.2   Specific Applications

14.3.2.1   Serologic Diagnosis of Parasitic Diseases

The typical approach to the diagnosis of most parasitic infections involves microscopic examination of body fluids or tissues for various stages of parasite development (e.g., eggs, larvae, cysts, and adult forms). This approach requires trained personnel and may not be possible in all cases due to the lack of suitable tissue specimen or it may in fact be contraindicated in certain infections (e.g., hydatid disease). As such, serologic testing including detection of parasite-specific antibody and parasite-specific antigen tests are available to aid in the diagnosis of a number of infections (Table 14.1).

The interpretation of serologic test results for parasitic infections must take into consideration a few key points. The detection of IgG antibodies is most commonly employed, with the exception of *Toxoplasma*, where detection of IgM and IgA

TABLE 14.1
Antibody and Antigen Detection Assays for Select Parasitic Infections

Infection (Reference)	Agent	Serologic Tests	
		Antigen	Antibody
Amebiasis [35]	*Entamoeba histolytica*	EIA	EIA
Babesiosis [17]	*Babesia microti*		IFA
Cryptosporidiosis [20]	*Cryptosporidium parvum*	EIA, DFA, IFA	
Giardiasis [12]	*Giardia lamblia*	EIA, DFA	
Toxoplasmosis [29]	*Toxoplasma gondii*		EIA, IFA, LA, other
Trichomoniasis [28]	*Trichomonas vaginalis*	DFA	
Cysticercosis [32]	*Taenia solium*		EIA, IB
Echinococosus [27]	*Echinococcus granulosis*		EIA, IB
Strongyloidiasis [7]	*Strongyloides stercoralis*		EIA

Note: EIA, enzyme immunoassay; DFA, direct fluorescent antibody; IFA, indirect fluorescent antibody; IB, immunoblot; LA, latex agglutination; and other (dye test, agglutination tests, and avidity).

antibodies is employed for diagnosis of infection in pregnancy. Consideration of travel history is also critical. For example, a positive IgG result for a parasitic infection in a person living in an endemic area is difficult to attribute to current infection since past infection(s) cannot be excluded. However, a positive IgG result in a person who has recently traveled to an endemic area for the first time is more likely to reflect a causal relationship between the parasite of interest and the current disease state [11,44].

14.3.2.2 Serologic Testing for Fungal Infections

As with other infectious diseases, the laboratory diagnosis of fungal infections is based on direct identification of organisms including culture and microscopic examination of tissues. Serologic detection of fungal antibodies and antigens in the assessment of fungal infections is an important component of the laboratory diagnostic approach. Table 14.2 lists common fungal infections for which serological antibody- and antigen-detection methods are available.

The detection of antibodies to fungal pathogens, although useful, is subject to several limitations. Antibodies to some fungal pathogens may not be detectable for a long period of time after initial infection. In addition, some fungal antibody assays are prone to cross-reactivity due to shared epitopes among different organisms. In these cases, serial monitoring of antibody levels to several organisms or the use of antigen-detection tests may help clarify the results [25].

14.3.2.3 Serologic Detection of Bacterial Infections

Table 14.3 lists common bacterial infections for which antibody testing is commonly employed. As is evident, the types of bacterial infections and the role of antibody testing in these infections are very diverse. Serological antibody detection systems are available for the diagnosis of postinfection sequelae [38]; detection of infection for which culture is not an optimal/efficient approach [40]; detection of infection with organisms that cannot be cultured [24]; and for detection of chronic infection for which suitable specimens cannot be readily obtained without invasive methods [43].

TABLE 14.2
Fungal Infections for Which Antibody/Antigen Testing Are Commonly Used

Agent (Reference)	Method	
	Antibody Detection	Antigen Detection
Aspergillus [13]	CF, ID	EIA, LA
Histoplasmosis [18]	CF, ID, LA	
Blastomycosis [41]	CF, ID	
Coccidioidomycosis [31]	CF, ID, LA, EIA	
Cryptococosis [39]	TA	LA, EIA

Note: CF, complement fixation; ID, immunodiffusion; EIA, enzyme immunoassay; LA, latex agglutination; and TA, tube agglutination.

TABLE 14.3
Antibody Methods and Applications for Select Bacterial Infections

Agent	Method	Application
S. pyogenes	ASO, DNase B, NT	Diagnosis of postinfection sequelae
H. pylori	EIA, IB	Provides evidence of infection at some time
T. pallidum	AG, IFA, EIA	Screening and confirmation of infection Monitoring success of therapy
B. burdorferi	IFA, EIA, IB	Screening and confirmation of infection
Rickettsia sp.	LA, IFA	Diagnosis
M. pneumoniae	EIA, LA, IFA	Diagnosis

Note: ASO, antistreptolysin O; DNase B, anti-DNase B; NT, neutralization; EIA, enzyme immunoassay; IB, immunoblot; AG, agglutination; and LA, latex agglutination.

TABLE 14.4
Antibody and Antigen Detection Assays for Select Viral Infections

Virus (Reference)	Serologic Method for Detection	
	Antibody	Antigen
HSV [2]	EIA, IB, IHA, LA, NT	
VZV [36]	IB, IFA, EIA, NT, AV	
EBV [5]	EIA, IFA, NT	
CMV [8]	EIA, LA, NT, AV, CF	DFA (pp65)
Parvovirus B19 [4]	EIA, IFA	
Measles	EIA, IFA, CF, NT, HI	
Mumps	EIA, IFA, CF, NT, HI	
Rubella	EIA, IFA, CF, HI, LA	
HAV [9]	EIA	
HBV [37]	EIA, RIA	EIA (HBsAg)
HCV [15]	EIA, RIBA	EIA
HDV [30]	EIA	
HEV [22]	EIA	
Arboviruses [23]	EIA, IFA, NT	
HTLV-1, 2 [21]	EIA	
HIV [3,19,42]	EIA, IB, IFA, RIPA	EIA (p24)

Note: EIA, enzyme immunoassay; IFA, indirect fluorescent antibody; CF, complement fixation; NT, neutralization; HI, hemagglutination inhibition; LA, latex agglutination; RIBA, recombinant immunoblot assay; RIPA, radioimmunoprecipitation assay; and RIA, radioimmunoassay.

14.3.2.4 Detection of Viral Infections

Antibody/antigen detection plays a critical role in the laboratory diagnosis of many viral infections. It is particularly important for those pathogens that cannot be or are difficult to culture. Certain groups of people are routinely screened for several past viral infections as a measure of immunity (e.g., pregnant women are screened for rubella antibodies). Several blood-borne viral pathogens are routinely screened for in blood and tissue donors (HIV, HBV, HCV, HTLV-I/II, CMV). Potentially, chronic viral infections serve as examples for the use of multiple antigens in serologic tests to define the stage of infection (e.g., EBV, HBV). Table 14.4 provides a list of viral infections for which antibody testing is commonly employed.

14.4 SUMMARY

Serologic methods for the diagnosis of infectious diseases have been employed for decades. Antibody and antigen detection methods have evolved and adapted to fill a variety of needs from diagnosis of acute versus chronic infections, and the assessment of immunity for bacterial, viral, fungal, and parasitic infections. Although many techniques have been developed over the years to enhance sensitivity, specificity, and ease of use, classic methods used decades ago are still employed. Advances in sources of antigens, technology, and instrumentation will continue to enhance the use of this diagnostic approach.

REFERENCES

1. Abbas, A. K. and A. H. Lichtman. 2003. *Cellular and Molecular Immunology*, 5th ed. Saunders, Philadelphia, PA.
2. Ashley, R. L. 2001. Sorting out the new HSV type specific antibody tests. *Sex Transm Infect* 77: 232–237.
3. Branson, B. M., H. H. Handsfield, M. A. Lampe, R. S. Janssen, A. W. Taylor, S. B. Lyss, and J. E. Clark. 2006. Revised recommendations for HIV testing of adults, adolescents, and pregnant women in health-care settings. *MMWR Recomm Rep* 55: 1–17; quiz CE1-4.
4. Brown, K. E. and N. S. Young. 1997. Parvovirus B19 in human disease. *Annu Rev Med* 48: 59–67.
5. Bruu, A. L., R. Hjetland, E. Holter, L. Mortensen, O. Natas, W. Petterson, A. G. Skar, T. Skarpaas, T. Tjade, and B. Asjo. 2000. Evaluation of 12 commercial tests for detection of Epstein–Barr virus-specific and heterophile antibodies. *Clin Diagn Lab Immunol* 7: 451–456.
6. Cao, Y., A. E. Rriedman-Kien, J. V. Chuba, M. Mirabile, and B. Hosein. 1988. IgG antibodies to HIV-1 in the urine of HIV-1 seropositive individuals. *Lancet* 1: 831–832.
7. Carroll, S. M., K. T. Karthigasu, and D. I. Grove. 1981. Serodiagnosis of human strongyloidiasis by an enzyme-linked immunosorbent assay. *Trans Roy Soc Trop Med Hyg* 75: 706–709.
8. Chou, S. 1990. Newer methods for diagnosis of cytomegalovirus infection. *Rev Infect Dis* 12(Suppl 7): S727–S736.
9. Cuthbert, J. A. 2001. Hepatitis A: old and new. *Clin Microbiol Rev* 14: 38–58.
10. Gallo, D., J. R. George, J. H. Fitchen, A. S. Goldstein, and M. S. Hindahl. 1997. Evaluation of a system using oral mucosal transudate for HIV-1 antibody screening and confirmatory testing. OraSure HIV clinical trials group. *JAMA* 277: 254–258.

11. Garcia, L. S. 2006. Antibody and antigen detection in parasitic infections. In L. Garcia (Ed.), *Diagnostic Medical Parasitology*, 4th ed. ASM Press, Washington, pp. 592–615.

12. Garcia, L. S. and R. Y. Shimizu. 1997. Evaluation of nine immunoassay kits (enzyme immunoassay and direct fluorescence) for detection of *Giardia lamblia* and *Cryptosporidium parvum* in human fecal specimens. *J Clin Microbiol* 35: 1526–1529.

13. Greenberger, P. A. 2002. Allergic bronchopulmonary aspergillosis. *J Allergy Clin Immunol* 110: 685–692.

14. Greensides, D. R., R. Berkelman, A. Landsky, and P. S. Sullivan. 2003. Alternative HIV testing methods among populations at high risk fo rHIV infection. *Public Health Rep* 118: 531–539.

15. Gretch, D. R. 1997. Diagnostic tests for hepatitis C. *Hepatology* 26: 43S–47S.

16. Hedman, K., M. Lappalainen, I. Seppaia, and O. Makela. 1989. Recent primary toxoplasma infection indicated by a low avidity of specific IgG. *J Inf Dis* 159: 736–740.

17. Homer, M. J., I. Aguilar-Delfin, S. R. Telford III, P. J. Krause, and D. H. Persing. 2000. Babesiosis. *Clin Microbiol Rev* 13: 451–469.

18. Kauffman, C. A. 2007. Histoplasmosis: a clinical and laboratory update. *Clin Microbiol Rev* 20: 115–132.

19. Keenan, P. A., J. M. Keenan, and B. M. Branson. 2005. Rapid HIV testing. Wait time reduced from days to minutes. *Postgrad Med* 117: 47–52.

20. Kehl, K. S., H. Cicirello, and P. L. Havens. 1995. Comparison of four different methods for detection of *Cryptosporidium* species. *J Clin Microbiol* 33: 416–418.

21. Kline, R. L., T. Brothers, N. Halsey, R. Boulos, M. D. Lairmore, and T. C. Quinn. 1991. Evaluation of enzyme immunoassays for antibody to human T-lymphotropic viruses type I/II. *Lancet* 337: 30–33.

22. Krawczynski, K., R. Aggarwal, and S. Kamili. 2000. Hepatitis E. *Infect Dis Clin North Am* 14: 669–687.

23. Lanciotti, R. S. and J. T. Roehrig. 2006. Arboviruses. In B. Detrick, R. G. Hamilton, and J. D. Folds (Eds.), *Manual of Molecular and Clinical Laboratory Immunology*, 7th ed. ASM Press, Washington, pp. 757–765.

24. Larsen, S. A., B. M. Steiner, and A. H. Rudolph. 1995. Laboratory diagnosis and interpretation of tests for syphilis. *Clin Microbiol Rev* 8: 1–21.

25. Lindsley, M. D., D. W. Warnock, and C. J. Morrison. 2006. Serological and molecular diagnosis of fungal infections. In B. Detrick, R. G. Hamilton, and J. D. Folds (Eds.), *Manual of Molecular and Clinical Laboratory Immunology*, 7th ed. ASM Press, Washington, pp. 569–605.

26. McCarron, B., R. Fox, K. Wilson, S. Cameron, J. McMenamin, G. McGregor, A. Pithie, and D. Goldberg. 1999. Hepatitis C antibody detection in dried blood spots. *J Viral Hepatitis* 6: 453–456.

27. McManus, D. P., W. Zhang, J. Li, and P. B. Bartley. 2003. Echinococcosis. *Lancet* 362: 1295–1304.

28. Miller, G. A., J. D. Klausner, T. J. Coates, R. Meza, C. A. Gaydos, J. Hardick, S. Leon, and C. F. Caceres. 2003. Assessment of a rapid antigen detection system for *Trichomonas vaginalis* infection. *Clin Diagn Lab Immunol* 10: 1157–1158.

29. Montoya, J. G. and O. Liesenfeld. 2004. Toxoplasmosis. *Lancet* 363: 1965–1976.

30. Negro, F. and M. Rizzetto. 1995. Diagnosis of hepatitis delta virus infection. *J Hepatol* 22: 136–139.

31. Pappagianis, D. and B. L. Zimmer. 1990. Serology of coccidioidomycosis. *Clin Microbiol Rev* 3: 247–268.

32. Proano-Narvaez, J. V., A. Meza-Lucas, O. Mata-Ruiz, R. C. Garcia-Jeronimo, and D. Correa. 2002. Laboratory diagnosis of human neurocysticercosis: double-blind comparison of enzyme-linked immunosorbent assay and electroimmunotransfer blot assay. *J Clin Microbiol* 40: 2115–2118.

33. Reiber, H. and P. Lange. 1991. Quantification of virus-specific antibodies in cerebrospinal fluid and serum: sensitive and specific detection of antibody synthesis in brain. *Clin Chem* 37: 1153–1160.
34. Rose, N. R., H. Friedman, and J. L. Fahey. 1986. *Manual of Clinical Laboratory Immunology*, 3rd ed. ASM Press, Washington.
35. Rosenblatt, J. E., L. M. Sloan, and J. E. Bestrom. 1995. Evaluation of an enzyme-linked immunoassay for the detection in serum of antibodies to *Entamoeba histolytica*. *Diagn Microbiol Infect Dis* 22: 275–278.
36. Schmid, D. S. and V. Loparev. 2006. Varicella-Zoster Virus. In B. Detrick, R. G. Hamilton, and J. D. Folds (Eds.), *Manual of Molecular and Clinical Laboratory Immunology*, 7th ed. ASM Press, Washington, pp. 631–636.
37. Servoss, J. C. and L. S. Friedman. 2006. Serologic and molecular diagnosis of hepatitis B virus. *Infect Dis Clin North Am* 20: 47–61.
38. Shet, A. and E. Kaplan. 2006. Diagnostic methods for group A streptococcal infections. In B. Detrick, R. G. Hamilton, and J. D. Folds (Eds.), *Manual of Molecular and Clinical Laboratoray Immunology*, 7th ed. ASM Press, Washington, pp. 428–433.
39. Tanner, D. C., M. P. Weinstein, B. Fedorciw, K. L. Joho, J. J. Thorpe, and L. Reller. 1994. Comparison of commercial kits for detection of cryptococcal antigen. *J Clin Microbiol* 32: 1680–1684.
40. Tugwell, P., D. T. Dennis, A. Weinstein, G. Wells, B. Shea, G. Nichol, R. Hayward, R. Lightfoot, P. Baker, and A. C. Steere. 1997. Laboratory evaluation in the diagnosis of Lyme disease. *Ann Intern Med* 127: 1109–1123.
41. Turner, S. and L. Kaufman. 1986. Immunodiagnosis of blastomycosis. *Semin Respir Infect* 1: 22–28.
42. West, K. 2004. HIV: test results in minutes: reviewing the prehospital occupational risk of acquiring HIV & the availability & value of rapid post-exposure testing. *JEMS* 29: 68–73.
43. Wilcox, M. H., T. H. Dent, J. O. Hunter, J. J. Gray, D. F. Brown, D. G. Wight, and E. P. Wraight. 1996. Accuracy of serology for the diagnosis of *Helicobacter pylori* infection—a comparison of eight kits. *J Clin Pathol* 49: 373–376.
44. Wilson, M., P. M. Schantz, and T. Nutman. 2006. Molecular and immunological approaches to the diagnosis of parasitic infections. In B. Detrick, R. G. Hamilton, and J. D. Folds (Eds.), *Manual of Molecular and Clinical Laboratory Immunology*, 7th ed. ASM Press, Washington, pp. 557–568.

15 Molecular Techniques Applied to Infectious Diseases

Jennifer S. Goodrich and Melissa B. Miller

CONTENTS

15.1   INTRODUCTION

As the complexity of diagnostic microbiology intensifies, so do the methods employed in the laboratory to detect infectious disease etiologies. The advent of molecular technology in the clinical laboratory has not only augmented traditional methods such as culture and serology that have historically been the "gold standard" for pathogen detection, but has also created a niche for itself. For routine bacteriology (i.e., blood cultures, urine cultures, respiratory cultures), culture has remained the gold standard primarily based on cost accounting and the potential complex nature of associated infections. However, there are circumstances where molecular methodologies would prove advantageous over standard methods; for example, there may be minute quantities of the pathogen present, the patient may have received antibiotics before specimen collection, or the etiologic agent may require unusual culture conditions.

The optimal use of molecular techniques in microbiology resides with specimens in which a limited number of pathogenic organisms are sought (i.e., detection of *Chlamydia trachomatis* [CT] and *Neisseria gonorrhoeae* [NG] from cervical specimens or detection of methicillin-resistant *Staphylococcus aureus* [MRSA] from nares for infection control purposes), and in cases where the enhanced sensitivity and faster turnaround time of molecular methods far outweighs the increased cost (i.e., detection of herpes simplex virus [HSV] in cerebrospinal fluid [CSF] or direct detection of *Mycobacterium tuberculosis* [MTB] in sputum samples). A particularly exciting use of molecular techniques involves response to new and emerging public health threats such as the severe acute respiratory syndrome coronavirus, avian influenza, or bioterrorism agents. A clinical laboratory with molecular expertise is at a higher level of preparedness for such unforeseen events than laboratories that rely on traditional methodologies.

This chapter aims to discuss the most common currently available molecular methodologies, select infectious diseases applications, and the advantages and disadvantages associated with molecular testing in the clinical laboratory.

15.2   NONAMPLIFIED DIRECT DETECTION

15.2.1   Probe Hybridization

Nucleic acid probes are used for culture confirmation as well as direct detection of organisms from clinical material. Although nucleic acid probes are more expensive than conventional culture and identification methods, their power lies in their moderately increased sensitivity and specificity and decreased turnaround time. This latter advantage is especially applicable for organisms that are slow growing or are difficult to grow, such as *Mycobacterium* spp. and fungi.

Probes are single-stranded oligonucleotides that can vary in size from 20 base pairs (bp) to a few kilobases, but are generally less than 50 bp. Probe specificity is defined by the nucleic acid sequence of the probe. Typical probe targets include genomic deoxyribonucleic acid (DNA) or ribonucleic acid (RNA)—messenger RNA (mRNA) or ribosomal RNA (rRNA). Bacterial identification using probes to 16S rRNA or 23S rRNA are most commonly employed due to the higher copy number of the genes encoding these rRNAs in the bacterial genome, thus increasing the sensitivity of direct detection. Further, rRNA sequences contain conserved regions in addition to hypervariable regions allowing for the level of identification to be varied depending on the sequence of the probe targets.

Probe detection occurs by hybridization, or annealing, of a labeled probe to a target sequence (Figure 15.1) that has been released by lysis of the organism. Hybridization generally occurs using stringent conditions (i.e., high temperature and low salt) to allow for the highest specificity; often even a single bp change can be detected. Following hybridization, the probe–target hybrids are isolated and detected. Detection of these hybrids is dependent on the reporter incorporated into the hybridization assay. Probe reporter molecules include enzymes, affinity labels, and fluorescent or chemiluminescent molecules.[1]

Table 15.1 outlines some of the most commonly used probes for culture confirmation and direct detection in the clinical laboratory and their reported sensitivities and specificities. For culture confirmation, probe hybridization assays generally offer greater sensitivity and specificity than routine biochemical methods.[2] In some instances, bacterial strains may possess unique polymorphisms that prevent hybridization and culture confirmation, thus rendering a false-negative result.[3,4] Though rare, there are also examples of cross-reactivity among the probes resulting in false-positive results.[5,6] Disadvantages include the limited number of different species that can be detected using commercial probes and the inability to probe clinical specimens directly. For direct detection of CT and NG from clinical specimens (Gen-Probe PACE 2C), the sensitivity and specificity is a factor of disease prevalence. This application is discussed further in Section 15.5.1.

Hybridization can occur in liquid phase or solid phase. In liquid-phase hybridization, both the probe and the target are free in solution to interact, allowing for more rapid annealing and thus shorter assay times. This format known as hybridization protection assay is commonly used in clinical microbiology laboratories for the identification of select bacterial, mycobacterial, and fungal species, as it is available in commercial kits used for culture confirmation (Gen-Probe Inc., San Diego, California). Disadvantages of liquid-phase hybridization include a requirement for relatively pure target nucleic acid; and interference from related but not identical target nucleic acid sequence can affect detection.[1]

In contrast, solid-phase hybridization uses solid media such as nitrocellulose or a nylon membrane to immobilize the target nucleic acids. Labeled probes are then added to the solid support allowing for detection of the nucleic acid sequence of interest. Researchers have used this approach (i.e., dot blots, southern blots, and

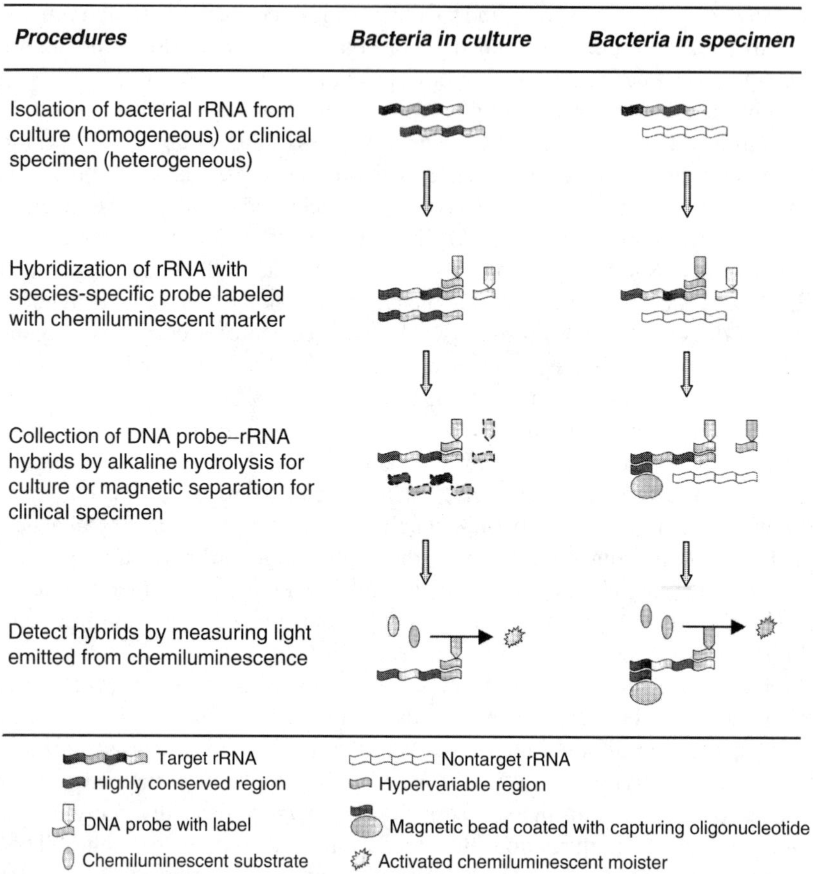

Procedures	Bacteria in culture	Bacteria in specimen
Isolation of bacterial rRNA from culture (homogeneous) or clinical specimen (heterogeneous)		
Hybridization of rRNA with species-specific probe labeled with chemiluminescent marker		
Collection of DNA probe–rRNA hybrids by alkaline hydrolysis for culture or magnetic separation for clinical specimen		
Detect hybrids by measuring light emitted from chemiluminescence		

Target rRNA — Nontarget rRNA
Highly conserved region — Hypervariable region
DNA probe with label — Magnetic bead coated with capturing oligonucleotide
Chemiluminescent substrate — Activated chemiluminescent moister

FIGURE 15.1 Probe-based nonamplified detection. DNA probe hybridization can be used for both culture confirmation and direct detection from clinical specimens. The procedures depicted were developed by Gen-Probe, Inc., San Diego, CA. (From Li, J. and Hanna, B. A., *Molecular Microbiology Diagnostic Principles and Practice*, Persing, ASM Press, Washington, DC, 2004. With permission from ASM Press.)

northern blots) for many years to identify cloned genes and analyze DNA and RNA fragments. The format can also be reversed such that the probe is immobilized on a solid surface and target sequences are added in solution. This reversed format allows for the detection of multiple analytes from a single specimen preparation or the detection of multiple polymorphisms in a single amplicon generated using consensus primers for amplification. One example of solid-phase hybridization is line probe technology that has been applied to detect polymorphisms associated with hepatitis C virus (HCV) and human papillomavirus (HPV) genotypes, and for resistance detection in MTB (discussed further in Sections 15.5.2 and 15.5.3). Two additional solid-phase hybridization techniques are microarrays and *in situ* hybridization (ISH), the latter of which is discussed in Section 15.2.2.

Capture probe technology uses a combination of solid- and liquid-phase hybridization through the use of glass or metal beads coated with probes. Bead-affixed

TABLE 15.1
Selected Commercially Prepared DNA Probes for the Detection of Bacterial and Fungal Pathogens

Organism	Specimen	Sensitivity (%)	Specificity (%)	Agreement (%)[a]
Campylobacter spp.[b]	Stool culture	100	99.7	99.8
Listeria monocytogenes	Cultured isolate	100	99.7	99.8
Streptococcus pyogenes	Throat swab culture	99.0	99.7	99.3
Haemophilus influenzae	CSF or throat swab culture	97.1	100	98.7
M. kansasii	Sputum culture	86.9	99.3	94.2
M. avium complex[c]	Sputum or BAL fluid culture	97.6	100	98.1
M. gordonae	Sputum or BAL fluid culture	98.8	99.7	99.3
MTB *complex*[d]	Sputum or BAL fluid culture	99.2	99.9	99.6
B. dermatitidis	Cultured isolate	98.1	99.7	99.2
C. immitis	Cultured isolate	98.8	100	99.6
NG	Urethral or cervical culture	95.4	99.8	99.0
H. capsulatum	Cultured isolate	100	100	100
CT	Urethral or cervical swabs	92.6	99.8	99.0
NG	Urethral or cervical swabs	95.4	99.8	99.0
S. pyogenes	Throat swab	91.7	99.3	97.4

Note: Summarized from information provided by Gen-Probe. CSF, cerebrospinal fluid; BAL, bronchoal-veoloar lavage.

[a] Agreement between probe identification and culture results.
[b] Including *C. jejuni, C. coli*, and *C. laridis*.
[c] The *M. avium* complex includes *M. avium* and *M. intarcellulare*.
[d] The MTB complex includes *M. tuberculosis, M. bovis, M. bovis BCG, M. africanum, M. microti*, and *M. canetti*.

probes act to capture potential sequences of interest with the subsequent addition of discriminatory and reporter probes. Capture probe technology can also be considered as a signal amplification assay (see Section 15.3). The signal is amplified due to the addition of anti-DNA (target)–RNA (probe) hybrid antibodies that are conjugated to multiple alkaline phosphatase molecules; this results in a 3000-fold increase in signal (i.e., Hybrid Capture [HC] system, Digene Corporation, Gaithersburg, Maryland). Analytes that can be detected directly from patient specimens using HC technology and are commercially available include HPV, cytomegalovirus (CMV), CT and NG (Digene Corporation). HC probe technology is currently the only Food and Drug Administration (FDA)-approved method for HPV detection and genotyping.

15.2.2 *In Situ* Hybridization

ISH allows for the detection of nucleic acid sequences in cells or tissues fixed to glass slides. This technology allows microorganisms to be identified in their natural, disease

state. Probes, which can be either DNA or RNA, are typically short (15–30 bp) allowing for easier penetration and access to the target site. Lysis of cellular membranes
and proteins must still occur to allow permeation and hybridization of the probe.
Both colorimetric and fluorescent ISH (FISH) probes have been described; common
fluorescent labels include fluorescein, rhodamine, their derivatives, and cyanine dyes.[1]
A powerful application of ISH is the use of multiple probes tagged with different fluo-
rophores for the simultaneous detection of multiple organisms in a single specimen,
or the simultaneous detection of an etiologic agent and immunohistologic markers.
Advantages of ISH applications in histopathology are that the host tissue response
can be evaluated and the exact cells displaying a specific morphotype can be probed
for suspected etiologic agent(s). In addition, organisms that are "nonculturable" or
difficult to culture can be probed, for example, *Tropheryma whipplei* for Whipple's
disease. Disadvantages include autofluorescence exhibited by some microorganisms
(including *Pseudomonas, Legionella*, and many yeasts and moulds), specificity and
reliability are highly probe sequence dependent, insufficient penetration of sample
material, secondary structure of target sequence, low target content, and photobleach-
ing.[7] Examples of the successful use of ISH in infectious disease testing include HPV
typing and the detection Epstein–Barr virus (EBV) transcripts in cellular material.[8,9]

For direct identification of microbial organisms in patient samples or cultures,
peptide nucleic acid–FISH (PNA–FISH) has been described. PNA probes have a
neutral peptidelike backbone, as opposed to a negatively charged sugar phosphate
backbone found in DNA probes (Figure 15.2).[10] However, like DNA probes, PNA
probes also hybridize to DNA and RNA in a sequence-specific manner and can
be labeled fluorescently for ease of detection. Reported advantages of PNA probes
include stronger and faster hybridization, ability to discriminate 1-bp differences,
resistance to nucleases and proteases, survival under stringent conditions that allow
for access to regions with secondary structure, and increased hydrophobicity that
allows for penetration of cell membranes during ISH.[10] PNA probes that are commercially available for culture identification include *S. aureus, Candida albicans*,

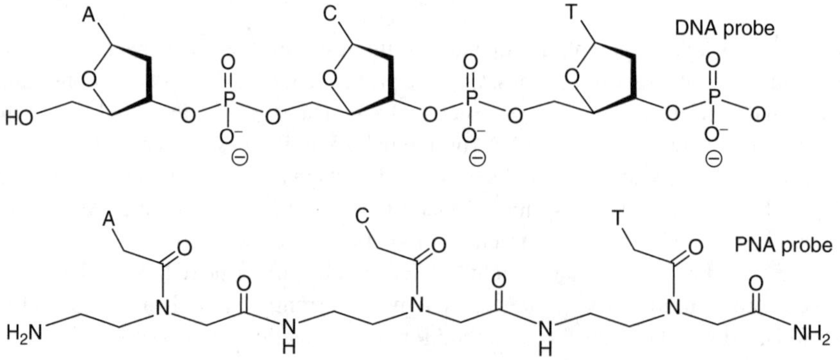

FIGURE 15.2 DNA and PNA probes. The DNA probe has a sugar phosphate backbone and
carries a negative charge, whereas the PNA probe has a peptide-like backbone and neutral
charge. Both probes hybridize in a sequence-specific manner to nucleic acids. (Images courtesy of AdvanDx, Inc., Woburn, MA.)

and *Enterococcus faecalis* (AdvanDx Inc., Woburn, Massachusetts). In addition, PNA–FISH probes have been used successfully by clinical microbiology labs in the identification of *S. aureus* and *C. albicans* directly from positive blood culture bottles within 2.5 h and the differentiation of MTB complex and nontuberculous mycobacteria from liquid cultures.[11–15]

15.3   SIGNAL AMPLIFICATION

Another application that is used for the direct detection of nucleic acid sequences in clinical specimens is signal amplification. This technique is unique in that the signal, as opposed to the target (as in the polymerase chain reaction [PCR]) is amplified. Owing to the amplification of the signal generated from hybridization, signal amplification is more sensitive than nonamplified direct detection but is generally not as sensitive as target amplification methods. However, there seems to be less risk for contamination with signal amplification than with target amplification techniques.

The most robust signal amplification method currently available is branched DNA (bDNA; Figure 15.3). Capture extenders, label extenders, and preamplifier and amplifier oligonucleotides act in concert to form a branched structure (i.e., bDNA)

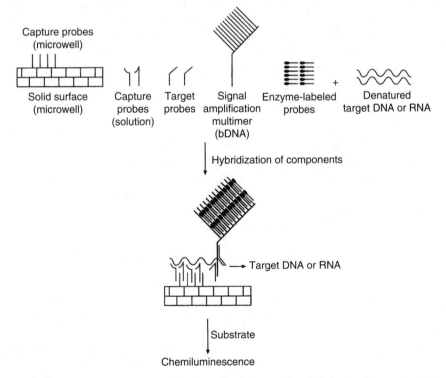

FIGURE 15.3   bDNA signal amplification. Capture extenders, label extenders, and preamplifier and amplifier oligonucleotides act in concert to form a branched structure (bDNA) on the target sequence to which multiple enzymatically labeled probes can hybridize, which results in signal amplification. (From Wolk, D., Mitchell, S., and Patel, R., *Infect. Dis. Clin. North Am.* 15(4), 1157, 2001. With permission from Elsevier.)

on the target sequence to which multiple enzymatically labeled probes can hybridize. Capture extenders hybridize to both the target sequence and to capture probes immobilized on a microtiter plate. Label extenders hybridize to the target sequence and provide complementary nucleotides for hybridization of preamplifier oligonucleotides. Once hybridized in the complex of oligonucleotides, the preamplifier serves as a template for the hybridization of amplifier oligonucleotides, which are then available for hybridization of labeled probes. The chemiluminescent signal is generated by the enzymatic reaction of the label (most commonly, alkaline phosphatase). In some assays, up to 3000 reporter probes are hybridized per target sequence. The signal generated is proportional to the amount of target sequence present in the clinical specimen, thus allowing for quantification of the analyte.

The first generation of bDNA assays lacked sensitivity compared to target amplification, but subsequent generations have improved on the signal-to-noise ratio, which allows for analytical sensitivity comparable to that of target amplification.[16] In addition, third-generation assays have improved specificity by increasing hybridization stringency of target and capture probes, and the incorporation of isoC- and isoG-containing probes have decreased nonspecific hybridization to nontarget sequences.[17] Further, amplification of the signal was improved by the introduction of preamplifiers. FDA-approved assays using bDNA signal amplification include quantification of human immunodeficiency virus (HIV)-1 and HCV (Bayer HealthCare, Tarrytown, New York), which are discussed further in Section 15.5.3. In addition, several user-defined assays have been described including the detection of hepatitis B virus (HBV), CMV, HPV, *Trypanosoma brucei*, and antibiotic-sensitive and resistant staphylococci.[17]

15.4 TARGET AMPLIFICATION

In contrast to the previously discussed platforms that represent variations of the detection format, target amplification relies on multiple rounds of thermocycling or polymerase activities to multiply the number of targets present for subsequent detection. Nucleic acid amplification (NAA) greatly increases the sensitivity of molecular techniques. This section summarizes the most commonly used target amplification systems, transcription-mediated amplification (TMA), nucleic acid sequence-based amplification (NASBA), strand displacement amplification (SDA), and PCR.

15.4.1 Transcription-Based Technologies: TMA and NASBA

TMA and NASBA are RNA-based amplification techniques that rely on multiple enzymes at isothermal conditions to amplify RNA targets rather than amplification based on a single enzyme (i.e., DNA polymerase) and repeated rounds of thermocycling (see Section 15.4.3). TMA and NASBA techniques are very similar with the exception of the number of enzymes used in each reaction: NASBA uses three enzymes, whereas TMA uses two enzymes. As illustrated in Figure 15.4, an RNA target sequence, usually rRNA because it is abundant, is converted into complementary DNA (cDNA) by reverse transcriptase (RT), which can then act as the template for multiple rounds of RNA transcription. TMA relies on the RNase activity of the RT enzyme, whereas NASBA uses a separate RNase H to degrade the DNA–RNA hybrids produced during cDNA synthesis.[18]

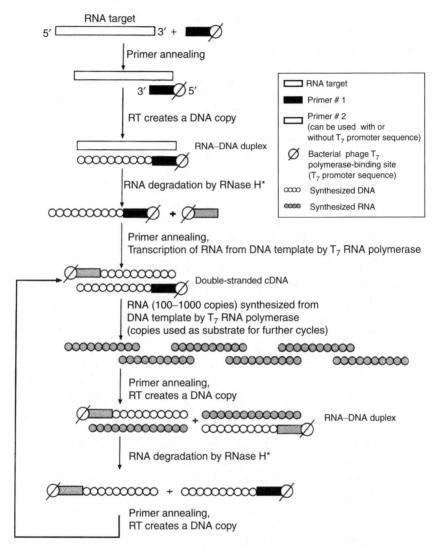

FIGURE 15.4 Transcription-based amplification systems (TAS). TAS includes TMA and NASBA. In an isothermal reaction, an RNA target is converted into cDNA by RT, which can then act as the template for multiple rounds of RNA transcription. NASBA uses three enzymes, whereas TMA uses two enzymes to complete the processes. (From Wolk, D., Mitchell, S., and Patel, R., *Infect. Dis. Clin. North Am.* 15(4), 1157, 2001. With permission from Elsevier.)

The advantages of TMA and NASBA are the robustness of the amplification and the fact that isothermal conditions do not require a thermocycler. TMA and NASBA assays are specific and sensitive, amplifying as few as 10–100 target molecules from samples. In addition, they detect the presence of RNA, which is suggestive of active infection of DNA viruses, rather than detecting any DNA present. The major disadvantage of TMA and NASBA is the relative instability of the RNA target.

The addition of RNase inhibitors and proper storage can help to overcome stability issues. Another disadvantage is the small optimal amplicon size (120–250 bp) for NASBA and TMA; longer products are amplified less efficiently.[19]

15.4.2 STRAND DISPLACEMENT AMPLIFICATION

SDA technology is based on the generation of single-stranded nicks by a restriction endonuclease (RE) before amplification (Figure 15.5). The reaction is isothermal because denaturation is enzyme driven as opposed to thermally driven. Primers that have both target-specific sequence and an RE site encoded in the primer sequence are used. The incorporation of α-thio-substituted nucleotides in the reaction produces hemiphosphorylated DNA that allows the RE to make a site-specific single-stranded nick as opposed to the double-stranded cleavage that occurs by wild-type endonucleases.[20] Because synthesized strands incorporate α-thiolated nucleotides, they are not cleaved by the RE. The DNA polymerase, which lacks 3′–5′ exonuclease activity, synthesizes a new DNA strand from the 3′ end of the nick displacing the nicked single strand.[18] The displaced strands are then available for repeated rounds of annealing, nicking, extension, and displacement of newly formed DNA strands. This results in exponential amplification of the targeted sequence.

Commercialization of SDA has allowed for the real-time detection of amplified products on an automated platform (BD ProbeTec ET system, BD Diagnostics, Sparks, Maryland). Detection occurs by the inclusion of probes that are labeled with a fluorescent dye in close proximity to a quencher molecule. When in proximity to the fluorescent label, the quencher prevents the emission of fluorescent signal. Because the probes contain target sequence, as the quantity of amplified product increases, the probes hybridize to the product and are then converted from stem–loop structures to oligonucleotides hybridized to the target. The probes also encode an RE site. When cleavage by the RE occurs, the fluorescent dye is separated from the quencher allowing for signal emission measured by a fluorometer; this is the energy transfer (ET) detection method. Applications that are FDA approved on this platform include CT, NG, and *L. pneumophila*. Assays available outside the United States include detection of MTB and agents of atypical pneumonia (*Chlamydiophila* and *Mycoplasma*).[21]

15.4.3 POLYMERASE CHAIN REACTION

PCR is the most commonly used target amplification technique, and its discovery in 1985 by Kary Mullis has revolutionized basic and translational research as well as clinical diagnostics. PCR utilizes two oligonucleotide sequences, or primers, specific to the target of interest. One primer is complementary to the sense strand, whereas the other complements the antisense strand of the DNA target. Generally, the primers are 150–3000 bp apart for conventional PCR and 50–100 bp for real-time PCR. PCR is a three-step temperature-dependent process as depicted in Figure 15.6. The process involves (1) denaturation of the two strands of the double-stranded DNA (dsDNA) template, (2) annealing of the primers to the denatured target DNA, and (3) extension by the addition of deoxyribonucleotides (dNTPs) by DNA polymerase. The temperature changes necessary to complete the three phases of PCR are accomplished using a controlled and programmable thermocycler. At the completion of each temperature

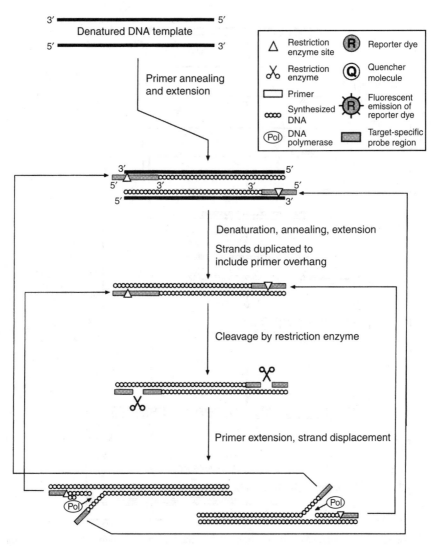

FIGURE 15.5 Stand displacement amplification. Single-stranded nicks are introduced by a restriction endonuclease prior to amplification. In an isothermal reaction, DNA polymerase synthesizes a new DNA strand from the 3' end of the nick displacing the nicked single strand. The displaced strands are then available for repeated rounds of annealing, nicking, extension, and displacement of newly formed DNA strands. (From Wolk, D., Mitchell, S., and Patel, R., *Infect. Dis. Clin. North Am.* 15(4), 1157, 2001. With permission from Elsevier.)

cycle, there is a theoretical doubling of the amplified product, and by using 35–50 cycles, there is over a million-fold amplification of the product. Although the primary target of PCR is DNA, the addition of RT and another temperature cycle can permit RNA to be used as the target nucleic acid in a method called RT PCR.

The results of PCR can be detected in two ways, end-point and real-time PCR detection. End-point detection involves the measurement of the amplified product

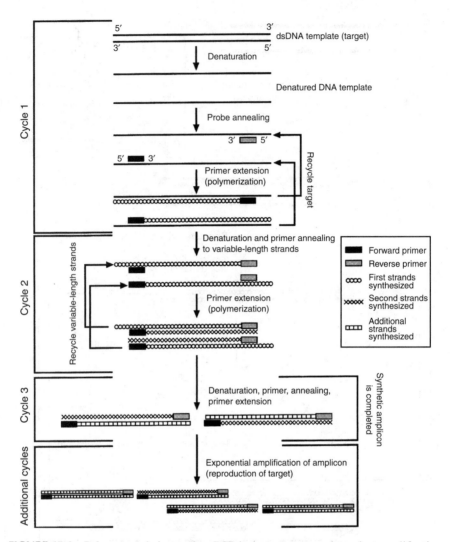

FIGURE 15.6 Polymerase chain reaction. PCR is the temperature-dependent amplification of a specific DNA template (target) using short oligonucleotide primers that are complimentary to the region of interest. The steps of PCR, denaturation, annealing, and extension are repeated for a specified number of cycles, which result in an exponential amplification of the product. (From Wolk, D., Mitchell, S., and Patel, R., *Infect. Dis. Clin. North Am.* 15(4), 1157, 2001. With permission from Elsevier.)

at the end of thermocycling. The primary methods of end-point detection are direct visualization by agarose gel electrophoresis and indirect detection by hybridization of a complementary probe. PCR followed by agarose gel electrophoresis is commonly called conventional PCR. Hybridization of probes to detect amplicons from NAA techniques has several different formats. A southern blot involves electrophoresing the amplified products on an agarose gel, transferring them to a nitrocellulose membrane, and detecting them with a labeled DNA probe complimentary to the target of interest.[22] Microtiter hybridization assays are suitable for high-volume tests and conducive to automation. This method, depicted in Figure 15.7, generates biotin-labeled PCR products that are captured onto streptavidin-coated microtiter wells, and after denaturation, the single-stranded DNA is hybridized to a target probe.[23] Each probe requires a separate well to interrogate the amplified products that are detected by the addition of a conjugate and substrate. Reverse hybridization methods have been developed to permit multiple probes to hybridize to a single PCR product. In this detection method, the probe is on a membrane or attached to beads and the labeled PCR product is added to the probe. The hybridized products are visualized by the addition of conjugate and substrate on the membrane or by a flow cytometer if fluorescent beads are used.[23] These end-point detection methods are inexpensive and adaptable, but labor-intensive and time-consuming.[22]

Real-time detection measures the accumulation of amplicons while they are being produced and eliminates postamplification manipulation, thus preventing potential carry-over contamination that can occur with end-point detection. While maintaining high sensitivity and specificity, real-time detection is accurate and reproducible with a greatly expanded linear dynamic range. Because of the decreased time to results, real-time assays are ideal for high-throughput assays.

Several different real-time detection platforms exist as seen in Table 15.2.[24] The primary differences in the platforms are highlighted in Table 15.2. The needs of each laboratory dictate the ideal instrument. The thermocycling of the large-capacity instruments is much slower than that of the smaller-capacity instruments that use the heating block principle. The lower-capacity instruments are more rapid and have workflow

Detection of PCR products in a microtiter plate

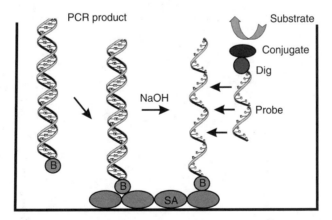

FIGURE 15.7 Microtiter hybridization assay. An end-point detection method where PCR products are linked with a biotin moiety (B) and captured by the streptavidin (SA) coated in the microtiter well. The double-stranded amplicon is denatured and the captured-single strand is detected by the addition of labeled oligonucleotide probe that is visualized by the addition of a conjugate and a colorimetric substrate reaction. (From van Doorn, L.-J., *Molecular Microbiology Diagnostic Principles and Practice*, ASM Press, Washington, DC, 2004, 147. With permission from ASM Press.)

TABLE 15.2
Real-Time PCR Platforms

Manufacturer	Instrument	Probe Chemistries Supported	Excitation (nm)	Detection (nm)	No. of Samples	Reaction Volume (µL)	Rapid Thermo-Cycling	Comments
Roche Applied Science (http://www.roche-applied-science.com/lightcycler-online/)	LightCycler 1.0	Hybridization probes, molecular beacons, TaqMan	LED470	530, 640, 710	32	10–20	Yes	Special sample containers
	LightCycler 2.0	Hybridization probes, molecular beacons, TaqMan	LED470	530, 555, 610, 640, 670, 710	32	10–100	Yes	Special sample containers
Cepheid (http://www.cepheid.com/pages/home.html)	SmartCycler II	Molecular beacons, TaqMan	LED450–495, LED500–550, LED565–590, LED630–650	510–527, 565–590, 606–650, 670–750	16	25–100	Yes	Special sample containers, independent modules
Corbett Research (http://www.corbettresearch.com/home.htm)	Rotor-Gene 3000	Molecular beacons, TaqMan	LED470, LED530, LED585, LED635	510, 550, 580, 610, 660	72	10–150, 5–50	Mid-range	Standard plastic tubes are used
ABI (http://www.appliedbiosystems.com/index.cfm)	Prism 7000	TaqMan, molecular beacons	Tungsten/halogen	Four-color multiplex	96		No	
	Prism 7300	TaqMan, molecular beacons	Tungsten/halogen	Three-color multiplex	96		No	
	Prism 7500	TaqMan, molecular beacons	Tungsten/halogen	Four-color multiplex	96		No	
	Prism 7900ht	TaqMan, molecular beacons	Laser 488	500–600	384		No	

Company	Instrument	Chemistry	Excitation	Detection	Wells	Volume (µL)		Notes
BioRad (http://www.bio-rad.com/)	MyiQ	TaqMan, molecular beacons	Tungsten/halogen	Single color	96	15–100	No	
	ICycler iQ	TaqMan, molecular beacons, hybridization probes	Tungsten/halogen	Four-color multiplex	96	50	No	
Stratagene (http://www.stratagene.com/homepage/)	Mx4000	TaqMan, molecular beacons,	Tungsten/halogen	Four-color multiplex	96	10–50	No	
	Mx3000p	TaqMan, molecular beacons	Tungsten/halogen	Four-color multiplex	96	25	No	
MJ Research (http://www.mjr.com/)	Chromo 4	TaqMan, molecular beacons	450–490, 500–535, 555–585, 620–730	515–530, 560–580, 610–650, 675–730	96	10–100	No	
	Opticon	TaqMan, molecular beacons	450–495	515–545	96	10–100	No	
	Opticon 2	TaqMan, molecular beacons	470–505	525–543, 540–700	96	10–100	No	
Genetic Discovery Technology (http://www.biogene.com/index.cfm)	SycChron	TaqMan, molecular beacons, hybridization probes	473 Laser	520–720	6	10–50	Yes	Electrically conducting polymer technology

flexibility that is not possible with the high-throughput machines, but these instruments are limited in the number of samples that can be run at one time.[24] Commercial availability of analyte-specific reagents (ASRs) and FDA-approved kits on specific instruments makes real-time assays feasible for many labs. Before commercial availability, the majority of assays were user-defined, or "homebrew" assays that required extensive expertise to develop and validate and were therefore limited to large reference laboratories and academic medical centers.

Real-time platforms have specific detection formats with a range of specificities. The most simple and nonspecific detection format for real-time PCR is the incorporation of SYBR® Green (Applied Biosystems, Foster City, California) into any dsDNA product. Although SYBR Green is sensitive, it is not specific. The use of melting curve analysis, defined by the length and %G–C content of the amplicons, allows detection of different amplification products produced by incorporation of SYBR Green, but this requires knowledge of the melting profile of the target amplicon of interest.[24] Due to the lack of specificity, SYBR Green assays are rarely used for clinical diagnosis, but are more commonly used as screening assays before more expensive, specific assays that are used for confirmation.

Specific detection is achieved by the use of fluorescent probes. Three nucleic acid probe technologies are most commonly used with real-time PCR, though the repertoire of available probe technologies is expanding.[25] As depicted in Figure 15.8, these probes include 5' nuclease (also known as TaqMan or hydrolysis) probes, molecular beacons, and fluorescence resonance ET (FRET) hybridization (also known as LightCycler) probes.[26] Each probe relies on the transfer of light between two dye molecules, a fluorophore and a quencher, a process referred to as FRET.[24,25] The distance between the fluorophore and the quencher determines if a signal is detected. 5' Nuclease probes (Figure 15.8A) and molecular beacons (Figure 15.8B) are two types of probes that contain a 5' fluorescent dye and a 3' quenching dye. To generate a signal from the 5' nuclease probe, the probe binds to the complementary target sequence and the 5' exonuclease activity of DNA polymerase cleaves the 5' end of the probe that separates the dye from the quencher and permits light emission. The accumulation of fluorescence (free dye) can be measured at any stage of PCR and is a direct measure of how much product has been amplified, which in turn is directly related to how much target was present in the original sample. To generate a signal from the molecular beacon, the probe binds to the complementary target sequence and releases the hairpin that is formed as a result of complementary sequence within the probe. Once bound to the target sequence, the fluorescent dye and the quencher are separated and fluorescence is emitted.[24,25] In addition to detecting PCR products, molecular beacons have been used for real-time detection in NASBA assays.[27,28] Unlike the previous two probes, FRET hybridization probes (Figure 15.8C) are composed of two oligonucleotide probes, each labeled with a different fluorescent dye, that target 40–50 bp of adjacent sequence, which provides increased specificity. The probes are designed to anneal next to one another in a head-to-tail fashion. When bound to the amplicon, the acceptor probe absorbs the emission light from the donor probe and emits fluorescence that is measured during the annealing phase. The main benefit to this probe technology is the ability to perform melting curve analysis, slowly raising the temperature of the reaction to measure the point at which half of

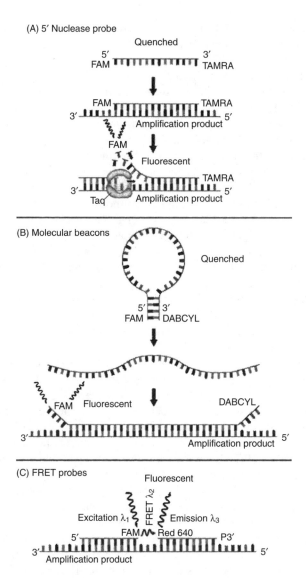

FIGURE 15.8 Real-time fluorescent probe technologies. (A) With 5′ nuclease probes the
reporter dye (FAM) and the quencher (TAMRA) are in close proximity. Once the probe
hybridizes with the target, the DNA polymerase exonuclease activity separates the reporter
dye from the quencher permitting detection of the product. (B) A molecular beacon has a
region of complimentarity within the probe that forms a hairpin that keeps the dye (FAM)
and quencher (DABCYL) in close proximity when the probe is not bound to the target. Once
the probe binds to the target, the dye and quencher are separated and signal is emitted.
(C) FRET hybridization probes contain two fluorescently labeled oligonucleotides that anneal
next to each other on the target of interest. Once annealed, the acceptor probe absorbs the
emission light from the donor probe and fluorescence is emitted. (From Cockerill, F. R., III
and J. R. Uhl, *Rapid cycle real-time PCR methods and applications*, Springer-Verlag, Berlin,
Germany, 2002, 3–27. With permission from Springer-Verlag GmbH.)

the FRET signal is lost. The melting temperature (T_m) is specific for the product and identifies the amplicon to which the probe is bound. Hybridization probes are useful when the targeted region is known to mutate frequently. The probes will still bind, but the T_m will be shifted revealing that presence of polymorphisms.[25]

For all three probe technologies, a positive result occurs when fluorescence is generated as a result of probes binding to the amplified target. The cycle number at which the fluorescence crosses the baseline is called the threshold cycle (C_t) and is indicative of a sample positive for the target. For many infectious diseases, a qualitative result is sufficient because the etiologic agent is not normally present, therefore, the presence of any amount is considered diagnostic. However, the florescent signal generated by real-time PCR is proportional to the amount of DNA in the reaction, therefore, calibration standards can be incorporated into the reaction to produce quantitative results. For quantification, the fluorescent readings must occur during the exponential phase of amplification, and algorithms incorporate linear-regression analysis to compare the C_t of the amplicon with the C_t of the known concentration of the standard to predict the quantity of nucleic acid present in the test samples.[29] Quantitative results are useful in determining the viral load, which is important to predict or monitor the response to therapy or to differentiate active from latent infection (see Section 15.5.3). Qualitative assays are generally more sensitive and thus serve to determine if the virus or infection is present and is the final assessment for treatment response.

The primary limitation of PCR is the availability of sequence data, which is required to design primers that target the nucleic acid of the organism of interest. In addition, bp mutations that arise in the targeted sequence of a PCR assay can cause false-negative reactions because the specificity of the reaction does not permit recognition of the altered sequence. The risk of false-positive results due to contamination from the environment or from another sample is also a major disadvantage of PCR assays. This can be prevented by strict working conditions that are optimized to reduce contamination. PCR can be too sensitive in that it is able to detect nonpathogenic levels of an organism in addition to not being able to distinguish between viable and nonviable organisms.[20] The disadvantages of real-time detection are the incompatibility of certain platforms with some fluorescent chemistries and the restricted multiplex capabilities due to a limited number of fluorescent chemistries and available detection channels. In addition, real-time PCR is initially very expensive to implement due to the cost of the instrumentation and reagents compared to those required for conventional PCR. The lack of interlaboratory standardization among assays and platforms is also problematic.[24] Nonetheless, the power of PCR has greatly enhanced the diagnosis of infectious diseases.

15.5 SPECIFIC APPLICATIONS OF DIRECT DETECTION AND AMPLIFICATION TECHNOLOGIES

15.5.1 Diagnosis of Bacterial Infections

15.5.1.1 *Chlamydia trachomatis* and *Neisseria gonorrhoeae*

The diagnosis of the sexually transmitted diseases (STDs) caused by CT and NG has been greatly enhanced by the application of molecular technology. The increased sensitivity offered by molecular detection is important not only for the diagnosis of symptomatic patients, but also for the asymptomatic individuals that account for more than 70% of positive cases. The Centers for Disease Control and Prevention (CDC) recommends routine screening for CT and NG due to the severe health consequences that may occur including pelvic inflammatory disease (PID), infertility, chronic pelvic pain, and pregnancy-related complications in women and urethritis in men.[30] In 2006, the CDC guidelines included screening all women under 25 years of age, pregnant women, and any woman with risk factors.[30] Implementation of CT screening programs has demonstrated a reduction in *Chlamydia* prevalence and the incidence of PID.[31]

Molecular detection of CT and NG was the first routinely used molecular diagnostic tool more than 10 years ago. Until then culture was the gold standard for CT and NG detection, which has subsequently been demonstrated to have only 60–75% sensitivity compared to NAA testing.[32] Cell culture for CT is relatively labor-intensive and requires technical expertise making interlaboratory standardization difficult. In addition, CT culture has a low sensitivity and a long turnaround time making this technique suboptimal. In contrast, GC culture has high sensitivity and specificity, and is inexpensive and easy to perform. Although a culture result will require up to 3 days, with GC culture, an isolate is available for further testing if needed for medicolegal purposes or therapeutic decisions. A disadvantage of culture for CT and NG is that organism viability must be preserved during transport for optimal sensitivity. Nonculture antigen detection tests are also available in which organism viability is not required. The latter tests include enzyme immunoassays (EIAs) for both CT and NG and direct fluorescent antibody (DFA) detection of CT.

Molecular technologies include both direct detection methods and NAA. Direct detection methods are more sensitive than culture, but not as sensitive as NAA. Commercially available direct detection assays include Gen-Probe PACE 2C (hybridization protection assay) and Digene HC II GC/CT test. Neither of the assays initially differentiates between CT and NG if positive, but subsequent testing can produce organism-specific results.[31] Compared to culture, these assays show sensitivities greater than 90%. In areas of moderate prevalence (5–20%), positive predictive values (PPV) of probe hybridization range from 96% to 99%.[31] The Digene assay is only approved for female specimens, and neither test is approved for urine specimens. The ease of specimen transport is a great advantage of hybridization tests over culture.

NAA is the most sensitive and is the recommended method for CT and NG detection.[30] Methods available include PCR, TMA, and SDA. NAA tests that are FDA cleared include AMPLICOR and COBAS AMPLICOR CT/NG (Roche Diagnostics, Indianapolis, Indiana), BD ProbeTec ET (BD Diagnostics) that use PCR and SDA, respectively, to amplify a CT-specific cryptic plasmid. Also FDA cleared, the Gen-Probe APTIMA COMBO 2 utilizes TMA to amplify and detect 23S rRNA sequences of CT. All three platforms are also FDA cleared to detect NG, targeting the cytosine methyltransferase gene, the multicopy pilin gene-inverting protein homologue, and 16S rRNA, respectively.[31] These methods are FDA cleared for use with endocervical swabs, urethral swabs, and urine. However, it should be noted that not all platforms are cleared for urine NG testing in women. Specimen types such as vaginal and conjunctival specimens have also been evaluated by independent users with NAA with satisfactory performance, but there are limited studies evaluating rectum and pharynx specimens.[31,32] Broader applications of NAA will increase detection from these alternative sites relative to culture. However, regulations require each laboratory to verify the performance characteristics of these noncleared specimens before offering results clinically.

Most reports of sensitivity and specificity of NAA tests include discrepant analysis which is thought to inflate the reported sensitivity.[31,32] However, a meta-analysis was performed using culture as the reference test method from these studies and NAA was 96.7% sensitive and nucleic acid hybridization tests were 92.2% sensitive.[33] Specificity is generally high; the primers used to detect CT are not known to cross-react with other non-CT DNA sequences. Some primers used for NG detection cross-react with saphrophytic *Neisseria* species, but the specimens that are FDA cleared for testing using these primers are unlikely to have these oropharyngeal commensals present.[31,32] Nonetheless, to increase the specificity of the assay, in addition to the positive–negative threshold, a gray zone may be established by the manufacturer. Commonly, both false-positives and -negatives hover around the threshold. Therefore, by utilizing a gray zone interpretation with subsequent confirmatory testing, the specificity and PPV can be increased for a particular NAA test. The implementation of routine confirmatory testing should be considered when the prevalence of either CT or NG is low, resulting in a PPV below 90%.[31] Methods of confirmatory testing include testing either the original specimen or a second specimen with a different assay that detects a different target, repeating the original assay by incorporating a blocking antibody or competitive probe, or, least favorable, repeating the original specimen with the original test.[31]

One disadvantage of CT and NG NAA is the presence of inhibitors in the specimen. Some manufacturers supply an internal control to monitor for the presence of inhibitors, but when inhibitors are present, the test is invalid. In addition, test-of-cure (only recommended post-CT treatment during pregnancy) should not be performed using NAA because results can remain positive for up to 3 weeks due to the presence of nucleic acids from nonviable organisms.[31] As with any NAA assay, contamination is a risk that could lead to false-positive results.

15.5.1.2 *Mycobacterium tuberculosis*

Detection of MTB has historically relied on conventional methods including decontamination and concentration of specimens followed by microscopic analysis of acid fast staining, culture, and biochemical identification. Although smears stained for acid fast bacilli (AFB) offer a same-day turnaround time, the associated sensitivity is very low, requiring at least 10^4 bacilli per milliliter to be positive.[34] Although conventional culture methods offer increased sensitivity compared to acid fast stains, the time of detection can be up to 8 weeks, and culture yield is affected in specimens requiring decontamination procedures. Advances in culture technology (i.e., automated liquid culture systems that measure the release of carbon dioxide during bacterial growth) have decreased the time of detection to 2–4 weeks and has offset yield lost by decontamination processes by offering increased sensitivity. However, these advances still do not provide a rapid laboratory diagnosis for MTB. Owing to the increasing frequency of isolation of mycobacterial species associated with immuno-compromised hosts and the increased incidence of multi-drug-resistant MTB, it has become imperative to offer accurate yet rapid diagnostic tools for the detection and identification of mycobacteria.

The introduction of the hybridization protection assay (AccuProbe; Gen-Probe Inc.) has dramatically decreased the time to identification for MTB complex, as well as *M. avium, M. intracellulare, M. gordonae,* and *M. kansasii.* Chemiluminescent probes to 16S rRNA sequences allow the accurate identification of the preceding mycobacterial species. Although the assay can be completed in 2 hours, it must be performed from an isolate, thereby limiting the rapidity with which a result can be obtained. Direct probe identification reportedly has a sensitivity and specificity of 88–100% and 99–100%, respectively, when compared to conventional biochemical identification.[3,4] Although AccuProbes are only FDA approved for culture confirmation, a number of reports have demonstrated the successful use of these probes with liquid media, further shortening the time to identification.[35,36]

Two FDA-approved assays are available for the detection of MTB complex directly from patient specimens: the COBAS/AMPLICOR MTB test (Roche Diagnostics) and the Gen-Probe Amplified MTB direct test (MTD). The AMPLICOR system employs PCR using genus-specific biotinylated primers with subsequent detection of the amplicon using an MTB-specific probe (Figure 15.7). This test is only approved for use in smear-positive respiratory specimens. The MTD uses TMA for amplification of tubercular rRNA transcripts followed by detection using the hybridization protection assay. An enhanced version of the MTD assay has been approved for use in both smear-positive and -negative respiratory specimens. Based on several studies in the literature using culture and clinical status as the reference method, the sensitivity of direct detection using the aforementioned two assays from smear-positive respiratory samples are higher (92–100%) than that of smear-negative respiratory specimens (40–93%) and specimens from extrapulmonary sites (27–98%).[37]

In addition to the two FDA-approved assays, there are commercially available assays that are not FDA approved for the direct detection of MTB in clinical samples. The BD ProbeTec ET Direct TB system (DTB; BD Diagnostics) utilizes SDA to simultaneously amplify IS6110 and 16S rRNA sequences. Smear-positive respiratory samples have reported sensitivities ranging from 90% to 100%, whereas smear-negative and extrapulmonary sensitivities range from 33% to 100%.[37] A study of 735 pulmonary specimens and 396 extrapulmonary specimens, including 125 culture-positive and 42 smear-positive specimens, determined the overall sensitivity, specificity, PPV, and negative predictive value (NPV) values for the ProbeTec assay to be 90%, 97%, 78%, and 99%, respectively.[34] The INNO-LiPA Mycobacteria (Innogenetics NV, Ghent, Belgium) identify mycobacteria using the intergenic transcribed spacer region between 16S and 23S rRNA genes, whereas the GenoType Mycobacteria (Hain Lifescience, Nehren, Germany) use the 23S rRNA sequence for identification.[34] These assays have been used to identify MTB from colony growth or from liquid culture systems, but have not been applied to direct detection from clinical specimens.[38] INNO-LiPA can identify 16 mycobacterial species and reportedly has 99% concordance when compared to AccuProbes or biochemical identification.[39] GenoType identifies 13 species and has demonstrated 89% correlation with AccuProbes and 16S rRNA gene sequencing.[34] Additional molecular techniques utilized for direct detection of MTB in clinical specimens include ISH,[34,40,41] NASBA,[42] and real-time PCR.[37]

Disadvantages of the molecular detection of MTB directly from clinical specimens include the inability to differentiate among the MTB complex. In addition, when testing is not performed from a cultured isolate, it is difficult to tell if the nucleic acid detected is from live or dead organisms. For this reason, amplification technologies should not be used on specimens collected from patients that have received antitubercular drugs for more than 7 days or have been treated for MTB within 2 months of collection.[34] Further, approximately 4% of pulmonary and 19% of extrapulmonary specimens have substances inhibitory to polymerase making negative test results invalid.[34] Because it is rare that all specimens from a given patient show inhibition, testing multiple samples can be advantageous.[43]

In 2000, the CDC published updated recommendations for the use of molecular testing in the detection of MTB from clinical specimens.[44] Recommendations were based on obtaining three separate sputum specimens for AFB microscopy and culture. If a specimen is smear positive, NAA should be performed on that specimen; but if all three AFB smears are negative, NAA should be performed on the first specimen collected. If the smear and NAA are both positive, it is presumed that the patient has tuberculosis (TB). If the specimen is smear positive and NAA negative, a test for inhibitors needs to be performed. If inhibitors are not detected on at least two smear-positive, NAA-negative samples, it is presumed that the patient has a nontuberculous mycobacterial infection. If at least two specimens are smear negative and NAA positive, it can be presumed that the patient has TB. If both the smear and NAA are negative in two specimens, it is presumed that the patient is not infectious with active pulmonary TB. Importantly, the CDC recommendations state that negative NAA results do not completely exclude active pulmonary TB, and clinical judgment is necessary for therapeutic decisions.

Although the specificity of NAA for TB diagnosis is very high, the sensitivity varies widely and depends both on the smear status of the specimen (positive or negative) and specimen type. The PPV and NPV further depend on the pretest probability of TB, or the degree of clinical suspicion.[43] NAA testing has a greater likelihood of influencing patient outcome in smear-negative patients in whom the clinical suspicion is high, as unnecessary treatment and further diagnostic procedures may be avoided.[34,43]

15.5.1.3 *Bordetella pertussis*

Conventional methods for the detection of *Bordetella pertussis* include culture, DFA staining, and serology. Although culture is very specific, its sensitivity suffers partially due to the fastidious nature of the organism, but primarily because the highest sensitivity for culture occurs before patients are symptomatic. Turnaround time for culture results is generally 5–7 days. Although DFA is rapid, offering same-day results, it has a high false-positive rate. Serologic testing can take up to 3–4 weeks as single, acute specimen results can be indeterminate and require a convalescent serum specimen. Because of the shortcomings of the preceding techniques, NAA has quickly become the new gold standard for the laboratory diagnosis of pertussis. NAA testing allows for same-day results and because erythromycin-resistant *B. pertussis* is still rare, a cultured isolate is rarely needed for antimicrobial susceptibility testing.[45]

The specimens of choice for *B. pertussis* NAA testing are nasopharyngeal swabs and aspirates, although calcium alginate swabs should not be used due to amplification inhibition.[46] Targets for *B. pertussis* detection include pertussis toxin promoter and the porin gene, but the most commonly amplified target is IS*481*.[46,47] The detection of a repetitive element such as IS*481* increases assay sensitivity due to multiple genomic copies. Multiplex assays that simultaneously detect the IS*481* element of *B. pertussis* and the IS*1001* element of *B. parapertussis* with analytic sensitivities as low as one organism per reaction also exist.[48,49] It should be noted that *B. holmseii* may cause a false-positive result with both targets.[47] Increasingly, commercially available non-FDA-approved real-time PCR assays are becoming available as ASRs. Although all reagents necessary for the amplification reaction can be purchased commercially, assay development and verification studies must be performed by individual laboratories. Unfortunately, there are no comparative studies between PCR procedures, including ASRs, limiting the application of such procedures in less-experienced laboratories.

Multiple studies have demonstrated significant increased detection of *B. pertussis* when comparing NAA to culture; PCR-positive, culture-negative samples reported range from 13 to 88%.[50] In a study by Chan et al.,[51] the sensitivity, specificity, PPV, and NPV were 100%, 97%, 88%, and 100% for real-time PCR and 12%, 100%, 100%, and 86% for culture, respectively. There are, however, both false-positives and -negatives with *B. pertussis* amplification procedures. Therefore, it is strongly recommended that the results be considered in context of patient's clinical presentation. Additionally, clinically inconsistent results should be confirmed by a second method. Because NAA is detecting DNA as opposed to live organisms, positive results can occur postantibiotic administration. This is advantageous when antibiotics have been administered before a confirmed diagnosis, but can be a disadvantage when a confirmed pertussis patient is not responding to therapy and follow-up testing is needed to differentiate persistent pertussis infection and anti-microbial-resistant *B. pertussis* from another source of symptoms. Studies have shown that on the fourth day of antimicrobial therapy 56% of specimens previously determined to be pertussis positive were still positive by culture, whereas 89% were positive by PCR; after 7 days, 0% was positive by culture and 56% were still positive by PCR.[52] PCR not only remains positive for longer period after therapy than culture, but it is also positive for a longer period after onset of symptoms.[46] Therefore, NAA is useful for patients presenting later in their illness. However, if symptoms have been present for weeks, serology is likely the most useful diagnostic test.[46]

15.5.2   IDENTIFICATION OF BACTERIAL RESISTANCE

Molecular detection is soon becoming the gold standard in clinical microbiology, particularly in cases where a specific pathogen or resistance gene is sought. Pertinent to this chapter, the molecular detection of antimicrobial resistance has been accomplished using probe hybridization and NAA technologies such as PCR and DNA sequencing. Once restricted to viral resistance detection, these methodologies are now being used for the detection of bacterial resistance.

Several caveats should be pointed out when discussing the use of molecular technology to detect antimicrobial resistance. Results can be difficult to interpret when (1) the resistance mechanism is multifactorial, (2) there is an uncertain cause–effect relationship between the genotypic and phenotypic resistance in clinical isolates, (3) the infection is polyclonal, or (4) there is synergism among multiple resistance phenotypes. Therefore, the molecular detection of antimicrobial resistance should be limited to organisms and infections in which the results regarding the genotypic relationship to clinical treatment and infection control precautions can be interpreted with confidence.

15.5.2.1 Vancomycin-Resistant *Enterococcus*

Vancomycin resistance was first detected in enterococci in 1988 (30 years after the introduction of vancomycin), and the past two decades have brought increasing resistance largely due to the increased use of vancomycin in the 1980s as oral therapy for *Clostridium difficile* colitis, treatment for MRSA, and in animal husbandry.[53,54] Vancomycin acts by binding to the D-Ala-D-Ala C-terminus of peptidoglycan precursors preventing their addition to the growing peptidoglycan chain and blocking the subsequent transglycosylation and transpeptidation steps of cell wall biosynthesis. The enterococci have developed two separate, yet similar, mechanisms for vancomycin resistance. The genes necessary for the resistance phenotype are encoded by the *van* genes, which encode for ligase enzymes that alter the acyl-D-Ala-D-Ala C-terminus of lipid II peptidoglycan precursors resulting in lower affinity of vancomycin for its target.[55] High-level resistance (MIC, ≥ 64 μg/mL), generally found in *E. faecium* and *E. faecalis*, is encoded by *vanA, vanB*, or *vanD*. The *vanA* and *vanB* operons are generally located on transposons, which are transferable through plasmids, or on large, mobile chromosomal elements,[56,57] whereas *vanD* appears to be located on the chromosome.[58] The transferability of *vanA* and *vanB* genes is the basis for infection control measures to monitor and prevent the spread of vancomycin-resistant *Enterococcus* (VRE). Low-level resistance (MIC, 2–32 μg/mL) found in *E. gallinarum, E. casseliflavus*, and *E. flavescens* is encoded by the *vanC* genes (*vanC1, vanC2*, and *vanC3*, respectively). Unlike *vanA* and *vanB, vanC* genes are chromosomally encoded and nontransferable. Therefore, *vanC*-containing species are not considered "true" VRE for infection control purposes and should be ruled out when screening for VRE.

To date, there is no FDA-approved assay for the molecular detection of VRE from rectal sites for infection control purposes, but there have been numerous user-defined NAA assays described in the literature, and commercial ASRs are also available.[24] Although VRE most commonly contains *vanA* in the United States and Europe, it is important to detect both *vanA* and *vanB* as prevalence varies with geography. It is debatable whether the assay requires detection of an *Enterococcus*-specific gene because *vanA* and *vanB* are rarely found in nonenterococcal species.[59,60] There are at least two advantages to the molecular detection of VRE: (1) amplification technology reportedly increases VRE detection by up to 120%, allowing better infection control, and (2) *vanC*-containing enterococci are accurately ruled out preventing unnecessary contact precautions and contributing to hospital savings.[24,61]

15.5.2.2 Methicillin-Resistant *Staphylococcus aureus*

Resistance to beta-lactams in *S. aureus* is due to both the production of beta-lactamase and an altered penicillin-binding protein (PBP).[62,63] The modified PBP is designated as PBP 2a and is encoded by *mecA*. Strains expressing *mecA* have low affinity for all beta-lactam antimicrobials and are designated as MRSA. MRSA is an important cause of health care-associated infections and is often resistant to additional antimicrobials including aminoglycosides, macrolides, and fluoroquinolones.[62] More recently, new strains of MRSA that contain *mecA* have appeared, but are more sensitive to other classes of antibiotics. These strains are associated with skin and soft tissue infections in outpatients and are called community-associated MRSA (CA-MRSA).[64] The incidence of CA-MRSA is increasing causing overall rates of MRSA to rise; therefore, it has become even more important to quickly and accurately identify resistant isolates.

Screening patients for MRSA carriage is a central strategy for preventing the spread of this organism in health care settings.[65] Conventional methods for the detection of MRSA include disk diffusion or MIC testing, screening plates containing oxacillin, and a latex agglutination test that detects PBP 2a (Oxoid, Basingstoke, Hampshire, United Kingdom). The reference method used to accurately detect resistance due to altered PBPs in *S. aureus* is NAA detection of the *mecA* gene. Conventional and real-time PCR have been used to detect *mecA* both on bacterial isolates and directly on patient specimens. However, it should be noted that direct specimen testing comes with limitations, often including a lower PPV than conventional methods.[66–68] For example, by simultaneously detecting *mecA* and a *S. aureus*-specific gene from nonsterile sites such as nares, one cannot rule out co-colonization of a methicillin-susceptible *S. aureus* and a methicillin-resistant *S. epidermidis*; such a patient does not need to receive contact isolation.[66,69] Currently, there is one FDA-approved real-time PCR assay (IDI-MRSA test, GeneOhm Sciences, San Diego, California) for the detection of MRSA directly from nasal swabs for infection control purposes.[68] In addition, there are commercially available ASRs and a number of user-defined assays that have been described for MRSA detection.[24,69] NAA detection of MRSA has been shown to be equal in sensitivity to culture-based methods, but has the advantage of offering a faster turnaround time, which may significantly decrease hospital costs.[24]

15.5.2.3 *Mycobacterium tuberculosis*

Although resistance to all first-line TB drugs has been reported, resistance to isoniazid (INH) and rifampin is most common.[70] The reference method for antimycobacterial susceptibility testing is agar proportion, but this method is extremely slow and labor-intensive. Most clinical laboratories have replaced this method with automated broth susceptibility testing because this method is faster, more convenient, and similar in accuracy to the reference method.[70] However, automated broth methods still take days to weeks to complete. The use of molecular technology for the detection of antitubercular resistance is promising that results could be available within 24 h.

INH has long been a central component in TB therapy. Mutation in the catalase gene, *katG*, results in INH resistance. Catalase is necessary for INH to be modified to its active form.[71] Around 65–80% of INH-resistant MTB are due to mutations in *katG*.[72,73] Rifampin has also long been used in combination with front-line therapy for TB. The cellular target of rifampin is the beta subunit of bacterial DNA-dependent RNA polymerase, which is encoded by *rpoB*. Point mutations in *rpoB* can render the organism resistant to rifampin due to decreased binding affinity and can result in high-level resistance.[74]

INH and rifampin resistance determinants, *katG* and *rpoB*, respectively, are the most common targets for mutation detection in MTB. The detection of *rpsL* and 16S rRNA mutations for streptomycin resistance and *embB* mutations for ethambutol resistance has also been reported.[75–77] Two commercially available assays, although not FDA approved, use a combination of NAA and a probe-based line blot for the detection of resistance in MTB. The INNO-LiPA Rif.TB kit (Innogenetics) detects rifampin resistance using a combination of five wild-type probes and four mutant probes to detect *rpoB* mutations.[78,79] The GenoType MTBDR assay (Hain Life-science GmbH) detects both rifampin and INH resistance by detecting four *rpoB* mutations and two common *katG* mutations.[80] These two assays were compared to routine susceptibility methods, and the MTBDR assay was 90.4% concordant for INH resistance, and both the MTBDR assay and the LiPA Rif.TB assay demonstrated 98.1% concordance for rifampin resistance.[81] Although a number of reports have described the molecular detection of rifampin and INH resistance utilizing line probe, real-time PCR or sequencing methodologies, it should be noted that conventional susceptibility testing is still the reference method due to incomplete data on all mutations and target genes associated with phenotypic resistance. Limited studies are available determining performance characteristics of line probe assays for the detection of MTB resistance directly from clinical specimens, but studies that include both smear-negative and extrapulmonary specimens demonstrate sensitivities of 59–98%.[43] The most appropriate use of molecular detection of antitubercular resistance is as an initial screening assay. This approach could potentially identify resistant isolates weeks earlier, allowing clinical care to be significantly impacted.

15.5.3 DIAGNOSIS OF VIRAL INFECTIONS

15.5.3.1 Human Immunodeficiency Virus

The diagnosis of HIV relies on an antibody screening test (EIA or rapid test) followed by confirmatory test (western blot or indirect immunofluorescence assay [IFA]) if positive.[30] Once a patient is determined to be HIV positive, prognosis and therapeutic response is determined by CD4+ T-cell counts and HIV quantification, or viral loads. Before the advent of HIV viral loads, therapeutic effectiveness was measured by CD4+ counts, p24 antigen levels, and HIV culture results. Although CD4+ cell count is an excellent prognostic factor for HIV-infected individuals, CD4+ counts can display significant variability and a limited dynamic range.[82] Measurement of p24 antigen levels in serum is also a good predictor of HIV-associated disease and response to antiretroviral therapy (ART), but sensitivity is generally low.[82] Quantitative HIV cultures are labor-intensive and expensive although only offering moderate sensitivity and poor reproducibility.[83]

FDA-approved tests for HIV quantification include the VERSANT HIV-1 RNA 3.0 (bDNA; Bayer HealthCare), NucliSens HIV-1 QT (NASBA; bioMerieux Inc., Durham, North Carolina), and AMPLICOR HIV-1 MONITOR test, v1.5 and COBAS AMPLICOR (RT-PCR; Roche Diagnostics). The linear ranges of the assays are bDNA, $50-5 \times 10^5$ copies/mL; NASBA, $51-5.39 \times 10^6$ copies/mL; and RT-PCR, COBAS $50-7.5 \times 10^5$ copies/mL. These assays are only approved for use to monitor disease progression or response to ART. Early generations of the Roche AMPLICOR assay were known to underquantify or undetect several HIV-1 non-B subtypes because of the heterogeneity of the *gag* gene target.[84] The current version of the assay (v1.5) has modified the primers and amplification reaction such that this underquantification is no longer an issue. All the three FDA-approved methods now detect subtypes A–H of HIV-1 group M. Of the three FDA-approved assays, the VERSANT bDNA assay demonstrated the least variation with a mean coefficient of variation (CV) of 12%, whereas NASBA and PCR had CVs of 42% and 52%, respectively;[84] although other reports claim similar intra-assay variability among the three assays.[82] Although the performance characteristics are similar among the three assays, it is recommended that patients be monitored using the same method throughout their treatment or clinical study to provide the most consistent and comparable results.

Many diagnostic companies are transitioning HIV detection to real-time detection formats, including the NucliSens EasyQ (bioMerieux) system that employs molecular beacon probes currently available as "research use only (RUO)." In addition, companies and independent users are implementing automated extraction procedures to improve reproducibility and throughput, for example, the COBAS AmpliPrep system (Roche Diagnostics). Additional molecular quantification assays for HIV include TMA (Procleix HIV-1/HCV nucleic acid test, Gen-Probe), COBAS Ampliscreen HIV-1 (Roche Diagnostics), and the UltraQual HIV-1 RT-PCR assay (National Genetics Institute), which are FDA approved for use in screening blood products.[83]

Department of Health and Human Services (DHHS) guidelines for ART of adults incorporate HIV viral load data. Initiation of ART is recommended when the $CD4^+$ count is <350 cells/mL regardless of HIV viral load and should also be considered when $CD4^+$ counts are >350 cells/mL, but HIV viral load is >100,000 copies/ mL.[83,85] HIV viral loads are used, in conjunction with $CD4^+$ cell counts, to monitor a patient's disease progression and response to therapy with the goal of "undetectable" HIV RNA by 16–24 weeks.[85] Recent data suggest that low-level intermittent viremia (50–500 copies/mL) occurs frequently and does not mark treatment failure or poor adherence, but rather assay and statistical variability.[83] In addition, acute illness and vaccinations may cause HIV RNA levels to increase for 2–4 weeks; therefore, it is recommended that HIV viral loads are not determined during this time.[85] Changes in viral loads are only statistically significant if threefold ($0.5 \log_{10}$) or greater, and RNA results should always be confirmed before initiating changes in therapy. DHHS guidelines recommend viral load determinations as follows:[85]

- Initial evaluation of new HIV diagnosis
- Every 3–4 months in patients not on therapy and patients on stable therapy
- 2–8 weeks following initiation or change in therapy
- Clinical event or decline in $CD4^+$ T cells
- Syndrome consistent with acute HIV infection

Because none of the HIV NAA tests are FDA approved for diagnostic purposes, patients who are HIV RNA positive and antibody negative, consistent with acute HIV infection, should have the diagnosis confirmed by subsequent antibody testing.[30,85] Reports from the public health sector indicate that screening for acute HIV infection using a pooling algorithm significantly interrupts the chain of transmission and allows patients to enter therapy sooner.[86] As new approaches to HIV diagnostics emerge and new technologies, such as real-time PCR, become available, HIV detection and viral load monitoring will be more affordable, accurate, and efficient allowing clinicians worldwide the benefit of HIV RNA results.

15.5.3.2   Hepatitis

HCV and HBV infections can be asymptomatic, or can present as acute or chronic hepatitis that can subsequently lead to cirrhosis or primary hepatocellular carcinoma.[82] Serological methods have long been used to diagnose viral hepatitis. However, similar to HIV, viral load monitoring has become valuable in the management of patients infected with HCV and HBV. Unlike HIV and HBV viral load monitoring, HCV viral loads are not predictive of prognosis or disease progression. Monitoring HCV and HBV serum quantities can help identify the patients that will benefit most from therapy and aid in identifying treatment failures and relapsing infections.[82] Genotyping methods have also been applied to assist in selecting appropriate treatment and duration of therapy.

15.5.3.2.1   Hepatitis C Virus

HCV serology results only identify the presence of anti-HCV antibody (and therefore exposure to the virus) and do not differentiate between acute, resolved, or chronic infection.[87] The detection of anti-HCV commonly utilizes third-generation EIAs that have increased sensitivity over previous generation tests. These assays detect a mixture of antibodies directed against a variety of epitopes in the core, nonstructural (NS) 3, NS4, and NS5 proteins.[88] However, EIA false-negative results can occur in immunocompromised patients and patients undergoing hemodialysis.[89] In addition, total HCV core antigenemia can be measured by EIA as an alternative to molecular methods for measuring viral replication and assessing therapeutic response (Ortho-Clinical Diagnostics, Raritan, New Jersey).[88,89] Although antigen titers (in picogram per milliliter) correlate well with HCV viral loads, core antigen levels are only detectable when HCV viral loads are greater than 10,000–20,000 international units (IU)/mL making it less sensitive than RNA testing.[88,90]

HCV NAA testing is central to the management of HCV infection. Viral load determinations assist with diagnosis, treatment decisions, and response to antiviral therapy. The recommended treatment for chronic HCV infection is combination therapy with polyethylene glycol conjugate-(pegylated) interferon α and ribavirin.[88,89] HCV NAA testing is indicated for the following situations:[87,91,92]

- Acute hepatitis before seroconversion
- Seronegative patients with immune deficiency or undergoing hemodialysis

- Indeterminate serological results
- Confirmation of positive as determined by an earlier generation of EIA for low-risk patients
- Determination of viremia before considering a patient for therapy
- Determination of whether or not infection has resolved in a treated patient
- Investigation of infection in newborns

In addition, HCV viral load monitoring helps provide a marker for appropriate endpoints for treatment.[82]

Qualitative HCV NAA assays are used to assist with diagnosis, whereas quantitative NAA tests and genotyping are used in treatment decisions of HCV-infected patients.[92] Qualitative NAA tests are the most sensitive test for the diagnosis of HCV infection. With lower limits of detection in the 10–50 IU/mL range, qualitative NAA tests can detect HCV viremia 1–3 weeks postexposure.[89] FDA-approved qualitative HCV tests available include AMPLICOR HCV Test v2.0 (RT-PCR; Roche Diagnostics) and VERSANT HCV RNA (TMA; Bayer HealthCare), which have limits of detection of 50 and 10 IU/mL, respectively. The specificity of these assays is 98–99%.[90] A single positive qualitative result confirms active HCV replication, whereas a single negative result does not exclusively rule out the presence of HCV, but rather may indicate a transient decline in viremia below the limit of detection.[89]

Although HCV viral load does not predict severity of disease or prognosis, it does correlate with the likelihood of response to therapy.[89] There are three commercially available assays for the quantitative detection of HCV RNA, although only one, VERSANT HCV RNA 3.0 Assay (bDNA; Bayer HealthCare), is FDA approved. Also available are the PCR-based tests (Roche Diagnostics) COBAS/AMPLICOR HCV MONITOR v2.0 (RUO) and COBAS TaqMan HCV Test (ASR), using endpoint and real-time detection, respectively. The detection limits of currently available assays range from 30 to 615 IU/mL, whereas the upper end of the linear range varies from 500,000 to 7,700,000 IU/mL.[90] Specificities of the assays are 98–99%, independent of genotype.[90] For viral load changes to be significant, differences must exceed 1 $\log_{10}$ because the biological variation of HCV viral load ranges from 0.5 to 0.75 $\log_{10}$.[93] The interrun precision of the AMPLICOR assay has a CV of 31%, whereas the semiautomated version on the COBAS has a 20–30% CV, and the VERSANT assay has a CV of 15–35%.[92] Note that an international HCV RNA standard has been used to calibrate these assays to permit normalization by reporting results in IU per milliliter; however, serial results on a single patient should only be compared when performed using the same platform.

The therapeutic goal is a sustained virologic response (SVR), which is defined as "undetectable" serum HCV RNA using a sensitive qualitative assay.[89] For patients who receive therapy and are infected with genotype 1, 4, 5, or 6 (see details for genotyping below), the recommendation is to treat for 48 weeks (as opposed to 24 weeks for genotype 2 or 3). However, if there is no likelihood of an SVR as determined by viral load monitoring, then treatment for 48 weeks can be avoided.[88] Patients infected with non-1 genotypes should have a qualitative assay performed at the end of treatment to assess SVR and 24 weeks after completion of therapy to see if the response is sustained.[88] For patients with genotype 1, the recommendations regarding the use of viral load determinations are as follows:[90]

- Initial determination to confirm baseline viremia (quantitative assay)
- 12 weeks into treatment to determine the likelihood of an SVR, defined as a 100-fold drop in viral load (quantitative assay)
- At the end of 48 weeks of treatment using a qualitiative assay to assess end-of-treatment and SVR
- 24 weeks after the completion of therapy using a qualitative assay

The presence of HCV RNA at the end of therapy is highly predictive of relapse, whereas the absence of HCV RNA indicates a SVR and clearance of infection for the majority of patients with a SVR.[88]

The primary factor impacting response to therapy is the HCV genotype; therefore, genotyping should be performed before initiation of therapy. The genotype not only affects the duration of treatment, but also therapeutic indications, dosing of ribavirin, and viral load monitoring schedule.[88] The SVR rate is 70–80% in patients infected with genotypes 2 or 3, whereas the SVR rate for those infected with genotype 1 is 40–50%.[88] The reference method for determining HCV genotype is sequencing the NS5B or E1 region, only available as user-defined assays. However, there are several non-FDA-approved commercial assays available for HCV genotyping: VERSANT HCV genotyping assay (line probe hybridization technology; Bayer HealthCare), TRUGENE 5′NC HCV (sequencing; Bayer HealthCare), Invader HCV ASR (fluorescent probe/cleavase technology; Third Wave Technologies, Madison, Wisconsin), and Abbott HCV ASR (real-time RT-PCR; Abbott Molecular, Des Plaines, Illinois).[94] Except the Abbott ASR, all commercially available assays depend on differentiation based on 5′ untranslated region (UTR) sequences; the Abbott assay uses NS5B sequences to identify 1a and 1b, but the 5′UTR region for genotypes 2–6.[94] Mistyping of the six genotypes is rare with the use of 5′UTR sequences, but subtyping errors occur 10–25% of the time due to the inherent variability of the 5′UTR of the HCV genome.[90] Interestingly, the Invader assay does not offer subtype results, which is deemed the appropriate use of genotyping assays directed at the 5′UTR of HCV.[94]

15.5.3.2.2 Hepatitis B Virus

The combination of serology, nucleic acid testing, and resistance testing is critical to HBV disease prevention and treatment. The mainstay of HBV diagnosis is detection of hepatitis B surface antigen (HBsAg); this serologic test in combination with the detection of hepatitis B e antigen (HBeAg), anti-HBsAg antibodies, anti-HBeAg antibodies, and anti-hepatitis B core antigen antibodies (total and IgM) help define a patient's infection status. Typically, HBsAg becomes detectable 6–10 weeks after exposure to the virus, whereas HBV DNA is possibly detected 21 days earlier.[95]

HBV viral load determinations can be beneficial in the following settings: (1) to establish acute HBV infection; (2) to distinguish between active and inactive disease; (3) to assess response to therapy; (4) to detect the development of antiviral resistance; (5) to measure infectivity, as with vertical transmission; and (6) to mediate atypical serology profiles.[95,96] HBV quantification, unlike serology results, have been shown to correlate with patient outcomes such as progression to cirrhosis or hepatocellular carcinoma.[96] There are four main commercial assays for HBV viral load testing, none of which are FDA approved: HC II (Digene Corporation), VERSANT HBV DNA 3.0 (bDNA; Bayer HealthCare), COBAS/AMPLICOR HBV MONITOR v2.0 (PCR; Roche Diagnostics), and COBAS TaqMan 48 HBV (real-time PCR; Roche Diagnostics). The dynamic ranges of the above assays are HC II (ultrasensitive), $4700-5.7 \times 10^7$ copies/mL; VERSANT, $2000-10^8$ copies/mL ($357-1.7 \times 10^7$ IU/mL); AMPLICOR, $1000-4.0 \times 10^7$ copies/mL; COBAS AMPLICOR, $200-2 \times 10^5$ copies/mL; COBAS TaqMan, $<50-8.5 \times 10^8$ copies/mL ($30-1.1 \times 10^8$ IU/mL).[97] Until the introduction of real-time PCR assays, no single HBV assay covered the entire range of HBV DNA values observed in untreated patients and in patients with chronic HBV infection.[97] A disadvantage of the nonamplification techniques is their inability to detect very low quantities of HBV. Similar to HCV biological variance, the fluctuation of HBV DNA in untreated patients tends to be less than 1 $\log_{10}$. Thus, a change in viral load is deemed significant if it is greater than 1 $\log_{10}$. Although the World Health Organization has established an international reference standard for HBV DNA testing using IU per milliliter, similar to HCV, it is still recommended to compare serial results on the same patient using only the same assay. As HBV viral load assays become more sensitive, the challenge will be to standardize the results and define clinically significant levels of HBV DNA.[95]

The HBV viral load threshold for treatment is arbitrarily set at 10^5 copies/mL (1.8×10^4 IU/mL) for HBeAg-positive patients and 10^4 copies/mL (1.8×10^3 IU/mL) for HBeAg-negative patients.[95,96] Currently, licensed medications for the treatment of HBV infections were approved using histology as the endpoint of therapy. However, an SVR (i.e., HBV DNA "undetectable") is evolving to become the primary endpoint of therapy.[96] The availability of new nucleoside/nucleotide analogue drugs active against HBV has prompted increased interest in viral load monitoring during therapy. The recommended frequency of HBV viral load monitoring during therapy is 2–4 times a year, or once a month in patients with cirrhosis or awaiting transplantation.[96] Patients not being treated should have a viral load performed once or twice a year. Viral load testing should be an adjunct to other diagnostic testing routinely performed such as serum alanine transaminase (ALT) levels. The goal is to achieve a 1 $\log_{10}$ or greater reduction in HBV viral load from a patient's baseline within 3 months of initiating therapy.[96]

Antiviral resistance in HBV has been described for the nucleoside/nucleotide analogues. Resistance testing is indicated when HBV DNA and ALT levels increase and systemic symptoms occur during therapy. Non-FDA-approved genotyping assays include direct sequencing of the POL/RT gene (TRUGENE; Bayer HealthCare) where all mutations described to date for HBV antiviral resistance can be identified, or INNO-LiPA HBV DR (Innogenetics) where only certain mutations are probed.[97]

15.5.3.3 Cytomegalovirus in Transplant Patients

CMV is the most frequent posttransplant infection and is a major cause of significant morbidity and mortality in transplant recipients. Both solid organ transplant (SOT) and hematopoeitic stem cell transplant (HSCT) recipients are at risk of CMV disease. Because CMV establishes a latent, persistent infection after primary exposure, the virus can reactivate in the transplant recipient. In naïve recipients, the virus can originate from donor tissue or primary infection. CMV replication can be identified in more than 50% of transplant recipients.[98] On reactivation, CMV can cause a wide extent of clinical effects including both local and systemic manifestations. Further, CMV infection further suppresses the immune system making patients more susceptible to other opportunistic infections.

The laboratory diagnosis of localized infections relies on tissue culture or user-defined NAA assays to detect CMV in tissue biopsies or fluids, whereas the diagnosis of systemic CMV infection depends on antigenemia or molecular determinations. Antigenemia tests quantify the number of peripheral blood leukocytes that express the nuclear pp65 antigen. Disadvantages of antigenemia tests include poor standardization between laboratories, the requirement for significant expertise, and they are labor-intensive.[99] Therefore, molecular assays have largely supplanted antigenemia assays for monitoring CMV replication in transplant recipients. Molecular assays commercially available to detect and monitor CMV include COBAS AMPLICOR CMV MONITOR (Roche Diagnostics), Murex HC version 2 (Digene Corporation), and a qualitative NASBA assay that detects pp67 transcripts (NucliSENS CMV pp67; bioMerieux). Several user-defined assays have also been described. Although molecular assays have greatly enhanced CMV diagnostics, the assays currently utilized do not provide comparable viral load results due to variations in sample type, sensitivity, lower limits of detection, target detected, and calibration standards used.[99] Therefore, there is a great need to establish an international standard. Whole blood is the specimen of choice as it yields CMV DNA levels 0.67 $\log_{10}$ higher than those observed when testing plasma or other blood compartments.[99]

An increase in both antigenemia and viral load are predictive of developing CMV disease.[99,100] Suggested thresholds predictive of disease for antigenemia and viral load, respectively, are as follows: 100 positive cells per 2×10^5 cells or 3×10^5 copies/mL in SOT patients, 1–2 positive cells per 2×10^5 cells or 10,000 copies/mL in HSCT patients.[101] However, no universal threshold has been established for initiation of preemptive treatment.[99] Although the periodicity at which blood should be assayed for viral load has yet to be defined,[99] it is currently recommended to test asymptomatic posttransplant patients weekly for 3 months followed by monthly testing until a year of posttransplantation.[101] Once a patient has a positive CMV viral load, it is suggested that twice weekly viral loads be performed to monitor the potential for disease progression and response to antiviral therapy.[101]

The primary applications of CMV viral load detection include (1) identification of those at risk for developing clinical CMV disease, (2) monitoring of antiviral therapy response to guide intensity and duration of treatment, and (3) optimization of preemptive therapy.[99] In addition, CMV viral loads can help identify patients who are at risk of recurrent CMV disease. High CMV viral loads before therapy and persistence of CMV DNA at therapy completion are markers for increased risk of CMV relapse.[99] Persistently high CMV viral loads during anti-CMV therapy could also be an indicator of antiviral resistance. Available methods for the detection of antiviral resistance in CMV include both phenotypic and genotypic assays. Genotyping has become the reference method due to increased specificity and faster turnaround time. Sequencing of the UL97 (phosphotransferase) and UL54 (DNA polymerase) genes of CMV can determine if mutations are present that confer ganciclovir and ganciclovir, foscarnet, or cidofovir resistance, respectively.

As more is learned regarding CMV viral load monitoring in transplant recipients, these lessons will be able to be applied to other herpes viruses. Similar approaches to quantitative virology are being used routinely in the monitoring of EBV viral loads in patients suspected of posttransplant lymphoproliferative disorder.[102] In addition, investigations into the utility of viral load monitoring of other viruses in the posttransplantation setting has begun, including human herpes virus 6 (HHV-6) and adenovirus detection in the pediatric HSCT setting and other transplant populations.[103,104]

15.5.3.4   Molecular Diagnosis of Meningitis and Encephalitis

Another successful clinical application of molecular technology in the past 10 years has been its use in the diagnosis of viral meningitis and encephalitis. NAA has become the gold standard for the detection of HSV and enteroviruses (EVs) in CSF. Further, the detection of EBV in the CSF of immunocompromised patients correlates strongly with the presence of a central nervous system (CNS) lymphoma, and the detection of JC virus in acquired immunodeficiency syndrome (AIDS) patients is a marker for progressive multifocal leukoencephalopathy.[105] NAA has also been successfully applied to other etiologies of viral CNS disease, such as CMV, varicella zoster virus (VZV), and HHV-6, but these assays have not been implemented as broadly as those for HSV and enterovirus (EV), therefore, they are still transitioning to become the method of choice. It should be noted that not all encephalitis viruses are readily detected by NAA; for example, due to the short period of viremia in many arboviral infections, CSF NAA has low sensitivity and serology remains the gold standard.[106] Historically, the gold standard for the detection of viral encephalitis has been either brain biopsy or cell culture accompanied by intrathecal serology results (the latter have poor sensitivity early in disease).[106] In contrast to those techniques, CSF NAA offers increased sensitivity and a more rapid turnaround, which significantly impacts clinical care and results in overall cost savings and a decrease in unnecessary antibacterial exposure.[105] Further, viral nucleic acid can be detected in the CSF for at least 5 days following antiviral therapy, allowing both immediate empiric antiviral therapy and accurate laboratory diagnostics.[105] However, CSF NAA is not without disadvantages including false-negative results due to collection of CSF very early or very late in illness, rapid viral clearance in immunocompetent hosts, and NAA inhibitors such as heme products.[105] False-positive CSF PCR results also occur primarily due to lack of data to suggest the detection of certain viral nucleic acids correlates with clinical CNS disease, but can also be caused by the presence of peripheral blood in the CSF due either to a break in the blood–brain barrier or a traumatic tap.[105] Unfortunately, the lack of standardized commercially available assays has made implementation of CSF NAA difficult in nonacademic settings. Although most laboratories offering CSF NAA use qualitative methods, investigations are beginning to indicate a role for quantitative CSF NAA in differentiating nonspecific presence of virus and virus-associated disease to aid in prognosis for improved patient management and in monitoring antiviral therapy.[106,107]

Clinically, the signs of CNS disease are very nonspecific necessitating the need for the detection of multiple viral etiologies, especially among the Herpesviridae family. CNS manifestations of Herpesviridae-related infections can be due to primary infection, reactivation, or reinfection. HSV is the most common cause of nonepidemic encephalitis in the United States and Western world with 90% of adult cases caused by HSV-1. CSF culture for HSV detects less than 2% of clinically determined adult HSV encephalitis cases and 40% of neonatal CNS disease, whereas HSV PCR is positive in most adult cases resulting in sensitivity and specificity of >95%.[108] In neonatal disease, HSV PCR has a sensitivity of 75%, specificity of 100%, a PPV of 100%, and an NPV of 98%.[105] There are numerous user-defined assays and ASRs designed to detect HSV-1 and -2 in CSF, but no FDA-cleared tests. An interesting approach to detect herpes viruses in CSF is the implementation of consensus or multiplex PCR that detects all members of the Herpesviridae family.[109] False-negative results have been known to occur if the CSF is obtained in the first 72 h of illness; therefore, it is recommended to retest these patients between days 4 and 7 and not discontinue acyclovir therapy due to an initial negative result.[105] HSV DNA can reportedly be found in the CSF of 47% of patients from 8 to 14 days after initiation of therapy, 21% after 14 days, and rarely after 30 days.[105] Therefore, repeat tests should be interpreted with caution. One value of repeat testing is the ability to differentiate between relapses after antiviral therapy due to residual active infection and postinfectious immune progresses.[110]

EV is the most common cause of aseptic meningitis in the summer and fall months in temperate climates. In addition, EV accounts for 10–20% of encephalitis cases.[111] Although EV is frequently isolated from CSF using culture techniques, molecular detection has increased detection rates.[106] Further, the turnaround time of EV culture can be 4–8 days for conventional culture and 2–3 days for shell vial cultures, which is often too long to be clinically effective.[111] The turnaround time of NAA can be as short as 4 h. Another advantage of NAA over culture is the ability to detect all EV types, with the possible exception of echovirus 22 (now reclassified as parechovirus type 1).[24] To recover the majority of EV types by culture an extensive array of cell lines must be employed, and culture sensitivity still remains approximately 70%, possibly due to host neutralizing antibody and low viral loads.[111] In contrast, both the sensitivity and specificity of CSF PCR for EV is estimated to be >95%, which is higher than sensitivities associated with PCR testing of serum (91–92%) or urine (62–77%).[105,108] NASBA assays reportedly have lower limits of detection of 1–10 copies depending on whether endpoint or real-time detection is used.[108] Most commonly the 5'UTR is the target of user-defined assays owing to its high conservation among all EV types.

15.5.4 Diagnosis of Parasitic Infections

The gold standard for diagnosis of parasites is microscopic examination, but the success of this technique is limited by the skill level of the person performing the examination, sample quality, and parasitic load. These conventional methods are inexpensive and do not require complex equipment, although still having a rapid turnaround time if skilled technologists are available. In addition, when a microscopic examination is performed, any and all parasites present can be identified. The development of NAA techniques has made identification of a small number of parasites in a specimen more sensitive and less subjective, but these techniques also have limitations. Only specified parasites will be identified when using molecular assays and testing for multiple parasites may be cost-prohibitive. Molecular assays are expensive to start-up, but once implemented, the reagents are generally afford-able, and molecular detection offers the additional benefit of eliminating interpreta-tive subjectivity. Further, the skills required for molecular assays are more definitive than those of microscopy.[24] But, problem is the lack of required equipment for molecular analysis in areas where malaria is endemic.

To date, the following parasites have been identified by NAA techniques: *Plasmodium, Babesia, Trypanosoma, Leishmania, Toxoplasma, Trichomonas, Cryptosporidium, Entamoeba*, and *Giardia*.[24] These assays are primarily PCR or real-time PCR assays, but user-defined assays using NASBA technology have also been developed.[112] The NASBA assays have the advantage of detecting viable organ-isms by RNA, whereas DNA detection (i.e., PCR) cannot distinguish between viable and dead organisms.[19] Detection of *Toxoplamsa gondii* and malaria is discussed in the following sections.

15.5.4.1 *Toxoplasma gondii*

Recently, much attention has been paid to the development of molecular diagnos-tics for the detection of *T. gondii* from blood, serum, CSF, and amniotic fluid, but no commercial kits are available. The conventional diagnosis of toxoplasmosis is made based on clinical presentation and serology, though this is not easy. Molecular diagnosis does not rely on immune response and can detect the parasite directly in the specimen for a definitive diagnosis.[113] NAA tests are especially useful for indi-viduals with immunodeficiencies or in prenatal cases. The targets of user-defined conventional PCR assays are most commonly the 18S rRNA gene with a sensitiv-ity and specificity of 47% and 100%, respectively; the 35-fold-repeated B1 gene with a sensitivity and specificity of 26% and 95%, respectively; and the AF146527 sequence—a 529 bp DNA fragment that is repeated 200- to 300-fold in the *T. gondii* genome, with a sensitivity and specificity of 42% and 100%, respectively, when using the gold standard of mouse inoculation in the context of ante- and neonatal diagno-sis.[114] In this study, the differences in the sensitivity and specificity of these three PCR techniques were not found to be statistically significant.

Real-time PCR assays for the detection of *T. gondii* have also been developed. These assays significantly shorten the turnaround time in addition to offering the ability to quantitate the parasite burden.[24] In one study, Contini et al.[115] found that the level of B1 DNA remained elevated irrespective of clinical and serologic resolution. Although the levels of the two bradyzoite genes investigated (SAG-4 and MAG-1) peaked during the symptomatic phase, they did not fall until 2 or 3 months of follow-up. This suggests that using a real-time PCR assay and alternative targets may have potential for providing molecular information regarding the state of infection.[115] Another study compared real-time PCR using the two previously described targets, the B1 gene, and the AF146527 target (529 bp DNA). They found that PCR in duplicate is required for diagnosis and that the 529 bp target was more sensitive for all samples (amniotic fluid, placenta, aqueous humor, whole blood, CSF, and bronchoalveolar fluids) tested.[116] Additional studies have found that real-time PCR assays are equivalent to PCR–enzyme-linked immunosorbent assay (ELISA) tests. In summary, NAA tests for the diagnosis of toxoplasmosis are useful in patients where serology is suboptimal (i.e., pregnant women, neonates, and immunocompromised) and in scenarios where culture or serology is not available.[24]

15.5.4.2 Malaria

The most vigorous application of real-time PCR for diagnosis of parasites has been to malaria. Although there are four species of *Plasmodium* that infect humans (*P. falciparum, P. vivax, P. malariae*, and *P. ovale*), *P. falciparum* causes the greatest morbidity and mortality. The laboratory diagnosis of malaria relies primarily on microscopic examination of blood smears and serologic methods. Recently, species-specific detection of *Plasmodium* has been made possible by PCR with unique primers to the rRNA of the parasites.[117–119]

There is a commercially available RUO kit (RealArt Malaria Assay; QIAGEN Hamburg GmbH, Hamburg, Germany) for detection of the 18S rRNA genes of the four *Plasmodium* species, but it is not able to speciate the infective parasite. This assay is 99.5% sensitive and 100% specific compared to a nested PCR assay.[120] Using the same target, several user-defined assays have also been developed that are equivalent to the microscopic detection and identification of *Plasmodium* species.[24] Only those that use melt curve analysis are able to speciate in a single assay.

15.5.5 Detection of Fungi

15.5.5.1 Dimorphic Fungi

The dimorphic fungi *Histoplasma capsulatum, Blastomyces dermatitidis*, and *Coccidioides immitis* are important human pathogens. Identification of these as the causative agent of an infection requires considerable expertise including recovery of the organism from the clinical specimen and direct examination of the isolate. Serological assays are available, but there is cross-reactivity and limited sensitivity and specificity.[121] One rapid and specific way to identify these species is the use of commercially available chemiluminescent-labeled nucleic acid probes that bind to fungal-specific rRNA sequences (Gen-Probe). This assay is a hybridization protection assay (Figure 15.1), and can be performed in less than 2 h after an isolate is recovered. The sensitivity and specificity of each probe was reported as 87.8% and 100% for *B. dermatitidis*, 99.2% and 100% for *C. immitis*, and 100% for both, respectively, for *H. capsulatum* compared to morphologic and microscopic examination.[122] Although these probes are relatively sensitive, specific, and rapid, an organism must still be recovered from culture before identification. The probes decrease the time of identification, but cannot be performed directly on the clinical specimen without a significant decrease in sensitivity.[123] Further, the probes must be tested individually on each isolate making the assay costly if the suspected fungus is unknown.

15.5.5.2 *Aspergillus*

Diagnosis of aspergillosis is critical as it causes a range of diseases that can ultimately be prevented if diagnosed and treated properly in a timely manner. There are two diagnostic strategies for aspergillosis: (1) The use of the galactomannan and validated PCR assays to screen blood from patients at high risk of developing invasive aspergillosis. Positive results will permit early targeting of antifungal therapy, whereas persistently negative results will permit withholding antifungal drugs. (2) The use of these assays only in situations when there is other evidence (clinical or radiological data) of aspergillosis to obtain a definitive diagnosis and target-specific anti-*Aspergillus* therapy.[124,125]

The galactomannan assay is an ELISA-based antigen detection assay. There are two commercially available kits, the Pastorex kit (Sanofi Diagnostics Pasteur, Marnes-La-Coquette, France) and the more commonly used Platelia ELISA (BioRad, Marnes-La-Coquette, France), which detect the heat-stable heteropolysaccharide that is present in most *Aspergillus* and *Penicillum* species from a variety of tissues or fluids. Galactomannan is composed of a nonimmunogenic mannan core with immunoreactive side chains of varying lengths containing galactofuranosyl units. The composition of galactomannan varies between genera and strains. The release and kinetics of circulating galactomannan are not defined but many factors such as growth phase, environment, host immune status, and pathology are influential.[124,126] The galactomannan assays use a monoclonal antibody derived from rats toward the $\beta(1,5)$-linked galactofuranoside side-chain residues of the galactomannan molecule that requires four or more epitopes for antibody binding. The analytical sensitivity of the assay is affected by the multiple epitopes required for binding, and antigens with fewer residues may escape detection. In addition, the acid-sensitive galactofuranoside residues may be degraded by the pretreatment acid step required for proper assay function. The lower limit of detection using the ELISA is significantly better (1 ng/L) than that achievable using latex agglutination (15 ng/L). Cross-reactivity with other filamentous fungi, bacteria, drugs (including piperacillin–tazobactam), and cotton swabs have been documented.[124,126] The clinical sensitivity of galactomannan ELISA is somewhat variable, with a range of 29–100% that is mostly affected by the patient population (immunocompromised versus chronic granulomatous disease or SOT), concomitant antifungal therapy, and poor specimen collection.[124,126] The clinical specificity of galactomannan is estimated to be greater than 90%.[124,126]

PCR has dominated the NAA techniques for diagnosis of aspergillosis, but the lack of standardization and commercially available kits have prevented it from becoming common practice. PCR offers the benefit of a rapid, sensitive, and possibly quantitative result, but the question of how to interpret PCR-positive, culture-negative results (contamination or subclinical organism burden) is problematic. *Aspergillus* nucleic acids can be detected from a variety of specimens including fluids, tissues, and even paraffin-embedded tissues, but the possibility of false-positives makes serum the optimal specimen for increased specificity.[127] *A. fumigatus* has been the primary species of interest with the 18S rRNA gene being the common target because of the conserved and variable sequences in addition to a significant number of sequences deposited into public databases. Other targets that have been used are the mitochondrial genes encoding transfer RNA (tRNA) and (apo)cytochrome b. These genes have been detected in a variety of formats including nested PCR to increase sensitivity and more recently real-time PCR using TaqMan, LightCycler, and molecular beacon probes.[124] The analytical sensitivity of the assays varies but detection limits range between 1 and 10 fg of DNA.[124] Comparison of real-time assays to the FDA-approved galactomannan ELISA test for *Aspergillus* detection revealed the sensitivity of PCR at 79–94% compared to ELISA at 50–90%, with specificities of 50–92% for PCR and 85–99% for galactomannan.[124,126,128] PCR results can be obtained more quickly than galactomannan results, and some studies suggest that the combination of PCR and galactomannan results would provide improved patient care through rapid diagnosis.[24]

15.6   SEQUENCING

Sequence amplification is quickly becoming a primary method to rapidly identify organisms that cause infection. This procedure can reduce the time of identification over the conventional methods, but initial growth of an isolate is still required before identification by sequencing in the clinical laboratory. The procedure of sequence identification includes DNA extraction from target agents grown on solid or liquid media, PCR amplification of the target gene, purification of the PCR product, cycle sequencing (similar to PCR in using DNA as template, but forward and reverse primers are used in separate reactions where the polymerase adds dye-terminated labeled and unlabeled nucleotides; see Figure 15.9), purification and size separation of the extension product, and finally data analysis.[18] Data analysis involves evaluating the quality of the sequence obtained and subsequent comparison of the sequence with known sequences through public and commercial databases such as National Center for Biotechnology Information (NCBI) GenBank or MicroSeq (Applied Biosystems), respectively. The process of sequencing is based on the incorporation of nucleotides (A, C, T, and G) that are each fluorescently labeled with a different dye. Synthesis is terminated after incorporation of the labeled nucleotide. The final product of the sequencing reaction consists of a collection of DNA fragments of different sizes, which when separated based on size produce fluorescent patterns that reveal the identity of the nucleotide in the terminal position (Figure 15.9).[18] Once an isolate is growing, the entire procedure can be done in approximately 1.5 working days. There are many targets commonly used for identification in the clinical labs that are discussed in the following sections.

Sequence results are more robust than conventional culture methods because they are less subjective given a comprehensive and accurate database for comparison. However, if an organism cannot be identified by sequence analysis, it likely represents a new organism; identification of novel organisms is a powerful benefit of sequencing.[129] Sequencing can also be used to identify organisms that cannot be cultured because they are inherently difficult to grow or a result of antibiotic therapy. In this situation, sequencing needs to be performed directly from the clinical specimen, but this practice must be used with caution and should only be performed on sterile sites. The major limitation of sequencing arises from the limited availability of sequence information in databases. Increased availability of commercial kits and comprehensive databases will make sequencing possible beyond academic laboratories in the future.[130,131]

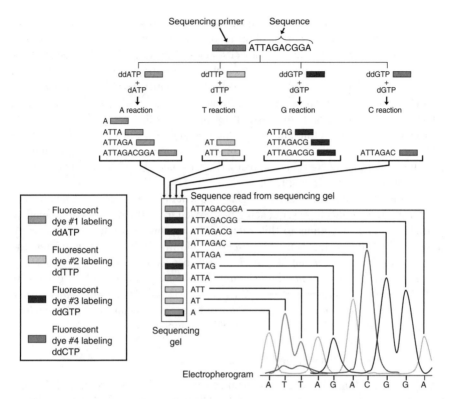

FIGURE 15.9   Sequencing. Dye-terminator cycle sequencing occurs after the target of interest is amplified by PCR and synthesis occurs using one primer as the starting point for the DNA polymerase that incorporates deoxynucleotides (dNTPs) and fluorescently labeled dideoxynucleotides (ddNTPs), which terminate the growing chain after addition. Combining the terminal ddNTP identity with the fragment size permits sequence identification as visualized by an electropherogram. (From Wolk, D., Mitchell, S., and Patel, R., *Infect. Dis. Clin. North Am.* 15(4), 1157, 2001. With permission from Elsevier.)

15.6.1   Bacterial Identification

15.6.1.1   16S rRNA

The 16S rRNA gene encodes for the highly conserved rRNA associated with the small subunit of the ribosome and is often used for taxonomic purposes and species identification. Although 16S rRNA is highly conserved among bacteria, there are nucleotide variations concentrated in specific regions that are unique to each species. The entire gene is 1550 bp long including the conserved and variable regions.[132] Universal primers complimentary to the conserved regions permit amplification of the gene from all bacterial species and the resulting amplicon contains unique sequence. Commercially available kits for sequencing of 16S rRNA labeled "for RUO" are available (MicroSeq; Applied Biosystems) and performed routinely in addition to conventional biochemical methods in laboratories that have the appropriate instrumentation.[133] Often used for the identification of common organisms with atypical biochemical profiles, 16S sequence results usually reveal that these species simply have an unusual phenotype of the taxon that they represent.[132] 16S rRNA gene sequencing can be used to identify novel species of bacteria and has been success-ful in differentiating between the species of viridans streptococci, which is useful for the diagnosis of endocarditis[134] as well as to identify *Bacillus anthracis* from patients with clinical symptoms and culture-negative samples.[135] Identification based on 16S rRNA sequencing is most useful for organisms that are difficult to classify or speciate based on routine phenotypic tests including species of *Nocardia*.[136] 16S rRNA gene sequencing can identify mycobateria that are typically slow-growing and difficult to identify by conventional methods. This has significant impact on patient care by decreasing the time of results.[132]

In studies comparing 16S rRNA gene sequencing to conventional methods of mycobacteria identification, sequencing has proven to be the most accurate method available.[132] The exceptions to this rule are the mycobacteria that cannot be dif-ferentiated by 16S sequencing: *M. abscessus* and *M. chelonae, M. marinum* and *M. ulcerans, M. kansasii* and *M. gastri*, the subspecies of *M. avium (M. avium* ssp. *avium, M. avium* ssp. *paratuberculosis*, and *M. avium* ssp. *silvaticum), M. genavense* and *M. simiae, M. peregrinum* and *M. septicum, M. murale* and *M. tokaiense*, and the members of the MTB complex.[132] Some of these species can be differentiated by sequencing the internal transcribed space (ITS) region.[137]

15.6.1.2 Internal Transcribed Spacer

The genes encoding the rRNAs are arranged in order of 16S-23S-5S and they are separated by two noncoding spacer regions. The region between the genes is known as the ITS. Most commonly, the region between 16S and 23S is the target of sequenc-ing. The 16S–23S ITS of mycobacteria exhibits variable length and sequence content that is sufficient to use as a target for sequence analysis for identification.[137] It is useful in discriminating between the closely related species such as *M. gastri* and *M. kansasii*, and *M. abscessus* and *M. chelonae*, but not useful in discriminating between *M. marinum* and *M. ulcerans* or among the members of the *M. avium* or MTB complexes.[137]

The advantages of ITS over 16S rRNA sequencing are the greater nucleotide variability and thus higher discriminatory power of ITS, and the target region is smaller in ITS resulting in a more efficient and sensitive target.[137] ITS sequence is often used in conjunction with 16S sequencing when 16S shows insufficient diversity between species to definitively identify an isolate. The disadvantage is that there are no commercially available reagents.

15.6.1.3 *hsp65*

The 65-kDa heat-shock protein gene (*hsp65*) is highly conserved among mycobac-terial species with hypervariable regions whose sequences may be used to identify these species.[138] Originally used for identification based on restriction fragment length polymorphism analysis, *hsp65* sequencing has now become an acceptable method of mycobacteria identification.[139] McNabb et al.[139] have performed an exten-sive analysis of the procedure and found that *hsp65* sequencing is cost-effective and more efficient at identifying most species of mycobacteria, but *hsp65* is neither able to distinguish between the members of the MTB complex nor is it able to identify the subspecies of *M. fortuitum* or *M. avium*. Using a more variable region, the 3′ region of the *hsp65* gene, the subspecies of the *M. avium* complex can be identified.[140]

Using *hsp65* sequence *in lieu* of 16S rRNA is advantageous for the identification of mycobacteria because it is more differential, uses a single set of reagents for identifying both rapid- and slow-growing mycobacteria, and there is no need to sequence a fragment larger than 401 bp to get discriminatory data.[139] The impediment to using *hsp65* sequencing routinely for identification of mycobacteria in the clinical laboratory is the lack of a comprehensive database of accurate *hsp65* sequences and identities.[139] Some *hsp65* sequences for valid species are available in public databases such as NCBI GenBank, but few entries exist for putative species or groups or for biochemically aberrant groups. As more laboratories use this methodology, the databases will become more comprehensive.

15.6.2 FUNGAL IDENTIFICATION

Invasive fungal disease plays a significant role in the morbidity and mortality of immunocompromised patients, in addition to becoming more common with the widespread use of antibiotics; thus, rapid identification of fungal etiologies is imperative. The poor sensitivity of fungal blood culture and histological practices has lead to the development of more sensitive techniques.[141] One technique is sequencing the ITS regions between the genes that encode the ribosomal subunits. Sequencing fungal isolates involves amplification using primers targeted toward the conserved regions of the 18S, 5.8S, and 28S rRNA genes and then sequencing the regions between them, the ITS1, between 16S and 5.8S, and ITS2, between the 5.8S and 28S rRNA genes. Although these regions of variable sequences, ITS1 and ITS2, are not translated into proteins, they have essential roles in generation of functional rRNA, and the sequence variations among species can be useful for identification.[141] This technique has been used to confirm or correctly identify fungi that have nonconventional biochemical results,[142,143] in addition to successfully identifying the fungal species that cause black-grain mycetomas.[144] The correct identification of medically important fungi by ITS1 and ITS2 sequence analysis have been reported at 96.8% and 98–99.7%, respectively, including both *Candida* and non-*Candida* species (e.g., *Aspergillus, Trichosporon, Cryptococcus, Geotrichum, Malassezia, Rhodotorula*, and others).[141,145] These studies reveal that sequencing of ITS2 is more species specific than ITS1, and it is a reliable, rapid (completed within 24 h of an isolated colony), and an accurate alternative to conventional identification methods as long as the ITS2 sequences of the unknown organism are in the NCBI GenBank database.[141] In addition to ITS sequencing, the use of the D1–D2 domain of the large subunit 28S rRNA gene for species identification has been reported for many pathogenic fungi[146] including *Malassezia*,[147] *Apsergillus*,[148] and the Zygomycetes,[149] but in general NCBI GenBank is more comprehensive for ITS sequence. There is a commercially available RUO kit for fungal identification using the D2 large rRNA subunit (MicroSeq; Applied Biosystems). Although fungal identification by sequencing is still in the early phases, improved targets and comprehensive databases will aid in its transition to routine analysis.

15.6.3   Viral Genotyping

Sequence analysis is used to determine the genotype of certain clinically important viruses including HIV, HCV, and CMV[150] (see also Section 15.5.3) Genotyping is performed on viruses to identify sequence variations (mutations) that give each viral strain unique phenotypic characteristics related to prognostic or therapeutic purposes and resistance to antiviral drugs. Multiple genes are the targets of sequencing for each virus. Genotyping of HIV has become a standard of care due to the rapid development of mutations in the antiretroviral targets, protease or RT genes, which confer drug resistance. The genome of HIV exists as a population of variants in infected individuals due to the absence of proofreading activity of the RT enzyme.[150] As a result, mutations appear randomly and are selected in the presence of pressures such as antiretroviral drugs. Thus, HIV genotyping is critical for therapeutic decisions. For HCV, genotyping is used to determine therapeutic regimen as discussed in Section 15.5.3. The targets of HCV genotyping by sequencing are the 5'UTR, the core gene (C), the envelope (E1) region, and the replicase (RNA dependent-RNA polymerase) (NS5B).[94] Another example of antiviral resistance detection by sequencing is the determination of mutations that confer resistance to nucleoside analogues in CMV (i.e., UL97 and UL54).[151] Because CMV grows very slowly, it is much more efficient to determine resistances based on sequence analysis than conventional phenotypic methods.[150]

15.7   SUMMARY

Advances in molecular biology in the past 10–15 years have led to significant improvements in the diagnosis of infectious diseases. Notably, TB can be confirmed in 24 h as opposed to 6–8 weeks, HSV encephalitis can be diagnosed rapidly preventing a brain biopsy and determining the need for acyclovir therapy, and sexually transmitted infections (STIs) such as those caused by *C. trachomatis* and *N. gonorrhoeae* can be rapidly and accurately identified improving treatment and prevention of transmission. Further, organisms with important infection control implications such as *B. pertussis*, MRSA, and VRE, can be quickly identified thus impacting appropriate therapy or precautions, and quantification of viral agents such as HIV, HCV, and HBV can augment current clinical recommendations regarding therapeutic monitoring.

The applications of molecular technology in clinical microbiology are endless, but disadvantages also abound. It is not without considerable cost and expertise that a molecular infectious disease laboratory is established. Further, we are still learning what many NAA results mean in terms of infectious etiology. For example, HHV-6 is detected in as many as 30% of "normal" CSF and can be found in a significant number of postmortem brain biopsies without evidence of CNS disease.[105] Therefore, it is difficult to ascertain the meaning of a positive HHV-6 NAA result from CSF. In such situations quantitative results may prove to be beneficial in identifying true infections. With the use of molecular technology to detect potential etiologic agents of disease, we are left to consider Koch's postulates.[152] Is the mere presence of an organism's nucleic acid convincing evidence of disease causation? Clearly, additional clinical scientific evidence is needed to make such a claim, and such evidence or lack thereof should be considered when interpreting molecular infectious disease results.

Although there is still much to be learned regarding the appropriate application and interpretation of molecular infectious disease testing, there are numerous exciting opportunities on the horizon. The implementation of real-time PCR in the clinical laboratory and associated user-defined, or "homebrew," testing has revolutionized clinical molecular infectious disease testing. (For an extensive review of the applications and performance of real-time PCR in clinical microbiology, see Espy et al.[24]) Recent advances in microarray technology have initiated its transition from pure research to the clinical setting. For example, Nanogen (San Diego, California) currently offers a respiratory virus assay capable of detecting influenza A and B, parainfluenza 1, 2, and 3, and respiratory syncytial virus using an electronic microarray format. As investigators refine molecular applications for infectious disease testing, diagnostic companies market such applications, quality control organizations standardize results; and as costs associated with implementation decrease and reimbursement increases, an extensive molecular menu will not only be available in academic medical centers and reference laboratories, but will also be available in community hospitals, thus more globally impacting patient care.

ABBREVIATIONS

AFB	Acid fast bacilli
AIDS	Acquired immunodeficiency syndrome
ALT	Alanine transaminase
ART	Antiretroviral therapy
ASR	Analyte-specific reagent
bDNA	Branched DNA
bp	Base pairs
CA-MRSA	Community-associated MRSA
CDC	Centers for Disease Control and Prevention
cDNA	Complementary DNA
CMV	Cytomegalovirus
CNS	Central nervous system
CSF	Cerebrospinal fluid
C_t	Threshold cycle
CT	*Chlamydia trachomatis*
CV	Coefficient of variation
DFA	Direct fluorescent antibody
DHHS	Department of Health and Human Services
DNA	Deoxyribonucleic acid
dNTP	Deoxyribonucleotide
dsDNA	Double-stranded DNA

EBV	Epstein–Barr virus
EIA	Enzyme immunoassay
ELISA	Enzyme-linked immunosorbent assay
ET	Energy transfer
EV	Enterovirus
FDA	Food and Drug Administration
FISH	Fluorescent ISH
FRET	Fluorescent resonance ET
HBeAg	Hepatitis B e antigen
HBsAg	Hepatitis B surface antigen
HBV	Hepatitis B virus
HC	Hybrid capture
HCV	Hepatitis C virus
HHV-6	Human herpes virus 6
HIV	Human immunodeficiency virus
HPV	Human papillomavirus
HSCT	Hematopoeitic stem cell transplant
HSV	Herpes simplex virus
IFA	Indirect immunofluorescence assay
INH	Isoniazid
ISH	*In situ* hybridization
ITS	Internal transcribed spacer
IU	International units
mRNA	Messenger RNA
MRSA	Methicillin-resistant *Staphylococcus aureus*
MTB	*Mycobacterium tuberculosis*
MTD	MTB direct test
NAA	Nucleic acid amplification
NASBA	Nucleic acid sequence-based amplification
NCBI	National Center for Biotechnology Information
NG	*Neisseria gonorrhoeae*
NPV	Negative predictive value
NS	Nonstructural
PBP	Penicillin-binding protein
PCR	Polymerase chain reaction
PID	Pelvic inflammatory disease
PNA	Peptide nucleic acid
PPV	Positive predictive value
RE	Restriction endonuclease
RNA	Ribonucleic acid
rRNA	Ribosomal RNA

RT	Reverse transcriptase
RT-PCR	Reverse transcriptase polymerase chain reaction
RUO	Research use only
SDA	Strand displacement amplification
SOT	Solid organ transplant
STI	Sexually transmitted infection
SVR	Sustained virologic response
TB	Tuberculosis
T_m	Melting temperature
TMA	Transcription-mediated amplification
tRNA	Transfer RNA
UTR	Untranslated region
VRE	Vancomycin-resistant *Enterococcus*
VZV	Varicella zoster virus

REFERENCES

1. Li, J. and Hanna, B. A., DNA probes for culture confirmation and direct detection of bacterial infections: a review of technology, in *Molecular Microbiology Diagnostic Principles and Practice*, Persing, D. H., Tenover, F. C., Versalovic, J., Tang, Y.-W., Unger, E. R., Relman, D. A., and White, T. J. (Eds.), ASM Press, Washington, 2004, pp. 19–26.
2. Tenover, F. C., DNA hybridization techniques and their application to the diagnosis of infectious diseases, *Infect. Dis. Clin. North Am.* 7(2), 171–181, 1993.
3. Lebrun, L., Espinasse, F., Poveda, J. D., and Vincent-Levy-Frebault, V., Evaluation of nonradioactive DNA probes for identification of mycobacteria, *J. Clin. Microbiol.* 30(9), 2476–2478, 1992.
4. Lumb, R., Lanser, J. A., and Lim, I. S., Rapid identification of mycobacteria by the Gen-Probe Accuprobe system, *Pathology* 25(3), 313–315, 1993.
5. Butler, W. R., O'Connor, S. P., Yakrus, M. A., and Gross, W. M., Cross-reactivity of genetic probe for detection of *Mycobacterium tuberculosis* with newly described species *Mycobacterium celatum*, *J. Clin. Microbiol.* 32(2), 536–538, 1994.
6. Ford, E. G., Snead, S. J., Todd, J., and Warren, N. G., Strains of *Mycobacterium terrae* complex which react with DNA probes for *M. tuberculosis* complex, *J. Clin. Microbiol.* 31(10), 2805–2806, 1993.
7. Moter, A. and Gobel, U. B., Fluorescence *in situ* hybridization (FISH) for direct visualization of microorganisms, *J. Microbiol. Meth.* 41(2), 85–112, 2000.
8. Amortegui, A. J. and Meyer, M. P., *In-situ* hybridization for the diagnosis and typing of human papillomavirus, *Clin. Biochem.* 23(4), 301–306, 1990.
9. Gulley, M. L., Molecular diagnosis of Epstein–Barr virus-related diseases, *J. Mol. Diagn.* 3(1), 1–10, 2001.
10. Perry-O'Keefe, H., Rigby, S., Oliveira, K., Sorensen, D., Stender, H., Coull, J., and Hyldig-Nielsen, J. J., Identification of indicator microorganisms using a standardized PNA FISH method, *J. Microbiol. Meth.* 47(3), 281–292, 2001.
11. Drobniewski, F. A., More, P. G., and Harris, G. S., Differentiation of *Mycobacterium tuberculosis* complex and nontuberculous mycobacterial liquid cultures by using peptide nucleic acid-fluorescence *in situ* hybridization probes, *J. Clin. Microbiol.* 38(1), 444–447, 2000.

12. Oliveira, K., Brecher, S. M., Durbin, A., Shapiro, D. S., Schwartz, D. R., De Girolami, P. C., Dakos, J., Procop, G. W., Wilson, D., Hanna, C. S., Haase, G., Peltroche-Llacsahuanga, H., Chapin, K. C., Musgnug, M. C., Levi, M. H., Shoemaker, C., and Stender, H., Direct identification of *Staphylococcus aureus* from positive blood culture bottles, *J. Clin. Microbiol.* 41(2), 889–891, 2003.

13. Oliveira, K., Procop, G. W., Wilson, D., Coull, J., and Stender, H., Rapid identification of *Staphylococcus aureus* directly from blood cultures by fluorescence *in situ* hybridization with peptide nucleic acid probes, *J. Clin. Microbiol.* 40(1), 247–251, 2002.

14. Rigby, S., Procop, G. W., Haase, G., Wilson, D., Hall, G., Kurtzman, C., Oliveira, K., Von Oy, S., Hyldig-Nielsen, J. J., Coull, J., and Stender, H., Fluorescence *in situ* hybridization with peptide nucleic acid probes for rapid identification of *Candida albicans* directly from blood culture bottles, *J. Clin. Microbiol.* 40(6), 2182–2186, 2002.

15. Wilson, D. A., Joyce, M. J., Hall, L. S., Reller, L. B., Roberts, G. D., Hall, G. S., Alexander, B. D., and Procop, G. W., Multicenter evaluation of a *Candida albicans* peptide nucleic acid fluorescent *in situ* hybridization probe for characterization of yeast isolates from blood cultures, *J. Clin. Microbiol.* 43(6), 2909–2912, 2005.

16. Nolte, F. S., Branched DNA signal amplification for direct quantification of nucleic acid sequences in clinical specimens, *Adv. Clin. Chem.* 33, 201–235, 1998.

17. Tsongalis, G. J., Branched DNA technology in molecular diagnostics, *Am. J. Clin. Pathol.* 126(3), 448–453, 2006.

18. Wolk, D., Mitchell, S., and Patel, R., Principles of molecular microbiology testing methods, *Infect. Dis. Clin. North Am.* 15(4), 1157–1204, 2001.

19. Compton, J., Nucleic acid sequence-based amplification, *Nature* 350(6313), 91–92, 1991.

20. Hayden, R. T., *In vitro* nucleic acid amplification techniques, in *Molecular Microbiology Diagnostic Principles and Practice*, Persing, D. H., Tenover, F. C., Versalovic, J., Tang, Y.-W., Unger, E. R., Relman, D. A., and White, T. J. (Eds.), ASM Press, Washington, 2004, pp. 43–69.

21. Kohler, L., Finn, S., Price, J., Joshi, S., Wang, S. S., and Hellyer, T., Detection of *Legionella pneumophila, Mycoplasma pneumoniae*, and the Chlamydiaceae family from a single throat swab using the BD ProbeTec™ ET system, *12th European Congress of Clinical Microbiology and Infectious Diseases*, Milan, Italy, 2002.

22. Podzorski, R. P., Gel electrophoresis, southern hybridization, and restriction fragment length polymorphism analysis, in *Molecular Microbiology Diagnostic Principles and Practice*, Persing, D. H., Tenover, F. C., Versalovic, J., Tang, Y.-W., Unger, E. R., Relman, D. A., and White, T. J. (Eds.), ASM Press, Washington, 2004, pp. 273–280.

23. van Doorn, L.-J., Mutation detection: PCR-based approaches, in *Molecular Microbiology Diagnostic Principles and Practice*, Persing, D. H., Tenover, F. C., Versalovic, J., Tang, Y.-W., Unger, E. R., Relman, D. A., and White, T. J. (Eds.), ASM Press, Washington, 2004, pp. 145–151.

24. Espy, M. J., Uhl, J. R., Sloan, L. M., Buckwalter, S. P., Jones, M. F., Vetter, E. A., Yao, J. D., Wengenack, N. L., Rosenblatt, J. E., Cockerill, F. R., III, and Smith, T. F., Real-time PCR in clinical microbiology: applications for routine laboratory testing, *Clin. Microbiol. Rev.* 19(1), 165–256, 2006.

25. Uhl, J. R. and Cockerill, F. R., The fluorescence resonance energy transfer system, in *Molecular Microbiology Diagnostic Principles and Practice*, Persing, D. H., Tenover, F. C., Versalovic, J., Tang, Y.-W., Unger, E. R., Relman, D. A., and White, T. J. (Eds.), ASM Press, Washington, 2004, pp. 295–306.

26. Cockerill, F. R., III and J. R., Uhl Applications and challenges of real-time PCR for the clinical microbiology laboratory, in *Rapid Cycle Real-Time PCR Methods and Applications*, U. Reischl, C. Wittwer, and Cockerill, F. R. (Eds.), Springer, Berlin, Germany, 2002, pp. 3–27.

27. Landry, M. L., Garner, R., and Ferguson, D., Real-time nucleic acid sequence-based amplification using molecular beacons for detection of enterovirus RNA in clinical specimens, *J. Clin. Microbiol.* 43(7), 3136–3139, 2005.

28. McClernon, D. R., Vavro, C., and St. Clair, M., Evaluation of a real-time nucleic acid sequence-based amplification assay using molecular beacons for detection of human immunodeficiency virus type 1, *J. Clin. Microbiol.* 44(6), 2280–2282, 2006.

29. Wittwer, C. T. and Kusukawa, N., Real-time PCR, in *Molecular Microbiology Diagnostic Principles and Practice*, Persing, D. H., Tenover, F. C., Versalovic, J., Tang, Y.-W., Unger, E. R., Relman, D. A., and White, T. J. (Eds.), ASM Press, Washington, 2004, pp. 71–84.

30. Centers for Disease Control and Prevention, Sexually Transmitted Diseases Treatment Guidelines, 2006, *MMWR* 55 (No. RR-11), 1–94, 2006.

31. Centers for Disease Control and Prevention, Screening Tests to Detect *Chlamydia trachomatis* and *Neisseria gonorrhoeae* Infections, 2002, *MMWR* 51 (No. RR-15), 1–38, 2002.

32. Chapin, K. C., Molecular tests for detection of the sexually-transmitted pathogens *Neisseria gonorrhoeae* and *Chlamydia trachomatis*, *Med. Health R I* 89(6), 202–204, 2006.

33. Koumans, E. H., Johnson, R. E., Knapp, J. S., and St. Louis, M. E., Laboratory testing for *Neisseria gonorrhoeae* by recently introduced nonculture tests: a performance review with clinical and public health considerations, *Clin. Infect. Dis.* 27(5), 1171–1180, 1998.

34. Cheng, V. C., Yew, W. W., and Yuen, K. Y., Molecular diagnostics in tuberculosis, *Eur. J. Clin. Microbiol. Infect. Dis.* 24(11), 711–720, 2005.

35. Badak, F. Z., Goksel, S., Sertoz, R., Nafile, B., Ermertcan, S., Cavusoglu, C., and Bilgic, A., Use of nucleic acid probes for identification of *Mycobacterium tuberculosis* directly from MB/BacT bottles, *J. Clin. Microbiol.* 37(5), 1602–1605, 1999.

36. Evans, K. D., Nakasone, A. S., Sutherland, P. A., de la Maza, L. M., and Peterson, E. M., Identification of *Mycobacterium tuberculosis* and *Mycobacterium avium-M. intracellulare* directly from primary BACTEC cultures by using acridinium-ester-labeled DNA probes, *J. Clin. Microbiol.* 30(9), 2427–2431, 1992.

37. Palomino, J. C., Newer diagnostics for tuberculosis and multi-drug resistant tuberculosis, *Curr. Opin. Pulm. Med.* 12(3), 172–178, 2006.

38. Scarparo, C., Piccoli, P., Rigon, A., Ruggiero, G., Nista, D., and Piersimoni, C., Direct identification of mycobacteria from MB/BacT alert 3D bottles: comparative evaluation of two commercial probe assays, *J. Clin. Microbiol.* 39(9), 3222–3227, 2001.

39. Lebrun, L., Gonullu, N., Boutros, N., Davoust, A., Guibert, M., Ingrand, D., Ghnassia, J. C., Vincent, V., and Doucet-Populaire, F., Use of INNO-LIPA assay for rapid identification of mycobacteria, *Diagn. Microbiol. Infect. Dis.* 46(2), 151–153, 2003.

40. Lefmann, M., Schweickert, B., Buchholz, P., Gobel, U. B., Ulrichs, T., Seiler, P., Theegarten, D., and Moter, A., Evaluation of peptide nucleic acid-fluorescence *in situ* hybridization for identification of clinically relevant mycobacteria in clinical specimens and tissue sections, *J. Clin. Microbiol.* 44(10), 3760–3767, 2006.

41. St Amand, A. L., Frank, D. N., De Groote, M. A., Basaraba, R. J., Orme, I. M., and Pace, N. R., Use of specific rRNA oligonucleotide probes for microscopic detection of *Mycobacterium tuberculosis* in culture and tissue specimens, *J. Clin. Microbiol.* 43(10), 5369–5371, 2005.

42. Gill, P., Ramezani, R., Amiri, M. V., Ghaemi, A., Hashempour, T., Eshraghi, N., Ghalami, M., and Tehrani, H. A., Enzyme-linked immunosorbent assay of nucleic acid sequence-based amplification for molecular detection of *M. tuberculosis*, *Biochem. Biophys. Res. Commun.* 347(4), 1151–1157, 2006.

43. Piersimoni, C. and Scarparo, C., Relevance of commercial amplification methods for direct detection of *Mycobacterium tuberculosis* complex in clinical samples, *J. Clin. Microbiol.* 41(12), 5355–5365, 2003.

44. Centers for Disease Control and Prevention, Update: Nucleic Acid Amplification Tests for Tuberculosis, *Morb. Mortal. Wkly. Rep.* 49(26), 593–594, 2000.

45. Hill, B. C., Baker, C. N., and Tenover, F. C., A simplified method for testing *Bordetella pertussis* for resistance to erythromycin and other antimicrobial agents, *J. Clin. Microbiol.* 38(3), 1151–1155, 2000.

46. Murdoch, D. R., Molecular genetic methods in the diagnosis of lower respiratory tract infections, *APMIS* 112(11–12), 713–727, 2004.

47. Reischl, U., Lehn, N., Sanden, G. N., and Loeffelholz, M. J., Real-time PCR assay targeting IS481 of *Bordetella pertussis* and molecular basis for detecting *Bordetella holmesii*, *J. Clin. Microbiol.* 39(5), 1963–1966, 2001.

48. Sloan, L. M., Hopkins, M. K., Mitchell, P. S., Vetter, E. A., Rosenblatt, J. E., Harmsen, W. S., Cockerill, F. R., and Patel, R., Multiplex LightCycler PCR assay for detection and differentiation of *Bordetella pertussis* and *Bordetella parapertussis* in nasopharyngeal specimens, *J. Clin. Microbiol.* 40(1), 96–100, 2002.

49. Templeton, K. E., Scheltinga, S. A., van der Zee, A., Diederen, B. M., van Kruijssen, A. M., Goossens, H., Kuijper, E., and Claas, E. C., Evaluation of real-time PCR for detection of and discrimination between *Bordetella pertussis*, *Bordetella parapertussis*, and *Bordetella holmesii* for clinical diagnosis, *J. Clin. Microbiol.* 41(9), 4121–4126, 2003.

50. Ieven, M. and Goossens, H., Relevance of nucleic acid amplification techniques for diagnosis of respiratory tract infections in the clinical laboratory, *Clin. Microbiol. Rev.* 10(2), 242–256, 1997.

51. Chan, E. L., Antonishyn, N., McDonald, R., Maksymiw, T., Pieroni, P., Nagle, E., and Horsman, G. B., The use of TaqMan PCR assay for detection of *Bordetella pertussis* infection from clinical specimens, *Arch. Pathol. Lab. Med.* 126(2), 173–176, 2002.

52. Edelman, K., Nikkari, S., Ruuskanen, O., He, Q., Viljanen, M., and Mertsola, J., Detection of *Bordetella pertussis* by polymerase chain reaction and culture in the nasopharynx of erythromycin-treated infants with pertussis, *Pediatr. Infect. Dis. J.* 15(1), 54–57, 1996.

53. Aubry-Damon, H., Soussy, C. J., and Courvalin, P., Characterization of mutations in the *rpoB* gene that confer rifampin resistance in *Staphylococcus aureus*, *Antimicrob. Agents Chemother.* 42(10), 2590–2594, 1998.

54. Luber, A. D., Jacobs, R. A., Jordan, M., and Guglielmo, B. J., Relative importance of oral versus intravenous vancomycin exposure in the development of vancomycin-resistant enterococci, *J. Infect. Dis.* 173(5), 1292–1294, 1996.

55. Bugg, T. D., Wright, G. D., Dutka-Malen, S., Arthur, M., Courvalin, P., and Walsh, C. T., Molecular basis for vancomycin resistance in *Enterococcus faecium* BM4147: biosynthesis of a depsipeptide peptidoglycan precursor by vancomycin resistance proteins VanH and VanA, *Biochemistry* 30(43), 10408–10415, 1991.

56. Cetinkaya, Y., Falk, P., and Mayhall, C. G., Vancomycin-resistant enterococci, *Clin. Microbiol. Rev.* 13(4), 686–707, 2000.

57. Courvalin, P., Vancomycin resistance in gram-positive cocci, *Clin. Infect. Dis.* 42 (Suppl 1), S25–S34, 2006.

58. Depardieu, F., Reynolds, P. E., and Courvalin, P., VanD-type vancomycin-resistant *Enterococcus faecium* 10/96A, *Antimicrob. Agents Chemother.* 47(1), 7–18, 2003.

59. Ballard, S. A., Pertile, K. K., Lim, M., Johnson, P. D., and Grayson, M. L., Molecular characterization of *vanB* elements in naturally occurring gut anaerobes, *Antimicrob. Agents Chemother.* 49(5), 1688–1694, 2005.

60. Sloan, L. M., Uhl, J. R., Vetter, E. A., Schleck, C. D., Harmsen, W. S., Manahan, J., Thompson, R. L., Rosenblatt, J. E., and Cockerill, F. R., III, Comparison of the Roche LightCycler *vanA/vanB* detection assay and culture for detection of vancomycin-resistant enterococci from perianal swabs, *J. Clin. Microbiol.* 42(6), 2636–2643, 2004.

61. Zirakzadeh, A. and Patel, R., Epidemiology and mechanisms of glycopeptide resistance in enterococci, *Curr. Opin. Infect. Dis.* 18(6), 507–512, 2005.

62. Rice, L. B., Antimicrobial resistance in gram-positive bacteria, *Am. J. Infect. Control* 34(5 Suppl 1), S11–S19; discussion S64–S73, 2006.

63. Tenover, F. C., Mechanisms of antimicrobial resistance in bacteria, *Am. J. Infect. Control* 34(5 Suppl 1), S3–S10; discussion S64–S73, 2006.

64. Naimi, T. S., LeDell, K. H., Como-Sabetti, K., Borchardt, S. M., Boxrud, D. J., Etienne, J., Johnson, S. K., Vandenesch, F., Fridkin, S., O'Boyle, C., Danila, R. N., and Lynfield, R., Comparison of community- and health care-associated methicillin-resistant *Staphylococcus aureus* infection, *JAMA* 290(22), 2976–2984, 2003.

65. Brown, D. F., Edwards, D. I., Hawkey, P. M., Morrison, D., Ridgway, G. L., Towner, K. J., and Wren, M. W., Guidelines for the laboratory diagnosis and susceptibility testing of methicillin-resistant *Staphylococcus aureus* (MRSA), *J. Antimicrob. Chemother.* 56(6), 1000–1018, 2005.

66. Becker, K., Pagnier, I., Schuhen, B., Wenzelburger, F., Friedrich, A. W., Kipp, F., Peters, G., and von Eiff, C., Does nasal cocolonization by methicillin-resistant coagulase-negative staphylococci and methicillin-susceptible *Staphylococcus aureus* strains occur frequently enough to represent a risk of false-positive methicillin-resistant *S. aureus* determinations by molecular methods? *J. Clin. Microbiol.* 44(1), 229–231, 2006.

67. Desjardins, M., Guibord, C., Lalonde, B., Toye, B., and Ramotar, K., Evaluation of the IDI-MRSA assay for detection of methicillin-resistant *Staphylococcus aureus* from nasal and rectal specimens pooled in a selective broth, *J. Clin. Microbiol.* 44(4), 1219–1223, 2006.

68. Warren, D. K., Liao, R. S., Merz, L. R., Eveland, M., and Dunne, W. M., Jr., Detection of methicillin-resistant *Staphylococcus aureus* directly from nasal swab specimens by a real-time PCR assay, *J. Clin. Microbiol.* 42(12), 5578–5581, 2004.

69. Huletsky, A., Giroux, R., Rossbach, V., Gagnon, M., Vaillancourt, M., Bernier, M., Gagnon, F., Truchon, K., Bastien, M., Picard, F. J., van Belkum, A., Ouellette, M., Roy, P. H., and Bergeron, M. G., New real-time PCR assay for rapid detection of methicillin-resistant *Staphylococcus aureus* directly from specimens containing a mixture of staphylococci, *J. Clin. Microbiol.* 42(5), 1875–1884, 2004.

70. Piersimoni, C., Olivieri, A., Benacchio, L., and Scarparo, C., Current perspectives on drug susceptibility testing of *Mycobacterium tuberculosis* complex: the automated non-radiometric systems, *J. Clin. Microbiol.* 44(1), 20–28, 2006.

71. Gillespie, S. H., Evolution of drug resistance in *Mycobacterium tuberculosis*: clinical and molecular perspective, *Antimicrob. Agents Chemother.* 46(2), 267–274, 2002.

72. Caws, M., Duy, P. M., Tho, D. Q., Lan, N. T., Hoa, D. V., and Farrar, J., Mutations prevalent among rifampin- and isoniazid-resistant *Mycobacterium tuberculosis* isolates from a hospital in Vietnam, *J. Clin. Microbiol.* 44(7), 2333–2337, 2006.

73. Lavender, C., Globan, M., Sievers, A., Billman-Jacobe, H., and Fyfe, J., Molecular characterization of isoniazid-resistant *Mycobacterium tuberculosis* isolates collected in Australia, *Antimicrob. Agents Chemother.* 49(10), 4068–4074, 2005.

74. Somoskovi, A., Parsons, L. M., and Salfinger, M., The molecular basis of resistance to isoniazid, rifampin, and pyrazinamide in *Mycobacterium tuberculosis*, *Respir. Res.* 2(3), 164–168, 2001.

75. Johnson, R., Jordaan, A. M., Pretorius, L., Engelke, E., van der Spuy, G., Kewley, C., Bosman, M., van Helden, P. D., Warren, R., and Victor, T. C., Ethambutol resistance testing by mutation detection, *Int. J. Tuberc. Lung Dis.* 10(1), 68–73, 2006.

76. Nachamkin, I., Kang, C., and Weinstein, M. P., Detection of resistance to isoniazid, rifampin, and streptomycin in clinical isolates of *Mycobacterium tuberculosis* by molecular methods, *Clin. Infect. Dis.* 24(5), 894–900, 1997.

77. Parsons, L. M., Salfinger, M., Clobridge, A., Dormandy, J., Mirabello, L., Polletta, V. L., Sanic, A., Sinyavskiy, O., Larsen, S. C., Driscoll, J., Zickas, G., and Taber, H. W., Phenotypic and molecular characterization of *Mycobacterium tuberculosis* isolates resistant to both isoniazid and ethambutol, *Antimicrob. Agents Chemother.* 49(6), 2218–2225, 2005.

78. Fluit, A. C., Visser, M. R., and Schmitz, F. J., Molecular detection of antimicrobial resistance, *Clin. Microbiol. Rev.* 14(4), 836–871, 2001.

79. Morgan, M., Kalantri, S., Flores, L., and Pai, M., A commercial line probe assay for the rapid detection of rifampicin resistance in *Mycobacterium tuberculosis*: a systematic review and meta-analysis, *BMC Infect. Dis.* 5, 62, 2005.

80. Hillemann, D., Weizenegger, M., Kubica, T., Richter, E., and Niemann, S., Use of the genotype MTBDR assay for rapid detection of rifampin and isoniazid resistance in *Mycobacterium tuberculosis* complex isolates, *J. Clin. Microbiol.* 43(8), 3699–3703, 2005.

81. Makinen, J., Marttila, H. J., Marjamaki, M., Viljanen, M. K., and Soini, H., Comparison of two commercially available DNA line probe assays for detection of multidrug-resistant *Mycobacterium tuberculosis*, *J. Clin. Microbiol.* 44(2), 350–352, 2006.

82. Hodinka, R. L., The clinical utility of viral quantification using molecular methods, *Clin. Diagn. Virol.* 10(1), 25–47, 1998.

83. Holodniy, M., HIV-1 load quantification: a 17-year perspective, *J. Infect. Dis.* 194(Suppl 1), S38–S44, 2006.

84. Peter, J. B. and Sevall, J. S., Molecular-based methods for quantifying HIV viral load, *AIDS Patient Care STDS* 18(2), 75–79, 2004.

85. Department of Health and Human Services, Guidelines for the Use of Antiretroviral Agents in HIV-1-Infected Adults and Adolescents, 1–113, 2006.

86. Pilcher, C. D., Fiscus, S. A., Nguyen, T. Q., Foust, E., Wolf, L., Williams, D., Ashby, R., O'Dowd, J. O., McPherson, J. T., Stalzer, B., Hightow, L., Miller, W. C., Eron, J. J., Jr., Cohen, M. S., and Leone, P. A., Detection of acute infections during HIV testing in North Carolina, *New Engl. J. Med.* 352(18), 1873–1883, 2005.

87. Fried, M. W., Diagnostic testing for hepatitis C: practical considerations, *Am. J. Med.* 107(6B), 31S–35S, 1999.

88. Chevaliez, S. and Pawlotsky, J. M., Use of virologic assays in the diagnosis and management of hepatitis C virus infection, *Clin. Liver Dis.* 9(3), 371–382, v, 2005.

89. National Institutes of Health, National Institutes of Health Consensus Development conference statement: management of hepatitis C: 2002—June 10–12, 2002, *Hepatology* 36(5 Suppl 1), S3–S20, 2002.

90. Pawlotsky, J. M., Use and interpretation of hepatitis C virus diagnostic assays, *Clin. Liver Dis.* 7(1), 127–137, 2003.

91. Erensoy, S., Diagnosis of hepatitis C virus (HCV) infection and laboratory monitoring of its therapy, *J. Clin. Virol.* 21(3), 271–281, 2001.

92. Podzorski, R. P., Molecular testing in the diagnosis and management of hepatitis C virus infection, *Arch. Pathol. Lab. Med.* 126(3), 285–290, 2002.

93. Halfon, P., Bourliere, M., Halimi, G., Khiri, H., Bertezene, P., Portal, I., Botta-Fridlund, D., Gauthier, A. P., Jullien, M., Feryn, J. M., Gerolami, V., and Cartouzou, G., Assessment of spontaneous fluctuations of viral load in untreated patients with chronic hepatitis C by two standardized quantification methods: branched DNA and Amplicor Monitor, *J. Clin. Microbiol.* 36(7), 2073–2075, 1998.

94. Weck, K., Molecular methods of hepatitis C genotyping, *Expert Rev. Mol. Diagn.* 5(4), 507–520, 2005.

95. Servoss, J. C. and Friedman, L. S., Serologic and molecular diagnosis of hepatitis B virus, *Clin. Liver Dis.* 8(2), 267–281, 2004.

96. Gish, R. G. and Locarnini, S. A., Chronic hepatitis B: current testing strategies, *Clin. Gastroenterol. Hepatol.* 4(6), 666–676, 2006.
97. Hatzakis, A., Magiorkinis, E., and Haida, C., HBV virological assessment, *J. Hepatol.* 44(1 Suppl), S71–S76, 2006.
98. Rubin, R. H., Cytomegalovirus in solid organ transplantation, *Transpl. Infect. Dis.* 3(Suppl 2), 1–5, 2001.
99. Razonable, R. R. and Emery, V. C., Management of CMV infection and disease in transplant patients, *Herpes* 11(3), 77–86, 2004.
100. Berger, A. and Preiser, W., Viral genome quantification as a tool for improving patient management: the example of HIV, HBV, HCV and CMV, *J. Antimicrob. Chemother.* 49(5), 713–721, 2002.
101. Gerna, G. and Lilleri, D., Monitoring transplant patients for human cytomegalovirus: diagnostic update, *Herpes* 13(1), 4–11, 2006.
102. Fan, H. and Gulley, M. L., Epstein–Barr viral load measurement as a marker of EBV-related disease, *Mol. Diagn.* 6(4), 279–289, 2001.
103. Seidemann, K., Heim, A., Pfister, E. D., Koditz, H., Beilken, A., Sander, A., Melter, M., Sykora, K. W., Sasse, M., and Wessel, A., Monitoring of adenovirus infection in pediatric transplant recipients by quantitative PCR: report of six cases and review of the literature, *Am. J. Transpl.* 4(12), 2102–2108, 2004.
104. Zerr, D. M., Corey, L., Kim, H. W., Huang, M. L., Nguy, L., and Boeckh, M., Clinical outcomes of human herpesvirus 6 reactivation after hematopoietic stem cell transplantation, *Clin. Infect. Dis.* 40(7), 932–940, 2005.
105. Debiasi, R. L. and Tyler, K. L., Molecular methods for diagnosis of viral encephalitis, *Clin. Microbiol. Rev.* 17(4), 903–925, 2004.
106. Cinque, P., Bossolasco, S., and Lundkvist, A., Molecular analysis of cerebrospinal fluid in viral diseases of the central nervous system, *J. Clin. Virol.* 26(1), 1–28, 2003.
107. Aberle, S. W. and Puchhammer-Stockl, E., Diagnosis of herpesvirus infections of the central nervous system, *J. Clin. Virol.* 25(Suppl 1), S79–S85, 2002.
108. Romero, J. R. and Kimberlin, D. W., Molecular diagnosis of viral infections of the central nervous system, *Clin. Lab. Med.* 23(4), 843–865, vi, 2003.
109. Calvario, A., Bozzi, A., Scarasciulli, M., Ventola, C., Seccia, R., Stomati, D., and Brancasi, B., Herpes consensus PCR test: a useful diagnostic approach to the screening of viral diseases of the central nervous system, *J. Clin. Virol.* 25(Suppl 1), S71–S78, 2002.
110. DeBiasi, R. L., Kleinschmidt-DeMasters, B. K., Weinberg, A., and Tyler, K. L., Use of PCR for the diagnosis of herpesvirus infections of the central nervous system, *J. Clin. Virol.* 25(Suppl 1), S5–S11, 2002.
111. Romero, J. R., Diagnosis and management of enteroviral infections of the central nervous system, *Curr. Infect. Dis. Rep.* 4(4), 309–316, 2002.
112. Schallig, H. D. and Oskam, L., Molecular biological applications in the diagnosis and control of leishmaniasis and parasite identification, *Trop. Med. Int. Health* 7(8), 641–651, 2002.
113. Switaj, K., Master, A., Skrzypczak, M., and Zaborowski, P., Recent trends in molecular diagnostics for *Toxoplasma gondii* infections, *Clin. Microbiol. Infect.* 11(3), 170–176, 2005.
114. Filisetti, D., Gorcii, M., Pernot-Marino, E., Villard, O., and Candolfi, E., Diagnosis of congenital toxoplasmosis: comparison of targets for detection of *Toxoplasma gondii* by PCR, *J. Clin. Microbiol.* 41(10), 4826–4828, 2003.
115. Contini, C., Giuliodori, M., Cultrera, R., and Seraceni, S., Detection of clinical-stage specific molecular *Toxoplasma gondii* gene patterns in patients with toxoplasmic lymphadenitis, *J. Med. Microbiol.* 55(Pt 6), 771–774, 2006.

116. Cassaing, S., Bessieres, M. H., Berry, A., Berrebi, A., Fabre, R., and Magnaval, J. F., Comparison between two amplification sets for molecular diagnosis of toxoplasmosis by real-time PCR, *J. Clin. Microbiol.* 44(3), 720–724, 2006.

117. McCutchan, T. F., de la Cruz, V. F., Lal, A. A., Gunderson, J. H., Elwood, H. J., and Sogin, M. L., Primary sequences of two small subunit ribosomal RNA genes from *Plasmodium falciparum*, *Mol. Biochem. Parasitol.* 28(1), 63–68, 1988.

118. Waters, A. P. and McCutchan, T. F., Rapid, sensitive diagnosis of malaria based on ribosomal RNA, *Lancet* 1(8651), 1343–1346, 1989.

119. Berry, A., Fabre, R., Benoit-Vical, F., Cassaing, S., and Magnaval, J. F., Contribution of PCR-based methods to diagnosis and management of imported malaria, *Med. Trop. (Mars)* 65(2), 176–183, 2005.

120. Farcas, G. A., Zhong, K. J., Mazzulli, T., and Kain, K. C., Evaluation of the RealArt Malaria LC real-time PCR assay for malaria diagnosis, *J. Clin. Microbiol.* 42(2), 636–638, 2004.

121. Lindsley, M. D., Hurst, S. F., Iqbal, N. J., and Morrison, C. J., Rapid identification of dimorphic and yeast-like fungal pathogens using specific DNA probes, *J. Clin. Microbiol.* 39(10), 3505–3511, 2001.

122. Stockman, L., Clark, K. A., Hunt, J. M., and Roberts, G. D., Evaluation of commercially available acridinium ester-labeled chemiluminescent DNA probes for culture identification of *Blastomyces dermatitidis, Coccidioides immitis, Cryptococcus neoformans*, and *Histoplasma capsulatum*, *J. Clin. Microbiol.* 31(4), 845–850, 1993.

123. Sandin, R. L. and Greene, J. N., Diagnostic molecular pathology and infectious disease, *Cancer Control* 2(3), 250–257, 1995.

124. Hope, W. W., Walsh, T. J., and Denning, D. W., Laboratory diagnosis of invasive aspergillosis, *Lancet Infect. Dis.* 5(10), 609–622, 2005.

125. Mennink-Kersten, M. A., Donnelly, J. P., and Verweij, P. E., Detection of circulating galactomannan for the diagnosis and management of invasive aspergillosis, *Lancet Infect. Dis.* 4(6), 349–357, 2004.

126. Wheat, L. J., Rapid diagnosis of invasive aspergillosis by antigen detection, *Transpl. Infect. Dis.* 5(4), 158–166, 2003.

127. Bretagne, S. and Costa, J. M., Towards a molecular diagnosis of invasive aspergillosis and disseminated candidosis, *FEMS Immunol. Med. Microbiol.* 45(3), 361–368, 2005.

128. White, P. L., Linton, C. J., Perry, M. D., Johnson, E. M., and Barnes, R. A., The evolution and evaluation of a whole blood polymerase chain reaction assay for the detection of invasive aspergillosis in hematology patients in a routine clinical setting, *Clin. Infect. Dis.* 42(4), 479–486, 2006.

129. Drancourt, M. and Raoult, D., Sequence-based identification of new bacteria: a proposition for creation of an orphan bacterium repository, *J. Clin. Microbiol.* 43(9), 4311–4315, 2005.

130. Hall, L., Doerr, K. A., Wohlfiel, S. L., and Roberts, G. D., Evaluation of the MicroSeq system for identification of mycobacteria by 16S ribosomal DNA sequencing and its integration into a routine clinical mycobacteriology laboratory, *J. Clin. Microbiol.* 41(4), 1447–1453, 2003.

131. Tortoli, E., Impact of genotypic studies on mycobacterial taxonomy: the new mycobacteria of the 1990s, *Clin. Microbiol. Rev.* 16(2), 319–354, 2003.

132. Clarridge, J. E., III, Impact of 16S rRNA gene sequence analysis for identification of bacteria on clinical microbiology and infectious diseases, *Clin. Microbiol. Rev.* 17(4), 840–862, table of contents, 2004.

133. Cloud, J. L., Neal, H., Rosenberry, R., Turenne, C. Y., Jama, M., Hillyard, D. R., and Carroll, K. C., Identification of *Mycobacterium* spp. by using a commercial 16S ribosomal DNA sequencing kit and additional sequencing libraries, *J. Clin. Microbiol.* 40(2), 400–406, 2002.

134. Millar, B. C. and Moore, J. E., Current trends in the molecular diagnosis of infective endocarditis, *Eur. J. Clin. Microbiol. Infect. Dis.* 23(5), 353–365, 2004.

135. Sacchi, C. T., Whitney, A. M., Mayer, L. W., Morey, R., Steigerwalt, A., Boras, A., Weyant, R. S., and Popovic, T., Sequencing of 16S rRNA gene: a rapid tool for identification of *Bacillus anthracis, Emerg. Infect. Dis.* 8(10), 1117–1123, 2002.

136. Brown-Elliott, B. A., Brown, J. M., Conville, P. S., and Wallace, R. J., Jr., Clinical and laboratory features of the *Nocardia* spp. based on current molecular taxonomy, *Clin. Microbiol. Rev.* 19(2), 259–282, 2006.

137. Roth, A., Fischer, M., Hamid, M. E., Michalke, S., Ludwig, W., and Mauch, H., Differentiation of phylogenetically related slowly growing mycobacteria based on 16S–23S rRNA gene internal transcribed spacer sequences, *J. Clin. Microbiol.* 36(1), 139–147, 1998.

138. Ringuet, H., Akoua-Koffi, C., Honore, S., Varnerot, A., Vincent, V., Berche, P., Gaillard, J. L., and Pierre-Audigier, C., *hsp65* sequencing for identification of rapidly growing mycobacteria, *J. Clin. Microbiol.* 37(3), 852–857, 1999.

139. McNabb, A., Eisler, D., Adie, K., Amos, M., Rodrigues, M., Stephens, G., Black, W. A., and Isaac-Renton, J., Assessment of partial sequencing of the 65-kilodalton heat shock protein gene (*hsp65*) for routine identification of *Mycobacterium* species isolated from clinical sources, *J. Clin. Microbiol.* 42(7), 3000–3011, 2004.

140. Turenne, C. Y., Semret, M., Cousins, D. V., Collins, D. M., and Behr, M. A., Sequencing of *hsp65* distinguishes among subsets of the *Mycobacterium avium* complex, *J. Clin. Microbiol.* 44(2), 433–440, 2006.

141. Leaw, S. N., Chang, H. C., Sun, H. F., Barton, R., Bouchara, J. P., and Chang, T. C., Identification of medically important yeast species by sequence analysis of the internal transcribed spacer regions, *J. Clin. Microbiol.* 44(3), 693–699, 2006.

142. Gutierrez, F., Masia, M., Ramos, J., Elia, M., Mellado, E., and Cuenca-Estrella, M., Pulmonary mycetoma caused by an atypical isolate of *Paecilomyces* species in an immunocompetent individual: case report and literature review of *Paecilomyces* lung infections, *Eur. J. Clin. Microbiol. Infect. Dis.* 24(9), 607–611, 2005.

143. Valenza, G., Valenza, R., Brederlau, J., Frosch, M., and Kurzai, O., Identification of *Candida fabianii* as a cause of lethal septicaemia, *Mycoses* 49(4), 331–334, 2006.

144. Desnos-Ollivier, M., Bretagne, S., Dromer, F., Lortholary, O., and Dannaoui, E., Molecular identification of black-grain mycetoma agents, *J. Clin. Microbiol.* 44(10), 3517–3523, 2006.

145. Ciardo, D. E., Schar, G., Bottger, E. C., Altwegg, M., and Bosshard, P. P., Internal transcribed spacer sequencing versus biochemical profiling for identification of medically important yeasts, *J. Clin. Microbiol.* 44(1), 77–84, 2006.

146. Rakeman, J. L., Bui, U., Lafe, K., Chen, Y. C., Honeycutt, R. J., and Cookson, B. T., Multilocus DNA sequence comparisons rapidly identify pathogenic molds, *J. Clin. Microbiol.* 43(7), 3324–3333, 2005.

147. Gupta, A. K., Boekhout, T., Theelen, B., Summerbell, R., and Batra, R., Identification and typing of *Malassezia* species by amplified fragment length polymorphism and sequence analyses of the internal transcribed spacer and large-subunit regions of ribosomal DNA, *J. Clin. Microbiol.* 42(9), 4253–4260, 2004.

148. Hinrikson, H. P., Hurst, S. F., De Aguirre, L., and Morrison, C. J., Molecular methods for the identification of *Aspergillus* species, *Med. Mycol.* 43(Suppl 1), S129–S137, 2005.

149. Schwarz, P., Bretagne, S., Gantier, J. C., Garcia-Hermoso, D., Lortholary, O., Dromer, F., and Dannaoui, E., Molecular identification of zygomycetes from culture and experimentally infected tissues, *J. Clin. Microbiol.* 44(2), 340–349, 2006.

150. Arens, M., Clinically relevant sequence-based genotyping of HBV, HCV, CMV, and HIV, *J. Clin. Virol.* 22(1), 11–29, 2001.

151. Drew, W. L., Paya, C. V., and Emery, V., Cytomegalovirus (CMV) resistance to antivirals, *Am. J. Transpl.* 1(4), 307–312, 2001.

152. Fredericks, D. N. and Relman, D. A., Sequence-based identification of microbial pathogens: a reconsideration of Koch's postulates, *Clin. Microbiol. Rev.* 9(1), 18–33, 1996.

16 Cytokines: Regulators of Immune Responses and Key Therapeutic Targets

Barbara Detrick, Chandrasekharam N. Nagineni, and John J. Hooks

CONTENTS

16.1   INTRODUCTION: OVERVIEW OF CYTOKINES

Over the last two decades, researchers have witnessed an explosion in cytokine biology. Biomedical researchers have tried to isolate and characterize selected molecules associated with specific biological activities and disease states. However, years of investigation have led to the description of mediators with a diverse range of actions on a variety of cell types. Cytokines are potent, low-molecular-weight protein cell regulators, produced transiently and locally by numerous cell types. Today, we recognize that cytokines are multifunctional proteins whose biological properties suggest a key role in hematopoiesis, immunity, infectious disease, tumorigenesis, homeostasis, tissue repair, cellular development, and growth.[1-4] This review will focus on the production and activities of cytokines with special emphasis on their clinical application.

Dramatic advances in recombinant deoxyribonucleic acid (DNA) and monoclonal antibody technology have greatly enhanced our knowledge of cytokines, their interactive role in health and disease, their activities in animal models, and their clinical utility. Because of their major participatory role in nearly all pathophysiologic processes and their therapeutic potential, there is a need to identify and measure cytokines. In the clinical laboratory, cytokine assessment has been used to monitor disease progression and activity. For example, the proinflammatory cytokines, tumor necrosis factor (TNF), interleukin-1 (IL-1), and interleukin-6 (IL-6) are frequently detected in the sera of patients with septic shock. These cytokines appear to play a critical role in the development of septic shock, and tracking their presence may be of prognostic value in severe sepsis. In addition, the increasing use of cytokines and cytokine antagonists as therapeutic modalities requires the measurement of cytokine levels to determine the pharmacokinetics of the administered molecules. In the research laboratory, the measurement of cytokine gene expression is currently being explored with the hope that this approach will offer clues to better define the mechanisms of cytokine action in disease processes. As a consequence of these expanding applications, the future use of cytokine monitoring in the clinical and research setting will undoubtedly increase. Dr. Maecker in Chapter 17 of this book describes the assays used to measure cytokines.

This chapter highlights the production and activities of cytokines and their receptors, as well as, spotlights their clinical applications, and has been divided into five areas. First, cytokines as biomarkers of disease is addressed and clues for mechanisms of disease are provided. Second, the measurement of cytokine production *in vitro* as a useful monitor of immune status is discussed. Third, the data supporting the use of recombinant cytokines as key therapeutic agents is reviewed. Fourth, how cytokines can be targets of therapeutics is highlighted. Here, the therapeutic utility of the cytokine receptor antagonist and anticytokines monoclonal antibodies that downregulate pathogenic responses to endogenous cytokine production is emphasized. Finally, the novel use of cytokines as key components of vaccine formulations is reviewed.

In this chapter, we will not attempt to identify all of the described cytokines and effector molecules. Rather, we have selected those cytokines that may have relevance in terms of clinical immunologic monitoring and which may be directed toward therapeutic strategies. A list of the current "interleukin cytokines" including a summary of their activities is presented in Table 16.1 (note: the normal ranges for the more common human serum cytokines are listed in Table 17.1, Chapter 17). Owing to the rapid developments in this area, cytokine lists will continually require updating. For more detailed descriptions of this diverse topic, see recent reviews in the literature.[3-7]

TABLE 16.1
Interleukins

Cytokine	Producers	Inducers	Effects
IL-1 (IL-1α and 1β)	Many cells, macrophages, dendritic cells, NK cells	LPS, TNF-α, infectious agents	Pleiotrophic factor; stimulates T cells; B cells; induces IL-2, -6, -8, GM-CSF, MCP-1
IL-2	T cells (CD4)	TCRs complex on APC, IL-1	Induces proliferation and maturation of T cells, B cells, NK cells, monocytes, and oligodendrocytes
IL-3	T cells, monocytes, NK cells, mast cells, endothelial cells	Antigen stimulation	Growth factor for hematopoietic cells; B cells; chemoattractant for eosinophils; activates monocytes
IL-4	T cells, mast cells, basophils, bone marrow stromal cells	Antigen stimulation	Growth factor for B cells; production of IgG and IgE; effects on monocytes, endothelial cells, and fibroblasts
IL-5	T cells, mast cells	Antigen stimulation	Growth and differentiation of eosinophils
IL-6	T cells, B cells, macrophages, fibroblasts, epithelial cells	LPS, IL-1, TNF-α, IFN-γ	B-cell stimulation, T cell and neuron growth, mediator of acute-phase reaction
IL-7	Bone marrow, thymic stromal cells, keratinocytes	Constitutively expressed	B-cell development, megakaryocyte maturation, proliferation and differentiation of T cells
IL-8	Many cell types, monocytes, fibroblasts, epithelial cells	LPS, TNF-α, IL-1	Neutrophil activation and chemotaxis, also attracts basophils and some T cells, angiogenic factor
IL-9	T cells, lymphoma cells	T-cell mitogens	Promotes mast cell and B-cell growth
IL-10	T cells, monocytes, macrophages	Antigen stimulation	Inhibits IFN-γ and IL-12 production, costimulator of thymocytes and mast cells
IL-11	Bone marrow stromal cells, mesenchymale cells	IL-1, TGF-β, TNF-α	Synergistic effects on hematopoiesis and thrombopoiesis, cytoprotective effects on epithelial cells
IL-12	Dendritic cells, macrophages, B cells		Induces production of IFN-γ and TNF-α and IL-2 by resting and activated T and NK cells
IL-13	CD4+ T cells, NK cells		Suppresses IL-1, -8, -12, and TNF-α by macrophages and endothelial cells; upregulates expression of CD71, CD72, and CD23 on B cells

(*continued*)

TABLE 16.1 (continued)
Interleukins

Cytokine	Producers	Inducers	Effects
IL-14	T cells		Mitogen for activated B cells
IL-15	T cells, atrocytes, microglia, fibroblasts, epithelial cells	IL-1, IFN-γ, TNF-α	Some biological activities similar to IL-2 induces proliferation of peripheral blood mononuclear cells, maturation of NK cells
IL-16	CD8+ T cells, epithelial cells		Chemotactic to CD4+ T cells, monocytes, eosinophils, suppresses HIV-1 replication
IL-17 (six members)	CD4+ T cells (Th17 T cells)		Stimulates epithelial, endothelial, and fibroblastic cells to produce IL-6, -8, G-CSF, and ICAM-1
IL-18	Liver (Kupffer cells) macrophages		Functional similarities to IL-2, induces IFN-γ in Th1 T cells and IL-2, GM-CSF in mononuclear cells
IL-19	Activated keratinocytes, monocytes		Induces IL-6 and TNF-α production by monocytes, T-helper cell differentiation to Th 2
IL-20	Skin and trachea, monocytes, keratinocytes		Induces the proliferation of multipotent hematopoietic progenitor cells, secretion of proinflammatory mediators in keratinocytes
IL-21	T cells		Related to IL-2, -4, and -15; plays a role in B- and T-cell proliferation and NK-cell maturation
IL-22	Normal T cells	Anti-CD3	Related to IL-10, activates STATs for acute-phase proteins in several cell lines
IL-23	Macrophages, dendritic cells		Similar to IL-12 (proliferation and IFN-γ production by T cells), distinct effects on memory T cells
IL-24 (mda-7)	B cells, CD4+ T cells NK cells, monocytes		Induces IL-1, -12, and TNF-α in monocytes; apoptosis in tumor cells; inhibits endothelial cell differentiation
IL-26	NK cells, T cells, endothelial cells		Induces secretion of IL-8, -10, expression of CD54
IL-28A and IL-29	Plasmacytoid, Dendritic cells	Viral infection, dsRNA	IL-28A, -28B, and -29 are also called IFN-λ2, IFN-λ3, and IFN-λ1, respectively; all three forms exhibit activities similar to type 1 IFNs; activate STAT 1 and 2 phosphorylation

(*continued*)

TABLE 16.1 (continued)
Interleukins

Cytokine	Producers	Inducers	Effects
IL-30	Antigen-presenting cells		Triggers expansion of CD+ T cells for Th1 response
IL-31	Th2 cells		Signals through JAK/STAT, and MAP kinase pathway; may contribute to pruritus and nonatopic dermatitis
IL-32	T cells and NK cells		Acts on macrophages; induces the production of TNF-α, IL-8, and MIP-2

Source: Thomson, A. and Lotze, M. *The Cytokine Handbook*, Academic Press, Amsterdam, 2003; Meager, A. *The Interferons: Characterization and Application*, Wiley, Weinheim, 2006; Santamaria, P. *Cytokines and Chemokines in Autoimmune Disease*, Kluwer Academic/Plenum Publishers, New York, 2003.

16.2 CYTOKINES: PRODUCTION AND ACTIVITIES

Before we enter into a description of the cytokines, there are certain general considerations that should be noted. First, the names used to describe the molecules are frequently erroneous, since the biological activities of the cytokines extend beyond the named activity. For example, interferon gamma (IFN-γ) was first described as an inhibitor of virus replication, but today it is also recognized as a potent immunoregulatory molecule. Moreover, the interleukins not only initiate communication among immune cells, but they can also induce profound effects on nonimmune cells.

Second, cytokines usually act as signaling molecules by binding to their own glycoprotein receptors on cell membranes.[8] This initial interaction is followed by a relay of the signal to the cell nucleus. Signal transduction is mediated as in many hormone-receptor systems by kinase-mediated phosphorylation of cytoplasmic proteins. In fact, tyrosine kinase activity is intrinsic to many cytokine receptors. For the purpose of this discussion, we will highlight two aspects of the receptor system: the production of soluble receptors and the potential role of receptor antagonists. Even when separated from the intact molecule, the receptor polypeptide domains can retain their functions. Receptors, for instance, released into the circulation can bind to their ligand. This can impact upon the interpretation of cytokine assays and cytokine therapy. Because receptor expression can be activated in a variety of disease states, approaches to block this activation and the specific cytokine–receptor interaction is actively under investigation and is discussed in the section on clinical applications of cytokines (Table 16.2).

Third, cytokine activity can be overlapping or redundant, and interactions among cytokines can occur through a cascading effect. Cytokines are potent regulatory molecules that modify inflammation, cell growth, and differentiation. The advent of cDNA cloning techniques has generated recombinant molecules that are pure

TABLE 16.2

Clinical Applications of Cytokines

Application	Example
Detection in diseases	IL-1, 2R, 6, TNF, IFN-γ in inflammatory and autoimmune diseases, graft rejection, and malignancy
Monitor immunocompetence	Cytokine (IL-2, IFN-γ) production *in vitro*
Cytokine therapy	
Ex vivo	
Stimulation of cells	CSF-treated bone marrow cells for transplantation of IL-2, IFN-γ-treated LAK cells in cancer
Gene insertion	TNF in malignancy
In vivo	IFNs in infectious disease, malignancy, immune dysfunction
	Cytokine combinations in bone marrow transplantation
Modalities for reducing cytokine activity	
Anticytokine antibody	Anti-IL-1, TNF, IFN-γ in inflammation and autoimmunity
Antireceptor antibody	Anti-IL-1R, anti IL-2R
Soluble receptor	sIL-2R
Receptor antagonists	IL-1ra
Enzyme inhibition	Inhibition of IL-1 convertases
Drugs	CyA, FK506, rapamycin
Vaccine therapy cytokine adjuvants	IFN-α, -β, -γ—infectious disease and tumor vaccines
	IL-2—infectious disease and tumor vaccines
	IL-7—infectious disease vaccines (HSV, HIV)
	TNF—cancer vaccines
	IL-12—parasitic vaccines (malaria)

preparations of individual cytokines. Studies using these recombinant molecules with transgenic and knockout mice have underscored the redundant and cascading or cyclic nature of these cytokine systems. It is now clear that cytokines share similar activities, and cytokines induce or augment the actions of other cytokines. An appreciation of these tenets has enriched our knowledge of the potential clinical applications of cytokines.

Production of cytokines by Th1 and Th2 subset of T cells is an example of cytokine interactions or cross-regulation.[9] Activation of the Th1 subset of T cells results in the production of IL-2, IFN-γ, and IL-12, which stimulates predominantly cell-mediated immune responses mediated by cytotoxic T cells, macrophages, and natural killer (NK) cells. In contrast, activation of Th2 subset of T cells results in the production of IL-4, IL-5, IL-6, and IL-10, which stimulates predominantly humoral immune responses mediated by antibodies. Opposing effects of cytokines are seen with IL-10 and IL-12. IFN-γ production is upregulated by IL-12 and downregulated by IL-10. An appreciation of the cytokine interactions is creating an impetus for investigators to try to modify pathologic responses by manipulating cytokine interactions within a particular tissue.

Fourth, it is now clear that cytokines are an integral component of the innate immune response and bridge innate and adaptive immunity. The body defends itself against microorganisms by turning on an innate immune response followed by an acquired immune response. (A general review of innate and acquired immunity is presented in Chapter 1.) The innate immune system can be activated within minutes after the invasion of a bacterium or virus. Recent studies identified that in the innate immune response, the body recognizes the invaders through toll-like receptors (TLRs) that are present on or within selected cells within the body. When the TLRs are engaged, genes important for an effective host defense, such as cytokines and chemokines, are activated.

The rapid emergence of new cytokines and the identification of new activities for established cytokines make a comprehensive comment on each individual cytokine difficult. Therefore, only selected cytokines are highlighted. For the purpose of this chapter, we have organized the cytokines into the following groups: IFNs, TNFs, interleukins, growth factors, chemokines, and adhesion molecules. Cytokines and their actions are listed in Tables 16.1 and 16.3.

TABLE 16.3
Cytokine Production and Action

Cytokine Family	Producers[a]	Effects
Interferons		
Alpha	Leukocytes	Antiviral, immunoregulatory, antiproliferative (enhance MHC class I, NK-cell activity)
Beta	Fibroblasts	Antiviral, immunoregulatory, antiproliferative
Gamma	T cells, NK cells	Antiviral, immunoregulatory, antiproliferative (enhance MHC class I and II, macrophage activation)
TNF		
Alpha	Macrophage, lymphocytes	Activate macrophages and cytotoxic cells; induce cachexia, acute-phase proteins, and cytokines (IL-1, -6)
Beta	T cells	Activate macrophages; induces cytokines (IL-1, -6)
Colony-Stimulating Factors		
M-CSF	Monocytes	Proliferation of macrophage precursors
G-CSF	Macrophages	Proliferation, differentiation and activation of neutrophilic granulocyte lineage
GM-CSF	T cells, macrophages	Proliferation of granulocyte and macrophage precursors
Stem cell factor	Bone marrow stromal cells, fibroblasts, fetal liver cells	Proliferation and differentiation of early myeloid and lymphoid cells (synergizes with other cytokines)

[a] This list is not inclusive; only the primary cells have been identified.

16.3 CYTOKINE GROUPS

16.3.1 INTERFERONS

Although the IFNs were first identified in 1957 as antiviral proteins, they are now also recognized as immunoregulatory proteins capable of altering a variety of cellular processes such as cell growth, differentiation, gene transcription, and translation.[10] The IFN family consists of three groups.[11] Type 1 IFNs comprise numerous genes and include IFN-α and IFN-β. Recently, IFN-ω, IFN-κ, and IFN-ε have been added to the group.[1] Type II IFN consists of a singe gene that produces IFN-γ. IFN-λ is a third group of IFN-like cytokines that have recently been described.[1] Our discussion is limited to IFN-α, IFN-β, and IFN-γ (Table 16.3).

The cell making the IFN and the substance triggering its production are important factors in determining the type of IFN produced. IFN-α is produced primarily by leukocytes in response to a variety of IFN inducers such as viruses, bacterial products, dsRNA, tumor cells, and allogeneic cells. The cell types responsible for synthesizing IFN-α include dendritic cells, B cells, T cells, macrophages, NK cells, and large granular cells. If any of these inducers should interact with fibroblasts, epithelial cells, or to a lesser extent, leukocytes, IFN-β is produced. As an integral part of the immune response, T cells are capable of manufacturing IFN-γ. Activated NK cells also produce IFN-γ. Moreover, the cytokines IL-2 and IL-12 can trigger T cells to produce IFN-γ.

The IFNs bind to their cellular receptor and activate the Janus-activated kinase (JAK) and signal transducers and activators of transcription (STAT) signaling pathways.[12] This process triggers activation of genes containing IFN-stimulated response elements or an IFN-γ-activated sequence. The different IFNs have overlapping biological activities such as antiviral actions, antiproliferative actions, and immunoregulatory actions. However, nonoverlapping functions also exist. For example, IFN-β is used to successfully treat patients with multiple sclerosis (MS), whereas IFN-γ has been shown to exacerbate the disease.

All of the IFNs can augment or depress a wide variety of immune reactions. Moreover, the IFNs are a key component of the innate immune response and regulate both innate and adaptive immunity. IFN proteins can modify immune reactivity by acting at the level of B cells, T cells, NK cells, dendritic cells, macrophages, basophils, or bone marrow stem cells. Immune responses altered by such interactions include antibody production, T-cell cytotoxicity, graft-versus-host reactions, mitogen and antigen stimulation, delayed-type hypersensitivity, NK-cell cytotoxicity, macrophage functions, IgE-mediated histamine release, and bone marrow stem cell maturation. It is possible that the actions of the IFNs on these cell types and the altered immune reactivity may be important in the pathophysiology of autoimmunity, immune deficiency, malignancy, and infectious diseases.

These potent actions of the IFNs and the advances in biotechnology are the underlying factors that have identified the clinical relevance of the IFNs. In fact, many of the IFNs have been FDA approved for the treatment of infections, malignancies, autoimmunity, and immunodeficiency (Table 16.4).

TABLE 16.4
Disease States That Respond to IFN Therapy

IFN	Infectious Diseases	Malignancy	Immune Dysfunctions	Vascular Proliferative Disorders
Alpha	Hepatitis B	Hairy cell leukemia		Angiomas
	Hepatitis C	Kaposi's sarcoma		Subretinal neovascular proliferative disease
	Genital warts	Chronic myelogenous leukemia		
	Herpes simplex virus	Non-Hodgkin's lymphoma		
	AIDS	Multiple myeloma		
	Laryngeal papillomatosis	Cutaneous T-cell lymphoma		
		Malignant melanoma		
		Basal cell carcinoma		
Beta			Multiple sclerosis	
Gamma	Intracellular parasites (leishmaniasis, toxoplasmosis)	Myeloid leukemia, cutaneous T-cell lymphoma metastatic renal cell carcinoma	Chronic granulomatosis disease	
	Leprosy	Kaposi's sarcoma		
		Basal cell carcinoma		
		Cervical intraepithelial neoplasia		

16.3.2 TUMOR NECROSIS FACTORS

TNFs are cytokines that participate in inflammatory and antitumor activity. At present, there are two types of TNF—TNF-α and TNF-β. TNF-α was originally detected in the circulation of mice. Endotoxin-treated mice contained a molecule, cachectin, which produced a wasting syndrome in mice.[13] In conjunction with the aforementioned study, Carswell and associates[14] discovered a serum factor (TNF) that induced tumor necrosis in LPS-treated mice. Cachectin and TNF were both identified to be the same molecule, now known as TNF-α. TNF-β, in contrast, was first isolated from activated T cells and was called lymphotoxin.[15,16] TNF-α and -β are structurally related, bind to the same cellular receptors, and produce similar biological changes in a variety of cells. TNF-α is produced by neutrophils, activated lymphocytes, macrophages, NK cells, and some nonlymphoid cells such as astrocytes, endothelial cells, and smooth muscle cells, whereas TNF-β appears to be produced solely by T cells.

Both the TNFs and the IFNs are highly pleiotropic factors in terms of their activities. In fact, TNF has been shown to induce or suppress the expression of a variety of genes resulting in the production of growth factors, other cytokines, inflammatory mediators, and acute-phase proteins. In addition to its ability to kill tumors, TNF is now recognized as a potent proinflammatory mediator, and this property is the primary focus for clinical applications. TNF modulates both lymphoid and nonlymphoid cells to synthesize and release other immunostimulatory cytokines such as IL-1 and -6. Furthermore, TNF induces fever, PGE2 and collagenase synthesis, bone and cartilage resorption, inhibition of lipoprotein lipase, and an increase in hepatic acute-phase proteins and complement components. Experimental administration of TNFs can result in hypotension, leucopenia, local tissue necrosis, and shock. Numerous clinical investigations have been initiated to evaluate agents that block TNF activity, particularly in septic shock and rheumatoid arthritis (RA). This is discussed in Section 16.4.

16.3.3 INTERLEUKINS

This term was originally coined in 1981 to describe the leukocyte-derived molecules that also acted on leukocytes. We now know that interleukins can be produced by and interact with a variety of cell types. The various interleukins (IL-1 through -33), the cells that produce these cytokines, and their diverse biological activities are listed in Table 16.1. In Section 16.4, IL-6, -2, and -15 are discussed in more detail.

16.3.4 GROWTH FACTORS

In addition to the hematopoietic growth factors, there are numerous additional cytokine growth factors that influence cellular growth and differentiation. One factor that has received widespread attention by both the research and clinical community is transforming growth factor beta (TGF-β). The transforming growth factors (TGF-β) are multifunctional agents that are involved in embryogenesis, tissue remodeling, and other diverse activities by regulating cell proliferation, differentiation, and migration.[17-19] Three isoforms—TGF-β1, TGF-β2, and TGF-β3—are expressed in mammals and exhibit many identical activities. However, these molecules also display some unique physiological and developmental functions. TGF-β1 and TGF-β2 are made as precursors of 390 and 414 amino acids, respectively. The mature polypeptide of 112 amino acids is produced by proteolytic cleavage at the C-terminal part of the precursor protein, known as latent TGF-β1. Mature TGF-β1 (active form) is a homodimer of two 12.5 kDa polypeptides linked by a disulfide bond. TGF-β1 and -2 synthesis, secretion, and activation are complex processes that are regulated at various steps for the tissue and cell-specific actions.[20]

Actions of all three isoforms of TGF-β are mediated predominantly through TGFβR-I (55 kDa) and -II (70 kDa) glycoprotein receptors that are ubiquitously expressed on most cell types.[18,19] TGF-β receptors belong to a family of transmembrane protein serine/threonine kinases. Binding of the active form of TGF-β to TGFβR-II results in a homodimeric complex, which then forms heterotetrameric receptor complexes by recruiting TGFβR-I homodimers. The affinity of TGF-β to TGFβR-II is very high and appears to be the initial event in the signal transduction pathways of TGF-β. TGFβR-I is not phosphorylated constitutively, and the formation of TGFβR-I and -II heterotetrameric complex leads to the phosphory-lation and activation of TGFβR-I kinase activity. TGFβR-I kinase phosphorylates Smad proteins, the only known intracellular mediators, of TGF-β actions. Smads are classified as regulatory Smads (R-Smads), coactivator Smads (C-Smads), and inhibitory Smads (I-Smads). Phosphorylated R-Smads (Smad-2, 3) complex with one C-Smad (Smad-4) to form the trimeric complex that translocates to the nucleus where it binds to specific DNA elements for transcriptional activation of the TGF-β responsive genes.[18,19]

TGF-β plays a major role in tissue repair and wound healing by promoting the synthesis of several extracellular matrix proteins such as collagens, fibronec-tin, thrombospondin, integrins, chondroitin sulfate, heparin sulfate proteoglycans, vascular endothelial growth factors, and platelet-derived growth factors.[17,18,21,22] Attempts have been made to use TGF-β for healing wounds in the skin and for the repair of retinal holes. Because of high fibrogenic potential of TGF-β, repeated administration resulted in generalized tissue fibrosis and other problems preclud-ing its use as a human therapeutic agent. Pathological production and activation of TGF-β result in a number of fibrotic disorders, notably affecting kidney, lung, liver, skin, and retina. This has been demonstrated in experimental animal mod-els as well as in human subjects, where progressive and excessive deposition of extracellular matrix was associated with tissue fibrosis. Currently, a number of investigators are working on identifying chemical inhibitors that can target TGF-β receptors and Smad proteins in an effort to block the actions of TGF-β.[23–25] Spe-cific inhibitor of Smad3 (SIS3) was shown to inhibit TGF-β-induced type I colla-gen expression in human dermal fibroblasts.[24] Small molecule screening resulted in identifying a number of TGFβR-I inhibitors (A-83-01, SB-431542, SB-505124) that block the serine/threonine kinase activity of the receptors.[23,25] These inhibi-tors were shown to inhibit TGF-β-induced epithelial–mesenchymal transition known to occur in cancer progression and apoptosis.[23,25] At this time, the clini-cal applications of TGF-β hold promise for treatment of various diseases such as fibrosis and cancer.

16.3.5 CHEMOKINES AND ADHESION MOLECULES

Inflammatory conditions are systemic (e.g., systemic lupus erythematosus) or restricted to target organs or tissue components.[26] The latter is observed in the skin in psoriasis and eczema, intestine in Crohn's disease and ulcerative colitis, eye in uveitis, central nervous system (CNS) in multiple sclerosis (MS), joints in RA, and pancreas in juvenile diabetes. These findings exemplify the critical role of selec-tive tissue infiltration by specific leukocytes. Chemokines are a superfamily of small proteins with a crucial role in immune and inflammatory reactions. Induction of leukocyte migration is the essential function of chemokines and their specific receptors but they also affect angiogenesis, collagen production, and proliferation of hematopoietic precursors. Adhesion molecules are also crucial molecules that facilitate leukocyte attachment to cells. There is a large repertoire of important chemokines/chemokine receptors and adhesion molecules that is beyond the scope of this chapter. A review of the latter can be found in recent publications.[7,26] Table 16.5 identifies three representative chemokines and adhesion molecules.

TABLE 16.5

Representative Chemokines and Adhesion Molecules

Family	Cell Type	Activity
Chemokines		
IL-8/CXCL8	Many cells	Activation of neutrophils and chemotactic activity for all migratory immune cells
RANTES/CCL5	Many cells	Chemotactic for T cells, monocytes, eosinophils, and basophils
CXCL9/10/11	Many cells	Chemotactic for Th1 cells
	IFN inducible	(CXCR3 positive T cells)
Adhesion Molecules		
sICAM-1	Endothelial cells	Adhesion and migration
sVCAM-1	Leukoyctes	
sP-selectin		
sE-selectin	Endothelial cells	Adhesion and migration

16.4 CLINICAL APPLICATIONS OF CYTOKINES

The broad involvement of cytokines and cytokine receptors in the pathogenesis and therapy of disease has made them valuable in the clinical arena.[27] In Table 16.2, some of the current clinical applications of cytokine molecules are outlined. First, the detection of cytokines can be useful as biomarkers of disease and provide clues for mechanisms of disease. Second, the measurement of cytokine production has provided us a useful monitor of immune status. Third, the increasing use of cytokines in therapy represents an exciting new class of therapeutic agents. Fourth, the emergence of cytokine antagonists offers yet another clinical tool for treating a variety of acute and chronic conditions. Anticytokine antibody, antireceptor antibody, soluble cytokine receptors, and receptor antagonists offer promise as selective immunosuppressive agents. Finally, the development of vaccines for infectious diseases and cancer frequently requires adjuvants to potentiate immunity. In a number of model systems, cytokines have been shown to be effective immunological adjuvants.

16.4.1 DETECTION OF CYTOKINES IN DISEASE

Cytokines can be useful as biomarkers of disease and provide clues for mechanisms of disease. However, it should be noted that the presence of cytokines in the circulation represents the host response to the pathologic process and may or may not be specific for the disease. Nevertheless, cytokine measurements in certain acute and chronic inflammatory states have been used as indicators of disease progression and activity. During acute inflammation and graft rejection, the proinflammatory cytokines, IL-1, -2R, -6, TNF-α, and IFN-γ are frequently detected. In this section, three cytokine groups that have been useful in some disease states are reviewed: the IFNs, IL-6, and the chemokine, CXCL10 (IP-10).

16.4.1.1 Detection of Interferons

IFN-α can be noted in sera of patients with active systemic lupus erythematosis (SLE).[28,29] During the past 5 years, there has been a resurgence of interest in this phenomenon since IFN-inducible genes have been shown to be upregulated in peripheral blood leukocytes from SLE patients. Recent studies have identified the inducer molecules and potential pathogenic mechanisms.[30,31] For example, DNA containing immune complexes purified from the serum of patients with SLE stimulate plasmacytoid dendritic cells to produce IFN-α in a TLR-9-dependent manner.[31]

Recently, we showed that the type 1 IFNs were detected in the sera from patients with retinal vasculitis,[32] which is a major component of ocular inflammation and plays a pivotal role in retinal tissue damage in patients with idiopathic uveitis and Behçet's disease. The predominant form of IFN observed was IFN-β that is produced by retinal vascular endothelial cells *in vitro*. Furthermore, the TLR-3 signaling pathway was shown to be operative in IFN-β production in these cells.[32] Further analysis of innate immune signaling may prove to be a novel target for future studies on pathogenic mechanisms and therapeutic approaches in retinal vasculitis.

16.4.1.2 Detection of Interleukin-6

Over the past decade, inflammation has been identified as a critical component of cardiovascular pathology. Furthermore, increasing evidence supports the involvement of inflammation in the various stages of the atherosclerotic process.[33] IL-6, a potent inflammatory cytokine that is produced by several cell types, including activated macrophages, endothelial cells, and smooth muscle cells, has been recognized as a potential marker linked to cardiovascular events.

IL-6 regulates the inflammatory system in a variety of ways, for example, by upregulating the hepatic synthesis of acute-phase reactants such as CRP. Numerous investigations have reported that increased levels of IL-6 and CRP are associated with increased cardiovascular disease, atherosclerosis, and myocardial infarction (MI).[34-36] In addition, enhanced levels of IL-6 in the circulation have been observed in unstable angina, acute MI, and chronic heart failure.[36-38] Moreover, IL-6 has been shown to be an independent risk factor for future MI and is associated with impaired short- and long-term prognosis in patients with acute coronary syndrome and has been found in atherosclerotic plaques.[39-41] Recent studies have revealed that IL-6 levels are associated with subclinical atherosclerotic lesions independent of traditional risk factors.[42] Data generated from a number of studies has identified that atheromatous lesions are active sites of inflammation and immune reactivity, and cytokines, in particular, appear to orchestrate the chronic development of atherosclerosis. IL-6 has been localized in macrophages within human atheroma, demonstrated to stimulate smooth muscle cell proliferation, and shown to accelerate atherosclerosis in murine models; however, the exact role of IL-6 in the human atherosclerotic process remains to be elucidated. During the past several years, it has become increasingly clear that inflammation is a critical component of cardiovascular disease and atherosclerosis. As we continue to decipher the mechanisms underlying these processes, the existing markers used to monitor heart disease may change. Based on the current data, IL-6 appears to be a reasonable marker of cardiovascular disease and perhaps even atherosclerosis. Whether or not IL-6 is eventually included into a panel of cardio-vascular markers remains to be determined. However, its inclusion into this group seems promising.

16.4.1.3 Detection of Chemokines

Immune cell migration into specific tissues is orchestrated by trafficking molecules such as chemokines and adhesion molecules.[26] Recent studies have shown that detection of chemokines in the sera may have prognostic values.[7] Chemokines have been shown to be key components of virus infections. In fact, in severe acute respiratory syndrome (SARS), the serum concentration of CXCL10 detected early after infections was shown to be an independent prognostic indicator of disease outcome.[43] Chemokines and their receptors are clearly important in inflammatory disorders such as atherosclerosis, asthma, RA, and transplantation.[7] In the future, clinical applications involving the measurement of these molecules will surely expand.

16.4.2 Cytokine Production as a Marker of Immune Status

The measurement of cytokine production *in vitro* has provided us a useful monitor of immune status. Incubation of peripheral blood leukocytes with specific antigens results in the activation of T cells that were previously sensitized to that antigen. The T cells proliferate and produce cytokines such as IL-2 and IFN-γ. The measurement of IFN-γ production in response to specific antigen stimulation is frequently used in experimental studies to evaluate immune competence. Recently, the application of this methodology has been introduced into the clinical laboratory. This test has been approved by the FDA to monitor immune reactivity to tuberculosis.[44]

16.4.3 Cytokines as Therapeutic Agents

The best example of the use of cytokines as therapeutic agents is reflected in the clinical applications of the IFNs. During the 1960s and 1970s, several limited clinical trails were performed to evaluate the efficacy of natural human leukocyte IFNs to treat virus infection and malignancy. The data generated provided us with some promising therapeutic options. Then, in the 1980s, recombinant IFN-α became available and proved efficacious for a variety of conditions. The IFNs were shown to be potent antiviral agents immunomodulators antineoplastic agents. In fact, the IFNs are licensed in more than 40 countries for therapeutic indications in infectious, immunologic, and neoplastic diseases (Table 16.4).[45] The U.S. FDA has approved the therapeutic use of the IFNs in hepatitis B and hepatitis C, chronic granulomatous disease (CGD), hairy cell leukemia, Kaposi's sarcoma, condyloma accuminatum (genital warts), and acquired immunodeficiency syndrome (AIDS). The varied diseases, which have responded positively to IFN therapy, are outlined in Table 16.4. These therapeutic indications for the IFNs are based on monotherapy approaches. Combinations of IFNs with chemotherapeutic agents are surely an integral part of the future of IFN therapy. As an example, IFN-α combined with the nucleoside analog ribavirin has proven to be more effective than IFN-α alone in the treatment of chronic hepatitis C.

Recombinant IFN-β is used for the treatment of patients with relapsing MS. Over the past 12 years, the IFN-β treatment of patients with MS has documented clearly the clinical efficacy of this drug.[46,47] The mechanism of action of IFN-β is believed to be through the immunosuppressive activity and stabilization of the blood–brain barrier. Serial magnetic resonance imaging (MRI) scans following IFN-β treatment demonstrated a reduction in gadolinium-enhancing lesions, suggestive of both immunosuppression and stabilization of the blood–brain barrier. Moreover, IFN-β treatment *in vitro* revealed a stabilizing effect on brain endothelial cells.[48,49]

In the 1990s, IFN-γ was approved for the treatment of CGD and severe malignant osteopetrosis. CGD is a rare primary immunodeficiency disorder that is characterized by mutations in genes encoding the nicotinamide adenine dinucleotide phosphate (NADPH) oxidase system leading to defective killing of engulfed pathogens (see Chapter 9 for details). CGD has been successfully treated by the administration of IFN-γ; however, the exact mechanisms leading to improved outcomes are not completely understood.[50]

16.4.4 CYTOKINES AS THERAPEUTIC TARGETS

A major clinical focus of cytokines is the concept that cytokines can be targets of therapeutics. Prime examples of this approach are the inhibitors of TNF-α in RA and Crohn's disease and inhibitors of IL-2 and -15 in transplantation and cancer.[45,51] This section highlights these two strategies that emphasize the therapeutic utility of cytokine receptor antagonists and anticytokine monoclonal antibodies that downregulate pathogenic responses to endogenous cytokine production. Table 16.6 identifies key examples of anticytokine therapies.

Key experimental studies helped to elucidate the role of cytokines in the pathogenic processes observed in RA. In the 1980s, several cytokines including IL-1, TNF-α, IL-6, -2, and GM-CSF were detected in synovial fluid of patients with RA.[52,53] It was then demonstrated that IL-1 mediated joint damage in an animal model and studies indicated that the IL-1 was regulated by TNF-α.[54] *In vitro* studies of cultured human synovial cells from RA patients demonstrated that these cells produced IL-1 and this could be blocked by the treatment of cells with anti-TNF antiserum.[55] These studies and others underscored the concept of cytokine cascades. For example, the production of TNF-α could trigger the production of several cytokines such as IL-1, -6, -8, and GM-CSF. A mouse model of collagen-induced arthritis provided evidence that anti-TNF-α monoclonal antibody treatment decreased disease severity.[56] Furthermore, overexpression of TNF-α in transgenic mice resulted in the onset of arthritis and this could be inhibited by treatment with anti-TNF-α monoclonal antibody.[57] Based on these studies, an open trial followed by a double-blind, randomized, placebo-controlled trial of anti-TNF-α monoclonal antibody treatment in RA were performed and demonstrated a therapeutic response.

Today, there are at least three major TNF-blocking agents licensed for clinical application. Remicade (infliximab) is a chimeric human–mouse monoclonal antibody that is approved for application in RA, Crohn's disease, and ankylosing spondylitis. Enbrel (etanercept) is a soluble TNF p75 receptor-IgG fusion protein that is approved for application in RA, juvenile RA, and psoriatic arthritis. Humira (adalimumab) is a human monoclonal antibody that is approved for application in RA. These novel

TABLE 16.6
Key Examples of Anticytokine Therapies

Cytokine	Drug	Trade Name/Code	Application	FDA Status
IL-1Rα	Recombinant IL-IRα antagonist	Kineret/ Anakinra	RA	Approved—2001
			Sepsis	Amgen
			Osteoarthritis	
IL-2	Antibody to IL-2Rα	Basiliximab/ Simulect	Renal transplantation	Approved—2001
				Novartis
IL-2	Humanized IL-2Rα-chain blocking mAb	Daclizumab/ Zenapax	Renal transplantation	Approved—2002
			Asthma	Hoffman-La Roche
			V(GVHD)	
			Multiple sclerosis	
			HIV	
			Psoriasis	
			Ulcerative uveitis	
TNF-α	Human anti-TNF-α mAb	Adalimumab/ Humira/D2E7	RA	Approved—2002
			Juvenile RA	Abbott Labs
			Ankylosing spondylitis	
			Psoriatic arthritis	
			Crohn's disease	
			Chronic plaque psoriasis	
TNF-α	Humanized anti-TNF-α mAb	Remicade/ Infliximab	RA	Approved—2005
			Crohn's disease	Centocor
			Ulcerative colitis	
			Psoriasis	
			Psoriatic arthritis	
			Ankylosing spondylitis	
			Juvenile idiopathic arthritis	
			Pediatric Crohn's disease	
TNF-α	Soluble p75 TNF receptor-Fc fusion	Etanercept/ Enbrel	RA	Approved—2003
			Juvenile RA	Immunex Corp
			Ankylosing spondylitis	
			Psoriatic arthritis	
IL-5	Humanized anti-IL-5	SCH-55700	Asthma/allergy	Clinical trials— Phase II
				Glaxo/SmithKline

TABLE 16.6 (continued)
Key Examples of Anticytokine Therapies

Cytokine	Drug	Trade Name/Code	Application	FDA Status
IL-5	Humanized anti-IL-5	Mepolizumab	Asthma	Clinical trials
			Hypereosinophilic syndrome	Glaxo/SmithKline
			Atopic dermatitis	
IL-6	Humanized anti-IL-6 receptor	MRA (Tocilizumab)	Crohn's disease	Clinical trials
			SLE	Phase I and -II
			RA	Chigai
			Myeloma	
			Systemic onset juvenile idiopathic arthritis	
IL-12	Human anti-IL 12	ABT874/J695	Crohn's disease	Clinical trials
			Multiple sclerosis	Abbott
IL-13	Human anti-IL-13	CAT 354	Asthma	Pending
IL-15	Humanized anti-IL-15	Humax IL-15/AMG-714	RA	Clinical trials
			Psoriasis	Amgen

treatment strategies have dramatically improved the prognosis of RA patients. Nevertheless, not all RA patients respond to this therapy. Promising new avenues of research have come forth as novel therapies for RA. Limiting IL-6 activity through monoclonal antibody blockage of IL-6 receptors is one such potential candidate.[58]

IL-2 and -15 and their receptors as therapeutic targets in transplantation and cancer provide another example of cytokine-based treatments. Thomas Waldmann of the NIH has been on the forefront of many of these experimental and clinical studies.[51,59,60] IL-2 and -15 share common receptor components and signaling pathways and both stimulate T-cell proliferation and the generation of cytotoxic T lymphocytes. The alpha chain of the IL-2 receptor (IL-2Rα, CD25) has been shown to be an effective target for immunotherapy. IL-2Rα is constitutively expressed in malignant T and B cells and by T cells involved in autoimmune diseases or in transplant rejection. Anti-IL-2Rα antibody treatment blocks the interaction between IL-2 and its receptor resulting in the death of IL-2-dependent T cells. A humanized antibody to IL-2Rα (Daclizumab, Zenapax) and chimeric antibody directed toward IL-2Rα (basiliximab) were shown to be efficacious in renal allograft recipients and have received FDA approval.[60–63] Moreover, daclizumab is useful for treatment of T-cell mediated autoimmune diseases. This therapy was effective for patients with noninfectious uveitis and in patients with MS.[64] Phase II/III clinical trials are ongoing in the latter conditions.

Studies targeting IL-15 and its receptor are in their infancy when compared to the promising record of IL-2 blockage. IL-15 is a proinflammatory cytokine that plays a key role in T-cell responses to pathogens by supporting CD8 memory T-cell survival. Blocking IL-15 or its receptor in animal models have shown efficacy in mouse models of collagen-induced arthritis, allograft rejection, and psoriasis. Waldmann's group has focused on a humanized antibody (Mikβ1) specific for IL-2/IL-15Rβ. Clinical trials are being initiated in patients with MS, RA, celiac disease, and HTLV-1-associated tropical spastic paraparesis.[51]

16.4.5 CYTOKINES IN VACCINES

Exciting advances introduced by molecular biology and peptide chemistry have led to the generation of a variety of recombinant proteins and synthetic peptides, which have revolutionized vaccine research. Unfortunately, these new candidate vaccines are often weak immunogens that require the addition of adjuvants to potentiate immune reactivity. The latter need has led to the search for new adjuvants, which can enhance vaccine effectiveness. In addition to adjuvants that induce a broad range of cytokines, cytokines themselves are now being tested as he adjuvants in vaccine formulations. Several cytokines have been shown to be effective adjuvants in a number of model systems, enhancing protection induced by viral, bacterial, and parasitic vaccines as well as stimulating immunity in tumor models. (Refer to more thorough reviews of cytokine adjuvants.)[65–67]

16.5 CONCLUDING REMARKS

Dramatic advances in the understanding of cytokine biology have paralleled the development of diagnostic, prognostic, and therapeutic applications. The administration of IFNs have clearly demonstrated the efficacy of cytokines as therapeutic agents. Numerous clinical trials are in progress to evaluate efficacies for a variety of cytokines, including IL-2, -4, -6, -10, -12, and -21.[27] Likewise, the inhibition of TNF-α clearly underscores strategies aimed at blocking cytokines. Numerous clinical trails involving the blockade of IL-1, -2, -4, -5, -6, -8, -9, -12, -13, -18, and TGF-β1 and -β2 are currently underway. Cytokines have emerged as regulators of the immune response and will continue to be key targets for new therapies. Continuing investigation of cytokines, their complex regulatory networks and their antagonists will take us toward the goal of manipulating these molecules to optimize their beneficial effects, while mitigating their deleterious effects.

REFERENCES

1. Meager, A. *The Interferons: Characterization and Application*, Wiley, Weinheim, Germany, 2006.
2. Santamaria, P. *Cytokines and Chemokines in Autoimmune Disease*, Kluwer Academic/Plenum Publishers, New York, 2003.
3. Thomson, A. and Lotze, M. *The Cytokine Handbook*, Academic Press, Amsterdam, 2003.
4. Oppenheim, J., Feldmann, M. and Durum, S. *Cytokine Reference: A Compendium of Cytokines and Other Mediators of Host Defense*, Academic Press, San Diego, CA, 2001.
5. Detrick, B., Hamilton, R.G. and Folds, J.D. *Manual of Molecular and Clinical Laboratory Immunology*, ASM Press, Washington, 2006.

6. Remick, D. Multiplex cytokine assays. In *Manual of Molecular and Clinical Laboratory Immunology* (eds. Detrick, B., Hamilton, R.G., and Folds, J.D.), ASM Press, Washington, 2006, pp. 340–352.

7. Medoff, B. and Luster, A. Chemokine and chemokine receptor analysis. In *Manual of Molecular and Clinical Laboratory Immunology* (eds. Detrick, B., Hamilton, R.G., and Folds, J.D.), ASM Press, Washington, 2006, pp. 371–384.

8. Sadowski, H.B., Shuai, K., Darnell, J.E., Jr. and Gilman, M.Z. A common nuclear signal transduction pathway activated by growth factor and cytokine receptors. *Science* 261, 1739–1744, 1993.

9. Paul, W.E. and Seder, R.A. Lymphocyte responses and cytokines. *Cell* 76, 241–251, 1994.

10. Isaacs, A. and Lindenmann, J. Virus interference. I. The interferon. *Proc R Soc Lond B Biol Sci* 147, 258–267, 1957.

11. Pestka, S., Krause, C.D. and Walter, M.R. Interferons, interferon-like cytokines, and their receptors. *Immunol Rev* 202, 8–32, 2004.

12. Levy, D.E. and Darnell, J.E., Jr. Stats: transcriptional control and biological impact. *Nat Rev Mol Cell Biol* 3, 651–662, 2002.

13. Cerami, A. Inflammatory cytokines. *Clin Immunol Immunopathol* 62, S3–S10, 1992.

14. Carswell, E.A., Old, L.J., Kassel, R.L., Green, S., Fiore, N. and Williamson, B. An endotoxin-induced serum factor that causes necrosis of tumors. *Proc Natl Acad Sci USA* 72, 3666–3670, 1975.

15. Williams, T.W. and Granger, G.A. Lymphocyte *in vitro* cytotoxicity: lymphotoxins of several mammalian species. *Nature* 219, 1076–1077, 1968.

16. Ruddle, N.H. and Waksman, B.H. Cytotoxicity mediated by soluble antigen and lymphocytes in delayed hypersensitivity. I. Characterization of the phenomenon. *J Exp Med* 128, 1237–1254, 1968.

17. Border, W.A. and Noble, N.A. Transforming growth factor beta in tissue fibrosis. *New Engl J Med* 331, 1286–1292, 1994.

18. Massague, J. TGF-beta signal transduction. *Annu Rev Biochem* 67, 753–791, 1998.

19. Feng, X.H. and Derynck, R. Specificity and versatility in TGF-beta signaling through Smads. *Annu Rev Cell Dev Biol* 21, 659–693, 2005.

20. Nagineni, C.N., Cherukuri, K.S., Kutty, V., Detrick, B. and Hooks, J.J. Interferon-gamma differentially regulates TGF-beta1 and TGF-beta2 expression in human retinal pigment epithelial cells through JAK-STAT pathway. *J Cell Phys* 210, 192–200, 2007.

21. Nagineni, C.N., Kutty, V., Detrick, B. and Hooks, J.J. Expression of PDGF and their receptors in human retinal pigment epithelial cells and fibroblasts: regulation by TGF-beta. *J Cell Phys* 203, 35–43, 2005.

22. Nagineni, C.N., Samuel, W., Nagineni, S., Pardhasaradhi, K., Wiggert, B., Detrick, B. and Hooks, J.J. Transforming growth factor-beta induces expression of vascular endothelial growth factor in human retinal pigment epithelial cells: involvement of mitogen-activated protein kinases. *J Cell Phys* 197, 453–462, 2003.

23. Tojo, M., Hamashima, Y., Hanyu, A., Kajimoto, T., Saitoh, M., Miyazono, K., Node, M. and Imamura, T. The ALK-5 inhibitor A-83-01 inhibits Smad signaling and epithelial-to-mesenchymal transition by transforming growth factor-beta. *Cancer Sci* 96, 791–800, 2005.

24. Jinnin, M., Ihn, H. and Tamaki, K. Characterization of SIS3, a novel specific inhibitor of Smad3, and its effect on transforming growth factor-beta1-induced extracellular matrix expression. *Mol Pharmacol* 69, 597–607, 2006.

25. DaCosta Byfield, S., Major, C., Laping, N.J. and Roberts, A.B. SB-505124 is a selective inhibitor of transforming growth factor-beta type I receptors ALK4, ALK5, and ALK7. *Mol Pharmacol* 65, 744–752, 2004.

26. Luster, A.D., Alon, R. and von Andrian, U.H. Immune cell migration in inflammation: present and future therapeutic targets. *Nat Immunol* 6, 1182–1190, 2005.

27. Cutler, A. and Brombacher, F. Cytokine therapy. *Ann N Y Acad Sci* 1056, 16–29, 2005.
28. Hooks, J.J., Moutsopoulos, H.M., Geis, S.A., Stahl, N.I., Decker, J.L. and Notkins, A.L. Immune interferon in the circulation of patients with autoimmune disease. *New Engl J Med* 301, 5–8, 1979.
29. Dall'era, M.C., Cardarelli, P.M., Preston, B.T., Witte, A. and Davis, J.C., Jr. Type I interferon correlates with serological and clinical manifestations of SLE. *Ann Rheum Dis* 64, 1692–1697, 2005.
30. Banchereau, J. and Pascual, V. Type I interferon in systemic lupus erythematosus and other autoimmune diseases. *Immunity* 25, 383–392, 2006.
31. Means, T.K. and Luster, A.D. Toll-like receptor activation in the pathogenesis of systemic lupus erythematosus. *Ann N Y Acad Sc* 1062, 242–251, 2005.
32. Lee, M.T., Hooper, L.C., Kump, L., Hayashi, K., Nussenblatt, R., Hooks, J.J. and Detrick, B. Interferon-beta and adhesion molecules (E-selectin and s-intracellular adhesion molecule-1) are detected in sera from patients with retinal vasculitis and are induced in retinal vascular endothelial cells by Toll-like receptor 3 signalling. *Clin Exp Immunol* 147, 71–80, 2007.
33. Libby, P. Inflammation in atherosclerosis. *Nature* 420, 868–874, 2002.
34. Erren, M., Reinecke, H., Junker, R., Fobker, M., Schulte, H., Schurek, J.O., Kropf, J., Kerber, S., Breithardt, G., Assmann, G. and Cullen, P. Systemic inflammatory parameters in patients with atherosclerosis of the coronary and peripheral arteries. *Arterioscler, Thromb Vasc Biol* 19, 2355–2363, 1999.
35. Volpato, S., Guralnik, J.M., Ferrucci, L., Balfour, J., Chaves, P., Fried, L.P. and Harris, T.B. Cardiovascular disease, interleukin-6, and risk of mortality in older women: the womens health and aging study. *Circulation* 103, 947–953, 2001.
36. Ridker, P.M., Rifai, N., Stampfer, M.J. and Hennekens, C.H. Plasma concentration of interleukin-6 and the risk of future myocardial infarction among apparently healthy men. *Circulation* 101, 1767–1772, 2000.
37. Biasucci, L.M., Vitelli, A., Liuzzo, G., Altamura, S., Caligiuri, G., Monaco, C., Rebuzzi, A.G., Ciliberto, G. and Maseri, A. Elevated levels of interleukin-6 in unstable angina. *Circulation* 94, 874–877, 1996.
38. McCarthy, M.J., Loftus, I.M., Thompson, M.M., Jones, L., London, N.J., Bell, P.R., Naylor, A.R. and Brindle, N.P. Angiogenesis and the atherosclerotic carotid plaque: an association between symptomatology and plaque morphology. *J Vasc Surg* 30, 261–268, 1999.
39. Fisman, E.Z., Benderly, M., Esper, R.J., Behar, S., Boyko, V., Adler, Y., Tanne, D., Matas, Z. and Tenenbaum, A. Interleukin-6 and the risk of future cardiovascular events in patients with angina pectoris and/or healed myocardial infarction. *The Am J Cardiol* 98, 14–18, 2006.
40. Lindahl, B., Toss, H., Siegbahn, A., Venge, P. and Wallentin, L. Markers of myocardial damage and inflammation in relation to long-term mortality in unstable coronary artery disease. FRISC Study Group. Fragmin during Instability in Coronary Artery Disease. *New Engl J Med* 343, 1139–1147, 2000.
41. Rus, H.G., Vlaicu, R. and Niculescu, F. Interleukin-6 and interleukin-8 protein and gene expression in human arterial atherosclerotic wall. *Atherosclerosis* 127, 263–271, 1996.
42. Amar, J., Fauvel, J., Drouet, L., Ruidavets, J.B., Perret, B., Chamontin, B., Boccalon, H. and Ferrieres, J. Interleukin 6 is associated with subclinical atherosclerosis: a link with soluble intercellular adhesion molecule 1. *J Hypertens* 24, 1083–1088, 2006.
43. Tang, N.L., Chan, P.K., Wong, C.K., To, K.F., Wu, A.K., Sung, Y.M., Hui, D.S., Sung, J.J. and Lam, C.W. Early enhanced expression of interferon-inducible protein-10 (CXCL-10) and other chemokines predicts adverse outcome in severe acute respiratory syndrome. *Clin Chem* 51, 2333–2340, 2005.

44. Ferrara, G., Losi, M., D'Amico, R., Roversi, P., Piro, R., Meacci, M., Meccugni, B., Dori, I.M., Andreani, A., Bergamini, B.M., Mussini, C., Rumpianesi, F., Fabbri, L.M. and Richeldi, L. Use in routine clinical practice of two commercial blood tests for diagnosis of infection with *Mycobacterium tuberculosis*: a prospective study. *Lancet* 367, 1328–1334, 2006.

45. Vilcek, J. and Feldmann, M. Historical review: Cytokines as therapeutics and targets of therapeutics. *Trends Pharmacol Sci* 25, 201–209, 2004.

46. Jacobs, L.D., Cookfair, D.L., Rudick, R.A., Herndon, R.M., Richert, J.R., Salazar, A.M., Fischer, J.S., Goodkin, D.E., Granger, C.V., Simon, J.H., Alam, J.J., Bartoszak, D.M., Bourdette, D.N., Braiman, J., Brownscheidle, C.M., Coats, M.E., Cohan, S.L., Dougherty, D.S., Kinkel, R.P., Mass, M.K., Munschauer, F.E. 3rd., Priore., R.L., Pullicino, P.M., Scherokman, B.J. and Whitham, R.H. Intramuscular interferon beta-1a for disease progression in relapsing multiple sclerosis. The Multiple Sclerosis Collaborative Research Group (MSCRG). *Ann Neurol* 39, 285–294, 1996.

47. Panitch, H., Goodin, D.S., Francis, G., Chang, P., Coyle, P.K., O'Connor, P., Monaghan, E., Li, D. and Weinshenker, B. EVIDENCE Study Group. Evidence of Interferon Dose-response: Europian North American Comparative Efficacy; University of British Columbia MS/MRI Research Group. Randomized, comparative study of interferon beta-1a treatment regimens in MS: The Evidence Trial. *Neurology* 59, 1496–1506, 2002.

48. Pozzilli, C., Bastianello, S., Koudriavtseva, T., Gasperini, C., Bozzao, A., Millefiorini, E., Galgani, S., Buttinelli, C., Perciaccante, G., Piazza, G., Bozzao, L. and Fieschi, C. Magnetic resonance imaging changes with recombinant human interferon-beta-1a: a short term study in relapsing-remitting multiple sclerosis. *J Neurol, Neurosurg, Psychiatr* 61, 251–258, 1996.

49. Kraus, J., Ling, A.K., Hamm, S., Voigt, K., Oschmann, P. and Engelhardt, B. Interferon-beta stabilizes barrier characteristics of brain endothelial cells *in vitro*. *Ann Neurol* 56, 192–205, 2004.

50. Marciano, B.E., Wesley, R., DeCarlo, E.S., Anderson, V.L., Barnhart, L.A., Darnell, D., Malech, H.L., Gallin, J.I. and Holland, S.M. Long-term interferon-gamma therapy for patients with chronic granulomatous disease. *Clin Infect Dis* 39, 692–699, 2004.

51. Waldmann, T.A. The biology of interleukin-2 and interleukin-15: implications for cancer therapy and vaccine design. *Nat Rev Immunol* 6, 595–601, 2006.

52. Fontana, A., Hengartner, H., Weber, E., Fehr, K., Grob, P.I. and Cohen, G. Interleukin 1 activity in the synovial fluid of patients with rheumatoid arthritis. *Rheum Int* 2, 49–53, 1982.

53. Buchan, G., Barrett, K., Turner, M., Chantry, D., Maini, R.N. and Feldmann, M. Interleukin-1 and tumour necrosis factor mRNA expression in rheumatoid arthritis: prolonged production of IL-1 alpha. *Clin Exp Immunol* 73, 449–455, 1988.

54. Chantry, D., DeMaggio, A.J., Brammer, H., Raport, C.J., Wood, C.L., Schweickart, V.L., Epp, A., Smith, A., Stine, J.T., Walton, K., Tjoelker, L., Godiska, R. and Gray, P.W. Profile of human macrophage transcripts: insights into macrophage biology and identification of novel chemokines. *J Leukocyte Biol* 64, 49–54, 1998.

55. Brennan, F.M., Chantry, D., Jackson, A., Maini, R. and Feldmann, M. Inhibitory effect of TNF alpha antibodies on synovial cell interleukin-1 production in rheumatoid arthritis. *Lancet* 2, 244–247, 1989.

56. Williams, R.O., Feldmann, M. and Maini, R.N. Anti-tumor necrosis factor ameliorates joint disease in murine collagen-induced arthritis. *Proc Natl Acad Sci USA* 89, 9784–9788, 1992.

57. Keffer, J., Probert, L., Cazlaris, H., Georgopoulos, S., Kaslaris, E., Kioussis, D. and Kollias, G. Transgenic mice expressing human tumour necrosis factor: a predictive genetic model of arthritis. *EMBO J* 10, 4025–4031, 1991.

58. Smolen, J.S. Expanding the frontiers of therapy. *Arthritis Res Ther* 8(Suppl 2), S1, 2006.

59. Waldmann, T.A. Immunotherapy: past, present and future. *Nat Med* 9, 269–277, 2003.

60. Waldmann, T.A. Anti-Tac (daclizumab, Zenapax) in the treatment of leukemia, autoimmune diseases, and in the prevention of allograft rejection: a 25-year personal odyssey. *J Clin Immunol*, 2007.

61. Nashan, B., Moore, R., Amlot, P., Schmidt, A.G., Abeywickrama, K. and Soulillou, J.P. Randomised trial of basiliximab versus placebo for control of acute cellular rejection in renal allograft recipients. CHIB 201 International Study Group. *Lancet* 350, 1193–1198, 1997.

62. Kirkman, R.L., Shapiro, M.E., Carpenter, C.B., McKay, D.B., Milford, E.L., Ramos, E.L., Tilney, N.L., Waldmann, T.A., Zimmerman, C.E. and Strom, T.B. A randomized prospective trial of anti-Tac monoclonal antibody in human renal transplantation. *Transplant Proc* 23, 1066–1067, 1991.

63. Vincenti, F., Kirkman, R., Light, S., Bumgardner, G., Pescovitz, M., Halloran, P., Neylan, J., Wilkinson, A., Ekberg, H., Gaston, R., Backman, L. and Burdick, J. Interleukin-2-receptor blockade with daclizumab to prevent acute rejection in renal transplantation. Daclizumab Triple Therapy Study Group. *The New England Journal of Medicine* 338, 161–165, 1998.

64. Nussenblatt, R.B., Fortin, E., Schiffman, R., Rizzo, L., Smith, J., Van Veldhuisen, P., Sran, P., Yaffe, A., Goldman, C.K., Waldmann, T.A. and Whitcup, S.M. Treatment of noninfectious intermediate and posterior uveitis with the humanized anti-Tac mAb: a phase I/II clinical trial. *Proc Natl Acad Sci USA* 96, 7462–7466, 1999.

65. Gonzalez, G., Crombet, T., Neninger, E., Viada, C. and Lage, A. Therapeutic vaccination with epidermal growth factor (EGF) in advanced lung cancer: analysis of pooled data from three clinical trials. *Hum Vaccin* 3, 8–13, 2007.

66. O'Hagan, D.T., MacKichan, M.L. and Singh, M. Recent developments in adjuvants for vaccines against infectious diseases. *Biomol Eng* 18, 69–85, 2001.

67. Lori, F., Weiner, D.B., Calarota, S.A., Kelly, L.M. and Lisziewicz, J. Cytokine-adjuvanted HIV-DNA vaccination strategies. *Springer Semin Immunopathol* 28, 231–238, 2006.

17 Measuring Human Cytokines

Holden T. Maecker

CONTENTS

17.1 INTRODUCTION

Cytokine measurement in clinical samples is an area of considerable and growing interest because of the importance of cytokines in mediating immunologic diseases (see chapter 16). Although most assays measuring cytokines are used in research settings rather than in clinical diagnosis or monitoring, this trend will likely change in the coming years. Already, clinical trials of vaccines for HIV and other diseases are using cytokine-based assays to measure the immunogenicity of candidate vaccines and to search for potential surrogate markers of protection from disease.[1] Licensed clinical assays measuring the cytokine interferon-gamma (IFN-γ) are currently available for diagnosis of tuberculosis (TB),[2] and other such clinical assays will likely follow. Here the major types of cytokine assays are discussed and compared, and sample data and reference ranges are provided for researchers using these assays on clinical samples.

Historically, three major methods were used for measuring cytokines: molecular assays, immunoassays, and bioassays.[3] Molecular assays measure cytokine messenger ribonucleic acid (mRNA), whereas immunoassays measure cytokine protein. Bioassays detect the actual bioactivity of the cytokine protein using a surrogate readout, such as proliferation, from cytokine-sensitive target cells.[4] In recent years, a new class of single-cell assays has emerged, which measure the number of cells that can produce a given cytokine, usually in response to *in vitro* stimulation with antigen or mitogen.[5]

Because they are time-consuming and difficult to standardize, bioassays have largely been replaced by molecular and immunoassay systems.[3,5] Molecular assays, in turn, are used almost exclusively in research settings and can be limited in their quantitative ability.[6] It should also be noted that positivity by a molecular assay might not strictly correlate with positivity by a protein immunoassay since cytokines can be regulated at a posttranscriptional level.[3] Finally, all three of these traditional assay types measure cytokine (or cytokine mRNA) from a bulk population of cells. Single-cell assays, by contrast, offer a new type of information by identifying and characterizing potentially rare subpopulations of cells that make particular cytokines.[5] The ability to simultaneously quantitate and phenotype specific cytokine-secreting cell populations make single-cell assays among the most popular and rapidly evolving ways of measuring cytokines in research and clinical trials.[7]

The most commonly accessed clinical specimen for measurement of cytokines is blood. Secreted cytokines can be measured in the serum, whereas blood leukocytes can be assessed for their ability to produce cytokines using cell-based assays. However, cytokines are most often secreted in specific lymphoid compartments or tissues rather than in blood.[3] Thus, informative assays of secreted cytokines can be done, for example, in cerebrospinal fluid,[8] tears,[9] or feces.[10] Cell-based cytokine assays can be performed on biopsies of gut or other mucosal tissues,[11] infected organs,[12] tumors,[13] etc. In many cases, measurement of cytokines in blood might be taken as a surrogate for processes occurring in regional compartments. However, direct access to those compartments usually results in a more definitive result in terms of a higher concentration of cytokines or of antigen-responsive cells.[12]

17.2 MOLECULAR ASSAYS

The traditional method for quantitating mRNA is the Northern blot. However, this technique is not in common use for measuring cytokine mRNAs, partially because of their low abundance[14] and also because of the desire to combine measurements of multiple markers in a single sample. For these reasons, newer techniques such as quantitative polymerase chain reaction (qPCR),[15,16] ribonucleic acid digesting enzyme (RNAse) protection assays (RPA),[17] and complementary deoxyribonucleic acid (cDNA) microarrays[18] have become more popular methods of detecting cytokine mRNAs. qPCR can be used to accurately detect minute quantities of mRNA in a specific and reproducible manner. RPA, although not as sensitive as qPCR, allows for semiquantitative detection of multiple mRNA species at once using probes of varying length. Finally, oligonucleotide or cDNA arrays can be used to detect a very large number of mRNA species in a single sample, but they are at best semiquantitative and can be subject to problems of specificity and sensitivity. A final alternative is *in situ* hybridization,[13,14] which is attractive in that it can show localized cytokine gene expression in tissue sections allowing possible identification of the cell types producing the cytokine and giving potential insight into pathological processes.[13] However, it is limited as a clinical laboratory tool because of its complexity. The techniques of qPCR, RPA, and cDNA arrays are discussed in more detail in Section 17.2.1.

17.2.1 QUANTITATIVE POLYMERASE CHAIN REACTION

PCR is an amplification technique that uses specific oligonucleotide primers to repeatedly amplify a sequence of interest from cDNA or genomic DNA to the point where that sequence can be visualized on an electrophoretic gel. qPCR, also known as real-time PCR, makes this technique quantitative by following the amplification process kinetically[15,16] (Figure 17.1). This process is done with a label that fluoresces upon binding to the amplified DNA. The amount of fluorescence is detected with each cycle of amplification avoiding the need for postamplification sample processing.

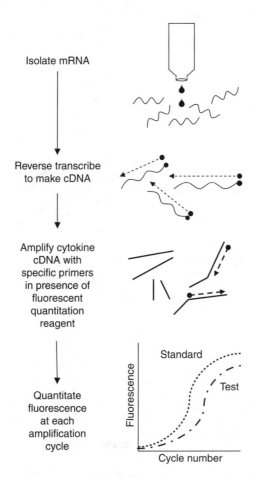

Isolate mRNA

Reverse transcribe to make cDNA

Amplify cytokine cDNA with specific primers in presence of fluorescent quantitation reagent

Quantitate fluorescence at each amplification cycle

Fluorescence

Standard

Test

Cycle number

FIGURE 17.1   Schematic representation of qPCR. (Figure courtesy of Laurel Nomura, BD Biosciences.)

With proper optimization and calibration, qPCR can be highly sensitive and quantitative.[19,20] Its main disadvantages include the relatively high cost of instrumentation and the technical skill required for assay setup. Ribonucleic acid (RNA) is highly labile and must be handled carefully to avoid degradation as well as contamination from genomic DNA. Without skilled handling and accurate pipetting, the assay can be quite variable. Also, the assay format does not easily lend itself to multiplexing. Nevertheless, it remains the most sensitive of the molecular methods for detecting cytokine mRNA.[14]

17.2.2 RNAse Protection Assays

RPA is similar in concept to a Northern blot in that it directly detects specific gene products separated by electrophoresis[17] (Figure 17.2). In contrast to Northern blotting, however, the labeled cDNA probe(s)[21] are annealed to the mRNA sample in solution rather than on a membrane blot. A single-stranded exonuclease (RNAse H) is used to digest away the nonhybridized mRNA as well as the single-stranded ends of

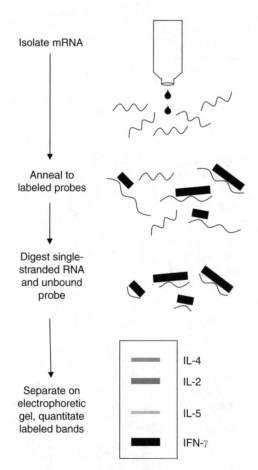

FIGURE 17.2 Schematic representation of RPA. (Figure courtesy of Laurel Nomura, BD Biosciences.)

the hybridized molecules. The remaining double-stranded material (probe + target mRNA) is then separated by electrophoresis resulting in labeled bands whose size corresponds to the size of the hybridized probe, and whose intensity is related to the amount of that mRNA species, which was available to bind the probe. By using probes of multiple lengths, the assay can detect a number of different target mRNAs at once. For example, commercial kits are available to detect a set of human cytokines secreted by T helper subset 1 (Th1) and T helper subset 2 (Th2) cells or a set of chemokines and related inflammatory cytokines.[22]

The main advantage of RPA is providing multiplex readouts for several cytokines while still maintaining reasonable quantitation of each.[17] The disadvantage is the relative complexity, including use of radiolabeled or chemiluminescent probes. The same considerations surrounding handling of mRNA in qPCR also apply to RPA. However, RPA is less sensitive than qPCR because it is not an amplification technique; and it is quantitative only up to the level of saturation of the detector. Still, it can be used to detect a small set of cytokine mRNAs in a single sample.[22]

17.2.3 cDNA MICROARRAYS

cDNA microarrays commonly called "gene arrays" consist of a set of nucleic acid probes (cDNA or oligonucleotides thereof) immobilized on a solid support[18] (Figure 17.3). Arrays that correspond to a set of genes of interest can be selected, for example, cytokine genes.[23] An mRNA sample is reverse transcribed to make labeled cDNA, which is then allowed to hybridize to the array. Detection is based on the degree of binding of labeled cDNA to the immobilized probes and can be done via fluorescence, chemiluminescence, or radioactivity.

Because hundreds or even thousands of different genes (or oligonucleotides) can be spotted onto a single array, these assays are highly multiplexed. However, they are not highly quantitative in that there are generally no controls by which to calibrate the intensity of spots for each target.[6] They can also suffer from specificity problems, especially when many targets are multiplexed in a single array. The specificity of these assays is only as good as that of the least specific probe in the array. Finally, like RPA, they require careful handling of the mRNA samples to avoid degradation. They are most useful as screening assays to get hints about what cytokines might be involved in a particular disease process. Follow-up measurement of candidate cytokines by other assays should then be done.

17.3 PROTEIN IMMUNOASSAYS

Protein immunoassays are called immunoassays because they rely on the specificity of immunoglobulins for detection of an analyte, these assays are workhorses of the immunology laboratory. The most common type of immunoassay is the enzyme-linked immunosorbent assay (ELISA), also known simply as enzyme immunoassay (EIA).[24] This immunoassay is a nonradioactive version of the older radioimmunoassay (RIA). It is generally performed in 96-well microtiter plates, which are coated with an antigen or antibody. Enzyme-labeled detector antibodies are used to measure the presence of captured analyte. Recently, multiplexed immunoassays have been

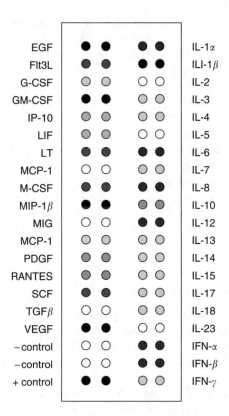

EGF	● ● ● ●	IL-1α
Flt3L	● ● ● ●	ILI-1β
G-CSF	○ ○ ○ ○	IL-2
GM-CSF	● ● ○ ○	IL-3
IP-10	○ ○ ○ ○	IL-4
LIF	○ ○ ○ ○	IL-5
LT	● ● ● ●	IL-6
MCP-1	○ ○ ○ ○	IL-7
M-CSF	● ● ● ●	IL-8
MIP-1β	● ● ○ ○	IL-10
MIG	○ ○ ● ●	IL-12
MCP-1	○ ○ ○ ○	IL-13
PDGF	○ ○ ○ ○	IL-14
RANTES	○ ○ ○ ○	IL-15
SCF	● ● ○ ○	IL-17
TGFβ	○ ○ ○ ○	IL-18
VEGF	● ● ○ ○	IL-23
−control	○ ○ ● ●	IFN-α
−control	○ ○ ● ●	IFN-β
+ control	● ● ○ ○	IFN-γ

FIGURE 17.3 Microarray technology. In a cDNA or oligonucleotide microarray, cDNAs or oligonucleotides are immobilized on a solid support such as a nitrocellulose filter. Hybridization of labeled cDNA is detected via radioactive, chemiluminescent, or fluorescent means. In an antibody microarray, monoclonal antibodies are spotted and detected by enzymatically labeled detector antibodies. Probes are usually spotted in duplicate as shown. (Figure courtesy of Laurel Nomura, BD Biosciences.)

developed using sets of beads that vary in size and fluorescence.[9,25–34] The beads are resolved by flow cytometry with one fluorescence channel used to quantitate the presence of bound analytes. Although these assays cannot be multiplexed to as high a degree as microarrays, they can accurately measure up to a dozen or more cytokines in a single sample of small volume.[9,25–34]

Recently, antibody arrays have been added to the list of immunoassays available for detection of cytokines.[35–37] Analogous to nucleic acid microarrays, these assays use large sets of antibodies spotted onto a solid support in order to achieve highly multiplexed detection of proteins. Although the specificity of monoclonal antibodies is usually well established, the antibody arrays still suffer from being less quantitative than the ELISA or multiplex bead assays; but they provide a convenient method to screen for the presence of many cytokines or other proteins.

In general, immunoassays do not necessarily detect bioactivity of a cytokine in contrast to the more traditional bioassays.[3–5] However, this drawback is rarely a concern in most biological samples. However, immunoassays do detect the presence of a cytokine protein directly as opposed to molecular assays, which only detect the cytokine mRNA.

Most commercially available immunoassays are robust and precise, with intra- and inter-assay coefficients of variation (CV) of less than 20% when used by experienced investigators.[26,36] However, optimization of a new immunoassay to achieve this level of precision is not always straightforward. The choice of antibody pair used for capture and detection is critical.[32] Incubation time for analyte, detector, and substrate needs to be optimized. Artifacts can occur including blocking factors present in serum as well as nonspecific binding factors in serum which can increase the background for particular cytokines.[3,5,25,29,31] Such factors can sometimes be mitigated not only by optimal dilution of the serum but also by the choice of buffer used for dilution.

The concentrations of cytokines in normal serum are extremely low (Table 17.1). This is because cytokines are mostly secreted locally between adjacent cells or in a specific microenvironment such as a lymph node.[3] Measurement of cytokines in serum can, however, detect their gross overproduction as a result of a diseased state.[38,39]

In some cases, testing the capacity of leukocytes to produce cytokines *in vitro* is desirable. In such cases, an immunoassay can be performed on a cell culture supernatant after several days of stimulation with mitogen, antigen, or allogeneic cells.[3] However, such assays are limited for several reasons. First, as bulk assays, they do not determine the cell type responsible for secreting the cytokines detected. Second, the level of cytokine produced is highly dependent on the starting cell concentration as well as length of culture and specific culture conditions. Small perturbations of the latter can greatly influence the results after a few days of stimulation. Finally, there is a dynamic balance between cytokine production and degradation or reuptake of cytokines during *in vitro* culture. As such, the optimum time of detection will vary

TABLE 17.1
Normal Ranges of Cytokines in Human Serum

	IL-1β[a]	IL-2[b]	IL-4[b]	IL-5[b]	IL-6[a]	IL-8[a]	IL-10[a]	IL-10[a]	IL-12[a]	IFN-γ[b]	TNF-α[b]	TNF-α[a]
5th percentile	3.6	0.0	0.0	0.0	3.4	9.5	0.0	0.0	1.7	0.0	0.0	2.5
Median	12.2	0.0	0.0	2.1	5.2	21.4	0.0	4.6	6.9	0.0	5.6	8.3
95th percentile	25.5	5.0	13.1	6.5	8.9	93.2	10.8	7.1	17.1	38.7	12.2	15.6

Note: Values are listed as pg/mL and were derived using a multiplex bead assay system from a set of 11 healthy adult donors. In some cases, the same cytokine was analyzed by two different kits and values for each are given. All values were derived using the kits' "serum buffer" to decrease nonspecific signals from serum-containing samples. Data courtesy of Roy Chen and Maria Jaimes, BD Biosciences.

[a] Data from BD CBA kit, inflammation panel (BD Biosciences).
[b] Data from BD CBA kit, Th1/Th2 panel (BD Biosciences).

by cytokine and possibly even by donor. All in all, such *in vitro* stimulation assays provide only qualitative hints about the immunocompetence of a particular donor's leukocytes or about any pathological processes that may be occurring.

Features of the ELISA, multiplex bead assays, and antibody arrays are described in more detail in Section 17.3.1.

17.3.1 Enzyme-Linked Immunosorbent Assay

In the so-called "sandwich ELISA," microtiter wells are coated with a capture antibody specific for a particular analyte (e.g., cytokine) (Figure 17.4). The wells are then incubated with serial dilutions of the sample containing the analyte (e.g., serum or cell culture supernatant). A second, nonblocking antibody specific for that analyte is used as a detector reagent (this second antibody being labeled with an enzyme such as horseradish peroxidase [HRP] or alkaline phosphatase [AP]). The amount of bound detector antibody is visualized by application of a substrate solution that creates a colored product in the presence of the respective enzyme with the amount of color being proportional to the amount of analyte. An alternative detection system uses electrochemiluminescence (ECL), whereby labeled antibodies emit light in the presence of an electrode-coated microtiter plate.[40,41] Commercial ELISA and ECL kits are available for a wide variety of cytokines. Such kits may simply consist of a pair of antibodies for capture and detection or they may come complete with precoated plates, detector antibody, substrate, standards, etc. Since the dynamic range

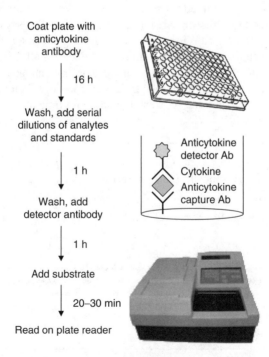

Coat plate with
anticytokine
antibody

↓ 16 h

Wash, add serial
dilutions of analytes
and standards

↓ 1 h

Wash, add
detector antibody

↓ 1 h

Add substrate

↓ 20–30 min

Read on plate reader

Anticytokine
detector Ab

Cytokine

Anticytokine
capture Ab

FIGURE 17.4 Schematic representation of ELISA.

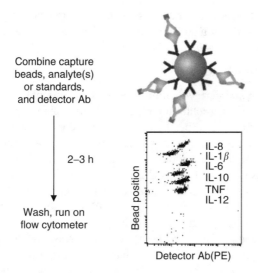

Combine capture
beads, analyte(s)
or standards,
and detector Ab

2–3 h

Wash, run on
flow cytometer

Bead position

IL-8
IL-1β
IL-6
IL-10
TNF
IL-12

Detector Ab(PE)

FIGURE 17.5   Schematic representation of multiplex bead assay.

of ELISA is limited, serial dilutions of the analyte solution are performed along with serial dilutions of a standard. Comparison of the optical density (OD) curves for the standard and test sample then allows the most accurate calculation of the analyte concentration. Such comparisons can be performed automatically using software that is built into many microtiter plate spectrophotometers.[24]

17.3.2   MULTIPLEXED BEAD ASSAYS

The adaptation of immunoassays to take advantage of the multiparametric readout of flow cytometry has resulted in multiplexed assays for detection of a wide range of cytokines from a small sample volume[9,25–34] (Figure 17.5). The principle is similar to that of an ELISA with a series of microparticles serving as the solid support rather than a microtiter well. Up to 30 or more sets of microparticles can be coated with individual capture antibodies for specific cytokines. The mixture of particles is applied to a biological sample and then further incubated with fluorescent-labeled detector antibodies. Finally, the sample is analyzed by flow cytometry that resolves the various bead populations (by size and differential fluorescence) and also quantitates the amount of fluorescent detector bound to each microparticle. Commercial kits are available ranging from do-it-yourself bead sets to fully optimized panels for codetection of a set of clinically related cytokines. For some systems, software is available to automatically calculate analyte concentrations.[25]

In addition to multiplexing, the main advantages of bead-based assays are their sensitivity (<20 pg/mL), wide dynamic range, and ability to work with very small sample volumes. For example, by using a cytometric bead array, investigators have been able to identify several Th2 cytokines in a 10 μL volume of tear fluid from allergic patients.[9] The main disadvantage to these assays is the requirement for a flow cytometer, which can be expensive and complex to set up and maintain. To address this, some manufacturers have made smaller and less expensive cytometers that are dedicated to bead-based assays or that run both bead-based and simple cell-based assays.

17.3.3 ANTIBODY MICROARRAYS

In contrast to cDNA microarrays, antibody microarrays immobilize specific monoclonal antibodies in spots on a solid support, usually a filter membrane[35–37] (Figure 17.3). Arrays with sets of cytokine-specific antibodies are commercially available in kits. The assay then works like a multiplexed ELISA. After incubation with the analyte containing sample (serum, cell culture supernatant, etc.), a mixture of enzyme-labeled detector antibodies is added. Detection is usually done by fluorescent or chemiluminescent means. Since this is more sensitive than colorimetric detection, the assays can have increased sensitivity compared to ELISA,[36,37] although this is not always the case.[42] Their dynamic range is also limited by the saturation of the detector. Performing serial dilutions of the sample in order to increase the dynamic range is not as simple or economical as for ELISA. Thus, these assays function best when the dynamic range of the expected result is known and the concentration of the sample can be adjusted accordingly. Still, they provide a powerful alternative to ELISA and multiplexed bead assays when a broad array of cytokines needs to be detected.

17.4 SINGLE-CELL ASSAYS

Although the measurement of cytokines in solution (for example, in serum or cell culture supernatant) is often quite useful, recent attention has focused increasingly on the enumeration of cytokine-producing cells. This process has been made possible by the development of enzyme-linked immunospot (ELISPOT) assays for cytokine-secreting cells[43–45] as well as flow cytometric detection of intracellular cytokine staining (ICS)[46–49] or of cytokine-secreting cells (cytokine secretion assay [CSA]).[50] Because the immune system is based on the actions of discrete cells, these assays answer fundamentally relevant types of questions, namely, what are the frequencies and functions of specific subsets of immune cells? In general, this is done via *in vitro* stimulation of a cell sample, as unstimulated levels of cytokine production are usually quite low.[51,52] Rather than quantifying the bulk level of cytokine produced from this stimulation, these assays count the number of responding cells that make a particular cytokine or set of cytokines (see Table 17.2 for typical results with model antigens). Assays based on flow cytometry (ICS and CSA) have the additional advantage of being able to resolve subsets of cells using antibodies to other markers in addition to cytokines.[53] This multiparametric readout affords the most in-depth view to date of the complexity and function of the immune system on a cellular level.[54]

Functional assays are intrinsically more complex and difficult to reproduce than simple immunoassays. Nevertheless, it is possible to achieve interlaboratory precision of around 20% CV with assays such as ICS[55] and ELISPOT.[56] Optimization of these assays with regard to reagents, protocols, and analysis are all important in arriving at sensitive and reproducible results. This optimization is especially true since the responding populations of antigen-specific T cells usually compromise minute percentages of the total T-cell population (Table 17.2).

TABLE 17.2

Normal Ranges of T-Cell Cytokine Responses to Model Antigens

		IFN-γ						IL-2						TNFα					
		Background	pp65 Peptide Mix	IE-1 Peptide Mix	CMV Lysate	Flu HA+M1 Peptide Mix	SEB	Background	pp65 Peptide Mix	IE-1 Peptide Mix	CMV Lysate	Flu HA+M1 Peptide Mix	SEB	Background	pp65 Peptide Mix	IE-1 Peptide Mix	CMV Lysate	Flu HA+M1 Peptide Mix	SEB
CD4	5th Percentile	0.000	0.08	0.00	0.60	0.00	2.39	0.000	0.06	0.01	0.25	0.00	5.23	0.005	0.07	0.01	0.25	0.01	4.85
	Median	0.010	0.81	0.05	2.10	0.03	5.87	0.015	0.48	0.03	0.89	0.04	10.50	0.030	0.63	0.04	2.44	0.05	10.52
	95th Percentile	0.067	4.81	0.82	12.56	0.11	20.53	0.047	3.20	0.42	6.42	0.09	21.99	0.087	4.87	0.72	10.83	0.10	27.96
CD8	5th Percentile	0.000	0.03	0.02	0.00	0.00	2.21	0.000	0.00	0.00	0.00	0.00	0.93	0.005	0.00	0.01	0.01	0.00	2.19
	Median	0.020	0.43	0.23	0.05	0.02	6.77	0.005	0.10	0.02	0.02	0.01	2.60	0.025	0.37	0.22	0.11	0.02	7.30
	95th Percentile	0.156	2.43	7.35	0.38	0.41	26.01	0.048	1.09	0.44	0.12	0.18	5.80	0.260	1.97	7.06	0.45	0.24	20.39

Note: Numbers are listed as percentage of CD4+ or CD8+ T cells and were derived from ICS assays performed on a set of 40 healthy adult donors (or a subset of 20 CMV seropositive donors for CMV-related antigens [pp65 and IE-1 peptide mixes and CMV lysate]). Peptide mixes are pools of 15 amino acid peptides overlapped by 11 amino acid residues each and spanning the stated protein(s).[102] SEB is a mitogen and commonly used positive control. pp65, phosphoprotein of 65 kilodaltons from human cytomegalovirus; IE-1, immediate early gene 1 from human cytomegalovirus; CMV, cytomegalovirus; HA, influenza hemagglutinin HA; M1, influenza matrix protein 1; SEB, staphylococcal enterotoxin B. Data courtesy of Margaret Inokuma, BD Biosciences.

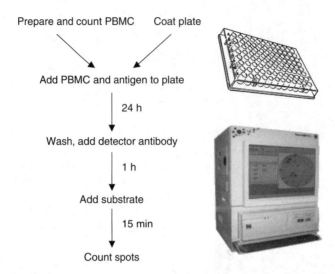

Prepare and count PBMC   Coat plate

Add PBMC and antigen to plate

24 h

Wash, add detector antibody

1 h

Add substrate

15 min

Count spots

FIGURE 17.6 Schematic representation of ELISPOT assay. (Figure courtesy of Laurel Nomura, BD Biosciences.)

17.4.1 ENZYME-LINKED IMMUNOSPOT

ELISPOT uses filter-bottom microtiter plates that are coated with a capture antibody specific for a cytokine, often IFN-γ or interleukin (IL)-2 (Figure 17.6). A sample of cells, for example, peripheral blood mononuclear cells (PBMC) is then placed in the wells along with a stimulus (specific antigen or mitogen). The plate is incubated 24–48 h to allow for secretion of cytokines, which are captured on the coated wells in the vicinity of the secreting cells. At the end of the incubation, the cells are washed out and the bound cytokines are visualized using an enzyme-labeled detector antibody and a colorimetric substrate. Discrete spots of color define the location of each cell that was secreting cytokine. These spots can be counted manually under a dissecting microscope or in an automated fashion using a dedicated ELISPOT reader.[57] The latter can also read spot sizes, which can be useful to discriminate various cellular sources of cytokine production.[58] Antibody pairs or kits containing precoated plates and other reagents are available to make the assay relatively simple to perform.

Although it is possible to detect two cytokines simultaneously in ELISPOT, this process requires additional complexity and optimization with a potential decrease in assay sensitivity and reproducibility.[59,60] Similarly, identification of the cell type responding in ELISPOT requires preselection of those cells,[46] for example, with immunomagnetic beads. This identification in turn introduces complexity and variability. Because of these limitations, ELISPOT is best used as a screening assay or for epitope mapping[61] of responses already characterized by more information-rich assays, such as ICS. In this regard, the ELISPOT has the advantage of built-in high throughput, being plate-based, and having an automated readout capability.[43–45]

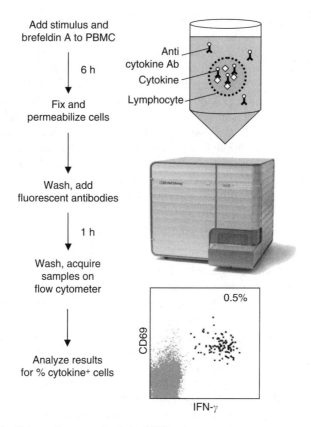

Add stimulus and
brefeldin A to PBMC

6 h

Fix and
permeabilize cells

Wash, add
fluorescent antibodies

1 h

Wash, acquire
samples on
flow cytometer

Analyze results
for % cytokine⁺ cells

Anti
cytokine Ab
Cytokine
Lymphocyte

0.5%

CD69

IFN-γ

FIGURE 17.7   Schematic representation of ICS assay.

17.4.2   INTRACELLULAR CYTOKINE STAINING

ICS, or cytokine flow cytometry (CFC),[46–49] is the most popular flow cytometric technique for detection of cytokine-producing cells (Figure 17.7). It is based on the ability of secretion inhibitors, such as brefeldin A or monensin, to allow intracellular accumulation of cytokines during *in vitro* stimulation. This accumulation can reach levels that are readily detected by flow cytometry in as little as 4–6 h giving these assays a faster turnaround time than ELISPOT. After incubation of cells with antigen or mitogen in the presence of the secretion inhibitor, the cells are fixed and then permeabilized (usually with a detergent such as saponin or Tween-20). They are then stained with fluorescently labeled antibodies to both surface and intracellular (cytokine) determinants. In some cases, antibodies to surface epitopes that are sensitive to fixation and permeabilization are applied to the cells prior to the fixation step so that their binding is not compromised. After washing, the cells are analyzed on a flow cytometer and commonly the percentage of cluster of differentiation (CD)4+ and/or CD8+ T cells producing a given cytokine is reported. When an absolute CD4 or CD8 count is run on the same sample, one can also report the number of cytokine-producing CD4+ or CD8+ T cells per milliliter of blood.[62]

The main advantage of the ICS assay is its information-rich readout. With newer digital cytometers capable of detecting up to 18 fluorescent labels simultaneously on a single cell, ICS can be used to precisely define cell subsets on the basis of both phenotype and function.[54] The assay is also highly sensitive, with generally greater efficiency of detection than ELISPOT.[63–67] However, its sensitivity is limited by the "background" frequency of spontaneous cytokine-secreting cells, which although low, can obscure responses of very rare cell populations (Table 17.2). Another disadvantage of ICS is its cost, particularly considering the need for a flow cytometer, which is both expensive and complex. In terms of throughput, ICS has been adapted to microtiter plates[68] and can even be performed with commercially available preconfigured lyophilized stimuli and staining plates.[55] Such advances together with the availability of plate-loading flow cytometers make ICS competitive to ELISPOT with regard to throughput.

17.4.3   Cytokine Secretion Assay

In some cases, assay of cytokine production in living cells is desirable so that the cells producing cytokine can be isolated and subjected to further analysis. This is not possible with ICS because the cells must be fixed and permeabilized for intracellular staining. However, CSA is useful in this regard, since it captures secreted cytokine on the surface of cells, allowing the cells to remain intact and alive for enrichment and further manipulation. The assay works as follows (Figure 17.8). A cell sample, for example, PBMC is stimulated as for ICS. After an initial incubation period, the cells are coated with a "catch reagent" consisting of a bispecific antibody binding to a ubiquitous cell-surface protein (CD45) as well as to a cytokine of interest. A final period of stimulation is done in the presence of this catch reagent allowing cytokine-producing cells to capture secreted cytokine on their cell surface. At the end of the incubation, a fluorescent-labeled anticytokine detector antibody is added along with antibodies to other cell-surface determinants as desired. The flow cytometric read-out is then similar to that for ICS.[50]

ICS and CSA assays can correlate well under optimized conditions, but not always.[69] There are additional technical complexities to the CSA assay that need to be taken into account (e.g., to avoid cross labeling of noncytokine-secreting cells).[70] CSA is also generally limited to detection of one cytokine at a time. Since ICS assays have been well standardized,[55] they are recommended for routine applications in which live cell isolation is not required.

17.4.4   Other Related Single-Cell Assays

Although the above three assays represent the major techniques used to measure cytokine-producing cells in the human system, there are related flow cytometry techniques that can be combined with ICS, for example, to extend the range of information gained from this method.

The recombinant production of major histocompatibility complex (MHC) molecules, made tetrameric via biotin–streptavidin interaction, and loaded with cognate peptide, has made possible a new level of detection of antigen-specific T cells.[71] These "tetramers," as they are commonly called, can be fluorescently labeled and used to identify rare subpopulations of T cells that recognize a particular

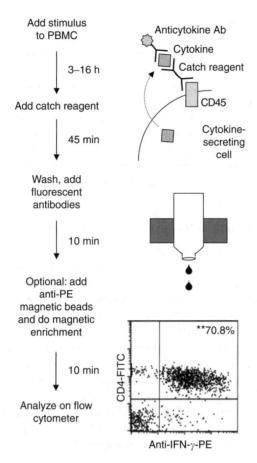

Add stimulus to PBMC

3–16 h

Add catch reagent

45 min

Wash, add fluorescent antibodies

10 min

Optional: add anti-PE magnetic beads and do magnetic enrichment

10 min

Analyze on flow cytometer

Anticytokine Ab

Cytokine

Catch reagent

CD45

Cytokine-secreting cell

**70.8%

CD4-FITC

Anti-IFN-γ-PE

FIGURE 17.8 Schematic representation of CSA.

MHC–peptide combination. Alternative versions of tetramers are also available, including MHC–peptide dimers fused to an immunoglobulin backbone,[72] MHC-peptide pentamers,[73] and others. These types of reagents can be used in conjunction with ICS to determine whether a population of T cells specific for a particular MHC-peptide combination is capable of producing cytokine(s).[74,75] The main disadvantage of the tetramer-type assays is that, due to their exquisite specificity for a particular peptide-binding T cell population, they are not useful for defining the total response to a protein or pathogen, particularly in an MHC-heterogeneous population.[76]

Recently, an assay measuring the degranulation capacity of T cells has been described.[77] This assay measures the cell-surface expression of CD107, a component of cytotoxic granules, which becomes transiently associated with the cell surface upon exocytosis and fusion of the granule membrane with the cell membrane. Although conditions must be carefully optimized to efficiently detect both CD107 and cytokines in the same sample,[7] such combined assays have proven useful in defining potentially clinically relevant "multifunctional" T cells in HIV+ individuals that do not progress to acquired immunodeficiency syndrome (AIDS).[78]

Assays of proliferation can also be carried out in combination with ICS using the incorporation of BrdU into DNA,[79] or the dilution of a cytophilic dye such as 5(6)-carboxylfluorescein diacetate succinimidyl ester (CFSE).[80] It should be stressed that the longer the period of *in vitro* culture, less is the functional capacity of the proliferated cells, which can be taken as a true measure of their functional capacity *in vivo*. Nevertheless, HIV viremia has been correlated with the suppression of HIV-specific CD4+ T-cell proliferation and reduced IL-2 production *in vitro*.[81] Conversely, long-term nonprogression of HIV has been correlated with maintenance of HIV-specific CD8+ T-cell proliferative capacity.[82] Thus, clinically relevant assays may use *in vitro* proliferation, cytokine production, or both, to measure properties of responding T cells that might then have prognostic implications.

17.5 STANDARDIZATION AND QUALITY CONTROL

Cytokine assays used in clinical diagnosis and monitoring obviously require a high level of standardization and quality control. They should achieve adequate accuracy, precision, linearity, and ruggedness/robustness in order to reliably differentiate patient classes and influence patient treatment decisions. Similarly, clinical labs using these assays need to validate their own ability to achieve expected levels of these parameters.[83] Such validation should be performed whenever a new assay is brought into the lab and should also be repeated, at least partially, whenever there are major changes in protocols and personnel performing the assays. Guidance on standard practices for many assays can be obtained from the Clinical and Laboratory Standards Institute (CLSI) (http://www.clsi.org) formerly known as the National Committee for Clinical Laboratory Standards (NCCLS). Laboratory performance of clinical tests is also regulated in the United States by the Clinical Laboratory Improvement Amendments (CLIA) (http://www.phppo.cdc.gov/clia/default.aspx).

In addition to true clinical assays, cytokine measurements made in clinical research and clinical trials are also in need of some level of standardization. Ideally, the results of one trial using a particular cytokine assay should be comparable to the results of another trial using the same assay. Unfortunately, the historic nature of cytokine measurement as a research tool has resulted in a proliferation of different protocols and reagents making this goal difficult. Recent efforts to standardize ELISPOT[56,84] and ICS[55] have begun to counter this trend. In addition, periodic proficiency testing programs are now underway for both ELISPOT (http://www.zellnet.com/cvceppannouncement) and ICS (http://bdicsqa.webbasix.com) allowing laboratories to test themselves against their peers to determine their level of performance. Expected performance characteristics of ELISPOT[67,85,86] and ICS[67,87] in clinical trial settings have also been published, such that laboratories doing their own validation have a standard for comparison.

Because single-cell assays rely on the quality of blood or PBMC specimens, attention must be paid to their handling. In particular, the time and temperature of storage of whole blood is important as antigen-presenting cell function can be lost in a matter of hours,[88] and platelet aggregation is enhanced at low temperatures.[89]

PBMC are cryopreserved and shipped to a central site for cytokine analysis a number of times. Standardized protocols for cryopreservation,[90] shipping,[91] and thawing[90] are important as viability and recovery of PBMC have an impact on their function.[67]

Assays that rely on complex readouts, such as flow cytometry, also require standardization of the instrument and the analysis routine in addition to standardized sample handling. In fact, gating can be a major source of variation in ICS assays performed across sites,[55] and use of a common gating template with "dynamic" gates can reduce such variation.[68] Instrument setup can be standardized and optimized by use of fluorescent particles and well-designed controls.[92] Criteria for the determination of positive results should be set[51,87] and can be guided by statistical tests[49] (http://maeckerlab.typepad.com/maeckerlab_weblog/2006/04/maecker_lab_too.html). Routine monitoring of microparticles run for the purpose of daily quality control is also invaluable for detection of potential instrument problems and variations.[93] These procedures need to be built in to the routine of the clinical or clinical research lab in order to assure consistent quality of flow cytometric results.

17.6 CONCLUSIONS AND FUTURE DIRECTIONS

The measurement of cytokines in biological samples, which was mostly a research endeavor, holds potential for clinical relevance in a number of diseases. Perhaps most immediately, the detection of cytokine-producing T cells responding to specific pathogens can be used diagnostically to determine prior exposure to that pathogen. This is currently done for TB using either an assay for secreted IFN-γ[94] or an IFN-γ ELISPOT.[95] In both cases, cells are stimulated with specific peptides present in the pathogenic *Mycobacterium tuberculosis* but not in the attenuated Bacillus Calmette–Guerin (BCG) vaccine for tuberculosis. Thus, these assays are able to distinguish exposure to the pathogen from vaccination with BCG.[2] Further refinements, perhaps involving phenotyping of the responsive cells in conjunction with ICS, may allow discrimination of active from latent TB or offer a surrogate of protection from disease resulting in an even more clinically useful assay.[96,97]

There is currently a massive "Global Enterprise" to discover a vaccine for HIV/AIDS.[98] A successful vaccine will most likely require the induction of a cellular immune response as well as neutralizing antibodies.[1] To detect such a cellular response, clinical trials for HIV vaccines are using ELISPOT and ICS assays to read out IFN-γ- and IL-2-producing T cells in vaccine recipients. Although the presence of T cells producing these two cytokines alone is unlikely to be a surrogate for protection from HIV disease, it is certainly a measure of immunogenicity and as such is probably a necessary but not sufficient property of a successful vaccine.[99] Additional refinements to the ICS assay, including other functions,[78] phenotypic markers,[100] or both,[101] may eventually uncover a true surrogate marker of protection.

Future clinical assays involving cytokines are sure to develop. Multiplexed bead assays and antibody microarrays are ideally suited to measure a complex milieu of cytokines that might be involved in the development of disease states such as sepsis.[34,36] Microarrays or multiparameter ICS assays will likely identify new cytokine signatures associated with good or poor prognosis in diseases such as cancer[70] and AIDS.[99] Although simplification and standardization might be required to make such assays clinically viable, there is clearly a future for cytokines and functional assays in the clinical laboratory.

REFERENCES

1. Pantaleo, G. and Koup, R. A., Correlates of immune protection in HIV-1 infection: what we know, what we don't know, what we should know, *Nat. Med.* 10(8), 806–810, 2004.
2. Ferrara, G., Losi, M., D'Amico, R., Roversi, P., Piro, R., Meacci, M., Meccugni, B., Dori, I. M., Andreani, A., Bergamini, B. M., Mussini, C., Rumpianesi, F., Fabbri, L. M., and Richeldi, L., Use in routine clinical practice of two commercial blood tests for diagnosis of infection with *Mycobacterium tuberculosis*: a prospective study, *Lancet* 367(9519), 1328–1334, 2006.
3. Whiteside, T. L., Cytokines and cytokine measurements in a clinical laboratory, *Clin. Diagn. Lab. Immunol.* 1(3), 257–260, 1994.
4. Meager, A., Measurement of cytokines by bioassays: theory and application, *Methods* 38(4), 237–252, 2006.
5. Bienvenu, J. A., Monneret, G., Gutowski, M. C., and Fabien, N., Cytokine assays in human sera and tissues, *Toxicology* 129(1), 55–61, 1998.
6. Fan, J., Tam, P., Woude, G. V., and Ren, Y., Normalization and analysis of cDNA micro-arrays using within-array replications applied to neuroblastoma cell response to a cyto-kine, *Proc. Natl. Acad. Sci. USA* 101(5), 1135–1140, 2004.
7. Suni, M. A., Maino, V. C., and Maecker, H. T., Ex vivo analysis of T-cell function, *Curr. Opin. Immunol.* 17(4), 434–440, 2005.
8. Baraczka, K., Nekam, K., Pozsonyi, T., Szuts, I., and Ormos, G., Investigation of cyto-kine (tumor necrosis factor-alpha, interleukin-6, interleukin-10) concentrations in the cerebrospinal fluid of female patients with multiple sclerosis and systemic lupus erythe-matosus, *Eur. J. Neurol.* 11(1), 37–42, 2004.
9. Cook, E. B., Stahl, J. L., Lowe, L., Chen, R., Morgan, E., Wilson, J., Varro, R., Chan, A., Graziano, F. M., and Barney, N. P., Simultaneous measurement of six cytokines in a single sample of human tears using microparticle-based flow cytometry: allergics vs. non-allergics, *J. Immunol. Methods* 254(1–2), 109–118, 2001.
10. Enocksson, A., Lundberg, J., Weitzberg, E., Norrby-Teglund, A., and Svenungsson, B., Rectal nitric oxide gas and stool cytokine levels during the course of infectious gastro-enteritis, *Clin. Diagn. Lab. Immunol.* 11(2), 250–254, 2004.
11. Shacklett, B. L., Yang, O., Hausner, M. A., Elliott, J., Hultin, L., Price, C., Fuerst, M., Matud, J., Hultin, P., Cox, C., Ibarrondo, J., Wong, J. T., Nixon, D. F., Anton, P. A., and Jamieson, B. D., Optimization of methods to assess human mucosal T-cell responses to HIV infection, *J. Immunol. Methods* 279(1–2), 17–31, 2003.
12. He, X. S., Rehermann, B., Lopez-Labrador, F. X., Boisvert, J., Cheung, R., Mumm, J., Wedemeyer, H., Berenguer, M., Wright, T. L., Davis, M. M., and Greenberg, H. B., Quan-titative analysis of hepatitis C virus-specific CD8(+) T cells in peripheral blood and liver using peptide-MHC tetramers, *Proc. Natl. Acad. Sci. USA* 96(10), 5692–5697, 1999.
13. Vitolo, D., Zerbe, T., Kanbour, A., Dahl, C., Herberman, R. B., and Whiteside, T. L., Expression of mRNA for cytokines in tumor-infiltrating mononuclear cells in ovarian adenocarcinoma and invasive breast cancer, *Int. J. Cancer* 51(4), 573–580, 1992.
14. Dallman, M. J., Montgomery, R. A., Larsen, C. P., Wanders, A., and Wells, A. F., Cytokine gene expression: analysis using northern blotting, polymerase chain reaction and in situ hybridization, *Immunol. Rev.* 119, 163–179, 1991.

15. Giulietti, A., Overbergh, L., Valckx, D., Decallonne, B., Bouillon, R., and Mathieu, C., An overview of real-time quantitative PCR: applications to quantify cytokine gene expression, *Methods* 25(4), 386–401, 2001.

16. Overbergh, L., Giulietti, A., Valckx, D., Decallonne, R., Bouillon, R., and Mathieu, C., The use of real-time reverse transcriptase PCR for the quantification of cytokine gene expression, *J. Biomol. Tech.* 14(1), 33–43, 2003.

17. Prediger, E. A., Detection and quantitation of mRNAs using ribonuclease protection assays, *Methods Mol. Biol.* 160, 495–505, 2001.

18. Hegde, P., Qi, R., Abernathy, K., Gay, C., Dharap, S., Gaspard, R., Hughes, J. E., Snesrud, E., Lee, N., and Quackenbush, J., A concise guide to cDNA microarray analysis, *Biotechniques* 29(3), 548–550, 552–554, 556 passim, 2000.

19. Boeuf, P., Vigan-Womas, I., Jublot, D., Loizon, S., Barale, J. C., Akanmori, B. D., Mercereau-Puijalon, O., and Behr, C., CyProQuant-PCR: a real time RT-PCR technique for profiling human cytokines, based on external RNA standards, readily automatable for clinical use, *BMC Immunol.* 6(1), 5, 2005.

20. Kuhne, B. S. and Oschmann, P., Quantitative real-time RT-PCR using hybridization probes and imported standard curves for cytokine gene expression analysis, *Biotechniques* 33(5), 1078, 1080–1082, 1084 passim, 2002.

21. von Wolff, M. and Tabibzadeh, S., Multiprobe RNase protection assay with internally labeled radioactive probes, generated by RT-PCR and nested PCR, *Front Biosci.* 4, C1–C3, 1999.

22. Krakauer, T., Chen, X., Howard, O. M., and Young, H. A., RNase protection assay for the study of the differential effects of therapeutic agents in suppressing staphylococcal enterotoxin B-induced cytokines in human peripheral blood mononuclear cells, *Methods Mol. Biol.* 214, 151–164, 2003.

23. Nikula, T., West, A., Katajamaa, M., Lonnberg, T., Sara, R., Aittokallio, T., Nevalainen, O. S., and Lahesmaa, R., A human ImmunoChip cDNA microarray provides a comprehensive tool to study immune responses, *J. Immunol. Methods* 303(1–2), 122–134, 2005.

24. Plested, J. S., Coull, P. A., and Gidney, M. A., Elisa, *Methods Mol. Med.* 71, 243–261, 2003.

25. Morgan, E., Varro, R., Sepulveda, H., Ember, J. A., Apgar, J., Wilson, J., Lowe, L., Chen, R., Shivraj, L., Agadir, A., Campos, R., Ernst, D., and Gaur, A., Cytometric bead array: a multiplexed assay platform with applications in various areas of biology, *Clin. Immunol.* 110(3), 252–266, 2004.

26. Ray, C. A., Bowsher, R. R., Smith, W. C., Devanarayan, V., Willey, M. B., Brandt, J. T., and Dean, R. A., Development, validation, and implementation of a multiplex immunoassay for the simultaneous determination of five cytokines in human serum, *J. Pharm. Biomed. Anal.* 36(5), 1037–1044, 2005.

27. Hill, H. R. and Martins, T. B., The flow cytometric analysis of cytokines using multianalyte fluorescence microarray technology, *Methods* 38(4), 312–316, 2006.

28. Khan, S. S., Smith, M. S., Reda, D., Suffredini, A. F., and McCoy, J. P., Jr., Multiplex bead array assays for detection of soluble cytokines: comparisons of sensitivity and quantitative values among kits from multiple manufacturers, *Cytometry B Clin. Cytom.* 61(1), 35–39, 2004.

29. Prabhakar, U., Eirikis, E., Miller, B. E., and Davis, H. M., Multiplexed cytokine sandwich immunoassays: clinical applications, *Methods Mol. Med.* 114, 223–232, 2005.

30. Kellar, K. L. and Douglass, J. P., Multiplexed microsphere-based flow cytometric immunoassays for human cytokines, *J. Immunol. Methods* 279(1–2), 277–285, 2003.

31. Prabhakar, U., Eirikis, E., Reddy, M., Silvestro, E., Spitz, S., Pendley, C., II, Davis, H. M., and Miller, B. E., Validation and comparative analysis of a multiplexed assay for the simultaneous quantitative measurement of Th1/Th2 cytokines in human serum and human peripheral blood mononuclear cell culture supernatants, *J. Immunol. Methods* 291(1–2), 27–38, 2004.

32. Elshal, M. F. and McCoy, J. P., Multiplex bead array assays: performance evaluation and comparison of sensitivity to ELISA, *Methods* 38(4), 317–323, 2006.
33. de Jager, W. and Rijkers, G. T., Solid-phase and bead-based cytokine immunoassay: a comparison, *Methods* 38(4), 294–303, 2006.
34. Tarnok, A., Hambsch, J., Chen, R., and Varro, R., Cytometric bead array to measure six cytokines in twenty-five microliters of serum, *Clin. Chem.* 49(6), 1000–1002, 2003.
35. Lin, Y., Huang, R., Cao, X., Wang, S. M., Shi, Q., and Huang, R. P., Detection of multiple cytokines by protein arrays from cell lysate and tissue lysate, *Clin. Chem. Lab. Med.* 41(2), 139–145, 2003.
36. Knight, P. R., Sreekumar, A., Siddiqui, J., Laxman, B., Copeland, S., Chinnaiyan, A., and Remick, D. G., Development of a sensitive microarray immunoassay and comparison with standard enzyme-linked immunoassay for cytokine analysis, *Shock* 21(1), 26–30, 2004.
37. Tam, S. W., Wiese, R., Lee, S., Gilmore, J., and Kumble, K. D., Simultaneous analysis of eight human Th1/Th2 cytokines using microarrays, *J. Immunol. Methods* 261(1–2), 157–165, 2002.
38. Hooks, J. J., Moutsopoulos, H. M., Geis, S. A., Stahl, N. I., Decker, J. L., and Notkins, A. L., Immune interferon in the circulation of patients with autoimmune disease, *N Engl. J. Med.* 301(1), 5–8, 1979.
39. Maury, C. P. and Teppo, A. M., Serum immunoreactive interleukin 1 in renal transplant recipients. Association of raised levels with graft rejection episodes, *Transplantation* 45(1), 143–147, 1988.
40. Rodriguez, R. M., Abdullah, R., Miller, R., Barry, L., Lungstras-Bufler, K., Bufler, P., Dinarello, C. A., and Abraham, E., A pilot study of cytokine levels and white blood cell counts in the diagnosis of necrotizing fasciitis, *Am. J. Emerg. Med.* 24(1), 58–61, 2006.
41. Sennikov, S. V., Krysov, S. V., Injelevskaya, T. V., Silkov, A. N., Grishina, L. V., and Kozlov, V. A., Quantitative analysis of human immunoregulatory cytokines by electrochemiluminescence method, *J. Immunol. Methods* 275(1–2), 81–88, 2003.
42. Copeland, S., Siddiqui, J., and Remick, D., Direct comparison of traditional ELISAs and membrane protein arrays for detection and quantification of human cytokines, *J. Immunol. Methods* 284(1–2), 99–106, 2004.
43. Mashishi, T. and Gray, C. M., The ELISPOT assay: an easily transferable method for measuring cellular responses and identifying T cell epitopes, *Clin. Chem. Lab. Med.* 40(9), 903–910, 2002.
44. Kalyuzhny, A. E., Chemistry and biology of the ELISPOT assay, *Methods Mol. Biol.* 302, 15–31, 2005.
45. Schmittel, A., Keilholz, U., Thiel, E., and Scheibenbogen, C., Quantification of tumor-specific T lymphocytes with the ELISPOT assay, *J. Immunother.* 23(3), 289–295, 2000.
46. Loza, M. J., Faust, J. S., and Perussia, B., Multiple color immunofluorescence for cytokine detection at the single-cell level, *Mol. Biotechnol.* 23(3), 245–258, 2003.
47. Ghanekar, S. A., Maecker, H. T., and Maino, V. C., Immune monitoring using cytokine flow cytometry, in Rose, N. R. (ed.), *Manual of Clinical Laboratory Immunology*, 7th ed., ASM Press, Washington, DC, 2005, pp. 353–360.
48. Maecker, H. T., Cytokine flow cytometry, in Hawley, T. S. and Hawley, R. G. (eds.), *Flow Cytometry Protocols*, 2nd ed., Humana Press, Totowa, NJ, 2004, pp. 95–107.
49. Maecker, H. T., Cytokine flow cytometry in the analysis of tumor-specific T cells, in Kieber-Emmons, T. (ed.), *Methods in Cancer Vaccine Development*, Humana Press, Totowa, NJ, 2005, in press.
50. Brosterhus, H., Brings, S., Leyendeckers, H., Manz, R. A., Miltenyi, S., Radbruch, A., Assenmacher, M., and Schmitz, J., Enrichment and detection of live antigen-specific CD4(+) and CD8(+) T cells based on cytokine secretion, *Eur. J. Immunol.* 29(12), 4053–4059, 1999.

51. Dunn, H. S., Haney, D. J., Ghanekar, S. A., Stepick-Biek, P., Lewis, D. B., and Maecker, H. T., Dynamics of CD4 and CD8 T cell responses to cytomegalovirus in healthy human donors, *J. Infect. Dis.* 186(1), 15–22, 2002.

52. Pittet, M. J., Zippelius, A., Speiser, D. E., Assenmacher, M., Guillaume, P., Valmori, D., Lienard, D., Lejeune, F., Cerottini, J. C., and Romero, P., Ex vivo IFN-gamma secretion by circulating CD8 T lymphocytes: implications of a novel approach for T cell monitoring in infectious and malignant diseases, *J. Immunol.* 166(12), 7634–7640, 2001.

53. Letsch, A. and Scheibenbogen, C., Quantification and characterization of specific T-cells by antigen-specific cytokine production using ELISPOT assay or intracellular cytokine staining, *Methods* 31(2), 143–149, 2003.

54. Roederer, M., Brenchley, J. M., Betts, M. R., and De Rosa, S. C., Flow cytometric analysis of vaccine responses: how many colors are enough? *Clin. Immunol.* 110(3), 199–205, 2004.

55. Maecker, H. T., Rinfret, A., D'Souza, P., Darden, J., Roig, E., Landry, C., Hayes, P., Birungi, J., Anzala, O., Garcia, M., Harari, A., Frank, I., Baydo, R., Baker, M., Holbrook, J., Ottinger, J., Lamoreaux, L., Epling, C. L., Sinclair, E., Suni, M. A., Punt, K., Calarota, S., El-Bahi, S., Alter, G., Maila, H., Kuta, E., Cox, J., Gray, C., Altfeld, M., Nougarede, N., Boyer, J., Tussey, L., Tobery, T., Bredt, B., Roederer, M., Koup, R., Maino, V. C., Weinhold, K., Pantaleo, G., Gilmour, J., Horton, H., and Sekaly, R. P., Standardization of cytokine flow cytometry assays, *BMC Immunol.* 6(1), 13, 2005.

56. Samri, A., Durier, C., Urrutia, A., Sanchez, I., Gahery-Segard, H., Imbart, S., Sinet, M., Tartour, E., Aboulker, J. P., Autran, B., and Venet, A., Evaluation of the interlaboratory concordance in quantification of human immunodeficiency virus-specific T cells with a gamma interferon enzyme-linked immunospot assay, *Clin. Vaccine Immunol.* 13(6), 684–697, 2006.

57. Lehmann, P. V., Image analysis and data management of ELISPOT assay results, *Methods Mol. Biol.* 302, 117–1132, 2005.

58. Guerkov, R. E., Targoni, O. S., Kreher, C. R., Boehm, B. O., Herrera, M. T., Tary-Lehmann, M., Lehmann, P. V., and Schwander, S. K., Detection of low-frequency antigen-specific IIL-10-producing CD4(+) T cells via ELISPOT in PBMC: cognate vs. nonspecific production of the cytokine, *J. Immunol. Methods* 279(1–2), 111–121, 2003.

59. Quast, S., Zhang, W., Shive, C., Kovalovski, D., Ott, P. A., Herzog, B. A., Boehm, B. O., Tary-Lehmann, M., Karulin, A. Y., and Lehmann, P. V., IL-2 absorption affects IFN-gamma and IL-5, but not IL-4 producing memory T cells in double color cytokine ELISPOT assays, *Cell Immunol.* 237(1), 28–36, 2005.

60. Palzer, S., Bailey, T., Hartnett, C., Grant, A., Tsang, M., and Kalyuzhny, A. E., Simultaneous detection of multiple cytokines in ELISPOT assays, *Methods Mol. Biol.* 302, 273–288, 2005.

61. Anthony, D. D. and Lehmann, P. V., T-cell epitope mapping using the ELISPOT approach, *Methods* 29(3), 260–269, 2003.

62. Jacobson, M. A., Maecker, H. T., Orr, P. L., D'Amico, R., Van Natta, M., Li, X. D., Pollard, R. B., and Bredt, B. M., Results of a cytomegalovirus (CMV)-specific CD8+/interferon- gamma + cytokine flow cytometry assay correlate with clinical evidence of protective immunity in patients with AIDS with CMV retinitis, *J. Infect. Dis.* 189(8), 1362–1373, 2004.

63. Tassignon, J., Burny, W., Dahmani, S., Zhou, L., Stordeur, P., Byl, B., and De Groote, D., Monitoring of cellular responses after vaccination against tetanus toxoid: comparison of the measurement of IFN-gamma production by ELISA, ELISPOT, flow cytometry and real-time PCR, *J. Immunol. Methods* 305(2), 188–198, 2005.

64. Whiteside, T. L., Zhao, Y., Tsukishiro, T., Elder, E. M., Gooding, W., and Baar, J., Enzyme-linked immunospot, cytokine flow cytometry, and tetramers in the detection of T-cell responses to a dendritic cell-based multipeptide vaccine in patients with melanoma, *Clin. Cancer Res.* 9(2), 641–649, 2003.

65. Moretto, W. J., Drohan, L. A., and Nixon, D. F., Rapid quantification of SIV-specific CD8 T cell responses with recombinant vaccinia virus ELISPOT or cytokine flow cytometry, *AIDS* 14(16), 2625–2627, 2000.

66. Karlsson, A. C., Martin, J. N., Younger, S. R., Bredt, B. M., Epling, L., Ronquillo, R., Varma, A., Deeks, S. G., McCune, J. M., Nixon, D. F., and Sinclair, E., Comparison of the ELISPOT and cytokine flow cytometry assays for the enumeration of antigen-specific T cells, *J. Immunol. Methods* 283(1–2), 141–153, 2003.

67. Maecker, H. T., Moon, J., Bhatia, S., Ghanekar, S. A., Maino, V. C., Payne, J. K., Kuus-Reichel, K., Chang, J. C., Summers, A., Clay, T. M., Morse, M. A., Lyerly, H. K., Delarosa, C., Ankerst, D. P., and Disis, M. L., Impact of cryopreservation on tetramer, cytokine flow cytometry, and ELISPOT, *BMC Immunol.* 6(1), 17, 2005.

68. Suni, M. A., Dunn, H. S., Orr, P. L., deLaat, R., Sinclair, E., Ghanekar, S. A., Bredt, B. M., Dunne, J. F., Maino, V. C., and Maecker, H. T., Performance of plate-based cytokine flow cytometry with automated data analysis, *BMC Immunol.* 4, 9, 2003.

69. Asemissen, A. M., Nagorsen, D., Keilholz, U., Letsch, A., Schmittel, A., Thiel, E., and Scheibenbogen, C., Flow cytometric determination of intracellular or secreted IFN-gamma for the quantification of antigen reactive T cells, *J. Immunol. Methods* 251(1–2), 101–108, 2001.

70. Maecker, H. T., The role of immune monitoring in evaluating cancer immunotherapy, in Disis, M. L. (ed.), *Cancer Drug Discovery and Development: Immunotherapy of Cancer*, Humana Press, Totowa, NJ, 2005, pp. 59–71.

71. Altman, J. D., Moss, P. A. H., Goulder, P. J. R., Barouch, D. H., McHeyzer-Williams, M. G., Bell, J. I., McMichael, A. J., and Davis, M. M., Phenotypic analysis of antigen-specific T lymphocytes, *Science* 274(5284), 94–96, 1996.

72. Greten, T. F., Slansky, J. E., Kubota, R., Soldan, S. S., Jaffee, E. M., Leist, T. P., Pardoll, D. M., Jacobson, S., and Schneck, J. P., Direct visualization of antigen-specific T cells: HTLV-1 Tax11-19- specific CD8(+) T cells are activated in peripheral blood and accumulate in cerebrospinal fluid from HAM/TSP patients, *Proc. Natl. Acad. Sci. USA* 95(13), 7568–7573, 1998.

73. Claassen, E. A., van der Kant, P. A., Rychnavska, Z. S., van Bleek, G. M., Easton, A. J., and van der Most, R. G., Activation and inactivation of antiviral CD8 T cell responses during murine pneumovirus infection, *J. Immunol.* 175(10), 6597–6604, 2005.

74. Appay, V. and Rowland-Jones, S. L., The assessment of antigen-specific CD8+ T cells through the combination of MHC class I tetramer and intracellular staining, *J. Immunol. Methods* 268(1), 9–19, 2002.

75. He, X. S., Rehermann, B., Boisvert, J., Mumm, J., Maecker, H. T., Roederer, M., Wright, T. L., Maino, V. C., Davis, M. M., and Greenberg, H. B., Direct functional analysis of epitope-specific CD8+ T cells in peripheral blood, *Viral Immunol.* 14(1), 59–69, 2001.

76. Betts, M. R., Casazza, J. P., Patterson, B. A., Waldrop, S., Trigona, W., Fu, T. M., Kern, F., Picker, L. J., and Koup, R. A., Putative immunodominant human immunodeficiency virus-specific CD8(+) T-cell responses cannot be predicted by major histocompatibility complex class I haplotype, *J. Virol.* 74(19), 9144–9151, 2000.

77. Betts, M. R., Brenchley, J. M., Price, D. A., De Rosa, S. C., Douek, D. C., Roederer, M., and Koup, R. A., Sensitive and viable identification of antigen-specific CD8+ T cells by a flow cytometric assay for degranulation, *J. Immunol. Methods* 281(1–2), 65–78, 2003.

78. Betts, M. R., Nason, M. C., West, S. M., De Rosa, S. C., Migueles, S. A., Abraham, J., Lederman, M. M., Benito, J. M., Goepfert, P. A., Connors, M., Roederer, M., and Koup, R. A., HIV nonprogressors preferentially maintain highly functional HIV-specific CD8+ T-cells, *Blood* 107, 4781–4789, 2006.

79. Mehta, B. A. and Maino, V. C., Simultaneous detection of DNA synthesis and cytokine production in staphylococcal enterotoxin B activated CD4+ T lymphocytes by flow cytometry, *J. Immunol. Methods* 208(1), 49–59, 1997.

80. Fulcher, D. and Wong, S., Carboxyfluorescein succinimidyl ester-based proliferative assays for assessment of T cell function in the diagnostic laboratory, *Immunol. Cell Biol.* 77(6), 559–564, 1999.

81. Iyasere, C., Tilton, J. C., Johnson, A. J., Younes, S., Yassine-Diab, B., Sekaly, R. P., Kwok, W. W., Migueles, S. A., Laborico, A. C., Shupert, W. L., Hallahan, C. W., Davey, R. T., Jr., Dybul, M., Vogel, S., Metcalf, J., and Connors, M., Diminished proliferation of human immunodeficiency virus-specific CD4+ T cells is associated with diminished interleukin-2 (IL-2) production and is recovered by exogenous IL-2, *J. Virol.* 77(20), 10900–10909, 2003.

82. Migueles, S. A., Laborico, A. C., Shupert, W. L., Sabbaghian, M. S., Rabin, R., Hallahan, C. W., Baarle, D. V., Kostense, S., Miedema, F., McLaughlin, M., Ehler, L., Metcalf, J., Liu, S., and Connors, M., HIV-specific CD8(+) T cell proliferation is coupled to perforin expression and is maintained in nonprogressors, *Nat. Immunol.* 3(11), 1061–1068, 2002.

83. Owens, M. A., Vall, H. G., Hurley, A. A., and Wormsley, S. B., Validation and quality control of immunophenotyping in clinical flow cytometry, *J. Immunol. Methods* 243(1–2), 33–50, 2000.

84. Cox, J. H., Ferrari, G., Kalams, S. A., Lopaczynski, W., Oden, N., and D'Souza M, P., Results of an ELISPOT proficiency panel conducted in 11 laboratories participating in international human immunodeficiency virus type 1 vaccine trials, *AIDS Res. Hum. Retroviruses* 21(1), 68–81, 2005.

85. Lathey, J., Preliminary steps toward validating a clinical bioassay: case study of the ELISpot assay, *Biopharm. Intl.* March, 42–50, 2003.

86. Lathey, J., Sathiyaseelan, J., Matijevic, M., and Hedley, M. L., Validation of pretrial ELISspot measurements, *BioProcess Intl.* Sept., 34–41, 2003.

87. Trigona, W. L., Clair, J. H., Persaud, N., Punt, K., Bachinsky, M., Sadasivan-Nair, U., Dubey, S., Tussey, L., Fu, T. M., and Shiver, J., Intracellular staining for HIV-specific IFN-gamma production: statistical analyses establish reproducibility and criteria for distinguishing positive responses, *J. Interferon. Cytokine Res.* 23(7), 369–377, 2003.

88. Hanekom, W. A., Hughes, J., Mavinkurve, M., Mendillo, M., Watkins, M., Gamieldien, H., Gelderbloem, S. J., Sidibana, M., Mansoor, N., Davids, V., Murray, R. A., Hawkridge, A., Haslett, P. A., Ress, S., Hussey, G. D., and Kaplan, G., Novel application of a whole blood intracellular cytokine detection assay to quantitate specific T-cell frequency in field studies, *J. Immunol. Methods* 291(1–2), 185–195, 2004.

89. Qi, R., Yatomi, Y., and Ozaki, Y., Effects of incubation time, temperature, and anticoagulants on platelet aggregation in whole blood, *Thromb. Res.* 101(3), 139–144, 2001.

90. Disis, M. L., dela Rosa, C., Goodell, V., Kuan, L. Y., Chang, J. C., Kuus-Reichel, K., Clay, T. M., Kim Lyerly, H., Bhatia, S., Ghanekar, S. A., Maino, V. C., and Maecker, H. T., Maximizing the retention of antigen specific lymphocyte function after cryopreservation, *J. Immunol. Methods* 308(1–2), 13–18, 2006.

91. Weinberg, A., Betensky, R. A., Zhang, L., and Ray, G., Effect of shipment, storage, anticoagulant, and cell separation on lymphocyte proliferation assays for human immunodeficiency virus-infected patients, *Clin. Diagn. Lab. Immunol.* 5(6), 804–807, 1998.

92. Maecker, H. T. and Trotter, J., Flow cytometry controls, instrument setup, and the determination of positivity, *Cytometry A*, 69(9), 1037–1042, 2006.

93. Agrawal, Y. P., Mahlamaki, E. K., and Penttila, I. M., Use of quality control standards in clinical flow cytometry, *Ann. Med.* 23(2), 127–133, 1991.

94. Todd, B., The QuantiFERON-TB Gold Test: a new blood assay offers a promising alternative in tuberculosis testing, *Am. J. Nurs.* 106(6), 33–34, 37, 2006.

95. Wagstaff, A. J. and Zellweger, J. P., T-SPOT.TB: an in vitro diagnostic assay measuring T-cell reaction to Mycobacterium tuberculosis-specific antigens, *Mol. Diagn. Ther.* 10(1), 57–63, discussion 64-5, 2006.

96. Kaufmann, S. H., Recent findings in immunology give tuberculosis vaccines a new boost, *Trends Immunol.* 26(12), 660–667, 2005.

97. Ellner, J. J., Hirsch, C. S., and Whalen, C. C., Correlates of protective immunity to *Mycobacterium tuberculosis* in humans, *Clin. Infect. Dis.* 30(Suppl 3), S279–S282, 2000.

98. The Global HIV/AIDS Vaccine Enterprise: Scientific Strategic Plan, *PLoS Med.* 2(2), 111–121, 2005.

99. Maecker, H. T. and Maino, V. C., T cell immunity to HIV: defining parameters of protection, *Curr. HIV Res.* 1, 249–259, 2003.

100. Letvin, N. L., Mascola, J. R., Sun, Y., Gorgone, D. A., Buzby, A. P., Xu, L., Yang, Z. Y., Chakrabarti, B., Rao, S. S., Schmitz, J. E., Montefiori, D. C., Barker, B. R., Bookstein, F. L., and Nabel, G. J., Preserved CD4+ central memory T cells and survival in vaccinated SIV-challenged monkeys, *Science* 312(5779), 1530–1533, 2006.

101. Nomura, L. E., Emu, B., Hoh, R., Haaland, P., Deeks, S. G., Martin, J. N., McCune, J. M., Nixon, D. F., and Maecker, H. T., IL-2 production correlates with effector cell differentiation in HIV-specific CD8+ T cells, *AIDS Res. Ther.* 3(1), 18, 2006.

102. Maecker, H. T., Dunn, H. S., Suni, M. A., Khatamzas, E., Pitcher, C. J., Bunde, T., Persaud, N., Trigona, W., Fu, T. M., Sinclair, E., Bredt, B. M., McCune, J. M., Maino, V. C., Kern, F., and Picker, L. J., Use of overlapping peptide mixtures as antigens for cytokine flow cytometry, *J. Immunol. Methods* 255(1–2), 27–40, 2001.

18 The Human Major Histocompatibility Complex and DNA-Based Typing of Human Leukocyte Antigens for Transplantation

Susana G. Marino, Andrés Jaramillo, and Marcelo A. Fernández-Viña

CONTENTS

18.1 INTRODUCTION

The Major Histocompatibility Complex (MHC) in humans (denominated the human leukocyte antigen [HLA] system), like in many other species, controls the acceptance or the rejection of transplanted foreign cells and tissues. Immunocompetent CD8+ and CD4+ T cells respond vigorously to foreign (allogeneic) HLA molecules (direct allorecognition) or to foreign HLA-derived peptides (indirect allorecognition), respectively. Allorecognition of HLA molecules may activate up to 10% of the total T-cell repertoire. The degree of allorecognition and alloactivation varies according to the extent of HLA (class I and class II) disparity between donor and recipient—the greater the extent of HLA disparity, the stronger the degree of the alloresponse.

The HLA class I antigens (-A, -B, and -C) are expressed on all nucleated cells and are recognized by CD8+ T cells. However, the HLA class II antigens (-DR, -DQ, and -DP) have a selected tissue distribution, are expressed on the cell surface of the antigen presenting cells, and are recognized by CD4+ T cells. The HLA class II antigens may also be expressed on a variety of nonimmune cell types on stimulation with IFN-γ. Each individual inherits six or seven HLA antigens (three class I and three or four class II) from each parent. All such antigens are codominantly expressed on the cell surface. The entire set of HLA-A, -B, -C, -DR(-DR51/52/53), -DQ, and -DP antigens encoded on chromosome 6 is called a haplotype.

Kidney, heart, liver, lung, skin, pancreas, cornea, bone marrow, and other hematopoietic stem cells can be transplanted from one human to another. Transplanted organs, tissues, and cells (except in the case of identical twins) are called allografts, indicating genetic differences between the donor and the recipient. When a transplant is performed, compatibility (matching) between donor and recipient HLA increases the chance for successful long-term allograft survival. For example, if the solid organ donor and recipient are not HLA-matched, the recipient's immune system (primarily mediated by T and B lymphocytes) recognizes the donor cells as nonself (foreign) and will mount an immune response against the transplanted organ resulting in its rejection and subsequent loss of function.

Because HLA differences lead to an alloreactive immune response, transplant physicians attempt to match donor and recipient HLA antigens and alleles. As indicated in Table 18.1, better-matched grafts have better survival. Thus, it is necessary to determine the HLA type in prospective donor–recipient pairs prior to transplantation. HLA matching between donors and recipients prevents graft rejection in solid organ transplantation and prevents both rejection and graft-versus-host disease (GvHD) in hematopoietic stem cell transplantation (HSCT). Clinical histocompatibility testing involves typing of the HLA antigens or the genes encoding these antigens and testing for the presence of HLA antibodies in the serum of a patient that are directed against the HLA antigens of the donor (see Chapter 19 for details of HLA antibody testing). HLA typing is performed in the laboratory using either serological or molecular techniques. The level of typing resolution (Table 18.2) depends on the type of transplant and type of donor. Traditionally, solid organ transplant patients and donors have been typed for the HLA-A, -B, and -DR antigens using serological methods or the more recent low-resolution molecular typing techniques. In contrast, HSCT with unrelated donors requires high-resolution typing of at least HLA-A, -B, -C, and -DRB1 loci.

In the last decade, a large number of laboratories have replaced serological typing methods with DNA-based typing techniques and currently perform

TABLE 18.1
Influence of HLA Matching on Kidney Graft Failure after 5 Years

HLA Mismatches	Deceased Donor Graft Failure Rate (%)	Living Donor Graft Failure Rate (%)
0	28.7 ± 1.0	13.8 ± 0.7
1	31.2 ± 1.2	22.8 ± 1.3
2	35.4 ± 0.7	24.7 ± 0.8
3	37.0 ± 0.5	25.5 ± 0.7
4	38.4 ± 0.5	22.0 ±1.6
5	41.1 ± 0.7	24.0 ± 1.9
6	43.4 ± 1.2	29.9 ± 3.1

Source: Data derived from the 2004 Annual report by UNOS (http://www.unos.org).

TABLE 18.2
Typing Resolution and Techniques

Level of Resolution	Technique	Example
Low	Serological	HLA-A2
Low	Molecular	HLA-A*02
High	Molecular	HLA-A*0201

low-resolution typing for the HLA-A, -B, -C, -DRB1, -DRB3/4/5, and DQB1 loci. New molecular techniques have significantly improved histocompatibility testing for clinical transplantation and histocompatibility laboratory's accuracy and efficiency. Recent studies have shown that approximately 12% of kidney allografts are lost during the first year posttransplantation due to rejection despite complete HLA matching by conventional methods [1]. This observation suggests that there could be undetected HLA mismatches in these recipient–donor pairs or mismatches in other non-HLA antigenic systems that may also play an important role in allograft rejection. Additional data suggests that low-resolution typing may no longer be considered adequate for solid organ transplant patients [2,3]. By the use of new emerging molecular techniques, the laboratory should be able to detect true HLA matches from a list of potentially related or unrelated donors in both solid organ and HSCT. Greater extent of HLA matching results in better graft survival, less GvHD, and requires less immunosuppression following transplantation.

HLA typing results play an integral role in the final medical decision for donor selection and evaluation of graft rejection in transplantation. An ongoing interactive communication between the transplant program and the HLA laboratory facilitates the correct analysis and interpretation of the testing results in the clinical context of specific patients.

18.2 THE MAJOR HISTOCOMPATIBILITY COMPLEX

The genetic contribution of the MHC to allograft rejection was first proposed by Bover [4], who observed that skin grafts between identical twins were not rejected like grafts were from genetically distinct individuals. The genes involved in the rejection process were first described in mice by Gorer [5]. Subsequently, Snell [6] used mouse cell lines to further define a genetic locus, which was called H for histocompatibility. Gorer [5] referred to the gene products of this locus as antigens II and the combined term H-2 was subsequently used for the mouse MHC system.

The MHC system in humans (HLA) was subsequently discovered in the early 1950s. Several investigators independently noted that blood from multiparous women or from previously transfused individuals contained antibodies that agglutinated leukocytes [7–9]. The latter discovery led to serological typing methods that originally identified a genetic system that was further split into two loci: HLA-A and -B. Initially, a variety of techniques were used for serologic testing and the microlymphocytotoxicity assay became the most widely used one [10,11]. Later it was observed that when cultured together in a mixed lymphocyte culture (MLC), lymphocytes from unrelated individuals identical in the HLA-A and -B antigens showed a vigorous proliferative response *in vitro* [12]. This observation led to the discovery of an additional locus initially called HLA-D [13]; it was later shown that mismatches in the HLA-DR and HLA-DQ gene products contributed to this stimulation. Detailed serologic analysis led to the discovery of HLA-C. The HLA-DP locus, initially called SB, was discovered while performing secondary stimulation tests that utilized primed lymphocytes (PLT) that revealed the recognition of other HLA antigens in addition to those recognized in primary MLCs.

The HLA complex is divided into three genetic regions containing genes of different classes (I, II, and III), all encoded by a gene complex located on the short arm of chromosome 6 (Figure 18.1). The class I and II regions of the HLA complex

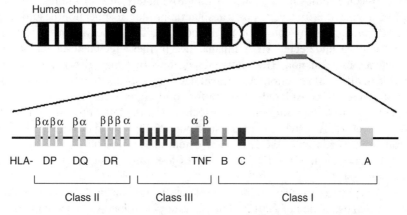

FIGURE 18.1 Diagram of the human major histocompatibility complex (HLA) on chromosome 6. HLA class I region is 3–6 kb long and HLA class II region is 4–11 kb long. HLA class III and TNF-α and TNF-β genes are not part of the polymorphic HLA system.

include genes coding for the HLA class I and II histocompatibility antigens. Between these two regions is found the class III region that contains genes encoding molecules involved in immune function that are not targets for allorecognition. Several cytokine genes such as TNF-α and -β are located inside the main HLA complex. In addition to the main HLA complex, the extended MHC complex covers 8 Mb and includes the hemachromatosis (*HFE*) gene, the farthest telomeric locus in the complex and the *Tapasin* gene (required for antigenic peptide processing), the farthest centromeric locus of the complex. In the center of the class II region there are genes that encode for intracellular proteins involved in antigen processing and presentation (TAP, subunits of the immune proteasome and HLA-DO).

As mentioned earlier, the gene products of the HLA class I, II, and III proteins are expressed differently on different tissues (Table 18.3). Despite the discovery of the HLA gene products as mediators of transplant rejection, recognition of allografts is not the main function of these proteins. HLA molecules appear on the surface of cells of the immune system allowing cell-to-cell communication during immune functions. As shown in Figure 18.2, HLA class I molecules consist of a long (heavy) glycoprotein of 45 kDa associated with a smaller protein of 12 kDa called the β2-microglobulin (β2m) encoded by a gene located in chromosome 15. The two chains are associated with one another on the cell surface by noncovalent bonds. The heavy chain has a transmembrane polypeptide anchoring the complex at the surface of the cell. HLA class II molecules are composed of two transmembrane proteins, the α chain of 33–35 kDa and the β chain of 26–28 kDa. The two proteins associate forming a groove, which will hold fragments of antigen that have been engulfed and processed by the cell (extracellular antigens). In contrast, antigens bound to HLA class I molecules are generated from macromolecules synthesized within the cell (intracellular antigens). HLA class I and II molecules present the antigen peptide fragments to CD8+ and CD4+ T lymphocytes, respectively (approximately 9 and

TABLE 18.3
Expression of MHC Genes

MHC Region	Gene Products	Tissue Location	Function
HLA class I	HLA-A, -B, -C	Nucleated cells	Recognition of tumor and virus-infected cells by CD8+ T lymphocytes
HLA class II	HLA-DR, -DQ, -DP	Antigen-presenting cells B lymphocytes Macrophages Dendritic cells Endothelial cells	Recognition of foreign antigens by CD4+ T lymphocytes
HLA class III	Complement C2, C4, B	Plasma	Lysis of extracellular pathogens

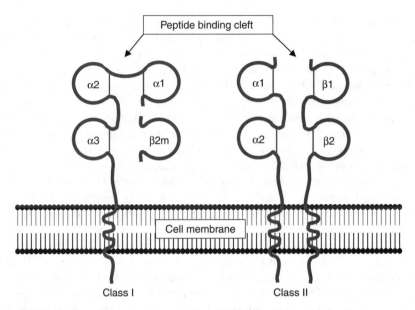

FIGURE 18.2 Structure of the HLA class I and class II molecules. HLA class I molecules are comprised of one polymorphic heavy chain (α) associated with a light chain called the β2-microglobulin (β2m). HLA class II molecules are composed of two polymorphic chains (α and β).

15 amino acids long, respectively). HLA class I and II molecules vary from one another (polymorphism), sometimes only by a single amino acid. Owing to these polymorphisms, different HLA molecules vary in their efficiency of binding antigen fragments resulting in a range of immune responses to a given antigen. This distinction can affect symptoms of disease, for example, the susceptibility of individuals of a particular HLA type infected with human immunodeficiency virus (HIV) to develop acquired immunodeficiency syndrome (AIDS) [14].

18.2.1 HUMAN LEUKOCYTE ANTIGEN POLYMORPHISMS

Genes of the HLA complex are the most polymorphic of the human genome. As mentioned earlier, polymorphisms in these loci were first defined phenotypically by acceptance or rejection of tissue or by reaction with defined antibodies (serological typing). Molecular typing methods reveal HLA polymorphisms that range from single nucleotide polymorphisms to loss or gain of entire genetic regions. The identification of HLA polymorphisms and typing of alloantigens was initially done by serologic and cell proliferation methodologies. These methodologies were successfully utilized to initially characterize the system; however, in spite of their broad application, there were limitations in terms of reproducibility, accuracy, and sample demands [15]. Alloantisera are usually in limited supply and both serologic and cellular assays require live cells to perform HLA typing. The main limitation of serologic typing is its inability to recognize microheterogeneity that may elicit T-cell allorecognition.

A general feature of the highly polymorphic HLA class I and II genes is that the distal membrane domains present a high degree of variability, whereas the proximal membrane domains as well as the transmembrane and cytoplasmic domains have limited or no polymorphism within each locus. The heavy chain of the HLA class I molecules comprises three extracellular domains and the α and β subunits of the HLA class II molecules each contain two extracellular domains. The genes encoding the HLA class I and II genes, like all eukaryotic genes, are discontinuous containing coding (exons) and noncoding (introns) segments. The class I genes contain eight exons whereas the class II genes contain six or seven exons.

The wide application of molecular methods allowed the characterization of numerous alleles in all HLA class I and II loci [16,17]. The analysis of the nucleotide sequences indicates that the vast majority of the nucleotide polymorphisms occur in exon 2 of the HLA class II genes and in exons 2 and 3 of the HLA class I genes. These exons encode for the distal membrane domains. It has been noted that most nucleotide polymorphisms in these exons involve changes that determine substitutions in the corresponding amino acids (nonsynonymous substitutions) and correlate well with phenotypic differences detected by cellular and serologic methods [15]. However, serological equivalents are not available for all described alleles. Furthermore, it is difficult to predict the serological specificities of alleles with polymorphisms corresponding to more than one antigenic group [18].

It has been observed that most of the distinguishing sequences between alleles are restricted to some segments of the gene (variable regions). Pairs of alleles associated to the same serotype differ only by a few nucleotide sequences, while distinguishing sequence motifs can be found in alleles of other serotype indicating the patchwork nature of the HLA polymorphism. The likely nature of HLA variation probably arose from the existence of a few allelic lineages followed by short segmental exchanges that significantly increased the number of alleles at a given locus. It appears that most of the polymorphism in the HLA region was generated by these mechanisms. Selected events have been necessary for novel alleles to reach significant population frequencies. However, it should not be ignored that some alleles have arisen from single-point mutations.

With the understanding of the DNA sequence variations and the ease of molecular techniques, several molecular methods were developed to type HLA alleles. These methodologies focused on the analysis of polymorphisms in exon 2 of the HLA class II genes and of exons 2 and 3 of HLA class I genes. The application of DNA-based procedures resulted in accurate, reproducible, and sensitive HLA typing [19]. The wide application of these procedures led to the identification of many novel alleles, some of them were undetectable with the preexisting serologic reagents [18]. The molecular methods are widely employed and take advantage of the simplicity of genomic DNA amplification by the polymerase chain reaction (PCR). The distinguishing sequences can be detected by sequence-specific oligonucleotide probe hybridization (SSOPH), amplification with sequence-specific priming (SSP), or determining the nucleotide sequence (sequence-based typing [SBT]).

18.2.2 Human Leukocyte Antigen Nomenclature

The polymorphic nature of the HLA system means that there are multiple alleles present in the human population. A standard nomenclature for expressing serologically defined antigens has been established by the World Health Organization (WHO) nomenclature committee [16,17]. HLA refers to the entire genetic region whereas A, B, C, DR, DQ, and DP each refer to a particular locus. A small w is included in HLA-C allele nomenclature. This connotation was originally a designation of alleles in workshop status and it is retained to distinguish it from the C designation of the complement genes.

A list of the serologically defined antigens accepted by the WHO is shown in Table 18.4. The WHO official nomenclature refers to serologically defined antigens by a number following the gene region name, for example, HLA-B51 denotes the HLA-B antigen 51. Subtypes from a broad specificity are followed by the number of the parent antigen in parentheses. For example, HLA-A24(9) denotes the HLA-A antigen 24 from parent, antigen 9. The derived antigens are called split specificities. Additional antigens have been defined by antiserum reactivity. The current number of class I and II antigens are shown in Table 18.5.

TABLE 18.4
Serologically Defined HLA Specificities

HLA-A	HLA-B	HLA-C	HLA-DR	HLA-DQ	HLA-DP
A1	B5	Cw1	DR1, DR103	DQ1	DPw1
A2, A203, A210	B51(5), B5102, B5103	Cw2	DR2	DQ5(1)	DPw2
A3	B52(5)	Cw3	DR15(2)	DQ6(1)	DPw3
A9	B7, B703	Cw9(w3)	DR16(2)	DQ2	DPw4
A23(9)	B8	Cw10(w3)	DR3	DQ3	DPw5
A24(9), A2403	B12	Cw4	DR17(3)	DQ7(3)	DPw6
A10	B44(12)	Cw5	DR18(3)	DQ8(3)	
A25(10)	B45(12)	Cw6	DR4	DQ9(3)	
A26(10)	B13	Cw7	DR5	DQ4	
A34(10)	B14	Cw8	DR11(5)		
A66(10)	B64(14)		DR12(5)		
A11	B65(14)		DR6		
A19	B15		DR13(6)		
A74(19)	B62(15)		DR14(6), DR1403, DR1404		
	B63(15)		DR7		
	B75(15)		DR8		
A29(19)	B76(15)		DR9		

TABLE 18.4   (continued)
Serologically Defined HLA Specificities

HLA-A	HLA-B	HLA-C	HLA-DR	HLA-DQ	HLA-DP
A30(19)	B77(15)		DR10		
A31(19)	B16		DR51		
A32(19)	B38(16)		DR52		
A33(19)	B39(16),		DR53		
A28	B3901,				
A68(28)	B3902				
A69(28)					
A36	B17				
A43	B57(17)				
A80	B58(17)				
	B18				
	B21				
	B49(12)				
	B50(12)				
	B22				
	B54(22)				
	B55(22)				
	B56(22)				
	B27, B2708				
	B35				
	B37				
	B40, B4005				
	B60(40)				
	B61(40)				
	B41				
	B42				
	B46				
	B47				
	B48				
	B53				
	B59				
	B67				
	B70				
	B71(70)				
	B72(70)				
	B73				
	B7801				
	B81				

Note: Broad antigen specificities are listed in parentheses. Associated antigens such as A2, A203, and A210 are listed together.

TABLE 18.5

HLA Specificities Identified by Serology Versus Molecular Methods (as of February 2007)

Gene	Serology	Molecular
HLA class I		
HLA-A	28	506
HLA-B	62	851
HLA-C	10	276
HLA class II		
HLA-DRA1	0	3
HLA-DRB1	25	476
HLA-DRB3	1	44
HLA-DRB4	1	13
HLA-DRB5	1	18
HLA-DQA1	0	34
HLA-DQB1	9	81
HLA-DPA1	0	23
HLA-DPB1	6	126

With the introduction of molecular biology techniques in the 1980s, HLA typing at the DNA level required nomenclature for specific DNA sequences. Many new alleles continue to be defined at the DNA level [20–23]. A revised nomenclature is used for denoting alleles at the DNA level. The gene name, such as *HLA-DRB1* is followed by an asterisk (*) and the allele family number (equivalent to the serological antigen) followed by a number for the specific allele (DNA sequence). For example, DRB1*1502 is the second specific allele 02 of the HLA-DRB1*15 family. The letter N following the specific allele number indicates lack of expression or null allele. For example, in the B*1307N type, the 07 allele of the 13 allele family in the HLA-B locus is not expressed. This null allele is due to a 15 base pair deletion in the gene at the site of the 07 allele. Null alleles can also be the result of nonsense, frameshift, splice site, or other premature stop mutations. The letters L and S indicate low expression or soluble molecules, respectively.

Silent mutations (changes in the DNA sequence that do not change the amino acid sequence), also called synonymous mutations, are designated by a number following the specific allele number. For example, A*020103(A*020103) indicates synonymous allele 03 of the first specific allele (A*020101) from the HLA-A*0201 family. Seventh and eighth number designations denote changes outside the coding regions (exons) of the genes.

The National Marrow Donor Program (NMDP) assigns alphabetical allele codes to allele combinations from submitted requests [24]. Generic codes can be used with several loci and allele families. For example, the combination of alleles 01/03 in any *HLA* gene is designated as AC so that B*1501/1503 = B*15AC. The lists of these codes and submissions for new codes are available at http://bioinformatics.nmdp.org/.

In this regard, the term ambiguity is the inability of the typing system to discriminate among several possible allele combinations as they are analyzed as a single reaction. Ambiguity is designated as/between the possible allele numbers. For example, if a typing test results in either B*0701 or B*0702, the notation is B*0701/0702. Ambiguity also arises from the inability of some typing methods to assign heterozygous alleles to one or the other chromosome. The term "resolution" is the level of detail with which the allele is determined (Table 18.2). Low resolution identifies broad allele types or groups of alleles using a two-digit nomenclature. A designation of A*26 is low resolution, which can be determined at the serological level (Table 18.2). Typing methods that detect specific alleles, in addition to identification of all serological types, are at medium resolution. Typing result A*2601/05/10 is medium resolution. High-resolution typing procedures can discriminate between almost all specific alleles. Thus, a typing of A*2601 is high resolution determined by DNA analysis. A range of methods from serological typing to direct DNA sequence analysis provides the laboratory with a choice of low-, medium-, or high-resolution typing (Table 18.2).

18.3 HUMAN LEUKOCYTE ANTIGEN TYPING

18.3.1 SEROLOGICAL TYPING

As mentioned earlier, HLA typing for organ transplantation has traditionally been performed serologically using alloantibodies of known HLA specificity to identify unknown cellular antigens. Although serological testing yields only low-resolution typing results, there are some advantages to this method. Serological typing is a relatively rapid method and reveals immunologically relevant epitopes. In addition, serological typing can be used to resolve some ambiguities or to confirm null alleles detected by molecular methods. Serological tests include HLA phenotype determination where patient cells are tested with known alloantisera.

18.3.2 MOLECULAR TYPING

Molecular methods are now available to define HLA alleles. The ability to amplify DNA segments by PCR has facilitated the application of these techniques. The PCR-based methods can be broadly classified into three categories according to the readout used. First, those that generate PCR products containing internally located polymorphisms that can be identified by a secondary technique, such as SSOP, SBT, or by other techniques involving digestion with restriction enzymes that yield characteristic restriction fragment length polymorphisms (RFLP). Second, those in which the polymorphisms are identified directly by the PCR process, without further steps, such as SSP. And third, methods in which the changes introduced by the nucleotide substitution result in detectable conformational changes in the physical characteristics of different alleles. These changes are identified by electrophoretic analysis such as hetroduplex analysis, single-strand conformational polymorphism (SSCP), denaturing gradient gel electrophoresis (DDGE), and temperature-gradient gel electrophoresis (TGGE).

The use of specific techniques depends on the laboratory requirements for resolution, clinical urgency, and sample volume. The different techniques have different requirements in terms of skills of laboratory workers, equipment, and costs. Currently, most laboratories utilize techniques involving detection by SSP or detection by either hybridization (SSOP) or nucleotide sequences (SBT). Many laboratories utilize combinations of these techniques to achieve the final results at the desired level of resolution. All of these methods should continually be evaluated to allow updates of newly described alleles and also to ensure that these alleles are consistently detected. The clinical applications of HLA typing will determine the level of resolution required, either high or low. In solid organ transplantation, the resolution required should at least parallel the resolution achieved by serologic typing, whereas in stem cell transplantation the resolution required needs to identify alleles with subtle differences that may elicit allorecognition by both B and T lymphocytes. In recent years, it has been recognized that alloantibodies may also recognize subspecificities; therefore higher resolution may also be required in histocompatibility testing for solid organ transplantation.

18.3.2.1 Steps for Molecular Typing

Molecular typing techniques involve three general steps: (1) The extraction of genomic DNA, (2) The amplification of segments of the gene(s) of interest, and (3) The detection of the sequence polymorphisms that define the alleles or allow the distinction of allele differences.

18.3.2.1.1 DNA Extraction

Genomic DNA is extracted from nucleated cells, typically using whole blood as the source of nucleated cells. Only a few micrograms of genomic DNA are sufficient to complete molecular typing. DNA purity is an important factor to achieve successful typing results. To amplify short DNA fragments, a salting out method is adequate. However, to amplify longer fragments, other DNA extraction methods that yield higher purity are usually required.

18.3.2.1.2 DNA Amplification

DNA is amplified by repeating thermal cycling of the PCR mixture. The PCR mixture (including dNTPs, primers, Taq DNA polymerase, and genomic DNA) is subjected to repeated cycles of heating to 94–96°C for double-stranded DNA denaturation, cooling down to the corresponding temperature for primer annealing, and lastly, warming up to 72°C for optimal activity to integrate the complementary nucleotide to the single-stranded DNA. After one amplification cycle, the DNA copies serve as templates to allow an exponential growth of the PCR product. Since both strands from the fragment need to be amplified, the primers for those strands should be designed and amplified to include intervening segments between the two. To ensure efficiency of the PCR reaction, the number of the PCR cycles and incubation time at each temperature should be based on the length and GC content from both the segments to be amplified and the primers used.

18.3.2.1.3   Detection of Sequence Polymorphisms That Define Alleles

18.3.2.1.3.1   Sequence-Specific Priming

SSP is a rapid method of typing that uses sets of primer pairs to amplify specific region of genomic DNA. The efficiency of the amplification reaction is controlled by the primers that amplify conserved sequences of a selected gene. The 3' end of the PCR primer must match the template for recognition by the DNA polymerase. By designing primers with polymorphic sequences at the 3' end, successful amplification (generation of a PCR product) can be used to type specific alleles (allele-level typing) or group of alleles (allele-group-level typing). The final results are interpreted by analyzing the amplification pattern, detected on an agarose gel electrophoresis, obtained with a particular sample.

PCR–SSP reactions could be set up in a 96 well plate format with different allele-specific primer sets in each well. Each PCR reaction mixture contains the sequence-specific primers and a set of amplification control primers. The amplification control primers should yield products for every specimen (except the negative control). The sequence-specific (allele-specific) primers should only yield products if the specimen has the specific allele matching the allele-specific primer sequences. The amplification control primers are designed to yield a PCR product of distinct size from the product of the allele-specific primers. The two amplicons can then be resolved by agarose gel electrophoresis (Figure 18.3). Specimens should yield two PCR products (amplification control and allele-specific product) only from those PCR reactions containing primers matching the specimen HLA alleles. PCR reactions containing primers that do not match the patient's HLA alleles should only amplify the amplification control. This method is easy to implement in the laboratory and it is adequate for low-volume laboratories.

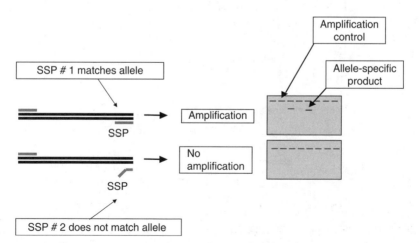

FIGURE 18.3   Principle of SSP. HLA alleles are amplified by PCR using SSPs. The PCR products are then detected by agarose gel electrophoresis. An amplification control is included with each reaction to detect false-negative results due to amplification failure.

18.3.2.1.3.2   Sequence-Specific Oligonucleotide Probe Hybridization

Specific PCR amplification of HLA alleles at a particular locus in the HLA region and subsequent probing of the product with probes immobilized on a nylon membrane is the method known as reverse SSOPH. As a result of conversion events generating polymorphism in the HLA system, a small nucleotide fragment of a particular allele is transferred to another allele. Therefore, many of the polymorphic region sequences are not allele-specific, that is, some regions tend to be shared by several different alleles. To identify a particular allele, several probes must be used. A battery of probes is required to differentiate the alleles, yet it is the pattern of reactivity with these probes that determines the HLA allele or allele-group type (Figure 18.4).

For this procedure, the HLA region under investigation is amplified using primers flanking the polymorphic sequences and labeled with biotin at the 5' end. Because the majority polymorphisms are located in exon 2 of the class II genes and exons 2 and 3 of the class I genes, the probes are designed to target these regions. The probes are short (19–20 bases) single-stranded DNA fragments designed to hybridize to specific HLA alleles. The probe sequences are based on sequence alignments available at http://www.ebi.ac.uk/imgt/hla/, with the polymorphic nucleotides located in the center of the probe sequence. Hybridization depends on the conditions of optimal binding of the probe matching a sequence in a particular allele. The amplified and labeled DNA bound to the immobilized probes is detected with a biotin-specific molecule,

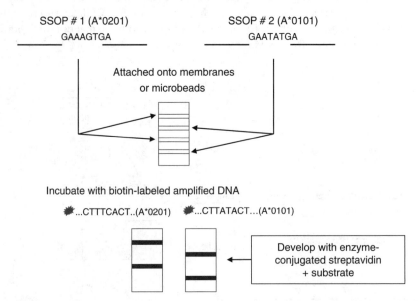

FIGURE 18.4 Principle of SSOPH. HLA genetic regions are amplified by PCR using generic primers covalently bound to biotin at the 5' end. The PCR product is then hybridized to panels of probes immobilized on a solid support, either membrane or beads. If the sequence of the amplified DNA matches and hybridizes to that of the probe, a secondary reaction with enzyme-conjugated streptavidin will produce a color when exposed to a substrate. If the sequence of the amplicon differs from that of the probe, no signal is generated.

streptavidin, conjugated with an enzyme. The enzyme will give a positive colorimetric signal on addition of the corresponding substrate. Panels of immobilized probes define specific alleles by the pattern with which probes bind to the amplified DNA under investigation. The number of probes used depends on the design of the assay. SSOPH is considered low to intermediate resolution depending on the number and types of probes used in the assay. For example, an intermediate resolution assay of the HLA-B locus might take 80–90 probes. Because some probes have multiple specificities, hybridization panels are complex and computer programs are used for accurate interpretation of the results. Recently, microbead array systems have been developed for the SSOPH strategy where fluorescently distinct beads carry the probes and the profile of beads with bound amplified DNA is detected with streptavidin conjugated with a fluorescent dye (phycoerythrin) in a bead microarray flow cytometry system (Luminex Corporation).

18.3.2.1.3.3 Sequence-Based Typing

The most accurate procedure and the gold standard for HLA typing is the direct identification of the complete nucleotide sequence of the HLA alleles carried by a DNA sample. The most widely used approach to detect the sequencing fragments is the dideoxy chain termination method [25] (Figure 18.5). The performance of

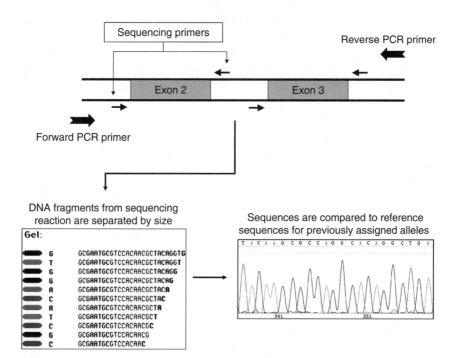

FIGURE 18.5 Principle of SBT. HLA genetic regions are amplified by PCR using locus-specific primers. The PCR products are then purified from unused PCR reaction components, sequencing reactions are performed using forward and reverse sequencing primers and these reactions are loaded onto the automated DNA sequencer to detect the nucleotide sequence of the targeted genes.

electrophoresis is usually assisted by the use of multiple dyes. When a single PCR reaction is performed for each HLA locus, simultaneous amplification and sequencing of both alleles carried by heterozygous samples is obtained, and heterozygous nucleotide assignments are observed at positions where both alleles have different nucleotides. Some heterozygous genotypes resulting in the same sequencing pattern result in ambiguous (alternative) sequencing types. Performance of additional tests targeting only one of the possible alleles either by sequencing primer or SSP usually resolve these ambiguous combinations, enabling physical separation of the two alleles of heterozygous samples.

SBT represents advantages over other procedures although analysis of the nucleotide sequences is more technically demanding than other methods.

18.4   ROLE OF HLA TYPING IN CLINICAL TRANSPLANTATION

18.4.1   Hematopoietic Stem Cell Transplantation

HSCT is an effective treatment for a variety of malignant and non-malignant hematological diseases. Matching of HLA-A, -B, -Cw, and -DRB1 is critical for the success of stem cell allografts. Mismatches of these HLA loci increase the risk of GvHD and graft rejection and reduce the probability of survival [26,27].

The first donor choice for a stem cell transplant patient is an HLA identical sibling. Individuals have 25% probability of being HLA identical to their siblings (Table 18.6 and Figure 18.6). Therefore, the first step on a donor search is screening all available siblings for the identification of a HLA identical sibling. If an HLA identical sibling is not available an unrelated donor search is initiated. The main sources of unrelated

TABLE 18.6
Family Segregation Analysis

HLA-	A*	B*	Cw*	DRB1*	DQB1*	Degree of Matching
Patient	0101	0702	0702	1302	0604	
	2501	1801	1203	0404	0302	
Father	0101	0702	0702	1302	0604	Haploidentical
	2402	3503	1203	1401	0503	
Mother	2301	4403	0401	0701	0202	Haploidentical
	2501	1801	1203	0404	0302	
Sibling 1	0101	0702	0702	1302	0604	Haploidentical
	2301	4403	0401	0701	0204	
Sibling 2	2402	3503	1203	1401	0503	Two-haplotype
	2301	4403	0401	0701	0202	mismatch
Sibling 3	0101	0702	0702	1302	0604	HLA identical
	2501	1801	1203	0404	0302	

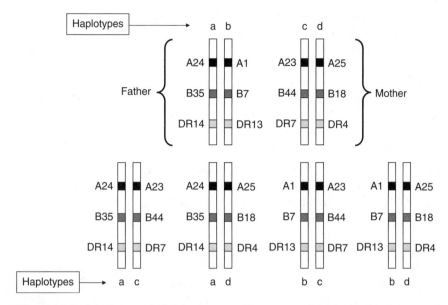

FIGURE 18.6   Inheritance of HLA haplotypes. *HLA* genes are inherited *en block* from each parent according to Mendelian laws.

donors are HLA typed volunteers participating in national and international bone marrow registries and umbilical cord blood units stored in national and international cord blood banks. Final donor confirmatory typing is performed either by the transplant program reference HLA laboratory or the donor registry reference laboratory.

Given the high level of polymorphism of the HLA system and the opportunity that even a single nucleotide difference between patient and donor alleles may initiate harmful immune reactions in the recipient, major challenges are faced in the HLA testing laboratory when performing high-resolution typing or allele-level typing. It is critical for the laboratory to utilize robust DNA-based typing technologies such as SBT to be able to accurately identify the HLA alleles. When a single allele is sequenced, this technique is capable of unambiguously differentiating the nucleotide bases of the amplified regions for that particular allele. However, when both alleles at a specific locus are sequenced on a single reaction, several alternative allele combinations may be obtained (ambiguities) as final results from the original locus-specific amplification. The laboratory may need to use additional testing strategies (SSP, sequencing) to resolve these original ambiguities.

Another challenge faced in the HLA testing laboratory is the identification of null alleles. Although null alleles have extensive DNA sequence similarity to expressed alleles, null alleles are not expressed on the cell surface and therefore are immunologically undetectable. Inability of the laboratory to identify null alleles of the donor or the recipient cells may lead to selection of a donor with a major antigen mismatch. For example, if the donor carries a Cw*0409N nonexpressed allele and the recipient carries a Cw*0401 allele, there is a major HLA-C mismatch in the graft-versus-host direction.

The minimum requirement for unrelated bone marrow donor typing is HLA-A, -B, -C, and -DRB1 at the high-resolution level (Table 18.2). The final HLA matching criteria is established by the transplant center and many transplant centers have even higher typing criteria including -DQB1 and -DPB1. Although many reports indicate that the level of matching required for successful unrelated umbilical cord blood transplants is lower than for bone marrow transplants, the minimum matching criteria have not clearly been established.

18.4.2   KIDNEY TRANSPLANTATION

Kidney transplantation is the treatment of choice for patients with end-stage renal disease. Major advances in this field were introduced in the 1980s with the introduction of calcineurin inhibitors (CI). Treatment with cyclosporine, tracolimus, and other CI has improved short-term (one-year) graft survival. However, no significant improvement in long-term graft survival have been observed in the past 25 years. A study of more than 97,000 kidney transplants performed in various centers in the United States comparing 10-year graft survival before and after the introduction of the CI drugs in transplantation shows that—although the overall survival has improved with these drugs—there is a consistent 18% difference in survival between six-antigen mismatches and zero-antigen mismatch [28]. This difference in survival has remained unchanged regardless of the therapy used. This study also shows the significant impact that a single HLA-A, -B, -DR mismatch exists on 10-year graft survival [28].

Overall, the rate of yearly graft loss remains at 2–5% per year. One of the main causes of long-term graft loss is chronic allograft nephropathy (CAN) as the end result of immunological and nonimmunological factors [29]. The immunological risk factors for the development of CAN are acute rejection episodes [30] and HLA mismatches between donor and recipient [28,31]. There is strong evidence that HLA matching improves graft survival and reduces the incidence of graft rejection in kidney transplants.

Standard complement-dependent cytotoxicity assay has been traditionally used for HLA-typing of kidney transplant patients. Advances in DNA-based testing technologies and the introduction of very sensitive solid-phase-based antibody detection assays have indicated that serological typing may no longer be sufficient to select the best donor or to improve long-term kidney graft survival [2,3].

18.4.3   HEART TRANSPLANTATION

Patients with congenital abnormalities, ischemic cardiomyopathy, and other causes of severe heart failure are candidates for heart transplantation. Long-term allograft and patient survival after heart transplantation have steadily improved due to advances in surgical techniques and immunosuppressive protocols [32]. Nevertheless, due to the development of chronic allograft rejection, long-term outcome of transplant recipients is still quite limited. Several studies have revealed significant benefits of HLA matching in heart transplantation [33–36]. A multicenter study of the UNOS registry yields clear results in favor of prospective HLA matching in cardiac transplantation [36]. An European multicenter study also shows that the extent of

HLA matching significantly influences allograft survival after heart transplantation independently of age, cold ischemic time, gender, type of underlying disease, and treatment [33]. Two or more HLA-A, -B, or -DR mismatches were associated with a 25% increased risk for graft failure within 3 years when compared with patients having zero to one mismatches.

Chronic heart allograft rejection or cardiac allograft vasculopathy (CAV) is a progressive obliterative form of atherosclerosis that represents the main cause of long-term morbidity and mortality after heart transplantation [37]. The main risk factors for CAV include the immune response to the allograft, ischemia-reperfusion injury, and viral infections [37]. Related studies have demonstrated that HLA matching significantly improves long-term patient survival after heart transplantation [38,39]. In patients with zero, one, and two HLA-DR mismatches, freedom from CAV after 5 years was 89%, 61%, and 54%, respectively [39]. Recent studies have also demonstrated that a significant proportion of allograft rejection is mediated by anti-HLA antibodies [40]. With regards to cardiac transplantation, it has been demonstrated that activation and deposition of the complement component C4d correlates with the presence of circulating anti-HLA antibodies leading to accelerated CAV and allograft failure [41,42].

The concept of prospective HLA compatibility is widely accepted in kidney transplantation. Thus, by prioritizing well-matched patients on the waiting lists, a further amelioration of long-term outcome in kidney transplant patients has become possible. However, despite our growing understanding about the immunological mechanisms of heart allograft rejection and the evidence for the beneficial effects of HLA matching, prospective HLA matching for heart transplantation has not been implemented mainly due to the short cold ischemic times allowed for thoracic organs.

18.4.4 LUNG TRANSPLANTATION

Lung transplant is an effective treatment option for patients with end-stage lung disease and a large number of lung transplants are performed worldwide each year. However, long-term allograft survival is limited by the high incidence of bronchiolitis obliterans syndrome (BOS), which affects up to 60% of patients at 5 years posttransplant and up to 90–100% at 9 years posttransplant [43–45].

It has been reported that HLA class I and -II mismatches, as well as acute rejection are the main risk factors for BOS development [46–48]. These data advocate for prospective HLA matching as an important determining factor in the organ allocation algorithm. As mentioned earlier for heart transplantation, due to time constraint and limited number of donors, HLA matching between donor and recipient is not performed in lung transplant patients.

18.4.5 LIVER TRANSPLANTATION

In contrast to the widely recognized positive effect of HLA matching on kidney, heart, lung, and hematopoietic stem cell transplant outcomes, no beneficial effects of HLA matching on liver transplants have been clearly demonstrated. Furthermore, a dualistic effect of HLA matching was reported in the late 1980s [49]. This study reported lower incidence of graft rejection as well as higher incidence of recurrence of disease in patients with HLA-matched liver transplants [49]. These observations were confirmed by subsequent studies [50,51]. A recent study of a large number of patients in the Organ Procurement and Transplantation Network (OPTN) database analyzed the effect of HLA compatibility on 5-year graft survival [52]. This study did not find a clinically significant effect of HLA compatibility on liver graft survival [52].

Extent and type of HLA matching between donor and liver transplant recipient have clinical significance in a number of situations. Increased risk of fatal GvHD has been reported in living related liver transplants with homozygous donors to haploidentical recipients [52–57]. Increased risk of graft loss has been reported in HLA-DR13 matched liver transplant patients with positive hepatitis B virus and primary sclerosing cholangitis [58]. Therefore, HLA typing of liver transplant recipients and potential donors should be considered for donor selection and risk evaluation.

18.5 CONCLUDING REMARKS

The preponderant role of HLA mismatching as a risk factor for early graft loss after solid organ transplantation, for lack of graft engraftment, and development of GvHD after HSCT has been firmly established. As a result of improvement in surgical techniques, immunosuppressive protocols, and management of peri- and postoperative complications, short-term allograft survival has greatly increased during the last years. However, long-term allograft survival is still limited by the development of chronic rejection. The current premise is that chronic rejection represents a common lesion in which different inflammatory insults such as ischemia-reperfusion, rejection, and infection can lead to similar histological and clinical outcomes. Clearly, the process of chronic rejection is a synergy of humoral and cellular immune components and soluble immune mediators as well as nonimmunologic factors [59]. However, a growing body of evidence suggests that chronic rejection is mainly the result of the immune response developed against mismatched HLA molecules of the allograft. This is supported by the observation that the main risk factors for the development of chronic rejection after solid organ transplantation and GvHD after HSCT include HLA class I or II mismatches between donor and recipient, and the severity and frequency of acute rejection episodes.

During allograft rejection, donor HLA class I and II molecules are directly recognized by CD8+ and CD4+ T cells, respectively. It has been suggested that acute rejection is predominantly mediated through direct allorecognition of donor antigen-presenting cells within the graft. However, the succeeding decline in the number of donor antigen-presenting cells within the graft suggests that chronic allograft rejection is predominantly mediated by CD4+ T cells activated through indirect allorecognition of donor HLA-derived peptides presented by recipient antigen-presenting cells. Because of the continuous supply of recipient antigen-presenting cells into the graft, alloreactive CD4+ T cells perpetuate the rejection process through an indolent delayed-type hypersensitivity response that provides the cytokines required for the expansion of alloreactive CD4+ T cells and, eventually, the development of chronic rejection.

Several studies have shown that development of anti-HLA class I and II antibodies are associated with the development of chronic rejection and graft loss after kidney, heart, and lung allograft transplantation [59–61]. Related studies have also shown that preexisting anti-HLA antibodies increase the rate of early kidney, heart, and lung allograft dysfunction [62,63]. Although it is possible that the *de novo* anti-HLA alloantibody production occurs as an epiphenomenon as a result of the activation of cellular immune mechanisms during the development of chronic rejection, there is now compelling evidence that the humoral immune response has a direct contribution to graft tissue injury. Several studies have demonstrated that anti-HLA class I antibodies induced intracellular signal transduction in endothelial and epithelial cells resulting in cellular proliferation, production of growth factors, and apoptosis [59,64].

Thus, all the manifestations of chronic rejection and GvHD derive from a fundamental failure of current immunosuppressive agents to effectively control an indolent immune response mainly to mismatched HLA class I and II antigens. In this regard, it has been suggested that CD4+ T cells primed by the indirect allorecognition pathway are less responsive to conventional immunosuppression than those primed by the direct allorecognition pathway [65]. To prevent chronic rejection, it is imperative that better methods and agents that block the indirect pathway be instituted and better prospective HLA matching be implemented to prevent further cycles of T cell activation, development of anti-HLA antibodies, and subsequent graft dysfunction or GvHD. As organ donation rates plateau, it is even more imperative to extend the usable life of each transplanted organ. This goal will be achieved only by understanding the process of chronic rejection and GvHD and improving immune regulation.

REFERENCES

1. Bresnahan, B.A. et al., Comparison between recipients receiving matched kidney and those receiving mismatched kidney from the same cadaver donor, *Am J Transplant*, 2, 366, 2002.
2. Ferraz, A.S. et al., Comparative study of HLA-DR typing by serology and sequence-specific primer analysis in a genetically highly diverse population of kidney transplant recipients, *Transplant Proc*, 34, 463, 2002.
3. Bray R.A. and Gebel H.M., Low-resolution HLA class II typing for solid organ transplantation is inadequate, *Transplant Proc,* 33, 500, 2001.
4. Bover, K., Homoisotransplantation van Epidermis bei eineigen Zwillingen, *Beitrage Zur Klinischen Chirurgica*, 141, 442, 1927.
5. Gorer, P., The genetic and antigenic basis for tumor transplantation, *J Pathol Bacteriol*, 44, 691, 1937.
6. Snell, G., Methods for the study of histocompatibility genes, *J Genet,* 49, 87, 1948.
7. Dausset, J., Iso-leuco-anticorps, *Acta Haematol (Basel)*, 20, 156, 1958.
8. van Rood, J.J., Leucocyte grouping. A method and its application, *Doctoral thesis.* Drukkerij Pasmans, Den Haag, 1962.
9. Payne, R., The development and persistence of leukoagglutinins in parous women, *Blood*, 19, 411, 1962.
10. Terasaki, P.I. and McClelland, J.D., Microdroplet assay of human serum cytotoxins, *Nature*, 204, 998, 1964.
11. Terasaki, P.I. and Rich, N.E., Quantitative determination of antibody and complement directed against lymphocytes, *J Immunol*, 92, 128, 1964.

12. Bach, F.H. and Amos, B., Hu-1: major histocompatibility locus in man, *Science*, 156, 1506, 1967.

13. Yunis, E.J. et al., Anomalous MLR responsiveness among siblings, *Transplant Proc*, 3, 118, 1971.

14. Heeney, J. et al., Origins of HIV and the evolution of resistance to AIDS, *Science*, 313, 462, 2006.

15. Noreen, H.J. et al., Validation of DNA-based HLA-A and HLA-B testing of volunteers for a bone marrow registry through testing with serology, *Tissue Antigens*, 57, 221, 2001.

16. Marsh, S.G.E. et al., Nomenclature for factors of the HLA system, *Tissue Antigens*, 65, 301, 2005.

17. http://www.ebi.ac.uk/imgt/hla/align.html.

18. Rodríguez, S.G. et al., Identification of a new allele, DRB1*1204, during routine PCR-SSOP typing of National Marrow Donor Program volunteers, *Tissue Antigens*, 48, 221, 1996.

19. Hurley, C.K. et al., Large-scale DNA-based typing of HLA-A and HLA-B at low resolution is highly accurate, specific, and reliable, *Tissue Antigens*, 55, 352, 2000.

20. Steiner, N. et al., Twenty nine new HLA-B alleles associated with antigens in the 5C CREG, *Tissue Antigens*, 57, 481, 2001.

21. Hurley, C.K. et al., Twelve novel HLA-B*15 alleles carrying previously observed sequence motifs are placed into B*15 subgroups, *Tissue Antigens*, 57, 474, 2001.

22. Steiner, N. et al., Novel HLA-B alleles associated with antigens in the 7C CREG, *Tissue Antigens*, 57, 486, 2001.

23. Rodríguez, S.G. et al., Molecular characterization of HLA-B71 from an African American individual, *Human Immunol*, 37, 192, 1993.

24. http://bioinformatics.nmdp.org/HLA/allele_code_lists.html.

25. Sanger, F. et al., DNA sequencing with chain-terminating inhibitors, *PNAS*, 74, 5463, 1977.

26. Petersdorf, E.W., HLA matching in allogeneic stem cell transplantation, *Curr Opin Hematol*, 11, 386, 2004.

27. Flomenberg, N. et al., Impact of HLA class I and II high-resolution matching on outcomes of unrelated donor bone marrow transplantation: HLA-C mismatching is associated with a strong adverse effect on transplantation outcome, *Blood*, 104, 1923, 2004.

28. Cecka, J.M., The effect of HLA mismatches on cadaver kidney graft survival-still with us after all these years, in *Visuals of the Clinical Histocompatibility Workshop 2003*, Terasaki, P.I. (Ed.), One Lambda, Canoga Park, CA, 2003, p. 84.

29. Pascual, M. et al., Strategies to improve long-term outcomes after renal transplantation, *N Engl J Med*, 346, 580, 2002.

30. Hariharan, S. et al., Improved graft survival after renal transplantation in the United States, 1988 to 1996, *N Engl J Med*, 342, 605, 2000.

31. Opelz, G., Factors influencing kidney long-term graft loss. The collaborative transplant study, *Transplant Proc*, 32, 647, 2000.

32. Trulock, E.P. et al., Registry of the international society for heart and lung transplantation: twenty-second official adult lung and heart lung transplant report-2005, *J Heart Lung Transplant*, 24, 956, 2005.

33. Opelz, G. and Wujciak, T., The influence of HLA compatibility on graft survival after heart transplantation, *N Engl J Med*, 330, 816, 1994.

34. Kerman, R.H. et al., The relationship among donor–recipient HLA mismatches, rejection, and death from coronary artery disease in cardiac transplant recipients, *Transplantation*, 57, 884, 1994.

35. Keogh, A. et al., HLA mismatching and outcome in heart, heart–lung, and single lung transplantation, *J Heart Lung Transplant*, 14, 444, 1995.

36. Hosenpud, J.D. et al., Influence of HLA matching on thoracic transplant outcomes. An analysis from the UNOS/ISHLT thoracic registry, *Circulation*, 94, 170, 1996.
37. Weis, M. and von Scheidt, W., Cardiac allograft vasculopathy: a review, *Circulation*, 96, 2069, 1997.
38. Cocanougher, B. et al., Degree of HLA mismatch as a predictor of death from allograft arteriopathy after heart transplantation, *Transplant Proc*, 25, 1383, 1993.
39. Kaczmarek, I. et al., HLA-DR matching improves survival after heart transplantation: is it time to change allocation policies? *J Heart Lung Transplant*, 25, 1057, 2006.
40. Colvin, R.B. and Smith, R.N. Antibody-mediated organ-allograft rejection, *Nat Rev Immunol*, 10, 807, 2005.
41. Diujvestijn, A.M. et al., Complement activation by anti-endothelial cell antibodies in MHC-mismatched and MHC-matched heart allograft rejection: anti-MHC-, but not anti non-MHC alloantibodies are effective in complement activation, *Transplant Int*, 13, 363, 2000.
42. Smith, R.N. et al., C4d deposition in cardiac allografts correlates with alloantibody, *J Heart Lung Transplant*, 24, 1202, 2005.
43. Smith, M.A. et al., Effect of development of antibodies to HLA and cytomegalovirus mismatch on lung transplantation survival and development of bronchiolitis obliterans syndrome, *J Thorac Cardiovasc Surg*, 116, 812, 1998.
44. Boehler, A. et al., Bronchiolitis obliterans after lung transplantation: a review, *Chest*, 114, 1411, 1998.
45. Estenne, M. and Hertz, M.I., Bronchiolitis obliterans after human lung transplantation, *Am J Respir Crit Care Med*, 166, 440, 2002.
46. Sundaresan, S. et al., HLA-A locus mismatches and development of antibodies to HLA after lung transplantation correlate with the development of bronchiolitis obliterans syndrome, *Transplantation*, 65, 648, 1998.
47. van den Berg, J.W. et al., Long-term outcome of lung transplantation is predicted by the number of HLA-DR mismatches, *Transplantation*, 71, 368, 2001.
48. Chalermskulrat, W. et al., Human leukocyte antigen mismatches predispose to the severity of bronchiolitis obliterans syndrome after lung transplantation, *Chest*, 123, 1825, 2003.
49. Markus, B.H. et al., Histocompatibility and liver transplant outcome. Does HLA exert a dualistic effect? *Transplantation*, 46, 372, 1988.
50. Donaldson, P. et al., Influence of human leukocyte antigen matching on liver allograft survival and rejection: the dualistic effect, *Hepatology*, 17, 1008, 1993.
51. Neumann, U.P. et al., Impact of human leukocyte antigen matching in liver transplantation. *Transplantation*, 75, 132, 2003.
52. Navarro, V. et al., The effect of HLA class I (A and B) and class II (DR) compatibility on liver transplantation outcomes: an analysis of the OPTN database, *Liver Transplantation*, 12, 652, 2006.
53. Nemoto, T. et al., Unusual onset of chronic graft-versus-host disease after adult living-related liver transplantation from a homozygous donor, *Transplantation*, 75, 733, 2003.
54. Soejima, Y. et al., Graft-versus-host disease following living donor liver transplantation, *Liver Transplantation*, 10, 460, 2004.
55. Whitington, P.F. et al., Complete lymphoid chimerism and chronic graft-versus-host disease in an infant recipient of a hepatic allograft from an HLA-homozygous parental living donor, *Transplantation*, 62, 1516, 1996.
56. Kiuchi, T. et al. One-way donor-recipient HLA-matching as a risk factor for graft-versus-host disease in living-related liver transplantation, *Transplant Int*, 11, S383, 1998.
57. Kamei, H. et al., Fatal graft-versus-host disease after living donor liver transplantation: differential impact of donor-dominant one-way HLA matching, *Liver Transplantation*, 12, 140, 2006.

58. Futagawa, Y., et al., The association of HLA-DR13 with lower graft survival rates in hepatitis B and primary sclerosing cholangitis Caucasian patients receiving a liver transplant, *Liver Transplantation*, 12, 600, 2006.
59. Jaramillo, A. et al., Immune mechanisms in the pathogenesis of bronchiolitis obliterans syndrome after lung transplantation, *Pediatric Transplantation*, 9, 84, 2005.
60. Davenport, A. et al., Development of cytotoxic antibodies following renal allograft transplantation is associated with reduced graft survival due to chronic vascular rejection, *Nephrol Dial Transplant*, 9, 1315, 1994.
61. Reed, E.F. et al., Monitoring of soluble HLA alloantigens and anti-HLA antibodies identifies heart allograft recipients at risk of transplant-associated coronary artery disease, *Transplantation*, 61, 566, 1996.
62. Terasaki, P.I. and Ozawa, M., Predicting kidney graft failure by HLA antibodies: a prospective trial, *Am J Transplant*, 4, 438, 2004.
63. Reinsmoen, N.L., et al., Anti-HLA antibody analysis and crossmatching in heart and lung transplantation, *Transplant Immunol*, 13, 63, 2004.
64. Jin, Y.P. et al., Anti-HLA class I antibodies activate endothelial cells and promote chronic rejection, *Transplantation*, 79, S19, 2005.
65. Sawyer, G.J., et al., Indirect T cell allorecognition: a cyclosporin A resistant pathway for T cell help for antibody production to donor MHC antigens, *Transplant Immunol*, 1, 77, 1993.

19 Relevance of Antibody Screening and Crossmatching in Solid Organ, Hematopoietic Stem Cell Transplantation, and Blood Transfusion

Chee L. Saw, Denise L. Heaney,
Howard M. Gebel, and Robert A. Bray

CONTENTS

19.1   INTRODUCTION

The importance of detecting antibody to human leukocyte antigens (HLA) has been known since the early years of clinical organ transplantation. The association of hyperacute rejection of kidney grafts with preexisting humoral antibody to donor cells represented a seminal finding in organ transplantation [1–4]. Since then, screening for donor-specific antibody (DSA) present in transplant patients has become one of the most important tests performed in the HLA laboratory. The total accumulation of data available today supports the utility of this practice and provides insight into the relevance of HLA antibody and the utility of multiple assays for both solid organ and stem cell transplantation. Most importantly, recent advances in antibody testing techniques have made antibody detection and identification easier for the transplant community. With the current techniques available for testing, detailed antibody studies for individual patients can be performed. Thus, a better understanding of patient antibody profiles can provide a better assessment of potential immunologic risk. Advances in immunosuppression and therapies for treating antibody-mediated rejection have shown that the presence of DSA is no longer an obligatory contraindication to transplantation but represents a "risk factor" that is taken into consideration when determining the appropriateness of transplantation. This chapter describes data on HLA antibody detection and its association with graft outcome. Additionally, we provide brief discussions on the methods used for HLA antibody detection and characterization.

19.2   BACKGROUND

Historically, complement-dependent cytotoxicity had been used to evaluate alloantibody reactions to antigens (i.e., HLA and non-HLA specificities) on donor cells. Recently, flow cytometry and solid-phase methods have become the new "gold standards" for evaluating pretransplant alloreactive antibodies. Many studies have demonstrated the clinical relevance of alloreactive antibodies in transplantation and furthermore, improvements in graft survival have correlated with improved antibody testing. However, as in any biological system, nothing is absolute and 100% correlative. Hence, some important issues related to alloantibody testing remain to be discussed: (1) crossmatch tests for renal transplantation; (2) crossmatch tests for nonrenal transplantation; (3) B-cell crossmatches (i.e., class II antibodies); (4) the immunoglobulin class and subclass of HLA antibodies; (5) *de novo* posttransplant antibodies and graft survival; and (6) methods for HLA antibody identification. The recent development and widespread use of new technologies has helped address some of these issues but has also raised new concerns and put forth new questions to resolve. Such concerns include (1) antibodies against HLA alleles; (2) low levels of DSA that do not produce a positive crossmatch. Nonetheless, the newer technologies have ushered in a new era in HLA testing. Even with these new technologies, important clinical questions remain to be addressed: (1) Do all donor-specific HLA antibodies (either present at the time of transplant or that develop and persist after transplantation) result in poor graft outcomes; (2) What is the role of HLA class II antigens (HLA-DR, -DQ, and -DP) in mediating graft dysfunction; (3) Can the risk represented by HLA antibodies be ameliorated by specific therapies such as intravenous immunoglobulin (IVIG) treatment with or without plasmapheresis or induction with Thymoglobulin® or Campath®. Many of the questions surrounding antibody detection focus on the aforementioned issues. Additional concerns revolve around crossmatching, and include topics such as the type(s) of cells that should be tested (T cell versus B cell), the techniques used for crossmatching, and clinical endpoints. Also, factors such as the type of organ transplanted, the status of the recipient (i.e., primary or a regraft), the titer or strength of preformed donor-directed antibody, the degree of HLA mismatch, and the immunosuppressive protocols were used for the transplant decision. Finally, HLA antibodies have also been linked to complications following transfusion, such as platelet transfusion refractoriness [5] and transfusion-related acute lung injury (TRALI).

19.2.1 Overview of Donor-Reactive Antibody in Solid Organ and Hematopoietic Stem Cell Transplantation

Using a simple crossmatch assay, an elegant study by Patel and Terasaki [4] was published in 1969, which correlated donor-reactive antibodies with poor graft survival. This retrospective analysis of renal allograft recipients revealed that kidneys were nonfunctional within 48 h after transplantation in 80% of recipients who exhibited a positive cytotoxicity crossmatch. In contrast, less than 5% of patients with undetectable antidonor antibodies exhibited such a so-called hyperacute rejection. Following this study, crossmatch assays became a mandatory pretransplant test [6]. Nonetheless, in the study there were patients who experienced graft loss but whose crossmatch results were negative as well as patients with a positive crossmatch but whose grafts remained functional. This raised questions regarding the sensitivity and specificity of the complement-dependent lymphocytotoxicity (CDC) assays.

To reduce the risk of rejection, there have been numerous attempts to assess the extent of risk represented by donor-reactive antibody. Some have examined the risk conferred by the presence of any lymphocytotoxic antibody in the recipient, as determined by the level of panel reactive antibodies (PRAs) [7–9]. PRA represents the percentage of the panel cells that are positive with a patient's serum. For example, one group [10] reported that the 1- and 3-year survival of renal allografts was impacted by PRA. In patients with PRA >50% (highly sensitized), the 1-year-graft survival was 56% compared with 77 and 93% in the groups with PRA 10–50% (sensitized) and <10% (nonsensitized), respectively. However, the 3-year survival was comparable. Lavee et al. [11] demonstrated a 28% decrease in the 5-year survival of heart transplant recipients who had PRA >25% compared to recipients with PRA <10%. Similarly, in studies by Zerbe et al. [12], lymphocytotoxic antibody status, reflected by percentage of PRA, was associated with increased rejection episodes as documented by histological biopsy. McKenzie et al. [13] also demonstrated that a positive PRA with a positive crossmatch can adversely affect the cardiac allograft survival.

Using only PRA screening, consistent prediction of outcome has not been achieved among patients with heart [14,15] and liver [16,17] transplants. Interestingly, PRA has been shown to be of little importance in liver transplantation. Studies by Gordon et al. [16] demonstrated that high PRA and preformed DSA was not associated with decreased patient or allograft survival for primary or retransplanted liver recipients.

One of the major drawbacks of such PRA testing is that previous methodologies could not accurately predict crossmatch results. Although cytotoxic PRA testing was useful at detecting relatively high-titer antibodies, in general, cytotoxicity testing lacked sensitivity and could not provide adequate specificity determinations. Subsequently, more sensitive techniques were employed in attempts not only to prevent all hyperacute rejections but also to improve long-term graft survival rates. The major improvements to the cytotoxicity test included variations in the number of wash steps, changes in incubation time(s), temperature(s), and the use of an antiglobulin reagent [18,19]. More importantly, the advent of flow cytometry launched a new phase of antibody detection [20–22]. Although these techniques provided greater sensitivity for testing, there was a growing need for better specificity as well. In the late 1990s, new state-of-the-art solid-phase techniques were developed, that is, enzyme-linked immunosorbent assay (ELISA) and microbead techniques began to enter the field. New studies that employed these more sensitive techniques have shown a better correlation with crossmatch prediction and that a negative crossmatch predicts better graft survival and decreased frequency and severity of rejection episodes. One major difference between these new solid-phase techniques and historic antibody screening techniques using viable cells is the source of HLA antigen. The newest techniques utilize recombinant HLA antigens produced by Epstein-Barr virus (EBV)-transformed cell lines. Through this technology, a variety of HLA antigens or alleles can be produced. These antigens will then be immobilized on microplates for ELISA assay preparation or coated on polystyrene beads for solid-phase microparticle assays. Assays using frozen cell panels are prone to viability issue and may have impact on the validity of results. The convenience of cell lines and the stability of antigens also contributed to the increased use of these techniques. Another advantage is that the purified antigens used in the solid-phase technique can exclude the false-positive reactions caused by nonspecific binding that could happen in cellular-based technique or crossmatch.

In one of the first studies to highlight the significance of flow cytometry, Cook et al. [23] found that flow cytometric crossmatch (FCXM) was better in predicting early (1 month) graft failure. Thirty-three percent of patients who had a positive FCXM had graft failure versus 8% among those who tested negative. Especially for the regrafts, 56% of patients who tested positive by FCXM lost the graft compared to only 21% of them who tested negative. Subsequently, other investigators have reported increased graft failure and incidence of early rejection episodes associated with a positive FCXM [24–29].

In contrast, some studies in renal [30–32], heart [11,33], and particularly liver transplantation [16,34–36] did not find the use of a more sensitive crossmatch technique to be beneficial. In these studies, the lack of correlation could be explained by lack of specificity of the crossmatch. Specifically, since the crossmatch is performed with cellular targets, a positive crossmatch cannot stand alone as proof of HLA-specific antibody. Hence, false-positive crossmatches (i.e., non-HLA antibodies) may have contributed to this discrepancy. As mentioned earlier, the recombinant HLA antigen-based solid-phase technique does not detect non-HLA antibodies. Thus, using this technique, Bray et al. [37] resolved some discrepancies and showed that patients with antibodies that were screened as non-HLA (i.e., crossmatch positive but solid-phase negative) could be transplanted with satisfactory graft survival at one or more years.

Similar to the role of donor reactive antibody in solid organ transplantation, detecting donor reactive antibody is equally important in facilitating bone marrow transplantation outcome. The reaction of donor reactive antibodies (both HLA and non-HLA) to the marrow graft takes a different route than that of graft rejection in solid organ transplantation. The impact has been associated with a higher frequency of engraftment failure. In a retrospective study, Anasetti et al. [38] showed that the crossmatch is of predictive value in bone marrow transplant outcome. Of 269 HLA nonidentical transplanted patients, 18 patients had positive crossmatches, and the graft failure rate was 39%. In patients who had negative crossmatches, the graft failure rate was only 10%. In a later update [39] of haploidentical bone marrow transplants in 522 patients surviving at least 21 days after the marrow infusion, the rate of graft failure was 62% in patients with a positive pretransplant crossmatch, compared with 7% in patients with a negative crossmatch ($p = 7.82\text{e-}10$). Anasetti concluded that before alloimmunization to donor histocompatibility, antigens has a profound effect on the probability of achieving sustained engraftment. Ottinger et al. [40] from Germany revealed similar observations of crossmatch-dependent graft failure. Their analysis of 60 cases attributed inferior overall survival and high incidence of graft failure to the subgroup of patients with HLA-mismatched, crossmatch-positive transplants ($p = 0.01$).

19.2.2 CLINICAL IMPACT OF HUMAN LEUKOCYTE ANTIGEN-SPECIFIC ANTIBODY

19.2.2.1 Pretransplant

In the pretransplant assessment of patients, two criteria should be addressed: (1) is there an HLA antibody present (sensitivity) and, if present, (2) what is the specificity. Recently, the quantity of antibody present is becoming increasingly important and is discussed in the following paragraphs.

As far as pretransplant assessment is concerned, every attempt should be made to identify antibodies present in patients. To accomplish this, sensitive and specific assays to detect HLA antibody must be employed. At present, the most sensitive method utilizes microparticles coated with HLA antigens that are analyzed by flow cytometry. FlowPRA® (One Lambda, Inc.) is a very sensitive and specific HLA antibody test. Once antibody has been identified, elucidating all HLA specificities is important. Despite the technical limitation that not every target for all known HLA antigens or alleles (now >2000) is available, the newer solid-phase techniques can assess a multitude of common HLA antigens or alleles. The lack of targets such as those to DP and DQ-alpha polymorphisms and non-HLA antibodies show the limitations of the current bead-based assays. Nonetheless, many studies have been able to determine the specificity of donor-reactive antibody with appropriate approaches to rule out non-HLA antibodies. These studies have shown that there is an increased level of risk that correlates with the presence of donor-specific HLA antibody.

It is generally accepted that B cells express higher number of HLA class I antigens than T cells, and B cells express HLA class II antigens that are not normally expressed on T cells. Thus, B-cell crossmatches were considered a more sensitive assay for evaluating transplant recipients. However, nonspecific binding of immunoglobulins to the Fc and complement receptors expressed on B cells produced many false-positive results, which led to significant controversies as to whether the B-cell crossmatch was of any clinical value. Ettenger et al. [41], Jeannet et al. [42], and Schäfer et al. [43], using different methods for B-cell separation (sheep red blood cell rosette or nylon wool adherence), observed no correlation between the development of B lymphocyte antibodies and graft outcome. The high incidence of positive B-cell crossmatch results were in conflict with the graft survival outcome for many more years. Until investigators had acquired improved test methods (i.e., flow cytometry), diminished graft outcome was not associated with a positive B-cell crossmatch. Kotb et al. [44] used flow cytometry to determine, retrospectively, the influence of a positive B-cell FCXM on the incidence of rejection. Of all 51 T-FCXM negative primary transplant recipients, 50% of those with positive B-FCXM had experienced at least one rejection episode within the first year. In contrast, only 29% of patients with a negative B-cell FCXM experienced rejection. There have also been studies that claimed that B-cell FCXM were irrelevant and not predictive for graft outcomes. At least one of the explanations for such results was that a B-cell FCXM may sometime be falsely positive due to nonspecific antibody binding of immunoglobulin to Fc receptors. In an effort to overcome this problem, Lobo et al. [45,46] performed flow cytometric crossmatches using donor B cells that were pretreated with pronase to further increase the specificity of the assay. Pronase, a mixture of proteolytic enzymes, was used to cleave Fc receptors on B lymphocytes, thus eliminating nonspecific binding of antibody. Their findings underscored the importance of improving the specificity of the FCXM as well as the significance of identifying weak anti-HLA class I antibodies that react only to B cells. In similar studies, Vaidya et al. [47] reported evaluations on three primary transplant recipients who lost their allografts to accelerated rejection. All pretransplant B-cell FCXMs of the patients were apparently whereas the T-cell FCXMs were apparently negative. These data were interpreted to mean that the antibodies were non-HLA (HLA antigens are either expressed on both T and B cells [class I] or only on B cells [class II]). However, after pronase treatment, both T- and B-cells FCXMs of each patient became strongly positive, and donor-specific anti-HLA class I antibody was identified in each case. The reason for the observed difference in the B-cell crossmatch results from the fact that pronase lowers nonspecific background fluorescence, particularly on B-cells;

thereby permitting better resolution of a true antibody. Pronase enhances signal-to-noise separation by reducing the background noise.

In cardiac transplantation, similar results were reported. The study led by Bunke et al. [48] found that heart transplant recipients who had a positive B-cell cross-match were linked to early, frequent graft rejection and therefore, a stronger immu-nosuppressive regime was given, which included OKT3 induction and replacing azathioprine with cytoxan for the first 6 months. In another study [49], investigators examining graft survival in 25 cardiac transplant recipients who had HLA-specific antibodies, found that two patients having DSA, not just HLA antibody, experienced acute fulminating rejections. The other 23 patients, who had HLA antibody but not DSA, had a survival rate similar to that of patients who had zero PRA. Other studies also demonstrated that antibody specific for donor HLA antigens are also deleterious to other types of transplanted organs. For example, in a combined pancreas–spleen transplantation case report, Peltenburg et al. [50] observed accelerated acute rejection of both organs in a patient with positive crossmatch. The patient was known to have neither T nor B-cell antibodies in sera obtained 3 months before transplantation; how-ever, the final crossmatch was performed retrospectively to shorten the cold preserva-tion time. The PRA tests show 0% at 3 months before transplant, but 75% at the time of transplant; the retrospective T-cell crossmatch turned out to be positive. Subsequent studies proved that the HLA antibodies were donor specific. From these data, it is clearly understood that the graft outcome has strong correlation with predetermined antibody specificity, and that the predetermined donor-specific HLA antibody is a def-inite contraindication to transplantation. However, interestingly, the transplant group from Berlin [51] has reported reduced incidence of acute rejection in 18 patients, who underwent combined kidney and liver transplantation. One important aspect of these transplants was that the liver is reported to protect the kidney from hyperacute rejec-tion despite a positive crossmatch, a shared hypothesis from other groups [52,53].

19.2.2.2   B-Cell (Class II) Antibody

Because donor-directed antibody specific for HLA class I antigens usually repre-sents a significant risk factor for transplantation, the relevance of class II-specific antibody has been controversial. Screening for HLA class II antibody has been dif-ficult because B cells, the primary target cell, possesses both class I and class II antigens. Patients with HLA antibodies may have antibodies against class I, -II, or both. Most often, patients have antibodies to both antigen groups making it dif-ficult to evaluate the contribution of class II antibodies. Nonetheless, investigators have reported class II–specific antibody to be a significant risk [54]. However, until the first solid-phase microbead-based HLA class I and -II assays were introduced [55,56], the identification of class II–only antibodies was not an easy task. Virtually, all antibody identification focused on class I HLA antibodies.

In addition to B cells possessing both classes of HLA antigens, nonspecific anti-body was easily bound to the B cell. There have been a few published reports of class II antibodies associated with hyperacute rejection. For example, Scornik et al. [57] has reported four cases of hyperacute rejection. Using the solid-phase assay, class II–specific antibody can be easily identified independent of other confounding immunoglobulins. Furthermore, the new solid-phase assays have made it possible to identify antibodies to other HLA loci that could not be previously evaluated. These include antigens such as HLA Cw, DQB, DQA, and DP antigens as well as DRB3, 4, and 5. Historically, such specificities were impossible to ascertain. Clinical reports are just beginning to emerge indicating the importance of other class II loci in graft outcomes and indicate that antibodies to these HLA antigens may very well be detrimental to graft survival. A similar finding has been associated with class I antigens as well. Specifically, anti-Cw locus antibodies (44%) were observed in 24 of 34 patients, who experienced renal allograft failure [58] when there was a high degree of Cw-locus mismatching (67%) [59]. An early case report also noted hyperacute rejection due to anti-Cw5 antibody [60]. Together, these data suggest that risk of rejection may be independent of the locus specificity.

19.2.2.3   Transfusion Medicine

HLA antibody has been implicated in transfusion-related issues such as platelet refractoriness and TRALI. In a study by Sato et al. [5], platelet refractoriness was attributed to HLA antibody present in the patients' serum. They used a flow-based technique to determine the antibody specificity and gave HLA-matched platelets to the patients to restore the platelet count. Because TRALI can be associated with HLA antibodies in blood components, it would be important to know their frequency in these components. To address this issue, Bray et al. [61], using new technologies, investigated the frequency of HLA antibodies present in normal blood donors. Their data revealed that approximately 22% of blood components tested contained HLA alloantibodies, 10-fold greater than reported earlier [62]. An earlier study [63] showed that volunteer platelet donors, especially female donors, contributed to the prevalence of HLA antibody in the donated blood products.

19.2.2.4   Immunoglobulin Class (IgG versus IgM)

There is consensus from several studies [64,65] showing that antibody of immunoglobulin isotype G represented a greater risk in transplant outcome. One-year renal graft survival was significantly lower for those patients who had positive crossmatches (due to IgG) than those who had crossmatches predominantly due to IgM. Kerman et al. [33] and Ratkovec et al. [66] used IgM-reduced reagent in the crossmatch tests to confirm the close association between cardiac graft rejection and IgG positive crossmatch. Katz et al. [67], with the same testing methodology, proved that IgG antibodies have similar impact in liver transplantation. Early studies revealed that IgM antibodies detected in a lymphocyte crossmatch are frequently autoantibodies that do not have a deleterious effect on transplant outcome [68–73]. However, one should not underestimate IgM antibody that is HLA-specific. Similar to IgG HLA-specific antibody, IgM HLA-specific antibody was associated with decreased survival in regraft patients (67%) compared to primary graft patients (88%) [74]. However, this study did not rule out the presence of underlying IgG antibodies. Although some investigators have observed correlation between increased rejection and any HLA-specific antibody, rejection episodes had been more frequent among those who have IgG antibody. The relevance of IgM in outcome of nonrenal transplant is unclear. In a study by Smith et al. [15] in cardiac transplantation, among 24 recipients who made antibodies posttransplant, 18 were found to be specific for HLA antigens (IgM, 6 cases and IgG, 12 cases). These antibodies and their occurrence were strongly correlated with rejection ($p < 0.001$). Using a solid-phase technique, Khan et al. [75] identified 13 patients with IgM HLA-specific antibodies from a group of 46 renal patients who tested positive for IgM autoantibodies. Among these 13 patients, 11 were regraft patients and 2 were primary patients. Five of these patients did not possess IgG class antibodies. This study highlights the importance of using solid-phase assays to verify the presence of HLA-specific IgG class antibody. In contrast, data on other classes of antibodies have been inconclusive or actually associated with better graft survival [76,77].

19.2.2.5 Historic Antibody

The timing of antibody response has always been a topic of interest. There are two aspects for consideration, namely, past positive and current negative antibodies and posttransplant DSAs. The latter will be discussed in the next session; the former are antibodies that have peaked at some period in history and subsequently diminished before transplantation. Several investigators [78,79] found no difference in graft survival between patients with a historic positive and current negative crossmatch and patients without antibody. In a study focused on retransplant patients, Barger et al. [80] found that past positive–current negative crossmatch patients have satisfactory graft survival when repeat mismatch antigens were avoided. However, in the same study, three out of four patients who exhibited HLA antibodies specific for previously encountered antigens lost their graft within 1.4 months. For many transplant programs, the presence of any historic DSA represents a contraindication to transplantation. Three additional studies have supported this concept [73,80,81]. Current practice of the Emory transplant group is to consider the previous year only when evaluating historic sera. In contrast, some transplant centers consider any historic antibody as a contraindication. Aspects that may affect transplant outcomes in these situations would include titer of antibody present, the nature of the sensitizing events that provoked the earlier antibody response, the immunoglobulin class and specificity of the historic antibody, and the type of immunosuppressive therapy utilized. Current trends with modern immunosuppression favor transplantation across a past positive–current negative crossmatch.

19.2.2.6 Posttransplant

Several studies have shown that transplant recipients, even with 0% PRA, can develop DSAs in the posttransplant period. However, these studies have shown that, for renal transplantation, posttransplant DSA is a harbinger of poor outcomes. As immunohistochemical techniques become available to identify antibody-mediated rejection by C4d staining, acute cellular and antibody-mediated rejection have been detected. In a long-term follow-up of 235 patients with mismatches at loci A, B, or DR enrolled within 1986 until 1998, Worthington [54] observed 57 patients who eventually lost their graft to HLA antibody. Of the latter, 17 had production of HLA class I antibody within 51–2798 days posttransplant and the grafts failed at a median of 634 days after the production of antibody. Three patients produced both HLA class I and -II antibodies at 88, 1098, and 4387 days after transplantation and these grafts also failed 720, 398, and 848 days after antibody production, respectively. For 14 patients, class II DSA was identified before graft failure. The mean time from transplantation to antibody production was 1541.8 days, whereas the mean time from antibody production to failure of the graft was 1409.1 days. It is clear that DSA in the posttransplant period can ultimately lead to graft failure weeks to years after transplantation. Recently, an international prospective trial on 4763 patients from 36 centers has suggested that chronic rejection may be predicted by the development of HLA-specific antibody following transplant [82]. Whether early detection and intervention can reverse this trend is yet to be determined. As a result, many programs have instituted posttransplant monitoring as a routine "standard-of-care" for transplant recipients.

Compared with chronic rejection, studies on acute rejection mediated by HLA antibody could be traced back to many years. Halloran et al. [83,84] investigated acute rejection mediated by class I–specific antibody in patients who developed the antibody following transplantation. All patients ($N = 13$) experienced rejection episodes, 80% of which were severe, and 5 (38%) ultimately lost their grafts. Of the remaining 51 patients with a 0% PRA, only 41% experienced rejection with only 17% experiencing graft failure. Recently, using two-color flow cytometry to measure antibody in diseased donor renal transplant recipients, Scornik et al. [85] found that 40% of patients experiencing acute rejection had posttransplant DSA (IgG or IgM), whereas only 9% of patients without rejection had developed antibody (exclusively IgM).

When evaluating the relevance of posttransplant DSA in other organ transplantation, one can find studies that support their clinical significance. Barr et al. [86] conducted a multiyear follow-up of heart transplant recipients evaluating the production of DSA. They found a significant difference in the 5-year graft survival among heart transplant patients, in whom DSA had developed or persisted after transplantation versus recipients without DSA (78 versus 91%). A positive correlation with DSA was also observed among patients who experienced acute cellular rejection episodes and developed atherosclerosis. In a different study using a cell panel of 70 people, Rose et al. [87] reported that the 4-year actuarial survival rate of patients who developed DSA (PRA > 10%) during the first 6 months after transplantation was 70%, whereas the survival rate of patients with low PRA (<10%) was 93%. George et al. [88] also reported reduced long-term survival among heart transplant recipients when posttransplant DSA developed. In lung transplantation, Schulman et al. [89] found that patients who developed HLA-specific antibody posttransplant correlated with the onset of rejection and led to a lower survival rate compared to those without antibody. Together, these studies suggest that DSA developed in the posttransplant period may be important in long-term graft survival and also warrant closer attention by posttransplant monitoring.

19.2.3 NON-HLA

Various investigators have reported that antibody to antigens other than HLA may play a role in transplant rejection. This is based, in part, on the 10–15% incidence of rejection (usually chronic) observed in HLA-identical kidney grafts, even with current immunosuppression [90,91]. One such new motif that has been implicated in allograft rejection is MHC class I chain-related gene A (MICA), which is expressed on vascular endothelial cells. The human MICA, located on chromosome 6 and close to HLA-B, encodes an inducible protein that is a ligand for the NKG2D receptor. In addition to its possible role in immune surveillance, data suggest that MICA may be involved in the immune response to solid organ transplants. Several studies [92,93] have reported MICA antigens on endothelial to be possible targets for antibody-mediated graft rejection, since antibody has been demonstrated in a significant number of transplant patients. Recently, data published by Terasaki et al. [94], from a 4-year follow-up study has provided evidence for the association of both HLA and MICA antibodies in allograft failure. However, since MICA proteins are not expressed on lymphocytes, traditional crossmatch tests are not adequate to detect these antibodies. Hence, recombinant MICA antigens adhered to solid matrices will be an invaluable tool for further testing.

19.3   ANTIBODY IDENTIFICATION: PAST TO PRESENT

19.3.1   ANTIGEN NONSPECIFIC ASSAYS

The presence of HLA-specific antibody has been shown to be associated with all realms of graft rejection, from hyperacute to chronic. Identification of such antibodies allows proper assessment of immunological risks and gives clinicians information that will assist in decision making for donor selection and patient care both pre- and posttransplant. The ability to accurately and efficiently detect antibody to HLA is therefore a very important part of patient workup and an essential function of the transplant laboratory. The technology for detecting antibody has evolved over the years and continues to evolve with a shift to more sensitive and specific methods. This section will discuss the chronology of method development for HLA antibody detection and identification over the past 35 years.

19.3.1.1   Cytotoxicity Assays

Initially, the laboratory gold standard for detecting HLA antibodies was the microlymphocytotoxicity assay, also known as the CDC assay [95,96]. This assay consists of combining and incubating isolated lymphocytes from a single individual or a defined cell panel with the patient's serum followed by the addition of rabbit complement. The binding of HLA antibody in the patient's sera to the donor (or cell panel) lymphocytes activates complement resulting in cell injury and death. The measurement of cell death is determined microscopically by the uptake of a vital dye (e.g., Eosin, Trypan blue, and Ethidium Bromide) (Figure 19.1).

Although the CDC assay was able to identify complement-binding HLA antibodies, patients with pretransplant HLA antibodies undetectable by this assay experienced early antibody-mediated rejection and subsequent graft loss [97]. These findings were the impetus to enhance and develop more sensitive assays to identify all clinically relevant antibodies. Various modifications were made to the CDC assay in attempting to make it more sensitive. First, extended incubation times of cells

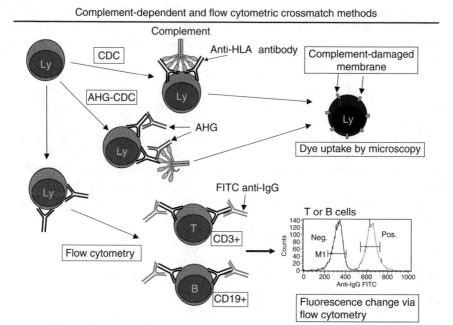

FIGURE 19.1   Schematic representation comparing methods used to detect HLA antibodies. CDC and AHG-CDC assays rely on antigen–antibody complexes to activate the complement. The AHG-CDC utilizes the addition of AHG (secondary antibody) to enhance the sensitivity. The complement activation in both assays results in cellular damage or death and is visualized microscopically by the uptake of a vital dye. In contrast, the FCXM, a complement-independent assay, utilizes a fluorochrome-labeled AHG (secondary antibody) to detect the presence of antibody bound to T and B cells. The increase in surface fluorescence of the lymphocytes indicates a positive result.

and serum along with increased complement incubation time were introduced to allow binding of low-titer and low-avidity antibodies. Then, additional washes were included to remove "anticomplementary" factors. Because IgM and IgG antibodies are both detected by CDC assays, the pretreatment of the patient's serum with a reducing agent, dithiothreitol (DTT) was also incorporated. DTT reduces the pentameric IgM antibody structure to IgM monomers that cannot bind complement. This was an important modification to the traditional protocol because IgM antibodies can interfere with the detection of clinically relevant IgG antibodies and may not also be as clinically relevant [4,18,19,98,99] (Table 19.1).

Finally, the use of antihuman globulin (AHG) was added (Figure 19.1). This particular modification allowed for the detection of low level of HLA antibodies as well as permitting the detection of noncomplement fixing antibodies, referred to as CYtotoxicity Negative, Adsorption Positive (CYNAP) antibodies [11], which has led to significantly improved graft survival [18,100–102]. This is still the most commonly used CDC-enhanced assay (Table 19.1).

Although the adjustments and additions to the CDC assay virtually eliminated hyperacute rejection and increased the ability to detect HLA antibody, there were

TABLE 19.1
Modifications to Standard CDC Assay

Extended Incubation	Increased Incubation Times of Cells or Serum and Complement
Amos	Three washes post cells or serum incubation before complement
Amos modified	One wash post cells or serum incubation before complement
DTT/dithioerythritol (DTE)	Pretreatment of serum to remove IgM antibodies
AHG-CDC	Amos modified with the addition of AHG before complement

TABLE 19.2
Methods for HLA Antibody Evaluation

Antigen Nonspecific	Antigen Specific
Cytotoxicity (cells)	ELISA
NIH	Yes or No
Variations	Percentage of PRA (I or II)
Washes	Specificity (I or II)
Extended incubation	Flow cytometry (beads)
Antiglobulin (AHG)	Percentage of PRA (I and II)
DTT/DTE	Specificity (I or II)
Flow cytometry (cells)	Multiplex
T cell or B cell	Suspension arrays
Pronase	Protein chips

patients who still had antibody-mediated early graft rejection and loss. Interestingly, majority of these patients had previous sensitizing events (e.g., retransplants, multi-transfused patients, and multiparous females). Again, this observation reiterated the need for an even more sensitive assay.

19.3.1.2 Flow Cytometry

In 1983, Garovoy et al. [20] developed a technique that utilized the flow cytometer and became known as the FCXM. This technique consists of incubating purified mononuclear cells (donor lymphocytes) with the patient's serum, washing off any unbound serum components and adding a fluorochrome-conjugated antihuman IgG. Antibody detection is observed by the change in surface fluorescence of the lympho-cytes. It was observed that this assay detected levels of HLA antibodies on lympho-cytes not detectable by other methods and soon it became a clinical laboratory test. Six years later, Bray et al. [22] refined the FCXM method by using dual color flow cytometry that allowed the simultaneous but separate evaluation of T- and B-cells reactivities (Figure 19.1).

Although more sensitive than the CDC assays, the FCXM had similar limitations and it too lacked antigen specificity (Table 19.2), which resulted in positive FCXMs not due to HLA antibodies. These false-positives can be due to (1) autoantibodies,

(2) the binding of other antibodies (non-HLA) to lymphocyte antigens, and (3) lymphocyte Fc receptors that bind immunoglobulin independent of antigen specificity. A positive FCXM due to any of these reasons are not considered a contraindication to transplantation. Laboratory methods to decrease the incidence of false-positive reactions have been introduced and implemented. These include ultracentrifugation of the patient's serum to clear or eliminate immune complexes that nonspecifically bind to donor lymphocytes along with pretreating donor cells with pronase, a proteolytic enzyme known to cleave Fc receptors on the cell surface (discussed earlier). Routine application of these methods to the FCXM significantly reduces false-positives, imparts higher sensitivity and specificity, and facilitates accurate interpretation of the results.

19.3.2 ANTIGEN-SPECIFIC ASSAYS

The CDC and FCXM assays are cellular-based assays and, although quite sensitive, lack antigen specificity. A positive result by either of these methods merely indicated the presence of "antilymphocyte" reactivity. What was now needed was a technology that conferred antigen specificity (Table 19.2), where true HLA-specific antibodies could be identified and quantified in a patient's serum. Recent technological advances have allowed for the purification of HLA antigens to be used as targets. Once purified, the HLA antigens can be adhered to a solid matrix claiming "membrane-independent" status because the purified antigens lack potentially interfering proteins found on cell membranes.

19.3.2.1   Enzyme-Linked ImmunoSorbent Assay

Kao et al. [103] developed the ELISA to assess the presence of HLA-specific antibodies. Briefly, the purified HLA proteins are adsorbed to the bottom of a 96-well, microtiter plate. The patient's serum is then added to the wells and incubated. After several washes to remove unbound serum proteins, a secondary IgG-specific antibody with an enzyme marker (i.e., alkaline phosphatase and peroxidase) is added and incubated. Finally, the enzyme substrate is added and a color reaction is produced in wells that have a positive reaction; hence, the presence of HLA antibody. Interpretation of the assay can be performed visually or by use of a spectrophotometer. The percentage of HLA antigens to react to the patient's sera can be calculated by the positive reactions and the number of wells and the patient assigned a PRA (percentage of PRA). Although the introduction of the ELISA in the 1990s allowed antibody screening for class I and -II independently, two separate procedures must be performed. In addition, the specificity of the positive sera could not be determined.

19.3.2.2   Flow Cytometry

In the late 1990s, the isolation of HLA class I and -II antigens from EBV-transformed cell lines and their coupling to microparticles, which could be analyzed by a standard flow cytometer was described [55,56,104]. Briefly, the HLA antigen-conjugated microparticles are incubated with the patient's serum, washed, and then stained with a fluoresceinated anti-IgG. The microparticles are then analyzed on the flow cytometer. The presence or absence of HLA antibodies in the patient's sera is determined by the increase or change in fluorescence and has been referred to as the FlowPRA (Figure 19.2). The FlowPRA has many advantages over both cytotoxic and ELISA assays. One advantage over the ELISA is the simultaneous detection of antibodies to class I and -II antigens. Other advantages of the FlowPRA assay include specificity for HLA antigens, comparable sensitivity to the FCXM, nonsubjectivity such as the cell-based assays, and capability of quantitatively measuring the strength of antibody.

Although the advantages are numerous, the FlowPRA is not without its limitations. One major limitation is the inability to determine individual HLA specificities that are present in a positive serum (i.e., someone with PRA). This limitation was addressed by modifying the FlowPRA procedure. Microparticles coated with either HLA haplotypes or individual HLA molecules were developed and the procedures defined as FlowPRA specificity and FlowPRA single antigen, respectively (Figure 19.3) [105]. FlowPRA single-antigen (Figure 19.3) assay is able to identify specific HLA antibodies—class I and -II—from FlowPRA positive sera.

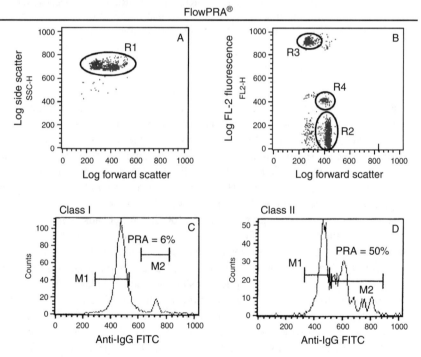

FIGURE 19.2 Illustration of FlowPRA I and II. Microparticles (R1) containing either HLA class I (R2) or class II (R3) antigens can be run simultaneously in the presence of control beads (R4) to detect the presence of anti-HLA antibody. Panels A and B show the scatter plots of the microparticles that are used for appropriate gating. Panels C and D show examples of positive results for both class I (6%) and class II (50%) antibodies determined by an increase in fluorescence and architectural change (M2) compared to the negative control position (M1). For the class I result, two beads would be considered positive (i.e., 2/30 = 6%).

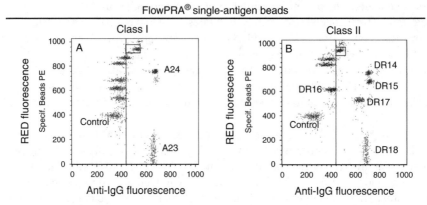

FIGURE 19.3 Example of FlowPRA single-antigen analysis with class I and II–coated microparticles. The background staining of the negative control is represented by the vertical line. Panel A shows two beads positive, one bead expressing HLA A23 and the other expressing HLA A24. Panel B shows the presence of class II antibody with four positive beads: DR14, 15, 17, and 18. The positive reaction for DR15 (split of DR2) and the negative reaction for DR 16 (split of DR2) illustrate the fine specificity that is gained by using this method. DR15 (DRB1*1501) differs from DR16 (DRB1*1601) by five amino acids.

19.3.2.3 Multiplex Platforms

The use of flow cytometry for the detection and identification of DSA antibodies has been a tremendous addition to clinical laboratory practice and patient assessment. However, classic flow cytometry is limited by the number of beads or microparticles that can be run simultaneously. Unfortunately, due to the high degree of polymorphism within the HLA complex, clinically relevant antigens may not be represented on the panel. Recently, HLA antibody detection and identification has been developed on the Luminex™ platform, a type of flow cytometer. The Luminex is a multiplex platform [106] that allows up to 100 different beads or microparticles to be analyzed simultaneously. The latter allows for a wide array of HLA antigens to be expressed on the beads, ultimately leading to better antibody identification. Two vendors (Tepnel, Stamford, Connecticut and One Lambda, Inc., Canoga park, California) have developed HLA applications on the Luminex platform.

Although Luminex represented the first introduction of protein arrays into clinical testing, there are several new platforms currently in development. Innogenetics, a Belgium-based company with offices in the United States (Innogenetics, Inc., Alpharetta, Georgia) has recently released a chip-based HLA antibody-typing platform called "4-MAT™." This system uses a porous microarray containing 400 individual spots to which molecular probes for DNA typing or proteins for HLA antibody identification can be attached. Invitrogen–Dynal (Carlsbad, California) has also recently released a microarray for HLA antibody testing, called the DynaChip™. This one-dimensional membrane contains individual spots coated with a constellation of HLA antigens (i.e., HLA class I or -II phenotype) or unique HLA alleles. Because both of these products are relatively new to the HLA testing market, no information is available yet on their performance compared to current assays or their clinical utility. Nonetheless, such products are clear steps toward increased automation in the HLA laboratory.

19.4   APPLICATIONS OF NEW METHODOLOGIES

19.4.1   Solid Organ Transplant

19.4.1.1   Pretransplant

The ability to appropriately and efficiently document the presence or absence of HLA antibodies in the patient's sera has been shown to be important in assessing the patient's immunological risks. The presence of donor-reactive antibody before transplantation is considered a risk and may lead to early graft rejection and loss. Histocompatibility laboratories are monitoring the patient's sera on a regular basis for the presence or absence of antibodies as well as changing antibody status. The sensitive techniques described previously to assess antibody status assist in interpreting whether a positive FCXM with donor cells is directed against HLA antigens. A positive FCXM with detectable donor-reactive HLA antibody should be considered a risk factor, whereas a positive FCXM not associated with HLA antibodies is not considered clinically relevant.

In addition to assisting in the interpretation of the FCXM, the proper assessment of antibody status and presence or absence of donor-reactive antibody permits laboratorians and clinicians the ability to determine which donors may or may not be suitable for the patient. This practice can effectively benefit the highly sensitized patients, who have a longer waiting time for donors and are transplanted at a lower rate. The determination of which HLA antibodies to avoid (i.e., unacceptable mismatches) and which antigens constitute acceptable mismatches will facilitate transplantation of the sensitized patient. Therefore, using the most sensitive techniques to detect and identify HLA antibodies is essential for proper risk assessment and providing the best information for clinical decision making.

19.4.1.2   Posttransplant

The existence of preformed antibodies and their potential impact on organ transplantation have been discussed; however, this only represents half of the immune equation with regard to transplant. Several groups have shown that antibodies that are formed in the posttransplant period also appear to have an effect on long-term graft survival [55,83,107–111]. Using the most sensitive technology (flow cytometry), both the specificity and quantity (i.e., titer) of the DSA can be determined. In addition, these quantitative technologies can be very useful in monitoring the levels of antibody during posttransplant therapies. Determining whether antibody levels are increasing or decreasing in response to therapy aids the clinician in the treatment of the patient.

19.4.2   Stem Cell Transplantation

Although matching a bone marrow transplant recipient with a donor is the most important aspect of bone marrow transplantation, antibody status is not without its importance. Studies show that the presence of antibody, more specifically, donor-reactive antibody can impair marrow engraftment [38]. If present, accurately detecting and identifying HLA antibodies can help assist in the selection of potential donors and improve the chances of successful engraftment.

19.4.3 TRANSFUSION MEDICINE

Platelet refractoriness has been attributed to the presence of HLA antibodies. It is therefore, essential to screen the potential recipients of platelets for HLA antibodies. With the advent of flow cytometric–based antibody screening and identification, the decision for which platelets may or may not be appropriate for a patient has become more efficient. This is also important because it reduces the patient's exposure to HLA antigens that they may already have antibodies to, rendering them less likely to develop refractoriness.

19.4.4 SUMMARY

To conclude, the presence of HLA antibodies has an impact on organ transplantation. The evolution of laboratory methods for antibody detection and identification over the past 35 years has assisted in improving patient and graft survival of organ transplantation. With the development of newer and more sensitive technologies, histocompatibility laboratorians and clinicians will continue to provide the best testing for patients awaiting transplantation.

REFERENCES

1. Kissmeyer-Nielsen, F., Olsen, S., Petersen, V.P., and Fjeldborg, O., Hyperacute rejection of kidney allografts, associated with pre-existing humoral antibodies against donor cells, *Lancet*, 2, 662, 1966.
2. Starzl, T.E., Lerner, R.A., Dixon, F.J., Groth, C.G., Brettschneider, L., and Terasaki, P.I., Shwartzman reaction after human renal homotransplantations, *N. Engl. J. Med.*, 278, 642, 1968.
3. Williams, G.M., Hume, D.M., Hudson, R.P. Jr., Morris, P.J., Kano, K., and Milgrom, F., "Hyperacute" renal-homograft rejection in man, *N. Engl. J. Med.*, 279, 611, 1968.
4. Patel, R. and Terasaki, P.I., Significance of a positive crossmatch test in kidney transplantation, *N. Engl. J. Med.*, 280, 735, 1969.
5. Sato, S., Sakurai, T., Yamamoto, Y., Mackawa, I., and Ikeda, H., Earlier detection of HLA alloimmunization in platelet transfusion refractoriness by flow cytometric analysis, *Transfusion*, 45(8), 1399, 2005.
6. Federal Register, DHHS, HCFA, PHS (42 CFR Part 493), February 28, 1992.
7. Iwaki, Y. and Terasaki, P.I., Sensitization effect, in *Clinical Transplants*, Terasaki, P.I., Ed., UCLA Tissue Typing Laboratory, Los Angeles, 1986, p. 257.
8. Opelz, G. for the Collaborative Transplant Study, Effect of HLA matching, blood transfusion, and presensitization in cyclosporine treated kidney transplant recipients, *Transplant. Proc.*, 17, 2179, 1985.
9. Iwaki, Y. and Terasaki, P.I., Effect of sensitization on kidney allografts, in *Clinical Kidney Transplants 1985*, Terasaki, P.I., Ed., UCLA Tissue Typing Laboratory, Los Angeles, 1985, p. 139.

10. Mjörnstedt, L., Konar, J., Nyberg, G., Olausson, M., Sandberg, L., and Karlberg, I., Renal transplantation in patients with lymphocytotoxic antibodies—a 5-year experience from a single centre, *Transplant. Proc.*, 24, 333, 1992.

11. Lavee, J., Kormos, R.L., Duquesnoy, R.J., Zerbe, T.R., Armitage, J.M., Vanek, M., Hardesty, R.L., and Griffith, B.P., Influence of panel-reactive antibody and lymphocytotoxic crossmatch on survival after heart transplantation, *J. Heart Lung Transplant.*, 10, 921, 1991.

12. Zerbe, T.R., Arena, V.C., Kormos, R.L., Griffith, B.P., Hardesty, R.L., and Duquesnoy, R.J., Histocompatibility and other risk factors for histological rejection of human cardiac allografts during the first three months following transplantation, *Transplantation*, 52, 485, 1991.

13. McKenzie, F.N., Tadros, N., Stiller, C., Keown, P., Sinclair, N., and Kostuk, W., Influence of donor-recipient lymphocyte crossmatch and ABO status on rejection risk in cardiac transplantation, *Transplant. Proc.*, 19, 3439, 1987.

14. McCloskey, D., Festenstein, H., Banner, N., Hawes, R., Holmes, J., Khaghani, A., Smith, J., and Yacoub, M., The effect of HLA lymphocytotoxic antibody status and crossmatch result on cardiac transplant survival, *Transplant. Proc.*, 21, 804, 1989.

15. Smith, J.D., Danskine, A.J., Rose, M.L., and Yacoub, M.H., Specificity of lymphocytotoxic antibodies formed after cardiac transplantation and correlation with rejection episodes, *Transplantation*, 53, 1358, 1992.

16. Gordon, R.D., Fung, J.J., Markus, B., Fox, I., Iwatsuki, S., Esquivel, C.O., Tzakis, A., Todo, S., and Starzl, T.E., The antibody crossmatch in liver transplantation, *Surgery*, 100, 705, 1986.

17. Donaldson, P.T., Thomson, L.J., Heads, A., Underhill, J.A., Vaughan, R.W., Rolando, N., and Williams, R., IgG donor-specific crossmatches are not associated with graft rejection or poor graft survival after liver transplantation, *Transplantation*, 60, 1016, 1995.

18. Cross, D.E., Whittier, F.C., Weaver, P., and Foxworth, J., A comparison of the antiglobulin versus extended incubation time crossmatch: results in 223 renal transplants, *Transplant. Proc.*, 9, 1803, 1977.

19. Zachary, A.A., Klingman, L., Thorne, N., Smerglia, A.R., and Teresi, G.A., Variations of the lymphocytotoxicity test: an evaluation of sensitivity and specificity, *Transplantation*, 60, 498, 1995.

20. Garovoy, M.R., Rheinschmidt, M.A., and Bigos, M., Flow cytometry analysis: a high technology crossmatch technique facilitating transplantation, *Transplant. Proc.*, 15, 1939, 1983.

21. Bray, R.A., Flow cytometry in the transplant laboratory, *Ann. NY Acad. Sci.*, 677, 138, 1996.

22. Bray, R.A., Lebeck, L.K., and Gebel, H.M., The flow cytometric crossmatch, *Transplantation*, 48, 834, 1989.

23. Cook, D.J., Terasaki, P.I., Iwaki, Y., Terashita, G.Y., and Lau, M., An approach to reducing early kidney transplant failure by flow cytometry crossmatching, *Clin. Transplant.*, 1, 253, 1987.

24. Ogura, K., Terasaki, P.I., Johnson, C., Mendez, R., Rosenthal, J.T., Ettenger, R., Martin, D.C., Dainko, E., Cohen, L., Mackett, T., Berne, T., Barba, L., and Lieberman, E., The significance of a positive flow cytometry crossmatch test in primary kidney transplantation, *Transplantation*, 56, 294, 1993.

25. Ogura, K., Koyama, H., Takemoto, S., Chia, J., Johnson, C., and Terasaki, P.I., Flow cytometry crossmatching for kidney transplantation, *Transplant. Proc.*, 25, 245, 1993.

26. Mahoney, R.J., Ault, K.A., Given, S.R., Adams, R.J., Breggia, A.C., Paris, P.A., Palomaki, G.E., Hitchcox, S.A., White, B.W., Himmelfarb, J., and Leeber, D.A., The flow cytometric crossmatch and early renal transplant loss, *Transplantation*, 49, 527, 1990.

27. Johnson, A., Hallman, J., Alijani, M.R., Melhorn, N., Lim, L.Y., Jenson, A.B., and Helfrich, G.B., A prospective study of the clinical relevance of the current serum antiglobulin-augmented T cell crossmatch in renal transplant recipients, *Transplant. Proc.*, 19, 792, 1987.
28. Lazda, V.A., Pollak, R., Mozes, M.F., and Jonasson, O., The relationship between flow cytometer crossmatch results and subsequent rejection episodes in cadaver renal allograft recipients, *Transplantation*, 45, 562, 1988.
29. Stratta, R.J., Mason, B., Lorentzen, D.F., Sollinger, H.W., D'Alessandreo, A.M., Pirsch, J.D., Kalayoglu, M., and Belzer, F.O., Cadaveric renal transplantation with quadruple immunosuppression in patients with a positive antiglobulin crossmatch, *Transplantation*, 47, 282, 1989.
30. Dafoe, D.C., Bromberg, J.S., Grossman, R.A., Tomaszewski, J.E., Zmijewski, C.M., Perloff, L.J., Naji, A., Asplund, M.W., Alfrey, E.J., Sack, M., Zellers, L., Kearns, J., and Barker, C.F., Renal transplantation despite a positive anti-globulin crossmatch with and without prophylactic OKT3, *Transplantation*, 51, 762, 1991.
31. Thistlethwaite, J.R.J., Heffron, T.G., Stevens, L., Buckingham, M., Stuart, J.K., and Stuart, F.P., The use of the T-cell flow cytometry crossmatch to evaluate the significance of positive B-cell serologic crossmatches in cadaveric donor renal transplantation, *Transplant. Proc.*, 22, 1897, 1990.
32. Kerman, R.H., Susskind, B., Buyse, I., Pryzbylowski, P., Ruth, J., Warnell, S., Gruber, S.A., Katz, S., Van Buren, C.T., and Kahan, B.D., Flow cytometry-detected IgG is not a contraindication to renal transplantation: IgM may be beneficial to outcome, *Transplantation*, 68, 1855, 1999.
33. Kerman, R.H., Kimball, P., Scheinen, S., Radovancevic, B., Van Buren, C.T., Kahan, B.D., and Frazier, O.H., The relationship among donor-recipient HLA mismatches, rejection, and death from coronary artery disease in cardiac transplant recipients, *Transplantation*, 57, 884, 1994.
34. Gordon, R.D., Fung, J.J., Iwatsuki, S., Duquesnoy, R.J., and Starzl, T.E., Immunological factors influencing liver graft survival, *Gastroenterol. Clin. North Am.*, 17, 53, 1988.
35. Iwatsuki, S., Ratsin, B.S., and Shaw, B.W.J., Liver transplantation against T cell-positive warm crossmatches, *Transplant. Proc.*, 16, 1427, 1984.
36. Lay, G., Schallon, D., Klein, A., Hopkins, K.A., Zachary, A.A., and Leffell, M.S., The effect of HLA match and antibody on liver transplant outcome, *Human Immunol.*, 44(S1), 110, 1995.
37. Bray, R.A., Nickerson, P.W., Kerman, R.H., and Gebel, H.M., Evolution of HLA antibody detection: technology emulating biology, *Immunol. Res.*, 29, 41, 2004.
38. Anasetti, C., Amos, D., Beatty, P.G., Appelbaum, F.R., Bensinger, W., Buckner, C.D., Clift, R., Doney, K., Martin, P.J., Mickelson, E., Nisperos, B., O'Quigley, J., Ramberg, R., Sanders, J.E., Stewart, P., Storb, R., Sullivan, K.M., Witherspoon, R.P., Thomas, E.D., and Hansen, J.A., Effect of HLA compatibility on engraftment of bone marrow transplants in patients with leukemia or lymphoma, *N. Engl. J. Med.*, 320, 197, 1989.
39. Anasetti, C. and Hansen, J., Bone marrow transplantation from HLA-partially matched related donors and unrelated volunteer donors, *Bone Marrow Transplantation*, Forman, S.J., Blume, K.G., and Thomas, E.D., Eds., Blackwell Scientific Publications, Boston, MA, Chapter 51, p. 665, 1994.
40. Ottinger, H.D., Rebmann, V., Pfeiffer, K.A., Beelen, D.W., Kremens, B., Runde, V., Schaefer, U.W., and Grosse-Wilde, H., Positive serum crossmatch as predictor for graft failure in HLA-mismatched allogeneic blood stem cell transplantation, *Transplantation*, 73(8), 1280, 2002.
41. Ettenger, R.B., Uittenbogaart, C.H., Pennisi, A.J., Malekzadeh, M.H., and Fine, R.N., Long-term cadaver allograft survival in the recipient with a positive B lymphocyte crossmatch, *Transplantation*, 27, 315, 1979.

42. Jeannet, M., Benzonana, G., and Arni, I., Donor-specific B and T lymphocyte antibodies and kidney graft survival, *Transplantation*, 31, 160, 1981.

43. Schäfer, A.J., Hasert, K., and Opelz, G., Collaborative Transplant Study crossmatch and antibody project, *Transplant. Proc.*, 17, 2469, 1985.

44. Kotb, M., Russell, W.C., Hathaway, D.K., Gaber, L.W., and Gaber, A.O., The use of positive B cell flow cytometry crossmatch in predicting rejection among renal transplant recipients, *Clin. Transplant.*, 13(1; Part 2), 83, 1999.

45. Lobo, P.I., Spencer, C.E., Stevenson, W.C., McCullough, C., and Pruett, T.L., The use of pronase-digested human leukocytes to improve specificity of the flow cytometric crossmatch. *Transpl. Int.*, 8(6), 472, 1995.

46. Lobo, P.I., Spencer, C.E., Isaacs, R.B., and McCullough, C., Hyperacute renal allograft rejection from anti-HLA class 1 antibody to B cells—antibody detection by two color FCXM was possible only after using pronase-digested donor lymphocytes. *Transpl. Int.*, 10(1), 69, 1997.

47. Vaidya, S., Cooper, T.Y., Avandsalehi, J., Barnes, T., Brooks, K., Hymel, P., Noor, M., Sellers, R., Thomas, A., Stewart, D., Daller, J., Fish, J.C., Gugliuzza, K.K., and Bray, R.A., Improved flow cytometric detection of HLA alloantibodies using pronase: potential implications in renal transplantation, *Transplantation*, 71, 422, 2001.

48. Bunke, M., Ganzel, B., Klein, J.B., and Oldfather, J., The effect if a positive B cell crossmatch on early rejection in cardiac transplant recipients, *Transplantation*, 56, 758, 1993.

49. Fenoglio, J., Ho, E., Reed, E., Rose, E., Smith, C., Reemstma, K., Marboe, C., and Suciu-Foca, N., Anti-HLA antibodies and heart allograft survival, *Transplant. Proc.*, 21, 807, 1989.

50. Peltenburg, H.G., Tiebosch, A., and van den Berg-Loonen, P.M., A positive T cell crossmatch and accelerated acute rejection of a pancreas–spleen allograft, *Transplantation*, 53, 226, 1992.

51. Lang, M., Kahl, A., Bechstein, W., Neumann, U., Knoop, M., Frei, U., and Neuhaus, P., Combined liver-kidney transplantation: long-term follow up in 18 patients, *Transpl. Int.*, 11, S155, 1998.

52. Kliem, V., Ringe, B., Frei, U., and Pichlmayr, R., Single-center experience of combined liver and kidney transplantation, *Clin. Transplant.*, 9, 39, 1995.

53. Demirci, G., Becker, T., Nyibata, M., Lueck, R., Bektas, H., Lehner, F., Tusch, G., Strassburg, C., Schwarz, A., Klempnauer, J., and Nashan, B., Results of combined and sequential liver-kidney transplantation, *Liver Transpl.*, 9(10), 1067, 2003.

54. Worthington, J.E., Martin, S., Al-Husseini, D.M., Dyer, P.A., and Johnson, R.W., Post-transplantation production of donor HLA-specific antibodies as a predictor of renal transplant outcome, *Transplantation*, 75(7), 1034, 2003.

55. Pei, R., Wang, G., Tarsitani, C., Rojo, S., Chen, T., Takemura, S., Liu, A., and Lee, J.H., Simultaneous HLA class I and class II antibody screening with flow cytometry, *Hum. Immunol.*, 59(5), 313, 1998.

56. Pei, R., Lee, J., Chen, T., Rojo, S., and Terasaki, P.I., Flow cytometric detection of HLA antibodies using a spectrum of microbeads, *Hum. Immunol.*, 60(12), 1293, 1999.

57. Scornik, J.C., LeFor, W.M., Cicciarelli, J.C., Brunson, M.E., Bogaard, T., Howard, R.J., Ackermann, J.R.W., Mendez, R., Shires, D.L.J., and Pfaff, W.W., Hyperacute and acute kidney graft rejection due to antibodies against B cells, *Transplantation*, 54, 61, 1992.

58. Uboldi de Capei, M., Pratico, L., and Curtoni, E.S., Comparison of different techniques for detection of anti-HLA antibodies in sera from patients awaiting kidney transplantation, *Eur. J. Immunogenet.*, 29(5), 379, 2002.

59. Rees, M.T. and Darke, C., HLA-A,B,C,DRB1,DQBQ matching heterogeneity in 'favourably matches' kidney recipients, *Transpl. Immunol.*, 12, 73, 2003.

60. Chapman, J.R., Taylor, C., Ting, A., and Morris, P.J., Hyperacute rejection of a renal allograft in the presence of anti-HLA-Cw5 antibody, *Transplantation*, 42(1), 91, 1986.
61. Bray, R.A., Harris, S.B., Josephson, C.D., Hillyer, C.D., and Gebel, H.M., Unappreciated risk factors for transplant patients: HLA antibodies in blood components, *Hum. Immunol.*, 65(3), 240, 2004.
62. Popovsky, M.A., Abel, M.D., and Moore, S.B., Transfusion-related acute lung injury associated with passive transfer of antileukocyte antibodies, *Am. Rev. Respir. Dis.*, 128, 185, 1983.
63. Kao, G.S., Wood, I.G., Dorfman, D.M., Milford, E.L., and Benjamin, R.J., Investigations into the role of anti-HLA class II antibodies in TRALI, *Transfusion*, 43, 2, 2003.
64. Kerman, R.H., Kimball, P.M., van Buren, C.T., Lewis, R.M., DeVera, V., Baghdahsarian, V., Heydari, A., and Kahan, B.D., AHG and DTE/AHG procedure identification of crossmatch-appropriate donor-recipient pairings that result in improved graft survival, *Transplantation*, 51, 316, 1991.
65. Martin, S., Liggett, H., Robson, A., Connolly, J., and Johnson, R.W., The association between a positive T and B cell flow cytometry crossmatch and renal transplant failure, *Transplant. Immunol.*, 1, 270, 1993.
66. Ratkovec, R.M., Hammond, E.H., O'Connell, J.B., Bristow, M.R., DeWitt, C.W., Richenbacher, W.E., Millar, R.C., and Renlund, D.G., Outcome of cardiac transplant recipients with a positive donor-specific crossmatch—preliminary results with plasmapheresis, *Transplantation*, 54, 651, 1992.
67. Katz, S.M., Kimball, P.M., Ozaki, C., Monsour, H., Clark, J., Cavazos, D., Kahan, B.D., Wood, R.P., and Kerman, R.H., Positive pretransplant crossmatches predict early graft loss in liver allograft recipients, *Transplantation*, 57, 616, 1994.
68. Barger, B., Shroyer, T.W., Hudson, S.L., Deierhoi, M.H., Barber, W.H., Curtis, J.J., Julian, B.A., Luke, R.G., and Diethelm, A.G., Successful renal allografts in recipients with crossmatch-positive, dithioerythritol-treated negative sera, *Transplantation*, 47, 240, 1989.
69. Cross, D.E., Greiner, R., and Whittier, F.C., Importance of the autocontrol crossmatch in human renal transplantation, *Transplantation*, 21, 307, 1976.
70. Reekers, P., Lucassen-Hermans, R., Koene, R.A.P., and Kunst, V.A.J.M., Autolymphocytotoxic antibodies and kidney transplantation, *Lancet*, 1, 1063, 1977.
71. Ting, A. and Morris, P.J., Renal transplantation and B-cell cross-matches with autoantibodies and alloantibodies, *Lancet*, 2, 1095, 1977.
72. Ting, A. and Morris, P.J., Successful transplantation with a positive T and B cell crossmatch due to autoreactive antibodies, *Tissue Antigens*, 21, 219, 1983.
73. Ettenger, R.B., Jordan, S.C., and Fine, R.N., Cadaver renal transplant outcome in recipients with autolymphocytotoxic antibodies, *Transplantation*, 35, 429, 1983.
74. Ten Hoor, G.M., Coopmans, M., and Allebes, W.A., Specificity and Ig class of preformed alloantibodies causing a positive crossmatch in renal transplantation, *Transplantation*, 56, 298, 1993.
75. Khan, N., Robson, A.J., Worthington, J.E., and Martin, S., The detection and definition of IgM alloantibodies in the presence of IgM autoantibodies using flowPRA beads, *Hum. Immunol.*, 64(6), 593, 2003.
76. Koka, P., Anti-HLA antibodies: detection and effect on renal transplant function, *Transplant. Proc.*, 25, 243, 1993.
77. Koka, P., Chia, D., Terasaki, P.I., Chan, H., Chia, J., and Ozawa, M., The role of IgA anti-HLA class I antibodies in kidney transplant survival, *Transplantation*, 56, 207, 1993.
78. Guerin, C., Pomier, G., Fleuru, H., Laverne, S., Le Petit, J.C., and Berthoux, F.C., Differential crossmatch in renal transplantation: allopositive historical crossmatch (B and/or T) but current T-cell negative crossmatch, *Transplant. Proc.*, 22, 2286, 1990.

79. Cristol, J.P., Mourad, G., Argiles, A., Seignalet, J., Iborra, F., Ramounau-Pigot, A., Chong, G., and Mion, C., Negative crossmatch on current sera and good HLA matching are the main requirements for renal transplantation in highly sensitized patients, *Transplant. Proc.*, 24, 2470, 1992.

80. Barger, B.O., Shroyer, T.W., Hudson, S.L., Deierhoi, M.H., Barber, W.H., Curtis, J.J., Julian, B.A., Luke, R.G., and Diethelm, A.G., Early graft loss in cyclosporine A-treated cadaveric renal allograft recipients receiving retransplants against previous mismatched HLA-A, -B, -DR donor antigens, *Transplant. Proc.*, 20, 170, 1988.

81. Chapman, J.R., Taylor, C.J., Ting, A., and Morris, P.J., Immunoglobulin class and specificity of antibodies causing positive T cell crossmatches: relationship to renal transplant outcome, *Transplantation*, 42, 608, 1986.

82. Terasaki, P.I. and Ozawa, M., Predicting kidney graft failure by HLA antibodies: a prospective trail, *Am. J. Transplant.*, 4, 438, 2004.

83. Halloran, P.F., Wadgymar, A., Ritchie, S., Falk, J., Solez, K., and Srinivasa, N.S., The significance of the anti-class I antibody response. I. Clincal and pathologic features of anti-class I-mediated rejection, *Transplantation*, 49, 85, 1990.

84. Halloran, P.F., Schlaut, J., Solez, K., and Srinivasa, N.S., The significance of the anti-class I response. II. Clinical and pathologic features of renal transplants with anti-class I-like antibody, *Transplantation*, 53, 550, 1992.

85. Scornik, J.C., Salomon, D.R., Lim, P.B., Howard, R.J., and Pfaff, W.W., Posttransplant antidonor antibodies and graft rejection. Evaluation by two-color flow cytometry, *Transplantation*, 47, 287, 1989.

86. Barr, M.L., Cohen, D.J., Benvenisty, A.I., Hardy, M., Reemtsma, K., Rose, E.A., Marboe, C.C., D'Agati, V., Suciu-Foca, N., and Reed, E., Effect of anti-HLA antibodies on the long-term survival of heart and kidney allografts, *Transplant. Proc.*, 25, 262, 1993.

87. Rose, E.A., Pepino, P., Barr, M.L., Smith, C.G., Ratner, A.J., Ho, E., and Berger, C., Relation of HLA antibodies and graft atherosclerosis in human cardiac allograft recipients, *J. Heart Lung Transplant.*, 11, S120, 1992.

88. George, J.F., Kirklin, J.K., Shroyer, T.W., Naftel, D.C., Bourge, R.C., McGiffin, D.C., White-Williams, C., and Noreuil, T., Utility of post-transplantation panel-reactive antibody measurements for the prediction of rejection frequency and survival of heart transplant recipients, *J. Heart Lung Transplant.*, 14, 856, 1995.

89. Schulman, L.L., Ho, E.K., Reed, E.F., McGregor, C., Smith, C.R., Rose, E.A., and Suciu-Foca, N.M., Immunologic monitoring in lung allograft recipients, *Transplantation*, 61, 252, 1996.

90. Braun, W.E. and Straffon, R.A., Long-term results in 35 HLA-identical sibling and 3 HLA-identical parent–child renal allograft recipients, *Nephron.*, 22(1), 232, 1978.

91. Opelz, G., Collaborative transplant Ssudy: non-HLA transplantation immunity revealed by lymphocytotoxic antibodies, *Lancet*, 365(9470), 1570, 2005.

92. Zwirner, N.W., Marcos, C.Y., Mirbaha, F., Zou, Y., and Stastny, P., Identification of MICA as a new polymorphic alloantigen recognized by antibodies in sera of organ transplant recipients, *Hum. Immunol.*, 61, 917, 2000.

93. Sumitran-Holgersson, S., Wilczek, H.E., Holgersson, J., and Soderstrom, K., Identification of the nonclassical HLA molecules, mica, as targets for humoral immunity associated with irreversible rejection of kidney allografts, *Transplantation*, 74(2), 268, 2002.

94. Terasaki, P.I., Ozawa, M., and Castro, R., Four-year follow up of a prospective trial of HLA and MICA antibodies on kidney graft survival, *Am. J. Transplant.*, 7, 408–415, 2007.

95. Terasaki, P.I. and McClelland, J.D., Micro-droplet assay of human serum cytotoxins, *Nature*, 204, 998–100, 1964.

96. Amos, B.D., Bashir, H., Bogle, W., et al., A simple microtoxicity test, *Transplantation*, 7, 220–223, 1970.

97. Johnson, A.H., Rossen, R.D., and Butler, W.T., Detection of alloantibodies using a sensitive antiglobulin microcytotoxicity test: identification of low levels of preformed antibodies in accelerated allograft rejections, *Tissue Antigens*, 2, 215–226, 1972.
98. Ting, A., Hasegawa, T., Ferrone, S., and Reisfeld, R.A., Presensitization detected by sensitive crossmatch tests, *Transplant. Proc.*, 5, 813–817, 1973.
99. Ross, J., Dickerson, T., and Perkins, H.A., Two techniques to make the lymphocytotoxic crossmatch more sensitive: prolonged incubation and the antiglobulin test, *Tissue Antigens*, 6(3), 129–136, 1975.
100. Fuller, T.C., Cosimi, A.B., and Russell, P.S., Use of antiglobulin-ATG reagent for detection of low levels of alloantibody—improvement of allograft survival in presensitized recipients, *Transplant. Proc.*, 10(2), 463–464, 1978.
101. Fuller, T.C., Phelan, D., Gebel, H.M., and Rodey, G.E., The antigenic specificity of antibody reactive in the antiglobulin-augmented lymphocyte-toxicity test, *Transplantation*, 34(1), 24–29, 1982.
102. Rodey, G.E. and Fuller, T.C., Public epitopes and the antigenic structure of HLA molecules, *Crit. Rev. Immunol.*, 7(3), 229–267, 1987.
103. Kao, K.J., Scornik, J.C., and Small, S.J., Enzyme-linked immunoassay for anti-HLA antibodies—an alternative to panel studies by lymphocytotoxicity, *Transplantation*, 55(1), 192–196, 1993.
104. Pei, R., Lee, J.H., Shih, N.J., Chen, M., and Terasaki, P.I., Single human leukocyte antigen flow cytometry beads for accurate identification of human leukocyte antigen antibody specificities, *Transplantation*, 75(1), 43–49, 2003.
105. Fulton, R.J., McDade, R.L., Smith, P.L., Kienker, L.J., and Kettman, J.R. Jr, Advanced multiplexed analysis with the FlowMetrix™ system, *Clin. Chem.*, 43(9), 1749–1756, 1997.
106. Christiaans, M.H.L., Overhof-de Roos, R., Nieman, F., et al., Donor-specific antibodies after transplantation by flow cytometry, *Transplantation*, 65, 427–433, 1998.
107. Mitzutani, K., Terasaki, P., Rosen, A., Esquenazi, V., Miller, J., Shih, R.N., Pei, R., Ozawa, M., and Lee, J., Serial ten-year follow-up of HLA and MICA antibody production prior to kidney graft failure, *Am. J. Transplant.*, 5(9), 2265–2272, 2005.
108. Hourmant, M., Cesbron-Gautier, A., Terasaki, P.I., Mizutani, K., Moreau, A., Meurette, A., Dantal, J., Giral, M., Blancho, G., Cantarovich, D., Karam, G., Follea, G., Soulillou, J.P., and Bignon, J.D., Frequency and clinical implications of development of donor-specific and non-donor-specific HLA antibodies after kidney transplantation, *J. Am. Soc. Nephrol.*, 16(9), 2804–2812, 2005.
109. Martin, L., Guignier, F., Mousson, C., Rageot, D., Justrabo, E., and Rifle, G., Detection of donor-specific anti-HLA antibodies with flow cytometry in eluates and sera from renal transplant recipients with chronic allograft nephropathy, *Transplantation*, 76, 395, 2003.
110. Piazza, A., Poggi, E., Borrelli, L., et al., Impact of donor-specific antibodies on chronic rejection occurrence and graft loss in renal transplantation: post-transplantation analysis using flow cytometric techniques, *Transplantation*, 71, 1106, 2001.
111. Kimball, P., Rhodes, C., King, A., et al., Flow crossmatching identifies patients at risk for postoperative elaboration of cytotoxic antibodies. *Transplantation*, 65, 444–446, 1998.

20 Functional Assessment of Immunosuppression: Monitoring Posttransplant Alloreactivity with Flow Cytometric Mixed Lymphocyte Cocultures

Chethan Ashok Kumar, Ali Abdullah,
Alison Logar, Patrick Wilson, Anjan Talukdar,
Nydia Chien, Mandal Singh, and Rakesh Sindhi

CONTENTS

20.1 INTRODUCTION

The daily management of antirejection drugs (immunosuppressants) in transplant recipients, including blood level monitoring, presents several challenges.

Immunosuppressant failures consisting of organ rejection or drug toxicity continue to affect roughly one-half of all patients [1]. Better assays are needed to measure immunosuppression in order to better manage it and assess individual risk/benefit ratios. Clinical necessities dictate that such an assay system replicate the host–graft interaction with minimal perturbations; be scalable to accommodate multiple samples in a clinical lab; delivers an output within a few hours to a day, and utilize a readily available source of recipient immune cells, especially if the recipient is a child. Because concerns for subject safety have unified all such efforts with the common purpose of developing a substitute for allograft biopsy, the peripheral blood lymphocyte has, by default, become the most studied representative of the human immune system.

The following sections describe the flow cytometric mixed lymphocyte response (MLR) in coculture experiments, as a measure of instantaneous risk of rejection in children who receive liver and intestine transplantation [2]. Inferences about the risk of rejection are essential, because early rejection, which often complicates the first 3 months after transplantation, is increasingly preventable with an enlarging repertoire of immunosuppressants. However, recurrent or late rejection, which results from mis-timed dose reductions, occurs unpredictably in up to 30% patients, because our immunosuppression management decisions are based on a "clinical" assessment of rejection risk. Multiple measurements in our pediatric recipients support this impression. Rejection risk, measured as persistence of donor-specific alloreactivity, was accompanied by a significantly greater risk of recurrent rejection following drug dose reductions based on clinical assessment alone [3].

20.2 PRELIMINARY STUDIES

Early work demonstrated that immunosuppressants inhibited mitogen-stimulated intracellular cytokine production and cell-surface marker expression in a concentration-dependent manner and that rejection-prone individuals were relatively resistant to immunosuppressants [4–6]. Parallel experiments showed that among rejection-prone children, this property was accompanied by enhanced donor-specific alloreactivity, measured as cell proliferation in a tritiated-thymidine (^{3}H-thymidine) MLR [6]. Because alloreactivity toward third-party (mismatched) peripheral blood lymphocyte (PBL) was similar among rejection-prone and rejection-free subjects, we inferred that donor-specific responses would better represent host immunoreactivity (IR) in any transplant simulation. Therefore, mitogen-stimulated responses have been provisionally abandoned. Our MLR studies incorporate carboxyfluorescein succinimidyl ester (CFSE) intravital dye dilution among prelabeled recipient cells, as a marker of cell proliferation in a 3-day coculture assay. By using fluorochrome-labeled tracer antibodies, we are able to measure the donor-induced proliferation of several T-cell subphenotypes. To minimize the known intrapatient variability of the MLR, all donor-induced events are expressed as a fraction of those induced using third party cells. This allows us to control for stresses on the immune system, which might introduce daily variability. It is assumed that such stresses influence both nonspecific and donor-specific alloreactivity comparably, on any given day. The resulting IR index can then be considered to reflect increased risk of rejection if >1 (enhanced donor-specific alloreactivity) or decreased risk, if <1.

20.3 METHODS

Three million recipient PBL are needed to perform the flow cytometric MLR, although we have been successful with half that amount. One million recipient PBL are incubated alone, with one million PBL from the donor and with one million PBL from a "third-party" (unmatched) donor. Third party PBL are obtained from normal human subjects, which are dissimilar to recipient and donor at all major human leukocyte antigen (HLA) loci (previously characterized by appropriate HLA typing methods, see Chapter 18).

Tissue typing usually provides characterization of alleles at the HLA-A, HLA-B, and HLA-DR loci. As is often the case, cells from the actual allograft donor are rarely available. In these situations, we use as "surrogate donors," PBL from normal humans, which are similar to the actual donor with respect to at least one allele at either HLA-A or HLA-B loci, and at least one allele at the HLA-DR locus.

20.3.1 PRELABELING

Before mixing of *recipient* PBL with either donor or third-party PBL, the recipient PBL are prelabeled with 1–5 μM CFSE (Molecular Probes, Invitrogen) using previously published techniques [7]. The amount of intravital dye needed varies, and should be adjusted such that the maximally stained recipient PBL generates a uniform fluorescence peak at the far right-hand side of a histogram (as shown in Figure 20.1a) using standardized instrument settings.

Donor and third-party PBL are prelabeled with anti-CD45-APC or anti-CD45-Pacific Blue. Other intravital dyes such as far red dye from molecular probes (DDAO) are unstable, if relied on to stain a cell for more than a few hours in coculture experiments. This prelabeling strategy allows for the clear separation of each cell type (as shown in Figure 20.1b) and the ability to analyze the functions within each cell type. It also eliminates the need to irradiate donor or third-party cells, as

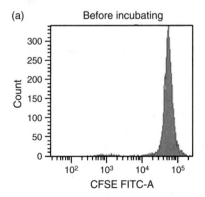 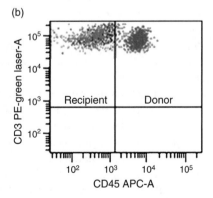

FIGURE 20.1 (a) CFSE intensity on recipient cell before incubation with donor or third party; CFSE concentration used in this assay was 4 μM and (b) shows the gating strategy used to distinguish between donor and third-party cells.

is the practice in conventional MLRs, since proliferation of recipient cells can be measured by the dilution of CFSE in daughter generations of recipient PBL, without being confounded by the inclusion of CFSE-negative, CD45-labeled PBL of donor or third-party origin. Live cells interacting with each other also better simulate the *in vivo* host–graft interaction, and can allow us to understand their effects on one another in a two-way interaction.

20.3.2 COMPENSATION

After 72 h of coculture, multiple fluorochrome-labeled antibodies are added, carefully compensating for spectral overlap. We have commonly used seven fluorochromes, which include Pacific Blue-labeled anti-CD45 as described previously, and CFSE, at the time of setting up the coculture. After 72 h, cells are harvested and labeled with five additional antibodies or fluorochromes. They are CD3, CD4, CD45RO (memory), CD25 (activation marker), and 7-AAD (to exclude dead cells). To avoid spectral overlap with this number of fluorochromes, Roederer's compensation scheme is very useful [8]. Compensation strategies are detailed in Chapter 6.

20.3.3 ACQUISITION

We use an LSRII® flow cytometer in a four-laser configuration as described by Perfetto et al. [9] and the FACSDiva software for acquisition and analysis (Becton Dickinson, San Jose, California).

20.4 ANALYSIS: CREATING SUBPHENOTYPES

Our choice of antibodies has been biased toward monitoring the T-regulatory phenotype, which reportedly expresses the memory and CD25 activation marker [10]. Although this has not identified a nonproliferating or "anergic" CD25+ cell with our current combination of markers, our classification scheme has helped us to understand how multiple subphenotypes can be identified for analytical purposes, and how they may contribute to the overall function of the parent phenotype, for example, the T helper cell.

Figure 20.2 demonstrates the derivation of the CD3+, CD3+CD4+ (T helper), and CD3+CD4− (T cytotoxic) populations, and thereafter, of the memory (CD45RO+), naïve (CD45RO−), activated (CD25+), and memory-activated (CD45RO+CD25+) subphenotypes, for each of these cell types. Thereafter, each subpopulation is characterized for CFSE-based fluorescence. In almost all cases, cells that have undergone proliferation and are represented by the CFSE^{low} fluorescent population of daughter cells can be easily separated from the CFSE^{high} noncycling parent cell population. These CFSE^{low} cells can be expressed either as cell counts per 10,000 total cells within that subphenotype, for example, the T helper memory cell, or as a proportion (%) of total cells within that subphenotype. For each subphenotype, CFSE^{low} cells are measured for each culture condition, that is, no stimulation (recipient cells only), donor-PBL stimulation, or third-party PBL stimulation.

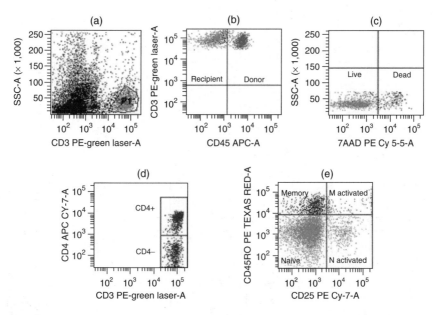

FIGURE 20.2 (a)–(d) The analysis of only the CD3+, CD3+CD4+ (T helper), and CD3+ CD4− (T cytotoxic) recipient population (d) by excluding the CD45-labeled donor or third-party cells (b). Live recipient CD3+CD4+ (T helper cells) and CD3+CD4− (T cytotoxic cells) are gated on by excluding dead cells (7AAD+; c), which are further divided into memory (CD45RO+), naïve (CD45RO−), memory-activated (CD45RO+CD25+), and naïve-activated (CD25+) subphenotypes as shown in (e).

20.5 CALCULATING AN IMMUNOREACTIVITY INDEX TOWARD THE DONOR

In the classical MLR, the donor stimulation indice (DSI) and third-party stimulation indice (TPSI) are raw numbers that represent multiples by which either donor-induced or third-party-induced proliferation exceed proliferation seen in unstimulated recipient PBL. The IR index that we have developed to express flow cytometric MLR data is the DSI/TPSI ratio. If this ratio exceeds 1, then the individual is felt to be at increased risk of rejection (due to increased reactivity to the donor). If the index is <1 then the individual's alloreactivity toward the donor is less than the alloresponse toward the third party. Such an individual is at decreased risk for rejection.

20.6 ILLUSTRATING ALLORESPONSES AMONG REJECTORS AND NONREJECTORS

Figures 20.3 (rejector) and 20.4 (nonrejector) demonstrate proliferation of recipient T cells (CD3+) and its subphenotypes, when cultured alone, with donor PBL and with third-party PBL.

Control culture of recipient cells alone (i.e., no coculture with no stimulator cells) results in very low proportions of proliferating cells (i.e., CSFE^{low} representing

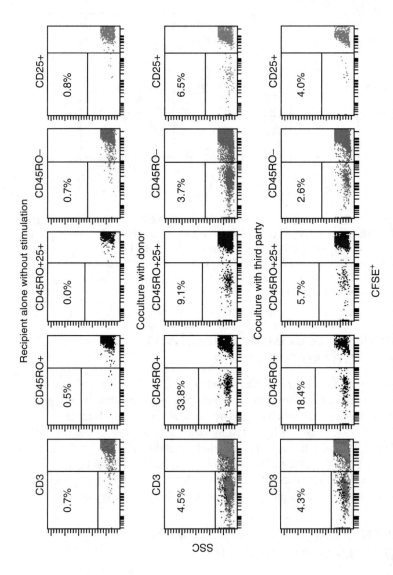

FIGURE 20.3 The IR MLR pattern in a 15-year-old subject with biopsy-proven liver allograft rejection. Proliferating (alloreactive, CFSE[low]) responder cells are shown in the left-hand side of each scatter plot. Each column of scatter plots represents from left- to right-hand side, the parent cell (T cell) and its memory, activated-memory, naïve, and activated subphenotypes. For each phenotype, proliferation in response to different culture conditions is represented in each row. These culture conditions are culture without stimulation, coculture with donor/surrogate donor, and third-party cells.

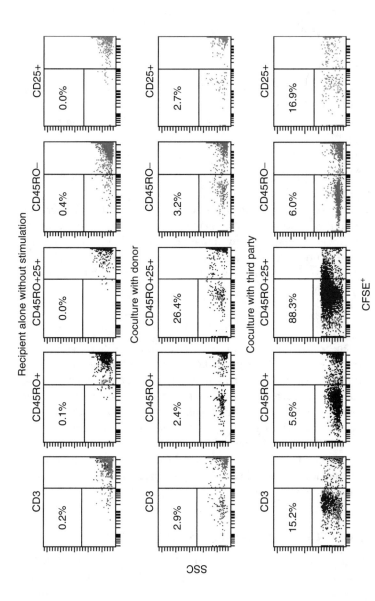

FIGURE 20.4 The scatter plots of a CFSE-MLR performed on PBL from a 6-year-old subject with a liver graft, whose biopsy at the time of CFSE-MLR did not reveal rejection. Each column represents proliferation of the T cell in response to no stimulation (alone) to donor and to third-party stimulation. Donor-induced proliferating cell frequencies (CFSE^{low}) were lower than the third-party-induced proliferating cell frequencies. The resulting DSI for each subphenotype was therefore less than the TPSI, yielding IR indices <1. The low likelihood of rejection suggested by these results was consistent with the absence of rejection.

the youngest daughter generations) in all the T-cell subphenotypes assessed (0–1%). In the example illustrated in Figure 20.3, coculture with third-party-PBL-induced stimulation, proliferating cells frequencies (CFSE^{low}) were markedly lower (TPSI) than the proliferating cell frequencies observed with donor-PBL-induced stimulation. As a result, the DSI exceeded the TPSI, and the IR index was >1 for all subphenotypes. This suggested increased likelihood of rejection, and was consistent with ongoing biopsy-proven rejection in this subject.

20.6.1 SUMMARY RESULTS

Results can be reported in three formats: number of CFSE^{low} cells (representing youngest generation of daughter cells)/10,000 cells within that phenotype, frequency/proportion (%) CFSE^{low} cells/10,000 within that phenotype, or the IR index for each cell type or subphenotype.

20.6.2 CLINICAL APPLICATION

IR indices measured for the T cell and its memory subphenotype provide the most significant correlation with clinical outcomes, suggesting to us that proliferation measurements within these two populations may be all that needs to be reported. This inference is biologically plausible, given that the memory T cell demonstrates donor specificity in rejection responses, and that responses of the parent cell type, for example, T cells represent a net summation of all pro- and antigraft forces at play within this cell population.

20.6.3 LIMITATIONS

Cell proliferation can be questioned as a marker of antigraft or donor specificity, given that cytokine markers such as IFN-γ, have become accepted markers of antigen-specific effector function [7]. However, the kinetics of cytokine expression in PBL stimulated by other PBL are such that after 3–5 day coculture, secondary mitogen restimulation or extensive additional cell manipulation may be needed to identify IFN-γ+ (donor-specific) proliferating cells by intracellular cytokine staining (ICS) procedures. The scant PBL obtained from low-volume blood samples from children, and the lymphocyte-depleting regimens used as induction immunosuppression in these subjects, compounded with cell losses during permeabilizing procedures for ICS, make cytokine assays challenging.

20.7 ANTIGEN-SPECIFIC ALLORESPONSE TO DONOR

Chattopadhyay et al.'s detection of intracellular CD154, without permeabilizing agents, presents us with a viable option to enumerate donor-specific, proinflammatory PBL [11]. In their report, CD154 was coexpressed with IFN-γ in T cells stimulated with viral peptides for a few hours. The transient and early surface expression of CD154 was detected by the inclusion of anti-CD154-PE (BD Pharmingen, San Diego, California) in the cell culture medium. Thereafter, antibody-bound surface CD154, which is normally internalized and degraded intracellularly, could be detected intracellularly, because of the inclusion of monensin in the cell culture medium. Monensin arrests intracellular degradation by inhibiting proteolytic intracellular enzymes. In a preliminary cohort of children with liver and intestine transplants, we have been able to demonstrate donor-specific, proinflammatory T cells (CD154+) using MLR cocultures with only 1.5 million total recipient PBL, and <24 h incubation. For example, liver graft rejectors demonstrated up to five times as many CD154+ memory T helper cells/10,000 memory T helper cells, in response to donor stimulation, compared with nonrejectors.

20.8 SUMMARY OVERVIEW

The adaptation of the MLR coculture to flow cytometry represents a significant advance over currently available techniques. Antigen-specific effector function can be localized to subphenotypes. The level of functional detail, for example, proinflammatory (CD154+) versus anti-inflammatory (CTLA4+, TGFβ), and its use to identify "fine" functional subsets is limited only by available antibodies, fluorochrome labels, and instrumentation. An important limitation is the number of PBL that can be isolated from clinical samples. In turn, this will determine the type of ICS procedure—permeabilization or nonpermeabilization—needed to identify functional markers with reliability.

Our assay system is scalable into a discovery tool, as lasers, photomultiplier tubes, custom-labeled antibodies, and new fluorochromes including quantum dots are brought on line. The ultimate acceptance of this and other function-based flow cytometric assays of immune response as standard clinical tools awaits rigorous validation studies, including the development of normal ranges of response.

REFERENCES

1. MacDonald AS, RAPAMUNE Global study group. A worldwide, phase III, randomized, controlled safety and efficacy study of sirolimus/cyclosporine regimen for prevention of acute rejection in recipients of primary mismatched renal allografts. *Transplantation* 71(2): 271–280, 2001.
2. Suchin EJ, Langmuir PB, Palmer E, Sayegh MH, Wells AD, Turka LA. Quantifying the frequency of alloreactive T cells *in vivo*: new answers to an old question. *J. Immunol.* 166(2): 973–981, 2001.
3. Khera N, Janosky J, Zeevi A, Mazariegos G, Marcos A, Sindhi R. Persistent donor-specific alloreactivity may portend delayed liver rejection during drug minimization in children. *FBS* 12: 660–663, 2007.
4. Sindhi R, Allaert J, Gladding D, Koppelman B, Dunne JF. Cytokines and cell-surface receptors as target endpoints of immunosuppression with cyclosporine A. *J. Interferon Cytokine Res.* 21(7): 507–514, 2001.
5. Sindhi R, LaVia MF, Pauling E, McMichael J, Burckart G, Shaw S, Sindhi LA, Livingston R, Sehgal S, Jaffe J. Stimulated response of peripheral lymphocytes may distinguish cyclosporine effect in renal transplant recipients on a cyclosporine+rapamycin regimen. *Transplantation* 69(3): 432–436, 2000.
6. Sindhi R, Magill A, Abdullah A, Seward J, Tresgaskes M, Bentlejewski C, Zeevi A. Enhanced donor-specific alloreactivity occurs independent of immunosuppression in children with early liver allograft rejection. *Am. J. Transplant.* 5: 96–102, 2005.

7. Tanaka Y, Ohdan H, Onoe T, Asahara T. Multiparameter flow cytometric approach for simultaneous evaluation of proliferation and cytokine-secreting activity in T cells responding to allo-stimulation. *Immunol. Invest.* 33(3): 309–324, 2004.
8. Roederer M. Spectral compensation for flow cytometry: visualization artifacts, limitations, and caveats. *Cytometry* 45(3): 194–205, 2001.
9. Perfetto SP, Chattopadhyay PK, Roederer, M. 17-Color flow cytometry: unraveling the immune system. *Nat. Rev. Immunol.* 4: 648–655, 2004.
10. Baecher-Allan C, Brown JA, Freeman GJ, Hafler DA, CD4+CD25 high regulatory cells in human peripheral blood. *J. Immunol.* 167(3): 1245–1253, 2001.
11. Chattopadhyay PK, Yu J, Roederer M. A live-cell assay to detect antigen-specific CD4+ T cells with diverse cytokine profiles. *Nat. Med.* 11(10): 1113–1117, 2005.

Index

A

absolute counts
flow cytometry, quality control, 206
primary immunodeficiency disease
immunophenotpying, 273–275
T-cell count single-platform detection,
209–212
accuracy, immunology statistics, 45–46
acid citrate dextrose (ACD), flow cytometry,
sample accession and processing,
193–195
acquired angioedema (AAE), classical
complement pathway deficiencies
and, 120
acquired immunodeficiency syndrome (AIDS),
flow cytometry investigation
CD4 T-cell measurements, 258–259
immune deficiency levels, 259–260
opportunistic infection prophylaxis and
antiretroviral treatment decisions,
261–263
overview, 257–258
prognosis, 260–261
acquisition process, flow cytometry,
187–188
actin, autoimmune disease and, 377
activation antigens, T-cell lineage differentiation
and, leukemia/lymphoma
markers, 234
acute leukemia, immunophenotyping, 235
acute lymphoblastic leukemia/lymphoma
(ALL/LBL),
immunophenotyping, 238–243
acute megakaryocytic leukemia,
immunophenotyping of, 238
acute myeloid leukemia (AML),
immunophenotyping,
236–238
acute promyelocytic leukemia (APL),
immunophenotyping of, 236–238
adaptive immune system
components, 138
cytokine production and, 501
inflammatory bowel disease, 410
adaptive immunity, overview, 18
adhesion molecules, classification and function,
505–506
adrenalitis, as autoimmune disease, 378
adult T-cell leukemia and lymphoma (ATLL),
immunophenotyping, 249
affinity maturation, antibodies, 423–424

age ranges
prevalence of autoantibodies and, 370–371
primary immunodeficiency disease
immunophenotpying, 274–275
age-related macular degeneration (AMD),
alternative complement pathway
deficiencies, 122
agglutination
antibody-antigen complex assays, 425–426
autoantibody detection, 378–379
AH50 assays
classical complement pathway deficiencies
and, 119–120
complement functional testing, 123–124
allele definition, human leukocyte antigen
polymorphisms, 553–556
allergic reactions, immunoglobulin E
antibodies, 72–73
alloantibody reactions, evaluation techniques,
566–567
alloantigens, human leukocyte antigen
polymorphisms and, 546–547
allograft rejection, major histocompatibility
complex and, 544–551
alloresponse assessment, transplant patients
antigen-specific donor response, 596–597
rejectors and nonrejectors, 593–596
αβ T-cell receptor, structure and properties,
314–315
alpha error, immunology statistics, 43
alternative complement pathway (AC)
control of activation in, 116
deficiencies of, 121–122
structure and function, 111–112
amplification loop, alternative complement
pathway, 112
analyte specific reagent (ASR) tests, absolute
T-cell counts, single-platform
detection, 209–212
anaplastic large cell lymphoma (ALCL),
immunophenotyping, 249
anhidrotic ectodermal dysplasias, innate
immune system defects, 302–303
animal models, autoimmune disease, 372–373
antagonists, cytokines, 506–507
antibodies. *See also* immunoglobulins
affinity maturation, 423–424
B cells and, 20
detection methods, 424–432
antibody-antigen complex assays,
425–426
avidity-based testing, 429–430